D0208760

LINEAR CONTROL SYSTEMS

McGraw-Hill Series in Electrical and Computer Engineering

Senior Consulting Editor

Stephen W. Director, *Carnegie Mellon University*

Circuits and Systems
Communications and Signal Processing
Computer Engineering
Control Theory
Electromagnetics
Electronics and VLSI Circuits
Introductory
Power and Energy
Radar and Antennas

Previous Consulting Editors

Ronald N. Bracewell, Colin Cherry, James F. Gibbons, Willis W. Harman, Hubert Heffner, Edward W. Herold, John G. Linvill, Simon Ramo, Ronald A. Rohrer, Anthony E. Siegman, Charles Susskind, Frederick E. Terman, John G. Truxal, Ernest Weber, and John R. Whinnery

Control Theory

Senior Consulting Editor

Stephen W. Director, *Carnegie Mellon University*

D'Azzo and Houpis: *Linear Control System Analysis and Design: Conventional and Modern*
Emanuel and Leff: *Introduction to Feedback Control Systems*
Friedland: *Control System Design: An Introduction to State Space Methods*
Houpis and Lamont: *Digital Control Systems: Theory, Hardware Software*
Raven: *Automatic Control Engineering*
Rohrs, Melsa, and Schultz: *Linear Control Systems*
Saadat: *Computational Aids in Control Systems Using MATLAB™*
Schultz and Melsa: *State Functions and Linear Control Systems*

LINEAR CONTROL SYSTEMS

Charles E. Rohrs
James L. Melsa
Donald G. Schultz

Based on the classical text by

James L. Melsa
Donald G. Schultz

McGraw-Hill, Inc.
New York St. Louis San Francisco Auckland Bogotá
Caracas Lisbon London Madrid Mexico Milan Montreal
New Delhi Paris San Juan Singapore Sydney Tokyo Toronto

This book was set in Times Roman by Electronic Technical Publishing Services.
The editors were Anne T. Brown and John M. Morriss;
the production supervisor was Denise L. Puryear.
The cover was designed by John Hite.
Project supervision was done by Electronic Technical Publishing Services.
Arcata Graphics/Halliday was printer and binder.

LINEAR CONTROL SYSTEMS

1 2 3 4 5 6 7 8 9 0 HAL HAL 9 0 9 8 7 6 5 4 3 2

ISBN 0-07-041525-0

Library of Congress Cataloging-in-Publication Data

Rohrs, Charles E.
Linear control systems / by Charles E. Rohrs, James L. Melsa, Donald G. Schultz; based on the classical text by James L. Melsa, Donald G. Schultz.
p. cm. — (McGraw-Hill series in electrical and computer engineering. Control theory.)
Rev. ed. of: Linear control systems / James L. Melsa. 1969.
Includes bibliographical references and index.
ISBN 0-07-041525-0
1. Control theory. 2. Automatic control. I. Melsa, James, L. II. Schultz, Donald G. III. Melsa, James L. Linear control systems. IV. Title. V. Series: McGraw-Hill series in electrical and computer engineering. VI. Series: McGraw-Hill series in electrical and computer engineering. Control theory.
QA402.3.R64 1993
629.8′312—dc20 92-35794

ABOUT THE AUTHORS

Charles E. Rohrs has taught the senior level course, Control Theory, at the University of Notre Dame seven times from 1982 through 1991. He has also taught the preceding course, Networks and Systems, twice and has taught more advanced courses in robust multivariable control, digital control, and adaptive control. He received his Masters, Engineers and Ph.D degrees from MIT during the late 1970's—a time and place where pioneering research and teaching were taking place in robust control theory and in the reconciliation of classical and modern control techniques. Dr. Rohrs' dissertation in 1982 and much of his subsequently published research is included in the significant early research efforts of applying robustness measures to adaptive control systems. Earlier, he received his B.S. degree from Notre Dame, graduating *summa cum laude*. He returned to his *alma mater* as an Assistant Professor of Electrical Engineering from 1982 to 1985. In 1985 he joined Tellabs Operations, Inc. of Lisle, IL as Manager of Adaptive System Research and led the research effort to create a state of the art echo canceller using adaptive filtering theory. Since 1986, he has held the position of Director of the Tellabs Research Center in South Bend, IN, where he is responsible for all research for Tellabs, a developer and manufacturer of telecommunications equipment. Throughout his tenure at Tellabs, he has also held a position at Notre Dame, where he is now an Adjunct Associate Professor.

James L. Melsa currently holds the position of Vice President and General Manager of the Data Communications Division at Tellabs Operations, Inc. in Lisle, IL. Dr. Melsa has previously held positions as Vice President of Strategic Planning and Advanced Technology, Vice President of Research and Development, and Vice President of Research for Tellabs. Dr. Melsa received his B.S.E.E. degree from Iowa State University, Ames, in 1960 and his M.S. and Ph.D. degrees from the University of Arizona, Tucson, in 1962 and 1965.

From 1973 to 1984, Dr. Melsa was professor and Chairman of Electrical Engineering at the Universtiy of Notre Dame, South Bend, IN. Dr. Melsa previously served as a Professor of Information and Control Sciences at Southern Methodist University and as an Assistant Professor of Electrical Engineering at the University of Arizona. He has done research in the areas of speech encoding and digital signal processing. He has published over one hundred papers and is the author or co-author of ten books in these areas. Dr. Melsa is a Fellow of the Institute of Electrical and Electronic Engineers. He has served in numerous positions in these professional organizations including President of the IEEE Control Systems Society.

Donald G. Schultz received his undergraduate degree from the University of Santa Clara and his graduate degrees from UCLA and Purdue University. He has been active in the control field for over 30 years and has former graduate students teaching in half a dozen universities in this country. He served as Professor in Electrical Engineering and later as professor and Head of the Systems and Industrial Engineering Department, both at the University of Arizona. Dr. Schultz retired from the U. of A. in 1988.

CONTENTS

PREFACE

I took my first course in Control Theory from the classical text on which this book is based. I learned the basic techniques of both modern and classical control theory. When I finished the book I could design a controller which would meet a set of specifications in both the time domain and frequency domain. However, I was left with a nagging question. "Why don't we make it better and faster?" I didn't understand what stopped me from making a system respond arbitrarily fast. In graduate school I learned that, for many plants, the most important factor that limits a control system's speed and performance is the inability of the linear time invariant model of the plant to accurately predict how a plant behaves.

When I began to teach control theory I could not find a book that both gave a modern presentation of the fundamental techniques of control theory and also discussed the inherent limitations present and tradeoffs necessary in the design of controllers based on imperfect models. *The single most important improvement in the rewriting of this book is the inclusion of the ramifications of model inaccuracies on the design of control systems.* The topics concerned with modeling problems are introduced early and stressed often. Section 2.7 discusses some typical manifestations of modeling error. In Sec. 5.6 a perturbation model is used to describe modeling uncertainty in the frequency domain. Robust stability, the maintenance of stability in the face of modeling error, is discussed in Sec. 6.6 while, in Sec. 6.7, robust performance is discussed. In Chap. 8, a new chapter on design, the impact of modeling inaccuracies on design strategy is stressed. It is also demonstrated that certain factors in the plant model such as right half-plane zeros and modeling uncertainty inherently limit the performance which can be achieved in a control system design.

There is a second point of emphasis in the rewriting of the design section of the book. Most first textbooks in control theory place the achievement of a specified transient response as the primary concern in the design of a control system. In this new edition the position is taken that, in a feedback control system design, *the major concern is to ensure stability and to reduce the sensitivity of the system to disturbances and plant perturbations.* If this sensitivity is made small, the transient response of the resulting closed loop system can be tuned by using a prefilter on the reference input. This view is in accord with the current view of control system researchers.[1]

At the time that the original book on which this book is based was written, state-space techniques for designing control systems were fairly new and rather imposing. The original book provided a vehicle for the study of the classical techniques of control theory in the context of state-space controller designs. In rewriting, we remain true to this concept; classical techniques and state-space techniques are not competing but complementary. However, with the emphasis on modeling uncertainty comes the increased importance of frequency domain concepts. Some notational changes have been adopted which make the exposition of classical transfer function techniques independent of whether the transfer function results from a state-space approach or not. Indeed, if desired, the material can be covered in two passes with the first pass covering classical techniques and the second pass adding the state-variable information. This is accomplished by skipping Secs. 2.3–2.6, 3.3–3.5, 4.5–4.6, and 8.11–8.12 on the first pass. The advantage of this approach is that it allows the exposure to design concepts earlier in the course; the disadvantage is that it breaks up the unifying treatment.

The speed in introducing design concepts can be further increased without continuity problems by using the following approach. Initially, Sec. 6.3 on the Routh-Hurwitz criterion can be skipped. More importantly, the detailed discussion of techniques for sketching the root locus in Chap. 7 can be replaced with the short introduction to root locus described in the appendix to Chap. 7 entitled, "Bypassing the root locus." It is a matter of some debate in the control education community whether details of the root locus and the Routh-Hurwitz criterion need to be learned due to the availability of computers to perform these functions. This book provides a path for those who wish to de-emphasize these subjects or those who wish to simply reorder the presentation from the more pedagogical ordering which is presented in the book to an ordering which allows design problems to be assigned sooner in the course. Covering at least the appendix to Chap. 7 is recommended since the concept of the root locus is used to add insight in Chap. 8 on design.

The book has also been updated in its use of Computer Aided Design (CAD) material. To demonstrate the capabilities of CAD programs which are now readily available, examples using a representative program, MATLAB™ by MathWorks, are provided throughout the book at the end of relevant sections under the heading of *CAD Notes*. In addition, problems which assume the availability of a CAD package have been added to give the students a chance to gain experience using these tools. I believe that CAD packages which are available today are quite similar and that the examples using MATLAB are representative of CAD programs in general and the actual code can be fairly easily translated to other CAD packages.

Some other changes have been made. The state-space material itself has been expanded with the inclusion of Sec. 3.5 on controllability and general pole placement.

---

[1] See the references to Doyle and Stein, Zames, Francis, and Doyle, Francis and Tannenbaum.

MATLAB is a registered trademark of the MathWorks, Inc.

Pole placement techniques using transfer functions and the Diophantine Equation have also been added in Secs. 3.2 and 8.11. Much of Chap. 4 has been rewritten to explain the effects of additional poles and zeros in systems which are dominantly first or second order. This replaces the previous chapter on specifications. Section 4.7 on steady state error constants has been added.

There has been one major notational change in the new book as it was decided that Laplace transforms should be represented with upper case symbols. Almost all the problems are new. However, the philosophy of placing exercises which illustrate and drill techniques at the end of each section while placing more challenging problems at the chapter's end remains. Many examples of the physical systems from which the mathematical models are derived have been added.

An emphasis on physical examples has been added. The use of the example of a dc motor positioning system has been retained to demonstrate modeling, state variable manipulation and pole placement control in Chaps. 2 and 3. The system can be equally considered to be representative of a system which positions an antenna, a telescope, or many other things. The example has been expanded and made more realistic in Sec. 2.7 to include unmodeled effects such as torsionally flexible shafts, saturation and hysteresis in the motor, and backlash in the gearing. Two examples are used through Chap. 8 on design. The theme problem for the development of the proportional, lag, and lead compensators in Secs. 8.3–8.5 is an automobile cruise control system. A realistic design on an attitude control system for the ATS-6 communications satellite is explored in detail in Secs. 8.7 and 8.8. The introduction of CAD tools allows the introduction of realistic design problems which the students can solve. The linearized dynamics of a chemical reaction process are developed and explored via simulation in Sec. 2.7 while the same is done for the longitudinal dynamics of a 747 airplane in the Appendix to Chap. 2. These two representations of physical systems appear throughout the end of chapter problems giving the students an opportunity to apply design skills to realistic problems as these skills are learned.

In summary, in rewriting *Linear Control Systems* much material has been added; terminology and philosophies have been brought up to date. Some effort has been made to explain concepts using the language that is used in more advanced control courses so that this study may lead naturally to the study of such topics as multivariable control. I hope that what I considered the strengths of the original book in the clear explanations of the basic concepts of control theory have been retained. The book should now be ready to provide an introduction to the basic concepts of control theory as they stand at the beginning of the 1990's.

I would like to thank Jim Melsa and Don Schultz for the opportunity to work with their text. It provides a core which was *user friendly* before the phrase became popular. I would also like to thank those people who brought me to the position where I was able to perform the task of rewriting this book. First, my parents, Charlie and Florence, who are responsible for the fundamentals of my education and gave me the words that make the book. While the words which have been added in this revision are mine, most of the ideas are a gift from Gunter Stein, who taught me not only the theory of control but also taught me how to think about problems and how to critically question the ideas with which I am presented. From Mike Athans I learned the importance of taking a stance on technical issues and believing in that stance. His example helped me to write a book that breaks the usual curriculum of control theory as I believe this book does. Jim Melsa not only granted me the authority to tamper with his words from the original book, he also helped bring me to this point by showing faith in me at a time when I had lost faith in myself.

Most of all, I thank my wife, Cathy, and daughters, Kelly, Ali, and J.J. Because I left academia at the beginning of this project, most of this book was written on my own time which, of course, is really *their* time. Without their support and understanding of my regularly disappearing into my den and temporarily out of their lives, this project would have been scrapped long ago. I also appreciate the indulgence of my superiors and coworkers at Tellabs Operations, Inc. for the part of me that this project inevitably stole from them.

I acknowledge the proofreading and suggestions from my coworkers at the Tellabs Research Center. Don Heard and his obviously dedicated group at Electronic Technical Publishing also deserve credit for the polish that they added to the manuscript. I thank the following reviewers for their many helpful comments and suggestions: James Palmer, Rochester Institute of Technology; William R. Perkins, University of Illinois; and Hal Tharp, University of Arizona. I thank Tony Michel and Dan Costello for the opportunity to continue my teaching of this material at Notre Dame and I thank the students with whom I have had the privilege of exploring this material.

I mentioned before that the words of this revision are mine. Unfortunately, the words usually emerge from my mind through my hand in only a barely legible form. The task of translating and formatting the words, equations, and figures fell mostly on Marilynn Anson and less so on Karen Spindler. They contributed greatly not only to the final form of the book but also to the preservation of my mental health. I hope that the reader might find the material of this book as challenging and interesting as I have found the journey that led me to produce it.

Charles E. Rohrs

LINEAR CONTROL SYSTEMS

CHAPTER 1

INTRODUCTION TO AUTOMATIC CONTROL SYSTEMS

1.1 INTRODUCTION

The term *automatic control system* is intended to be somewhat self-explanatory. The word *system* implies not just one component but a number of components that work together in a prescribed fashion to achieve a particular goal. This goal is the control of some physical quantity, and the control is to be achieved in an automatic fashion, often without the aid of human supervision.

The topic of automatic control is certainly a romantic one. In this age of unmanned ballistic missiles, nuclear power plants, manufacturing robots, and high performance aircraft and spacecraft, the need for the control of such dynamic systems is not only evident, it is dramatized repeatedly in the mass communication media of newspaper, radio and television. As an unmanned missile hits its target with incredible accuracy, the country and the world marvel. Similarly, as a satellite tumbles out of control or a nuclear power plant goes into automatic shutdown, millions are instantly aware of the fact and wait in fear that control may not be regained.

In the more prosaic area of civilian and industrial endeavors, the subject of automatic control is romanticized by the word *automation*. Automation means the automatic production of processed material. The end product of an automated plant may vary between such wide extremes as high-octane gasoline or a computer chip where the margin of error is measured in microns. However, be it robotically welded cars or the creation of pure and perfect silicon crystals, the unifying feature is that the process is under control every step of the way and often by automatic means, without which the needed accuracy could not be attained.

Unfortunately, the study of automatic control systems is not as romantic as its applications. Because the systems and processes that must be controlled are dynamic rather than static, their behavior is described by differential rather then algebraic equations. In wrestling with the complexities inherent in such controlled dynamic systems, students have been known to sweat and to swear. Yet the rewards to be gained from such a study are immense. Satellites and missiles and automatic factories do not just happen; they evolve from the ideas and understanding of men and women. And a surprisingly short time ago those presently in charge of exotic control applications were in the same situation as are the present readers of this book. They were students, and they were struggling; their victory was understanding. One object of this book is to make the reader's own victory in the understanding of automatic control systems as painless as possible.

The purpose of the introductory material in this first chapter is to give the reader an intuitive feeling for what is meant by an automatic control system and to outline the approach used in this book for the systematic study of this subject.

1.2 CLOSED-LOOP VS. OPEN-LOOP CONTROL

This book deals almost exclusively with *closed-loop automatic control* systems, and it is therefore necessary at the outset to distinguish between open-loop and closed-loop systems. This is most easily done by means of a familiar example. Consider the control of the temperature of a fluid within a tank filled by fluids at different temperatures. As an everyday example, one might think of a person trying to adjust the water in the bathtub to a desired temperature.

One method of realizing the desired temperature is to open the hot water tap a specified amount, open the cold water tap a specified amount, and let the water run for a fixed time or until the water has reached a given level. This is the way in which a person in a rush would fill the tub, and if she had been rushed a sufficient number of times in the past, she might know rather well the necessary settings of the hot and cold water faucets to realize the desired water temperature. This is an example of an *open-loop* control system.

An *open-loop control system* is one in which neither the output nor any of the other system variables has any effect on the control of the output. In this example the output is the temperature of the water in the tub, and control is exercised by setting the valves on the hot and cold water lines. If, for any reason, the water in the tub is not at the desired temperature, this fact is not known and no control is exercised to force the actual water temperature toward the desired water temperature.

In this familiar example there are a number of factors that might affect the final water temperature, the most obvious of which is the amount of hot water available. Suppose that the normal hot water supply were depleted, because someone else had just finished a bath or because the washing machine had just completed its cycle. Then, of course, when the water reached its final level, it would be too cold.

Someone in less of a rush might feel the water in the tub at several intervals while the tub is filling. If the water were not at the right temperature, adjustments could then be made to either the hot or the cold water faucet, as required. This is an approach toward *closed-loop control*. In a *closed-loop control system*, the system output and other system variables affect the control of the system. In this case the loop would be closed intermittently by the person making measurements and taking corrective control action.

A person with lots of time might proceed in yet a different fashion. Instead of feeling the water in the tub only at intervals, he might continuously stir the contents of the tub and continuously measure the average water temperature, at the same time making the necessary valve adjustments. Since it is assumed that the desired water temperature is higher than that of the tub, the incoming water must initially be at a higher temperature than the desired temperature to overcome the thermal capacity of the tub itself. As the tub heats up to the desired water temperature, the temperature of the incoming water must be decreased. Thus, the person exercising the control would find that continuous adjustment is necessary to maintain a constant water temperature as the tub is filling. This system is a closed-loop system, where the loop is actually closed by the human operator. Because a human operator is needed to make the system function properly, such a system is not classified as automatic.

Even in the latter situation, where the person is devoting full time to controlling the water temperature, a less than ideal situation may result. Assume, for example, that the supply of hot water is nearly depleted at the start of the operation. As the temperature of the hot water being supplied begins to diminish, the person stirring eventually notices that the temperature of the water in the tub is decreasing, and she might well increase the demand for hot water. If indeed the hot water were by this time cold, such action would produce the opposite of the desired result. A more astute tub filler might measure not just the temperature of the water in the tub but the temperature of the water coming from the tap as well. In other words, she might measure more than one variable and therefore be able to perform the task of control more effectively. As a further refinement, she might even profit by measuring the rate at which the temperature of the hot water supply is decreasing as the hot water is used up.

Consider briefly now the automatic control of the water temperature in the tub. In view of the discussion above, the hardware requirements of such a control operation are evident. First, we need to measure the temperature of the water in the tub and, if we wish to do a good job of control, it might be appropriate to measure the temperature of the incoming water and the rate of change of the temperature of the hot water. *Sensors* are used to measure the required variables. Next, a power element of some sort is necessary so that the control valves, the hot and cold water faucets, are positioned automatically. Assume that an electric motor is used as the power element. The power element is often referred to as an *actuator* by control engineers. Since an electric motor is inherently a high-speed device and the valves move at relatively slow speeds, a gear train is necessary to make the velocities compatible. To ensure that the temperature sensor in the tub measures the average water temperature, a stirring device might be clamped over the side of the tub. To accomplish the actual control, the measured water temperature in the tub and the other measured state-variables must be combined and compared with the desired tub temperature. This is accomplished in the controller and the resulting control signal is then amplified to a sufficient level to enable it to drive the actuators, which in turn position the valves.

The open-loop and the automatic closed-loop control systems for controlling the water temperature in the tub are illustrated in Fig. 1.2-1. Recall that the open-loop control requires only the initial positioning of the two control valves, the water faucets. This example is typical of open-loop control situations, where success depends upon two things:

1. The accuracy of the system model
2. The repeatability of related events over an extended period of time, that is, the lack of external disturbances

In this example, for each desired water temperature there is a corresponding setting of the control valves and an underlying assumption that the water temperatures of the hot and cold water are known. If successful open-loop control is to be achieved, this calibration and the source water temperatures must remain fixed. Open-loop control is not successful unless all parameters associated with the system remain fixed over an extended period of time. If either the hot or the cold water or the tub itself is

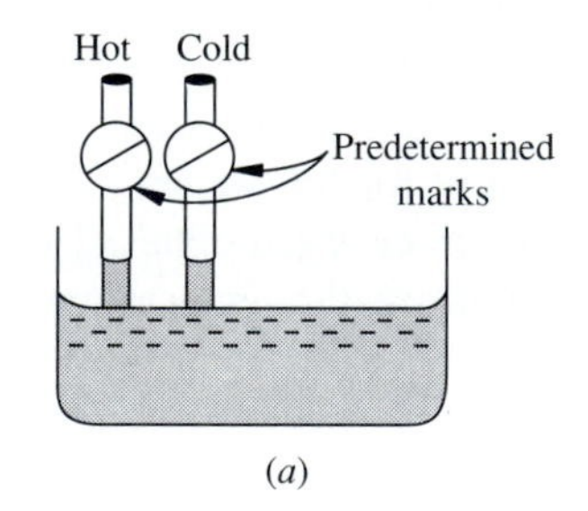

(*a*)

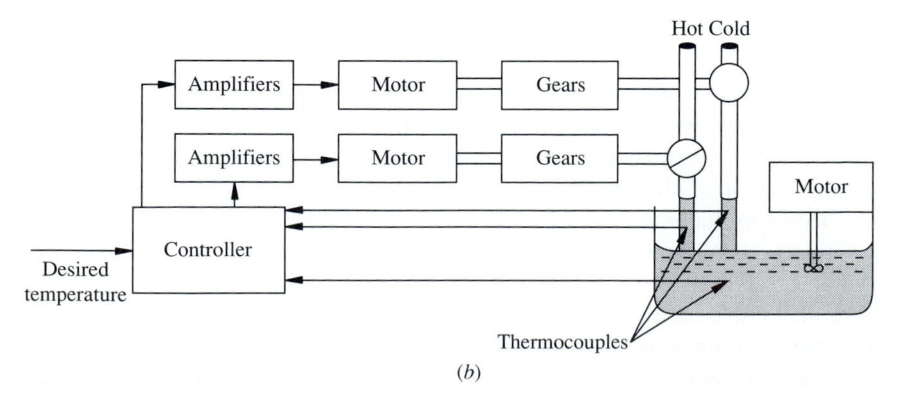

(*b*)

FIGURE 1.2-1
Liquid temperature control system. (*a*) Open loop; (*b*) closed loop.

at a different temperature than it was on the day of calibration, the end result is not predictable.

It is obvious from Fig. 1.2-1 that the closed-loop control system is considerably more complicated than the open-loop system. The trade off here is between complexity and performance. This is most easily seen if one considers the simplest case in which only the output, the tub water temperature, is fed into the controller. In this case assume that the controller is just a subtracting circuit, so that the control signal is proportional to the error between the required water temperature and the actual water temperature. If the required and the actual temperatures are the same, no error is present to drive the valve settings to new positions. The ultimate behavior of the system is not dependent upon the hot or cold water temperatures or the tub temperature.

The distinguishing feature between the open-loop and closed-loop systems of Fig. 1.2-1 is the use of the state of the output to affect the input in the closed-loop control case. Such use is termed *feedback*. The final error may be further affected if other system variables are fed back in addition to the output. This is often the case, and the feedback of the other system variables in addition to the output is of considerable aid in ensuring that the system under consideration can be properly controlled.

This last remark brings up the important question of stability and instability. The notion of *stability* implies that the system operates within a fairly narrow range of behavior. *Instability* implies that some variables of the system grow to unacceptably large values. In the open-loop case the settings of the control elements are programmed according to the desired output. In the open-loop example, the control elements are set once and for all, and there is no question as to whether the system is stable or not. In a closed-loop system, this need not be the case. In fact, one of the major difficulties in the design and synthesis of high performance closed-loop control systems is the necessary compromise between desired performance and required stability. On the basis of the simple example that has been discussed, it may be difficult for the reader to appreciate that a stability problem might exist.

To illustrate the possible occurrence of such instability, let us suppose that we control only the hot water tap, with the cold water rate set at some predetermined value. Let us assume, in addition, that we are slow to make a correction but, when we do decide to make the correction, we make a very large one by turning the tap either fully on or fully off. If the water in the tub is initially warmer than desired, we turn the hot water completely off. As the water begins to cool, because of our slow response time, we might let it become too cool before we turn the hot water full on again, and so forth, so that we oscillate between water that is too hot and water that is too cold.

This example illustrates the two most common causes of instability in automatic control systems: delay and high gain. Delay in this case was caused by our slow response speed, and high gain is the result of our plan to turn the faucet fully on or fully off. Each of these items will be examined in detail in later chapters.

In this bathtub example, it is very clear how the system responds to inputs. Increasing the hot water supply causes the tub water to slowly warm up. Increasing the cold causes a slow cooling. In most control applications, capturing the way the system reacts to inputs results in a complicated mathematical model. If the model does not predict the system response properly, the controller may not take proper actions. Since no system can be perfectly modeled, designers must be careful that the controllers they design based on mathematical models behave properly when they are implemented on the real system.

In this section, we have attempted to give the reader an intuitive feeling for just what a closed-loop control system is by comparing it with an open-loop case. On the surface, it may appear that the open-loop system is indeed the better. Not only is it less complex, but no stability problem exists in open-loop systems. The only negative feature is that performance may be degraded if the system parameters deviate from their expected values. Actually, the comparison made on the basis of complexity, performance, and stability is somewhat of an academic question and is useful only to demonstrate the nature of a closed-loop system. In many instances, open-loop control cannot even be considered. The only cases in which open-loop control is possible are those in which the desired performance is known in advance. In a great many instances the desired output is not known in advance. Suppose that in our tub example the temperature of the water is required to be a function of some other variable, such as the weather or the temperature of a liquid in some adjacent tank. Since these variables are unpredictable and beyond our control, it would be impossible to program the desired output in advance to control the valve settings. A closed-loop system is necessary to produce the desired response.

An anti-aircraft gun control system is another obvious situation in which open-loop control simply does not work. In such a gun control system, a gun is made to follow a moving target such as an aircraft. Since the flight path of the target aircraft is unknown and may even be intentionally evasive, it is evident that no open-loop control system can be programmed in advance to follow the plane's trajectory. If a human operator were used to close the loop, his response time would usually be much too slow. Thus, in spite of the complexity and the stability problems associated with automatic closed-loop control, there is often no alternative. The design portions of this book are concerned with ensuring adequate performance and at the same time minimizing complexity and maintaining stability.

Exercise

1.2-1. List five examples of feedback control systems which you encounter in your day-to-day existence.

1.3 HISTORICAL AND MATHEMATICAL BACKGROUND

The preceding section served to distinguish between open-loop and closed-loop control systems on the basis of feedback. Feedback is a concept or principle that seems to be fundamental in nature and not necessarily peculiar to engineering. In human social and political organizations, for example, a leader remains the leader only as long as she is successful in realizing the desires of the group. If she fails, another is elected or by other means obtains the effective support of the group. The system output in this case is the success of the group in realizing its desires. The actual success is measured against the desired success, and if the two are not closely aligned, that is, if the error is not small, steps are taken to ensure that the error becomes small. In this case, control may be accomplished by deposing the leader. Individuals act in much the same way. Studying for this course involves feedback. If your study habits do not produce the desired understanding and grades, you change your study habits so that actual results become the desired results.

Because feedback is so evident in both nature and humanity, it is impossible to determine when feedback was first intentionally used. Newton, Gould, and Kaiser[1] cite the use of feedback in water clocks built by the Arabs as early as the beginning of the Christian era, but their next reference is not dated until 1750. In that year Meikle invented a device for automatically steering windmills into the wind, and this was followed in 1788 by Watt's invention of the flyball governor for regulation of the steam engine.

However, these isolated inventions cannot be construed as reflecting the application of any automatic control theory. There simply was no theory although at roughly the same time as Watt was perfecting the flyball governor both Laplace and Fourier were developing the two transform methods that are now so important in electrical engineering and in control theory in particular. The final mathematical background was laid by Cauchy, with his theory of the complex variable. It is unfortunate that the readers of this text cannot be expected to have completed a course in complex variables, although some may be taking this course at present. It is expected, however, that the reader is versed in the use of the Laplace transform. Note the word *use*. Present practice is to begin the use of Laplace transform methods early in the engineering curriculum so that, by the senior year, the student is able to use the Laplace transform in solving linear, ordinary differential equations with constant coefficients. But not until complex variables are mastered does a student actually appreciate how and why the Laplace transform is so effective. In this text we assume that the reader does not have any knowledge of complex variables but does have a working knowledge of Laplace transform methods. A short summary of the more commonly used Laplace transform theorems is included in Appendix A, and Appendix B is a table of direct and inverse Laplace transforms. Although the Laplace transform is the mathematical language of the control engineer, in using this book the reader will not find it necessary to use more transform theory than appears in these two appendices.

Although the mathematical background for control engineering was laid by Cauchy (1789–1857), it was not until about 75 years after his death that an actual control theory began to evolve. Important early papers were "Regeneration Theory," by Nyquist, 1932, and "Theory of Servomechanisms," by Hazen, 1934. World War II produced an ever-increasing need for working automatic control systems and thus did much to stimulate the development of a cohesive control theory. Following the war a

[1] G. C. Newton, L. A. Gould, and J. F. Kaiser, *Analytical Design of Linear Feedback Controls*, John Wiley & Sons, Inc., New York, 1957.

large number of linear control theory books began to appear, although the theory was not yet complete. As recently as 1958 the author of a widely used control text stated in his preface that "feedback control systems are designed by trial and error."

In the early 1960s a new control design method referred to as *modern* control theory appeared. The basis of much of this theory is highly mathematical in nature and almost completely oriented to the time domain. A key idea is the use of state-variable system representation and feedback, with matrix methods used extensively to shorten notation. Originally it was thought that these state-variable methods would lead to a well-defined control design procedure that could be mechanized. Later, it was recognized that mechanized designs are only as good as the model and specifications used to develop them. Many issues dealing with modeling uncertainty still need to be analyzed using standard frequency domain methods. In addition, frequency domain methods have been developed to produce mechanized designs equivalent to the state-space designs when used on the same output feedback problem.

In this book the frequency domain and the state-space or time domain methods are developed side by side. Each method has its advantages and proponents. The well-educated young control engineer should become conversant in both methods. We note here, however, that the material of this book can be absorbed in two passes. All the state-space material can be skipped on the first pass without loss of continuity. This material can then be assimilated on the second pass. The state-space material which can be skipped on the first pass is Sections 2.4–2.6, 3.3–3.5, 4.5–4.6, and 8.11–8.12. Also, Chapter 7 on root locus can be covered in less depth by using the Root Locus Bypass of Section 7.A. While such an approach may be less cohesive and unified, it has the advantage of encountering design issues much earlier in the course.

1.4 OUTLINE OF THE BOOK

Broadly speaking, this book is divided into three sections: modeling, analysis, and synthesis. The section on modeling is made up of the two chapters immediately following this introductory chapter. These chapters treat, respectively, plant representation and system representation. The *plant* means the physical hardware associated with the quantity being controlled. Usually the plant is considered unalterable. The elements added to effect control are designated as the *controller*, and the plant and the controller taken together constitute the entire closed-loop control system, or simply the *system*.

The discussion of plant representation emphasizes the different means by which the same plant can be described in terms of differential equations or transfer functions or state-variable equations. It is assumed that the student is already familiar with the use of Laplace transforms and elementary matrix manipulations. In all cases, it is assumed that the plant can be adequately represented by a linear, time-invariant system. The limitations of the low-order linear models generally used in control design are discussed.

The separate discussion of system representation in Chapter 3 serves to emphasize the effects of feedback. Two common controller configurations, which are used for the remainder of the book, are defined. Feedback properties are explored in the context of these configurations. The development of controllers to achieve desired closed-loop poles is addressed for both the transfer function and the state-variable approaches. The concept of controllability is addressed and defined for the state-variable approach.

Chapters 4 to 7 constitute the analysis portion of the book. Chapter 4 explores the time response of systems given the system transfer function or state-variable representation. Since much of control theory is concerned with poles and zeros of systems, the material of Chapter 4 provides the bridge needed to understand how the actual behavior of a system in the time domain is related to the pole and zero formations of the system.

In Chapter 5 another bridge is built. This time the poles and zeros of a system are related to the system's frequency response using Bode diagrams. Using the system's poles and zeros as intermediaries, the material of Chapters 4 and 5 relates time domain behavior to frequency domain information. This is a very valuable concept since the frequency domain properties of a system can be fairly easily manipulated in a design procedure. Also included in Chapter 5 are discussions of identifying a plant model in the frequency domain and quantifying the uncertainty connected with such models.

The important question of stability is addressed in Chapter 6, using both frequency domain methods and the roots of the characteristic equation. The powerful Nyquist theory is developed and it is shown how closed-loop frequency domain characteristics are attainable from open-loop frequency domain characteristics. The robustness problem of maintaining stability and performance in the face of model uncertainties is discussed. In Chapter 7 the root locus methods for transient response analysis of control systems are explored.

Chapter 8 develops synthesis or design issues. In Chapter 8 it is shown how the analysis techniques of Chapters 4 and 7 can point the way to improved systems by manipulating the open-loop transfer function using a series compensator. The long-recognized techniques of PID and lead-lag control are explored. A full-blown iterative design example is performed and evaluated. Included in the design criteria is the important goal of tolerance to model uncertainty. The problems associated with controlling plants with poles and zeros in the right half-plane are discussed. It is shown that the right half-plane zeros cause problems for controller designs. Right half-plane zeros can arise physically from time delays or inadequate measurements. It is shown that if all state-variables can be measured, problems arising from right half-plane zeros are alleviated. Designs that place closed-loop poles arbitrarily are investigated using state-variable feedback and output feedback. It is shown that the designs must be evaluated using frequency domain techniques.

By the time a student finishes this book, the student should be equipped with the basic knowledge required to understand and design linear, single-input, single-output control systems in the face of practical uncertainties. That knowledge will have been presented in such a way that it leads naturally to the theory of multi-input, multi-output controller design, which is saved for another course.

CHAPTER 2

PLANT REPRESENTATION

2.1 INTRODUCTION

In this chapter we consider the most fundamental part of any control problem, namely the representation of the *plant* by appropriate *mathematical models*. The plant means the physical hardware associated with the quantity being controlled. Usually, the plant is considered to be unalterable. The concept of a mathematical model is not new to the reader. Before any engineering analysis or design may be undertaken, it is necessary to abstract from the physical object in question a description in terms of mathematical formulas. For example, the mathematical model for an electrical circuit is the set of loop or node equations that describe the circuit.

Although we deal almost exclusively with the mathematical model for a system, one must not forget that the ultimate interest must be in the control of real, physical systems. One does not control a set of equations; rather, one controls a radar antenna or a rocket engine. A mathematical model gives us only an approximation to the actual behavior of a real, physical system. In this book, after developing techniques to control the behavior of mathematical models, we develop additional techniques to quantify how closely the mathematical model must approximate the actual behavior of the physical system to ensure that the behavior of the physical system is adequately controlled. In addition, throughout this book various practical features of control system design, which are not always obvious from the mathematical description, are illustrated with examples or exercises for the reader.

Often more than one mathematical model may be found for the same physical object. Consider, for example, the loop and node representations of electrical circuits.

In such cases, the various models tend to complement each other, with one model better suited for one use and another model better suited for another use.

In control theory, two different methods of representation are of interest. These two methods differ more in degree than in nature. To be more precise, the first method, known as the *input-output relation method*, provides less detail and therefore is less complete than the second method, known as the *state-variable method*. As with electrical circuits, these two methods complement rather than conflict with each other.

This last statement may seem strange since it may appear that the more complete state-variable method would be superior in all situations. That this is not true may be traced to the fact that one simply does not always need and often cannot profitably use the more complete representation. Therefore, there is considerable need for both methods. This chapter contains not only a development of each of these two methods of representation but also a considerable discussion of the relationship between the two methods. The input-output relation method of plant representation is discussed first, since it is easier and probably more familiar to the reader. Following this development, a convenient graphical representation of the input-output relation information is achieved by the introduction of the block diagram concept.

The state-variable representation of a plant is then introduced and various properties of this representation are developed. The relationship among the transfer function, various state-variable representations, and various block diagrams of a plant are investigated. Finally, the limitations of all the above methods of modeling physical systems are discussed.

This chapter is restricted to plant representation, but *not* because the methods discussed apply only to that case. On the contrary, the methods of this chapter form the entire basis of the subject of closed-loop system representation presented in the next chapter. The restriction has been made here for the sake of simplicity and to emphasize the important and natural distinction between the uncontrolled plant and the controlled plant, i.e., the closed-loop system. By restricting our attention in this fashion, we hope that the reader will be able to grasp the fundamental concepts of representation with the least chance of confusion.

2.2 TRANSFER FUNCTIONS AND BLOCK DIAGRAMS

2.2.1 Transfer Functions

One of the principal tools of the input-output representation is the transfer function. The idea of using transfer functions to represent physical systems is a natural outgrowth of the use of Laplace transform methods to solve linear differential equations. Since these methods have been so successful in simplifying and systematizing the problem of obtaining the time response of a system, it appears reasonable that they should also be valuable in system representation.

The general configuration of a closed-loop control system is shown in Fig. 2.2-1. The system consists of two basic elements, the plant and the controller.

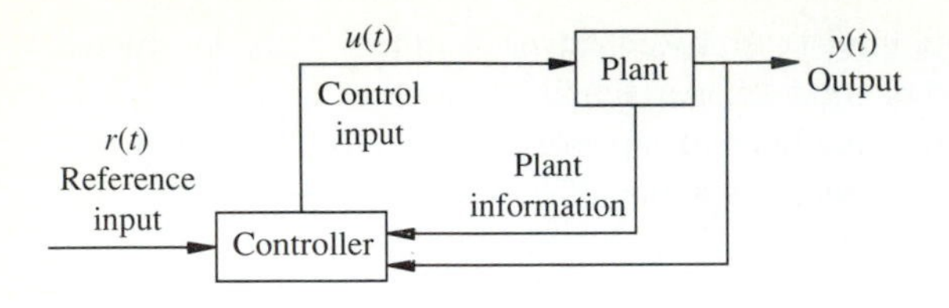

FIGURE 2.2-1
General closed-loop system.

The plant encompasses the unalterable portion of the system. The controller is added by the designer to achieve proper performance of the overall system. In this chapter we restrict our attention to the description of the plant, for simplicity; we consider the entire system in the next chapter.

To understand how transfer functions are used in plant representation, let us suppose that we have a general nth-order plant with a control input $u(t)$ and an output $y(t)$, as shown in Fig. 2.2-1. The behavior of many physical systems we wish to control may be closely approximated by the behavior of an nth-order linear time-invariant differential equation relating the input to the output. The general form of such an equation is given by

$$\frac{d^n y(t)}{dt^n} + a_{n-1}\frac{d^{n-1}y(t)}{dt^{n-1}} + \cdots + a_o y(t)$$
$$= c_m \frac{d^m u(t)}{dt^m} + c_{m-1}\frac{d^{m-1}u(t)}{dt^{m-1}} + \cdots + c_o u(t) \qquad (2.2\text{-}1)$$

Here all the a_i's and c_i's are assumed to be constant.

This differential equation provides a complete description of the plant, in the sense that, for any given input and initial conditions, the output may be determined. Actually, the differential equation (2.2-1) is already a mathematical model that only approximates the behavior of the physical plant. This model, however, is rather unwieldy and therefore is seldom used in this form.

A more basic starting point for our development of transfer functions is the set of differential equations that describe the physical laws governing the behavior of the individual components of the plant. Such an approach tends to obscure the basic concept of transfer functions with an extensive development of plant components and their describing equations. It is assumed that the reader is somewhat familiar with the modeling of electromechanical systems, and therefore a separate development on components is not included here. On the other hand, several of the examples and exercises begin with real, physical systems to acquaint the reader with the formulation of some of the more common control system elements.

If we take the Laplace transform of Eq. (2.2-1), assuming that all initial conditions are zero, we obtain

$$s^n Y(s) + a_{n-1}s^{n-1}Y(s) + \cdots + a_o Y(s) = c_m s^m U(s) + \cdots + c_o U(s)$$

or

$$\left(s^n + a_{n-1}s^{n-1} + \cdots + a_o\right) Y(s) = \left(c_m s^m + \cdots + c_o\right) U(s) \qquad (2.2\text{-}2)$$

The *transfer function* $G_p(s)$ of a plant is defined as the ratio of the Laplace transform of the output $Y(s)$ to the Laplace transform of the input $U(s)$, assuming that *all initial conditions are zero*. For the general example above, this becomes

$$\frac{Y(s)}{U(s)} = G_p(s) = \frac{c_m s^m + c_{m-1}s^{m-1} + \cdots + c_o}{s^n + a_{n-1}s^{n-1} + \cdots + a_1 s + a_o} \tag{2.2-3}$$

Note that the transfer function is the ratio of two polynomials in s, known as the numerator polynomial and the denominator polynomial. As we shall see in Chap. 4, the denominator polynomial plays a key role in determining the character of the behavior of the plant.

Several comments are in order at this time. First, the transfer function $G_p(s)$ completely characterizes the plant, since it contains all the information concerning the coefficients of Eq. (2.2-1), the original differential equation describing the plant. In other words, given the transfer function, it is possible to reconstruct the differential equation description. In fact, this reconstruction may be done by inspection, by cross-multiplying and letting $s^k Y(s)$ be replaced by the kth derivative of $y(t)$.

Second, the transfer function is dependent only on the plant and not on the input or the initial conditions. The input does not enter into the transfer function, since the transfer function is defined as the ratio of the Laplace transform of the output to the Laplace transform of the input. Note that the input does not appear on the right-hand side of Eq. (2.2-3). The statement that $G_p(s)$ is independent of the initial conditions is somewhat misleading since the definition of a transfer function requires that all initial conditions are zero. If the initial conditions are not zero, one must either return to the differential equation description (Eq. (2.2-1)) or make use of the state-variable representation discussed later in this chapter. Since many properties of interest are completely determined by the plant, the lack of initial-condition information is not a serious problem, and transfer functions find a wide range of use. The reader is reminded, however, never to forget the basic assumption of zero initial conditions whenever working with transfer functions.

The transfer function of Eq. (2.2-3) is said to have n poles and m zeros. The reason for this statement is more obvious if the transfer function $G_p(s)$ is represented in factored form as

$$G_p(s) = c_m \frac{(s+\delta_1)(s+\delta_2)\cdots(s+\delta_m)}{(s+\lambda_1)(s+\lambda_2)\cdots(s+\lambda_n)} = \frac{K_p N_p(s)}{D_p(s)}$$

where

$$N_p(s) = (s+\delta_1)(s+\delta_2)\cdots(s+\delta_m)$$

$$D_p(s) = (s+\lambda_1)(s+\lambda_2)\cdots(s+\lambda_m)$$

$$K_p = c_m$$

Here c_m is changed to K_p to emphasize that this is a gain inherently associated with the plant. The m values of s, namely, $-\delta_1, -\delta_2, \cdots, -\delta_m$, which make the numerator polynomial of $G_p(s)$ zero, are known as the *zeros* of $G_p(s)$. The n values of s, namely, $-\lambda_1, -\lambda_2, \cdots, -\lambda_n$, which make the denominator polynomial of $G_p(s)$ zero, or make the overall transfer function infinite, are known as the *poles* of $G_p(s)$. Hence, we say that the transfer function has n poles and m zeros.

A transfer function that has strictly more poles than zeros is called *strictly proper*. It can be argued that no physical system can react instantaneously to a change in input although certain electrical systems can react very quickly. A system that does not react instantaneously can be modeled by a strictly proper transfer function. We usually assume that physical systems are modeled by strictly proper transfer functions. We allow electrical components of a controller to be modeled with a transfer function that is simply *proper*, that is, a transfer function with the same number of zeros as poles. No physically realizable system will be modeled with a transfer function with a denominator of lesser degree than its numerator, i.e., an *improper* transfer function.

2.2.2 Block Diagrams

It is often helpful to make a pictorial representation of a transfer function using a technique known as a *block diagram*. Block diagrams of the plant described by the transfer function in Eq. (2.2-3) are shown in Fig. 2.2-2. There we see that the quantity contained in a block is the transfer function relating the input and output of the block. It is common practice to assume that a block diagram represents information transmission in one direction only, from the input to the output, as indicated by the arrows in Fig. 2.2-2. In other words, if the output $Y(s)$ is forced to have some behavior, from a block diagram the input is assumed to be unaffected. On the other hand, if the input is known, the output may be found by making use of the transfer function.

Although the block diagrams of Fig. 2.2-2 are nothing more than pictorial representations of Eq. (2.2-3), the block-diagram representation often proves useful in gaining a clearer understanding of a control problem by making it possible to visualize the interrelationships of the various parts of the problem.

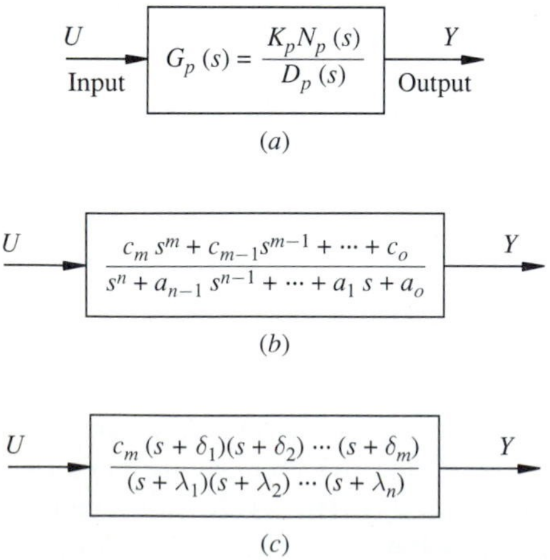

FIGURE 2.2-2
Block diagrams. (*a*) Compact form; (*b*) expanded form; (*c*) factored or pole-zero form.

At this point, the reader is cautioned against the following pitfall. Although the equation

$$\frac{Y(s)}{U(s)} = G_p(s) \tag{2.2-4}$$

is correct and unambiguous, the "analogous" time domain statement, namely,

$$\frac{y(t)}{u(t)} = G_p(t)$$

is completely meaningless. Clearly the two are not analogous statements. The input-output relationship as expressed by the transfer-function relationship of Eq. (2.2-4) has meaning only for transformed quantities. This is, in fact, the basis for the definition of a transfer function: a ratio of the Laplace transform of the output to the Laplace transform of the input, with zero initial conditions.

Hence we see that the describing differential equation (2.2-1), the transfer function (2.2-3), and the block diagrams of Fig. 2.2-2 are entirely equivalent methods of representation. One may, by inspection, determine from any one of the representations the others. These forms of representation constitute what we have loosely referred to as the *input-output method* of plant representation. To illustrate these various methods of plant representation, two examples are presented.

Example 2.2-1. As the first example, let us consider the simple electrical circuit shown in Fig. 2.2-3. If we write a Kirchhoff loop equation for this circuit, we find that

$$L\frac{di(t)}{dt} + (R_1 + R_2)\,i(t) = u(t)$$

The output voltage $y(t)$ is given by

$$y(t) = R_2 i(t)$$

If these two equations are combined and $i(t)$ is suppressed, the resulting differential equation is[1]

$$\frac{L}{R_2}\dot{y}(t) + \frac{R_1 + R_2}{R_2}y(t) = u(t)$$

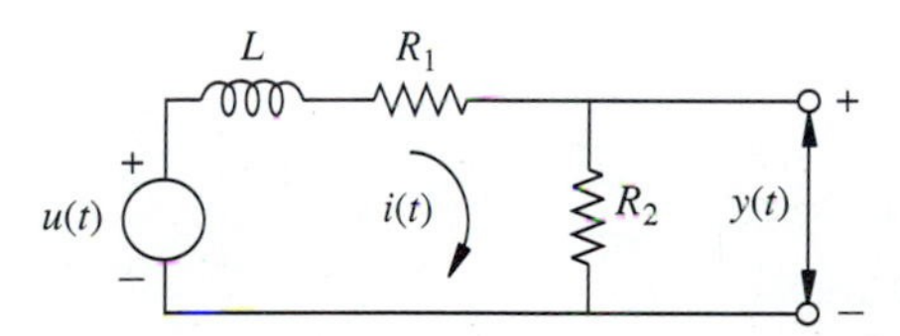

FIGURE 2.2-3
Simple electrical circuit.

[1]The dot convention for time derivatives is used whenever the order of the differentiation is low. Hence $\dot{y}(t) = dy(t)/dt$ and $\ddot{y}(t) = d^2y(t)/dt^2$, etc.

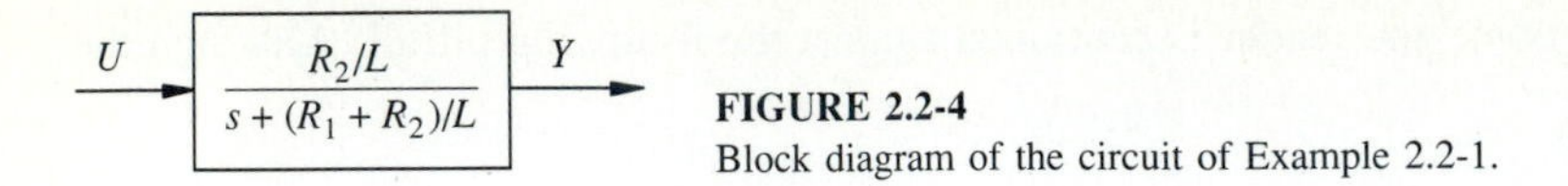

FIGURE 2.2-4
Block diagram of the circuit of Example 2.2-1.

Then the transfer function for this circuit is determined by taking the Laplace transform of this equation (assuming zero initial conditions) and solving for the ratio of $Y(s)$ to $U(s)$, with the result

$$G_p(s) = \frac{Y(s)}{U(s)} = \frac{R_2}{Ls + (R_1 + R_2)} = \frac{R_2/L}{s + (R_1 + R_2)/L}$$

The block diagram for this system is shown in Fig. 2.2-4.

Note that in this example a cause-and-effect relationship is indicated by the arrows in the block diagram. Here $u(t)$ is the input and $y(t)$ is the output. If a current were given as flowing in the circuit, we could not predict the effect on $u(t)$. This is true in general. The block diagram presupposes the input as indicated, with the flow of information as shown by the arrow.

Example 2.2-2. A dc motor may be used to position the angular attitude of a platform. The platform may be pointing a telescope at a star, an antenna at a satellite, or a gun at a target. Since the motor in these cases provides the power to actuate the desired response, the motor in this role may be called the actuator. There are two ways to configure a dc motor. In this example (see Fig. 2.2-5), a field-controlled dc motor is examined. The armature current is assumed to be constant. The input is the applied field voltage $e(t)$, the output is the shaft's angular position $\theta_o(t)$, and

$$\begin{aligned} i_f(t) &= \text{field current} \\ I_a &= \text{constant armature current} \\ \tau(t) &= \text{torque} \\ R_f &= \text{field resistance} \\ R_a &= \text{armature resistance} \\ L_f &= \text{field inductance} \\ \beta &= \text{viscous damping coefficient} \\ J &= \text{moment of inertia of motor and load} \\ K_\tau &= \text{torque constant} \end{aligned}$$

The Kirchhoff equation for the field circuit is

$$L_f \dot{i}_f(f) + R_f i_f(t) = e(t)$$

The Newtonian equation for the mechanical load is

$$J\ddot{\theta}_o(t) + \beta\dot{\theta}_o(t) = \tau(t)$$

and the torque field-current relation is

$$\tau(t) = K_\tau i_f(t)$$

since the armature current is constant.

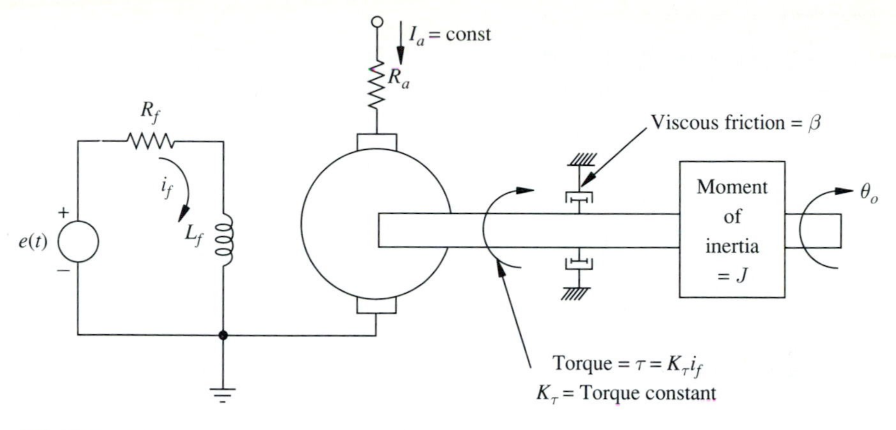

FIGURE 2.2-5
Schematic diagram of a field-controlled dc motor.

Let us transform each of these three equations to obtain

$$\left(L_f s + R_f\right) I_f(s) = E(s)$$

$$\left(Js^2 + \beta s\right) \Theta_o(s) = T(s)$$

and

$$T(s) = K_\tau I_f(s)$$

These three algebraic frequency domain equations can be solved simultaneously to obtain the desired transfer function $\Theta_o(s)/E(s)$

$$\frac{\Theta_o(s)}{E(s)} = \frac{K_\tau}{\left(Js^2 + \beta s\right)\left(L_f s + R_f\right)} = \frac{K_\tau/JL_f}{s\,(s + \beta/J)\left(s + R_f/L_f\right)}$$

or

$$\frac{\Theta_o(s)}{E(s)} = \frac{K_\tau}{JL_f s^3 + \left(L_f\beta + R_f J\right)s^2 + \beta R_f s}$$

The describing differential equation can now be written by inspection as

$$JL_f\dddot{\theta}_o(t) + \left(L_f\beta + R_f J\right)\ddot{\theta}_o(t) + \beta R_f\dot{\theta}_o(t) = K_\tau e(t)$$

or the block diagram can be drawn in either polynomial or pole-zero form as shown in Fig. 2.2-6.

2.2.3 Block-Diagram Algebra

Closely associated with the use of block diagrams to represent the input-output characteristics of a plant is a set of procedures for block-diagram manipulation commonly referred to as block-diagram algebra. In essence these manipulations are nothing more than a graphical procedure for manipulating algebraic equations, such as the ones determined in Example 2.2-2. Consequently, these procedures are often helpful in

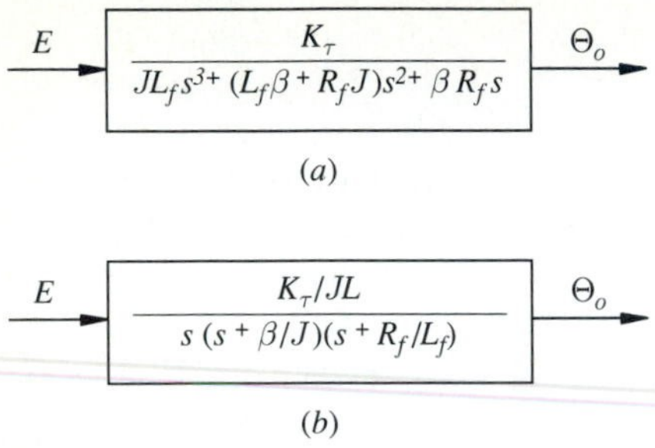

$E \rightarrow \boxed{\dfrac{K_T}{JL_f s^3 + (L_f\beta + R_f J)s^2 + \beta R_f s}} \rightarrow \Theta_o$

(*a*)

$E \rightarrow \boxed{\dfrac{K_T/JL}{s(s+\beta/J)(s+R_f/L_f)}} \rightarrow \Theta_o$

(*b*)

FIGURE 2.2-6
Block diagram of the field-controlled dc motor. (*a*) Polynomial form; (*b*) pole-zero form.

finding the transfer function of a complex plant by combining the block diagrams of various parts of the plant to find the overall input-output block diagram and hence the transfer function. In addition, this block-diagram algebra may be used in the design and analysis phases of a control problem to arrange the system into some particularly advantageous form.

The simplest block-diagram manipulation involves the reduction of two blocks in series or cascade, as shown in Fig. 2.2-7*a*, to one overall block, as shown in Fig. 2.2-7*b*. To verify this reduction, we simply write the two algebraic frequency domain equations represented by the two blocks

$$Y(s) = G_2(s)X(s) \qquad X(s) = G_1(s)U(s)$$

and then combine them so that

$$Y(s) = G_2(s)X(s) = G_2(s)G_1(s)U(s) = G_1(s)G_2(s)U(s)$$

The input-output transfer function becomes

$$\frac{Y(s)}{U(s)} = G_1(s)G_2(s)$$

which is the desired result.

It should be noted that the order in which the two blocks occur does not alter the input-output result, so that the series combination of Fig. 2.2-7*c* also reduces to the single block of Fig. 2.2-7*b*. However, although the order of the blocks is unimportant

$U \rightarrow \boxed{G_1(s)} \xrightarrow{X} \boxed{G_2(s)} \rightarrow Y$

(*a*)

$U \rightarrow \boxed{G_1(s)G_2(s)} \rightarrow Y$

(*b*)

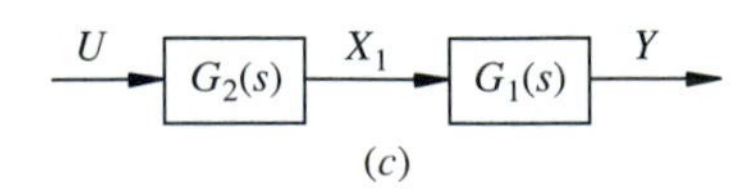

(*c*)

FIGURE 2.2-7
Block diagram manipulation of two blocks in series. (*a*) Two transfer functions in cascade; (*b*) and (*c*) equivalent input-output forms.

from an input-output view, it does affect the internal variable found between the blocks. This important distinction is discussed in detail in Sec. 2.3 on state-variable representation.

This block-diagram reduction may also be reversed in the sense that one may use it to divide one block into two blocks in series. In other words, the block-diagram equivalents indicated in Fig. 2.2-7 are reciprocal, and one may proceed from any one to any other.

It is obvious that by considering groups of two blocks at a time any number of blocks in series may be reduced to a single block. Similarly, any single block may be broken into any number of blocks in series.

In addition to this simple block-diagram identity, there are many others. Some of the more useful ones are summarized in Fig. 2.2-8. In addition to the familiar diagrams, this figure also contains a new symbol known as the error detector, the summing junction, or just the summer. Its properties are displayed in Fig. 2.2-9. This element plays an important role in almost all practical control systems.

All the identities shown in Fig. 2.2-8 may be verified by writing the expressions for the outputs of the two plants. Consider the identity of Fig. 2.2-8*f*, for example. This particular interconnection of blocks is often called the *feedback interconnection*. It is of fundamental importance in the study of control systems. The output of the left-hand representation is

$$Y(s) = G(s)U(s) \pm G(s)H(s)Y(s)$$

or

$$[1 \mp G(s)H(s)]Y(s) = G(s)U(s)$$

The output is then

$$Y(s) = \frac{G(s)}{1 \mp G(s)H(s)} U(s)$$

which is seen to be equivalent to the output of the representation on the right. Verification of the remaining identities is left to the reader as an exercise (see Exercise 2.2-1).

Since extensive use is made of this block-diagram algebra throughout this book, the reader is urged to study Fig. 2.2-8 carefully.

Example 2.2-3. To illustrate the usefulness of the block-diagram identities of Fig. 2.2-8 in the representation of a plant, let us consider once again the field-controlled dc motor of Example 2.2-2. In that example three algebraic transformed equations were found for various portions of the plant. Those equations are repeated here for convenience:

$$\left(L_f s + R_f\right) I_f(s) = E(s)$$

$$\left(Js^2 + \beta s\right) \Theta_o(s) = T(s)$$

and

$$T(s) = K_\tau I_f(s)$$

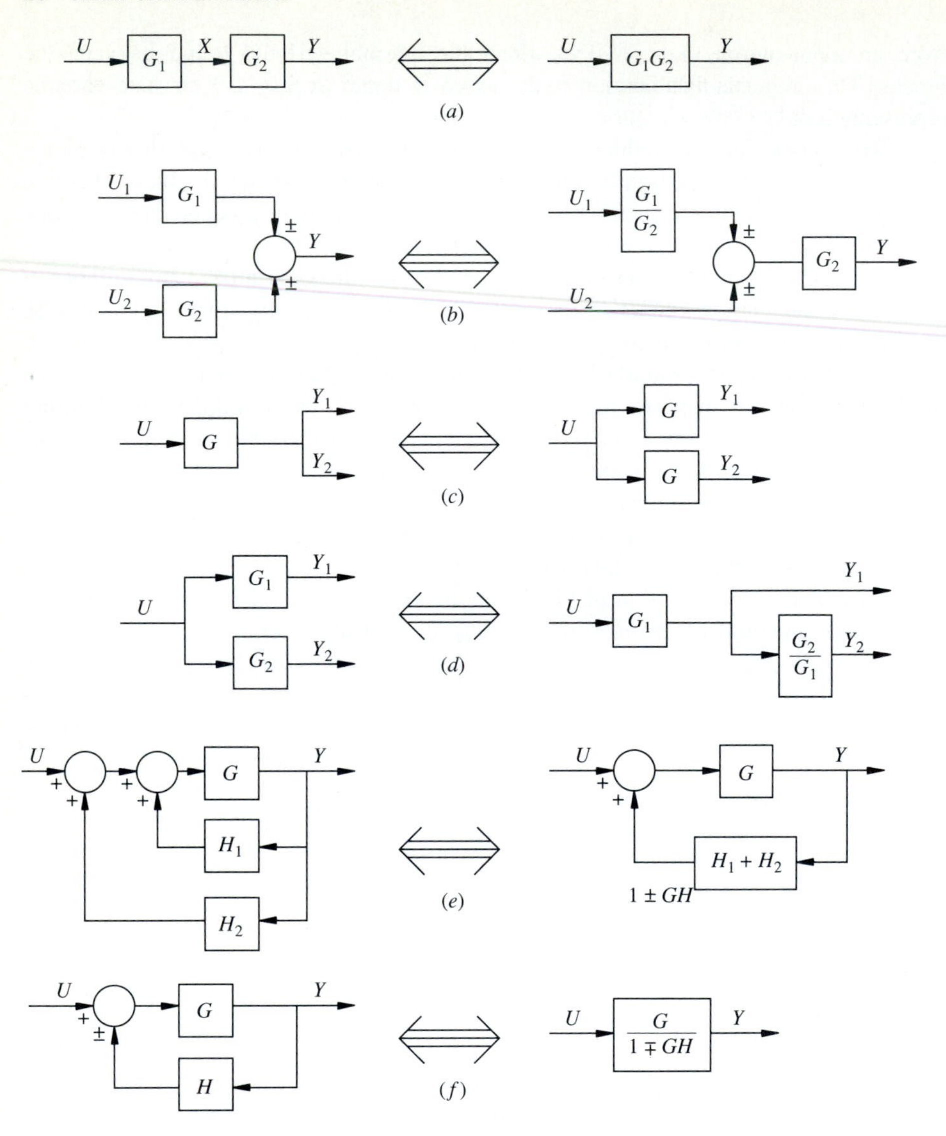

FIGURE 2.2-8
Block-diagram identities.

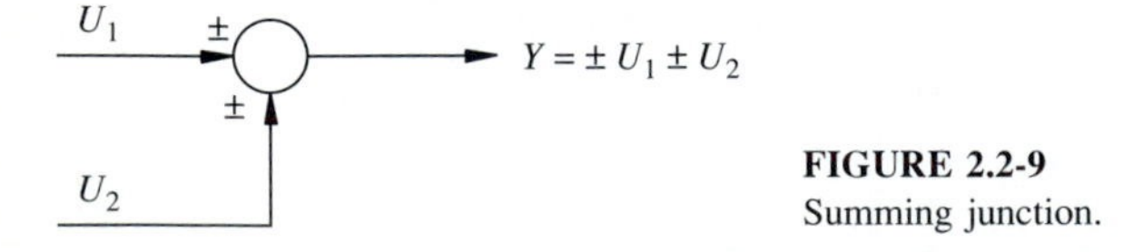

FIGURE 2.2-9
Summing junction.

These three equations may be represented by three block diagrams, as shown in Fig. 2.2-10*a*. If the equivalent input and output variables of these three blocks are connected, the block diagram shown in Fig. 2.2-10*b* results. One may now make use of the series-combination rule of Fig. 2.2-8*a* to obtain the block diagram shown in Fig. 2.2-10*c*. Note that in this final block diagram the electrical and mechanical portions of the plant may still be easily separated.

Example 2.2-4. As a second example, let us consider the representation of the armature-controlled dc motor shown schematically in Fig. 2.2-11. Again the load is inertia plus friction; in this case, the field current is held constant. (This is often a much easier task to accomplish in practice than holding the armature current constant.)

Kirchhoff's equation for the armature circuit is

$$(L_a s + R_a)\, I_a(s) = E(s) - E_c(s)$$

where $E_c(s)$ is the back emf of the motor:

$$E_c(s) = K_v s \Theta_o(s)$$

In the mechanical portion of the plant we have

$$\left(Js^2 + \beta s\right) \Theta_o(s) = T(s)$$

and

$$T(s) = K'_\tau I_a(s)$$

If each of these four equations is represented by a block diagram, we have the result shown in Fig. 2.2-12. As the next step, we again connect the equivalent variables as shown in Fig. 2.2-13*a*. This result may be further reduced to the final form shown in Fig. 2.2-13*c* by making use of the identities given in Fig. 2.2-8*a* and *f*. From this final block diagram, we can easily write either the transfer function of the plant or the describing differential equation.

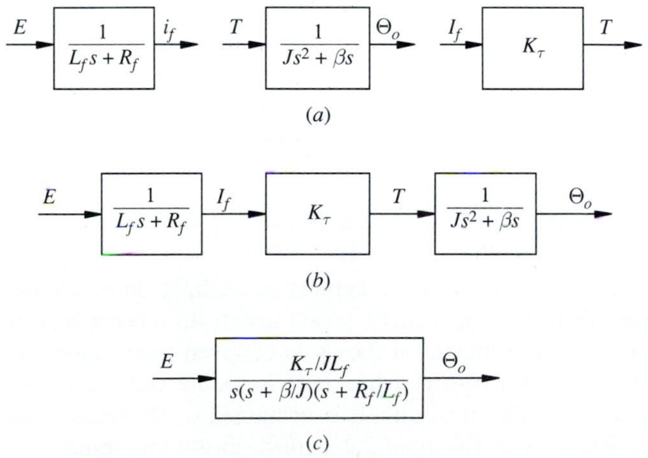

(*a*)

(*b*)

(*c*)

FIGURE 2.2-10
Use of Block-diagram algebra to find the transfer function for the field-controlled motor. (*a*) Subsystems; (*b*) connected subsystems; (*c*) overall block diagram.

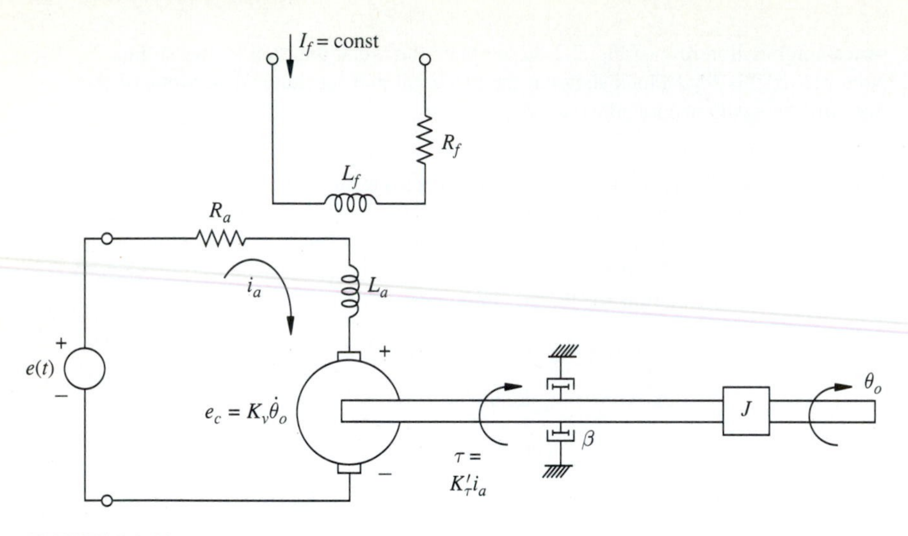

FIGURE 2.2-11
Schematic diagram of an armature-controlled dc motor.

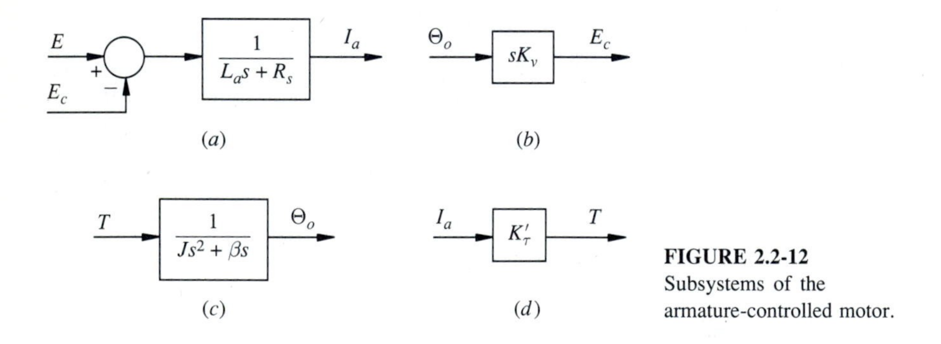

FIGURE 2.2-12
Subsystems of the armature-controlled motor.

If the final block diagram of Fig. 2.2-13*c* is compared with the final block diagram of the field-controlled dc motor (Fig. 2.2-10*c*), it is seen that in the present case the electrical and mechanical portions of the plant cannot be easily separated. In practice the fact that the electrical and mechanical portions of the plant cannot be separated is not important and both field-controlled and armature-controlled motors are used in applications.

To be able to connect the subsystem block diagrams as we have done in these two examples, it is necessary that the "impedance" levels match. In other words, if the subsystem block diagrams are determined on the basis of given termination and input conditions, these conditions must be met when the blocks are connected. For example, if the block diagram of an electrical circuit is computed on the basis of an open-circuit termination, as was done in Example 2.2-1, then, to use this result, it is necessary to ensure that the termination is effectively an open circuit.

Before leaving this section it should be reemphasized that all the block-diagram identities presented in Fig. 2.2-8 are valid only in an input-output sense. Because of

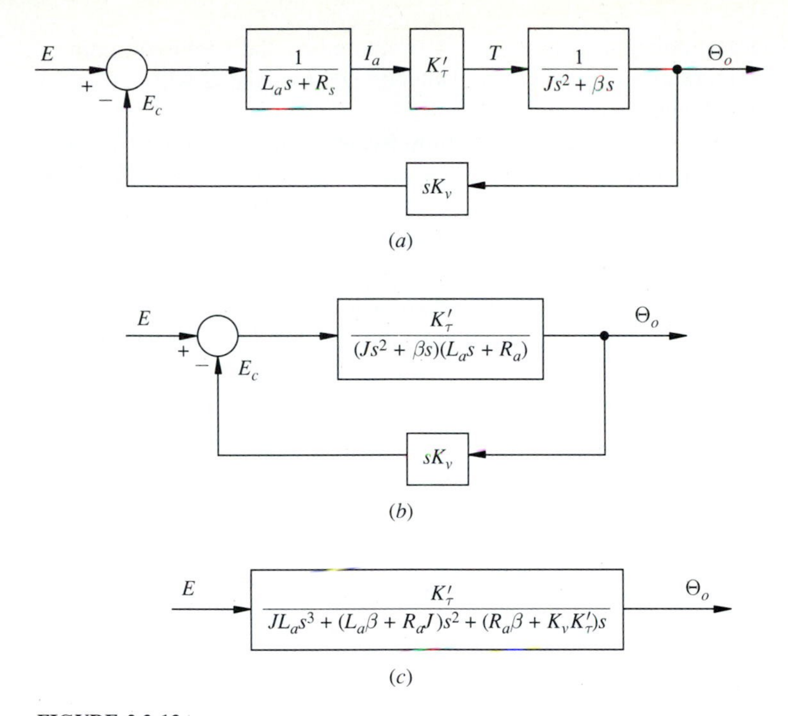

FIGURE 2.2-13
Block diagrams of the armature-controlled motor. (*a*) Connected subsystems; (*b*) first reduction; (*c*) final block diagram.

this, internal variables may be and often are distorted or even obliterated. Consider, for example, the three block-diagram representations of the armature-controlled motor shown in Fig. 2.2-13. In Fig. 2.2-13*a*, we can identify three meaningful physical, internal variables: I_a, E_c, and T. In Fig. 2.2-13*c*, however, none of these variables appear.

This complete lack of regard for internal variables is a characteristic of the input-output representation of a plant. By its very name, this method is labeled as being strongly related to input-output properties. Such an approach is often sufficient, but there are also many situations in which a more complete description including the internal behavior of the plant is necessary. In these cases, one may use the state-variable representation.

2.2.4 CAD Notes[1]

MATLAB®, like many other CAD programs, manipulates transfer functions by manipulating polynomials. A polynomial is represented by a vector comprising the polynomial coefficients starting with the coefficient of the highest power of s and working

[1]See Appendix D for an introduction to the *CAD Notes* sections of this text. MATLAB is a registered trademark of The MathWorks, Inc.

down. For example, the polynomial $s^2 + 2s + 4$ is entered by the vector assignment `poly=[1 2 4]`. Numerator and denominator polynomials associated with a transfer function are maintained separately.

Two transfer functions are multiplied by multiplying the numerator and denominator polynomials separately. Polynomials `poly1` and `poly2` are multiplied to form `poly3` with the statement `poly3=conv(poly1,poly2)`. The name `conv` may seem strange. It arises as follows. When two blocks of polynomials in s are placed in series the blocks are combined by multiplying the two polynomials or equivalently by convolving the two associated impulse responses. The name `conv` arises from the convolution interpretation. Roots of the polynomial `poly3` can be found and placed in the vector `rp3` by using the statement `rp3=roots1(poly3)`.

Example 2.2-5. Suppose

$$G_1(s) = \frac{s+3}{(s+1)(s+4)}$$

$$G_2(s) = \frac{4(s+2)}{\left(s^2+4s+5\right)}$$

Find $G(s) = G_1(s)G_2(s)$

```
>>g1num=[1 3];
g1den1=[1 1];
g1den2=[1 4];
g1den=conv(g1den1,g1den2);
g2num=4*[1 2];
g2den=[1 4 5];
gnum=conv(g1num,g2num)
gnum=
  4  20  24
gden=conv(g1den,g2den)
gden =
  1  9  29  41  20
zeros=roots1(gnum)
zeros =
  -2.0000
  -3.0000
poles=roots1(gden)
poles =
 -1.0000 + 0.0000i
 -2.0000 - 1.0000i
 -2.0000 + 1.0000i
 -4.0000 + 0.0000i
```

The answer is

$$G(s) = \frac{4s^2 + 20s + 24}{s^4 + 9s^3 + 29s^2 + 41s + 20} = \frac{4(s+2)(s+3)}{(s+1)(s+4)\left(s^2+4s+5\right)}$$

Adding and subtracting transfer functions is a bit trickier since MATLAB works with polynomials so that adding two transfer functions must be accomplished with the formula

$$\frac{N_{G1}(s)}{D_{G1}(s)}+\frac{N_{G2}(s)}{D_{G2}(s)}=\frac{N_{G1}(s)D_{G2}(s)+N_{G2}(s)D_{G1}(s)}{D_{G1}(s)D_{G2}(s)}$$

In addition, since MATLAB adds polynomials by adding the vectors representing the polynomial, it requires that the vectors added be the same length. This often creates the need to pad some vectors with leading zeros to represent zero coefficients of high powers of s. The closed-loop transfer function of a negative feedback system as shown in Fig. 2.2-8*f* can be computed using the identity

$$\frac{G(s)}{1+G(s)H(s)}=\frac{N_G(s)D_H(s)}{D_G(s)D_H(s)+N_G(s)N_H(s)}$$

Example 2.2-5 cont. Find the closed-loop transfer function $G_{CL}(s)$ if the $G(s)$ given above is placed in feedback with $H(s)=1$. Find $G_{\text{sum}}(s)=G_1(s)+G_2(s)$.

```
%pad gnum with zeros so additions can be performed
  gnum=[0 0 gnum];
%find closed-loop num and den
  gclnum=gnum
gclnum =
  0  0  4  20  24
  gclden=gden+gnum
gclden =
  1  9  33  61  44
%find g1+g2
  gsumnum=conv(g1num,g2den)+conv(g2num,g1den)
gsumnum =
  5  35  73  47
  gsumden=conv(g1den,g2den)
gsumden =
  1  9  29  41  20
```

The answers are found by reading off the polynomials:

$$G_{CL}(s)=\frac{4s^2+20s+24}{s^4+9s^3+33s^2+61s+44}$$

$$G_{\text{sum}}(s)=\frac{5s^3+35s^2+73s+47}{s^4+9s^3+29s^2+41s+20}$$

Exercises 2.2

2.2-1. Verify the block-diagram identities of Fig. 2.2-8 and use them to represent the plant shown in Fig. 2.2-14*a* in the form given in Fig. 2.2-14*b* and *c*.
Answers:

$$H(s)=4s+1 \qquad G(s)=\frac{5}{s(s+22)}$$

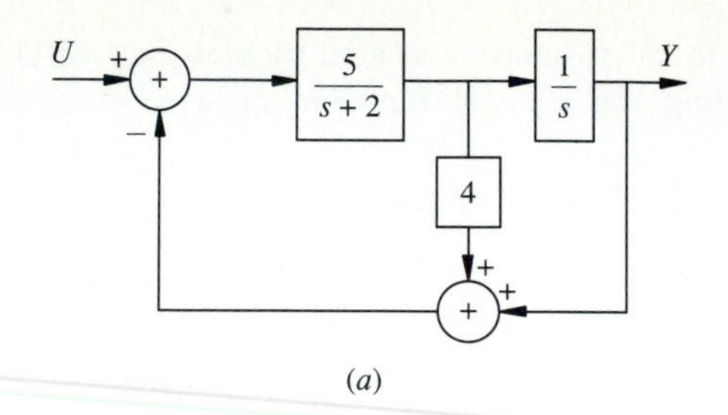

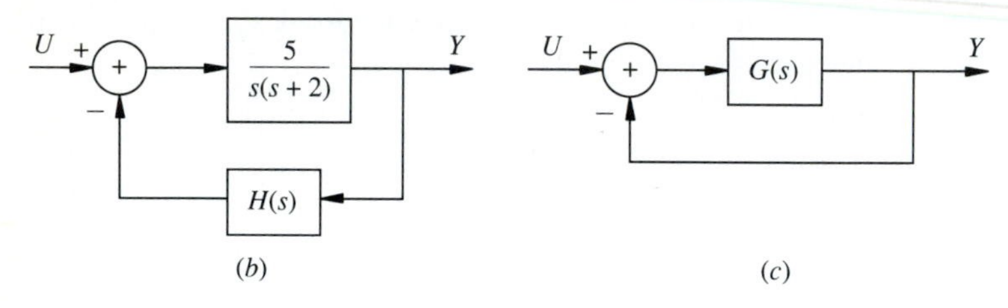

FIGURE 2.2-14
Exercise 2.2-1.

2.2-2. Starting from the block diagram of Fig. 2.2-15*a* use block diagram manipulations to find $P_1(s)$ and $G_{C1}(s)$ to make Fig. 2.2-15*b* equivalent to Fig. 2.2-15*a*. Find $P_2(s)$ and $H_2(s)$ to make Fig. 2.2-15*c* equivalent to Fig. 2.2-15*a*.

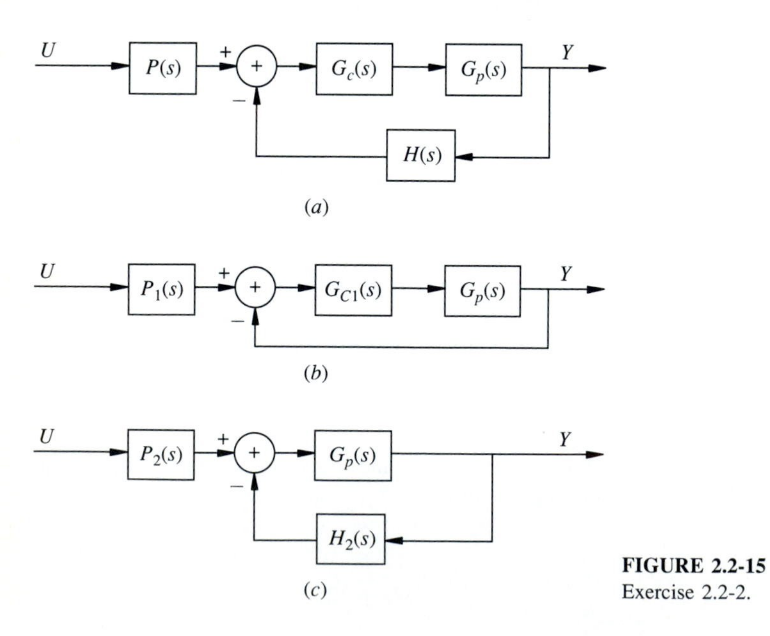

FIGURE 2.2-15
Exercise 2.2-2.

Answers:

$$G_{C1}(s) = H(s)G_C(s) \qquad P_1(s) = P(s)/H(s)$$

$$H_2(s) = G_C(s)H(s) \qquad P_2(s) = P(s)G_C(s)$$

2.2-3. For the plant shown in Fig. 2.2-16 show that $Y(s)/U(s)$ is

$$\frac{Y(s)}{U(s)} = \frac{G_1G_2G_3}{H_1G_2G_3 + H_3G_1G_2G_3 + G_1G_2H_2 + 1}$$

2.2-4. The plant shown in Fig. 2.2-17 consists of a field-controlled motor and a constant-field generator. The motor's torque constant is K_τ newtons-m/amp. The generator constant is K_v volts/rad/sec. Assume that the torque required to turn the generator under load τ_g is $K_v i_{ag}$. The connecting shaft has an inertia of J and viscous friction β. Draw a block diagram for each equation in the following set of differential equations. Connect its input and output as given. Connect the blocks to form a block diagram for the entire plant; find $Y(s)/U(s)$.

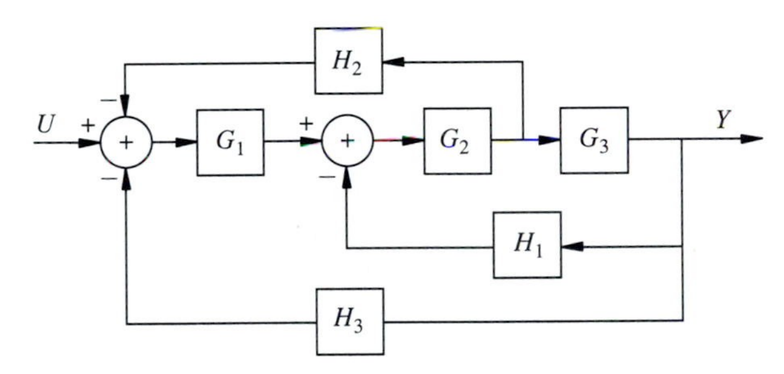

FIGURE 2.2-16
Exercise 2.2-3.

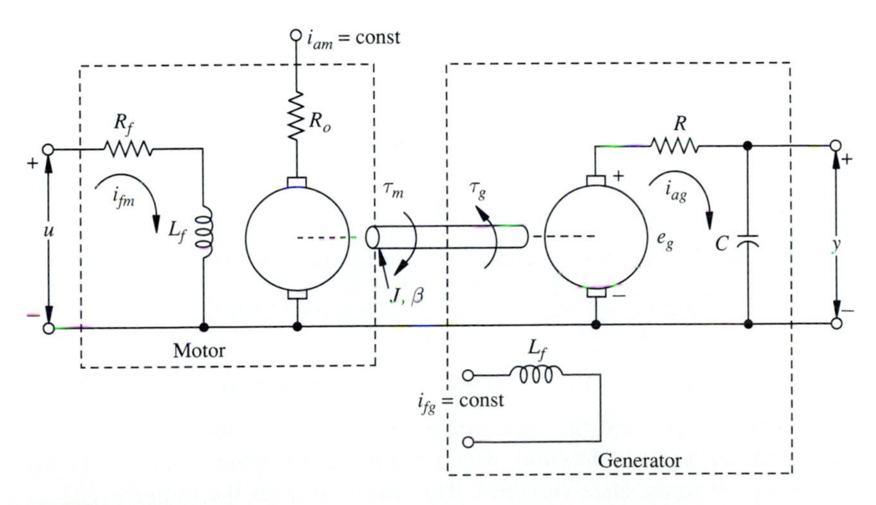

FIGURE 2.2-17
Exercise 2.2-4.

Equation	Input	Output
$L_f \dfrac{di_{fm}(t)}{dt} + R_f i_{fm}(t) = u(t)$	U	I_{fm}
$\tau_m(t) = K_\tau i_{fm}(t)$	I_{fm}	T_m
$J \dfrac{d^2\theta(t)}{dt^2} + \beta \dfrac{d\theta(t)}{dt} = \tau_m(t) - \tau_g(t)$	$T_m - T_g$	$s\Theta$
$e_g(t) = K_v \dfrac{d\,\theta(t)}{dt}$	$s\Theta$	E_g
$e_g(t) = RC \dfrac{dy(t)}{dt} + y(t)$	E_g	Y
$i_{ag}(t) = C \dfrac{dy(t)}{dt}$	Y	I_{ag}
$\tau_g(t) = K_g i_{ag}(t)$	I_{ag}	T_g

Answer:

$$\frac{Y(s)}{U(s)} = \frac{K_\tau K_v}{\left(sL_f + R_f\right)\left[(Js + \beta)(sRC + 1) + sK_v K_g C\right]}$$

2.3 STATE-VARIABLE REPRESENTATION

The state-variable method of plant representation is concerned not only with the input-output properties of the plant but also with its *complete internal behavior*. This is the feature that distinguishes the state-variable representation from the input-output representation.

2.3.1 State-Variable Notation

We begin our study of the state-variable representation of a plant by establishing the standard notation used in this approach. A more complete discussion of the equations introduced here begins in Sec. 2.3.2. For the moment, be content to become familiar with the notation.

Plant representation in terms of state-variables is accomplished by means of a set of n linear first-order differential equations known as the *plant equations*. These equations take the general form

$$\begin{aligned}
\dot{x}_1(t) &= a_{11}x_1(t) + a_{12}x_2(t) + \cdots + a_{1n}x_n(t) + b_1 u(t) \\
\dot{x}_2(t) &= a_{21}x_1(t) + a_{22}x_2(t) + \cdots + a_{2n}x_n(t) + b_2 u(t) \\
&\vdots \\
\dot{x}_n(t) &= a_{n1}x_1(t) + a_{n2}x_2(t) + \cdots + a_{nn}x_n(t) + b_n u(t)
\end{aligned} \tag{2.3-1}$$

where the n variables $x_i(t)$ are the *state-variables* and $u(t)$ is the *plant input*. In addition, one needs an *output expression*, which is a linear algebraic equation relating the output of the plant to the state-variables. This equation takes the form

$$y(t) = c_1 x_1(t) + c_2 x_2(t) + \cdots + c_n x_n(t) \tag{2.3-2}$$

The use of matrix notation permits the plant equations and the output expression to be written in a particularly simple and convenient form. The plant equation may be written as

$$\dot{\boldsymbol{x}}(t) = \boldsymbol{A}\boldsymbol{x}(t) + \boldsymbol{b}u(t) \tag{$\boldsymbol{Ab}$}$$

Here $\boldsymbol{x}(t)$ is the n-dimensional vector known as the *state vector*; its elements are called *state-variables*.

$$\boldsymbol{x}(t) = \text{col}\,(x_1(t), x_2(t), \ldots, x_n(t)) \tag{2.3-3}$$

The vector $\dot{\boldsymbol{x}}(t)$ is the derivative of the state vector, and its elements are therefore the derivatives of the state-variables, or

$$\begin{aligned}\dot{\boldsymbol{x}}(t) &= \frac{d}{dt}\boldsymbol{x}(t) = \frac{d}{dt}\text{col}\,(x_1(t), x_2(t), \ldots, x_n(t)) \\ &= \text{col}\,(\dot{x}_1(t), \dot{x}_2(t), \ldots, \dot{x}_n(t))\end{aligned}$$

The $n \times n$ matrix $\boldsymbol{A}$, which is defined as usual by

$$\boldsymbol{A} = \begin{bmatrix} a_{11} & a_{12} & \cdots & a_{1n} \\ a_{21} & a_{22} & \cdots & a_{2n} \\ \vdots & \vdots & \ddots & \vdots \\ a_{n1} & a_{n2} & \cdots & a_{nn} \end{bmatrix}$$

is known as the *plant matrix*. The n-dimensional vector

$$\boldsymbol{b} = \text{col}\,(b_1, b_2, \ldots, b_n)$$

is referred to as the *input vector* since it describes how the control input $u(t)$ affects the plant. If all these results are combined, the expanded form of Eq. ($\boldsymbol{Ab}$) becomes

$$\begin{bmatrix} \dot{x}_1(t) \\ \dot{x}_2(t) \\ \vdots \\ \dot{x}_n(t) \end{bmatrix} = \begin{bmatrix} a_{11} & a_{12} & \cdots & a_{1n} \\ a_{21} & a_{22} & \cdots & a_{2n} \\ \vdots & \vdots & \ddots & \vdots \\ a_{n1} & a_{n2} & \cdots & a_{nn} \end{bmatrix} \begin{bmatrix} x_1(t) \\ x_2(t) \\ \vdots \\ x_n(t) \end{bmatrix} + \begin{bmatrix} b_1 \\ b_2 \\ \vdots \\ b_n \end{bmatrix} u(t) \tag{$\boldsymbol{Ab}$}$$

which, if the indicated matrix operations are completed, is equal to the initial plant equations (2.3-1).

In a similar fashion, the output expression may be written as

$$y(t) = \boldsymbol{c}^T\boldsymbol{x}(t) \tag{$\boldsymbol{c}$}$$

The n-dimensional vector $\boldsymbol{c} = \text{col}(c_1, c_2, \ldots, c_n)$ is known as the *output vector*. In expanded form this equation becomes

$$y(t) = [\,c_1 \quad c_2 \quad \cdots \quad c_n\,] \begin{bmatrix} x_1(t) \\ x_2(t) \\ \vdots \\ x_n(t) \end{bmatrix} = c_1x_1(t) + c_2x_2(t) + \cdots + c_nx_n(t) \tag{$\boldsymbol{c}$}$$

Once again we see that this result is identical with the output expression of Eq. (2.3-2). Often we let $y = x_1$ so that all the elements of $\boldsymbol{c}$ are zero except c_1, which is equal to 1.

Since Eqs. (**Ab**) and (**c**) play fundamental roles and therefore appear frequently in this book, they are designated by special and common symbols for the convenience of the reader. In addition, the equation symbols have been selected to assist the reader in remembering the form of the equation.

The use of the vector-matrix formulation therefore allows us to write the state-variable representation in terms of two simple equations

$$\dot{\boldsymbol{x}}(t) = \boldsymbol{A}\boldsymbol{x}(t) + \boldsymbol{b}u(t) \qquad (\boldsymbol{Ab})$$

and

$$y(t) = \boldsymbol{c}^T\boldsymbol{x}(t) \qquad (\boldsymbol{c})$$

The simplicity of these equations in comparison with the original form is obvious. Of course, these equations only symbolize Eqs. (2.3-1) and (2.3-2). Whenever an actual calculation is made for a specific plant, either the original Eqs. (2.3-1) and (2.3-2) or the expanded form of the above matrix equations must be used, since only they contain the detailed information regarding the given plant. On the other hand, the compact symbolic representation is useful for general derivations and discussions.

2.3.2 The Concept of State

As mentioned previously, the variables $x_1(t), x_2(t), \ldots, x_n(t)$ are known as state-variables. At any time t_0, their values, taken collectively, comprise the state of the plant.

It is well known that to solve such an nth-order differential equation one must know, in addition to the input for $t \geq t_0$, a set of n initial conditions. Because of the method of solution, these initial conditions are usually chosen as the value of the output y and its $n-1$ derivatives at time t_0, that is, $y(t_0), \dot{y}(t_0), \ldots, d^{n-1}y(t_0)/dt^{n-1}$. However, this is not the only set of initial conditions that may be used. The values at t_0 of any set of n linearly independent variables of the plant, $x_1(t_0), x_2(t_0), \ldots, x_n(t_0)$ are also sufficient. The meaning of linear independence will be discussed later.

In precise mathematical language, the definition of state takes the following form. The *state* at t_0 of an nth-order plant is described by a set of n numbers, $x_1(t_0), x_2(t_0), \ldots, x_n(t_0)$, which, along with the input to the plant for $t \geq t_0$, is sufficient to determine the behavior of the plant for all $t \geq t_0$.

In other words, the state of the plant represents a sufficient amount of information about the plant at t_0 to determine its future behavior without reference to the input before t_0. At the same time, the state of the plant at t_0 represents a complete description of the plant in the sense that no other information except the input is needed to determine its response. In addition, any other plant variables may be determined from a knowledge of the state.

Consider, for example, games of checkers or chess; although these are discrete processes and hence not exactly analogous to the processes we are discussing, they illustrate the concept of state. Here the state at the end of each move is simply the position of the pieces on the board. If the game is interrupted, one could resume play with this knowledge; it would not be necessary to know how the game had been played up to that point.

It may appear that the state of a plant at any time is not unique. This, however, is not true; only the method of representing the state information is not unique. Any set of state-variables we choose provides exactly the same information about the plant. This situation is similar to specifying the size of a physical object in different units or in a different coordinate system. Assuredly the units or the coordinate system used do not change the physical object, although it may alter the description of the object.

2.3.3 Block Diagrams for State Equations

The frequency-domain form of the state-variable representation of Eqs. (***Ab***) and (***c***) may be obtained by taking the Laplace transform of these equations. Since we will use the results to find a transfer function for the plant, let us assume that all initial conditions are zero. The transformation yields

$$s\boldsymbol{X}(s) = \boldsymbol{A}\boldsymbol{X}(s) + \boldsymbol{b}U(s) \tag{2.3-4}$$

$$Y(s) = \boldsymbol{c}^T\boldsymbol{X}(s) \tag{2.3-5}$$

For frequency domain form, we may draw a block diagram of the state-variable representation as shown in Fig. 2.3-1. This result may be compared with the above representation if both sides of Eq. (2.3-4) are divided by s so that

$$\boldsymbol{X}(s) = \frac{1}{s}\boldsymbol{I}\,[\boldsymbol{A}\boldsymbol{X}(s) + \boldsymbol{b}U(s)] \tag{2.3-6}$$

where $\boldsymbol{I}$ is the identity matrix which is needed to preserve the proper dimensioning. The broad arrows in Fig. 2.3-1 are used to distinguish vectorial quantities, such as $\boldsymbol{X}$, from scalars, such as U.

In addition to the general symbolic form of the block-diagram representation in Fig. 2.3-1, it is often useful to make a detailed block-diagram representation of the state-variable equations. Such a block diagram may be easily formed by beginning with n blocks containing the term $(1/s)$, each of whose input and output is labeled, respectively, sX_i and X_i, $i = 1, 2, \ldots, n$. These blocks are then connected by making use of the state-variable equations to generate each sX_i. The output is formed from the output equation as the last step. A block diagram of this form, that is, one containing

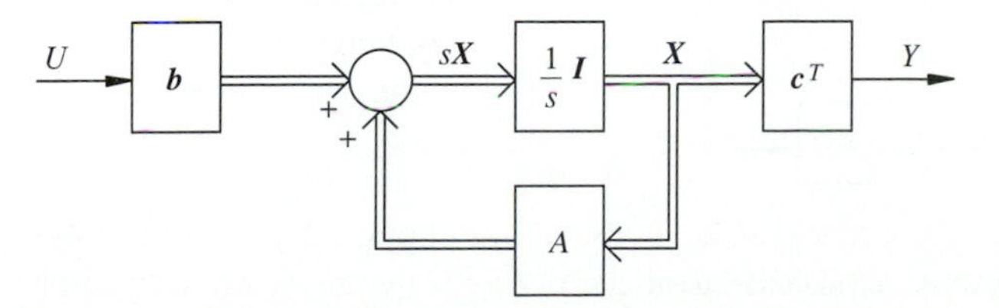

FIGURE 2.3-1
General block diagram for the state-variable representation.

n blocks labeled $(1/s)$, is referred to as the *elementary block diagram* of the state-variable representation.

As an illustration, Fig. 2.3-2 shows the elementary block diagram for the most general second-order plant. The state-variable representation of this plant is

$$\dot{\boldsymbol{x}} = \begin{bmatrix} a_{11} & a_{12} \\ a_{21} & a_{22} \end{bmatrix} \boldsymbol{x} + \begin{bmatrix} b_1 \\ b_2 \end{bmatrix} u$$

and

$$y = [\, c_1 \quad c_2 \,] \boldsymbol{x}$$

In most practical problems, many of the elements of $\boldsymbol{A}$, $\boldsymbol{b}$, and $\boldsymbol{c}$ are zero, and many of the interconnections are not needed.

The exact form of the plant representation, that is, the elements of $\boldsymbol{A}$, $\boldsymbol{b}$, and $\boldsymbol{c}$, depends on the set of state-variables chosen for the plant. Again it must be emphasized that the plant does not change, only its representation does. There is, in fact, an infinite number of state-variable representations for a given plant, depending on which set of state-variables has been selected for its representation.

Example 2.3-1. To illustrate the state-variable method of representation presented above, let us consider once again the field-controlled motor. We have seen the transfer func-

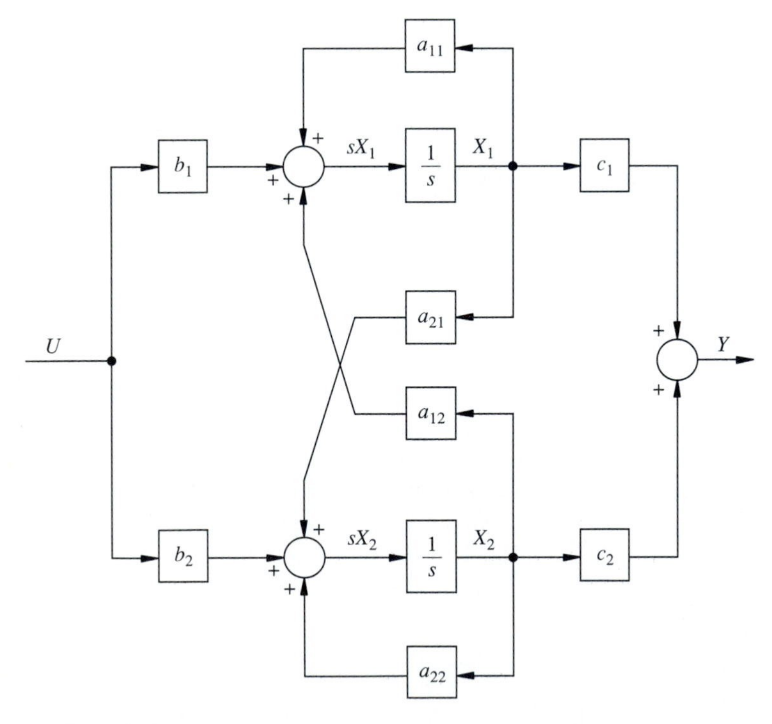

FIGURE 2.3-2
Elementary block diagram of the general second-order plant.

tion representation of this plant in Example 2.2-2. The choice of state-variables will be discussed in Secs. 2.4 and 2.5.

For now, let us consider the representation of this plant using the state-variables $\theta_o(t)$, $\dot{\theta}_o(t)$, and $i_f(t)$. Here we define

$$x_1(t) = \theta_o(t)$$

$$x_2(t) = \dot{\theta}_o(t)$$

$$x_3(t) = i_f(t)$$

$$u(t) = e(t)$$

From these definitions, one plant equation is known, namely,

$$\dot{x}_1(t) = x_2(t)$$

To find the other two plant equations, we must consider the two original equations determined for this plant. These are repeated here for reference as

$$L_f \frac{di_f(t)}{dt} + R_f i_f(t) = e(t)$$

and

$$J\ddot{\theta}_o(t) + \beta\dot{\theta}_o(t) = \tau(t)$$

Substituting the state-variable definitions into these equations yields

$$L_f \dot{x}_3(t) + R_f x_3(t) = u(t)$$

$$J\dot{x}_2(t) + \beta x_2(t) = \tau(t)$$

Since $\tau(t) = K_\tau i_f(t)$, the last equation becomes

$$J\dot{x}_2(t) + \beta x_2(t) = K_\tau x_3(t)$$

so that the three plant equations are

$$\dot{x}_1(t) = x_2(t)$$

$$\dot{x}_2(t) = -\frac{\beta}{J} x_2(t) + \frac{K_\tau}{J} x_3(t)$$

$$\dot{x}_3(t) = -\frac{R_f}{L_f} x_3(t) + \frac{1}{L_f} u(t)$$

The output expression is

$$y(t) = x_1(t)$$

The $\boldsymbol{A}$, $\boldsymbol{b}$, and $\boldsymbol{c}$ matrices are

$$\boldsymbol{A} = \begin{bmatrix} 0 & 1 & 0 \\ 0 & -\beta/J & K_\tau/J \\ 0 & 0 & -R_f/L_f \end{bmatrix} \qquad \boldsymbol{b} = \begin{bmatrix} 0 \\ 0 \\ 1/L_f \end{bmatrix} \qquad \boldsymbol{c} = \begin{bmatrix} 1 \\ 0 \\ 0 \end{bmatrix}$$

The elementary block diagram of this state-variable representation is shown in Fig. 2.3-3*a*. In Fig. 2.3-3*b*, some internal loops have been removed to make the result

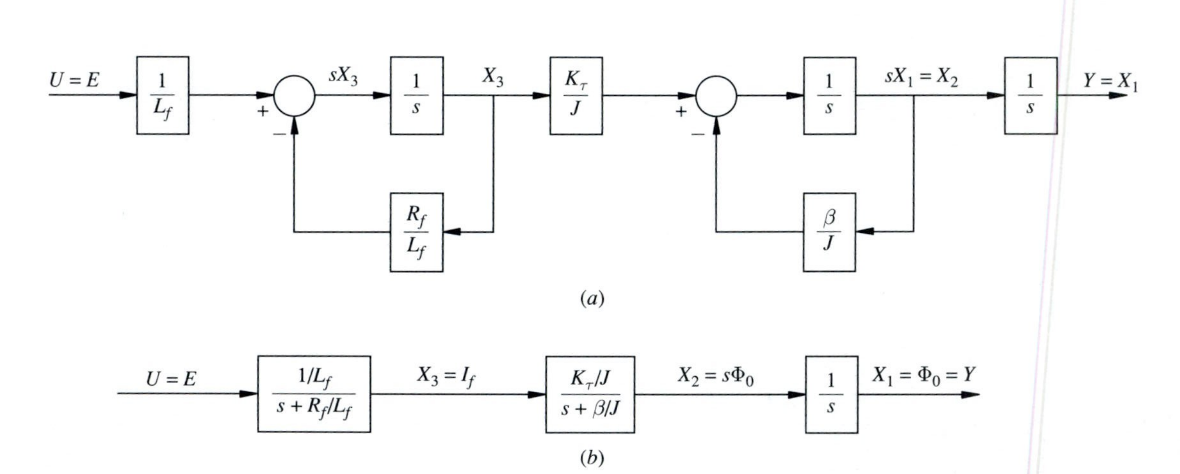

FIGURE 2.3-3
Block diagrams of the field-controlled motor, using the transforms of the state-variables $\theta_o(t)$, $\dot{\theta}_o(t)$ and $i_t(t)$. (*a*) Elementary block diagram; (*b*) block diagram with some internal loops eliminated.

more convenient. Note, however, that the state-variables are still clearly labeled and have not been suppressed.

2.3.4 Relation to Transfer Function

The question that must naturally arise in the mind of the reader is: How are the state-variable representation and the input-output representation related? In particular, one would like to know how to find the transfer function of a plant, given its state-variable representation, and how to find its state-variable representation, given its transfer-function representation.

Since the transfer-function representation specifies only the input-output behavior, one can always make an arbitrary selection of state-variables for a plant specified only by a transfer function. This means that, in general, an infinite number of state-variable representations exist for a given transfer function. The next two sections discuss two of the more common methods of choosing the state-variables, beginning with a transfer-function representation.

On the other hand, if a state-variable representation of a plant is known, the transfer function of the plant is completely and uniquely specified. This fact is just one more manifestation that the state-variable representation is a more complete description. To determine the transfer function of the plant $Y(s)/U(s)$ from the state-variable representation, we begin with the frequency domain form of the state-variable representation as given by Eqs. (2.3-4) and (2.3-5). These equations are repeated here for reference.

$$s\boldsymbol{X}(s) = \boldsymbol{A}\boldsymbol{X}(s) + \boldsymbol{b}U(s) \tag{2.3-4}$$

$$Y(s) = \boldsymbol{c}^T\boldsymbol{X}(s) \tag{2.3-5}$$

It must be remembered that, when writing these equations, we assume that all initial conditions are zero. This assumption is proper in this situation since we are searching for a transfer function.

Grouping the two $\boldsymbol{X}(s)$ terms in Eq. (2.3-4) we have

$$(s\boldsymbol{I} - \boldsymbol{A})\boldsymbol{X}(s) = \boldsymbol{b}U(s) \tag{2.3-7}$$

where the identity matrix has been introduced to maintain dimensionality and to allow the indicated factoring. If both sides of this equation are premultiplied by the matrix $(s\boldsymbol{I} - \boldsymbol{A})^{-1}$, then Eq. (2.3-7) becomes

$$\boldsymbol{X}(s) = (s\boldsymbol{I} - \boldsymbol{A})^{-1}\boldsymbol{b}U(s) \tag{2.3-8}$$

The matrix $(s\boldsymbol{I} - \boldsymbol{A})^{-1}$ is referred to as the *resolvent matrix* and is designated by $\boldsymbol{\Phi}(s)$

$$\boldsymbol{\Phi}(s) = (s\boldsymbol{I} - \boldsymbol{A})^{-1} \tag{2.3-9}$$

Using this definition, Eq. (2.3-8) becomes

$$\boldsymbol{X}(s) = \boldsymbol{\Phi}(s)\boldsymbol{b}U(s)$$

If this result is substituted into Eq. (2.3-5), $Y(s)$ is given by

$$Y(s) = \boldsymbol{c}^T \boldsymbol{\Phi}(s)\boldsymbol{b}U(s)$$

so that the transfer function $Y(s)/U(s)$ is

$$\frac{Y(s)}{U(s)} = G_p(s) = \boldsymbol{c}^T \boldsymbol{\Phi}(s)\boldsymbol{b} \tag{2.3-10}$$

Since the resolvent matrix is[1]

$$\boldsymbol{\Phi}(s) = (s\boldsymbol{I} - \boldsymbol{A})^{-1} = \frac{\text{adj}(s\boldsymbol{I} - \boldsymbol{A})}{\det(s\boldsymbol{I} - \boldsymbol{A})}$$

$G_p(s)$ becomes

$$G_p(s) = \frac{\boldsymbol{c}^T \text{adj}(s\boldsymbol{I} - \boldsymbol{A})\boldsymbol{b}}{\det(s\boldsymbol{I} - \boldsymbol{A})} \tag{2.3-11}$$

Here the scalar quantity $\boldsymbol{c}^T[\text{adj}(s\boldsymbol{I} - \boldsymbol{A})]\boldsymbol{b}$ is simply a polynomial in s and forms the numerator polynomial of $G_p(s)$. The scalar quantity $\det(s\boldsymbol{I} - \boldsymbol{A})$, which is also a polynomial in s, forms the denominator polynomial of $G_p(s)$. In other words, if $G_p(s)$ is written as the ratio of a gain times a numerator polynomial $K_pN_p(s)$ to a denominator polynomial $D_p(s)$, that is,

$$G_p(s) = \frac{K_pN_p(s)}{D_p(s)} \tag{2.3-12}$$

then

$$K_pN_p(s) = \boldsymbol{c}^T \text{adj}(s\boldsymbol{I} - \boldsymbol{A})\boldsymbol{b} \tag{2.3-13}$$

and

$$D_p(s) = \det(s\boldsymbol{I} - \boldsymbol{A}) \tag{2.3-14}$$

From the above result, we see that the values of s that satisfy the equation

$$\det(s\boldsymbol{I} - \boldsymbol{A}) = 0 \tag{2.3-15}$$

also satisfy the equation

$$D_p(s) = 0$$

and therefore are the poles of $G_p(s)$. In matrix terminology, these values of s are known as *eigenvalues* of the matrix $\boldsymbol{A}$, and Eq. (2.3-15) is referred to as the *characteristic equation* of the matrix $\boldsymbol{A}$. Hence we see that the eigenvalues of $\boldsymbol{A}$ correspond to the poles of $G_p(s)$.

---

[1]The operators adj and det stand for the adjoint and the determinant of a matrix, respectively. See Appendix C.

Example 2.3-2. To illustrate the use of Eq. (2.3-10), let us consider the representation of the field-controlled motor in terms of the variables $\theta_o(t)$, $\dot{\theta}_o(t)$, and $i_f(t)$ found in Example 2.3-1. That representation is repeated here for reference.

$$\dot{\boldsymbol{x}}(t) = \begin{bmatrix} 0 & 1 & 0 \\ 0 & -\beta/J & K_\tau/J \\ 0 & 0 & -R_f/L_f \end{bmatrix} \boldsymbol{x}(t) + \begin{bmatrix} 0 \\ 0 \\ 1/L_f \end{bmatrix} u(t)$$

$$y(t) = [1 \quad 0 \quad 0]\boldsymbol{x}(t)$$

For this problem, the matrix $(s\boldsymbol{I} - \boldsymbol{A})$ becomes

$$(s\boldsymbol{I} - \boldsymbol{A}) = \begin{bmatrix} s & -1 & 0 \\ 0 & s+\beta/J & -K_\tau/J \\ 0 & 0 & s+R_f/L_f \end{bmatrix}$$

and its inverse is

$$\boldsymbol{\Phi}(s) = (s\boldsymbol{I} - \boldsymbol{A})^{-1} = \begin{bmatrix} 1/s & 1/s(s+\beta/J) & K_\tau/s(Js+\beta)\left(s+R_f/L_f\right) \\ 0 & 1/(s+\beta/J) & K_\tau/(Js+\beta)\left(s+R_f/L_f\right) \\ 0 & 0 & 1/\left(s+R_f/L_f\right) \end{bmatrix}$$

Therefore the transfer function is given by

$$G_p(s) = [1 \quad 0 \quad 0] \begin{bmatrix} 1/s & 1/s(s+\beta/J) & K_\tau/s(Js+\beta)\left(s+R_f/L_f\right) \\ 0 & 1/(s+\beta/J) & K_\tau/(Js+\beta)\left(s+R_f/L_f\right) \\ 0 & 0 & 1/\left(s+R_f/L_f\right) \end{bmatrix} \begin{bmatrix} 0 \\ 0 \\ 1/L_f \end{bmatrix}$$

$$= \frac{K_\tau/L_f}{s(Js+\beta)\left(s+R_f/L_f\right)} = \frac{K_\tau/JL_f}{s(s+\beta/J)\left(s+R_f/L_f\right)}$$

which is identical to the previous result of Example 2.2-2.

Although Eq. (2.3-10) provides a direct method for determining the transfer function of a plant from a state-variable representation of the plant, it is not an easy method, since it requires the inversion of the matrix $(s\boldsymbol{I} - \boldsymbol{A})$. The inversion of a matrix is never an easy task, and the job is even more difficult in this case since the elements are functions of s. Because of this problem, it is sometimes easier to obtain the transfer function by carrying out block-diagram reductions on the state-variable representation. In fact, by using this approach, for certain types of state-variable representations the transfer function may be obtained by inspection. For high-order systems, the use of a digital computer is probably the best approach. Indeed, it is in conjunction with the computational power of computers that the state-variable approach shows its greatest utility.

Example 2.3-3. Consider, for example, the field-controlled dc motor problem of Example 2.3-1. The block diagram of the state-variable representation is shown in Fig. 2.3-3*b*. A simple combination of the three series blocks provides the desired transfer function. Note, however, that in recovering the transfer functions the information relating the internal variables $x_3 = i_f$ and $x_2 = \dot{\theta}_o$ to the input and output is lost.

Before proceeding with the problem of attaining a state-variable realization from a transfer function description, we stop to discuss the degrees of the polynomials which result from applying Eq. (2.3-11) to obtain a transfer function from a state-variable description. Equation (2.3-11) is repeated here for reference.

$$G_p(s) = \frac{\boldsymbol{c}^T \operatorname{adj}(s\boldsymbol{I} - \boldsymbol{A})\boldsymbol{b}}{\det(s\boldsymbol{I} - \boldsymbol{A})} \tag{2.3-11}$$

The properties of the determinant dictate that the degree of the denominator polynomial of $G_p(s)$ is n, where n is the number of state-variables and the order of $\boldsymbol{A}$. The numerator polynomial of $G_p(s)$ has degree $n - 1$ or less.

Thus a state-variable description as we have defined it always gives rise to a transfer function with the property that the degree of its denominator polynomial is strictly greater than the degree of its numerator polynomial, i.e., a *strictly proper transfer function.* It would appear that an nth-order state-variable description always gives rise to an nth-order transfer function. However, when the transfer function $G_p(s)$ computed from Eq. (2.3-11) is factored to display its poles and zeros, the numerator and denominator may contain one or more common factors. Clearly, these common factors may be canceled, just as one cancels common factors in fractions. Such cancellations have no effect upon the value of the transfer function when considered as a function of a complex variable, but cancellations change the order of the transfer function.

Polynomials which contain no common factors are called *relatively prime* or *coprime.* The order of a transfer function is taken as the degree of the denominator polynomial of the transfer function after all common factors have been canceled so that the numerator polynomial and denominator polynomial of the transfer function are relatively prime. Therefore, an nth-order state-variable description can give rise to a transfer function that has order less than n. Remember, a state-variable description may be more complete than a transfer function description of a system.

2.3.5 Linear Independence and the Concept of State

One situation in which common factors appear in the numerator and denominator of Eq. (2.3-11) is when there are state-variables in the state-variable description that provide only redundant information. To avoid useless states one should always choose state-variables to be linearly independent. A set of n variables $x_1(t), x_2(t), \ldots, x_n(t)$ is said to be *linearly independent* if there is no set of constants, $\alpha_1, \alpha_2, \ldots, \alpha_n$ other than all zero that satisfy the following equation for all $t \geq t_0$

$$\alpha_1 x_1(t) + \alpha_2 x_2(t) + \cdots + \alpha_n x_n(t) = 0 \tag{2.3-16}$$

Otherwise the set is said to be *linearly dependent.* If a set of variables is linearly dependent then any variable $x_i(t)$ associated with a non-zero α_i can be expressed as a *linear combination* of the remaining variables, that is

$$x_i(t) = \beta_1 x_1(t) + \cdots + \beta_{i-1} x_{i-1}(t) + \beta_{i+1} x_{i+1}(t) + \cdots + \beta_n x_n(t) \tag{2.3-17}$$

with the constants $\beta_1 = -\alpha_1/a_i, \beta_2 = -\alpha_2/\alpha_i, \ldots$. Notice that if $x_i(t)$ can be expressed as a linear combination of other states there is no reason to retain the equation that describes the evolution of $x_i(t)$

$$\dot{x}_i(t) = a_{i1}x_1(t) + a_{i2}x_2(t) + \cdots + a_{in}x_n(t) + b_i u(t)$$

because $x_i(t)$ may be obtained from the other states through Eq. (2.3-17). In addition, the right hand side of Eq. (2.3-17) can be substituted anywhere $x_i(t)$ appears in the remaining equations describing the system. Thus any state-variable that is linearly dependent on other state-variables can be removed from the state-variable description with no loss of information.

Example 2.3-4. Consider the system of equations

$$\dot{\boldsymbol{x}}(t) = \begin{bmatrix} -1 & 1 & 1 \\ 0 & -1 & 0 \\ -1 & -1 & 1 \end{bmatrix} \boldsymbol{x}(t) + \begin{bmatrix} 0 \\ 1 \\ 2 \end{bmatrix} u(t)$$

$$y(t) = [0 \quad 0 \quad 1]\boldsymbol{x}(t)$$

and assume that

$$x_3(t) = x_1(t) + 2x_2(t)$$

Then $x_1(t)$, $x_2(t)$, and $x_3(t)$ are linearly dependent since

$$x_1(t) + 2x_2(t) - x_3(t) = 0$$

Note that the state equation is consistent with the linear dependence of $x_3(t)$ on $x_1(t)$ and $x_2(t)$ since

$$\dot{x}_3(t) = \dot{x}_1(t) + 2\dot{x}_2(t)$$

The transfer function for this system is computed from Eq. (2.3-11) as

$$G_p(s) = \frac{\boldsymbol{c}^T \operatorname{adj}(s\boldsymbol{I} - \boldsymbol{A})\boldsymbol{b}}{\det(s\boldsymbol{I} - \boldsymbol{A})} = \frac{(2s+3)s}{(s+1)s^2} = \frac{2s+3}{s(s+1)}$$

As predicted, there is a common factor s in the numerator and denominator of $G_p(s)$. The system can be rewritten using only two state-variables as

$$\dot{x}_1(t) = -x_1(t) + x_2(t) + (x_1(t) + 2x_2(t))$$

$$\dot{x}_2(t) = -x_2(t) + u(t)$$

$$y(t) = x_1(t) + 2x_2(t)$$

Or, in matrix notation, we write

$$\dot{\boldsymbol{x}}(t) = \begin{bmatrix} 0 & 3 \\ 0 & -1 \end{bmatrix} \boldsymbol{x}(t) + \begin{bmatrix} 0 \\ 1 \end{bmatrix} u(t)$$

$$y(t) = [1 \quad 2]\boldsymbol{x}(t)$$

The reader should check that, by using the new state-variable description, the reduced form of $G_p(s)$ is attained directly.

Since linearly dependent state-variables do not add to our information about the plant we will henceforth assume that all state-variables are chosen to form a linearly independent set. It is important to realize that the assumption of linear independence does not ensure that the numerator and denominator of $G_p(s)$ as computed by

Eq. (2.3-11) will be relatively prime. Additional situations in which nth-order state-variable representations can give rise to transfer functions with order less than n will be studied in Sec. 3.5.

In this section, we have discussed the state-variable representation of a plant and we have seen how to derive the transfer function representation from the state-variable representation. The problem of determining a state-variable representation of a plant whose transfer function is known is more complicated than the above problems suggest, since there is an infinite number of state-variable representations for the same transfer function. Sections 2.4 and 2.5 discuss two of the more common means of selecting the state-variables.

2.3.6 CAD Notes

CAD programs can be very useful in manipulating state-variable descriptions when such descriptions are expressed using numerical values. MATLAB has a function to help compute Eqs. (2.3-9) and (2.3-10); this allows a conversion of a plant description from state-space form to transfer function form. The pertinent MATLAB statement has the form

```
[num,den]=ss2tf(a,b,c,d,iu)
```

The function returns the numerator and denominator polynomials of the transfer function. The arguments `a,b,` and `c` provide the state-space description. The argument `d` is available for more general state-space descriptions than are considered in this book. For our purposes `d` should be a vector of zeros whose row dimension equals the number of inputs (usually one) and whose column dimension equals the number of outputs (usually one). The argument `iu` indicates which input we are considering. In our case, to get a transfer function for a system with one input and one output, `d=[0]` and `iu=1`.

Example 2.3-5. Here a state-space description is entered and the corresponding transfer function is computed.

```
>>A=[-4 0 1;1 -3 0;1 1 -5]
A =
 -4   0   1
  1  -3   0
  1   1  -5

b=[1;0;1]
b =
  1
  0
  1
```

```
c=[1 1 0]
c =
  1  1  0

d=[0]
d =
  0

[num,den]=ss2tf(A,b,c,d,1)
num =
     0  1.0000  10.0000  24.0000

den =
  1.0000  12.0000  46.0000  56.0000
```

The resulting transfer function is

$$G_p(s) = \frac{s^2 + 10s + 24}{s^3 + 12s^2 + 46s + 56}$$

The resolvent matrix $\mathbf{\Phi}(s)$ can be computed one column at a time by creating a related system that depends only on the $\boldsymbol{A}$ matrix and has three inputs and three outputs. The $\boldsymbol{b}$ and $\boldsymbol{c}$ matrices are 3×3 identity matrices and the $\boldsymbol{d}$ matrix is a 3×3 matrix of zeros. The last argument in `ss2tf` is used to establish which column of $\mathbf{\Phi}(s)$ is determined.

Example 2.3-5 cont.

```
% A remains the same

c=eye(3)
c =
  1  0  0
  0  1  0
  0  0  1

b=eye(3)
b =
  1  0  0
  0  1  0
  0  0  1

d=zeros(3)
d =
  0  0  0
  0  0  0
  0  0  0
```

```
%find the first column of the resolvent
[num1,den1]=ss2tf(A,b,c,d,1)
num1 =
     0   1.0000   8.0000  15.0000
     0        0   1.0000   5.0000
     0        0   1.0000   4.0000
den1 =
   1.0000    12.0000    46.0000    56.0000

%find the second column of the resolvent
[num2,den2]=ss2tf(A,b,c,d,2)
num2 =
     0        0   0.0000   1.0000
     0   1.0000   9.0000  19.0000
     0        0   1.0000   4.0000
den2 =
   1.0000  12.0000  46.0000  56.0000

%find the third column of the resolvent
[num3,den3]=ss2tf(A,b,c,d,3)
num3 =
     0   0.0000   1.0000   3.0000
     0        0   0.0000   1.0000
     0   1.0000   7.0000  12.0000
den3 =
   1.0000  12.0000  46.0000  56.0000
```

Of course, all the denominators are the same since each is just the characteristic equation of $\boldsymbol{A}$. The resolvent matrix can now be read off as

$$\boldsymbol{\Phi}(s) = \frac{\begin{bmatrix} s^2+8s+15 & 1 & s+3 \\ s+5 & s^2+9s+19 & 1 \\ s+4 & s+4 & s^2+7s+12 \end{bmatrix}}{s^3+12s^2+46s+56}$$

Exercises 2.3

2.3-1. For the state-variable description

$$\dot{\boldsymbol{x}} = \begin{bmatrix} 0 & 1 & 1 \\ 0 & -2 & 2 \\ -1 & -1 & -4 \end{bmatrix} \boldsymbol{x} + \begin{bmatrix} 0 \\ 0 \\ 2 \end{bmatrix} u$$

$$y = [1 \quad 0 \quad 0]\boldsymbol{x}$$

Find the elementary block diagram of the system and use block diagram reductions to find the transfer function $Y(s)/U(s)$.
Answer:

$$\frac{Y(s)}{U(s)} = \frac{2(s+4)}{s^3+6s^2+11s+4}$$

2.3-2. For the state-variable description given in Exercise 2.3-1, find the resolvent matrix $\mathbf{\Phi}(s)$ and the transfer function $Y(s)/U(s)$ using the matrix Eq. (2.3-11).
Answer:

$$\mathbf{\Phi}(s) = \frac{1}{s^3 + 6s^2 + 11s + 4}\begin{bmatrix} s^2 + 6s + 10 & s + 3 & s + 4 \\ -2 & s^2 + 4s + 1 & 2s \\ -s - 2 & -s - 1 & s^2 + 2s \end{bmatrix}$$

$$\frac{Y(s)}{U(s)} = \frac{2(s + 4)}{s^3 + 6s^2 + 11s + 4}$$

2.3-3. Write the state equation for the combination motor and generator system of Exercise 2.2-4. The equations governing the system are given in Exercise 2.2-4. Use the state-variables i_{fm}, $\frac{d\theta}{dt}$, and y. The tricky part is to express $\tau_g(t)$ in terms of the state-variables. The intermediate steps involve expressing $\tau_g(t)$ in terms of $i_{ag}(t)$, then in terms of $\frac{dy}{dt}$, then in terms of $y(t)$ and $e_g(t)$, and finally in terms of $y(t)$ and $\frac{d\theta(t)}{dt}$.
Answer:

$$\dot{\mathbf{x}}(t) = \begin{bmatrix} \frac{-R_f}{L_f} & 0 & 0 \\ \frac{K_\tau}{J} & \left(-\frac{\beta}{J} - \frac{K_v K_g}{JR}\right) & \frac{K_g}{JR} \\ 0 & \frac{K_v}{RC} & \frac{-1}{RC} \end{bmatrix}\mathbf{x}(t) + \begin{bmatrix} \frac{1}{L_f} \\ 0 \\ 0 \end{bmatrix}u(t)$$

$$y(t) = [0 \quad 0 \quad 1]\mathbf{x}(t)$$

2.3-4. Compute the transfer function for the state-space system shown as the answer of Exercise 2.3-3. Notice that you need only compute one element of adj$(s\mathbf{I} - \mathbf{A})$. The correct answer is the same as the answer to Exercise 2.2-4.

2.3-5. Given the state-variable description

$$\dot{\mathbf{x}} = \begin{bmatrix} -1 & 0 & 1 \\ -1 & -2 & 1 \\ 0 & 2 & 0 \end{bmatrix}\mathbf{x} + \begin{bmatrix} 1 \\ 0 \\ 1 \end{bmatrix}u \qquad \mathbf{x}(0) = \begin{bmatrix} 3 \\ 2 \\ 1 \end{bmatrix}$$

$$y = [0 \quad 1 \quad 0]\mathbf{x}$$

Give a two-dimensional state-space description by eliminating a linearly dependent state.
Answer: Any state can be eliminated. If you eliminate x_2 you obtain

$$\dot{\mathbf{x}} = \begin{bmatrix} -1 & 1 \\ 2 & -2 \end{bmatrix}\mathbf{x} + \begin{bmatrix} 1 \\ 1 \end{bmatrix}u \qquad \mathbf{x}(0) = \begin{bmatrix} 3 \\ 1 \end{bmatrix}$$

$$y = [1 \quad -1]\mathbf{x}$$

2.4 PHASE VARIABLES

Although there are an infinite number of ways of selecting the state-variables for any plant, and hence an infinite number of state-variable representations, only a limited number are in common use. These representations have either mathematical advantage or physical meaning. The method of phase variables presented in this section falls into the category of representations that possess mathematical advantage.

Phase variables are defined as the particular set of state-variables that consists of one variable and its $n - 1$ derivatives. To introduce the method, let us consider the phase-variable representation of the nth-order plant with unity numerator whose block diagram is shown in Fig. 2.4-1. From the block diagram, the transfer function is seen to be

$$G_p(s) = \frac{Y(s)}{U(s)} = \frac{1}{s^n + a_{n-1}s^{n-1} + \cdots + a_1 s + a_0}$$

This transfer function represents the nth-order differential equation

$$\frac{d^n y(t)}{dt^n} + a_{n-1}\frac{d^{n-1}y(t)}{dt^{n-1}} + \cdots + a_1\frac{dy(t)}{dt} + a_0 y(t) = u(t) \tag{2.4-1}$$

To express this equation in phase-variable form, let $x_1(t) = y(t)$, and then, according to the definition of phase variables,

$$x_2(t) = \dot{y}(t) = \dot{x}_1(t), \qquad x_3(t) = \ddot{y}(t) = \dot{x}_2(t), \ldots, x_n(t) = \frac{d^{n-1}y(t)}{dt^{n-1}} = \dot{x}_{n-1}(t)$$

If these definitions are substituted into Eq. (2.4-1), the result is

$$\dot{x}_n(t) + a_{n-1}x_n(t) + a_{n-2}x_{n-1}(t) + \cdots + a_1 x_2(t) + a_0 x_1(t) = u(t) \tag{2.4-2}$$

In addition, we have the $n-1$ defining equations

$$\begin{aligned} \dot{x}_1(t) &= x_2(t) \\ \dot{x}_2(t) &= x_3(t) \\ &\vdots \\ \dot{x}_{n-1}(t) &= x_n(t) \end{aligned}$$

and the output expression

$$y(t) = x_1(t)$$

These equations are of the state-variable form ($\boldsymbol{Ab}$) and ($\boldsymbol{c}$), where

$$\boldsymbol{A} = \begin{bmatrix} 0 & 1 & 0 & \cdots & 0 & 0 \\ 0 & 0 & 1 & \cdots & 0 & 0 \\ \vdots & \vdots & \vdots & \ddots & \vdots & \vdots \\ 0 & 0 & 0 & \cdots & 0 & 1 \\ -a_0 & -a_1 & -a_2 & \cdots & -a_{n-2} & -a_{n-1} \end{bmatrix} \quad \boldsymbol{b} = \begin{bmatrix} 0 \\ 0 \\ \vdots \\ 0 \\ 1 \end{bmatrix} \quad \boldsymbol{c} = \begin{bmatrix} 1 \\ 0 \\ \vdots \\ 0 \\ 0 \end{bmatrix} \tag{2.4-3}$$

For example, the phase-variable representation for the third-order plant

$$G_p(s) = \frac{Y(s)}{U(s)} = \frac{1}{s^3 + a_2 s^2 + a_1 s + a_0}$$

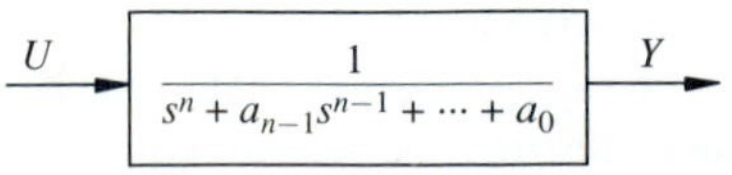

$U \rightarrow \boxed{\dfrac{1}{s^n + a_{n-1}s^{n-1} + \cdots + a_0}} \rightarrow Y$

FIGURE 2.4-1
An nth-order plant with unity numerator.

is

$$\dot{x}_1(t) = x_2(t)$$
$$\dot{x}_2(t) = x_3(t)$$
$$\dot{x}_3(t) = -a_1 x_2(t) - a_2 x_3(t) + u(t) \tag{2.4-4}$$

with

$$y(t) = x_1(t) \tag{2.4-5}$$

The matrices $\boldsymbol{A}$, $\boldsymbol{b}$, and $\boldsymbol{c}$ are therefore

$$\boldsymbol{A} = \begin{bmatrix} 0 & 1 & 0 \\ 0 & 0 & 1 \\ -a_0 & -a_1 & -a_2 \end{bmatrix} \qquad \boldsymbol{b} = \begin{bmatrix} 0 \\ 0 \\ 1 \end{bmatrix} \qquad \boldsymbol{c} = \begin{bmatrix} 1 \\ 0 \\ 0 \end{bmatrix}$$

Possible block diagrams for the phase-variable representation of both an nth-order plant and the above third-order example are shown in Fig. 2.4-2. A simple examination of these block diagrams, of the original transfer function, of the nth-order differential equation, and of the phase-variable representation reveals that any of these methods of representation may be determined from the other *by inspection*. This is one of the important features of the phase-variable representation and makes it easy to transfer from one method of representation to the other.

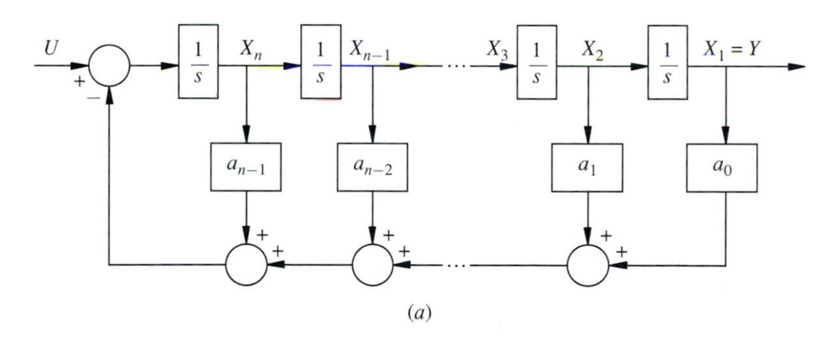

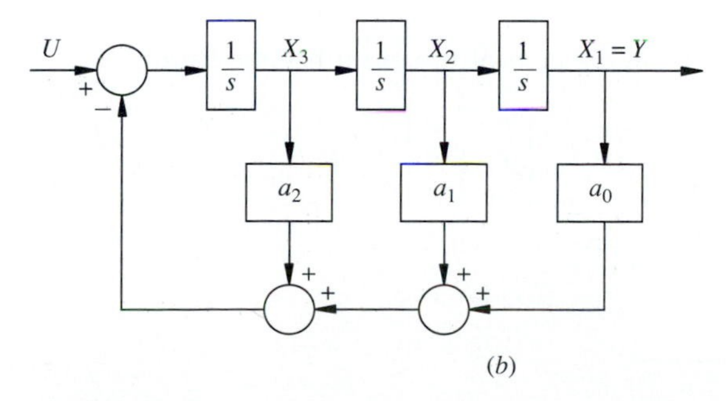

FIGURE 2.4-2
Block diagrams of the phase-variable representation. (*a*) An nth-order plant; (*b*) a third-order plant.

So far we have considered only the case where the transfer function $G_p(s)$ has a unity numerator. To include the general case, a slight modification must be made in this approach. To see why this is necessary, let us consider the same third-order example except with a zero added, so that

$$G_p(s) = \frac{c_1 s + c_0}{s^3 + a_2 s^2 + a_1 s + a_0}$$

If we proceed as before by letting $x_1(t) = y(t)$, $x_2(t) = \dot{y}(t)$, and $x_3(t) = \ddot{y}(t)$, then the equivalent differential equation can be written as

$$\dot{x}_3(t) + a_2 x_3(t) + a_1 x_2(t) + a_0 x_1(t) = c_1 \dot{u}(t) + c_0 u(t)$$

The phase-variable representation would now contain a $\dot{u}(t)$ term on the right-hand side, which is a violation of the assumed form of Eq. (***Ab***).

To avoid this problem, the transfer function is divided into two parts, as shown in Fig. 2.4-3, and written as

$$G_p(s) = \frac{Y(s)}{U(s)} = \frac{X_1(s)}{U(s)} \frac{Y(s)}{X_1(s)}$$

where

$$\frac{X_1(s)}{U(s)} = \frac{1}{s^3 + a_2 s^2 + a_1 s + a_0}$$

and

$$\frac{Y(s)}{X_1(s)} = c_1 s + c_0$$

The transfer function $X_1(s)/U(s)$ is identical to the original transfer function without the added zero, and therefore its phase-variable representation is still given by Eq. (2.4-4). From the second transfer function, however, we see that $y(t)$ is no longer equal to $x_1(t)$ but is now

$$y(t) = c_1 \dot{x}_1(t) + c_0 x_1(t) = c_1 x_2(t) + c_0 x_1(t)$$

where the second expression has been written by making use of the fact that $\dot{x}_1(t) = x_2(t)$.

The complete phase-variable representation of this plant has the form

$$\dot{\boldsymbol{x}}(t) = \begin{bmatrix} 0 & 1 & 0 \\ 0 & 0 & 1 \\ -a_0 & -a_1 & -a_2 \end{bmatrix} \boldsymbol{x}(t) + \begin{bmatrix} 0 \\ 0 \\ 1 \end{bmatrix} u(t)$$

$$\boldsymbol{y}(t) = [c_0 \quad c_1 \quad 0] \boldsymbol{x}(t)$$

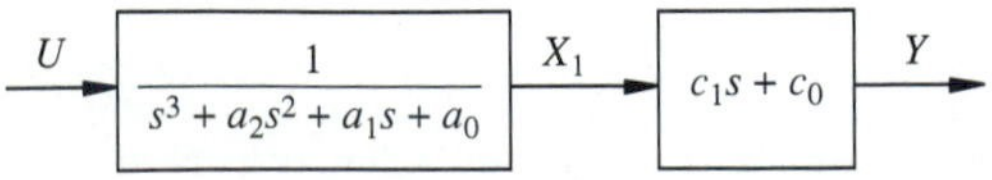

FIGURE 2.4-3
Method of handling zeros.

A comparison of this result with the representation of the unity numerator system indicates that the only change is in the output expression. This is, in fact, always the case. For the general nth-order system with m zeros and $m < n$, the transfer function

$$G_p(s) = \frac{c_m s^m + c_{m-1}s^{m-1} + \cdots + c_1 s + c_0}{s^n + a_{n-1}s^{n-1} + \cdots + a_1 s + a_0} \tag{2.4-6}$$

is expressed in phase-variable representation as

$$\dot{\boldsymbol{x}}(t) = \begin{bmatrix} 0 & 1 & 0 & \cdots & 0 & 0 \\ 0 & 0 & 1 & \cdots & 0 & 0 \\ \vdots & \vdots & \vdots & \ddots & \vdots & \vdots \\ 0 & 0 & 0 & \cdots & 0 & 1 \\ -a_0 & -a_1 & -a_2 & \cdots & -a_{n-2} & -a_{n-1} \end{bmatrix} \boldsymbol{x}(t) + \begin{bmatrix} 0 \\ 0 \\ \vdots \\ 0 \\ 1 \end{bmatrix} u(t) \tag{2.4-7}$$

$$y(t) = [\, c_0 \quad c_1 \quad \cdots \quad c_{m-1} \quad c_m \quad 0 \quad \cdots \quad 0 \,]\, \boldsymbol{x}(t) \tag{2.4-8}$$

The block diagrams for the third-order example and the general nth-order case are shown in Fig. 2.4-4. Note that the phase-variable representation is still easily determined from the transfer function by inspection, and vice versa. Note also that the elements of $\boldsymbol{b}$ are all zero except for the last element, which is always equal to 1, and that the first $n-1$ rows of $\boldsymbol{A}$ are always of the same form.

We assume that $G_p(s)$ as given in Eq. (2.4-6) is in a reduced form with its numerator and denominator relatively prime so that it is truly an nth-order transfer function. Notice that the phase-variable method produces an nth-order state-variable description, the minimal order state-variable description that can produce an nth-order transfer function.

The resolvent matrix $\boldsymbol{\Phi}(s)$ associated with a general phase-variable plant matrix, $\boldsymbol{A}$, is fairly complicated. However, the control vector $\boldsymbol{b}$ contains zeros in all but the last entry. For the phase-variable form

$$\boldsymbol{\Phi}(s)\boldsymbol{b} = (s\boldsymbol{I} - \boldsymbol{A})^{-1}\,\boldsymbol{b} = \frac{\begin{bmatrix} 1 \\ s \\ \vdots \\ s^{n-2} \\ s^{n-1} \end{bmatrix}}{s^n + a_{n-1}s^{n-1} + \cdots + a_1 s + a_0} \tag{2.4-9}$$

From this point it is easy to see that the transfer function of Eq. (2.4-6) can be recovered simply from the phase-variable representation.

Although the use of phase variables provides a simple means of representing a plant in state-variable form, the method has the disadvantage that the phase variables are not, in general, meaningful physical variables of the plant and therefore not available for measurement or manipulation. If the transfer function has a unity numerator, the phase variables are the output and its $n-1$ derivatives. If n is greater than 2 or 3, it is physically difficult to obtain these derivatives. If, on the other hand, $G_p(s)$ has

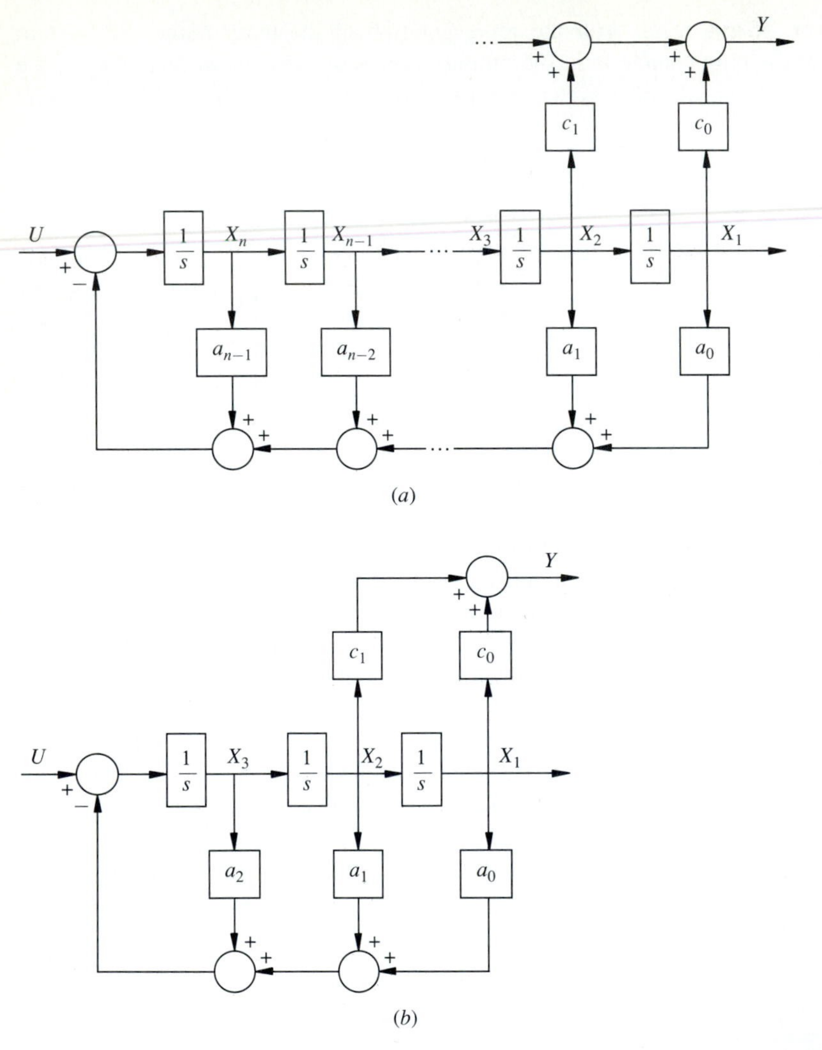

FIGURE 2.4-4
General phase-variable representations. (*a*) An *n*th-order plant; (*b*) a third-order plant with one zero.

one or more zeros, the phase variables bear little or no resemblance to real, physical quantities in the plant.

Thus, although the phase-variable representation provides mathematical advantages, it is not a practical set of state-variables from a control point of view, since we eventually wish to measure all the state-variables. This does not mean that phase variables are not a useful and even valuable method of representation. On the contrary, we shall make extensive use of them throughout this book. In Sec. 2.5 we consider a method of state-variable representation that is closely tied to measurable physical variables.

2.4.1 CAD Notes

To go from a transfer function to a phase-variable representation, MATLAB provides a function of the form

```
[a,b,c,d]=tf2ss(num,den)
```

Unfortunately, MATLAB produces a slightly different phase-variable form that arises from taking the phase-variable form presented here and renumbering the states in reverse order.

Example 2.4-1. This example shows the MATLAB form of the phase-variable representation for the transfer function that was the result in the CAD Notes of the last section. The quantity $(s\boldsymbol{I} - \boldsymbol{A})^{-1}\boldsymbol{b}$ is then computed and shown to be of the form of Eq. (2.4-9) with the numbering of the states reversed. The denominator polynomial is recovered.

```
>>num=[1 10 24];
den=[1 12 46 56];
[Ap,bp,cp,dp]=tf2ss(num,den)
Ap =
 -12  -46  -56
   1    0    0
   0    1    0
bp =
   1
   0
   0
cp =
   1   10   24
dp =
   0

%find (inv(sI-A))b for phase variables
c=eye(3);
d=zeros(3,1);
[phinum,phiden]=ss2tf(Ap,bp,c,d,1)
phinum =
   0   1.0000  -0.0000  -0.0000
   0  -0.0000   1.0000   0.0000
   0  -0.0000  -0.0000   1.0000
phiden =

   1.0000  12.0000  46.0000  56.0000
```

One can read off that

$$(s\boldsymbol{I} - \boldsymbol{A})^{-1}\boldsymbol{b} = \frac{\begin{bmatrix} s^2 \\ s \\ 1 \end{bmatrix}}{s^3 + 12s^2 + 46s + 56}$$

Exercises 2.4

2.4-1. Represent the following plants using phase variables:

$$(a)\quad G_p(s) = \frac{1}{s^2 + 4s + 1} \qquad (b)\quad G_p(s) = \frac{s+3}{s^3 + s^2 + 10}$$

$$(c)\quad G_p(s) = \frac{10(s+3)(s+1)}{s(s+2)(s+5)}$$

Answers:

$$(a)\quad \dot{\boldsymbol{x}} = \begin{bmatrix} 0 & 1 \\ -1 & -4 \end{bmatrix}\boldsymbol{x} + \begin{bmatrix} 0 \\ 1 \end{bmatrix}u \qquad y = [1 \quad 0]\boldsymbol{x}$$

$$(b)\quad \dot{\boldsymbol{x}} = \begin{bmatrix} 0 & 1 & 0 \\ 0 & 0 & 1 \\ -10 & 0 & -1 \end{bmatrix}\boldsymbol{x} + \begin{bmatrix} 0 \\ 0 \\ 1 \end{bmatrix}u \qquad y = [3 \quad 1 \quad 0]\boldsymbol{x}$$

$$(c)\quad \dot{\boldsymbol{x}} = \begin{bmatrix} 0 & 1 & 0 \\ 0 & 0 & 1 \\ 0 & -10 & -7 \end{bmatrix}\boldsymbol{x} + \begin{bmatrix} 0 \\ 0 \\ 1 \end{bmatrix}u \qquad y = [30 \quad 40 \quad 10]\boldsymbol{x}$$

2.4-2. Find the transfer functions for the following plants:

$$(a)\quad \dot{\boldsymbol{x}}(t) = \begin{bmatrix} 0 & 1 \\ -6 & -2 \end{bmatrix}\boldsymbol{x}(t) + \begin{bmatrix} 0 \\ 1 \end{bmatrix}u(t) \qquad y(t) = [3 \quad 0]\boldsymbol{x}(t)$$

$$(b)\quad \dot{\boldsymbol{x}}(t) = \begin{bmatrix} 0 & 1 & 0 \\ 0 & 0 & 1 \\ -5 & -3 & -2 \end{bmatrix}\boldsymbol{x}(t) + \begin{bmatrix} 0 \\ 0 \\ 1 \end{bmatrix}u(t) \qquad y(t) = [1 \quad 2 \quad 0]\boldsymbol{x}(t)$$

$$(c)\quad \dot{\boldsymbol{x}}(t) = \begin{bmatrix} 0 & 1 & 0 \\ 0 & 0 & 1 \\ 0 & -1 & -3 \end{bmatrix}\boldsymbol{x}(t) + \begin{bmatrix} 0 \\ 0 \\ 1 \end{bmatrix}u(t) \qquad y(t) = [3 \quad 8 \quad 2]\boldsymbol{x}(t)$$

Answers:

$$(a)\quad G_p(s) = \frac{3}{s^2 + 2s + 6} \qquad (b)\quad G_p(s) = \frac{2s+1}{s^3 + 2s^2 + 3s + 5}$$

$$(c)\quad G_p(s) = \frac{2s^2 + 8s + 3}{s^3 + 3s^2 + s}$$

2.4-3. Find the phase-variable representation of the armature-controlled motor of Example 2.2-4.

Answer:

$$\dot{\boldsymbol{x}} = \begin{bmatrix} 0 & 1 & 0 \\ 0 & 0 & 1 \\ 0 & -\left(R_a\beta + K_v K_\tau'\right)/JL_a & -(L_a\beta + R_a J)/JL_a \end{bmatrix}\boldsymbol{x} + \begin{bmatrix} 0 \\ 0 \\ 1 \end{bmatrix}u$$

$$y = [K_\tau'/JL_a \quad 0 \quad 0]\boldsymbol{x}$$

2.4-4. Given

$$\boldsymbol{A} = \begin{bmatrix} -1 & 2 \\ -2 & -2 \end{bmatrix} \qquad \boldsymbol{b} = \begin{bmatrix} 1 \\ 0 \end{bmatrix} \qquad \boldsymbol{c}^T = [1 \quad 0]$$

find the transfer function and phase-variable representation.

Answer:

$$\dot{x} = \begin{bmatrix} 0 & 1 \\ -6 & -3 \end{bmatrix} x + \begin{bmatrix} 0 \\ 1 \end{bmatrix} u$$

$$y = [2 \quad 1]x$$

2.5 PHYSICAL VARIABLES

The method of plant representation in terms of real, physical variables relies heavily on an understanding of the physical character of the plant. It is an engineer's approach to the problem of state-variable representation, as opposed to the mathematician's approach of the preceding section. In the preceding section, no attempt was made to relate the state-variables to real, physical quantities. Here, the method is almost intuitive, and the reader may feel that the approach is so straightforward that it should not be called a method. However, this approach is often the starting point to creating the transfer functions of complicated plants with many interconnections.

Let us begin our discussion of physical variables by considering a third-order plant whose transfer function is

$$G_p(s) = \frac{c_0}{s^3 + a_2 s^2 + a_1 s + a_0}$$

This transfer function might represent either a field-controlled or an armature-controlled dc motor or, in fact, any other third-order plant with no zeros.

This is exactly the strength as well as the weakness of the transfer function representation. All plants with the same input-output characteristics, that is, the same describing differential equations, are assigned the same transfer function. Therefore, if we are given only the transfer function, we may only speculate on the physical origins of the problem. Obviously a field-controlled motor is not the same physical device as an armature-controlled motor, yet the transfer function makes no distinction. Therefore, to represent a plant in terms of a set of meaningful physical variables, we must know more than just the transfer function; we must know the physical origins of the problem.

In Example 2.3-1 we used a set of physical variables, namely, $\theta_o(t)$, $\dot{\theta}_o(t)$, $i_f(t)$, to represent the field-controlled motor, let us, therefore, assume that the above transfer function represents an armature-controlled motor of the nature described in Example 2.2-4. There, a set of four transformed equations was found to describe the plant. These equations are presented here by means of their differential equation counterparts.

$$L_a \frac{di_a(t)}{dt} + R_a i_a(t) = e(t) - e_c(t)$$

$$e_c(t) = K_v \dot{\theta}_o(t)$$

$$J\ddot{\theta}_o(t) + \beta\dot{\theta}_o(t) = \tau(t)$$

$$\tau(t) = K'_\tau i_a(t)$$

In addition, the expanded block diagram of Fig. 2.2-13*a* is repeated here as Fig. 2.5-1.

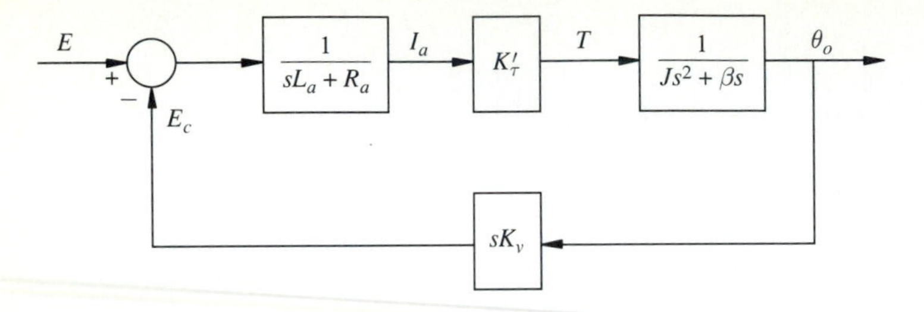

FIGURE 2.5-1
Armature-controlled motor.

The preceding equations employ five variables, $i_a(t)$, $\theta_o(t)$, $\dot{\theta}_o(t)$, $\tau(t)$, and $e_c(t)$, and at first glance it might appear that the plant should have five state-variables. However, we always assume that the plant is specified with a set of linearly independent variables. We see that both $\tau(t)$ and $i_a(t)$ may not be used, since they are not linearly independent. Since $i_a(t)$ is perhaps the easier variable to measure, let us use it as one state-variable. The output $\theta_o(t)$ is a logical choice for a second variable. Although we could choose $e_c(t)$ as the third state-variable, we see that it is linearly dependent on $\dot{\theta}_o(t)$ and therefore $\dot{\theta}_o(t)$ could also be used. Since the back emf does not exist as a separate, measurable quantity, and since $\dot{\theta}_o(t)$ can be easily measured with a tachometer, we choose the physical variables $i_a(t)$, $\theta_o(t)$, and $\dot{\theta}_o(t)$ as the state-variables for the system. Therefore, let

$$x_1(t) = i_a(t)$$

$$x_2(t) = \theta_o(t)$$

$$x_3(t) = \dot{\theta}_o(t)$$

and

$$u(t) = e(t)$$

If these definitions are substituted into the differential equations for this plant, we have

$$L_a\dot{x}_1(t) + R_a x_1(t) = u(t) - e_c(t)$$

$$e_c(t) = K_v x_3(t)$$

$$J\dot{x}_3(t) + \beta x_3(t) = \tau(t)$$

$$\tau(t) = K'_\tau x_1(t)$$

Combining the two algebraic equations with the differential equations, we obtain

$$L_a\dot{x}_1(t) + R_a x_1(t) = u(t) - K_v x_3(t)$$

$$J\dot{x}_3(t) + \beta x_3(t) = K'_\tau x_1(t)$$

or

$$\dot{x}_1(t) = -\frac{R_a}{L_a}x_1(t) - \frac{K_v}{L_a}x_3(t) + \frac{1}{L_a}u(t)$$

$$\dot{x}_3(t) = \frac{K'_\tau}{J}x_1(t) - \frac{\beta}{J}x_3(t)$$

In addition, we have the defining equation

$$\dot{x}_2(t) = x_3(t)$$

In the form of the matrix equation ($\boldsymbol{Ab}$), this result becomes

$$\dot{\boldsymbol{x}}(t) = \begin{bmatrix} -R_a/L_a & 0 & -K_v/L_a \\ 0 & 0 & 1 \\ K'_\tau/J & 0 & -\beta/J \end{bmatrix} \boldsymbol{x}(t) + \begin{bmatrix} 1/L_a \\ 0 \\ 0 \end{bmatrix} u(t)$$

Since the output $y(t)$ is equivalent to $\theta_o(t)$ and $\theta_o(t) = x_2(t)$, the output expression $y(t) = x_2(t)$ can be expanded in matrix form as

$$y(t) = [0 \quad 1 \quad 0]\boldsymbol{x}(t)$$

To make the relationship $\dot{\theta}_o(t) = x_3(t)$ more obvious, the block diagram of Fig. 2.5-1 has been redrawn as Fig. 2.5-2*a*. This block diagram has been further redrawn as Fig. 2.5-2*b* by joining the two places where $x_3(t)$ appears or, equivalently, by using the block-diagram identity of Fig. 2.2-8*d*. Note that the above physical-variable representation can be easily written from this block-diagram representation. Of course, Fig. 2.5-2*b* could be reduced to the overall block diagram of Fig. 2.2-13*c*.

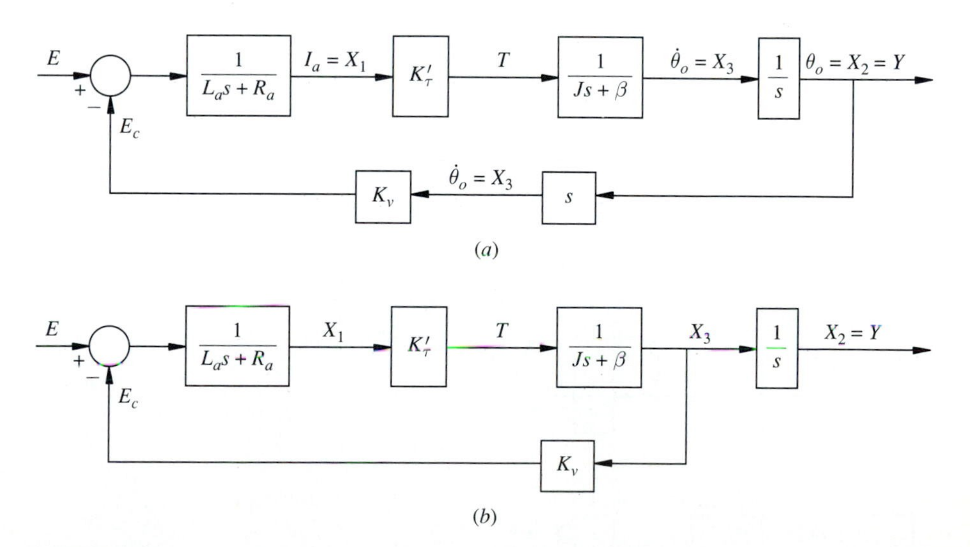

FIGURE 2.5-2
Physical-variable representation of the armature-controlled dc motor. (*a*) Initial block diagram; (*b*) reduced block diagram.

Obviously, physically meaningful state-variables could not be chosen from that block diagram.

Note that in the block diagram of Fig. 2.5-2*b* the output variable is labeled x_2 and the variables are not in any particular numerical order. Previously we have always designated the state-variable that appears on the right in the block diagram as x_1, and the state-variables always appeared in numerical order. This ordering of state-variables occurs for phase-variable descriptions. Here we wish to emphasize that no such ordering is necessary. The state-variables may be arbitrarily interchanged. In this example, for instance, had θ_o been designated as x_1, $\dot{\theta}_o$ as x_2, and i_a as x_3, the state equations would have included the matrices

$$\boldsymbol{A}^* = \begin{bmatrix} 0 & 1 & 0 \\ 0 & -\beta/J & K'_\tau/J \\ 0 & -K_v/L_a & -R_a/L_a \end{bmatrix} \quad \boldsymbol{b}^* = \begin{bmatrix} 0 \\ 0 \\ 1/L_a \end{bmatrix} \quad \boldsymbol{c}^* = \begin{bmatrix} 1 \\ 0 \\ 0 \end{bmatrix}$$

Although $\boldsymbol{A}^*$, $\boldsymbol{b}^*$, and $\boldsymbol{c}^*$ are different from the matrices associated with the original choice of state-variables, the transfer function does not change; that is,

$$G_p(s) = \boldsymbol{c}^T\,(s\boldsymbol{I} - \boldsymbol{A})^{-1}\,\boldsymbol{b} = \boldsymbol{c}^{*T}\,(s\boldsymbol{I} - \boldsymbol{A})^{-1}\,\boldsymbol{b}^*$$

Thus far we have considered only cases in which the numerator of $G_p(s)$ has been a constant. The treatment when $G_p(s)$ has zeros calls for slight modifications, much as in the phase-variable case. To initiate the discussion, we consider the two third-order plants of Fig. 2.5-3*a* and *b*. (Here, although the transfer functions are the same, the two figures do not represent the same plant.) In Fig. 2.5-3*a* the zero does not appear in the left-most block, and no difficulty arises in writing the state equations in the form of Eq. (***Ab***). The state equations may be written directly from that block diagram as

$$\frac{X_1(s)}{X_2(s)} = \frac{1}{s} \qquad \rightarrow \dot{x}_1 = x_2$$

$$\frac{X_2(s)}{X_3(s)} = \frac{s+3}{s+1} \rightarrow \dot{x}_2 = -x_2 + \dot{x}_3 + 3x_3$$

$$\frac{X_3(s)}{U(s)} = \frac{1}{s+5} \rightarrow \dot{x}_3 = -5x_3 + u$$

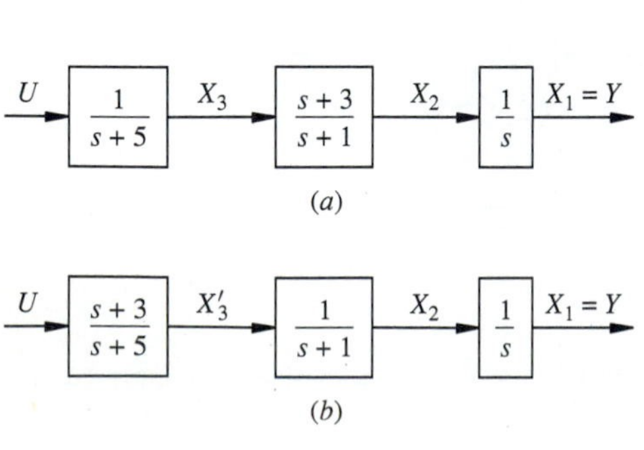

FIGURE 2.5-3
Two cases of a zero in third-order plants described in physical variables.

After the $\dot{x}_3$ term is eliminated from the second equation, the result is

$$\dot{x}_1 = x_2$$

$$\dot{x}_2 = -x_2 - 2x_3 + u$$

$$\dot{x}_3 = -5x_3 + u$$

Note here that the vector $\boldsymbol{b}$ now has two nonzero elements since

$$\boldsymbol{b} = \begin{bmatrix} 0 \\ 1 \\ 1 \end{bmatrix}$$

The reader may recall that, in phase variables, the effect of zeros is to add additional nonzero terms to the output vector $\boldsymbol{c}$. In physical variables, additional nonzero terms are normally added in $\boldsymbol{b}$, whereas $\boldsymbol{c}$ usually remains unaffected. The $\boldsymbol{A}$ matrix is also different.

If the state equations are written for the plant of Fig. 2.5-3b, the resulting equations contain a $\dot{u}$ term, and hence they are not of the required form of Eq. ($\boldsymbol{Ab}$). The $\dot{u}$ term arises from the block containing the zero, as may be seen by writing the differential equation corresponding to the transfer function $X_3'(s)/U(s)$. This equation is

$$\dot{x}_3' = -5x_3' + \dot{u} + 3u$$

This difficulty may be avoided by redrawing Fig. 2.5-3b, using the feedforward form shown in Fig. 2.5-4. This configuration is suggested by dividing the denominator into the numerator of $X_3'(s)/U(s)$, leading to the expression

$$\frac{s+3}{s+5} = 1 - \frac{2}{s+5}$$

The consequences of this division are shown in Fig. 2.5-4, where it has become necessary to indicate a new state-variable x_3. If the state equations are written in terms of x_1, x_2, and x_3, with x_3' equal to $u - x_3$, the result is

$$\dot{x}_1 = x_2$$

$$\dot{x}_2 = -x_2 + x_3' = -x_2 - x_3 + u$$

$$\dot{x}_3 = -5x_3 + 2u$$

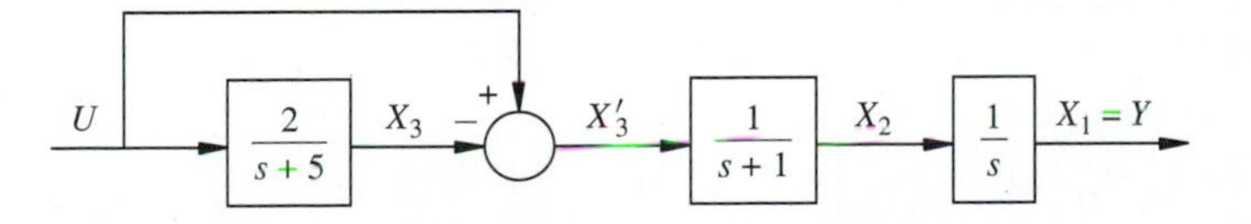

FIGURE 2.5-4
Redrawing of Fig. 2.5-3b to eliminate the $\dot{u}$ term in the describing equations.

Here $\boldsymbol{b}$ again has an additional nonzero term, as $\boldsymbol{b}$ is now

$$\boldsymbol{b} = \begin{bmatrix} 0 \\ 1 \\ 2 \end{bmatrix}$$

It should be noted that, if x_3' is a real, physical variable, then, in all probability, x_3 is not. (However, the configuration of Fig. 2.5-4 and unprimed variables are used hereafter to ensure that the describing equations are of the form ($\boldsymbol{Ab}$) and ($\boldsymbol{c}$).) Answers expressed in terms of x_3 therefore must be converted to x_3' for physical interpretation using the equation $x_3' = u - x_3$.

Before proceeding to the topic of linear changes of variables, let us make the following observations concerning phase variables versus physical variables. The phase-variable method provides a simple, direct, systematic, and unique method of translating the transfer function information into state-variable form. At the same time, phase variables in general lack physical significance, particularly if the plant transfer function has zeros.

In contrast, the physical-variable representation, because of its close relation to the physical plant, does not produce a unique form for the resulting state-variable representation; that is, the same transfer function may yield two or more different representations, depending on the physical plant involved. On the other hand, the state-variables used in this approach are, by their very definition, real, physically meaningful variables.

Although we shall make considerable use of both these approaches to state-variable representation throughout this book, we must never forget that our basic interest is in controlling real plants, and therefore we must ultimately deal with real, physical variables.

Exercises 2.5

2.5-1. Some of the simplest practical physical models are second-order systems arising from Newton's Law, Force equals mass times acceleration ($\boldsymbol{F} = m\boldsymbol{a}$) or its equivalent in rotational terms, Torque equals moment of inertia times angular acceleration ($\boldsymbol{\tau} = I\ddot{\boldsymbol{\theta}}$). In these systems the state-variables of interest are usually the (angular) position and the (angular) velocity of a physical component of the plant. In this exercise we build some simple physical state-variable models of simple systems one may want to control.

(a) Firing a rocket in space produces a thrust that exerts a force on the rocket. Assume that this force is an input and the only force acting on the rocket. (This may be a good approximation in space.) Use the physical state-variables x_1 for position and x_2 for velocity, and form the state-variable description for the differential equation $m\ddot{x}_1(t) = u(t)$ with the force $u(t)$ as the input and the position $x_1(t)$ as the output. Find the transfer function. Such a transfer function is referred to as a double integrator.

Answer:

$$\dot{\boldsymbol{x}}(t) = \begin{bmatrix} 0 & 1 \\ 0 & 0 \end{bmatrix} \boldsymbol{x}(t) + \begin{bmatrix} 0 \\ \frac{1}{m} \end{bmatrix} u(t) \qquad y(t) = [1 \quad 0]\,\boldsymbol{x}(t)$$

$$\frac{Y(s)}{U(s)} = \frac{1}{ms^2}$$

(*b*) If a movable nozzle is used on the rocket of Exercise 2.5-1*a*, some fraction of the thrust is turned into a torque about the center of mass of the rocket. Let this torque τ be the input, $u(t)$, let $x_1(t)$ be the angle of the rocket, $\theta(t)$, and let $x_2(t)$ be the angular velocity of the rocket. Find the state-variable description using $u(t)$ as the input and $y(t) = x_1(t)$ as the output. Find the transfer function. The differential equation describing the system is $\tau = I\ddot{\theta}$.

Answer: Same as (*a*) with I replacing m.

(*c*) On earth, accelerating forces are usually opposed by dissipative forces due to friction, wind resistance, etc. The dissipative force, or damping is generally proportional to the velocity of the mass. In an automobile cruise control system, wind resistance and friction increase with velocity, and the differential equation describing this relationship is

$$m\ddot{p}(t) = u(t) - \beta\dot{p}(t)$$

where $p(t)$ is the automobile's position.

Using $u(t)$ as the input, $x_1(t) = p(t)$ and $x_2(t) = \dot{p}(t)$ as the state-variables, and $y(t) = p(t)$ as the output, give the state-variable description and the transfer function for this system. We have seen the equivalent representation for a rotational system as part of our position control system of Example 2.2-2.

Answer:

$$\dot{\mathbf{x}}(t) = \begin{bmatrix} 0 & 1 \\ 0 & -\frac{\beta}{m} \end{bmatrix}\mathbf{x}(t) + \begin{bmatrix} 0 \\ \frac{1}{m} \end{bmatrix}u(t) \qquad y(t) = [\,1 \quad 0\,]\mathbf{x}(t)$$

$$\frac{Y(s)}{U(s)} = \frac{1}{ms^2 - \beta s}$$

2.5-2. More complicated second-order dynamics result when there is a force such as a spring that attempts to restore a system to a particular position. Such restorative forces may come from an active feedback control system or may occur as part of the plant. Restorative forces often are proportional to the position of a system component, and often produce oscillations or instability in a system. Here is an example of a rotating physical system with a restorative force.

(*a*) The missile shown in Fig. 2.5-5 is traveling through the atmosphere with an input torque $u(t)$ acting on it from a movable nozzle. The missile trajectory is vertically upward but the missile has rotated by an angle $\theta(t)$ around its center.

Assuming that the missile is traveling upward through the atmosphere at a constant speed, the air pressure on the fin creates a restorative torque proportional to $\theta(t)$. There is also a damping force from the atmosphere opposing any rotation. The differential equation for this system is

$$I\ddot{\theta}(t) = u(t) - \beta\dot{\theta}(t) - K\theta(t)$$

Find the state-space description for this system using $u(t)$ as the input, $x_1(t) = \theta(t)$, $x_2(t) = \dot{\theta}(t)$ and $y(t) = \theta(t)$. Find the transfer function $Y(s)/U(s)$.

Answer:

$$\dot{\mathbf{x}}(t) = \begin{bmatrix} 0 & 1 \\ -\frac{K}{I} & -\frac{\beta}{I} \end{bmatrix}\mathbf{x}(t) + [\,0 \quad \tfrac{1}{I}\,]u(t) \qquad y(t) = [\,1 \quad 0\,]\mathbf{x}(t)$$

$$\frac{Y(s)}{U(s)} = \frac{1}{Is^2 - \beta s + K}$$

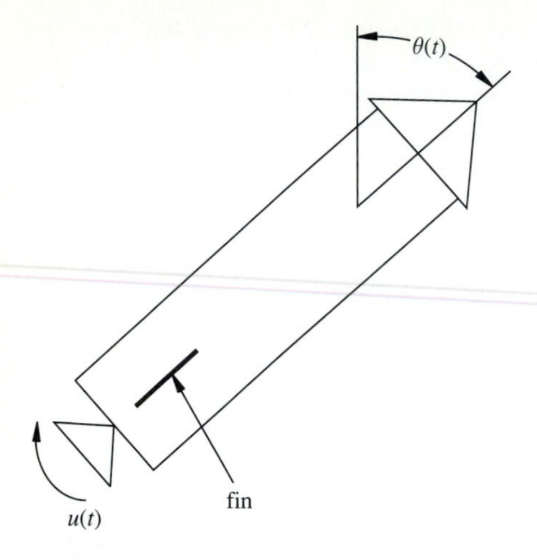

FIGURE 2.5-5
Diagram of a missile for Exercise 2.5-2.

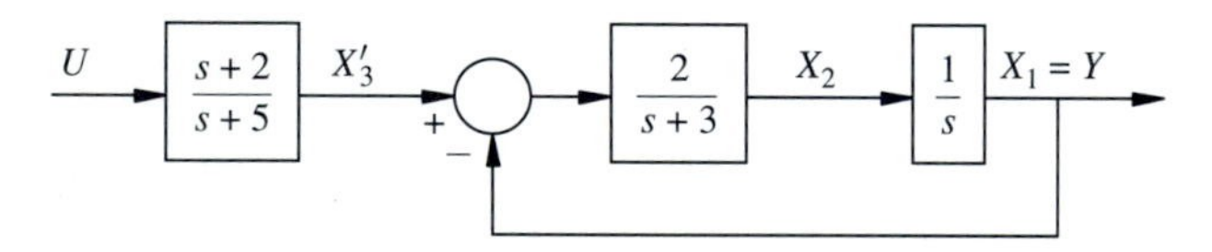

FIGURE 2.5-6
Exercise 2.5-3.

(*b*) When $\beta = 0$ (i.e., no damping) the system is called an oscillator. Show that the poles of this system are purely imaginary.

2.5-3. Find a state-variable representation of the system shown in Fig. 2.5-6 in the form $(\boldsymbol{A}\boldsymbol{b})$ and $(\boldsymbol{c})$ by defining an artificial state-variable $x_3 = x_3' - u$.
Answer:

$$\dot{x}_1 = x_2$$

$$\dot{x}_2 = 2x_1 - 3x_2 + 2x_3 + 2u$$

$$\dot{x}_3 = 5x_3 - 3u$$

$$y = x_1$$

2.5-4. Show that the two block diagrams in Fig. 2.5-7 have the same input-output transfer function.

2.6 LINEAR TRANSFORMATION OF VARIABLES

In the three preceding sections we have discussed the fact that a given plant has an infinite number of state-variable representations. In this section, we examine more

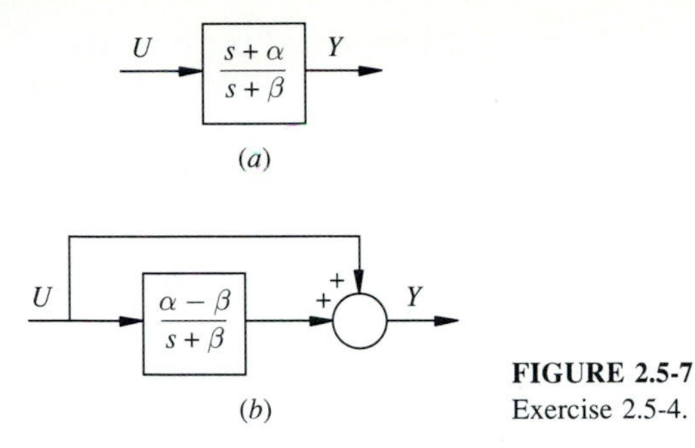

FIGURE 2.5-7
Exercise 2.5-4.

closely the relationship between these various state-variable representations using a technique known as a linear transformation.

To introduce this technique, let us consider once again the field-controlled motor. In Example 2.2-2, the transfer function of a field-controlled motor is developed. From the transfer function we can immediately write down the phase-variable representation by inspection as

$$\dot{\boldsymbol{x}}(t) = \begin{bmatrix} 0 & 1 & 0 \\ 0 & 0 & 1 \\ 0 & -\beta R_f/JL_f & -\left(L_f\beta + R_f J\right)/JL_f \end{bmatrix} \boldsymbol{x}(t) + \begin{bmatrix} 0 \\ 0 \\ 1 \end{bmatrix} u(t)$$

$$y(t) = [\,K_\tau/JL_f \quad 0 \quad 0\,]\,\boldsymbol{x}(t)$$

where

$$x_1(t) = \left(JL_f/K_\tau\right)\theta_o(t), \qquad x_2(t) = \left(JL_f/K_\tau\right)\dot{\theta}_o(t), \qquad x_3(t) = \left(JL_f/K_\tau\right)\ddot{\theta}_o(t),$$

and

$$u(t) = e(t)$$

A second representation is developed in terms of physical variables in Example 2.3-1. This representation takes the form

$$\dot{\boldsymbol{x}}^*(t) = \begin{bmatrix} 0 & 1 & 0 \\ 0 & -\beta/J & K_T/J \\ 0 & 0 & -R_f/L_f \end{bmatrix} \boldsymbol{x}^*(t) + \begin{bmatrix} 0 \\ 0 \\ 1/L_f \end{bmatrix} u(t)$$

$$y(t) = [\,1 \quad 0 \quad 0\,]\boldsymbol{x}^*(t)$$

where

$$x_1^*(t) = \theta_o(t), \qquad x_2^*(t) = \dot{\theta}_o(t), \qquad x_3^*(t) = i_f(t), \qquad \text{and } u(t) = e(t)$$

Here we have starred the physical variables to distinguish them from the phase variables.

For these two representations, there are two questions we investigate. First, how are the two sets of state-variables related? Second, how are the two plant representations related? Both questions have basically the same answer.

Let us begin with the first question. From the definitions of the two sets of state-variables, we easily see that

$$x_1(t) = \left(JL_f/K_\tau\right) x_1^*(t)$$

$$x_2(t) = \left(JL_f/K_\tau\right) x_2^*(t)$$

The relationship between the third variable in each set is not so obvious. To find this relationship, let us consider the second equation of the physical-variable representation

$$\dot{x}_2^*(t) = -\frac{\beta}{J} x_2^*(t) + \frac{K_\tau}{J} x_3^*(t)$$

Since $x_2^*(t) = \left(K_\tau/JL_f\right) x_2(t)$, then $\dot{x}_2^*(t) = \left(K_\tau/JL_f\right) \dot{x}_2(t) = \left(K_\tau/JL_f\right) x_3(t)$, and this equation may be rewritten as

$$\left(K_\tau/JL_f\right) x_3(t) = -\frac{\beta}{J} x_2^*(t) + \frac{K_\tau}{J} x_3^*(t)$$

Therefore, we have the three equations that relate the two sets of variables, and the first question has been answered.

$$x_1(t) = \left(JL_f/K_\tau\right) x_1^*(t)$$

$$x_2(t) = \left(JL_f/K_\tau\right) x_2^*(t)$$

$$x_3(t) = \frac{\beta L_f}{K_\tau} x_2^*(t) + L_f x_3^*(t)$$

Note, however, that each of these three equations is a linear equation, and therefore this result can be put into matrix form as

$$\boldsymbol{x}(t) = \boldsymbol{P}\boldsymbol{x}^*(t) \tag{2.6-1}$$

where

$$\boldsymbol{P} = \begin{bmatrix} JL_f/K_\tau & 0 & 0 \\ 0 & JL_f/K_\tau & 0 \\ 0 & -\beta L_f/K_\tau & L_f \end{bmatrix} \tag{2.6-2}$$

The relationship between the two sets of variables represented by Eq. (2.6-1) is known as a *linear transformation*, and the matrix $\boldsymbol{P}$ is referred to as the *transformation matrix*. Any two sets of state-variables for a linear plant must be related by a linear transformation.

To answer the second question concerning the relationship of the two plant representations, let us consider a general problem. Suppose that we have two sets of state-variables: a set $\boldsymbol{x}(t)$ whose plant representation is

$$\dot{\boldsymbol{x}}(t) = \boldsymbol{A}\boldsymbol{x}(t) + \boldsymbol{b}u(t) \tag{$\boldsymbol{Ab}$}$$

$$y(t) = \boldsymbol{c}^T\boldsymbol{x}(t) \tag{$\boldsymbol{c}$}$$

and a set $\boldsymbol{x}^*(t)$ with the associated representation

$$\dot{\boldsymbol{x}}^*(t) = \boldsymbol{A}^*\boldsymbol{x}^*(t) + \boldsymbol{b}^*u(t) \tag{$\boldsymbol{A}^*\boldsymbol{b}^*$}$$

$$y(t) = \boldsymbol{c}^{*T}\boldsymbol{x}^*(t) \tag{$\boldsymbol{c}^*$}$$

The variables $\boldsymbol{x}(t)$ and $\boldsymbol{x}^*(t)$ are related by the linear transformation

$$\boldsymbol{x}(t) = \boldsymbol{P}\boldsymbol{x}^*(t) \tag{2.6-1}$$

where $\boldsymbol{P}$ is a *constant nonsingular matrix*. The problem is to find the relationship between the matrices $\boldsymbol{A}$, $\boldsymbol{b}$, and $\boldsymbol{c}$ and the matrices $\boldsymbol{A}^*$, $\boldsymbol{b}^*$, and $\boldsymbol{c}^*$.

Let us begin by taking the derivative of both sides of Eq. (2.6-1). Since $\boldsymbol{P}$ has constant elements, it is not involved in the differentiation, and we can write

$$\dot{\boldsymbol{x}}(t) = \boldsymbol{P}\dot{\boldsymbol{x}}^*(t) \tag{2.6-3}$$

The substitution of Eqs. (2.6-1) and (2.6-3) into Eqs. ($\boldsymbol{Ab}$) and ($\boldsymbol{c}$) yields

$$\boldsymbol{P}\dot{\boldsymbol{x}}^*(t) = \boldsymbol{A}\boldsymbol{P}\boldsymbol{x}^*(t) + \boldsymbol{b}u(t) \tag{2.6-4a}$$

$$y(t) = \boldsymbol{c}^T\boldsymbol{P}\boldsymbol{x}^*(t) \tag{2.6-4b}$$

The premultiplication of Eq. (2.6-4a) by $\boldsymbol{P}^{-1}$ yields:

$$\dot{\boldsymbol{x}}^*(t) = \boldsymbol{P}^{-1}\boldsymbol{A}\boldsymbol{P}\boldsymbol{x}^*(t) + \boldsymbol{P}^{-1}\boldsymbol{b}u(t)$$

$$y(t) = \boldsymbol{c}^T\boldsymbol{P}\boldsymbol{x}^*(t)$$

A comparison of these equations with Eqs. ($\boldsymbol{A}^*\boldsymbol{b}^*$) and ($\boldsymbol{c}^*$) above reveals that

$$\boldsymbol{A}^* = \boldsymbol{P}^{-1}\boldsymbol{A}\boldsymbol{P} \qquad \boldsymbol{b}^* = \boldsymbol{P}^{-1}\boldsymbol{b} \qquad \text{and} \qquad \boldsymbol{c}^* = \boldsymbol{P}^T\boldsymbol{c} \tag{2.6-5}$$

Hence we see that the matrix $\boldsymbol{P}$ not only relates the two sets of state-variables by means of Eq. (2.6-1) but also relates their associated representation by means of Eqs. (2.6-5).

The above multiplication by $\boldsymbol{P}^{-1}$ indicates why the $\boldsymbol{P}$ matrix must be nonsingular. (Note that the $\boldsymbol{P}$ matrix given by Eq. (2.6-2) is nonsingular.) However, $\boldsymbol{P}$ may be any nonsingular matrix. Therefore a plant initially expressed by one set of state-variables may be transformed into an infinite number of alternative representations.

To use Eqs. (2.6-5) to find $\boldsymbol{P}$ if $\boldsymbol{A}$, $\boldsymbol{b}$, $\boldsymbol{c}$, $\boldsymbol{A}^*$, $\boldsymbol{b}^*$, and $\boldsymbol{c}^*$ are known, one premultiplies the first two parts of Eq. (2.6-5) by $\boldsymbol{P}$ to obtain

$$\boldsymbol{P}\boldsymbol{A}^* = \boldsymbol{A}\boldsymbol{P} \qquad \boldsymbol{P}\boldsymbol{b}^* = \boldsymbol{b} \qquad \text{and} \qquad \boldsymbol{c}^* = \boldsymbol{P}^T\boldsymbol{c} \tag{2.6-6}$$

These three matrix equations generate a set of $n^2 + 2n$ linear equations in the elements of $\boldsymbol{P}$, which may be solved to determine $\boldsymbol{P}$. Since there are only n^2 elements in the matrix $\boldsymbol{P}$, the reader may wonder how it is possible to satisfy $n^2 + 2n$ equations. The simple answer is that exactly $2n$ of these equations are redundant. Therefore for any two compatible state-variable representations, that is, representations that truly describe the same plant, these equations may always be solved to determine $\boldsymbol{P}$.

Although the above text indicates the use of Eqs. (2.6-6) to obtain the transformation matrix $\boldsymbol{P}$, this is not necessarily the recommended procedure. The method always works, but the algebra is often tedious. The transformation matrix can often be found more directly as it was in Eq. (2.6-2). A transformation is needed when the problem is worked in variables other than the physical variables and it is desired to transform the results to physical variables for implementation.

The use of Eqs. (2.6-5) to transform one state-variable representation into another is simply a matter of selecting a nonsingular $\boldsymbol{P}$ matrix and then substituting $\boldsymbol{A}$, $\boldsymbol{b}$, $\boldsymbol{c}$, and $\boldsymbol{P}$ into Eqs. (2.6-5) to obtain $\boldsymbol{A}^*$, $\boldsymbol{b}^*$, and $\boldsymbol{c}^*$. The only difficult part of this operation is the inversion of the matrix $\boldsymbol{P}$.

Example 2.6-1. To illustrate the use of Eqs. (2.6-5) to transform a state-variable representation, let us assume that the physical-variable representation of the field-controlled motor is not known but that the transformation matrix $\boldsymbol{P}$ from Eq. (2.6-2) and the representation in terms of the phase variables are both known. As a first step we obtain the inverse of $\boldsymbol{P}$

$$\boldsymbol{P}^{-1} = \begin{bmatrix} K_\tau/JL_f & 0 & 0 \\ 0 & K_\tau/JL_f & 0 \\ 0 & \beta/JL_f & 1/L_f \end{bmatrix}$$

Then $\boldsymbol{A}^*$ can be determined through substitution to yield

$$\boldsymbol{A}^* = \begin{bmatrix} K_\tau/JL_f & 0 & 0 \\ 0 & K_\tau/JL_f & 0 \\ 0 & \beta/JL_f & 1/L_f \end{bmatrix} \begin{bmatrix} 0 & 1 & 0 \\ 0 & 0 & 1 \\ 0 & -\beta R_f/JL_f & -\left(L_f\beta + R_f J\right)/JL_f \end{bmatrix} \begin{bmatrix} JL_f/K_\tau & 0 & 0 \\ 0 & JL_f/K_\tau & 0 \\ 0 & -\beta L_f/K_\tau & L_f \end{bmatrix}$$

$$= \begin{bmatrix} 0 & 1 & 0 \\ 0 & -\beta/J & K_\tau/J \\ 0 & 0 & -R_f/L_f \end{bmatrix}$$

The matrix $\boldsymbol{b}^*$ is given by

$$\boldsymbol{b}^* = \begin{bmatrix} K_\tau/JL_f & 0 & 0 \\ 0 & K_\tau/JL_f & 0 \\ 0 & \beta/JL_f & 1/L_f \end{bmatrix} \begin{bmatrix} 0 \\ 0 \\ 1 \end{bmatrix} = \begin{bmatrix} 0 \\ 0 \\ 1/L_f \end{bmatrix}$$

and the matrix $\boldsymbol{c}^*$ becomes

$$\boldsymbol{c}^* = \begin{bmatrix} JL_f/K_\tau & 0 & 0 \\ 0 & JL_f/K_\tau & 0 \\ 0 & -\beta L_f/K_\tau & L_f \end{bmatrix} \begin{bmatrix} K_\tau/JL_f \\ 0 \\ 0 \end{bmatrix} = \begin{bmatrix} 1 \\ 0 \\ 0 \end{bmatrix}$$

If these results are compared with the physical-variable representation given at the beginning of this section, the answers for $\boldsymbol{A}^*$, $\boldsymbol{b}^*$, and $\boldsymbol{c}^*$ are seen to be correct.

The transformation of one state-variable representation into another representation may also be accomplished with block diagrams and block-diagram algebra. However, the proper manipulations to achieve the block-diagram transformations may be quite obscure.

Before concluding our discussion of linear transformation, let us consider briefly two of its properties:

1. Since the input-output transfer function is insensitive to the state-variables used to represent the plant, it must be true that

$$\boldsymbol{c}^{*T}\left(s\boldsymbol{I} - \boldsymbol{A}^*\right)^{-1}\boldsymbol{b}^* = \boldsymbol{c}^T\left(s\boldsymbol{I} - \boldsymbol{A}\right)^{-1}\boldsymbol{b} \tag{2.6-7}$$

2. Since $\det(s\boldsymbol{I} - \boldsymbol{A})$ is equal to the denominator $G_p(s)$, which is also insensitive to the choice of state-variables, then

$$\det(s\boldsymbol{I} - \boldsymbol{A}^*) = \det(s\boldsymbol{I} - \boldsymbol{A}) \tag{2.6-8}$$

This result also indicates that the eigenvalues of the plant are not affected by a linear transformation of the state-variables that represent the plant.

We can show Eq. (2.6-8) is true by using Eq. (2.6-5)

$$\det(s\boldsymbol{I} - \boldsymbol{A}^*) = \det\left(s\boldsymbol{I} - \boldsymbol{P}^{-1}\boldsymbol{A}\boldsymbol{P}\right)$$

Use $\boldsymbol{I} = \boldsymbol{P}^{-1}\boldsymbol{P}$ to obtain

$$\det(s\boldsymbol{I} - \boldsymbol{A}^*) = \det\left(\boldsymbol{P}^{-1}(s\boldsymbol{I} - \boldsymbol{A})\boldsymbol{P}\right)$$

Now, using the fact that the determinant of a product of square matrices is equal to the product of the determinants of the matrices

$$\begin{aligned}
\det(s\boldsymbol{I} - \boldsymbol{A}^*) &= \det\left(\boldsymbol{P}^{-1}\right)\det(s\boldsymbol{I} - \boldsymbol{A})\det\boldsymbol{P} \\
&= \det(s\boldsymbol{I} - \boldsymbol{A})\det\left(\boldsymbol{P}^{-1}\right)\det\boldsymbol{P} \\
&= \det(s\boldsymbol{I} - \boldsymbol{A})\det\left(\boldsymbol{P}^{-1}\boldsymbol{P}\right) \\
&= \det(s\boldsymbol{I} - \boldsymbol{A})\det(\boldsymbol{I}) \\
&= \det(s\boldsymbol{I} - \boldsymbol{A})
\end{aligned}$$

The last equality follows from the fact that the determinant of the identify matrix is unity.

Equation (2.6-7) can be proven with similar substitutions.

$$\begin{aligned}
\boldsymbol{c}^{*T}(s\boldsymbol{I} - \boldsymbol{A}^*)^{-1}\boldsymbol{b}* &= \boldsymbol{c}^T\boldsymbol{P}\left(s\boldsymbol{I} - \boldsymbol{P}^{-1}\boldsymbol{A}\boldsymbol{P}\right)^{-1}\boldsymbol{P}^{-1}\boldsymbol{b} \\
&= \boldsymbol{c}^T\boldsymbol{P}\left(\boldsymbol{P}^{-1}(s\boldsymbol{I} - \boldsymbol{A})\boldsymbol{P}\right)^{-1}\boldsymbol{P}^{-1}\boldsymbol{b} \\
&= \boldsymbol{c}^T\boldsymbol{P}\,\boldsymbol{P}^{-1}(s\boldsymbol{I} - \boldsymbol{A})^{-1}\boldsymbol{P}\,\boldsymbol{P}^{-1}\boldsymbol{b} \\
&= \boldsymbol{c}^T(s\boldsymbol{I} - \boldsymbol{A})^{-1}\boldsymbol{b}
\end{aligned}$$

2.6.1 CAD Notes

Linear transformations of state-variables are carried out through matrix multiplication and inversion. The operations are performed easily with MATLAB.

Example 2.6-2. In the CAD Notes of Sec. 2.4 it is mentioned that the phase-variable description of Sec. 2.4 and the phase-variable description used by MATLAB differ by a reverse ordering of the state-variables. Thus, these two descriptions are related by a linear transformation. Let the MATLAB version be $\boldsymbol{x}$ and the version of Sec. 2.4 be $\boldsymbol{x}^*$, so that

$$x_1 = x_3^*$$

$$x_2 = x_2^*$$

$$x_3 = x_1^*$$

or

$$\boldsymbol{P} = \begin{bmatrix} 0 & 0 & 1 \\ 0 & 1 & 0 \\ 1 & 0 & 0 \end{bmatrix}$$

Find the **A***, **b***, and **c*** that represent the phase-variable description of Sec. 2.4.

```
>>%create MATLAB phase-variable description
num=[1 10 24];
den=[1 12 46 56];
[Ap,bp,cp,dp]=tf2ss(num,den)
Ap =
  -12   -46   -56
    1     0     0
    0     1     0
bp =
    1
    0
    0
cp =
    1   10   24
dp =
    0

%enter P
P=[0 0 1;0 1 0;1 0 0]
P =
   0   0   1
   0   1   0
   1   0   0

%compute inverse of P
Pinv=inv(P)
Pinv =
     0   0   1
     0   1   0
     1   0   0

%compute As,bs,cs
As=Pinv*Ap*P
As =
    0     1     0
    0     0     1
  -56   -46   -12
```

```
bs=Pinv*bp
bs =

   0
   0
   1

%Matlab returns a 1 x 3 matrix for cp so cp must
%be transposed to be like c in the book's notation
cs=P'*cp';
cs'
ans =
  24  10  1
```

Exercises 2.6

2.6-1. For the armature-controlled motor find the transformation matrix $\boldsymbol{P}$ that relates the phase-variable representation found in Exercise 2.4-3 to the physical-variable representation of Sec. 2.5 using physical reasoning rather than Eqs. (2.6-6). Verify that this $\boldsymbol{P}$ matrix satisfies Eqs. (2.6-6).

Answer:

If $\boldsymbol{x}$ = physical variables and $\boldsymbol{x}^*$ = phase variables, then

$$\boldsymbol{x} = \begin{bmatrix} 0 & \beta/JL_a & 1/L_a \\ K_\tau'/JL_a & 0 & 0 \\ 0 & K_\tau'/JL_a & 0 \end{bmatrix} \boldsymbol{x}^*$$

2.6-2. A plant has the state-variable representation

$$\dot{\boldsymbol{x}}(t) = \begin{bmatrix} -1 & 1 & 1 \\ 1 & 0 & 1 \\ 0 & 0 & 1 \end{bmatrix} \boldsymbol{x}(t) + \begin{bmatrix} 0 \\ 0 \\ 2 \end{bmatrix} u(t) \qquad y(t) = [\,1 \quad 0 \quad 1\,]\boldsymbol{x}(t)$$

(*a*) Find the phase-variable representation.

(*b*) Let $\boldsymbol{x}^*$ represent the phase variables. Find the linear transformation $\boldsymbol{P}$ so that

$$\boldsymbol{x} = \boldsymbol{P}\boldsymbol{x}^*$$

2.6-3. In an example in the next section we perform a common linear transformation by taking the sum and difference of the state-variables. Starting with the equations

$$\tau_m = J/2\,\ddot{\theta}_m + \beta/2\,\dot{\theta}_m + K_s\left(\theta_m - \theta_p\right)$$

$$K_s\left(\theta_m - \theta_p\right) = J/2\,\ddot{\theta}_p + \beta/2\,\dot{\theta}_p$$

find the state-variable realization using θ_m, $\dot{\theta}_m$, θ_p, and $\dot{\theta}_p$. Find the linear transformation into the state-variables $\theta_{\text{ave}} = \left(\theta_m + \theta_p\right)/2$, $\theta_{\text{diff}} = \left(\theta_m + \theta_p\right)/2$ and their derivatives and apply the transformation.

Answer:

$$\boldsymbol{A} = \begin{bmatrix} 0 & 1 & 0 & 0 \\ 0 & -\beta/J & 0 & 0 \\ 0 & 0 & 0 & 1 \\ 0 & 0 & -K/J & -\beta/J \end{bmatrix} \qquad \boldsymbol{b} = \begin{bmatrix} 0 \\ 1/J \\ 0 \\ 1/J \end{bmatrix}$$

2.7 LIMITATIONS OF MATHEMATICAL MODELS

Any mathematical representation, whether it be a state-variable representation or a transfer function representation, is, at best, an approximation of the behavior of a real, physical system. This point cannot be overemphasized. By necessity, the material in this book is devoted to methods of manipulating and controlling mathematical entities. The final goal, however, is to be able to control real, physical systems. It is therefore important to know how the behavior of physical systems differs from the mathematically predicted behavior. We can then use this knowledge to alter our mathematical manipulations so that the controllers designed in the mathematical world are likely to behave properly when implemented in the real, physical world. In this section, we investigate and classify some common limitations that models may have in describing physical systems. We shall use the example of the field-controlled dc motor whose mathematical model was developed in Example 2.2-2 in transfer function form and later in state-variable form.

2.7.1 Parameter Inaccuracies

The first problem with mathematical models of physical systems is that the parameters used in the models cannot be determined with absolute accuracy. Inaccurate parameters can arise from many different factors. Experiments performed to establish parameters are subject to inaccuracies in data collection. Experimental conditions may not reproduce operating conditions exactly. The values of parameters may change with time due to wear. Finally, we would like to design one controller for a number of systems that are supposed to be identical but which inevitably differ in many ways including different parameter values. Parts cannot be manufactured identically. There will always be some differences.

Example 2.7-1. In the field-controlled motor of Example 2.2-2, the electrical parameters R_f and L_f may be manufactured within a 10 percent tolerance; for example, the actual parameter value will be within the range given by the advertised value plus or minus 0.10 times the advertised value for every motor manufactured. The mechanical constants β and J may vary even more greatly as the operating conditions of the plant change.

Assume that the motor is used to position a large antenna platform at the correct angle to receive the signal from a particular satellite. The moment of inertia J will change as components of the telescope are changed or moved. The damping β will change as parts wear and friction is increased. We would like a controller that will perform well over a range of parameter values.

In another application, assume that a motor is used to control the daisy wheel of a computer's printer. Clearly, not every printer wheel will come off the manufacturing line with exactly the same parameters. Still, we would like to design one control law that will perform well for each printer produced.

In the remainder of this book we learn to design controllers that create an input such that the output of a mathematical model behaves as we desire. We find that by using feedback we can reduce the sensitivity of the output to changes in plant parameters. From the above discussion we can see that low sensitivity to plant parameter variation is vital if the controller is to behave properly when attached to a real, physical system, which invariably exhibits different parameters than the mathematical model.

2.7.2 Unmodeled Dynamics

In creating a mathematical model of a physical system, it is often desirable to keep the model as simple as possible while retaining enough information to produce an effective controller. Thus, high-order effects that are possible but difficult to model are often ignored. We must be sure to remember that such high-order effects are present even if they don't appear explicitly in our model.

We must be sure that our controller behaves properly in the presence of unmodeled dynamics. We can assume that the unmodeled dynamics have certain characteristics. The most important characteristic assumed about the unmodeled dynamics is that they have little effect on the plant output as long as the input to the plant contains the bulk of its energy at low frequencies. This characteristic enables us to design controllers that behave well in the presence of unmodeled dynamics by assuring that the controllers produce only low frequency inputs. The need for a tradeoff between using high frequency inputs for quick response and avoiding high frequency inputs for safety against unmodeled dynamics is further developed in Sec. 6.6 and thereafter.

Effects that often are not included in low-order models used for controller designs include small time delays in process control systems and bending modes in elements of mechanical systems. The example of a bending mode in the angular positioning system that follows shows some typical characteristics of unmodeled dynamics.

Example 2.7-2. Assume again that the field-controlled motor of Example 2.2-2 is used to position a large antenna platform. In Example 2.2-2 we assumed that the torque produced by the motor appeared directly on the load, in this case, the antenna platform. In reality there is a shaft that connects the motor to the platform and this shaft twists when a torque is applied to one end. To get a better model of the dynamics associated with the twisting shaft, we can model the shaft with two segments connected by a torsional spring, as in Fig. 2.7-1.

The equations for this system are

$$\tau_m = J_m \ddot{\theta}_m + \beta_m \dot{\theta}_m + K_s \left(\theta_m - \theta_p\right)$$

$$K_s \left(\theta_m - \theta_p\right) = J_p \ddot{\theta}_p + \beta_p \dot{\theta}_p$$

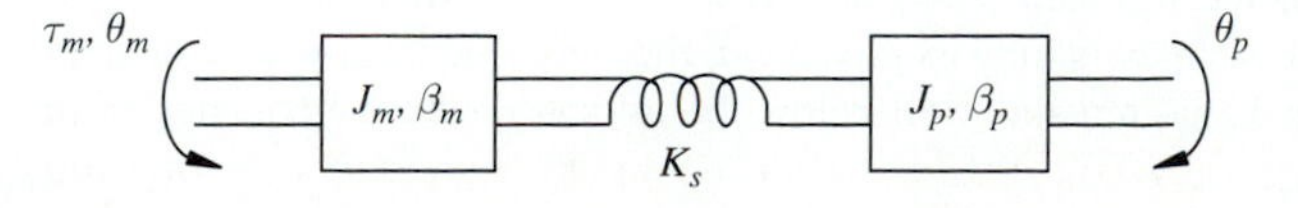

FIGURE 2.7-1
Example 2.7-2. A flexible shaft modeled with two segments.

where J_m is the moment of inertia of the motor
β_m is the damping in the motor
J_p is the moment of inertia of the platform
β_p is the damping of the platform
K_s is the spring constant indicating the torsional stiffness of the shaft

These equations can be put in a more convenient form by taking the sum and the difference of the equations. (Note that these operations are the same as performing a linear transformation on the state-variable representation of these equations. See Exercise 2.6-3.)

$$\tau_m = J_m\ddot{\theta}_m + J_p\ddot{\theta}_p + \beta_m\dot{\theta}_m + \beta_p\dot{\theta}_p$$

$$\tau_m = J_m\ddot{\theta}_m - J_p\ddot{\theta}_p + \beta_m\dot{\theta}_m - \beta_p\dot{\theta}_p + 2K_s\left(\theta_m - \theta_p\right)$$

To understand the qualitative behavior of this system, let us assume that

$$J_m = J_p = J/2$$

$$\beta_m = \beta_p = \beta/2$$

We also define

$$\theta_{\text{ave}} = \frac{\theta_m + \theta_p}{2}$$

$$\theta_{\text{diff}} = \frac{\theta_m - \theta_p}{2}$$

The equations then become

$$\tau_m = J\ddot{\theta}_{\text{ave}} + \beta\dot{\theta}_{\text{ave}}$$

$$\tau_m = J\ddot{\theta}_{\text{diff}} + \beta\dot{\theta}_{\text{diff}} + K_s\theta_{\text{diff}}$$

In transfer function form

$$\Theta_{\text{ave}}(s) = \frac{T_m(s)}{Js^2 + \beta s}$$

$$\Theta_{\text{diff}}(s) = \frac{T_m(s)}{Js^2 + \beta s + K_s}$$

$$\Theta_p(s) = \Theta_{\text{ave}}(s) - \Theta_{\text{diff}}(s) = \left[\frac{1/J}{s(s+\beta/J)} - \frac{1/J}{s^2 + \beta/Js + K_s/J}\right] T_m(s)$$

Thus, the original block diagram for the field-controlled motor position system given in Figure 2.7-2*a* is replaced by the block diagram of Fig. 2.7-2*b* with the twisting of the shaft is taken into account.

The only difference between these two block diagrams is the additional box containing the dynamics associated with the *bending mode*, $\Theta_m - \Theta_p$. We learn more about classifying the behavior of such dynamics in Chap. 4. However, at this point we can see that if the shaft is stiff, so that K_s/J is large, the steady-state response of the bending mode dynamics to low frequency inputs is small compared to the response, Θ_{ave}. However, if there is little damping, the steady-state response of the bending mode dynamics to a sinusoidal input at the resonant frequency $\omega = \sqrt{K_s/J}$ is quite large. Thus, the bending mode dynamics can be safely neglected if the controller can ensure that the frequency of the torque signal remains well below the critical resonant frequency of the bending mode.

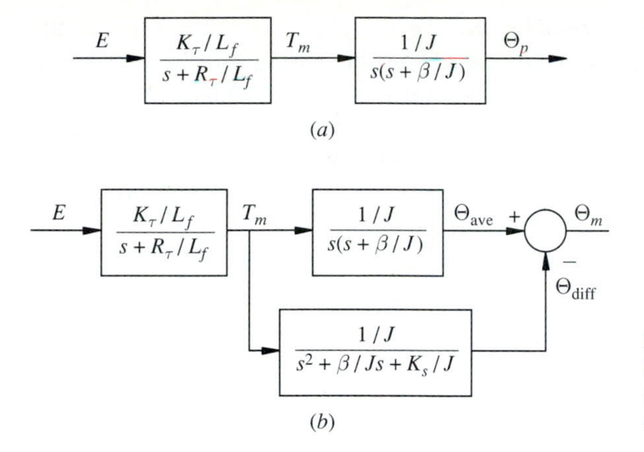

FIGURE 2.7-2
Block diagram for motor controlling a shaft. (*a*) Stiff shaft; (*b*) shaft with a bending mode.

The plant model used to design the controller is called the *nominal model*. All unmodeled dynamics such as the bending mode dynamics of Example 2.7-2 can be grouped under the heading of *plant perturbations*. When designing a controller using a nominal model, one must always remain aware of the possible plant perturbations that have been left out of the model.

There is a question that naturally arises here. Why not include everything that can be modeled in the nominal model and not worry about perturbations? This question can be answered in two stages. First, it is impossible to completely model a physical system with a transfer function or state-variable representation since the physical system has distributed parameters rather than lumped parameters and has infinite dimensionality. In Example 2.7-2 the shaft was modeled with two segments connected by a torsional spring. It could have easily been modeled with three, four, or any number of segments producing higher-order, more accurate models.

No matter how well a system is nominally modeled, there will always be unmodeled dynamics to produce plant perturbations.

Once it is realized that the physical dynamics are never modeled completely, the engineer must then decide how accurate a model is adequate. Large models are cumbersome and may obscure the fundamental aspects of a problem. Parameter inaccuracies in higher dimensional models may render the additional dimensions useless. The control designer needs a model which describes the physical system accurately enough to enable control objectives to be met when using the controller both on the model and on the physical system modeled. If more stringent requirements are made on the control system, better models are required. If better models are unavailable, control requirements must be relaxed. It is important that the control designer realize that there will always be perturbations to the nominal model and that the controller to be designed must meet its objectives in the face of those perturbations.

2.7.3 Nonlinearities

All of the models of physical systems used in this book are linear and time-invariant. The property of linearity is required to use either the transfer function model or the state-variable model as we have defined them. We use linear modeling because the analysis techniques we develop for linear models are far more powerful and useful than any techniques which have been developed for nonlinear models.

While physical systems are in general nonlinear in their responses to inputs, most physical systems can be accurately modeled by linear systems as long as all variables associated with the system are constrained to remain within a restricted range. The linear system may be derived using the technique of *linearization*, which uses the Taylor series expansion of a function. We shall demonstrate the technique on a first-order system. Suppose

$$\dot{y} = -\sin y + u$$

This system is nonlinear and we do not take the Laplace transform of this equation. We can linearize this system by first choosing an *operating point*. The operating point is made up of one value of each variable in a problem. For the model obtained by linearization to remain valid, each variable must stay near its operating point value. The functions in the problem can then be expanded in a Taylor series around the operating point. The Taylor series expansion for the function $f(x)$ around the point x_o is given by

$$f(x) = f(x_o) + \left.\frac{df}{dx}\right|_{x=x_o} (x - x_o) + \frac{1}{2!}\left.\frac{d^2 f}{dx^2}\right|_{x=x_o} (x - x_o)^2 + \cdots$$

If all terms except the first two are dropped, the result is a function which has the form of a linear function plus a constant. The resulting function approximates the original function as long as x stays close to x_o so that the values of the dropped terms remain small. In our example, if we take the operating point to be $y = 0$, the Taylor expansion of $\sin y$ is given by

$$\sin y = y - \frac{y^3}{3!} + \frac{y^5}{5!} + \cdots$$

The linearization of the system is given by

$$\dot{y} = -y + u$$

As long as y stays near the operating point, $y = 0$, the linear system's behavior closely approximates the nonlinear system's behavior. Note that if we are controlling this system we must take care that the control signal u never drives y too far from the operating region or our model becomes invalid.

Example 2.7-3 Linearization of a chemical reaction process. Control of large chemical reactions is a vitally important part of industrial production. The mathematical models for such reactions are usually nonlinear. To create a model that is useful for control, an operating point is chosen and the system of equations is linearized around the nominal operating point.

Consider the process depicted schematically in Fig. 2.7-3. A reaction tank of volume $V = 5000$ gallons accepts a feed of reactant which contains a substance A in concentration $C_{A,0}$. The feed enters at a rate of F gallons per hour and at a temperature T_0. In the tank some of the reactant A is turned into the desired product B. The output

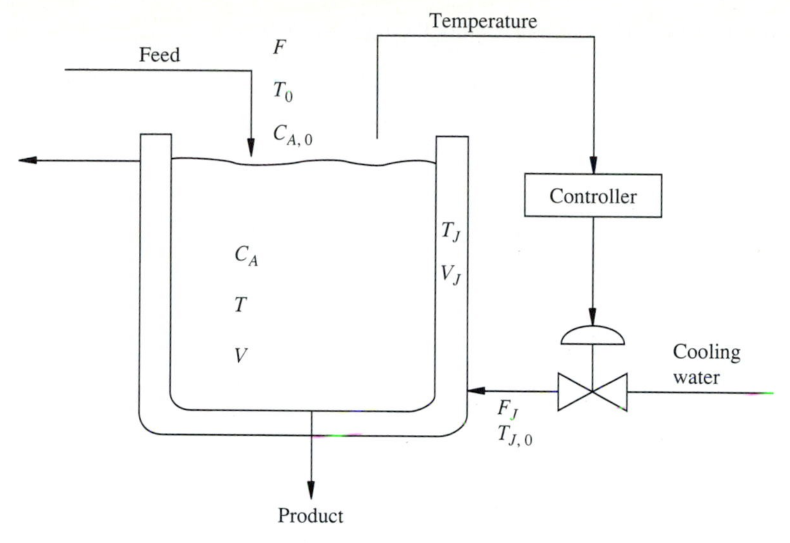

FIGURE 2.7-3
Schematic view of a chemical reaction tank.

product mixture is removed from the tank at the same rate that the feed enters the tank. The contents of the tank are mixed continuously.

Perfect mixing is assumed (i.e., mixing dynamics are ignored) so that the mixture in the tank is assumed to have a uniform concentration of A, C_A, and a uniform temperature, T.

The chemical reaction is exothermic so that if the tank is not properly cooled, the temperature in the tank rises, the speed of reaction increases and creates even more heat, which further increases the speed of reaction in an unstable manner. The tank is cooled by a flow of water around the jacket of the tank. Heat is transferred from the tank to the the water in the jacket. The temperature of the water in the jacket is assumed uniform at T_J. The temperature of the water flowing into the jacket is $T_{J,0}$. The system is controlled by measuring the temperature, T, in the tank and controlling the flow of water in the jacket, F_J, by actuating a valve. If the cooling water flows too slowly the system heats up. If the cooling water flows too quickly, the reaction slows down and poor product yield results.

There are three equations describing the evolution of C_A, concentration of A in the tank, T, the temperature in the tank, and T_J, the temperature in the cooling jacket. The change in the concentration of A arises from three terms: the amount of A that is added with the feed, the amount of A that leaves with the product flow, and the amount of A that is used up in the reaction. The higher the concentration of A and the higher the temperature, the faster A is used up in the reaction. This relationship is expressed mathematically as

$$\dot{C}_A = \frac{F}{V}C_{A,0} - \frac{F}{V}C_A - k_1 C_A e^{-\frac{k_2}{T}}$$

Inserting correct values for the constants and assuming that $C_{A,0} = 0.5$ yields

$$\dot{C}_A = (0.83)(0.5) - 0.83C_A - 7.08 \times 10^{10} C_A e^{-\frac{15098}{T}}$$

This equation is linearized by approximating the right-hand side

$$f(C_A, T) = (0.83)(0.5) - 0.83C_A - 7.08 \times 10^{10} C_A e^{-\frac{15098}{T}}$$

with

$$f(C_A, T) \approx f\left(\overline{C}_A, \overline{T}\right) + \left.\frac{\partial f(C_A, T)}{\partial C_A}\right|_{C_A=\overline{C}_A, T=\overline{T}} \delta C_A + \left.\frac{\partial f(C_A, T)}{\partial T}\right|_{C_A=\overline{C}_A, T=\overline{T}} \delta T$$

where

$$C_A \approx \overline{C}_A + \delta C_A \qquad \text{and} \qquad T \approx \overline{T} + \delta T$$

In this case,

$$\left.\frac{\partial f(C_A, T)}{\partial C_A}\right|_{C_A=\overline{C}_A, T=\overline{T}} = 7.08 \times 10^{10} e^{-\frac{15098}{\overline{T}}} - 0.83$$

and

$$\left.\frac{\partial f(C_A, T)}{\partial T}\right|_{C_A=\overline{C}_A, T=\overline{T}} = \left(-\frac{15098}{\overline{T}^2}\right) 7.08 \times 10^{10} \overline{C}_A e^{-\frac{15098}{\overline{T}}}$$

The left-hand side of the nonlinear differential equation is simplified by observing that

$$\dot{C}_A \approx \dot{\overline{C}}_A + \delta\dot{C}_A = \delta\dot{C}_A \qquad \text{since } \overline{C}_A \text{ is constant}$$

The resulting linearized equation is

$$\delta\dot{C}_A \approx (0.83)(0.5) - 0.83\overline{C}_A - 7.08 \times 10^{10} \overline{C}_A e^{-\frac{15098}{\overline{T}}} + \left(7.08 \times 10^{10} e^{-\frac{15098}{\overline{T}}} - 0.83\right) \delta C_A + \left(-\frac{15098}{\overline{T}^2}\right) 7.08 \times 10^{10} \overline{C}_A e^{-\frac{15098}{\overline{T}}} \delta T$$

A linearized equation for the evolution of the temperature of the fluid in the reaction tank is derived using similar logic. There are four terms describing the change in temperature: a term for the heat that enters with the feed flow, a term for the heat that leaves with the product flow, a term for the heat created by the reaction and a term for the heat that is transferred to the cooling jacket. The heat generated by the reaction depends on the speed of the reaction, which in turn depends on the concentration of A and the current temperature in the reaction tank. The speed with which heat is transferred to the cooling jacket depends on the difference in temperature between the fluid in the reaction tank and the fluid in the cooling jacket. This relationship is expressed mathematically as

$$\dot{T} = \frac{F}{V} T_0 - \frac{F}{V} T - k_3 k_1 C_A e^{-\frac{k_2}{T}} - k_4 (T - T_J)$$

Inserting correct values for the constants and assuming that $T_0 = 70°$F yields

$$\dot{T} = (0.83)(70) - 0.83T - (800)\left(7.08 \times 10^{10}\right) C_A e^{-\frac{15098}{T}} - 2.4(T - T_J)$$

This equation is linearized in the same way as is the concentration equation except that this equation must be linearized with respect to three state-variables. The result is

$$\delta\dot{T} = (0.83)(70) - 0.83\overline{T} - (800)\left(7.08 \times 10^{10}\right) C_A e^{-\frac{15098}{\overline{T}}} - 2.4\left(\overline{T} - \overline{T}_J\right)$$
$$+ (800)\left(7.08 \times 10^{10}\right) e^{-\frac{15098}{\overline{T}}}\delta C_A$$
$$+ \left(\left(-\frac{15098}{\overline{T}^2}\right)(800)\left(7.08 \times 10^{10}\right)\overline{C}_A e^{-\frac{15098}{\overline{T}}} - 0.83 - 2.4\right)\delta T + 2.4\delta T_J$$

There are three terms associated with the changes of the temperature of the fluid in the jacket: one term representing the heat entering the jacket with the cooling fluid flow, one term representing the heat leaving the jacket with the outflow of cooling liquid, and one term representing the heat transferred from the fluid in the reaction tank to the fluid in the jacket. This relationship is expressed as

$$\dot{T}_J = \frac{F_J}{V_J}T_{J,0} - \frac{F_J}{V_J}T_J + k_5\left(T - T_J\right)$$

Inserting correct values for the constants and assuming that $F_J/V_J = 13$ and $T_{J,0} = 70°\text{F}$ yields

$$\dot{T}_J = (13)(70) - 13T_J + 148\left(T - T_J\right)$$

This equation is also linearized around the nominal jacket flow, F_J, which functions as the control variable with a nominal value of $F_J = 1240$ gallons per hour. After linearization this equation becomes

$$\delta\dot{T}_J = (13)(70) - 13\overline{T}_J + 148\left(\overline{T} - \overline{T}_J\right) - 13\delta T_J + 148\left(\delta T - \delta T_J\right) - 0.16\delta F_J$$

To find the equilibrium state $(\overline{C}_A, \overline{T}, \overline{T}_J)$ one sets the incremental variables $(\delta C_A, \delta T, \delta T_J)$ and their derivatives equal to zero. This indicates that the state-variables do not change when the system is at the operating point. This procedure creates three equations with the operating point variables as the three unknowns. One can solve this system of equations and find

$$\overline{C}_A = 0.245, \qquad \overline{T} = 140, \qquad \overline{T}_J = 93.3$$

By substituting the equilibrium values into the linearized equations, the state-variable description is attained

$$\begin{bmatrix}\delta\dot{C}_A \\ \delta\dot{T} \\ \delta\dot{T}_J\end{bmatrix} = \begin{bmatrix}-1.7 & -2.13 \times 10^{-4} & 0 \\ 696 & 2.9 & 2.4 \\ 0 & 6.5 & -19.5\end{bmatrix}\begin{bmatrix}\delta C_A \\ \delta T \\ \delta T_J\end{bmatrix} + \begin{bmatrix}0 \\ 0 \\ -.16\end{bmatrix}\delta F_J$$

$$\delta T = [0 \quad 1 \quad 0]\begin{bmatrix}\delta C_A \\ \delta T \\ \delta T_J\end{bmatrix}$$

This state-space system describes the behavior of the actual system with reasonable accuracy only when the incremental variables remain small. One can check how differently the system would respond if the variables were allowed to change substantially from their operating values by trying different equilibrium values in the linearized equations. Often different control laws are designed for different operating points and the control laws are changed depending on the value of certain measured variables. Such a technique of control design is called *gain scheduling*.

It should be noted that besides the model imperfections attributable to the approximations made in the linearization process, there are other modeling imperfections present. The dynamics of the stirring process have been neglected; also, it is assumed that a change in temperature is measured instantaneously. The dynamics of the chemical reaction have been greatly simplified. In all, this model is typical in that it provides a good enough description to design a control system as long as it is not extended too far.

Even though a model may not have been derived by linearization techniques, every physical system is likely to contain some nonlinear behavior. The next example looks at possible nonlinearities in the field-controlled motor positioning system. The example also shows how nonlinear effects can be partially accounted for within the context of linear systems.

Example 2.7-4. In the derivation of the field-controlled motor, it was assumed that the torque of the motor is proportional to the field current

$$\tau = K_\tau i_f$$

In reality the torque is proportional to the magnetic flux created by the field current. The magnetic flux and thus the torque is related to the field current by the relationship shown in Fig. 2.7-4.

This curve has two nonlinear effects. The first effect is saturation. If the field current is increased past a certain value, the torque will not increase but will level off or saturate due to the saturation of the magnetic flux.

Saturation is a common nonlinear phenomenon. Control surfaces on airplanes can be moved only so far; valves in process control systems cannot be opened further than fully open. Saturation may be dealt with by assuring that the variables causing saturation are held within the region for which the linear model is applicable. This is similar to the measures taken when a formal linearization procedure is followed.

The second nonlinear effect exhibited by the torque versus field current curve of Fig. 2.7-4 is the effect of hysteresis. As the current is increased the right-hand curve is followed, and as the current is decreased the left-hand curve is followed. When the current changes from increasing to decreasing or from decreasing to increasing, the flux

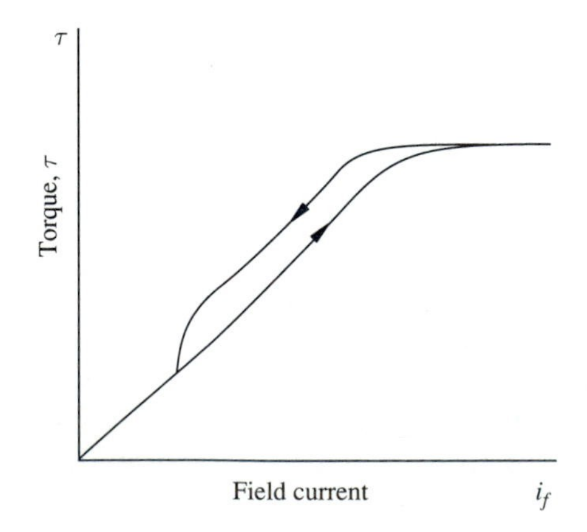

FIGURE 2.7-4
Torque versus field current showing saturation and hysteresis in a field-controlled dc motor.

and the torque remain constant while the relationship switches from one curve to the other.

A similar effect may occur in the gearing that connects a motor to a platform. In gears the effect is known as *backlash*. Backlash occurs when the gears do not mesh precisely. A gap of length δ may exist between gear teeth as shown schematically in Fig. 2.7-5.[1] Let the position of one gear be given by x and the position of the other gear be given by y. As x goes through a cyclic variation the relationship of y to x is as depicted in Fig. 2.7-6. This is similar to the situation of magnetic hysteresis.

The nonlinear effect of backlash or hysteresis may be partially captured in the context of a linear system model by examining the output $y(t)$ if the input $x(t)$ is a sinusoid. The curves of $x(t)$ and $y(t)$ versus time are depicted in Fig. 2.7-7.

The most important aspect of the relationship between x and y for control system design is that the major frequency component of $y(t)$ lags the sinusoid $x(t)$ in phase. In linear control applications systems improperly handled phase lags can cause unwanted oscillations and possibly instability. Even though we cannot deal with the

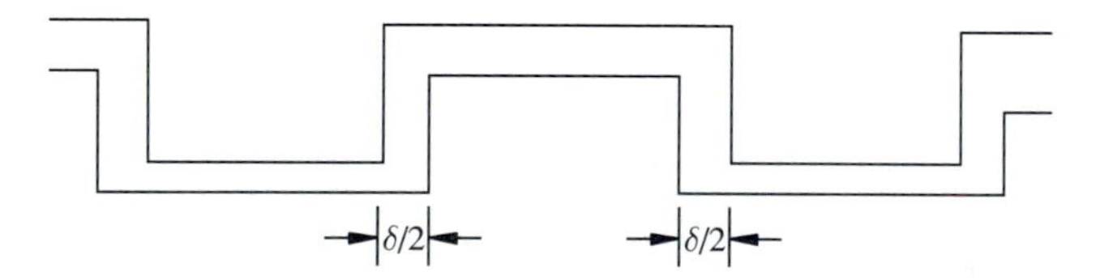

FIGURE 2.7-5
Schematic representation of a gap in gear teeth creating backlash.

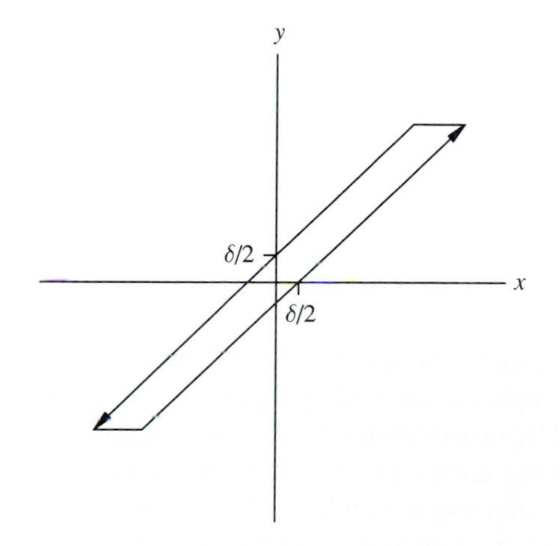

FIGURE 2.7-6
Graph of the position of one gear versus another during cyclic variation when backlash is present.

[1]The drawing of Fig. 2.7-5 is a crude schematic drawing whose purpose is not to show how gears look but rather to help define the problem. In particular, gear teeth are not rectangular.

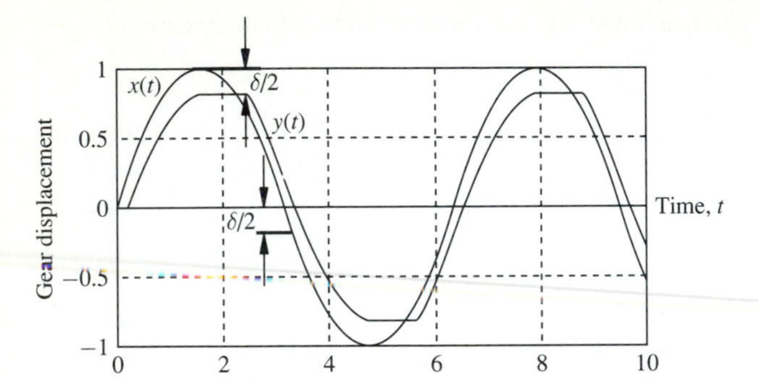

FIGURE 2.7-7
The time plots of two gears when $x(t)$ is sinusoidal and there is backlash between the two gears.

nonlinear model of backlash or hysteresis directly while using the techniques of linear control theory, it is important to know that these effects may exist and to allow for the extra phase lag that may be produced.

In the remainder of this book, we spend the bulk of our time learning techniques to analyze and design controllers that achieve control objectives when used with the nominal plant model. However, we must always remain aware that our design should be able to withstand plant perturbations without a serious degradation in performance. In Chaps. 5 and 6 we develop a method of including the effects of plant perturbations in our analysis.

2.7.4 CAD Notes

CAD tools usually provide the ability to simulate a set of differential equations that describe a system. Usually, the tool calls on a numerical integration routine to provide the solution of the differential equations at sample points. The MATLAB function for simulating a state-space system has the form

```
y=lsim(a,b,c,d,u,t,x0)
```

The matrices `a,b,c,` and `d` describe the system; the vector `t` provides the time instants of interest; the matrix *u* provides the inputs at the times given by `t`; `x0` gives the initial conditions for the state-variables. The matrix *u* should have one dimension that matches the dimension of `t` and the other dimension equal to the number of input channels in *b*. The output *y* has one dimension equal to the dimension of `t` and the other dimension equal to the number of output channels in `c`. Systems in transfer function form can be simulated assuming zero initial conditions by the command

```
y=lsim(num,den,u,t)
```

More extensive simulation capabilities including the simulations of nonlinear systems are available in a separate package for MATLAB.

Example 2.7-5 Simulation of the chemical reactor system. In this example the chemical reactor system developed in Example 2.7-3 is simulated. The first simulation explores what happens when the flow of the cooling fluid is increased.

```
>>% enter state space matrices
a=[-1.7 -.000213 0; 696 2.9 2.4;0 6.5 -19.5];
b=[0; 0; -0.16];
% let c be the identity matrix so that
% all the states may be inspected
c=eye(3);
d=[0; 0; 0];
% let the initial conditions be zero
x0=[0; 0; 0];
% set the time vector
t=[0.0:0.01:2];
% let the input be a unit step function indicating that
% the cooling flow is increased and held at the new value
u=ones(1,201);
% run the simulation
% y will have three columns for the three state-variables
y=lsim(a,b,c,d,u,t,x0);
% plot the state-variables versus time; label the axes.
plot(t,y),xlabel('time(hours)'),ylabel('state-variables')
% pause stops the program until a key is struck
pause
```

The plot produced by MATLAB appears as Fig.2.7-8*a*. MATLAB plots all three components of y. The first component, the concentration of A, is plotted using a solid line; however, the values are so small compared to the size of the values of the other variables plotted that the solid line appears as a line with zero ordinate. The second component of y, the temperature in the reaction tank, is plotted with a dashed line and can be seen to be decreasing exponentially, i.e., the variable δT grows exponentially with a negative sign. If the simulation is allowed to run longer, the temperature continues to decrease through zero and through unboundedly large negative values. Of course, unboundedly large negative temperatures don't make sense. The problem is that the linearized model makes sense only for values of the variables near their nominal values. The third component of y, the temperature in the cooling jacket, is plotted with dotted line and can be seen to also be decreasing exponentially.

The problem of not being able to see the response of the concentration of A can be solved either by rescaling the first state-variable or by creating a separate plot of the first output, as is done here.

```
% plot only y1, the concentration of A
% place grid lines on plot
plot(t,y(:,1)),xlabel('time(hours)'),
               ylabel('concentration of A'),grid
pause
```

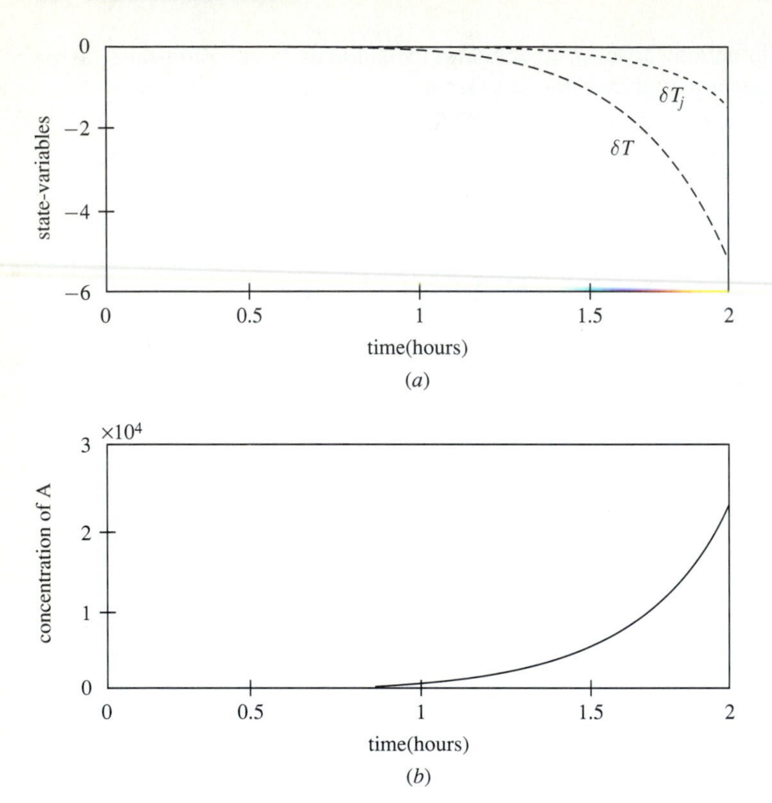

FIGURE 2.7-8*a,b*
Simulations of a chemical reaction system. (*a*) Increased cooling flow; (*b*) concentration of A.

The plot appears in Fig. 2.7-8*b*. Note the scale factor on the ordinate. The concentration of A increases exponentially, indicating that less product is produced. This variable will continue to increase through the point where the linearized model becomes meaningless.

Sometimes, it is desirable to observe how a system evolves after a state-variable has been disturbed from its equilibrium value when there is no change in input. Such a desire is handled by simulating from a nonzero initial condition. Assume that the concentration of A in the reaction tank becomes somewhat larger than its equilibrium value. (Maybe a perturbation of the strength of the feed occurs and is then corrected, leaving the feed correct but the tank mixture off from equilibrium.) The following simulation shows the result.

```
% simulate the system with no input but with an initial
% concentration of A slightly larger than the equilibrium value
u=zeros(1,201);
x0=[.01; 0; 0];
y=lsim(a,b,c,d,u,t,x0);
plot(t,y),xlabel('time(hours)'),ylabel('state-variables')
```

The resulting plot appears in Fig. 2.7-8*c*. All three state-variables grow exponentially (although the first state's growth cannot be seen due to scale). Clearly this system

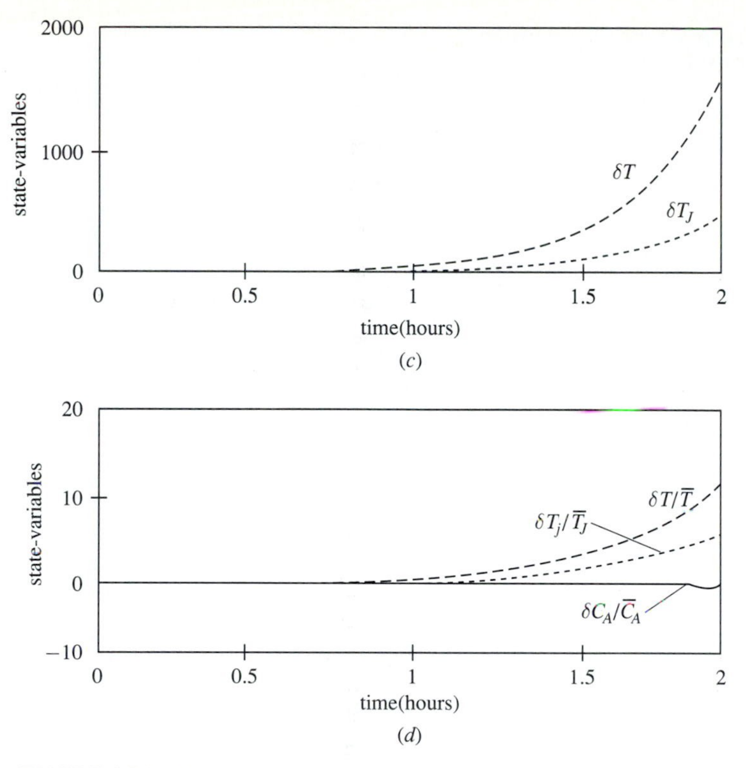

FIGURE 2.7-8*c,d*
Simulations of a chemical reaction system. (*c*) Initial offset in concentration; (*d*) initial offset in concentration with scaled state-variables.

is in need of control. The equilibrium of this system is said to be unstable since any small perturbation away from the equilibrium results in movement of the state-variables further away from equilibrium.

The problem of the various state-variables being on different scales can be alleviated by scaling the state-variables. A logical choice of scale factors is to scale each linearized variable by its equilibrium value. Such a scaling has added benefits in that it usually results in a state-variable description which is more conducive to numerical manipulation. It also provides a simpler interpretation of when the state-variables are too large for the linearized model to be valid since the new state-variables are interpreted as a percentage change from equilibrium. The scaling is performed by a simple transformation of state. In the notation of Sec. 2.6, letting the new variables be the starred variables means that P^{-1} is chosen as follows:

```
>>%Scale the state-variables using a linear transformation
%enter Pinv
Pinv=[1/.245 0 0;0 1/140 0;0 0 1/93.3]
Pinv =
  4.0816       0       0
       0  0.0071       0
       0       0  0.0107
```

```
%compute P as the inverse of Pinv
P=inv(Pinv)
P =
   0.2450          0          0
        0   140.0000          0
        0          0    93.3000

%compute as,bs,cs
as=Pinv*a*P
as =
  -1.7000  -0.1217          0
   1.2180   2.9000     1.5994
        0   9.7535   -19.5000
bs=Pinv*b


bs =
       0
       0
 -0.0017

% Since the c matrix used here is just the identity matrix used
% to see the states, this c won't be changed for the simulation
y=lsim(as,bs,c,d,u,t,x0);
plot(t,y),xlabel('time(hours)'),ylabel('state-variables')
```

The resulting plot appears in Fig. 2.7-8*d*. Here the ending value of the temperature near 10 means that the temperature is about ten times its nominal value of 140°, or 1400°. This of course agrees with the value in Fig.2.7-8*c*.

The linear transformation is completed by computing

```
c=[0  1/140  0]
```

The resulting state-space description uses δF_J as an input and T as an output. However, it is sensible to also normalize the input and output variables by their nominal values. The resulting state-space description we use in the remainder of the book for the chemical reaction system is

$$\begin{bmatrix} \delta\dot{C}_A/\overline{C}_A \\ \delta\dot{T}/\overline{T} \\ \delta\dot{T}_J/\overline{T}_J \end{bmatrix} = \begin{bmatrix} -1.7 & -0.12 & 0 \\ 1.218 & 2.9 & 1.6 \\ 0 & 9.75 & -19.5 \end{bmatrix} \begin{bmatrix} \delta C_A/\overline{C}_A \\ \delta T/\overline{T} \\ \delta T_J/\overline{T}_J \end{bmatrix} + \begin{bmatrix} 0 \\ 0 \\ -2.1 \end{bmatrix} \delta F_J/\overline{F}_J$$

$$\delta T/\overline{T} = [0 \quad 1 \quad 0] \begin{bmatrix} \delta C_A/\overline{C}_A \\ \delta T/\overline{T} \\ \delta T_J/\overline{T}_J \end{bmatrix}$$

It is interesting to note that the unstable dynamics shown here for an exothermic chemical reaction process are very similar to the dynamics associated with the reaction process involved in some nuclear power plants under certain operating conditions. However, in the nuclear power plant there are control rods available to control the rate of the reaction itself.

Exercises 2.7

2.7-1. Consider the nonlinearity of backlash in gearing. Assume that two gears have a gap of δ as shown in Figure 2.7-5. The major effect in response to a sinusoidal input is that the output $y(t)$ lags the input by the distance $\delta/2$ as shown in Figure 2.7-7. If the input $x(t)$ is a sinusoid of amplitude A, approximate the phase lag ϕ in radians between $x(t)$ and $y(t)$ as a function of δ and A
Answer:

$$\phi \approx \sin^{-1} \frac{\delta}{2A}$$

2.7-2. A saturation nonlinearity is particularly bothersome when there is an integrator in the feedback loop. In the system of Figure 2.7-9 assume that the maximum output of the saturation box is 1 volt. If $r(t)$ is a constant 2 volts, what happens to the output of the integrator over time?
Answer: The integrator output increases without bound. This effect is given the name *integrator windup*.

2.7-3. A common set of neglected dynamics arises from small time delays in the system. If the input to a time delay box is $x(t)$, the output is $x(t-T)$, the original signal delayed by an amount T. Find the transfer function for a time delay, and give the magnitude and phase response as a function of frequency $s = j\omega$.
Answer:

$$G(s) = e^{-sT}$$

$$|G(j\omega)| = 1 \text{ for all } \omega$$

$$\arg(G(j\omega)) = -\omega T$$

2.7-4. The equation of motion for a simple pendulum is given by

$$u(t) - mgl \sin\theta(t) = ml^2 \frac{d^2\theta(t)}{dt^2}$$

where $u(t)$ is an input force on the pendulum and $\theta(t)$ is the angle of the pendulum with $\theta(t) = 0$ when the pendulum hangs straight down. The constants are g, the acceleration due to gravity, m, the mass at the end of the pendulum, and l, the length of the pendulum. Linearize the system around $\theta(t) = 0$ and give the resulting transfer function from $U(s)$ to $\Theta(s)$. Now, linearize the system around $\theta(t) = \pi$. Let $\Delta\theta(t) = \theta(t) - \pi$ and give the resulting transfer function from $U(s)$ to $\Delta\Theta(s)$.

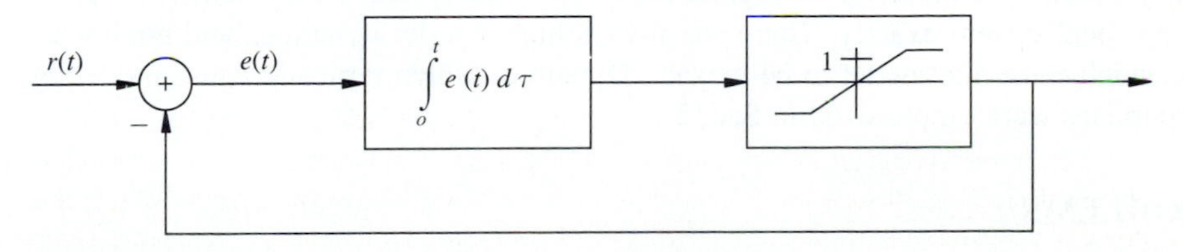

FIGURE 2.7-9
Loop with integrator and saturation nonlinearity for Exercise 2.7-2.

Answers:

$$\frac{\Theta(s)}{U(s)} = \frac{1}{ml^2s^2 + mgl}$$

$$\frac{\Delta\Theta(s)}{U(s)} = \frac{1}{ml^2s^2 - mgl}$$

2.8 CONCLUSIONS

In this chapter the concepts and techniques of plant representation have been developed and discussed. Two basic methods of plant representation were presented. The first, known as the input-output representation, deals only with the terminal characteristics of the plant and ignores all internal behavior. The second, the state-variable representation, provides a description of the internal as well as the terminal behavior of the plant.

In addition to the introduction of these two methods, considerable effort has been made to provide the reader with manipulative techniques appropriate and useful for each method, such as the block diagram and matrix algebra as well as the linear transformation. The interrelations of the two means of plant representation have also been emphasized, often by means of the block diagram.

The block diagram was introduced at the same time that the concept of the transfer function was introduced. However, the careful reader no doubt observed that we continued to use the block diagram to picture a variety of different plants described in state-variable form. This is possible because the block diagram is a picture of transformed differential equations, regardless of the form in which they appear. Block diagrams will continue to be used throughout the book.

In the third-order examples considered in this section, the original describing differential equations were neither one third-order equation nor three first-order equations. In models of electrical, mechanical, and electromechanical plants, it often happens that the original plant is described by sets of coupled first- and second-order equations. As convenient, these coupled equations are arranged as one nth-order differential equation, associated with the transfer function, or n first-order equations, associated with the state-variable representation, as shown graphically in Fig. 2.8-1. Regardless of how the plant is represented, a block diagram may always be used to display the relationship of the plant variables.

All the representations discussed in this chapter begin from a small set of linear differential equations with constant coefficients. While such equations can fairly closely predict the behavior of a great many physical systems, they cannot describe any physical system exactly. There are always higher order dynamics and nonlinearities which cause the model to be inexact. Dynamics which typically cause modeling inaccuracies were emphasized in Sec. 2.7.

PROBLEMS

2.1. A summing and inverting integrator can be realized with a feedback circuit around an operational amplifier or op amp as shown in Fig. P2.1. Assume that the op amp has

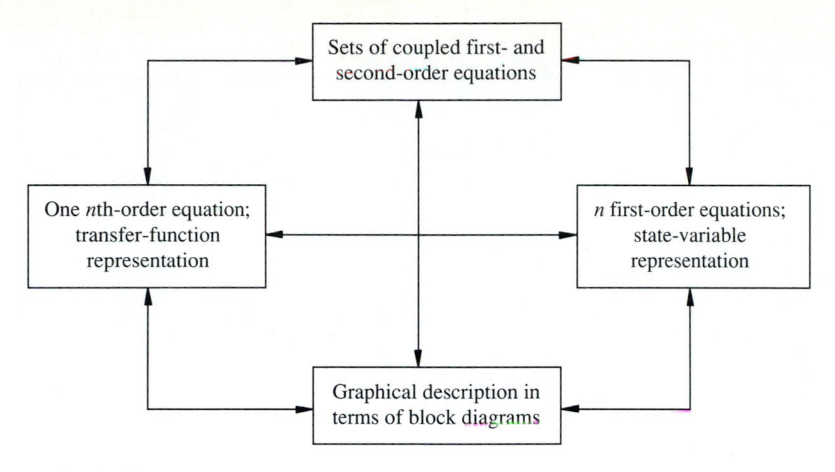

FIGURE 2.8-1
Methods of plant representation.

infinite input impedance so that $i_A(t) = 0$. Assume that the op amp amplifies the voltage by a large amount A so that

$$y(t) = Av_1(t)$$

Write an expression for the sum of the currents leaving the node at the top of the op amp; then find $\dot{y}(t)$ in terms of $y(t)$, $u_1(t)$, and $u_2(t)$. Because the amplification is very large, let $A \to \infty$ and look at the resulting equation.

The circuit given can provide a realization for the summer and the $1/s$ block in a block diagram. Repeat the problem with the capacitor C replaced by a resistor R but this time find $y(t)$ in terms $u_1(t)$ and $u_2(t)$. What do you get?

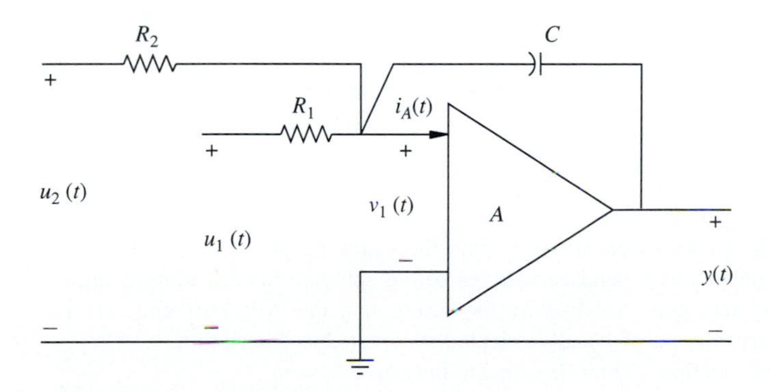

FIGURE P2.1
Circuit for Problem 2.1.

2.2. Use the results of Problem 2.1 to replace the op amp circuits of Fig. P2.2 with summers, blocks of $1/s$, or inverters to produce a block diagram of the circuit. What is the transfer function $Y(s)/U(s)$?

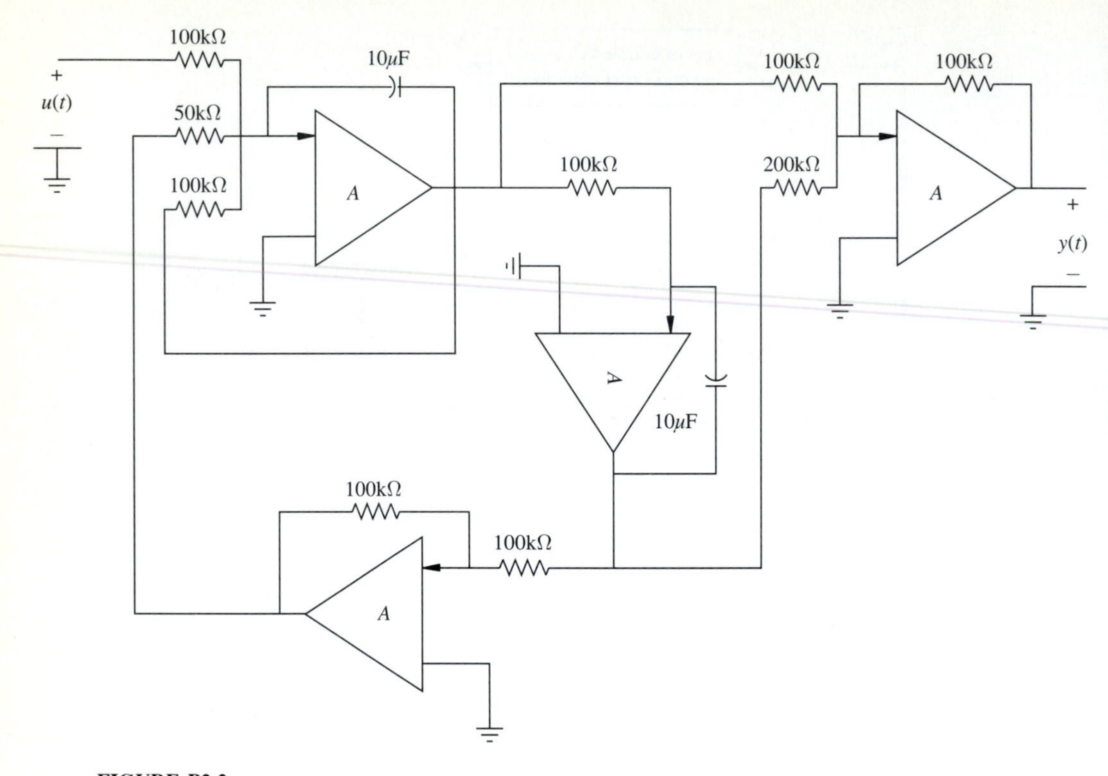

FIGURE P2.2
Circuit for Problem 2.2.

2.3. An often-used academic control experiment consists of an inverted pendulum on a cart as in Fig. P2.3. Controlling the angle of the pendulum without measuring this angle is similar to trying to balance a broom on your hand while you are blindfolded. The equations of motion for this problem are given by

$$u(t) = (M + m)\frac{d^2 z(t)}{dt^2} + ml\frac{d^2 \sin\theta(t)}{dt^2}$$

$$mg\sin\theta(t) = m\frac{d^2(z(t)\cos\theta(t))}{dt^2} + ml\frac{d^2\theta(t)}{dt^2}$$

(a) To linearize this system of equations around the operating point $z(t) = 0$, $\theta = 0$, $d\theta(t)/dt = 0$ you should expand derivatives before applying linearization identities. Linearize this system and explain why the assumption that $d\dot{\theta}(t)/dt$ and $z(t)$ are small is necessary. Give a physically-oriented state-variable description.

(b) Find the transfer function $Z(s)/U(s)$ for the linearized system.

2.4. An automotive cruise control system demonstrates how additional dynamics can be easily appended in a physical-variable description of a system. Assume initially for a cruise control system that the force, F_T, produced to accelerate a car of mass M is proportional to throttle position δ_T, which is the plant input, that is, $F_T = K_T\delta_T$. There is a dissipative or damping force caused by friction and aerodynamic drag that is proportional to the velocity of the car and opposes the motion, that is, $F_d = -K_d V$.

(a) Write a state-variable description using Newton's law, $\boldsymbol{F} = m\boldsymbol{a}$, with δ_T as the input

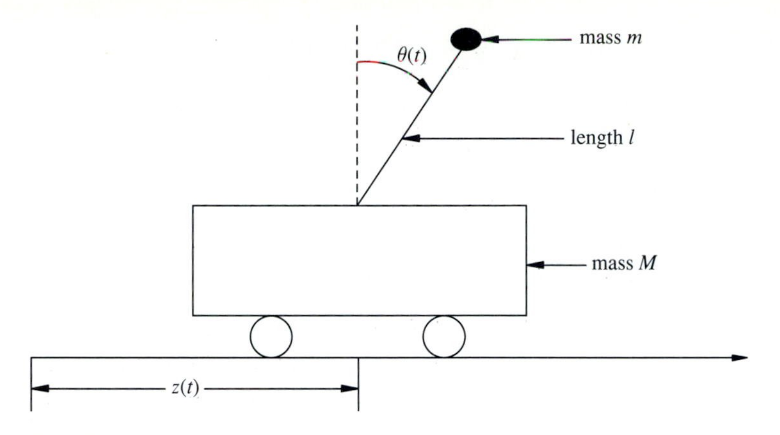

FIGURE P2.3
The inverted pendulum on a cart for Problem 2.3.

and the velocity V as the output.

(*b*) Of course, the force F_T on the car is not directly proportional to the throttle position but is the output of the motor dynamics. These dynamics may be very complicated but we will represent them as a simple first-order system.

$$\dot{F}_T = -\frac{1}{\tau_T}F_T + \frac{K_T}{\tau_T}\delta_T$$

Now write the state-space equations relating the input δ_T and the output V. What are the state-variables?

2.5. Series and parallel connections for state-space descriptions can be solved in general.

Let
$$\dot{\boldsymbol{x}}_1(t) = \boldsymbol{A}_1\boldsymbol{x}_1(t) + \boldsymbol{b}_1 u_1(t) \qquad y_1(t) = \boldsymbol{c}_1^T\boldsymbol{x}_1(t)$$
$$\dot{\boldsymbol{x}}_2(t) = \boldsymbol{A}_2\boldsymbol{x}_2(t) + \boldsymbol{b}_2 u_2(t) \qquad y_2(t) = \boldsymbol{c}_2^T\boldsymbol{x}_2(t)$$

be state-space descriptions of two systems.

(*a*) Find the state-space description of the series connection of these two systems as shown in Fig. P2.5*a*.

(*b*) Find the state-space description of the parallel connection of the two systems shown in Fig. P2.5*b*.

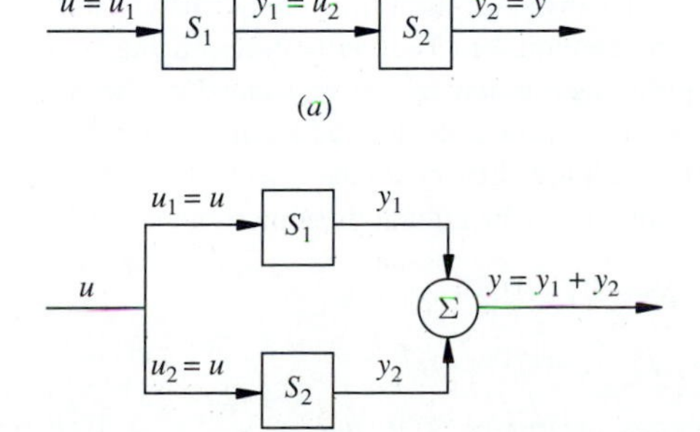

FIGURE P2.5
Series and parallel connections for Problem 2.5.

2.6. The A matrix in a physical-variable description often has most of its non-zero elements on or near the diagonal of the matrix. The eigenvalues of such matrices are usually less sensitive to small changes in the matrix elements than the matrices that result from phase-variable descriptions. (The sensitivity of the phase-variable description also demonstrates the sensitivity of the roots of a transfer function to small changes in the coefficients of the transfer function. Why?)

(*a*) Use the state-variables given to find the A matrix of a physical-variable type description of the system of Fig. P2.6. Find the A matrix of the phase-variable description of the system. Find the eigenvalues of both A matrices.

(*b*) Find the eigenvalues of the phase-variable A matrix if the a_{31} element is changed by 1/8 to 0. What happens to the eigenvalues of the physical-variable type description if the a_{31} element is changed by 1/8? What if the a_{32} element is changed by 1/8? What if the a_{33} element is changed by 1/8?

(*c*) This problem with phase variables becomes worse as the number of variables becomes larger. Start with the phase-variable description of an nth order system with pole polynomial $D(s) = s^n$. This system has n eigenvalues at zero. Find the eigenvalues if the a_{n1} element is changed to 2^{-n}. How much does the magnitude of each eigenvalue change with this small change to one element when n is large?

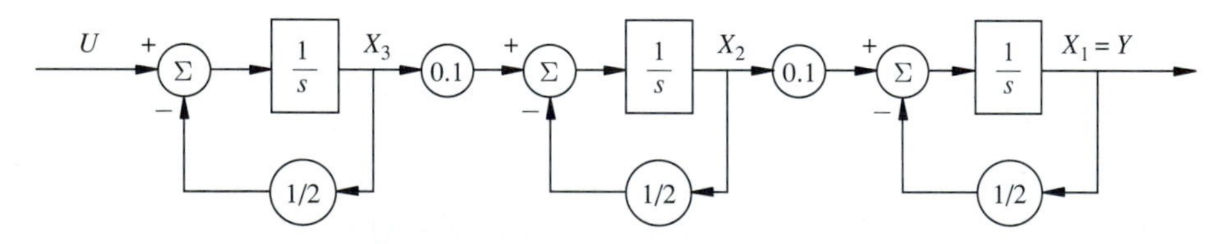

FIGURE P2.6
System for Problem 2.6.

2.7. Dynamics due to bending and flexing often occur in elements that are usually thought to be rigid. An extreme case is the behavior of a building during an earthquake or heavy winds. Control systems are being studied to actively control the flexibility of buildings during earthquakes. In this problem we examine the model of a one story structure, shown schematically in Fig. P2.7. The building is modeled as a foundation connected by walls to the mass of the first floor, m_1. Let x_1 be the distance the first floor is offset from the center due to being flexed by the wind or earthquake. The walls between the two floors produce a restorative force and damping is provided by the resistance of the air surrounding the building. A large mass, m_2, with a spring, a damper and a motor attached may be used to control the oscillations of the building. Let x_2 be the offset of the mass m_2 from the same reference point used to measure m_1. The equations of motion are given by

$$m_1\ddot{x}_1(t) = -\beta_1\dot{x}_1(t) - K_1x_1(t) + K_2\left(x_2(t) - x_1(t)\right)$$

$$m_2\ddot{x}_2(t) = -\beta_2\dot{x}_2(t) - K_2\left(x_2(t) - x_1(t)\right) + u(t)$$

Representative values for the parameters are $m_1 = 100$, $m_2 = 1$, $K_1 = 100$, $K_2 = 10000$, $\beta_1 = 1$, and $\beta_2 = 200$. Use these values throughout the problem.

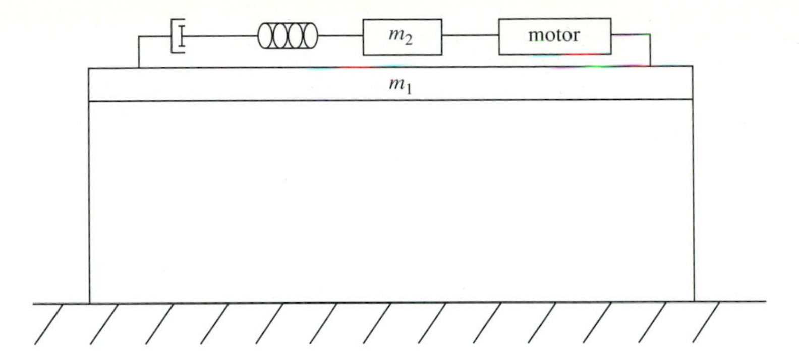

FIGURE P2.7
Building with earthquake control.

(*a*) Using the numerical values given except for K_2, write the state equations for this problem using x_1, $\dot{x}_1$, x_2, and $\dot{x}_2$ as state-variables, $y = x_1$ as the output and u the force produced by the motor, as the input. Find the eigenvalues of the **A** matrix when $K_2 = 0$. The pair of complex eigenvalues with very small real part indicate that the building left on its own oscillates badly. Now find the eigenvalues when $K_2 = 10000$. (A root finding or an eigenvalue program is needed for this part.) The fact that all the eigenvalues are real indicates that the oscillation has been stopped by the counterweight even when the motor is inactive. Proper control of the motor can further improve the response. For the rest of the problem use $K_2 = 10000$.

(*b*) A more easily measurable set of physical variables is

$$x_1' = x_1 \qquad \dot{x}_1' = \dot{x}_1$$

$$x_2' = x_2 - x_1 \qquad \dot{x}_2' = \dot{x}_2 - \dot{x}_1$$

Use a linear transformation to find the state-variable description using the new state-variables.

(*c*) Draw the elementary block diagram for the system with the new state-variables. Find the transfer function using block-diagram manipulations and verify it using the matrix equation (2.3-11). Give the phase-variable representation for the system.

2.8. One place where delay enters in control systems is in the digital implementation of a controller. Some fairly crude approximations capture the main effect of the digital implementation with a single delay of half the sampling period. A block view of a digitally-implemented control system is shown in Fig. P2.8-1.

The digital-to-analog converter (D/A) is conceptually broken into two pieces: a continuous-signal reconstructor which forms an analog signal from the digital signal and a zero-order-hold that holds the value of the signal constant between sample times. The combination of the analog-to-digital converter (A/D), the digital controller and the signal reconstructor can be designed to closely match over a specified frequency range a desired analog transfer function $G_C(s)$ times a scale factor $1/T$, where T is the time between samples. (The quantity $1/T$ is called the sampling rate.) The transfer function of the zero-order-hold can be lumped into the plant as unmodeled dynamics. In this problem we derive the effect of this unmodeled portion. The zero-order-hold turns the impulse coming out of the signal reconstructor into a staircase function as shown in Fig. P2.8-2. The zero-order-hold is therefore a linear time-invariant system with the response to a

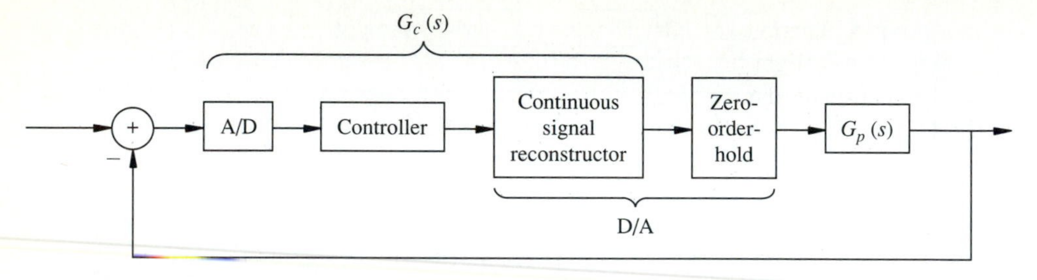

FIGURE P2.8-1
Block diagram for a digitally-implemented controller.

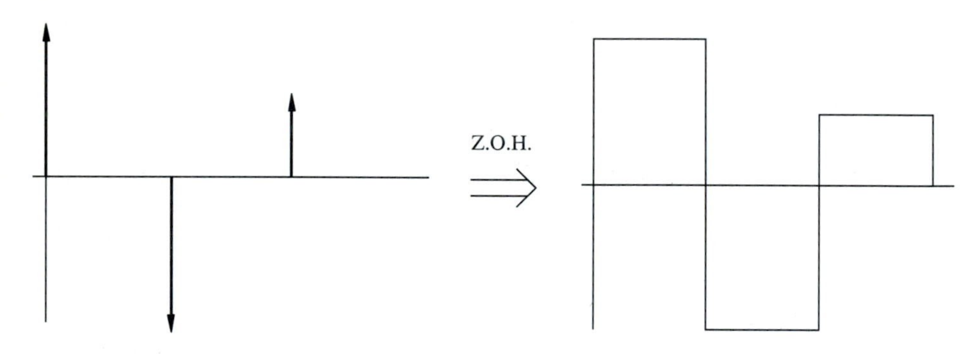

FIGURE P2.8-2
The effect of a zero-order-hold on a string of impulses.

single unit impulse being a pulse of duration T.

$$h(t) = u(t) - u(t - T)$$

Find the transfer function of the zero-order-hold. Show that the frequency response $H(j\omega)$ of the zero-order-hold can be approximated by

$$H(j\omega) = Te^{-\frac{j\omega T}{2}} \qquad \text{for} \qquad \omega \ll 2/T$$

The frequency response $e^{-\frac{j\omega T}{2}}$ is just the frequency response of a time delay of one-half of the sampling period.

Hint:

$$1 - e^{-j\omega T} = 2je^{-\frac{j\omega T}{2}} \frac{\left(e^{\frac{j\omega T}{2}} - e^{-\frac{j\omega T}{2}}\right)}{2j} = 2je^{-\frac{j\omega T}{2}} \sin\left(\frac{\omega T}{2}\right)$$

2.A APPENDIX TO CHAPTER 2—AIRPLANE DYNAMICS

The knowledge assimilated in learning control theory is applicable to problems in many different fields. Often general knowledge of system behavior makes it easier to understand and appreciate the dynamics of a particular system. One of the more interesting applications of control theory is the control of airplanes. The dynamics associated with flight are very complicated. In this appendix to Chap. 2 we examine some airplane dynamics. It can be fun to use system knowledge to understand something that is part of our everyday existence but that few people understand. This section is written in the form of problems but it can be read with benefit even if the problems are not worked. For the second problem some CAD tool is needed to perform simulations.

Problem 2.A-1 Linearization of the longitudinal dynamics of an airplane. Through linearization, fairly simple dynamic models of airplanes can be obtained. These linearized equations predict the airplane behavior well near steady-state flight conditions. The motion of an airplane is usually described using axes aligned with the airplane's axis as shown in Fig. 2.A-1. In this problem we investigate the longitudinal motion of the airplane, that is, the forward motion and particularly the control of altitude. Side and turning motions are called lateral motions and are not considered here.

Intimately connected with a plane's altitude is the plane's pitch angle, which measures whether the plane's nose is pointing up or down. The pitch angle, θ, is the angle between the horizon and the plane's x-axis. The direction of an airplane's motion is usually not along its x-axis. Usually an airplane is pitched up slightly so that the aerodynamics of its wings develop enough lift to overcome gravity. An angle, α, called the angle of attack, is defined as the angle between the velocity vector of the airplane and the airplane body-oriented x-axis. Under most flight conditions the amount of lift increases when α increases. Level flight occurs not when $\theta = 0$ but when $\theta = \alpha$. In Fig. 2.A-1 the plane is climbing since $\theta > \alpha$. The state-variables of the longitudinal dynamics are usually taken as

V	the magnitude of the velocity vector
α	the angle of attack, which gives the direction of the velocity vector
θ	the pitch angle
$q = \dot{\theta}$	the pitch rate

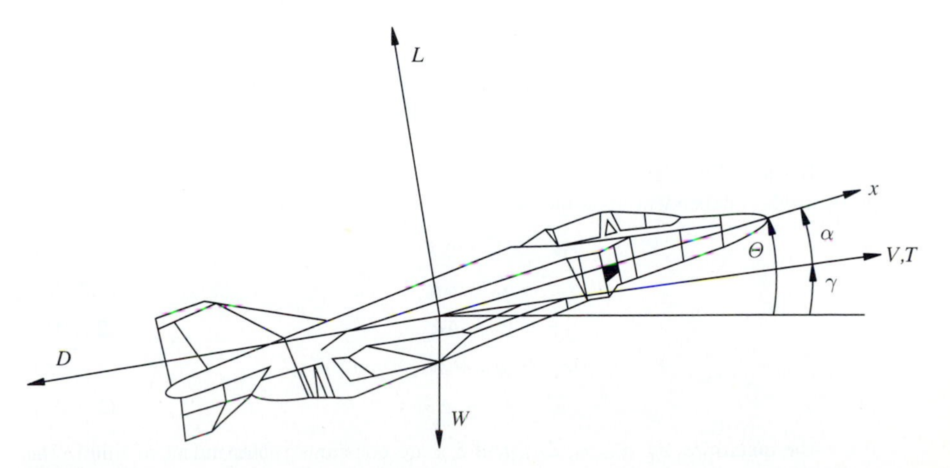

FIGURE 2.A-1
Forces and angles for airplane longitudinal dynamics.

Control is effected by changing the throttle position or the angle of the elevators, which are pieces of the wing that can tilt so that the aerodynamic forces rotate the plane's orientation nose up or nose down.
Define

δ_e deflection of the elevators
δ_T throttle position

The forces on a plane are also shown in Fig. 2.A-1. They are

L lift caused by the aerodynamics of the wing acting perpendicular to the velocity
T thrust produced by the plane's engines in the direction of the velocity vector
D drag, the aerodynamic dissipative forces opposing thrust, usually proportional to velocity
W weight, the force of gravity, acting perpendicular to the horizon

Writing $\boldsymbol{F} = m\boldsymbol{a}$ in the direction of the velocity vector produces the Drag Equation

$$T - D - W\sin(\theta - \alpha) = m\dot{V} \tag{2.A-1}$$

where $W\sin(\theta - \alpha)$ is the component of the gravity force acting against velocity.
The lift force, L, is perpendicular to velocity and does not enter this equation.
Writing $\boldsymbol{F} = m\boldsymbol{a}$ perpendicular to the velocity vector produces the Lift Equation

$$L - W\cos(\theta - \alpha) = mV\frac{d}{dt}\sin(\theta - \alpha) \tag{2.A-2}$$

The right side represents the mass times the acceleration that results from changing the direction of the velocity vector.
Finally there is an equation for the rotational pitching moments

$$M = I\ddot{\theta} \tag{2.A-3}$$

where M is the sum of the pitching torques or moments
I is the moment of inertia around the center of mass or gravity
The initial linearization of the Drag Equation, Eq. (2.A-1), is accomplished by a simple expansion of the sin function using $\sin x \approx x$ and $\cos x \approx 1$ under the assumption that all the angles are small. In the initial linearization of the Lift Equation, Eq. (2.A-2) the time derivative of the $V\sin(\theta - \alpha)$ term should be taken before expanding functions and dropping terms.
The total linearization is complicated by the fact that the forces in the equations are themselves dependent upon the state variables. An incremental model is built. Let

$$V = V_o + \Delta V \tag{2.A-4a}$$

$$\alpha = \alpha_o + \Delta\alpha \tag{2.A-4b}$$

$$\theta = \theta_o + \Delta\theta \tag{2.A-4c}$$

$$\delta_T = \delta_{To} + \Delta\delta_T \tag{2.A-4d}$$

$$\delta_e = \delta_{eo} + \Delta\delta_e \tag{2.A-4e}$$

The quantities V_o, α_o, θ_o, δ_{To}, and δ_{eo} are constants (independent of time). The system is said to be linearized around these nominal quantities. The state variables become ΔV, $\Delta\alpha$, $\Delta\theta$, $\Delta\delta_t$, and $\Delta\delta_e$.

Also, the forces are given first-order expansions about the nominal quantities.

$$T = T_o + T_v \Delta V + T_{\delta_T} \Delta \delta_T \tag{2.A-5a}$$

where

$$T_v = \left. \frac{\partial T}{\partial V} \right|_{V=V_o}$$

is the first partial derivative of thrust with respect to velocity, evaluated at the nominal velocity, and similarly

$$T_{\delta_T} = \left. \frac{\partial T}{\partial \delta_T} \right|_{\delta_T = \delta_{T_o}}$$

Physical intuition for these partial derivatives is developed further on in this Appendix. Thrust is independent of α, θ, q, and δ_e. Other first partial derivatives are defined analogously. Partial derivatives equal to zero are omitted.

$$D = D_o + D_v \Delta V + D_\alpha \Delta \alpha \tag{2.A-5b}$$

$$L = L_o + L_v \Delta V + L_\alpha \Delta \alpha \tag{2.A-5c}$$

$$\begin{aligned} M = {} & M_o + M_v \Delta V + M_\alpha \Delta \alpha + M_{\dot{\alpha}} \Delta \dot{\alpha} \\ & + M_{\dot{\theta}} \Delta \dot{\theta} + M_{\delta_e} \Delta \delta_e \end{aligned} \tag{2.A-5d}$$

We also assume that the nominal parameters represent steady-state flight so that

$$\ddot{\theta}_o = \dot{V}_o = \dot{\alpha}_o = \dot{\theta}_o = 0 \tag{2.A-6}$$

$$T_o - D_o - W_o \sin(\theta_o - \alpha_o) = 0 \tag{2.A-7a}$$

$$L_o - W_o \cos(\theta_o - \alpha_o) = 0 \tag{2.A-7b}$$

$$M_o = 0 \tag{2.A-7c}$$

Complete the linearization of the system using Eqs. (2.A-4)–(2.A-7) in the initial linearization of Eqs. (2.A-1)–(2.A-3) and dropping all second order terms, assuming that all incremental state variables and their first derivatives with respect to time are small. The first derivatives with respect to time are small if the incremental variables are small since the derivatives are, to a first-order approximation, linear combinations of the variables. Show that the resulting state equations are given by

$$\frac{d}{dt}\begin{bmatrix} \Delta V \\ \Delta \alpha \\ q \\ \Delta \theta \end{bmatrix} = \begin{bmatrix} \dfrac{T_v - D_v}{m} & \dfrac{W - D_\alpha}{m} & 0 & \dfrac{-W}{m} \\ \dfrac{-L_v}{m V_o} & \dfrac{-L_\alpha}{m V_o} & 1 & 0 \\ \left(\dfrac{M_v}{I} - \dfrac{M_{\dot{\alpha}} L_v}{\text{Im}\ V_o}\right) & \left(\dfrac{M_\alpha}{I} - \dfrac{M_{\dot{\alpha}} L_\alpha}{\text{Im}\ V_o}\right) & \left(\dfrac{M_{\dot{\theta}} + M_{\dot{\alpha}}}{I}\right) & 0 \\ 0 & 0 & 1 & 0 \end{bmatrix}$$

$$\times \begin{bmatrix} \Delta V \\ \Delta \alpha \\ q \\ \Delta \theta \end{bmatrix} + \begin{bmatrix} 0 & \dfrac{T_{\delta_T}}{m} \\ 0 & 0 \\ \dfrac{M_{\delta_e}}{I} & 0 \\ 0 & 0 \end{bmatrix} \begin{bmatrix} \Delta \delta_e \\ \Delta \delta_T \end{bmatrix} \tag{2.A-8}$$

where $q = \Delta\dot{\theta} = \dot{\theta}$.

The system has two possible inputs. Each input can be viewed individually by taking as the $\boldsymbol{b}$ vector the appropriate column of the matrix multiplying the inputs.

Problem 2.A-2 Simulation of the longitudinal dynamics of a Boeing 747. The incremental state-variable description for the longitudinal dynamics of an airplane was developed in Example 2.A-1. In this problem we use computer simulation to investigate the longitudinal motions of a Boeing 747. The first-order derivatives defined in Example 2.A-1 may take on different values depending on the design of an airplane. We discuss the meaning of these variables here to give some feel for the dynamics.

T_v change of thrust with velocity. This term arises because different engines react differently to the air intake properties at different speeds. This is usually zero for turbojets, negative for propeller-driven airplanes, and may be positive for ramjets.

D_v change of drag with velocity. This is always positive. It is the damping of the aerodynamics due to a velocity change.

$\frac{W}{m} = g$ the acceleration due to gravity.

D_α change in drag with angle of attack. This is positive due to added wind resistance associated with the higher profile angle.

L_α change in lift with angle of attack. This is positive. It is also called the vertical damping parameter. A positive change in α with constant θ changes the velocity vector to point further down. The increased lift due to L_α opposes this change in vertical speed as a damping force would.

M_v moment due to a change in velocity. This might be positive or negative. Effort is made to keep it near zero so that changing speed doesn't produce pitching.

M_α moment due to change in angle of attack. An increase in angle of attack means that the airplane is moving down more vertically in its own axis coordinate system. It is desirable to have M_α be negative to oppose the downward motion by pitching the nose of the airplane down (reducing θ), helping to restore the original angle of attack. M_α is sometimes called the angle of attack stability.

$M_{\dot{\alpha}}$ moment due to the rate of change of angle of attack or angle of attack damping. It is usually negative because under downward vertical accelerations the air moving past the wing causes the tail to gain more lift, pitching the airplane down.

M_q moment due to change in pitch rate. This is usually negative, producing pitch damping via aerodynamic forces.

M_{δ_e} moment due to change in elevator or elevator effectiveness. This is usually negative, by convention, pushing the nose down (reducing θ) for forward stick movement (positive δ_e).

T_{δ_T} change in thrust due to a changed throttle position. This is positive by convention, indicating increased thrust with increased throttle position.

Given this description, typical numbers for the state-space description of the longitudinal dynamics are given in Eq. (2.A-9). The dynamics given correspond to a Boeing 747 jumbo jet flying near sea level at a speed of 190 miles per hour.

$$\frac{d}{dt}\begin{bmatrix} \Delta V \\ \Delta\alpha \\ q \\ \Delta\theta \end{bmatrix} = \begin{bmatrix} -0.0188 & 11.5959 & 0 & -32.2 \\ -0.0007 & -0.5357 & 1 & 0 \\ 0.000048 & -0.4944 & -0.4935 & 0 \\ 0 & 0 & 1 & 0 \end{bmatrix}\begin{bmatrix} \Delta V \\ \Delta\alpha \\ q \\ \Delta\theta \end{bmatrix} + \begin{bmatrix} 0 & 1 \\ 0 & 0 \\ -0.5632 & 0 \\ 0 & 0 \end{bmatrix}\begin{bmatrix} \Delta\delta_e \\ \Delta\delta_T \end{bmatrix} \tag{2.A-9}$$

We can now examine characteristic motions using simulations. When the pilot pulls back on the stick to raise the nose of the airplane, the elevator is moved and the change in aerodynamics causes a pitching moment on the plane. The plane begins to rotate with the nose rising. The elevator must be restored to its original position when the desired new climbing angle is reached or the plane keeps rotating. The maneuver can be represented in a simulation by a $\Delta\delta_e$ signal which is -1 at $t = 0$ and zero afterwards. (The minus sign by convention represents pulling the stick back.)

(*a*) First, let's look at a simplified system that results if it is assumed that the velocity doesn't change during the maneuver. When we assume that $\Delta V = 0$ throughout then the top row of Eq. (2.A-9) can be removed. The last row can also be removed as $\Delta\theta$ does not affect any other state-variable except ΔV. Write the second-order state-variable description for this situation using $\Delta\alpha$ and q. Start with $\Delta\alpha = q = 0$. Input the negative pulse described above for $\Delta\delta_e$ and simulate the system for 20 secs. Plot the response of $\Delta\alpha$ and q. Your response should show that these variables oscillate and settle quickly back to their original values. This characteristic oscillatory response is associated with what is called the short period dynamics of a plane's longitudinal response.

The situation can be seen more clearly by adding some physical states which do not drive other dynamics but are of interest on their own. First, return to the fourth state equation, which gives θ as the integral of q, to the state equations. Also it is interesting to monitor the change in vertical height, Δh, of the airplane. The change in height is determined by the vertical component of the velocity vector.

$$\Delta\dot{h} = V\sin(\theta - \alpha) \approx V_o(\theta - \alpha) \tag{2.A-10}$$

Divide both sides of this equation by $10V_o$ and add this linearized equation for $\Delta\dot{h}/10V_o$ to your state equation. Rerun your simulation, and plot both the old and new state-variables. (Scaling down Δh allows for nicer composite plots.)

Your simulation should look like Fig. 2.A-2 and should support this scenario. The change in elevator causes $\Delta\theta$ and $\Delta\alpha$ to increase as the airplane pitches up. The increased angle of attack causes increased lift through L_α. The airplane rises, meaning $\Delta\alpha$ is greater than it was but less than $\Delta\theta$. The plane is now climbing faster than when it started. Meanwhile there is a restorative force pitching the plane down. This force arises due to the increased α through M_α. The two opposing forces of the elevator moment and the restorative force may cause some oscillation but there is damping through $M_{\dot{\alpha}}$ and M_q. The opposing force developed aerodynamically through M_α keeps the angle of the attack and pitch angle from growing unboundedly and thus M_α is referred to as the angle of attack stability coefficient.

This angle of attack stability moment arises as follows. An increase in α produces increased lift, which can be thought of as acting in the aerodynamic center of the wing. If the aerodynamic center of the wing is aft of the center of gravity of the plane the

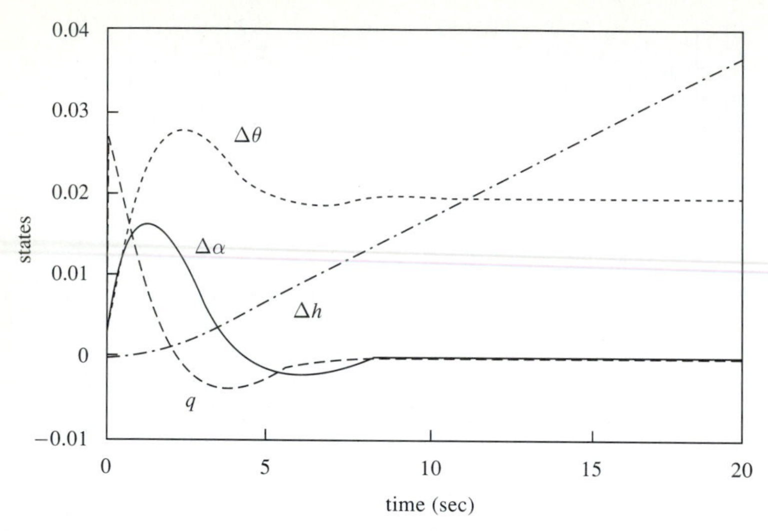

FIGURE 2.A-2
Simulation of simplified short period dynamics.

extra lift causes a moment about the center of gravity, restoring the nose downward. The aerodynamic center of the wing can be slightly forward of the center of gravity and the plane can still have a negative M_α for stability if enough lift is developed by the tail to create the needed restorative moment.

Try a simulation with the following system where M_α has been made positive. Let A(2,2) in the new system (A(3,3) in the original system) be changed from -0.4935 to $+0.85$. Run the simulation and report the results.

Clearly if $M_\alpha > 0$, the external control system becomes very important. Why would anyone build an airplane with $M_\alpha > 0$? In supersonic flight the aerodynamic center of an airplane shifts aft, significantly decreasing M_α. For high performance fighter planes to perform well at supersonic speeds they are sometimes designed with $M_\alpha > 0$ at subsonic speeds. A control system which stabilizes the longitudinal dynamics is required for the plane to fly. If the control system does not perform properly both the plane and pilot could be lost.

(*b*) Actually, the longitudinal dynamics are more complicated than the dynamics explored in (*a*) since the velocity does change in response to the other longitudinal variables. Return to the complete set of the state-variable equations given by Eq. (2.A-9) and add the equation for $\Delta h/10V_o$ as was done in (*a*). Run the simulation of the response to the pulse in the elevator deflection as described in (*a*). Run the simulation for 200 seconds and plot all the state-variables except q. (Scale ΔV by 0.001 to make the coincident plots readable. Do this in the output equation because the correct ΔV is needed in the state equation to drive the other state-variables.)

The result is surprising. Your results should be as in Fig. 2.A-3. The oscillatory motion seen in this simulation has a period much longer than the short period dynamics seen in (*a*). These oscillations have been given the colorful but apparently untraceable name of *phugoid dynamics*. The period of the oscillations of the phugoid mode for this example is about 50 sec., compared with a period of less than 10 sec. for the short period dynamics. The phugoid oscillations also decay much more slowly. In the phugoid oscillations the plane climbs and dives. The change in angle of attack, $\Delta\alpha$, is small compared to the change in pitching angle, $\Delta\theta$, so that the plane keeps almost a constant orientation with respect to its direction. The plane pitches up when it climbs and down while it drops. The plane is almost level when it is at its maximum and minimum altitudes.

The phugoid motion is related to the change in the speed of the plane as its pitch changes. The plane pitches up and begins to climb. Since it must climb against gravity the plane slows down. The plane then loses lift, pitches down, and loses altitude. Now, the plane's speed is aided by gravity. The plane speeds up, gains lift, pitches up and the cycle repeats.

It is interesting to see what happens to the full dynamics when the elevator is reset to a new position in the hope of pitching the plane up and climbing. Run a simulation of the complete system for 500 sec. using a negative unit step as the elevator input. Plot the same variables as in Fig. 2.A-3.

The plane actually winds up steadily losing altitude in an oscillatory trajectory after its initial move up. The problem is that the larger angle of attack creates more drag, causing the plane to slow and lose lift. The steady-state pitch angle θ returns to its original position; the increased α indicates a downward component of velocity.

Climbing maneuvers are made in planes not by using the elevator to tip the plane up but by increasing the thrust to increase the lift. Run a simulation for 500 sec. using the second column of the $\boldsymbol{B}$ matrix in Eq. (2.A-9) as the $\boldsymbol{b}$ vector. Now the state equations describe the system when the engine throttle is the input. Use a positive step of height five as the throttle input. Scale back the height in the output matrix by an additional factor of 0.1; otherwise, plot the same outputs. Your results should look like Fig. 2.A-4.

The increased lift makes the angle of attack decrease, which means the plane begins to climb. The decreased angle of attack excites the short period dynamics, which

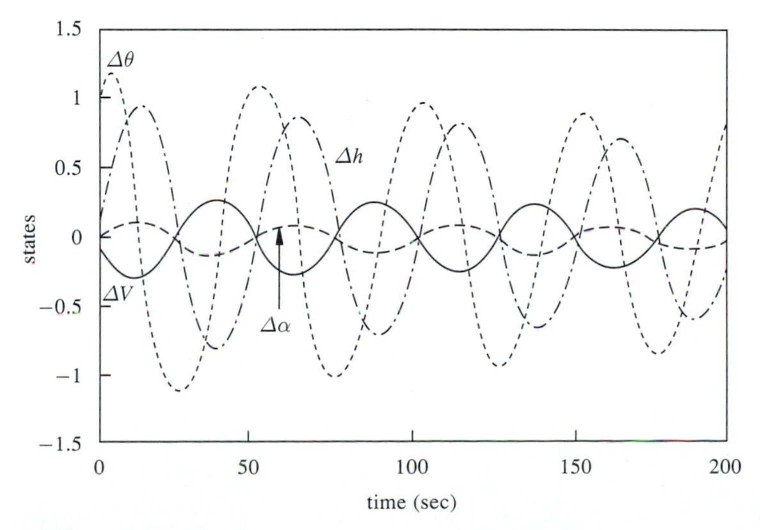

FIGURE 2.A-3
Simulation of phugoid dynamics.

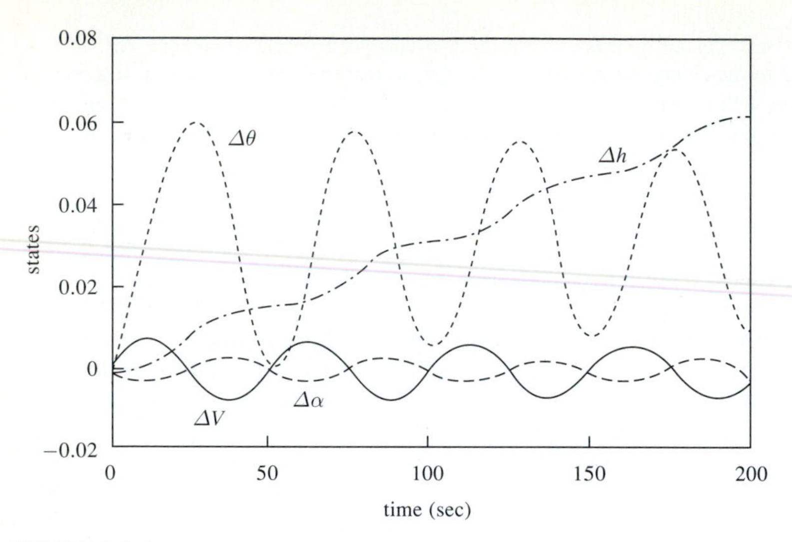

FIGURE 2.A-4
Simulation of the response to a change in throttle.

respond by pitching the airplane quickly nose up, further adding to the climb. The phugoid oscillation is still present but now the plane pitches around a steeper steady-state pitch angle and the height oscillates around a climbing trend.

CHAPTER 3

CLOSED-LOOP-SYSTEM REPRESENTATION

3.1 INTRODUCTION

In Chapter 2 we discussed the representation of a plant, under the assumption that the unalterable plant is described by linear differential equations with constant coefficients. Often the describing equations appear as sets of simultaneous equations of first and second order. With a little manipulation, these equations may be expressed as either one nth-order equation or as a set of n first-order equations. But regardless of how the plant is described, it always proves convenient to picture the governing equations by drawing a block diagram.

Recall that the block diagram involves not the differential equations themselves but the Laplace transforms of the differential equations. This is an important point. Pictures prove useful, both academically and in practice, and hence in this chapter we continue to utilize the Laplace transform, block-diagram approach. Refer to Fig. 2.2-1 which represents a general closed-loop system. In this chapter, we introduce two specific controller configurations, each of which divides into two classes depending upon the plant information passed to the controller. In the *output feedback problem*, the signal the controller receives from the plant is the output, $y(t)$. In the *state feedback problem*, it is assumed that all states of the plant can be measured, and used by the controller. Both controller configurations can be analyzed by the same methods.

It is important at the outset to recognize the difference between the terms *plant* and *control system*. In the state feedback problem, the plant is described by Eqs. ($\boldsymbol{Ab}$)

and ($\boldsymbol{c}$)

$$\dot{x} = \boldsymbol{A}\boldsymbol{x} + \boldsymbol{b}u \tag{$\boldsymbol{Ab}$}$$

$$y = \boldsymbol{c}^T\boldsymbol{x} \tag{$\boldsymbol{c}$}$$

Here the input u is termed the *control input*. The *closed-loop control system*, which is the concern of this chapter, is described by Eqs. ($\boldsymbol{Ab}$) and ($\boldsymbol{c}$), plus an additional equation that specifies the control signal. The control signal considered in this book for the state feedback problem is given as Eq. ($\boldsymbol{k}$)

$$u = K\left(r - \boldsymbol{k}^T\boldsymbol{x}\right) = K\left(r - k_1x_1 - k_2x_2 - \cdots - k_nx_n\right) \tag{$\boldsymbol{k}$}$$

The input r is the *reference input*, the input the control system is attempting to follow. In the general case each of the state-variables x_i is weighted by a linear gain or attenuation factor k_i and subtracted from the reference input.

In the output feedback problem, the control input u is related to the reference input r and the plant output y by the transfer function.

We begin our treatment of closed-loop-system representation in Sec. 3.2 by discussing some of the more common effects of feedback within the context of the output feedback problem. The next section presents the methods and techniques of state-variable feedback using a specific example system. In Sec. 3.4 the results and methods developed in Sec. 3.3 are generalized, based on the plant resolvent matrix, previously mentioned in Chap. 2. In Sec. 3.5 we discuss the concept of the controllability in a state-variable representation. Controllability is the property which allows arbitrary placement of closed-loop poles by state-variable feedback.

3.2 THE EFFECTS OF FEEDBACK IN THE OUTPUT FEEDBACK PROBLEM

The basic concept of feedback was introduced in Chap. 1. There we mentioned some of the effects of feedback and considered systems for which feedback could be used advantageously. As we discuss the representation of closed-loop feedback systems, we examine the effects of feedback in more detail.

Although the effects of feedback are too numerous to list in their entirety, five effects are usually of primary interest:

1. The reduction of sensitivity to plant-parameter variations
2. The reduction of sensitivity to output disturbances
3. The ability to control the system bandwidth
4. The stabilization of an unstable system
5. The ability to control the system transient response

Although these results are not universally obtained, one or more of these effects is usually the primary goal of a feedback control system. Let us consider each of these items in more detail using the output feedback structure.

3.2.1 Reduction of Sensitivity to Plant-Parameter Variations

To illustrate the effect of feedback on plant-parameter sensitivity most clearly, let us consider the simple cascade system in Fig. 3.2-1*a*. Each block represents a positive-gain, *frequency-independent*, linear amplifier. This same system, with the addition of a feedback loop around the second amplifier, is shown in Fig. 3.2-1*b*. The system of Fig. 3.2-1 corresponds to a frequency-independent version of the output feedback equation, Eq. (OF).

The transfer function (gain) of these two systems may be easily found by use of the block-diagram reductions of Chap. 2. For the open-loop system (Fig. 3.2-1*a*) the result is

$$\frac{Y(s)}{R(s)} = M_O(s) = M_O = G_1 G_p \tag{3.2-1}$$

and for the closed-loop system (Fig. 3.2-1*b*) we have

$$\frac{Y(s)}{R(s)} = M_C(s) = M_C = \frac{G_1 G_p}{1 + GH} \tag{3.2-2}$$

Notice that the input-output transfer functions are not dependent on s, since G_p, G_1, and H are assumed to be frequency independent.

As a preliminary indication of the effect of feedback on sensitivity, let us examine the closed-loop gain M_C when the amplifier gain G_p becomes large. To do this, let us rewrite M_C in the form

$$M_C = \frac{G_1}{1/G_p + H}$$

Now if G_p becomes very large while H remains fixed, eventually the $1/G_p$ term becomes small compared with H, and M_C can be approximated by the expression

$$M_C \approx \frac{G_1}{H}$$

In this case the closed-loop gain has become effectively independent of the amplifier gain G_p as long as G_p is large, i.e., the closed-loop gain has become insensitive to

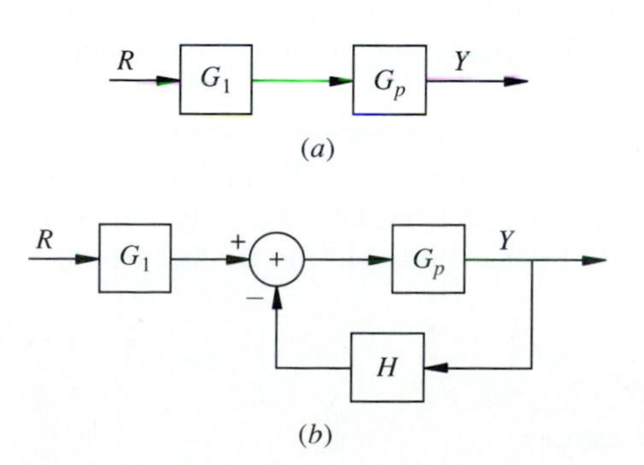

FIGURE 3.2-1
Open-loop and closed-loop amplifiers.

variations in G_p. It is clearly not possible to obtain this result in the case of the open-loop gain function M_O.

If we assume, for the moment, that $G_1 = 1$, then the closed-loop gain M_C effectively depends only on H whenever $1/G_p \ll H$ or $G_p H \gg 1$. The quantity $G_p H$ is referred to as the *loop gain*. The loop gain is taken as the negative of the gain around the loop by convention. If the gain H can be fixed precisely, as when H is formed by a precision electrical network, for example, then the closed-loop gain is independent of G_p as long as G_p is large enough. This is exactly the approach used to stabilize an amplifier gain in many practical situations.

To place the discussion of sensitivity on a more mathematical, quantitative basis, it is necessary to introduce the concept of a *sensitivity function*. Although there are many ways to define a sensitivity function, one normally refers to the *percentage* variation of some specific system quantity, such as gain, with respect to a *percentage* variation of the system parameter in question.[1] Therefore the sensitivity function S_a^M, which indicates the sensitivity of the gain M with respect to variation in the parameter α, is written as

$$S_\alpha^M = \frac{\%\text{ change in } M}{\%\text{ change in } \alpha} = \frac{dM/M}{d\alpha/\alpha} = \frac{\alpha}{M}\frac{dM}{d\alpha} \tag{3.2-3}$$

The effect of changes in the parameter α on M is minimized if the sensitivity function S_α^M is minimized.

Let us use this sensitivity function to examine the effect of feedback on the gain sensitivity for the systems shown in Fig. 3.2-1. In particular, let us consider first the sensitivity of the system's response with regard to the amplifier gain G. For the open-loop system in Eq. (3.2-1), the sensitivity function is given by

$$S_{G_p}^{M_O} = \frac{G_p}{M_O}\frac{d\,M_O}{d\,G_p} = \frac{1}{G_1}G_1 = 1$$

For the closed-loop system in Eq. (3.2-2), the sensitivity function is

$$S_{G_p}^{M_C} = \frac{G_p}{M_C}\frac{dM_c}{dG} = \frac{1}{1 + G_p H} \tag{3.2-4}$$

The results confirm the conclusions in the previous discussion. The sensitivity of the gain of the closed-loop system M_C to small changes in G_p is less than the sensitivity of open-loop gain to small changes in G_p. In addition, the sensitivity of the closed-loop system can be made as small as desired by increasing the loop gain function $G_p H$.

The sensitivity functions for the amplifier gains with respect to other variables may be found in similar fashion. The results are given Table 3.2-1. From this table, we note that the addition of the feedback loop has no effect on the sensitivity functions related to G_1. This is not surprising, since G_1 is not included in the feedback loop.

[1]In economics, such a sensitivity function is called an elasticity.

TABLE 3.2-1
Sensitivity functions

Parameter	Open-loop system	Closed-loop system
G_p	$S_{G_p}^{M_0} = 1$	$S_{G_p}^{M_C} = \dfrac{1}{1+G_pH}$
G_1	$S_{G_1}^{M_O} = 1$	$S_{G_1}^{M_C} = 1$
H	. . .	$S_H^{M_C} = \dfrac{-G_pH}{1+G_pH}$

Although it is not possible to discuss the effect of feedback on the sensitivity function related to H, since H does not exist in the open-loop system, it is interesting to note that the magnitude of sensitivity of M_C to H increases to onc as the loop gain G_pH increases. This result, which is just the opposite of the result obtained for the gain G_p, is not difficult to explain. As the loop gain G_pH increases, the closed-loop gain M_C begins to depend more and more on H and less and less on G_p. Hence, as the sensitivity to variations in G_p is reduced, the sensitivity to variations in H is increased.

In effect, then, one can conclude that the addition of feedback has not really reduced the sensitivity but has merely allowed the sensitivity to be transferred, in part, to the feedback elements. This effect, however, is of significant practical importance since the feedback-path elements are often under the complete control of the designer and may therefore be rigidly specified.

On the other hand, the forward-path elements may be beyond the direct control of the designer, since they include the plant being controlled, which is assumed to be unalterable. In addition, the plant may not be known exactly and may vary. Feedback allows the designer to transfer the sensitivity dependence to different elements in the feedback path, an advantage that may be very helpful in a practical situation. This general topic was also discussed in Chap. 1, where it was indicated that the use of feedback does not eliminate the need for calibration but simply allows the calibration to be done at a more convenient and acceptable point in the system. The need for such a transference of calibration and the existence of an acceptable alternative location for the calibration are two of the basic factors in deciding the utility of a feedback system.

From this simple example, we can draw the following tentative conclusions regarding the effects of feedback on sensitivity:

1. Feedback has no effect on elements that are not included in the feedback loop.
2. Feedback may be used to reduce the sensitivity of the closed-loop transfer function to small changes in forward-path elements by transferring the sensitivity to feedback-path elements. The sensitivity to small changes in forward path elements decreases as the loop gain increases.

Although based only on our simple example, these results remain essentially correct for even the most general case.

Of course, if the elements involved are frequency-dependent transfer functions, as they generally are, these effects of feedback on sensitivity are also frequency-dependent. In particular, when G_p is actually $G_p(s)$, G_1 is actually $G_1(s)$, and H is actually $H(s)$, Eq. (3.2-4) remains unaltered, since differentiation is not with respect to s. To indicate the resulting dependence on s, Eq. (3.2-4) is rewritten as

$$S_{G_p(s)}^{M_C(s)} = \frac{1}{1 + G_p(s)H(s)} \tag{3.2-5}$$

For this sensitivity function to remain less than unity, so that feedback reduces the sensitivity, the magnitude of the denominator of Eq. (3.2-5) must remain greater than one for the range of interest of s, i.e., it must satisfy

$$|1 + G_p(s)H(s)| > 1 \tag{3.2-6}$$

In later chapters on design, considerable effort will be expended to achieve this result.

The quantity $1+G_p(s)H(s)$, called the *return difference*, is clearly an important one. The return difference represents a fictitious transfer function that uses as input the reference input, $R(s)$, and as output the difference between $R(s)$ and the signal returned by the loop, $-G_p(s)H(s)R(s)$. The return difference is always equal to one plus the loop gain.

The sensitivity function of Eq. (3.2-5), which relates changes in the closed-loop transfer function to changes in the plant transfer function, is the most important of the various sensitivity functions that can be defined. The function in Eq. (3.2-5) is often referred to as simply *the sensitivity function*. As Eq. (3.2-5) shows, the sensitivity function is the inverse of the return difference function.

3.2.2 Reduction of Sensitivity to Output Disturbances

As a basis for this discussion, we consider the output feedback control system pictured in Fig. 3.2-2. In addition to the usual reference input R, this figure includes an additional input D, which is a disturbance input. As a practical example, we might consider the control of the position of a radar antenna. The plant, $G_p(s)$ of the positioning system includes the motor and gearing necessary to control the antenna, the block $H(s)$ includes any components added in the feedback path to accomplish control, and the block $G_1(s)$ includes the prefiltering of the reference signal, which may also be used for control. The disturbance D, represents any influence that tends to disturb the output, such as wind blowing on the antenna. To maintain control, we wish to minimize the effect of disturbances.

If, for the moment, the reference input R is allowed to be zero, the block diagram of Fig. 3.2-2*a* may be redrawn with the disturbance as the input. This is done in Fig. 3.2-2*b*. The transfer function $Y(s)/D(s)$ then becomes

$$\frac{Y(s)}{D(s)} = \frac{1}{1 + G_p(s)H(s)} \tag{3.2-7}$$

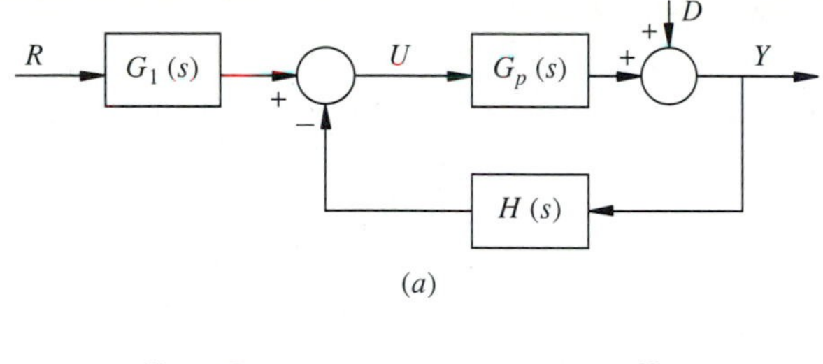

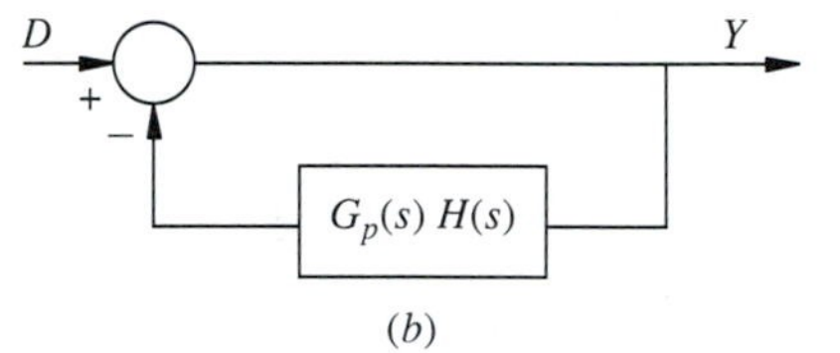

FIGURE 3.2-2
Illustration of the effect of feedback on output disturbances. (*a*) Typical control system including an output disturbance; (*b*) closed-loop system redrawn with the disturbance as the input and $R = 0$.

This result is very similar to that expressed in Eq. (3.2-5), and we see that, *to reduce the effect of an output disturbance at any frequency, it is necessary to ensure that the return difference for that frequency is much greater than one.*

3.2.3 Control of Bandwidth

The bandwidth of a system is usually defined in terms of the response of the system to sinusoidal inputs. Here, we assume that the closed-loop response takes the shape of a low pass filter, i.e., we assume that the magnitude of the closed-loop transfer function is larger at low frequencies and decreases as frequency increases. Let us define *bandwidth* as the frequency at which the magnitude of the output is $1/\sqrt{2}$ or approximately 0.707 times the magnitude of the output at very low frequency. It is shown in Chap. 5 in the discussion of frequency response that the magnitude of the output in response to a sinusoidal input may be determined by evaluating the system transfer function at the frequency of interest. Let us use the two simple first-order examples of Fig. 3.2-3 to illustrate the effect of feedback on system bandwidth.

For the open-loop system (Fig. 3.2-3a) the transfer function $Y(s)/R(s)$ is given by

$$\frac{Y(s)}{R(s)} = M_O(s) = \frac{1}{s+1}$$

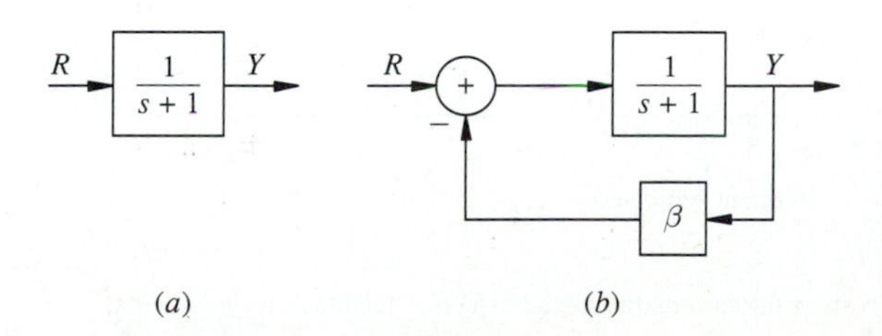

FIGURE 3.2-3
Two systems illustrating the effect of feedback on bandwidth. (*a*) Open-loop system; (*b*) closed-loop system.

When $s = j\omega = j1$, the magnitude of $M_O(s)$ is

$$|M_O(j1)| = \frac{1}{|j1+1|} = \frac{1}{\sqrt{2}}$$

and this is 0.707 times the magnitude of $M_O(s)$ at $s = 0$. The bandwidth of this system is 1 rad/sec.

In the closed-loop system of Fig. 3.2-3*b*, $M_C(s)$ is $1/(s+1+\beta)$, and the low-frequency value of $|M_C(j\omega)|$ is found by setting $s = j0$ so that

$$|M_C(j0)| = \frac{1}{1+\beta}$$

At $\omega = 1+\beta$ the magnitude of $M_C(j\omega)$ is

$$|M_C\,(j(1+\beta))| = \frac{1}{|j(1+\beta)+1+\beta)|} = \frac{1}{(1+\beta)\sqrt{2}}$$

i.e., the value is 0.707 times the dc value. The bandwidth of the closed-loop system is $(1+\beta)$ rad/sec. Hence the bandwidth of the closed-loop system is greater than that of the open-loop system as long as β is positive (negative feedback).

However, in addition to increasing the bandwidth, the use of feedback has *reduced* the low-frequency gain since the low-frequency gain of the open-loop system is one, whereas that of the closed-loop system is $1/(1+\beta)$. A graphical representation of this response is shown in Fig. 3.2-4 for $\beta = 0$ (open-loop system), $\beta = 1$, and $\beta = 2$.

From this illustration, it is obvious that the increase in bandwidth is achieved only at the expense of reducing the low-frequency gain. Here we see that feedback does not provide a magic wand for increasing bandwidth but provides only a means of trading gain for bandwidth.

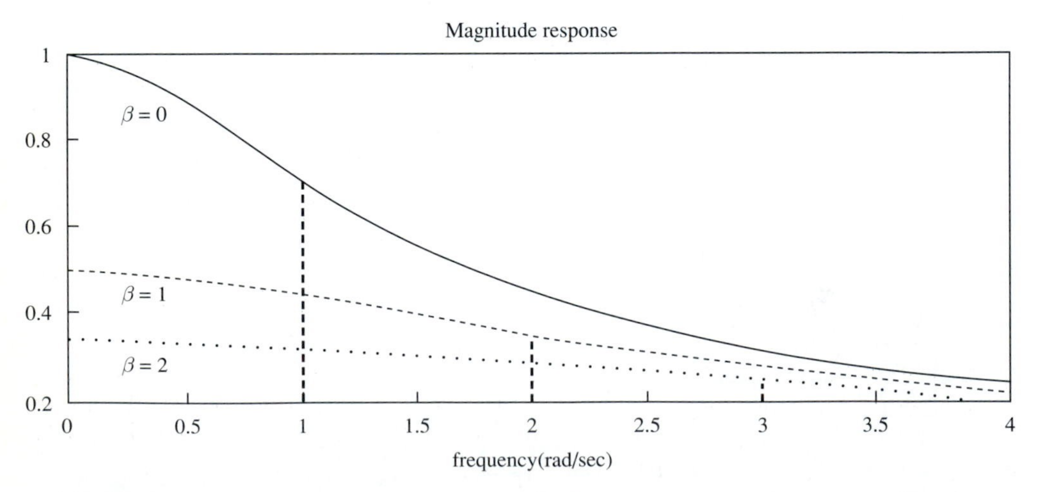

FIGURE 3.2-4
Magnitude of the closed-loop frequency response, demonstrating the effect of feedback on bandwidth.

3.2.4 Stabilizing Unstable Systems

Although a complete discussion of stability issues in control systems will wait until Chap. 6, it is important to mention here that the ability to stabilize systems is an important advantage of feedback control. If a plant is unstable it cannot be stabilized using an open-loop configuration as in Fig. 3.2-1*a*. Even in the closed-loop configuration of Fig. 3.2-1*b*, the element G_1, which is not in the loop, has no effect in stabilizing the system. However, by correctly using the feedback element, H in Fig. 3.2-1*b*, the unstable plant can be turned into a stable closed-loop system.

It should also be mentioned here that feedback control can cause a stable system to become unstable. Indeed, much of the control engineer's task is to ensure adequate performance with the advantageous effects of feedback and at the same time to maintain the necessary assurance of stability.

3.2.5 Control of the System Transient Response

A primary purpose in designing a control system is often to achieve a desired response to reference inputs. As a mathematical abstraction, any desired transfer function $G_d(s)$, relating $R(s)$ to $Y(s)$, can be created using the open-loop configuration of Fig. 3.2-1*a* by using

$$G_1(s) = \frac{G_d(s)}{G_p(s)} \tag{3.2-8}$$

so that

$$\frac{Y(s)}{R(s)} = M_O(s) = \frac{G_d(s)}{G_p(s)} G_p(s) = G_d(s) \tag{3.2-9}$$

There are some problems with this scheme of canceling the plant dynamics and inserting the desired dynamics. First of all, we will see when we are studying stability in Chap. 6 that not all plant dynamics can be canceled. The only plant dynamics we will be allowed to cancel are those dynamics that are originally well-behaved. The second problem with the open-loop scheme is that all the benefits of feedback discussed in this section are lost in an open-loop scheme. In particular, the open-loop system would be extremely sensitive to modeling inaccuracies and disturbances.

Thus, we turn our attention to the closed-loop scheme of Fig. 3.2-1*b*. The closed-loop system transfer function is given by

$$\frac{Y(s)}{R(s)} = M_C(s) = \frac{G_p(s)}{1 + G_p(s)H(s)} G_1(s) \tag{3.2-10}$$

Let's examine the purpose of $G_1(s)$ in this control configuration. Remember that $G_1(s)$ had no role in achieving the advantages of reduced sensitivity to plant modeling inaccuracies and output disturbances. The transfer function $G_1(s)$ also has no effect on the stability of the system. From Eq. (3.2-10), it is seen that $G_1(s)$ enters into the system transfer function in the manner of a simple open-loop component.

This observation suggests the following strategy for the design of an output feedback controller. Letting $G_1(s) = 1$, a controller $H(s)$ that reduces sensitivity to

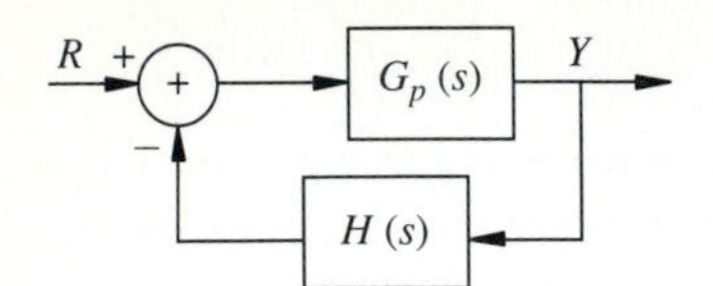

FIGURE 3.2-5
A feedback configuration.

plant modeling inaccuracies and output disturbances, provides adequate stability, and attains a reasonably good system transfer function is designed. The transfer function $G_1(s)$ can then be used to fine tune the shape of the system's transfer function and transient response. The transfer function $G_1(s)$ used for this purpose is referred to as a *command shaping filter* or *prefilter*. We will spend much time exploring the first part of the design process, i.e., designing the feedback element in the system configuration given in Fig. 3.2-5 to achieve the advantages of feedback. A command shaping filter may then be included in the final design to fine tune the system's transient response.

3.2.6 Sensor Noise

The use of closed-loop control is not without risk. There are some ways in which a closed-loop control system can degrade performance. A designer would like to feed back the exact plant output; however, only a measurement of the plant output is available, and that measurement is corrupted by noise. The noise is also fed back and disrupts the control system. Part of a designer's job is to minimize the impact of sensor noise. In Fig. 3.2-6*a* the manner in which sensor noise enters into a control loop is displayed. Unlike a disturbance, the sensor noise $N(s)$ is not part of the output itself. It is part of the signal fed back to form the plant input. In Fig. 3.2-6*b* the reference input is taken to be zero and the system is redrawn to more clearly display the transfer function from the sensor noise to the output.

$$\frac{Y(s)}{N(s)} = \frac{-G_p(s)H(s)}{1 + G_p(s)H(s)} \tag{3.2-11}$$

From Eq. (3.2-11) we see that, *to attenuate sensor noise, the loop gain must be kept small.* The loop gain needs to be large to achieve control objectives such as

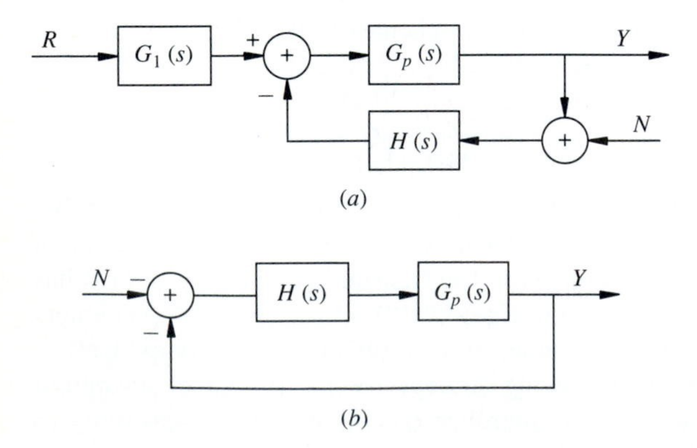

FIGURE 3.2-6
Illustration of the effects of sensor noise on feedback. (*a*) Typical control system including sensor noise; (*b*) system redrawn with the sensor noise as input and $R = 0$.

sensitivity reduction and disturbance rejection. Often, sensor noise is small enough that the sensor noise attenuation becomes a factor only when there are no other overriding concerns. The conflicting objectives of large loop gain and small loop gain require making a tradeoff. The usual situation is that the concerns requiring large loop gain are associated with low frequency phenomena, and the loop gain is made small at high frequencies. Notice that, for sensor noise attenuation, within the constraint of small loop gain, the return difference should still be as large as possible since the return difference appears in the denominator of Eq. (3.2-11). Of course, when the loop gain is small the largest value that the return difference can achieve is 1.

3.2.7 Closed-loop Configurations

So far in this section we have studied the closed-loop configuration of Fig. 3.2-1*b*. We found that, to reduce the sensitivity of the output to plant perturbations and output disturbances, $H(s)$ should be maximized. For large $H(s)$ we derive the approximate expression for the closed-loop transfer function

$$\frac{Y(s)}{R(s)} = \frac{G_p(s)G_1(s)}{1 + G_p(s)H(s)} \approx \frac{G_1(s)}{H(s)}$$

Often it is desirable to have the output $Y(s)$ follow the command signal of the reference input $R(s)$ as closely as possible. In such a situation it is natural to choose $G_1(s)$ equal to $H(s)$. Defining $G_c(s)$ as the common transfer function of $G_1(s)$ and $H(s)$, i.e.,

$$G_c(s) = G_1(s) = H(s)$$

and using the block diagram manipulations of Fig. 2.2-8*b*, Fig. 3.2-1*b* can be redrawn as Fig. 3.2-7.

The configuration of Fig. 3.2-7 is important both historically and in current control designs. It is often referred to as a *unity feedback* control system because there is unity gain in the feedback portion of the loop. It is also referred to as a *series compensation* control loop, since the controller is placed in series with the plant. An interesting aspect of this configuration is that the error between the reference input and the actual output appears as the signal E at the point indicated in Fig. 3.2-7.

There is an advantage to the configuration of Fig. 3.2-7 over the configuration of Fig. 3.2-1*b*. In the configuration of Fig. 3.2-1*b* the sensitivity of the closed-loop transfer function to changes in the controller is given by

$$S_{H(s)}^{M_C(s)} = \frac{-G(s)H(s)}{1 + G(s)H(s)}$$

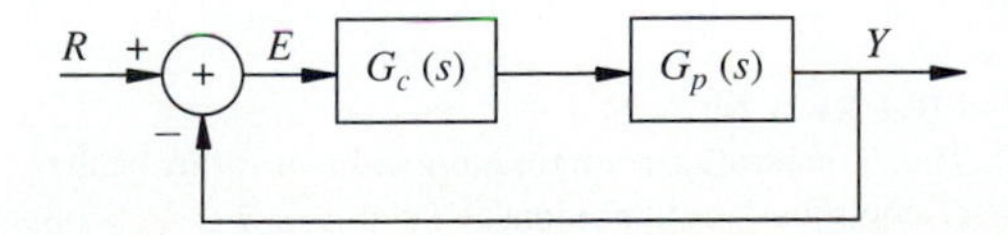

FIGURE 3.2-7
The series compensation or unity feedback configuration.

In the configuration of Fig. 3.2-7, the sensitivity of the closed-loop transfer function to changes in the controller is given by

$$S_{G_c(s)}^{M_C(s)} = \frac{1}{1 + G_c(s)G_p(s)}$$

The reduced sensitivity to inaccuracies in the controller when large loop gain is used in the series compensation scheme can be helpful when controllers consist of parts other than high precision electrical components or digital algorithms. The sensitivity to changes in the plant is the same for both configurations if $H(s)$ equals $G_c(s)$. Finally, note that a closed-loop response achieved by the design of the series compensator of Fig. 3.2-7 can still be fine tuned or more generally altered by the addition of a command shaping prefilter, as shown in Fig. 3.2-8.

In the remainder of the book, two closed-loop control configurations will be studied. The two configurations are given in Fig. 3.2-9. The configuration of Fig. 3.2-9*a* is called the *H configuration*. It is similar to the configuration of Fig. 3.2-1*b* but a constant gain K is allowed in the forward path.

The second configuration of interest is the series compensator of Fig. 3.2-7, which is repeated in Fig. 3.2-9*b*. Many of the properties of this configuration depend

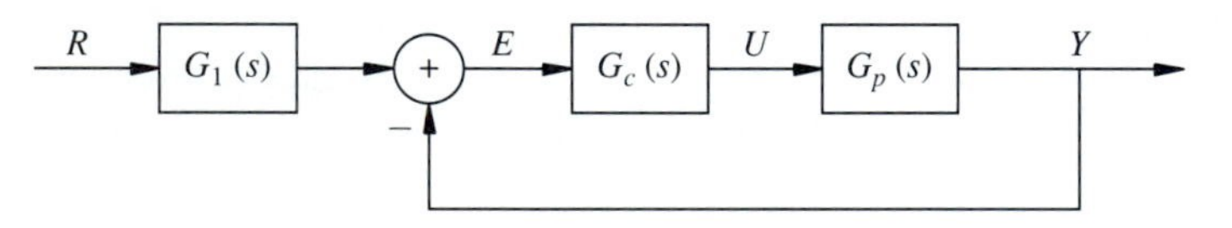

FIGURE 3.2-8
The series compensator configuration with a command shaping prefilter.

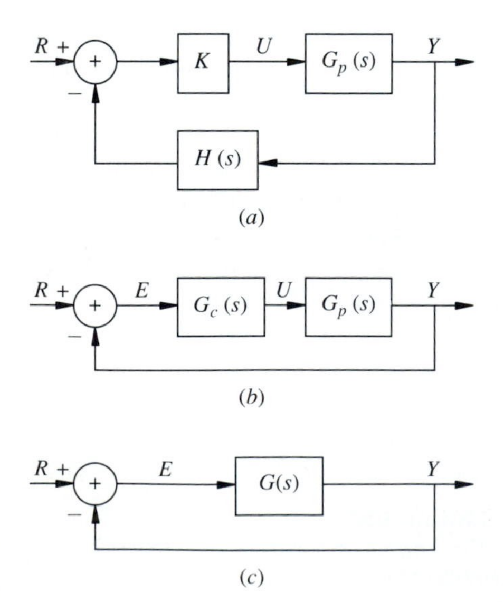

FIGURE 3.2-9
The H controller configuration and two forms of the G controller configurations.

upon the product of $G_p(s)$ and $G_c(s)$ rather than the individual transfer functions. Define

$$G(s) = G_p(s)G_c(s) \tag{3.2-12}$$

A system equivalent to the series compensator of Fig. 3.2-9*b* is given in Fig. 3.2-9*c*. The two forms of Fig. 3.2-9*b* and Fig. 3.2-9*c* are considered as one controller configuration, called the *G configuration*. The form used is a question of convenience. The defining Eq. (3.2-12) is implicit in each form. The G configuration is also referred to as the *series compensator* or the *unity feedback controller*.

We will design controllers using either the G configuration or the H configuration to achieve the desired closed-loop properties while remaining mindful that a command shaping prefilter may be added to either configuration to alter the final closed-loop system transfer function. With the addition of a prefilter, a system designed using the H configuration can be realized using the G configuration, and vice versa. Block diagram manipulations are used to convert from one configuration to another.

3.2.8 Control Effort

A control designer must also ensure that the plant input signals generated by a feedback controller are not too large. Large control signals can cause actuators to saturate and nonlinearities in a plant to come into effect. The plant will then not behave as the linear model predicts and severe problems may result. In Fig. 3.2-9*b* a control system in G configuration is shown. The transfer function from the reference input $R(s)$ to the plant input $U(s)$ is given by

$$\frac{U(s)}{R(s)} = \frac{G_c(s)}{1 + G_c(s)G_p(s)} \tag{3.2-13}$$

Again, a large return difference is desirable this time in conjunction with a small controller transfer function. Since the only part of the loop gain under the influence of the designer is the controller transfer function, a small loop gain is preferable. The usual tradeoff is that large loop gain is used at low frequencies to attain desired objectives and small loop gain is used at higher frequencies, partially because large amplitude, high frequency signals at the plant input are especially troublesome.

3.2.9 Closed-loop Poles and Zeros

It is interesting both for historical significance and for fundamental insight to ask how the closed-loop transfer function can be altered using feedback without the benefit of a prefilter. In particular, we demonstrate a fundamental limitation of feedback control. While feedback control can often move the poles of the closed-loop system to desired locations, feedback control cannot move the zeros of the plant. The original plant zeros either appear in the closed-loop transfer function or they are canceled by poles.

Let's first investigate the G configuration of Fig. 3.2-9*b*. To investigate their poles and zeros, the transfer functions $G_c(s)$ and $G_p(s)$ are represented in terms of their numerator and denominator polynomials:

$$G_p(s) = \frac{N_p(s)}{D_p(s)} \qquad G_c(s) = \frac{N_c(s)}{D_c(s)} \tag{3.2-14}$$

The closed-loop transfer function is computed as:

$$\frac{Y(s)}{R(s)} = \frac{G_c(s)G_p(s)}{1 + G_c(s)G_p(s)} = \frac{N_c(s)N_p(s)}{D_p(s)D_c(s) + N_c(s)N_p(s)} \tag{3.2-15}$$

As long as there are no pole-zero cancellations in the plant, there are no common zero factors in $D_p(s)$ and $N_p(s)$. This is important because if there were a common zero factor in $D_p(s)$ and $N_p(s)$ this factor would necessarily appear in the denominator of the closed-loop transfer function and be a pole of the closed-loop system. In the absence of such a common factor, a desired pole polynomial can be achieved by setting the expression for the denominator polynomial in Eq. (3.2-15) equal to the desired pole polynomial and solving for the controller's numerator $N_c(s)$ and denominator $D_c(s)$. For example, letting $D_d(s)$ be the desired pole polynomial, we can write

$$D_p(s)D_c(s) + N_c(s)N_p(s) = D_d(s) \tag{3.2-16}$$

Solving Eq. (3.2-16) for the controller parameters in $N_c(s)$ and $D_c(s)$, the resulting closed-loop system will have the desired poles defined by the zeros of $D_d(s)$. Equation (3.2-16) is called a *Diophantine equation.* It can always be solved; however, constraints must be placed on the minimum order of $D_d(s)$ if the resulting controller is proper and thus realizable. A control design based on achieving a particular set of closed-loop poles is called a *pole placement* design.

In the following discussion it is assumed that the coefficient of the highest power of s in the denominator polynomials is one. The highest power of s in a polynomial is the *order* of the polynomial. The coefficient associated with the highest power of s is called the *leading coefficient*. A polynomial with a leading coefficient equal to one is called *monic*.

3.2.10 Conditions for a Proper Pole Placement Controller

Theorem. Suppose that one is given a strictly proper nth order plant where the degree of the monic polynomial $D_p(s)$ is n and the degree of $N_p(s)$ is strictly less than n. An mth order proper compensator is to be found, i.e., the degree of the monic polynomial $D_c(s)$ is m and the degree of $N_c(s)$ is less than or equal to m. The resulting closed-loop polynomial is a monic polynomial of degree $n + m$. The non-leading coefficients of the closed-loop polynomial that determine the poles of the closed-loop system can be chosen arbitrarily if and only if there are no pole-zero cancellations between $D_p(s)$ and $N_p(s)$ and if $m \geq n - 1$.[1]

---

[1] A proof of this theorem can be found in Kailath, T., *Linear Systems*, Prentice-Hall, Englewood Cliffs, N.J., 1980, p. 306f.

Example 3.2-1. Let

$$G_p(s) = \frac{N_p(s)}{D_p(s)} = \frac{1}{s^3 + s}$$

and let $G_c(s)$ be a first order controller, so that

$$D_c(s) = s + d_0$$

$$N_c(s) = c_1 s + c_0$$

Substituting into Eq. (3.2-16) gives

$$\begin{aligned} D_p(s)D_c(s) + N_p(s)N_c(s) &= \left(s^3 + s\right)(s + d_0) + 1\,(c_1 s + c_0) \\ &= s^4 + d_0 s^3 + s^2 + d_0 s + c_1 s + c_0 \\ &= s^4 + d_0 s^3 + s^2 + (d_0 + c_1)\,s + c_0 \end{aligned}$$

The closed-loop pole polynomial cannot be chosen arbitrarily by choosing the c's and d because the coefficient of s^2 is unaffected by the choices of the c's and d.

In this problem $n = 3$. The theorem states that for arbitrary pole placement by a proper transfer function we must choose $m \geq n - 1 = 2$. The first-order compensator chosen above is insufficient. Now let

$$D_c(s) = s^2 + d_1 s + d_0$$

$$N_c(s) = c_2 s^2 + c_1 s + c_0$$

Now

$$\begin{aligned} D_p(s)D_c(s) + N_p(s)N_c(s) &= s^5 + d_1 s^4 + (1 + d_0)s^3 + d_1 s^2 + d_0 s + c_2 s^2 + c_1 s + c_0 \\ &= s^5 + d_1 s^4 + (1 + d_0)s^3 + (d_1 + c_2)s^2 + (d_0 + c_1)s + c_0 \end{aligned}$$

The correct coefficients may now be chosen to set the closed-loop polynomial equal to any desired monic 5th order polynomial, thus allowing arbitrary placement of five-closed-loop poles. Suppose, for example, that it is desired to place closed-loop poles at $s = -3 + j3$, $s = -3 - j3$, $s = -5$, $s = -5$ and $s = -10$.

$$\begin{aligned} D_d(s) &= (s + 3 - j3)(s + 3 + j3)(s + 5)(s + 5)(s + 10) \\ &= s^5 + 26s^4 + 263s^3 + 1360s^2 + 3750s + 4500 \end{aligned}$$

Equating $D_p(s)D_c(s) + N_p(s)N_c(s)$ to $D_d(s)$ gives

$$d_1 = 26$$

$$d_0 = 262$$

$$c_2 = 1334$$

$$c_1 = 3488$$

$$c_0 = 4500$$

The controller

$$G_c(s) = \frac{1334s^2 + 3488s + 4500}{s^2 + 26s + 262}$$

produces the desired closed-loop poles.

We can see that feedback control is a powerful tool to rearrange the poles of a system. The zeros, however, cannot be rearranged. Equation (3.2-15) shows that the zeros of the closed-loop transfer function are the union of the zeros of the plant and the zeros of the series compensator. Zeros can be added to the closed-loop system by adding zeros to the series compensator. Zeros can be removed only by cancellation with closed-loop poles. (We will discover that cancellation of right half-plane zeros is not allowed.) Zeros cannot be rearranged or moved by feedback.

Similar results occur for the H configuration of Fig. 3.2-9*a*. Let $H(s)$ be expressed in terms of its numerator and denominator:

$$H(s) = \frac{N_H(s)}{D_H(s)} \tag{3.2-17}$$

The closed-loop transfer function is then computed as:

$$\frac{Y(s)}{R(s)} = \frac{KG_p(s)}{1 + KG_p(s)H(s)} = \frac{KN_p(s)D_H(s)}{D_p(s)D_H(s) + KN_p(s)N_H(s)} \tag{3.2-18}$$

The zeros of the closed-loop system are formed by the union of the zeros of the plant and the *poles* of the feedback controller, $H(s)$. Zeros cannot be moved by feedback; they can only be augmented or canceled.

A desired pole polynomial can be realized by the solution of the Diophantine equation

$$D_p(s)D_H(s) + KN_p(s)N_H(s) = D_d(s) \tag{3.2-19}$$

The constraints on the order of the polynomials to obtain a proper controller are the same as the constraints for the G configuration.

3.2.11 Summary

In this section, we explored some useful properties of closed-loop control system configurations. We review those properties here and relate them to the configurations.

Two important quantities defined for any closed-loop control system are the loop gain transfer function and the return difference transfer function. The return difference is equal to one plus the loop gain. Many properties depend on these quantities. In particular, the sensitivity to plant perturbations and the response to output disturbances are both given by the inverse of the return difference transfer function. The inverse of the return difference is called the sensitivity function. It is desirable to keep the loop gain large for as wide a frequency range as possible. These results are summarized for the G and H configurations in Table 3.2-2.

It was shown that the closed-loop system configuration can be useful for controlling the bandwidth of a system. A closed-loop system may also be used to stabilize an unstable plant; however, caution must be used since a closed-loop system can become unstable even when the plant is stable.

Closed-loop transfer functions were explored. It was found that, through feedback, closed-loop poles can be placed in desired locations. The closed-loop zeros, however, must include the zeros of the plant unless these zeros are canceled. If a reasonably well-behaved closed-loop system is achieved through feedback, the closed-loop transient response can be fine tuned through the use of a prefilter.

Problems that can be caused by feedback were also discussed. We have already mentioned that feedback can cause instability if improperly designed. Also, feedback can introduce variations due to sensor noise. Finally, it was noted that if the control signals are too large, the plant may not respond as expected and problems can result. The expression for the amount of sensor noise appearing at the output and the expression for the size of the control signal are included in Table 3.2-2.

In exploring feedback properties, it was found that keeping the return difference as large as possible provides maximum control benefits. Some benefits, such as reduced sensitivity to plant perturbations and added disturbances, require large loop gains. Some problems, such as sensor noise and large control signals, are inhibited with small loop gain. We will find in Chap. 6 that stability considerations require small loop gains, at least at high frequencies. The usual compromise is to create high loop gain at low frequency and force small loop gain at high frequency. The power of the frequency domain methods of control design we will learn is their ability to manage this compromise.

In this section, we learned about properties of closed-loop systems using transfer function analysis on the output feedback problem with two system configurations, the G and the H configurations. In Secs. 3.3 and 3.4, we will derive controllers using linear state-variable feedback. To better understand the properties of the resulting closed-loop systems we will convert the state-variable representation of the closed-loop system to transfer function representations that are equivalent to the G and the H configurations.

TABLE 3.2-2

	***G* Configuration**	***H* Configuration**
Loop gain	$G(s) = G_c(s)G_p(s)$	$K\, G_p(s)H(s)$
Return difference	$1 + G(s)$	$1 + K\, G_p(s)H(s)$
Sensitivity to plant perturbations	$S_{G_p}^{M_C} = \dfrac{1}{1+G(s)}$	$S_{G_p}^{M_C} = \dfrac{1}{1+K\, G_p(s)H(s)}$
Disturbance rejection	$\dfrac{Y(s)}{D(s)} = \dfrac{1}{1+G(s)}$	$\dfrac{Y(s)}{D(s)} = \dfrac{1}{1+KG_p(s)H(s)}$
Sensor noise	$\dfrac{Y(s)}{N(s)} = \dfrac{G(s)}{1+G(s)}$	$\dfrac{Y(s)}{N(s)} = \dfrac{G_p(s)H(s)}{1+G_p(s)H(s)}$
Control effort	$\dfrac{U(s)}{R(s)} = \dfrac{G_c(s)}{1+G_c(s)G_p(s)}$	$\dfrac{U(s)}{R(s)} = \dfrac{1}{1+G_p(s)H(s)}$

3.2.12 CAD Notes

The solution of the Diophantine equation can be mechanized into a set of matrix manipulations. First, the multiplication of two polynomial can be represented in a specific matrix form. For example, the polynomial multiplication

$$\left(a_3 s^3 + a_2 s^2 + a_1 s + a_0\right)\left(b_2 s^2 + b_1 s + b_0\right) = \left(d_5 s^5 + d_4 s^4 + d_3 s^3 + d_2 s^2 + d_1 s + d_0\right)$$

can be represented as

$$\begin{bmatrix} a_3 & 0 & 0 \\ a_2 & a_3 & 0 \\ a_1 & a_2 & a_3 \\ a_0 & a_1 & a_2 \\ 0 & a_0 & a_1 \\ 0 & 0 & a_0 \end{bmatrix} \begin{bmatrix} b_2 \\ b_1 \\ b_0 \end{bmatrix} = \begin{bmatrix} d_5 \\ d_4 \\ d_3 \\ d_2 \\ d_1 \\ d_0 \end{bmatrix}$$

The Diophantine equation

$$\left(s^3 + dp_2 s^2 + dp_1 s + dp_0\right)\left(s^2 + dc_1 s + dc_0\right)$$
$$+ \left(np_2 s^2 + np_1 s + np_0\right)\left(nc_2 s^2 + nc_1 s + nc_0\right) = \left(s^5 + d_4 s^4 + d_3 s^2 + d_1 s + d_0\right)$$

becomes, in matrix notation

$$\begin{bmatrix} 1 & 0 & 0 & 0 & 0 & 0 \\ dp_2 & 1 & 0 & np_2 & 0 & 0 \\ dp_1 & dp_2 & 1 & np_1 & np_2 & 0 \\ dp_0 & dp_1 & dp_2 & np_0 & np_1 & np_2 \\ 0 & dp_0 & dp_1 & 0 & np_0 & np_1 \\ 0 & 0 & dp_0 & 0 & 0 & np_0 \end{bmatrix} \begin{bmatrix} 1 \\ dc_1 \\ dc_0 \\ nc_2 \\ nc_1 \\ nc_0 \end{bmatrix} = \begin{bmatrix} 1 \\ d_4 \\ d_3 \\ d_2 \\ d_1 \\ d_0 \end{bmatrix}$$

The matrix on the left is called the *Sylvester matrix* associated with the transfer function $N_p(s)/D_p(s)$. The theorem that guarantees solutions to the Diophantine equation is proved using known properties of this matrix. It is invertible whenever there are no pole-zero cancellations in $N_p(s)/D_p(s)$. By setting up the pole placement problem this way it can be solved using MATLAB by matrix inversion and multiplication.

Example 3.2-2. Solve the problem of Example 3.2-1 using MATLAB.

```
>>%setup the required matrix and vectors
%den of P is [1 0 1 0]
%num of P is [0 0 1]
%Sylvester matrix is S
S=[1 0 0 0 0 0; 0 1 0 0 0 0; 1 0 1 0 0 0;0 1 0 1 0 0;
   0 0 1 0 1 0;0 0 0 0 0 1]
S =
  1  0  0  0  0  0
  0  1  0  0  0  0
  1  0  1  0  0  0
  0  1  0  1  0  0
  0  0  1  0  1  0
  0  0  0  0  0  1
```

```
%the vector from the desired polynomial is d
d=[1; 26; 263; 1360; 3750; 4500]
d =
     1
    26
   263
  1360
  3750
  4500

%solve for the required compensator polynomials
c=inv(S)*d
c =
     1
    26
   262
  1334
  3488
  4500
```

The answer is interpreted as

$$\frac{N_c(s)}{D_c(s)} = \frac{1334s^2 + 3488s + 4500}{s^2 + 26s + 262}$$

Exercises 3.2

3.2-1. One way to achieve a desired transfer function $M_d(s)$ is to simply invert the plant transfer function and insert the desired dynamics in an open-loop configuration as shown in Fig. 3.2-10.

(*a*) Let $G_c(s) = M_d(s)G_p^{-1}(s)$ and find the sensitivity of the resulting total transfer function to a change in the plant itself. ($G_c(s)$ doesn't change.)

(*b*) The closed-loop system essentially inverts the plant in a preferable way. Find the transfer function $U(s)/R(s)$ for the system of Fig. 3.2-11 when $G_c(s)$ gets large.

(*c*) Find the sensitivity of the closed-loop transfer function, to a change in the plant for the system in (*b*).

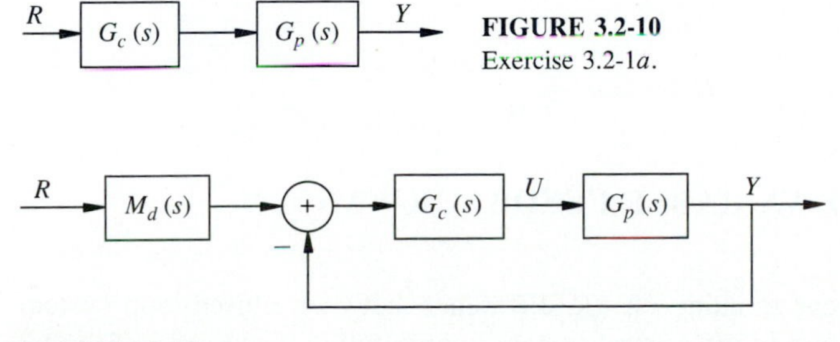

FIGURE 3.2-10
Exercise 3.2-1*a*.

FIGURE 3.2-11
Exercise 3.2-1*b*.

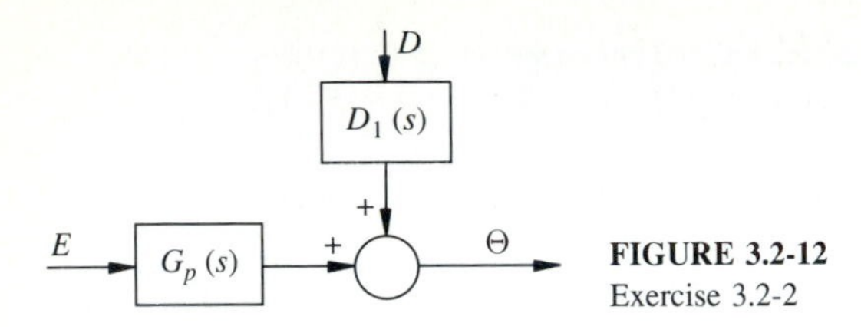

FIGURE 3.2-12
Exercise 3.2-2

Answers:

(*a*) $S_{G_p(s)}^{M_c(s)} = 1$

(*b*) $\dfrac{U(s)}{R(s)} \approx M_d(s)G_p^{-1}(s)$

(*c*) $S_{G_p(s)}^{M_c(s)} = \dfrac{1}{1 + G_c(s)G_p(s)} \approx 0$

3.2-2. We have analyzed the effect of a disturbance on the output of a plant. Often a disturbance occurs at the plant input or in the middle of the dynamics of a plant. We can reflect this disturbance to the output by defining a fictitious disturbance at the output that has the same effect as the original disturbance. Refer to the armature-controlled motor block diagram of Fig. 2.5-1. Assume there is a disturbance $D_1(s)$ added to the torque T. (Such a disturbance may come from a wind gust in a positioning system). Use block diagram manipulations and solve for $D_1(s)$ and $G_p(s)$ so that the block diagram of Fig. 3.2-12 properly represents the system and disturbance on an input-output basis.
Answer:

$$G_p(s) = \frac{K'_\tau}{JL_a s^3 + (L_a\beta + R_a J)\, s^2 + \left(R_a\beta + K_v K'_\tau\right) s}$$

$$D_1(s) = \frac{R_a + sL_a}{JL_a s^3 + (L_a\beta + R_a J)\, s^2 + \left(R_a\beta + K_v K'_\tau\right) s}$$

3.2-3. In this exercise, we arbitrarily place the poles of a system using a series compensator and the Diophantine equation. Assume a plant is given by

$$G_p(s) = \frac{2(s+1)}{(s+3)(s+4)}$$

Find a series compensator $G_c(s)$ which places closed-loop poles at $s = -3$ and $s = -6$. Add as many factors of $(s + 10)$ as you need to the desired pole polynomial to find a proper controller.
Answer:

$$G_c(s) = \frac{-2(s+3)}{s+16}$$

3.3 LINEAR STATE-VARIABLE FEEDBACK—AN EXAMPLE

As an aid in orienting our thinking on the difference between closed-loop system representations, let us return to the general system configuration discussed in Sec. 2.2 and illustrated in Fig. 2.2-1. At that point, the idea of state-variables had not been introduced. Let us now redraw Fig. 2.2-1 as Fig. 3.3-1, this time on the basis of state-variable equations. The plant is described by Eqs. (**Ab**) and (**c**); hence the plant portion of the closed-loop system is identical to that portrayed in Fig. 2.2-1. Again, the broad arrows indicate vector quantities, and single lines indicate the scalar quantities Y and U.

The closed-loop nature of the system of Fig. 3.3-1 is indicated by the presence of the controller to generate the control signal U from the knowledge of the state-variables. In contrast to Fig. 2.2-1, input to the controller is specifically indicated here as $\boldsymbol{X}$, the state of the plant, rather than the previously used vague statement, "information about the plant."

It is important to note that, except for the reference input the state of the plant is the only information needed by the controller. This must obviously be true since the state, by definition, provides a complete summary of the plant's past, and any other information concerning the plant's future can be synthesized from the state. The output y, for example, can always be obtained from a knowledge of the state using the output expression

$$y = \boldsymbol{c}^T \boldsymbol{x}$$

and therefore we need not show the controller as being dependent on the output.

There is an important difference between the state feedback problem discussed in this and the following sections and the output feedback problem discussed in the previous section. In the state feedback problem we assume that the entire state vector is measured and available for feedback while in the output feedback problem only the plant output is available for feedback. We will see that the characteristics of feedback studied in the previous section in the context of the output feedback problem also hold in the state feedback problem.

The output of the controller, namely, the control input u, is shown in this general state-variable feedback system as a general function $f(\boldsymbol{x}, r)$ of the state and

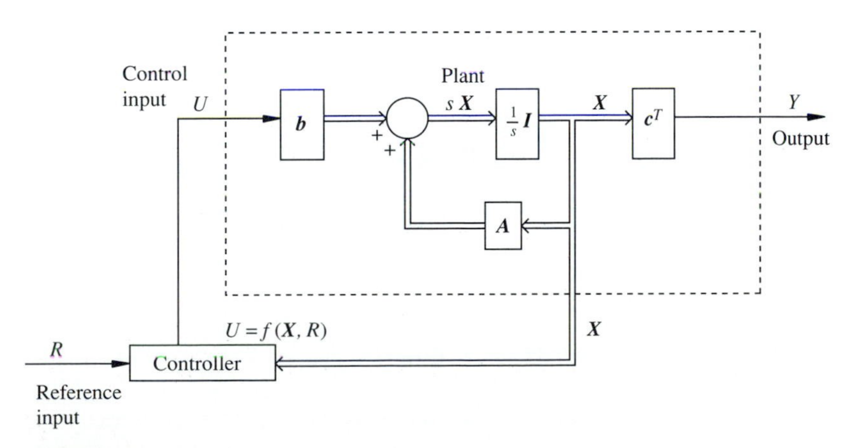

FIGURE 3.3-1
General closed-loop system with state-variable feedback.

the reference input. However, rather than this general, vague form of controller, we consider a more specific case known as *linear state-variable feedback*.

In the linear state-variable feedback case, the control input is computed by multiplying by a gain K the difference between the reference input and a linearly weighted sum of the state-variables. As a mathematical expression, this controller relation takes the form

$$u = f(\boldsymbol{x}, r) = K\,[r - (k_1 x_1 + k_2 x_2 + \cdots + k_n x_n)] \qquad (\boldsymbol{k})$$

Here the k_i's are referred to as *feedback coefficients*, since they are the coefficients of the various state-variables in our linear feedback expression for the control input. The gain K is referred to as the *controller gain*.

Equation ($\boldsymbol{k}$) may be simplified by use of matrix notation to

$$u = K(r - \boldsymbol{k}^T \boldsymbol{x}) \qquad (\boldsymbol{k})$$

where

$$\boldsymbol{k} = \text{col}\,(k_1, k_2, \ldots, k_n) \qquad (3.3\text{-}1)$$

A graphical representation of the linear state-variable feedback system configuration is shown in Fig. 3.3-2.

The linear state-variable feedback system has been chosen for consideration for three basic reasons. First, the configuration results in a linear system, thereby allowing us to make use of the powerful transform techniques for analysis and synthesis. Second, this configuration is sufficiently general to obtain satisfactory performance in many practical control problems. Third, the linear state-variable feedback approach serves as a suitable introduction to a broad array of topics usually referred to as

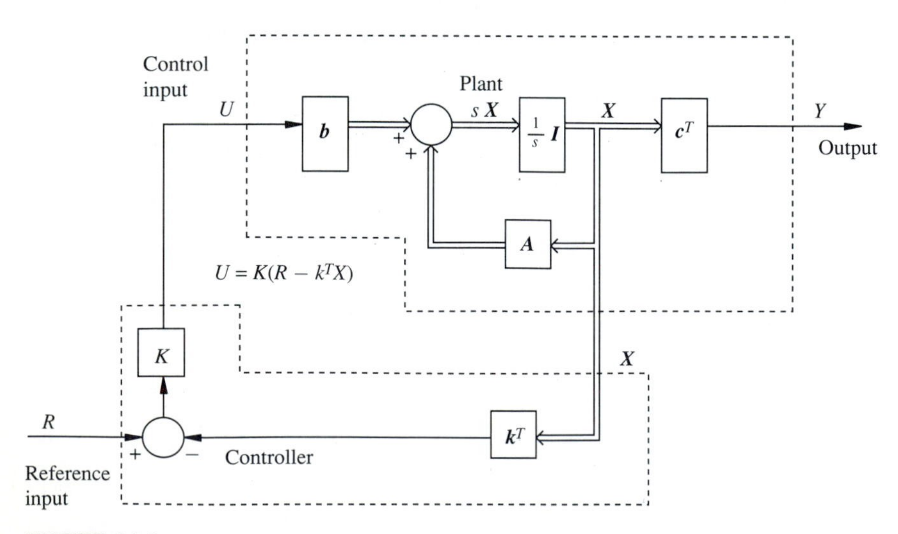

FIGURE 3.3-2
Linear state-variable-feedback system.

modern control theory. In fact, the use of state-variable feedback is one of the most important practical contributions of this modern approach.

The above statements and conclusions may not be completely obvious at this time; this should not be a cause for concern. As the reader gains a greater understanding of and appreciation for the linear state-variable feedback configuration in this and the following chapters, the above comments will become more meaningful.

Instead of dwelling further on generalities concerning linear state-variable feedback, let us consider a specific example of a simple positioning system to illustrate its practical implications. The next section provides the generalization of results obtained from this specific example.

Let us consider the field-controlled dc motor discussed in Chap. 2. This dc motor is used to supply the torque to position an inertial load, such as a telescope. Control of the plant, that is, the motor and the load, is to be achieved through state-variable feedback, as in Fig. 3.3-3. The state variables used to describe the system are the output angular position, $\theta_o = y$, the angular velocity, $x_2 = \dot{\theta}_o$, and the motor field current, $x_3 = i_f$. The output is converted into a voltage signal using a potentiometer, and a tachometer is used to measure the angular velocity. (Many control motors contain a tachometer as an integral part of the motor.) The motor field current is obtained by measuring the voltage across a small series resistor.

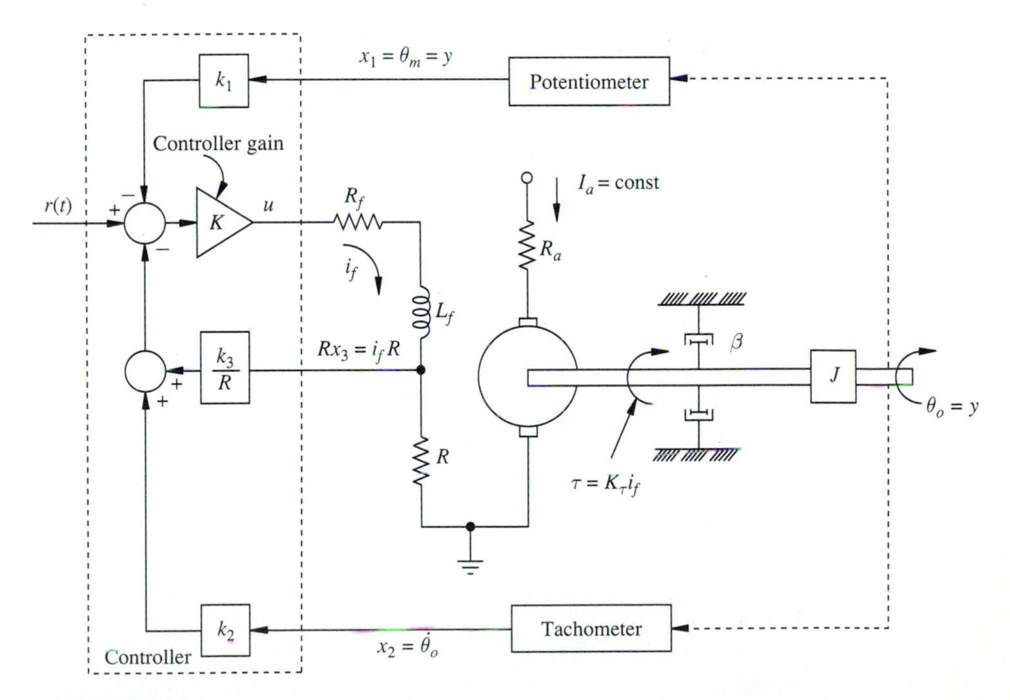

FIGURE 3.3-3
Simple positioning system using a field-controlled dc motor.

A block diagram of this simple positioning system is shown in Fig. 3.3-4*a*, with the controller and the plant indicated. To simplify our further discussion of this example, let us assume specific numerical values for the parameters so that the system shown in Fig. 3.3-4*b* results. Here we have not specifically isolated the controller from the plant. This is the convention we adopt for future use. The summers, the controller gain K, and the feedback coefficients $\boldsymbol{k}$ are always part of the controller but are not necessarily located together in the physical system. As indicated in Fig. 3.3-4*b*, there are four parameters to be chosen in the system: the three feedback coefficients k_1, k_2, k_3 and the controller gain K.

The state-variable representation of the *plant* may be easily obtained from Fig. 3.3-4*b* as

$$\dot{\boldsymbol{x}} = \begin{bmatrix} 0 & 1 & 0 \\ 0 & -1 & 2 \\ 0 & 0 & -4 \end{bmatrix} \boldsymbol{x} + \begin{bmatrix} 0 \\ 0 \\ 2 \end{bmatrix} \boldsymbol{u} \tag{3.3-2}$$

$$y = [1 \quad 0 \quad 0]\boldsymbol{x} \tag{3.3-3}$$

Similarly, the input-output transfer function of the plant may be obtained directly from Fig. 3.3-4*b* or from the state-variable representation by use of Eq. (2.3-10). The result in either case is

$$\frac{Y(s)}{U(s)} = G_p(s) = \frac{4}{s(s+1)(s+4)} \tag{3.3-4}$$

To find the state-variable representation of the entire closed-loop system, it is necessary only to substitute the expression ($\boldsymbol{k}$) for u in the state-variable representation of the plant to obtain

$$\dot{\boldsymbol{x}} = \begin{bmatrix} 0 & 1 & 0 \\ 0 & -1 & 2 \\ 0 & 0 & -4 \end{bmatrix} \boldsymbol{x} + \begin{bmatrix} 0 \\ 0 \\ 2 \end{bmatrix} K\,(\boldsymbol{r} - [\,k_1 \quad k_2 \quad k_3\,]\,\boldsymbol{x})$$

or

$$\dot{\boldsymbol{x}} = \begin{bmatrix} 0 & 1 & 0 \\ 0 & -1 & 2 \\ 0 & 0 & -4 \end{bmatrix} \boldsymbol{x} + \begin{bmatrix} 0 \\ 0 \\ 2 \end{bmatrix} K\boldsymbol{r} - \begin{bmatrix} 0 \\ 0 \\ 2 \end{bmatrix} K\,([\,k_1 \quad k_2 \quad k_3\,]\,\boldsymbol{x})$$

If the indicated matrix multiplication is carried out and the two terms involving $\boldsymbol{x}$ are grouped, the result is

$$\dot{\boldsymbol{x}} = \left(\begin{bmatrix} 0 & 1 & 0 \\ 0 & -1 & 2 \\ 0 & 0 & -4 \end{bmatrix} + \begin{bmatrix} 0 & 0 & 0 \\ 0 & 0 & 0 \\ -2Kk_1 & -2Kk_2 & -2Kk_3 \end{bmatrix} \right) \boldsymbol{x} + \begin{bmatrix} 0 \\ 0 \\ 2 \end{bmatrix} K\boldsymbol{r}$$

which may be reduced to

$$\dot{\boldsymbol{x}} = \begin{bmatrix} 0 & 1 & 0 \\ 0 & -1 & 2 \\ -2Kk_1 & -2Kk_2 & -4-2Kk_3 \end{bmatrix} \boldsymbol{x} + \begin{bmatrix} 0 \\ 0 \\ 2 \end{bmatrix} K\boldsymbol{r} \tag{3.3-5}$$

The output expression is still

$$y = [1 \quad 0 \quad 0]\boldsymbol{x} \tag{3.3-6}$$

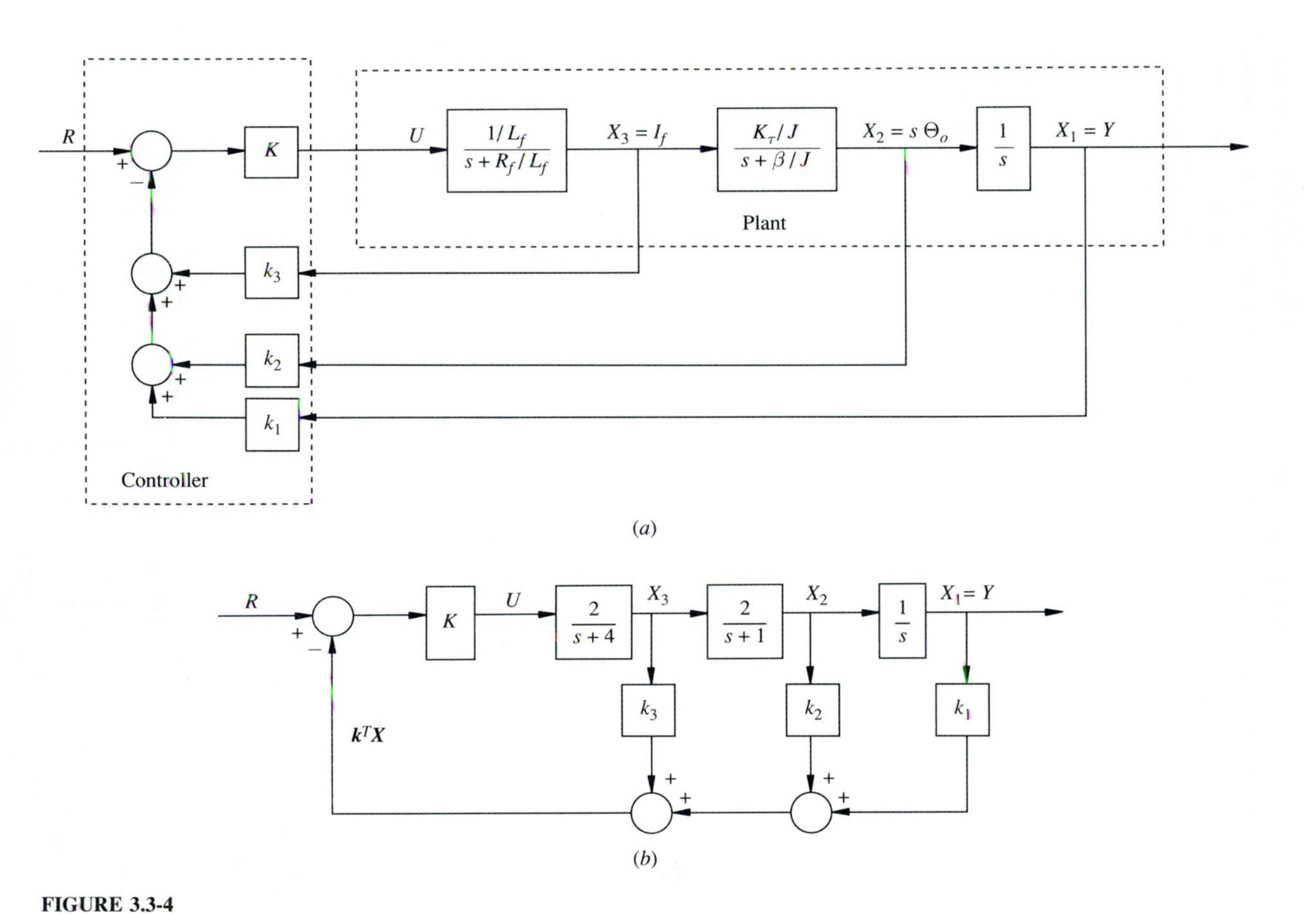

FIGURE 3.3-4
Block diagram of the system of Fig. 3.3-3. (*a*) General system; (*b*) the system defined with numerical values for the parameters: $L_f = 0.5$, $R_f = 2$, $K_\tau/J = 2$, and $\beta/J = 1$.

Equations (3.3-5) and (3.3-6) form the desired state-variable representation of the closed-loop system. Note that there are two differences between this system representation and the original plant representation of Eqs. (3.3-2) and (3.3-3). First, $K\boldsymbol{r}$, the reference input times K, has replaced the control input u, and second, the $\boldsymbol{A}$ matrix has been modified by the feedback of the state-variables. The $\boldsymbol{b}$ and $\boldsymbol{c}$ vectors have remained unchanged. Although we could obtain the closed-loop transfer function $Y(s)/R(s)$ from the above state-variable representation, let us approach the problem in a different manner.

Our approach is to force the system into one of the two configurations shown in Fig. 3.3-5. These configurations are referred to as the H_{eq} and the G_{eq} forms. The transfer functions H_{eq} and G_{eq} are read as *H equivalent* and *G equivalent*. This terminology was chosen because, in the linear state-variable feedback problem, neither H_{eq} nor G_{eq} is a physical transfer function. But, by artificially representing this state-variable feedback problem in the H_{eq} or G_{eq} form, the state-variable feedback problem can be analyzed using the same techniques developed with transfer function techniques and the G and H configurations in the output feedback problem of the previous section.

If the system of Fig. 3.3-4*b* is reduced to either configuration of Fig. 3.3-5, the closed-loop transfer function may be easily determined as it was in the previous section.

$$\frac{Y(s)}{R(s)} = \frac{KG_p(s)}{1 + KG_p(s)H_{eq}(s)} = \frac{G_{eq}(s)}{1 + G_{eq}(s)} \tag{3.3-7}$$

Let us consider the reduction of the system of Fig. 3.3-4*b* to each of these two forms. We begin with the H_{eq} configuration of Fig. 3.3-5*a*.

The H_{eq} configuration may be achieved by alternating use of the identities of Fig. 2.2-8*c* and *e* to move the inner state-variable feedback loops outward. This procedure begins with the innermost loop and is successively applied until the desired H_{eq} form is obtained. This procedure is illustrated in a step-by-step fashion in Fig. 3.3-6 for the system of Fig. 3.3-4*b*.

In Fig. 3.3-6*a* the path containing the k_3 feedback coefficient has been moved to the right past the blocks $2/(s+1)$ and $1/s$ and combined with the k_1 feedback coefficient. Next, the k_2 feedback coefficient is moved to the right and also combined with the k_1 block to achieve the desired H_{eq} configuration shown in Fig. 3.3-6*b*. From

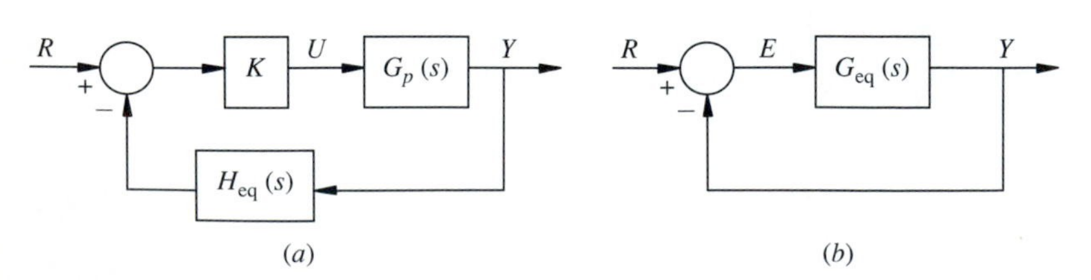

FIGURE 3.3-5
Two basic configurations for the closed-loop system. (*a*) The H equivalent form; (*b*) the G equivalent form.

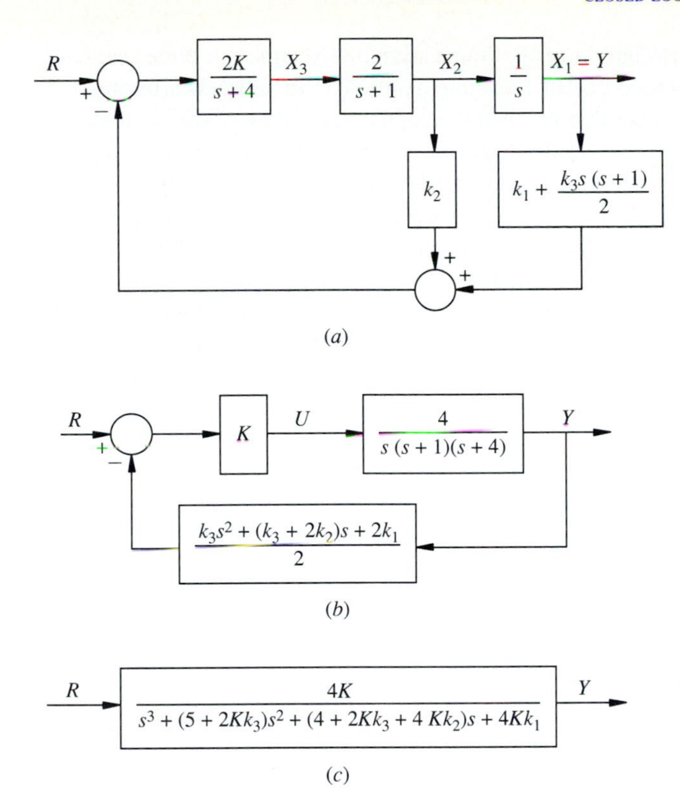

FIGURE 3.3-6
Step-by-step reduction to the H_{eq} form. (*a*) Movement of k_3 forward to combine with k_1; (*b*) H_{eq} form; (*c*) resulting closed-loop transfer function.

Fig. 3.3-6*b* it is easily seen that $H_{eq}(s)$ is given by

$$H_{eq}(s) = \frac{k_3 s^2 + (k_3 + 2k_2)s + 2k_1}{2} \tag{3.3-8}$$

Note that the numerator of $H_{eq}(s)$ is a quadratic in s, of the form $As^2 + Bs + C$, and that the location of the zeros may be determined as desired by appropriate choices for the values of k_1, k_2, and k_3. It is interesting to note also that these zero locations are independent of the controller gain K.

The closed-loop transfer function of the system may now be obtained by use of Eq. (3.3-7). The result, as shown in Fig. 3.3-6*c*, is

$$Y(s) = \frac{4K}{s^3 + (5 + 2Kk_3)s^2 + (4 + 2Kk_3 + 4Kk_2)s + 4Kk_1} R(s) \tag{3.3-9}$$

The denominator of $Y(s)/R(s)$ is a cubic in s, and the three coefficients may be adjusted by varying the feedback coefficients and the gain K. Thus the poles of the closed-loop system may be placed at any desired location. By selecting the values of the three coefficients in the closed-loop denominator, only three equations are established for the determination of the four variable elements: the three feedback coefficients and the gain. Hence one of these elements may be selected arbitrarily.

It is important to remember that H_{eq} is not actually realized by physical elements. Thus, it is not important that the resulting H_{eq} is an improper transfer function. An advantage of measuring the state-variables is that a controller of the configuration of Fig. 3.3-4 can realize the equivalent of an improper transfer function and place the poles of the closed-loop system without the constraints imposed by the need to create a proper H_{eq} for the controller to be realizable.

Let us consider next the reduction of the system of Fig. 3.3-4*b* to the G_{eq} form of Fig. 3.3-5*b*. This may be accomplished by beginning with the inner loop and applying the reduction of Fig. 2.2-8*f* on each successive feedback loop. The outermost loop must be broken into two loops containing $k_1 - 1$ and 1 as feedback coefficients, so that the desired unity-gain feedback form may be achieved. This procedure is illustrated in a step-by-step manner in Fig. 3.3-7.

In Fig. 3.3-7*a* the k_1 feedback loop has been reduced. Note that k_3 affects only the position of the pole that had been at $s = -4$. If the transfer function of the first

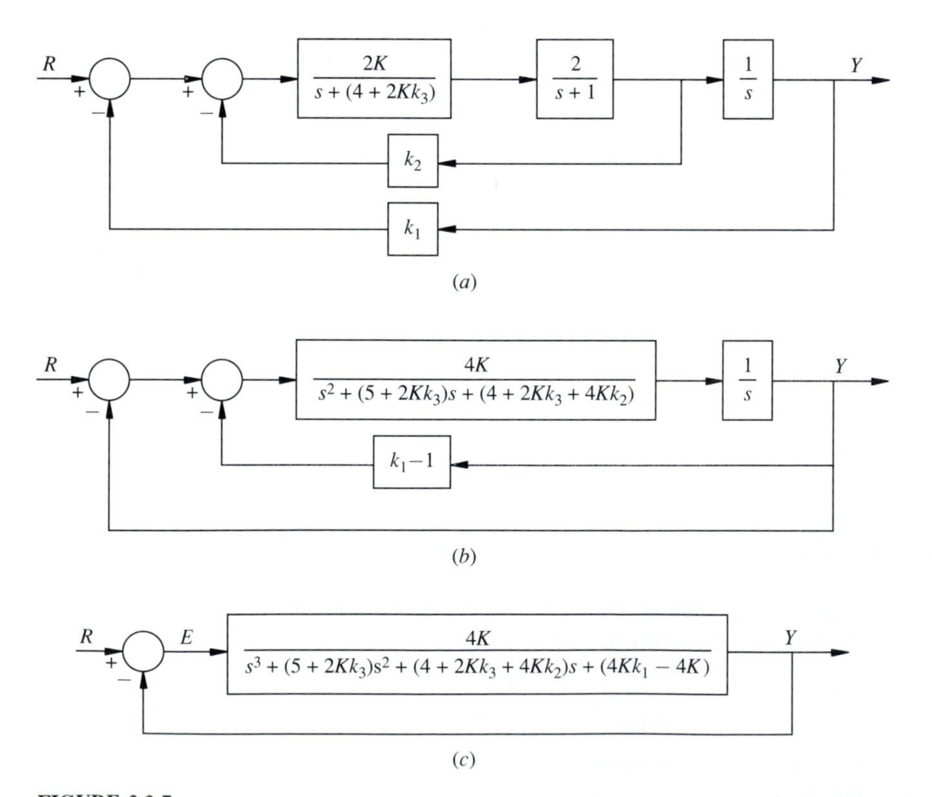

FIGURE 3.3-7
Step-by-step reduction to the G_{eq} form. (*a*) Absorption of the k_3 path; (*b*) absorption of k_2 and the splitting of k_1; (*c*) G_{eq} form.

block had been $K(s+a)/(s+4)$ rather than just $2K/(s+4)$, the zero at $s = -a$ would be unaffected. Thus the effect of feedback through pure-gain elements is to alter pole locations but to leave zeros, if any exist, unaltered. *No new zeros are ever created.* These statements are true regardless of the complexity of the plant transfer function, as we shall show in the next section. The statements are, of course, consistent with the lessons learned in the previous section considering the G and H configurations.

The reduction of the k_2 loop is shown in Fig. 3.3-7*b* along with the splitting of the k_1 loop into two parts as required. The final reduction of the $k_1 - 1$ loop yields the single desired unity-gain feedback loop in G_{eq} form as shown in Fig. 3.3-7*c*, where G_{eq} is easily recognized as

$$G_{eq}(s) = \frac{4K}{s^3 + (5 + 2Kk_3)s^2 + (4 + 2Kk_3 + 4Kk_2)s + (4Kk_1 - 4K)} \tag{3.3-10}$$

The closed-loop transfer function may be determined by applying Eq. (3.3-7) and is identical to the result obtained by using the H_{eq} approach. Note that the input applied to $G_{eq}(s)$ is the error signal $E(s) = R(s) - Y(s)$, that is, the difference between the input $R(s)$ and the output $Y(s)$, as expected from our results on the G configuration.

On the basis of this third-order example, it seems a rather large step to assume that all single-input, single-output linear control systems can be represented in either the $G_{eq}(s)$ or $H_{eq}(s)$ form. However, this is the case, and these two forms of system representation can be quite useful. In the next section the quantities $G_{eq}(s)$ and $H_{eq}(s)$ are expressed in terms of the matrices that make up the system description, that is, in terms of $\boldsymbol{A}$, $\boldsymbol{b}$, $\boldsymbol{c}$, and $\boldsymbol{k}$.

With regard to this third-order example, we are also able to conclude that the zeros of the closed-loop system are identical to those of the open-loop plant. Perhaps more important is the fact that the closed-loop system poles may be positioned anywhere. This result is also true more generally as is indicated in the next section.

Exercises 3.3

3.3-1. Find the $H_{eq}(s)$ and $G_{eq}(s)$ representations of the feedback control system of Fig. 3.3-8.

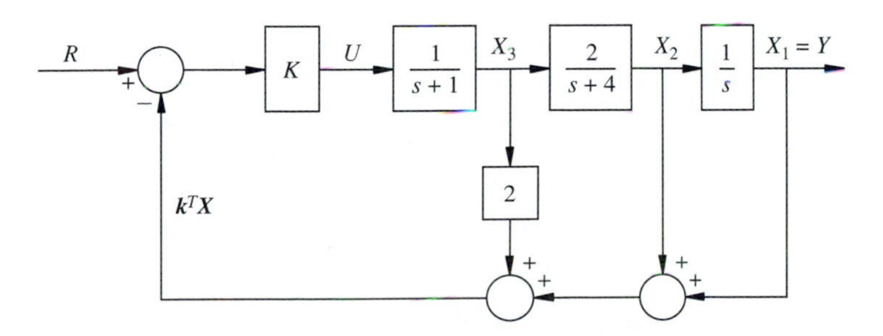

FIGURE 3.3-8
Exercise 3.3-1 and Exercise 3.3-2.

Answers:

$$H_{eq}(s) = s^2 + 5s + 1$$

$$G_{eq}(s) = \frac{2K}{s^3 + (5+2K)s^2 + (4+10K)s}$$

3.3-2. Determine the closed-loop transfer function $Y(s)/R(s)$ for the system of Exercise 3.3-1 (Fig. 3.3-8) using both forms of Eq. (3.3-7).

Answer:

$$\frac{Y(s)}{R(s)} = \frac{2K}{s^3 + (5+2K)s^2 + (4+10K)s + 2K}$$

3.3-3. Given

$$\dot{\boldsymbol{x}} = \begin{bmatrix} -2 & 0 & 0 \\ 0 & 1 & 2 \\ 1 & 0 & -4 \end{bmatrix} \boldsymbol{x} + \begin{bmatrix} 1 \\ 0 \\ 0 \end{bmatrix} \boldsymbol{u}$$

let

$$\boldsymbol{u} = -[k_1 \quad k_2 \quad k_3]\boldsymbol{x}$$

and find the values of $\boldsymbol{k} = [k_1 \quad k_2 \quad k_3]^T$ so that the eigenvalues of the closed-loop system matrix $\boldsymbol{A} - \boldsymbol{b}\boldsymbol{k}^T$ are -2, -3, and -4.

Answer:

$$\boldsymbol{k} = [4 \quad 30 \quad 12]^T$$

3.4 LINEAR STATE-VARIABLE FEEDBACK—GENERAL CASE

In many situations the block-diagram methods of the preceding section provide an easy method of reducing a given closed-loop system to either the $G_{eq}(s)$ or the $H_{eq}(s)$ form. In other situations a matrix approach is more useful. The matrix approach is needed to solve general problems where no specific block diagrams are available. General properties of state-variable feedback can then be established. Matrix methods are also useful when working on specific problems when numerical values are available. Computer programs developed for general systems can be used to compute specific characteristics of a specific problem. One of the major contributions of state-space methods is that these methods lend themselves to more reliable computational procedures than do transfer function methods.

By starting with the system equations it is a relatively simple matter to determine general expressions for both $G_{eq}(s)$ and $H_{eq}(s)$. In addition, expressions for the loop gain transfer function $KG_p(s)H_{eq}(s)$ and the closed-loop transfer function $Y(s)/R(s)$ are also derived. The reader may recall that in Chap. 2 we established that the plant transfer function can be written as

$$G_p(s) = \frac{Y(s)}{U(s)} = \boldsymbol{c}^T\boldsymbol{\Phi}(s)\boldsymbol{b}$$

where $\boldsymbol{\Phi}(s)$ is the plant resolvent matrix given by $\boldsymbol{\Phi}(s) = (s\boldsymbol{I} - \boldsymbol{A})^{-1}$.

To treat the most general case, let us assume an nth-order plant is described as always by

$$\dot{\boldsymbol{x}} = \boldsymbol{A}\boldsymbol{x} + \boldsymbol{b}u \qquad (\boldsymbol{A}\boldsymbol{b})$$

$$y = \boldsymbol{c}^T\boldsymbol{x} \qquad (\boldsymbol{c})$$

Next we assume that u is generated by a linear state-variable feedback relationship of the form

$$u = K\left(r - \boldsymbol{k}^T\boldsymbol{x}\right) \qquad (\boldsymbol{k})$$

If this expression is substituted into Eq. $(\boldsymbol{A}\boldsymbol{b})$, we obtain the desired closed-loop representation

$$\dot{\boldsymbol{x}} = \boldsymbol{A}\boldsymbol{x} - K\boldsymbol{b}\boldsymbol{k}^T\boldsymbol{x} + K\boldsymbol{b}r$$

Grouping the two terms involving $\boldsymbol{x}$, we have a complete description of the closed-loop system.

$$\dot{\boldsymbol{x}} = \left(\boldsymbol{A} - K\boldsymbol{b}\boldsymbol{k}^T\right)\boldsymbol{x} + K\boldsymbol{b}r$$

$$y = \boldsymbol{c}^T\boldsymbol{x} \qquad (3.4\text{-}1)$$

To put this result into a more compact form, let us define

$$\boldsymbol{A}_k = \boldsymbol{A} - K\boldsymbol{b}\boldsymbol{k}^T \qquad (3.4\text{-}2)$$

Then the closed-loop system representation becomes

$$\dot{\boldsymbol{x}} = \boldsymbol{A}_k\boldsymbol{x} + K\boldsymbol{b}r \qquad (\boldsymbol{A}_k\boldsymbol{b})$$

$$y = \boldsymbol{c}^T\boldsymbol{x} \qquad (\boldsymbol{c})$$

Again we note that the closed-loop system representation is identical to the original plant representation except that the $\boldsymbol{A}$ matrix has been changed to the *closed-loop system matrix* $\boldsymbol{A}_k$, and u becomes Kr.

Before considering the G_{eq} and H_{eq} approach to closed-loop system representation, let us determine the closed-loop transfer function directly from the above state-variable representation. This result may be obtained by employing the same procedure used in Sec. 2.3 to find the plant transfer function. We begin by taking the Laplace transform of Eqs. $(\boldsymbol{A}_k\boldsymbol{b})$ and $(\boldsymbol{c})$, assuming all initial conditions are zero.

$$s\boldsymbol{X}(s) = \boldsymbol{A}_k\boldsymbol{X}(s) + K\boldsymbol{b}R(s) \qquad (3.4\text{-}3)$$

$$Y(s) = \boldsymbol{c}^T\boldsymbol{X}(s) \qquad (3.4\text{-}4)$$

If we group the two terms involving $\boldsymbol{X}(s)$ in Eq. (3.4-3) and solve for $\boldsymbol{X}(s)$, the result is

$$\boldsymbol{X}(s) = K(s\boldsymbol{I} - \boldsymbol{A}_k)^{-1}\boldsymbol{b}R(s) \qquad (3.4\text{-}5)$$

Following the previous definition of $\mathbf{\Phi}(s)$, we define the $\mathbf{\Phi}_k(s)$ *closed-loop resolvent matrix*

$$\mathbf{\Phi}_k(s) = (s\boldsymbol{I} - \boldsymbol{A}_k)^{-1} = \left(s\boldsymbol{I} - \boldsymbol{A} + K\boldsymbol{b}\boldsymbol{k}^T\right)^{-1} \tag{3.4-6}$$

Using this definition, Eq. (3.4-5) becomes

$$\boldsymbol{X}(s) = K\mathbf{\Phi}_k(s)\boldsymbol{b}R(s) \tag{3.4-7}$$

If this result is substituted into Eq. (3.4-4), $Y(s)$ is given by

$$Y(s) = K\boldsymbol{c}^T\mathbf{\Phi}_k(s)\boldsymbol{b}R(s)$$

so that the closed-loop transfer function is

$$\frac{Y(s)}{R(s)} = K\boldsymbol{c}^T\mathbf{\Phi}_k(s)\boldsymbol{b} \tag{3.4-8}$$

This is the desired result.

Using the relationship

$$\mathbf{\Phi}_k(s) = (s\boldsymbol{I} - \boldsymbol{A}_k)^{-1} = \frac{\operatorname{adj}(s\boldsymbol{I} - \boldsymbol{A}_k)}{\det(s\boldsymbol{I} - \boldsymbol{A}_k)}$$

the above result may be written as

$$\frac{Y(s)}{R(s)} = K\frac{\boldsymbol{c}^T\operatorname{adj}(s\boldsymbol{I} - \boldsymbol{A}_k)\boldsymbol{b}}{\det(s\boldsymbol{I} - \boldsymbol{A}_k)} \tag{3.4-9}$$

Here $\boldsymbol{c}^T[\operatorname{adj}(s\boldsymbol{I} - \boldsymbol{A}_k)]\boldsymbol{b}$ forms the numerator polynomial of the closed-loop transfer function, and det $(s\boldsymbol{I} - \boldsymbol{A}_k)$ forms the denominator, or closed-loop characteristic polynomial. The closed-loop characteristic equation is formed by setting the closed-loop characteristic polynomial equal to zero.

$$\det(s\boldsymbol{I} - \boldsymbol{A}_k) = 0$$

The values of s that satisfy this characteristic equation are then the closed-loop poles which are just the *eigenvalues of the closed-loop system matrix*. It is important to distinguish between the open-loop and closed-loop poles. The open-loop poles are associated with the plant and the designed compensator. The poles of the closed-loop system are a consequence of the effects of feedback, and they must be found by factoring the closed-loop characteristic polynomial.

Now let us find the matrix expressions for $G_{\text{eq}}(s)$ and $H_{\text{eq}}(s)$. For easy reference, the H_{eq} and G_{eq} configurations are repeated in Fig. 3.4-1. Let us begin with the determination of $H_{\text{eq}}(s)$. Comparing Fig. 3.4-1*a* with the Laplace transform of Eq. (**k**) we see that $H_{\text{eq}}(s)$ is the transfer function from $Y(s)$ to $\boldsymbol{k}^T\boldsymbol{X}(s)$, the linear combination of states used to form the control signal. Thus

$$H_{\text{eq}}(s) = \frac{\boldsymbol{k}^T\boldsymbol{X}(s)}{Y(s)}$$

Since $Y(s) = \boldsymbol{c}^T\boldsymbol{X}(s)$, this expression becomes

$$H_{\text{eq}}(s) = \frac{\boldsymbol{k}^T\boldsymbol{X}(s)}{\boldsymbol{c}^T\boldsymbol{X}(s)}$$

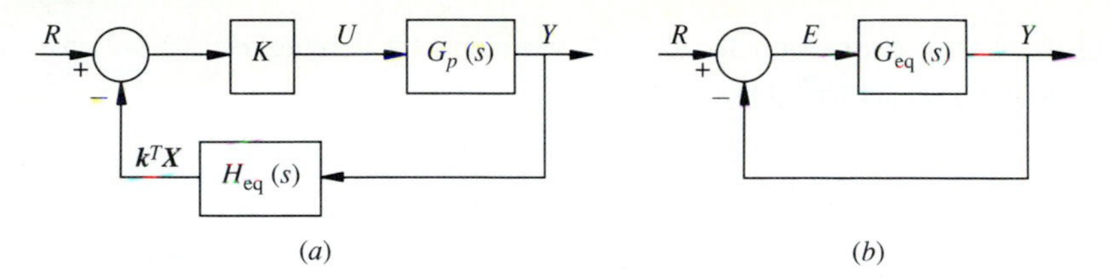

FIGURE 3.4-1
Two basic configurations for the linear state-variable-feedback system. (*a*) H_{eq} form; (*b*) G_{eq} form.

Notice that since $\boldsymbol{k}$, $\boldsymbol{c}$ and $\boldsymbol{X}$ are all vectors, the $\boldsymbol{X}$ factors in the numerator and denominator of this expression cannot be canceled.

If we next use Eqs. (2.3-8) and (2.3-9) to write $\boldsymbol{X}(s)$

$$\boldsymbol{X}(s) = \boldsymbol{\Phi}(s)\boldsymbol{b}U(s)$$

the expression for $H_{eq}(s)$ becomes

$$H_{eq}(s) = \frac{\boldsymbol{k}^T\boldsymbol{\Phi}(s)\boldsymbol{b}}{\boldsymbol{c}^T\boldsymbol{\Phi}(s)\boldsymbol{b}} \tag{3.4-10}$$

(Note that the scalar $U(s)$ may be canceled.) The reader is urged to determine $H_{eq}(s)$ for the system discussed in Sec. 3.3 by using this expression (see Exercise 3.4-1).

Using this result, the loop gain transfer function $KG_p(s)H_{eq}(s)$ takes on a particularly simple form since, as the reader will remember, $G_p(s)$ is given by

$$G_p(s) = \boldsymbol{c}^T\boldsymbol{\Phi}(s)\boldsymbol{b} \tag{3.4-11}$$

and the forward transfer function $KG_p(s)$ is therefore

$$KG_p(s) = K\boldsymbol{c}^T\boldsymbol{\Phi}(s)\boldsymbol{b} \tag{3.4-12}$$

The loop gain transfer function $KG_p(s)H_{eq}(s)$ becomes simply

$$KG_p(s)H_{eq}(s) = K\boldsymbol{c}^T\boldsymbol{\Phi}(s)\boldsymbol{b}\frac{\boldsymbol{k}^T\boldsymbol{\Phi}(s)\boldsymbol{b}}{\boldsymbol{c}^T\boldsymbol{\Phi}(s)\boldsymbol{b}} = K\boldsymbol{k}^T\boldsymbol{\Phi}(s)\boldsymbol{b} \tag{3.4-13}$$

where the cancelation is possible because the quantities involved are scalars.

Since the closed-loop transfer function is given by Eq. (3.3-7) as

$$\frac{Y(s)}{R(s)} = \frac{KG_p(s)}{1 + KG_p(s)H_{eq}(s)}$$

the above results may be used to write $Y(s)/R(s)$

$$\frac{Y(s)}{R(s)} = \frac{K\boldsymbol{c}^T\boldsymbol{\Phi}(s)\boldsymbol{b}}{1 + K\boldsymbol{k}^T\boldsymbol{\Phi}(s)\boldsymbol{b}} \tag{3.4-14}$$

This expression affords an alternative method of determining $Y(s)/R(s)$ by matrix methods. Note that this result depends on the plant resolvent matrix $\boldsymbol{\Phi}(s)$ rather than the closed-loop resolvent matrix $\boldsymbol{\Phi}_{\boldsymbol{k}}(s)$.

To find the matrix expression for $G_{eq}(s)$, we begin by solving Eq. (3.3-7) for $G_{eq}(s)$ in terms of $G_p(s)$ and $H_{eq}(s)$.

$$G_{eq}(s) = \frac{KG_p(s)}{1 + KG_p(s)H_{eq}(s) - KG_p(s)}$$

If we substitute the above expressions for $KG_p(s)$ and $KG_p(s)H_{eq}(s)$, then $G_{eq}(s)$ becomes

$$G_{eq}(s) = \frac{K\mathbf{c}^T\mathbf{\Phi}(s)\mathbf{b}}{1 + K(\mathbf{k} - \mathbf{c})^T\mathbf{\Phi}(s)\mathbf{b}} \tag{3.4-15}$$

These matrix expressions are summarized for easy reference in Table 3.4-1.

Example 3.4-1. As an example of the use of the expressions of Table 3.4-1, let us return to the field-controlled motor of Sec. 3.3, as given in the block diagram of Fig. 3.3-4. To make the example more specific, let us choose $k_1 = 1$, $k_2 = \frac{13}{40}$, and $k_3 = \frac{7}{20}$.

Before using Table 3.4-1, the equations describing the plant must be written. From these equations, $\mathbf{A}$ may be used to find $\mathbf{\Phi}(s)$.

$$\dot{x}_1 = x_2$$

$$\dot{x}_2 = -x_2 + 2x_3$$

$$\dot{x}_3 = -4x_3 + 2u$$

$\mathbf{A}$ is

$$\mathbf{A} = \begin{bmatrix} 0 & 1 & 0 \\ 0 & -1 & 2 \\ 0 & 0 & -4 \end{bmatrix}$$

and $\mathbf{b}$ and $\mathbf{c}$ are

$$\mathbf{b} = \begin{bmatrix} 0 \\ 0 \\ 2 \end{bmatrix} \quad \text{and} \quad \mathbf{c} = \begin{bmatrix} 1 \\ 0 \\ 0 \end{bmatrix}$$

TABLE 3.4-1
Summary of matrix-relation results

		$\mathbf{\Phi}(s)$ **Expression**	
$G_p(s)$	Plant transfer function	$\mathbf{c}^T\mathbf{\Phi}(s)\mathbf{b}$	Eq. (3.4-11)
$KG_p(s)$	Forward transfer function	$K\mathbf{c}^T\mathbf{\Phi}(s)\mathbf{b}$	Eq. (3.4-12)
$H_{eq}(s)$	Equivalent feedback transfer function	$\dfrac{\mathbf{k}^T\mathbf{\Phi}(s)\mathbf{b}}{\mathbf{c}^T\mathbf{\Phi}(s)\mathbf{b}}$	Eq. (3.4-10)
$KG_p(s)H_{eq}(s)$	Open-loop transfer function	$K\mathbf{k}^T\mathbf{\Phi}(s)\mathbf{b}$	Eq. (3.4-13)
$G_{eq}(s)$	Equivalent forward transfer function	$\dfrac{K\mathbf{c}^T\mathbf{\Phi}(s)\mathbf{b}}{1 + K(\mathbf{k} - \mathbf{c})^T\mathbf{\Phi}(s)\mathbf{b}}$	Eq. (3.4-15)
$Y(s)/R(s)$	Closed-loop transfer function	$\dfrac{K\mathbf{c}^T\mathbf{\Phi}(s)\mathbf{b}}{1 + K\mathbf{k}^T\mathbf{\Phi}(s)\mathbf{c}}$	Eq. (3.4-14)

Use of the formula $\boldsymbol{\Phi}(s) = (s\boldsymbol{I} - \boldsymbol{A})^{-1}$ yields

$$\boldsymbol{\Phi}(s) = (s\boldsymbol{I} - \boldsymbol{A})^{-1} = \frac{1}{s(s+1)(s+4)} \begin{bmatrix} (s+1)(s+4) & s+4 & 2 \\ 0 & s(s+4) & 2s \\ 0 & 0 & s(s+1) \end{bmatrix}$$

and we may now use Table 3.4-1 to calculate the various transfer functions. From Fig. 3.3-4 it is evident that the plant transfer function $G_p(s) = Y(s)/U(s)$ is

$$G_p(s) = \frac{4}{s(s+1)(s+4)}$$

This same result may be obtained from the first entry of Table 3.4-1, Eq. (3.4-11), which states

$$G_p(s) = \boldsymbol{c}^T \boldsymbol{\Phi}(s) \boldsymbol{b}$$

In this example,

$$G_p(s) = [1 \quad 0 \quad 0] \left(\frac{1}{s(s+1)(s+4)} \begin{bmatrix} (s+1)(s+4) & s+4 & 2 \\ 0 & s(s+4) & 2s \\ 0 & 0 & s(s+1) \end{bmatrix} \right) \begin{bmatrix} 0 \\ 0 \\ 2 \end{bmatrix}$$

$$= 2\phi_{13}(s) = \frac{4}{s(s+1)(s+4)}$$

where $\phi_{13}(s)$ represents the transfer function in the first row and third column of the matrix $\boldsymbol{\Phi}(s)$.

Let us also calculate $H_{\text{eq}}(s)$, given in Table 3.4-1 as

$$H_{\text{eq}}(s) = \frac{\boldsymbol{k}^T \boldsymbol{\Phi}(s) \boldsymbol{b}}{\boldsymbol{c}^T \boldsymbol{\Phi}(s) \boldsymbol{b}}$$

The denominator transfer function of $H_{\text{eq}}(s)$ is already known, as it is just $G_p(s)$. The numerator is

$$\boldsymbol{k}^T \boldsymbol{\Phi}(s) \boldsymbol{b} = [1 \quad \tfrac{13}{40} \quad \tfrac{7}{20}] \begin{bmatrix} \phi_{11} & \phi_{12} & \phi_{13} \\ \phi_{21} & \phi_{22} & \phi_{23} \\ \phi_{31} & \phi_{32} & \phi_{33} \end{bmatrix} \begin{bmatrix} 0 \\ 0 \\ 2 \end{bmatrix}$$

$$= [1 \quad \tfrac{13}{40} \quad \tfrac{7}{20}] \begin{bmatrix} 2\phi_{13} \\ 2\phi_{23} \\ 2\phi_{33} \end{bmatrix} = \frac{4 + 13s/10 + (7s/10)(s+1)}{s(s+1)(s+4)}$$

The feedback transfer function $H_{\text{eq}}(s)$ is found by dividing the above by $\boldsymbol{c}^T \boldsymbol{\Phi}(s) \boldsymbol{b}$ to yield

$$H_{\text{eq}}(s) = \frac{4 + 13s/10 + (7s/10)(s+1)}{4}$$

or equivalently

$$H_{\text{eq}}(s) = \frac{7s^2 + 20s + 40}{40}$$

The previous expression for $H_{\text{eq}}(s)$ from Eq. (3.3-8) is

$$H_{\text{eq}}(s) = \frac{k_3 s^2 + (k_3 + 2k_2)\, s + 2k_1}{2}$$

and the two results are seen to agree exactly for the given feedback coefficients.

It may seem somewhat ridiculous in the above example to use the matrix expressions to find $G_p(s)$ and $H_{\text{eq}}(s)$. They may be determined almost by inspection from Fig. 3.3-4, whereas the matrix approach requires the inversion of a 3×3 matrix. The point of this section is generality and a demonstration of the equivalence of the block-diagram and matrix approaches. In an actual application, the easier method is used. In situations where the plant being controlled is not simply a group of transfer functions in series, the matrix approach may be easier. The matrix approach may be solved by computer. Also, the generality of the matrix approach is convenient to illustrate the properties of linear state-variable feedback.

Exercise 3.4-1 requires the completion of Example 3.4-1 to find all the transfer functions indicated in Table 3.4-1.

3.4.1 Properties of Linear State-variable Feedback

To establish some of the properties of linear state-variable feedback, let us consider a plant having n poles and m zeros whose transfer function is therefore

$$G_p(s) = \frac{c_m s^m + c_{m-1} s^{m-1} + \cdots + c_0}{s^n + a_{n-1} s^{n-1} + \cdots + a_0} \tag{3.4-16}$$

Since we are interested in properties of transfer functions, the specific state-variable representation is immaterial. Therefore, for simplicity, we make use of phase variables so that the plant representation becomes

$$\dot{\boldsymbol{x}} = \begin{bmatrix} 0 & 1 & \cdots & 0 \\ 0 & 0 & \cdots & 0 \\ \vdots & \vdots & \ddots & \vdots \\ -a_0 & -a_1 & \cdots & -a_{n-1} \end{bmatrix} \boldsymbol{x} + \begin{bmatrix} 0 \\ 0 \\ \vdots \\ 1 \end{bmatrix} u \tag{3.4-17}$$

and

$$y = [\, c_0 \quad c_1 \quad \cdots \quad c_m \quad 0 \quad \cdots \quad 0\,]\boldsymbol{x} \tag{3.4-18}$$

If we start with a transfer function, the plant can always be expressed in phase variable form. If we start with a physical-state variable realization, we can usually provide a linear transformation to phase variable form, analyze or design our system, then transform the problem back into physical state-variable form for implementation. This procedure is explored in Sec. 3.5. Note however that using such a transformation does not always result in the most numerically reliable algorithm and more direct methods are available.

Let us examine $H_{\text{eq}}(s)$ first by considering Eq. (3.4-10). Since $\boldsymbol{\Phi}(s)$ is given by

$$\boldsymbol{\Phi}(s) = (s\boldsymbol{I} - \boldsymbol{A})^{-1} = \frac{\operatorname{adj}(s\boldsymbol{I} - \boldsymbol{A})}{\det(s\boldsymbol{I} - \boldsymbol{A})} \tag{3.4-19}$$

Eq. (3.4-10) may be written

$$H_{\text{eq}}(s) = \frac{\dfrac{\boldsymbol{k}^T \operatorname{adj}(s\boldsymbol{I} - \boldsymbol{A})\,\boldsymbol{b}}{\det(s\boldsymbol{I} - \boldsymbol{A})}}{\dfrac{\boldsymbol{c}^T \operatorname{adj}(s\boldsymbol{I} - \boldsymbol{A})\,\boldsymbol{b}}{\det(s\boldsymbol{I} - \boldsymbol{A})}} = \frac{\boldsymbol{k}^T \operatorname{adj}(s\boldsymbol{I} - \boldsymbol{A})\,\boldsymbol{b}}{\boldsymbol{c}^T \operatorname{adj}(s\boldsymbol{I} - \boldsymbol{A})\,\boldsymbol{b}} \tag{3.4-20}$$

If we recall that $G_p(s)$ can be expressed by Eq. (2.3-10) or Eq. (3.4-11) as

$$G_p(s) = \boldsymbol{c}^T \boldsymbol{\Phi}(s)\boldsymbol{b} = \frac{\boldsymbol{c}^T \operatorname{adj}(s\boldsymbol{I} - \boldsymbol{A})\,\boldsymbol{b}}{\det(s\boldsymbol{I} - \boldsymbol{A})}$$

We know that

$$\det(s\boldsymbol{I} - \boldsymbol{A}) = s^n + a_{n-1}s^{n-1} + \cdots + a_0 \tag{3.4-21}$$

and, using Eq. (2.4-9) for the phase variable form, we can see that

$$\boldsymbol{c}^T \operatorname{adj}(s\boldsymbol{I} - \boldsymbol{A})\,\boldsymbol{b} = c_m s^m + c_{m-1}s^{m-1} + \cdots + c_0 \tag{3.4-22}$$

By analogous reasoning, $\boldsymbol{k}^T \operatorname{adj}(s\boldsymbol{I} - \boldsymbol{A})\,\boldsymbol{b}$ must therefore be

$$\boldsymbol{k}^T \operatorname{adj}(s\boldsymbol{I} - \boldsymbol{A})\,\boldsymbol{b} = k_{n-1}s^{n-1} + k_{n-2}s^{n-2} + \cdots + k_0 \tag{3.4-23}$$

since the only change from Eq. (3.4-22) is that $\boldsymbol{k}$ has been substituted for $\boldsymbol{c}$. Note, however, that c_i is zero for $i > m$, whereas normally this is not true for the elements of $\boldsymbol{k}$.

If Eqs. (3.4-22) and (3.4-23) are substituted into Eq. (3.4-20) for $H_{\text{eq}}(s)$, we have

$$H_{\text{eq}}(s) = \frac{k_{n-1}s^{n-1} + k_{n-2}s^{n-2} + \cdots + k_0}{c_m s^m + c_{m-1}s^{m-1} + \cdots + c_0} \tag{3.4-24}$$

Here it is seen that the numerator of $H_{\text{eq}}(s)$ is a polynomial in s of order $n-1$ and that all the coefficients are adjustable by proper selection of $\boldsymbol{k}$. In order words, $H_{\text{eq}}(s)$ has $n-1$ arbitrarily-placed zeros whose locations are under the direct control of the designer.

In addition, we note the poles of $H_{\text{eq}}(s)$ are exactly equal to the zeros of $G_p(s)$; that is, the denominator polynomial of $H_{\text{eq}}(s)$ is equal to the numerator polynomial $G_p(s)$.

The loop gain transfer function $KG_p(s)H_{\text{eq}}(s)$ is therefore given by

$$KG_p(s)H_{\text{eq}}(s) = \frac{K\left(k_{n-1}s^{n-1} + k_{n-2}s^{n-2} + \cdots + k_0\right)}{s^n + a_{n-1}s^{n-1} + \cdots + a_1 s + a_0} \tag{3.4-25}$$

The arbitrarily-placed zeros of $H_{\text{eq}}(s)$ are the only zeros of the loop gain transfer function, and the poles of $G_p(s)$ are the only poles. In other words, the loop transfer function has poles where $G_p(s)$ has poles and $n-1$ zeros whose locations are determined by $\boldsymbol{k}$.

Let us consider next the closed-loop transfer function $Y(s)/R(s)$ by using Eq. (3.4-14) to write $Y(s)/R(s)$ as

$$\frac{Y(s)}{R(s)} = K\frac{\boldsymbol{c}^T \boldsymbol{\Phi}(s)\boldsymbol{b}}{1 + K\boldsymbol{k}^T \boldsymbol{\Phi}(s)\boldsymbol{b}}$$

If we now substitute for $\Phi(s)$ by using Eq. (3.4-19), $Y(s)/R(s)$ becomes

$$\frac{Y(s)}{R(s)} = \frac{K\boldsymbol{c}^T \operatorname{adj}(s\boldsymbol{I} - \boldsymbol{A})\,\boldsymbol{b}/\det(s\boldsymbol{I} - \boldsymbol{A})}{1 + K\boldsymbol{k}^T \operatorname{adj}(s\boldsymbol{I} - \boldsymbol{A})\,\boldsymbol{b}/\det(s\boldsymbol{I} - \boldsymbol{A})}$$

$$= \frac{K\boldsymbol{c}^T \operatorname{adj}(s\boldsymbol{I} - \boldsymbol{A})\,\boldsymbol{b}}{\det(s\boldsymbol{I} - \boldsymbol{A}) + K\boldsymbol{k}^T \operatorname{adj}(s\boldsymbol{I} - \boldsymbol{A})\,\boldsymbol{b}} \tag{3.4-26}$$

Using Eqs. (3.4-21) to (3.4-23), this result may be rewritten as

$$\frac{Y(s)}{R(s)} = \frac{K\left(c_m s^m + c_{m-1}s^{m-1} + \cdots + c_0\right)}{\left(s^n + a_{n-1}s^{n-1} + \cdots + a_0\right) + K\left(k_{n-1}s^{n-1} + k_{n-2}s^{n-2} + \cdots + k_0\right)}$$

Grouping the like powers of s in the denominator, we have

$$\frac{Y(s)}{R(s)} = \frac{K\left(c_m s^m + c_{m-1}s^{m-1} + \cdots + c_0\right)}{s^n + (a_{n-1} + Kk_{n-1})\,s^{n-1} + \cdots + (a_0 + Kk_0)} \tag{3.4-27}$$

This result could also have been obtained by writing the phase-variable representation of the closed-loop system

$$\dot{\boldsymbol{x}} = \begin{bmatrix} 0 & 1 & \cdots & 0 \\ 0 & 0 & \cdots & 0 \\ \vdots & \vdots & \ddots & \vdots \\ -(a_0 + Kk_0) & -(a_1 + Kk_1) & \cdots & -(a_{n-1} + Kk_{n-1}) \end{bmatrix} \boldsymbol{x} + \begin{bmatrix} 0 \\ 0 \\ \vdots \\ 1 \end{bmatrix} K\boldsymbol{r}$$

$$y = [\,c_1 \quad c_2 \quad \cdots \quad c_m \quad 0 \quad \cdots \quad 0\,]\boldsymbol{x} \tag{3.4-28}$$

From this result, the closed-loop transfer function may be determined by inspection. The result is identical to Eq. (3.4-27).

From Eq. (3.4-27) we see that the coefficients of the denominator polynomial of $Y(s)/R(s)$ may be adjusted at will by proper selection of $\boldsymbol{k}$ and K. Once again, since there are only n coefficients and $n + 1$ adjustable parameters, it is possible to fix one of the parameters arbitrarily.

From either Eq. (3.4-27) or Eq. (3.4-28), it can be seen that the denominator of a closed-loop transfer function that arises from the linear state-variable feedback of a plant described with phase variables is given by

$$D_M(s) = s^n + (a_{n-1} + Kk_{n-1})\,s^{n-1} + (a_{n-2} + Kk_{n-2})\,s^{n-2}$$
$$+ \cdots + (a_1 + Kk_1)\,s + (a_0 + Kk_0)$$

Let $D_d(s)$ be any nth-order polynomial that is desired as the closed-loop pole polynomial

$$D_d(s) = s^n + d_{n-1}s^{n-1} + d_{n-2}s^{n-2} + \cdots + d_1 s + d_0$$

The poles of the closed-loop system can then be set to the desired location by setting the feedback coefficients of the phase variables according to

$$\boldsymbol{k}_p = \frac{\boldsymbol{d} - \boldsymbol{a}}{K} \tag{3.4-29}$$

where $\boldsymbol{d}$ and $\boldsymbol{a}$ are vectors formed from the d_i and a_i terms. (This choice of $\boldsymbol{k}$ is correct only if the phase variables are the variables measured. The correct $\boldsymbol{k}$ for other state-variable representations is explored in Sec. 3.5.)

Note also that except for K the numerator of $Y(s)/R(s)$ is identical to the numerator of $G_p(s)$; that is, the closed-loop zeros are equal to the open-loop zeros. In other words, linear state-variable feedback has no effect on the zeros of $Y(s)/R(s)$. However, the freedom of choice of the poles of $Y(s)/R(s)$ means that closed-loop poles may be positioned anywhere. A method of removing an unwanted zero is to place a closed-loop pole at the same location as the zero, effectively canceling it. However, we will see in Chap. 6 that zeros in the right half-plane cannot be canceled. The freedom in state-variable feedback to place poles arbitrarily while not being able to change zeros was also found for output feedback using the transfer function in Sec. 3.2. However, the output feedback pole placement may be subject to restrictions to ensure realizability of the resulting controller. The state-variable feedback problem automatically meets realizability constraints as long as all state-variable are measurable.

Example 3.4-2. Assume that, as in Example 3.3-1, the plant $G_p(s)$ is given by

$$G_p(s) = \frac{4}{s(s+1)(s+4)}$$

The phase variable form of this plant is available by inspection

$$\boldsymbol{A_p} = \begin{bmatrix} 0 & 1 & 0 \\ 0 & 0 & 1 \\ 0 & -4 & -5 \end{bmatrix} \qquad \boldsymbol{b_p} = \begin{bmatrix} 0 \\ 0 \\ 1 \end{bmatrix} \qquad \boldsymbol{c}_p^T = [4 \quad 0 \quad 0]$$

We wish to place the poles of the closed-loop system at $s = -2$, $s = -2 - j2$ and $s = -2 + j2$. Thus

$$D_d(s) = s^3 + 6s^2 + 16s + 16$$

The vectors needed to apply Eq. (3.4-29) are

$$\boldsymbol{a} = \begin{bmatrix} 0 \\ 4 \\ 5 \end{bmatrix} \qquad \boldsymbol{d} = \begin{bmatrix} 16 \\ 16 \\ 6 \end{bmatrix}$$

Arbitrarily setting $K = 1$, the resulting feedback gains necessary to place the poles in the desired locations are given by Eq. (3.4-29)

$$\boldsymbol{k} = \begin{bmatrix} 16 \\ 12 \\ 1 \end{bmatrix}$$

This $\boldsymbol{k}$ can be used to form $H_{\text{eq}}(s)$.

$$H_{\text{eq}}(s) = \frac{\boldsymbol{k}^T \boldsymbol{\Phi}(s)\boldsymbol{b}}{\boldsymbol{c}^T \boldsymbol{\Phi}(s)\boldsymbol{b}} = \frac{\dfrac{s^2 + 12s + 16}{s^3 + 5s^2 + 4s}}{\dfrac{4}{s^3 + 5s^2 + 4s}} = \frac{1}{4}s^2 + 3s + 4$$

Note that, as a transfer function, $H_{\text{eq}}(s)$ is improper and unrealizable. In the present context we see that the *effect* of this $H_{\text{eq}}(s)$ can be realized by state-variable feedback if the needed phase variables can be measured. In Sec. 3.5 we explore the more realistic problem of determining the feedback gains necessary to realize this $H_{eq}(s)$ using physical variables.

To summarize, the following conclusions concerning the effects of linear state-variable feedback may be reached.

1. Feedback transfer function $H_{eq}(s)$
   (*a*) The poles of $H_{eq}(s)$ are the zeros of $G_p(s)$.
   (*b*) $H_{eq}(s)$ has $n-1$ arbitrarily-placed zeros.

2. Loop gain transfer function $KG_p(s)H_{eq}(s)$
   (*a*) The poles of $KG_p(s)H_{eq}(s)$ are the poles of $G_p(s)$.
   (*b*) The zeros of $KG_p(s)H_{eq}(s)$ are the zeros of $H_{eq}(s)$, that is, there are $n-1$ arbitrarily-placed zeros.

3. Closed-loop transfer function $Y(s)/R(s)$
   (*a*) The poles of $Y(s)/R(s)$ may be arbitrarily positioned by proper selection of $\boldsymbol{k}$ and K. For plant descriptions in phase variable form the appropriate $\boldsymbol{k}$ is easily found using Eq. (3.4-29).
   (*b*) The zeros of $Y(s)/R(s)$ are the zeros of $G_p(s)$.

One might view the above conclusions jointly and say that the ability to select the locations of the loop gain transfer function zeros, which are the zeros of $H_{eq}(s)$, ensures that the closed-loop poles can be located at will.

If one is not careful, situations can arise that appear to violate the properties of state-variable feedback developed above. First, if the system description is not in the form of Eq. ($\boldsymbol{Ab}$), the properties stated above are usually violated. This situation most often occurs because of a zero in the input block, which causes a $\dot{u}$ term to appear in the equations, as discussed in Sec. 2.5. The solution to this problem is to redefine the state-variables, using, for example, the feedforward scheme suggested in Sec. 2.5 so that the system is represented by Eqs. ($\boldsymbol{Ab}$) and ($\boldsymbol{c}$).

The second situation is a more insidious one. If $H_{eq}(s)$ is found by block-diagram methods, it is possible that poles and zeros of $H_{eq}(s)$ may be canceled, resulting in an $H_{eq}(s)$ that does not contain as poles all the zeros of $G_p(s)$. Consider, for example, the system shown in Fig. 3.4-2. Using block-diagram manipulations, one may easily find that

$$H_{eq}(s) = k_1 \qquad \text{and} \qquad G_p(s) = \frac{s+4}{(s+1)(s+2)(s+5)}$$

Note that the zero of $G_p(s)$ at $s=-4$ does *not* appear as a pole of $H_{eq}(s)$, as required. However, if the matrix equation (3.4-10) is used, the correct $H_{eq}(s)$ is found to be

$$H_{eq}(s) = \frac{k_1(s+4)}{s+4}$$

To avoid this problem whenever block-diagram methods are used to find $H_{eq}(s)$, one should always check to see if all the zeros of $G_p(s)$ appear as poles of $H_{eq}(s)$. If there

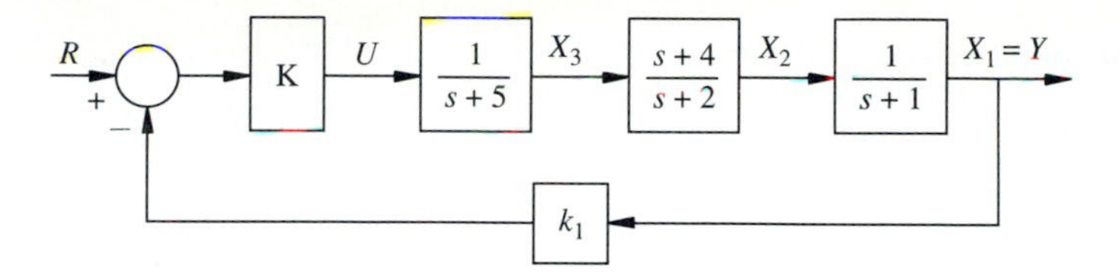

FIGURE 3.4-2
Example illustrating how $H_{\mathrm{eq}}(s)$ can be incorrectly found by block-diagram methods.

are some zeros that do not appear, then $H_{\mathrm{eq}}(s)$ should be multiplied in the numerator and denominator by these factors to ensure that all the zeros of $G_p(s)$ appear as poles of $H_{\mathrm{eq}}(s)$. This may seem an unnecessary complication of the problem; however, by ensuring that the properties $H_{\mathrm{eq}}(s)$ are always satisfied, the later developments may be considerably simplified by excluding special cases. After gaining familiarity with the methods, the reader will find several shortcuts. But, at the beginning, the reader is urged to follow the rules carefully to avoid difficulty.

One point the reader should note from this discussion is that the poles and zeros of $H_{\mathrm{eq}}(s)$ should not be canceled lest the properties of $H_{\mathrm{eq}}(s)$ be destroyed. In addition, the poles of $G_p(s)$ should not be canceled by the zeros of $H_{\mathrm{eq}}(s)$ in forming the loop gain transfer function although the poles of $H_{\mathrm{eq}}(s)$ and the zeros of $G_p(s)$, which are always identical, may be canceled.

A case where the properties of state feedback are actually modified arises from the physical variable description of certain systems. With certain systems not all closed-loop poles can be placed or controlled. Such *uncontrollable* systems are considered in Sec. 3.5.

3.4.2 CAD Notes

The quantities summarized in Table 3.4-1 can be computed using the MATLAB function `ss2tf` with appropriate arguments and some further manipulations.

Example 3.4-3. Find $H_{\mathrm{eq}}(s)$, $G_{\mathrm{eq}}(s)$, and the closed-loop transfer function for the system given in Example 3.4-2.

```
>>a=[0 1 0; 0 0 1; 0 -4 -5];
b=[0; 0; 1];
c=[4; 0; 0];
d=[0];
k=[16; 12; 1];
K=[1];
%find the numerator of Heq
%(and the characteristic polynomial of A)
[numHeq,charA]=ss2tf(a,b,k',d,1)
numHeq =
   0 1.0000 12.0000 16.0000
charA =
  1  5  4  0
```

```
%find the denominator of Heq
[denHeq,charA]=ss2tf(a,b,c',d,1)

denHeq =
    0     0     0  4.0000
charA =
   1  5  4  0

%find the numerator of Geq
numGeq=K*denHeq
numGeq =
    0     0     0  4.0000

%find the denominator of Geq
[preden,charA]=ss2tf(a,b,k'-c',d,1)
preden =
     0  1.0000 12.0000 12.0000
charA =
   1  5  4  0

denGeq=charA+K*preden
denGeq =
   1.0000   6.0000   16.0000   12.0000

%numerator of the closed-loop transfer function
%is the same as the numerator of Geq
%find the denominator of the closed-loop transfer function
denAk=charA+K*numHeq
denAk =
   1.0000   6.0000   16.0000   16.0000
```

Exercises 3.4

3.4-1. Complete Example 3.4-1 using the matrix expressions given in Table 3.4-1 to find $G_{\text{eq}}(s)$, $KG_p(s)H_{\text{eq}}(s)$, and $Y(s)/R(s)$ for the field-controlled motor shown in Fig. 3.3-4. As in Example 3.4-1, let $k_1 = 1, k_2 = \frac{13}{40}$, and $k_3 = \frac{7}{20}$. Compare the answers with those obtained by block-diagram manipulation in Sec. 3.3.

3.4-2. Find the state-variable representation and the value of $\boldsymbol{k}$ for the block diagram of Fig. 3.3-8 of Exercise 3.3-1. Use the expressions from Table 3.4-1 to determine $G_p(s)$, $G_{\text{eq}}(s)$, $H_{\text{eq}}(s)$, the loop gain transfer function, and the closed-loop transfer function $Y(s)/R(s)$. Compare your answers to those found in Exercises 3.3-1 and 3.3-2.

3.4-3. Indicate each correct answer. More than one answer may be correct.

(*a*) The poles of the closed-loop system $Y(s)/R(s)$ are:

(*i*) The poles of $G_p(s)$

(*ii*) The zeros of $H_{\text{eq}}(s)$

(*iii*) Arbitrarily located by K and $\boldsymbol{k}$

(*iv*) The poles of $G_{\text{eq}}(s)$

(*v*) None of the above

(*b*) The zeros of the closed-loop system $Y(s)/R(s)$ are:

(*i*) The zeros of $H_{\text{eq}}(s)$

(*ii*) The poles of $G_{\text{eq}}(s)$

(*iii*) The poles of $H_{eq}(s)$
(*iv*) Arbitrarily located by K and $\boldsymbol{k}$
(*v*) None of the above

(*c*) The loop gain $KG_p(s)H_{eq}(s)$ has:
(*i*) $n - 1$ arbitrary poles
(*ii*) n zeros
(*iii*) Arbitrary poles and zeros
(*iv*) Zeros identical to those of $G_p(s)$
(*v*) None of the above

(*d*) A linear transformation of variables:
(*i*) Does not alter the eigenvalues
(*ii*) Changes the transfer function
(*iii*) Does not change the $\boldsymbol{c}$ matrix
(*iv*) Requires a nonsingular transformation matrix
(*v*) None of the above

Answers:
(*a*) *iii* (*b*) *iii* (*c*) *v* (*d*) *i, iv*

3.5 CONTROLLABILITY AND POLE PLACEMENT

In the previous section, it was determined that, if a plant is expressed in phase variable form, any nth-order polynomial can be achieved as the closed-loop pole polynomial by using the state variable feedback gains that satisfy Eq. (3.4-29), repeated here.

$$\boldsymbol{k}_p = \frac{\boldsymbol{d} - \boldsymbol{a}}{K} \tag{3.4-29}$$

The vectors $\boldsymbol{d}$ and $\boldsymbol{a}$ are formed from the coefficients of the desired closed-loop pole polynomial and the plant pole polynomial respectively.

In this section, we see that the closed-loop poles of many other state-variable descriptions can be placed arbitrarily. All state-variable descriptions whose closed-loop poles can be placed arbitrarily share a property called *controllability*. All controllable representations can be transformed into phase-variable descriptions by linear transformations; the pole placement problem can then be solved and the description transformed back to the original physical variables, which can be measured and used for feedback. Systems that are not controllable cannot have their closed-loop poles arbitrarily set.

For a state-variable description given by Eq. ($\boldsymbol{Ab}$) and Eq. ($\boldsymbol{c}$), the *controllability matrix*, $\boldsymbol{C_{Ab}}$, is a square matrix whose first column is given by $\boldsymbol{b}$, second column is given by the product $\boldsymbol{Ab}$, third column is given by $\boldsymbol{A}^2\boldsymbol{b}$ and so on until the nth column is given by $\boldsymbol{A}^{n-1}\boldsymbol{b}$. For notation, we write

$$\boldsymbol{C_{Ab}} = [\,\boldsymbol{b} \quad \boldsymbol{Ab} \quad \boldsymbol{A}^2\boldsymbol{b} \quad \cdots \quad \boldsymbol{A}^{n-1}\boldsymbol{b}\,] \tag{3.5-1}$$

Note that the output vector $\boldsymbol{c}^T$ does not enter into the controllability matrix; it will not enter into the issue of controllability.

A state-variable description given by Eq. ($\boldsymbol{Ab}$) is said to be *controllable* if its associated controllability matrix is invertible.

We first note that controllability does not change when the system undergoes a linear transformation of state-variables. Recall from Sec. 2.6 that if two sets of state-variables are related by an invertible linear transformation

$$\boldsymbol{x}(t) = \boldsymbol{P}\,\boldsymbol{x}^*(t) \tag{2.6-1}$$

then the matrices of the state-variable descriptions are related by

$$\boldsymbol{A}^* = \boldsymbol{P}^{-1}\boldsymbol{A}\,\boldsymbol{P},\ \ \boldsymbol{b}^* = \boldsymbol{P}^{-1}\boldsymbol{b} \text{ and } \boldsymbol{c}^* = \boldsymbol{P}^T\boldsymbol{c} \tag{2.6-5}$$

The controllability matrix of the $\boldsymbol{x}^*$ description is given by

$$\begin{aligned}
\boldsymbol{C}_{A^*b^*} &= \left[\boldsymbol{b}^* \vdots \boldsymbol{A}^*\boldsymbol{b}^* \vdots \ldots \vdots \boldsymbol{A}^{*n-1}\boldsymbol{b}^*\right] \\
&= \left[\boldsymbol{P}^{-1}\boldsymbol{b} \vdots \boldsymbol{P}^{-1}\boldsymbol{APP}^{-1}\boldsymbol{b} \vdots \left(\boldsymbol{P}^{-1}\boldsymbol{AP}\right)^2\boldsymbol{P}^{-1}\boldsymbol{b} \vdots \ldots \vdots \left(\boldsymbol{P}^{-1}\boldsymbol{AP}\right)^{n-1}\boldsymbol{P}^{-1}\boldsymbol{b}\right] \\
&= [\boldsymbol{P}^{-1}\boldsymbol{b} \vdots \boldsymbol{P}^{-1}\boldsymbol{Ab} \vdots \boldsymbol{P}^{-1}\boldsymbol{A}^2\boldsymbol{b} \vdots \ldots \vdots \boldsymbol{P}^{-1}\boldsymbol{A}^{n-1}\boldsymbol{b}] \\
&= \boldsymbol{P}^{-1}\left[\boldsymbol{b} \vdots \boldsymbol{Ab} \vdots \ \boldsymbol{A}^2\boldsymbol{b} \vdots \ldots \vdots \boldsymbol{A}^{n-1}\boldsymbol{b}\right] \\
&= \boldsymbol{P}^{-1}\boldsymbol{C}_{Ab}
\end{aligned}$$

Since $\boldsymbol{P}$ is invertible, $\boldsymbol{C}_{A^*b^*}$ is invertible if and only if $\boldsymbol{C}_{Ab}$ is invertible. Therefore, the linear transformation of a controllable representation is controllable and the linear transformation of an uncontrollable representation is uncontrollable.

Next, we establish that the phase variable representation of a system is always controllable. The phase variable representation of a general nth-order system was given in Eq. (2.4-7) and Eq. (2.4-8). The representation matrices are

$$\boldsymbol{A_p} = \begin{bmatrix} 0 & 1 & 0 & \cdots & 0 & 0 \\ 0 & 0 & 1 & \cdots & 0 & 0 \\ \vdots & \vdots & \vdots & \ddots & \vdots & \vdots \\ 0 & 0 & 0 & \cdots & 0 & 1 \\ -a_0 & -a_1 & -a_2 & \cdots & -a_{n-1} & -a_{n-1} \end{bmatrix} \qquad \boldsymbol{b_p} = \begin{bmatrix} 0 \\ 0 \\ \vdots \\ 0 \\ 1 \end{bmatrix}$$

$$\boldsymbol{c_p}^T = [\,c_0 \quad c_1 \quad \cdots \quad c_{m-1} \quad c_m \quad 0 \quad \cdots \quad 0\,] \tag{3.5-2}$$

The first few columns of the controllability matrix for this representation are

$$\boldsymbol{b} = \begin{bmatrix} 0 \\ 0 \\ \vdots \\ 0 \\ 1 \end{bmatrix} \qquad \boldsymbol{Ab} = \begin{bmatrix} 0 \\ 0 \\ \vdots \\ 1 \\ -a_{n-1} \end{bmatrix} \qquad \boldsymbol{A}^2\boldsymbol{b} = \begin{bmatrix} 0 \\ 0 \\ \vdots \\ 0 \\ 1 \\ -a_{n-1} \\ a_{n-1}^2 - a_{n-1} \end{bmatrix}$$

The important pattern is that the ith column of the controllability matrix has zeros in the first $n-i$ rows, a one in row $n-i+1$ and other entries in lower rows. The controllability matrix for the phase variable representation has the form

$$C_{A_p b_p} = \begin{bmatrix} 0 & 0 & \dots & 0 & 1 \\ 0 & 0 & \dots & 1 & * \\ \vdots & \vdots & \ddots & \vdots & \vdots \\ 0 & 1 & \dots & * & * \\ 1 & * & \dots & * & * \end{bmatrix} \tag{3.5-3}$$

where each * represents an element of undetermined value.

Since the determinant of the matrix of Eq. (3.5-3) is unity the matrix is invertible and the phase variable representation is controllable.

Since the phase variable representation is controllable, the state-variable representations that can be transformed into a phase variable representation are controllable. We next look at the problem of finding the linear transformation needed to transform an arbitrary controllable representation into a phase variable representation. We solve this problem by first transforming the system into a new form called the *controllability canonical form* and then transforming the representation from the controllability canonical form into the phase variable form. The $\boldsymbol{A}_{cc}$ and $\boldsymbol{b}_{cc}$ elements of the controllability canonical form are

$$\boldsymbol{A}_{cc} = \begin{bmatrix} 0 & 0 & \cdots & 0 & 0 & -a_0 \\ 1 & 0 & \cdots & 0 & 0 & -a_1 \\ 0 & 1 & \cdots & 0 & 0 & -a_2 \\ \vdots & \vdots & \ddots & \vdots & \vdots & \vdots \\ 0 & 0 & \cdots & 1 & 0 & -a_{n-2} \\ 0 & 0 & \cdots & 0 & 1 & -a_{n-1} \end{bmatrix} \qquad \boldsymbol{b}_{cc} = \begin{bmatrix} 1 \\ 0 \\ 0 \\ \vdots \\ 0 \\ 0 \end{bmatrix} \tag{3.5-4}$$

where, as usual, the a_i terms are the coefficients of the pole polynomial of the plant. The output vector $\boldsymbol{c}^T$ is not as directly related to the transfer function as the output vector of the phase variable form. The output vector will be determined later using linear transformation relationships. Note that the $\boldsymbol{A}_{cc}$ matrix is simply the transpose of the phase variable representation $\boldsymbol{A}_p$ matrix. The controllability canonical form derives its name from the fact that its associated controllability matrix is always equal to the identity matrix, i.e.,

$$C_{A_{cc} b_{cc}} = \boldsymbol{I} \tag{3.5-5}$$

This fact is easily verifiable by directly calculating the controllability matrix from Eq. (3.5-4).

An important property of the controllability canonical form is that there is a simple expression which provides the linear transformation between the controllability canonical form and any other controllable state-variable representation. Let us derive this relationship by finding the $\boldsymbol{P}$ matrix in the expression

$$\boldsymbol{x} = \boldsymbol{P}\,\boldsymbol{x}_{cc} \tag{3.5-6}$$

where $\boldsymbol{x}$ is from any controllable state-variable representation and $\boldsymbol{x}_{cc}$ is the state vector of the controllability canonical form. From Eq. (2.6-5) we have

$$\boldsymbol{P}\,\boldsymbol{b}_{cc} = \boldsymbol{b} \tag{3.5-7}$$

and

$$\boldsymbol{P}\,\boldsymbol{A}_{cc} = \boldsymbol{A}\boldsymbol{P} \tag{3.5-8}$$

Let $\boldsymbol{p}_i$ be the ith column of $\boldsymbol{P}$ so that

$$\boldsymbol{P} = [\boldsymbol{p}_1 \quad \boldsymbol{p}_2 \quad \cdots \quad \boldsymbol{p}_n\,] \tag{3.5-9}$$

From the form of $\boldsymbol{b}_{cc}$, Eq. (3.5-7) becomes

$$\boldsymbol{p}_1 = \boldsymbol{b} \tag{3.5-10}$$

We then express the product on each side of Eq. (3.5-8) in terms of its columns and use the form of $\boldsymbol{A}_{cc}$ to obtain

$$\boldsymbol{P}\,\boldsymbol{A}_{cc} = \left[\boldsymbol{p}_2 \vdots \boldsymbol{p}_3 \vdots \ \ldots \ \vdots \boldsymbol{p}_n \vdots * \right] = \left[\boldsymbol{A}\boldsymbol{p}_1 \vdots \boldsymbol{A}\boldsymbol{p}_2 \vdots \ \ldots \ \vdots \boldsymbol{A}\boldsymbol{p}_{n-1} \vdots \boldsymbol{A}\boldsymbol{p}_n\right] \tag{3.5-11}$$

where the last column of the left hand product is left undetermined. Thus, we have

$$\boldsymbol{p}_i = \boldsymbol{A}\,\boldsymbol{p}_{i-1} \qquad i = 2, 3, \ldots, n \tag{3.5-12}$$

Starting with Eq. (3.5-10) and successively substituting previously determined columns of $\boldsymbol{P}$ in Eq. (3.5-12) the entire $\boldsymbol{P}$ matrix is obtained.

$$\boldsymbol{P} = \left[\boldsymbol{b} \vdots \boldsymbol{A}\,\boldsymbol{b} \vdots \boldsymbol{A}^2\boldsymbol{b} \vdots \ \ldots \ \vdots \boldsymbol{A}^{n-1}\boldsymbol{b}\right] = \boldsymbol{C}_{Ab} \tag{3.5-13}$$

We see that the needed transformation matrix is just the controllability matrix of the general controllable form $\boldsymbol{Ab}$. The need for $\boldsymbol{Ab}$ to be controllable is reflected in the need for the transformation $\boldsymbol{P}$ to be invertible.

Any controllable representation can be transformed into a controllability canonical form by the equation

$$\boldsymbol{x}_{cc} = \boldsymbol{C}_{Ab}^{-1}\boldsymbol{x} \tag{3.5-14}$$

A controllability canonical representation can be transformed into any other controllable representation by Eq. (3.5-6) with $\boldsymbol{P}$ as in Eq. (3.5-13). In particular, a controllability canonical representation can be transformed into a phase variable representation by the equation

$$\boldsymbol{x}_p = \boldsymbol{C}_{A_p b_p}\boldsymbol{x}_{cc} \tag{3.5-15}$$

Combining Eq. (3.5-14) and Eq. (3.5-15) we obtain the transformation needed between any controllable representation and the phase variable representation

$$\boldsymbol{x}_p = \boldsymbol{P}_p\boldsymbol{x} \tag{3.5-16}$$

where

$$\boldsymbol{P}_p = \boldsymbol{C}_{A_p b_p}\boldsymbol{C}_{Ab}^{-1} \tag{3.5-17}$$

Example 3.5-1. Start with the description of the field-controlled dc motor from Eq. (3.3-2) and Eq. (3.3-3).

$$\boldsymbol{A} = \begin{bmatrix} 0 & 1 & 0 \\ 0 & -1 & 2 \\ 0 & 0 & -4 \end{bmatrix} \qquad \boldsymbol{b} = \begin{bmatrix} 0 \\ 0 \\ 2 \end{bmatrix} \qquad \boldsymbol{c}^T = [1 \quad 0 \quad 0]$$

The controllability matrix for this representation is given by

$$\boldsymbol{C_{Ab}} = \begin{bmatrix} 0 & 0 & 4 \\ 0 & 4 & -20 \\ 2 & -8 & 32 \end{bmatrix}$$

Because this matrix is invertible the system is controllable and can be transformed into phase variable form.

$$\boldsymbol{C}_{\boldsymbol{Ab}}^{-1} = \begin{bmatrix} 1 & 1 & 0.5 \\ 1.25 & 0.25 & 0 \\ 0.25 & 0 & 0 \end{bmatrix}$$

The transfer function of the plant is

$$\boldsymbol{G_P}(s) = \frac{4}{s^3 + 5s^2 + 4s}$$

The phase variable representation matrices are

$$\boldsymbol{A_p} = \begin{bmatrix} 0 & 1 & 0 \\ 0 & 0 & 1 \\ 0 & -4 & -5 \end{bmatrix} \qquad \boldsymbol{b_p} = \begin{bmatrix} 0 \\ 0 \\ 1 \end{bmatrix} \qquad \boldsymbol{c}_p^T = [4 \quad 0 \quad 0]$$

The controllability matrix of the phase variable representation is

$$\boldsymbol{C}_{\boldsymbol{A_p b_p}} = \begin{bmatrix} 0 & 0 & 1 \\ 0 & 1 & -5 \\ 1 & -5 & 21 \end{bmatrix}$$

Therefore, the transformation matrix between the physical variables and the phase variables is obtained by applying Eq. (3.5-17)

$$\boldsymbol{P_p} = \begin{bmatrix} 0.25 & 0 & 0 \\ 0 & 0.25 & 0 \\ 0 & -0.25 & 0.5 \end{bmatrix}$$

The reader should check that

$$\boldsymbol{A_p} = \boldsymbol{P_p A P}_p^{-1} \qquad \boldsymbol{b_p} = \boldsymbol{P_p b} \qquad \boldsymbol{c_p} = \boldsymbol{P}_p^T \boldsymbol{c}$$

where

$$\boldsymbol{P}_p^{-1} = \begin{bmatrix} 4 & 0 & 0 \\ 0 & 4 & 0 \\ 0 & 2 & 2 \end{bmatrix}$$

The last step in solving the pole placement problem for state-variable feedback using any controllable state-variable realization involves solving for the feedback gains $\boldsymbol{k}$ needed for the measurable physical variables in terms of the feedback gains $\boldsymbol{k_P}$ computed for the phase variables by Eq. (3.4-29). The appropriate plant input formed by the controller is given by Eq. ($\boldsymbol{k}$) for the phase variable description

$$u = K\left(r - \boldsymbol{k}_p^T \boldsymbol{x}_p\right) \tag{3.5-18}$$

By substituting Eq. (3.5-16) relating $\boldsymbol{x}_p$ with the physical variables $\boldsymbol{x}$ we obtain

$$u = K\left(r - \boldsymbol{k}_p^T \boldsymbol{P}_p \boldsymbol{x}\right) \tag{3.5-19}$$

so that, for use in equation $\boldsymbol{k}$

$$\boldsymbol{k}^T = \boldsymbol{k}_p^T \boldsymbol{P}_p \tag{3.5-20}$$

By substituting Eq. (3.5-17) and Eq. (3.4-29) into Eq. (3.5-20), we obtain a general equation for the feedback gain required to place the poles of an arbitrary controllable state-variable realization.

$$\boldsymbol{k}^T = \frac{\boldsymbol{d}^T - \boldsymbol{a}^T}{K} \boldsymbol{C}_{A_p b_p} \boldsymbol{C}_{Ab}^{-1} \tag{3.5-21}$$

The constant K may be set arbitrarily. The parameters needed in Eq. (3.5-21) come from the desired closed-loop pole polynomial and the pole polynomial of the original controllable state-variable realization. Although the particular technique of a transformation to phase variable form was used to derive Eq. (3.5-21), the equation may be used directly to place the poles of any controllable form. The need for the physical representation to be controllable is reflected in the appearance of the inverse of the controllability matrix in Eq. (3.5-21). If a state representation is not controllable, then Eq. (3.5-21) cannot be solved for an arbitrary set of closed-loop polynomial coefficients $\boldsymbol{d}$. The poles of an uncontrollable plant description cannot be placed arbitrarily.

There are other observations to be made about Eq. (3.5-21). If $\boldsymbol{d} - \boldsymbol{a}$ is large, that is, if the desired closed-loop poles are very different from the plant poles, the feedback gains $\boldsymbol{k}$ are large and the loop gain, given by $\boldsymbol{k}^T \boldsymbol{\Phi}(s) \boldsymbol{b}$, is also large. Also, if the physical variable representation is nearly uncontrollable so that $\boldsymbol{C}_{Ab}^{-1}$ has large elements, a large loop gain is required to move some poles even a small distance. We will see in Chap. 6 that a large loop gain can cause stability problems.

Example 3.5-2. In this example, we find the appropriate feedback gains to place the poles of the field-controlled motor. The physical variable description is given in Example 3.5-1. The desired poles and the appropriate feedback gains for the phase variable description of the system are given in Example 3.4-2. From Example 3.4-2, with $K = 1$, we have

$$\boldsymbol{k}_p = \frac{\boldsymbol{d} - \boldsymbol{a}}{K} = \begin{bmatrix} 16 \\ 12 \\ 1 \end{bmatrix}$$

From Example 3.5-1, we have

$$\boldsymbol{C}_{A_p b_p} \boldsymbol{C}_{Ab}^{-1} = \begin{bmatrix} 0.25 & 0 & 0 \\ 0 & 0.25 & 0 \\ 0 & -0.25 & 0.5 \end{bmatrix}$$

Therefore, $\boldsymbol{k}$ is given by Eq. (3.5-21) as

$$\boldsymbol{k} = \begin{bmatrix} 4 \\ 2.75 \\ 0.5 \end{bmatrix}$$

A natural question arises. What happens if a system is uncontrollable? To answer this question we develop another characterization of an uncontrollable system using the eigenvalues and eigenvectors of a plant. For the rest of this section we assume that the plant has distinct eigenvalues. The results hold more generally but they are more difficult to derive when the eigenvalues are not distinct. In the following text we use results concerning the eigenvalues of matrices from Appendix C.

The poles of the plant are given by eigenvalues of the plant matrix $\boldsymbol{A}$. Associated with each eigenvalue λ_i are a right eigenvector $\boldsymbol{v}_i$ and a left eigenvector $\boldsymbol{w}_i$ which satisfy

$$\boldsymbol{A}\boldsymbol{v}_i = \lambda_i \boldsymbol{v}_i$$

$$\boldsymbol{w}_i^T \boldsymbol{A} = \lambda_i \boldsymbol{w}_i^T$$

The poles of the closed-loop system are given by the eigenvalues of the closed-loop system matrix, $\boldsymbol{A} - K\boldsymbol{b}\boldsymbol{k}^T$. Now assume that, for the ith left eigenvector of $\boldsymbol{A}$

$$\boldsymbol{w}_i^T \boldsymbol{b} = 0 \tag{3.5-22}$$

Then

$$\boldsymbol{w}_i^T \left(\boldsymbol{A} - K\boldsymbol{b}\boldsymbol{k}^T\right) = \boldsymbol{w}_i^T \boldsymbol{A} - K\boldsymbol{w}_i^T \boldsymbol{b}\boldsymbol{k}^T = \lambda_i \boldsymbol{w}_i^T \tag{3.5-23}$$

Equation (3.5-23) shows that λ_i is an eigenvalue of the closed-loop system matrix; therefore, λ_i is a pole of both the closed-loop system and the open-loop system Clearly, this system is uncontrollable since the plant pole at λ_i cannot be moved. If Eq. (3.5-22) holds for the ith left eigenvector of a plant with distinct eigenvalues the plant is uncontrollable and it is said that the eigenvalue λ_i represents an *uncontrollable mode*. That the system is uncontrollable can be verified, assuming Eq. (3.5-22) is true using the relationship:

$$\begin{aligned}\boldsymbol{w}_i^T [\boldsymbol{b} \quad \boldsymbol{A}\boldsymbol{b} \quad \boldsymbol{A}^2\boldsymbol{b} \quad \cdots \quad \boldsymbol{A}^{n-1}\boldsymbol{b}] &= [\boldsymbol{w}_i^T \boldsymbol{b} \quad \lambda_i \boldsymbol{w}_i^T \boldsymbol{b} \quad \lambda_i^2 \boldsymbol{w}_i^T \boldsymbol{b} \quad \cdots \quad \lambda_i^{n-1} \boldsymbol{w}_i^T \boldsymbol{b}] \\ &= \boldsymbol{0}^T\end{aligned}$$

Since $\boldsymbol{w}_i \neq \boldsymbol{0}$, the controllability matrix is not invertible. It can also be shown that if Eq. (3.5-22) does not hold for each of the left eigenvectors of a system matrix with distinct eigenvalues, the system is controllable.[1] Thus, for system matrices with distinct eigenvalues, we have another characterization of controllability that associates the concept of uncontrollability with a particular pole or mode that cannot be moved.

Example 3.5-3. Let

$$\boldsymbol{A} = \begin{bmatrix} -3 & -1 & -1 \\ -2 & -3 & -2 \\ 2 & 1 & 0 \end{bmatrix} \qquad \boldsymbol{b} = \begin{bmatrix} 1 \\ -1 \\ 0 \end{bmatrix}$$

---

[1] The eigenvectors of a matrix with distinct eigenvalues are linearly independent. Since $\boldsymbol{w}_i^T \boldsymbol{C}_{Ab} \neq 0$ for n linearly independent vectors $\boldsymbol{w}_i$, $i = 1, \ldots, n$, then $\boldsymbol{C}_{Ab}$ is invertible.

The eigenvalues and the eigenvectors of $\boldsymbol{A}$ are

$$\boldsymbol{\Lambda} = \begin{bmatrix} -1 & 0 & 0 \\ 0 & -2 & 0 \\ 0 & 0 & -3 \end{bmatrix} \qquad \boldsymbol{V} = \begin{bmatrix} 0 & -1 & 1 \\ -1 & 0 & 1 \\ 1 & 1 & -1 \end{bmatrix} \qquad \boldsymbol{W} = \begin{bmatrix} 1 & 0 & 1 \\ 0 & 1 & 1 \\ 1 & 1 & 1 \end{bmatrix}$$

where the ith column of $\boldsymbol{V}$ contains the right eigenvector of eigenvalue $\boldsymbol{\Lambda}_{ii}$ and the ith row of $\boldsymbol{W}$ contains the transpose of the left eigenvector of eigenvalue $\boldsymbol{\Lambda}_{ii}$.

In particular, $\quad \boldsymbol{w}_3^T = [1 \quad 1 \quad 1] \qquad \lambda_3 = -3$

and $\quad \boldsymbol{w}_3^T \boldsymbol{A} = \lambda_3 \boldsymbol{w}_3^T \qquad \boldsymbol{w}_3^T \boldsymbol{b} = 0$

Therefore, the eigenvalue $\lambda_3 = -3$ represents an uncontrollable mode and must be a closed-loop pole if state-variable feedback is used. Let

$$\boldsymbol{k}^T = [k_1 \quad k_2 \quad k_3]$$

Then

$$\boldsymbol{A} - \boldsymbol{b}\boldsymbol{k}^T = \begin{bmatrix} -3-k_1 & -1-k_2 & -1-k_3 \\ -2+k_1 & -3+k_2 & -2+k_3 \\ 2 & 1 & 0 \end{bmatrix}$$

$$s\boldsymbol{I} - \left(\boldsymbol{A} - \boldsymbol{b}\boldsymbol{k}^T\right) = \begin{bmatrix} s+3+k_1 & 1+k_2 & 1+k_3 \\ 2-k_1 & s+3-k_2 & 2-k_3 \\ -2 & -1 & s \end{bmatrix}$$

$$\det\left(s\boldsymbol{I} - \left(\boldsymbol{A} - \boldsymbol{b}\boldsymbol{k}^T\right)\right) = s^3 + (6 + k_1 - k_2)s^2 +$$

$$(11 + 4k_1 - 5k_2 + k_3)\, s + 6 + 3k_1 - 6k_2 + 3k_3$$

$$= (s+3)\left(s^2 + (3 + k_1 - k_2)s + (2 + k_1 - 2k_2 + k_3)\right)$$

There is, indeed, a closed-loop pole at $s = -3$. The reader should check that the controllability matrix for this example is not invertible.

There is another interesting characteristic of an uncontrollable mode. Where there is an uncontrollable pole it is always canceled by a zero in the plant, a zero in the closed-loop transfer function, and a zero in the loop gain transfer function. The zero appears in every entry of the vector $(s\boldsymbol{I} - \boldsymbol{A})^{-1}\boldsymbol{b}$ from which those transfer functions are formed.

To see that there must always be a zero at the same location as an uncontrollable pole in each entry of the vector $(s\boldsymbol{I} - \boldsymbol{A})^{-1}\boldsymbol{b}$, we assume the eigenvectors of $\boldsymbol{A}$ are distinct. First, multiply the eigenvalue expansion for $(s\boldsymbol{I} - \boldsymbol{A})^{-1}$, taken from Eq. (C-20) in Appendix C, by $\boldsymbol{b}$.

$$(s\boldsymbol{I} - \boldsymbol{A})^{-1}\boldsymbol{b} = \sum_{i=1}^{n} \frac{\boldsymbol{v}_i \boldsymbol{w}_i^T \boldsymbol{b}}{s - \lambda_i} \tag{3.5-24}$$

If there is an uncontrollable mode, λ_j, then $\boldsymbol{w}_j^T \boldsymbol{b} = 0$ and $s - \lambda_j$ does not appear in the denominator of any element in the vector $(s\boldsymbol{I} - \boldsymbol{A})^{-1}\boldsymbol{b}$ of Eq. (3.5-24). However we know another expression for this vector.

$$(s\boldsymbol{I} - \boldsymbol{A})^{-1}\boldsymbol{b} = \frac{\text{adj}(s\boldsymbol{I} - \boldsymbol{A})\boldsymbol{b}}{\det(s\boldsymbol{I} - \boldsymbol{A})}$$

Since we also know that $s - \lambda_j$ is a factor of $\det(s\boldsymbol{I} - \boldsymbol{A})$, $s - \lambda_j$ must be a factor of every element of $\text{adj}(s\boldsymbol{I} - \boldsymbol{A})\boldsymbol{b}$ for it to cancel and disappear from Eq. (3.5-24).

Since $s - \lambda_j$ is a factor of every element of $\text{adj}(s\boldsymbol{I} - \boldsymbol{A})\boldsymbol{b}$, we can discover from Table 3.4-1 that $s - \lambda_j$ is also a factor of the numerator polynomial of the plant and closed-loop transfer function

$$N_p(s) = \boldsymbol{c}^T \text{adj}(s\boldsymbol{I} - \boldsymbol{A})\boldsymbol{b}$$

and the numerator polynomial of the loop gain

$$N_k(s) = K\boldsymbol{k}^T \text{adj}(s\boldsymbol{I} - \boldsymbol{A})\boldsymbol{b}$$

no matter what the values of $\boldsymbol{c}$ and $\boldsymbol{k}$. From Eq. (3.4-14) of Table 3.4-1, we have the expression for the closed-loop transfer function

$$\frac{Y(s)}{R(s)} = \frac{\text{K}\boldsymbol{c}^T\boldsymbol{\Phi}(s)\boldsymbol{b}}{1 + K\boldsymbol{k}^T\boldsymbol{\Phi}(s)\boldsymbol{b}} = \frac{\dfrac{N_p(s)}{D_p(s)}}{1 + \dfrac{N_k(s)}{D_p(s)}} = \frac{N_p(s)}{D_p(s) + N_k(s)} \tag{3.5-25}$$

where $D_p(s)$ is the plant denominator, $\det(s\boldsymbol{I} - \boldsymbol{A})$. If λ_j is an uncontrollable plant pole, $s - \lambda_j$ is a factor of $D_p(s)$ and it is also a factor of $N_p(s)$. The denominator of the right side of Eq. (3.5-25) cannot be set to an arbitrary desired closed-loop pole polynomial since the resulting Diophantine equation may not have a solution. Instead, $s - \lambda_j$, the common factor of $D_p(s)$ and $N_k(s)$, must be a factor of the closed-loop pole polynomial, indicating again that λ_j is a closed-loop pole. Since $s - \lambda_j$ is also a factor of $N_p(s)$, λ_j is also a closed-loop zero and cancels the uncontrollable pole in the closed-loop transfer function

Example 3.5-4. Take the $\boldsymbol{A}$ and $\boldsymbol{b}$ matrix from Example 3.5-3.

$$\boldsymbol{A} = \begin{bmatrix} -3 & -1 & -1 \\ -2 & -3 & -2 \\ 2 & 1 & 0 \end{bmatrix} \qquad \boldsymbol{b} = \begin{bmatrix} 1 \\ -1 \\ 0 \end{bmatrix}$$

Then

$$\det(s\boldsymbol{I} - \boldsymbol{A}) = (s+1)(s+2)(s+3)$$

$$\text{adj}(s\boldsymbol{I} - \boldsymbol{A}) = \begin{bmatrix} (s+2)(s+1) & -(s+1) & -(s+1) \\ -2(s+2) & (s+1)(s+2) & -2(s+2) \\ 2(s+2) & (s+1) & (s+4.4)(s+1.6) \end{bmatrix}$$

$$\text{adj}(s\boldsymbol{I} - \boldsymbol{A})\boldsymbol{b} = \begin{bmatrix} (s+3)(s+1) \\ -(s+3)(s+2) \\ s+3 \end{bmatrix}$$

The uncontrollable mode represented by the factor $s + 3$ appears in each element of $\text{adj}(s\boldsymbol{I} - \boldsymbol{A})\boldsymbol{b}$.

If we let

$$c^T = [\, c_1 \quad c_2 \quad c_3 \,]$$

$$k^T = [\, k_1 \quad k_2 \quad k_3 \,]$$

we find that

$$G_p(s) = \frac{(s+3)[(c_1 - c_2)s + (c_1 - 2c_2 + c_3)]}{(s+1)(s+2)(s+3)}$$

$$H_{\text{eq}}(s) = \frac{(s+3)[(k_1 - k_2)s + (k_1 - 2k_2 + k_3)]}{(s+1)(s+2)(s+3)}$$

and

$$\frac{Y(s)}{R(s)} = \frac{G_p(s)}{1 + G_p(s)H_{\text{eq}}(s)} = \frac{(s+3)[(c_1 - c_2)s + (c_1 - 2c_2 + c_3)]}{(s+3)\,[(s+2)(s+1) + [(k_1 - k_2)s + (k_1 - 2k)]]}$$

The factor $s + 3$ associated with the uncontrollable mode appears everywhere it is expected. The closed-loop pole polynomial can be computed from the above expression; it, of course, matches the expression found in Example 3.5-3.

In summary, we found that there are two ways to characterize the controllability of a system. A system is controllable if its controllability matrix is invertible. A system with distinct eigenvalues is also controllable if it has no uncontrollable mode. A mode or eigenvalue is uncontrollable if its associated left eigenvector satisfies Eq. (3.5-22).

If a system is controllable, its n poles can be arbitrarily placed with state feedback gains given by Eq. (3.5-21). If a mode is uncontrollable, the uncontrollable mode will appear as a closed-loop mode or pole. Uncontrollable poles are always canceled by zeros in the plant, and in the closed-loop and loop gain transfer functions.

3.5.1 CAD Notes

The feedback gains necessary for pole placement can be computed using two functions included in the MATLAB Control Toolbox. One function, *acker*, uses a method called the Ackerman method to compute the feedback gain vector. The other function, *place*, uses another method that is better conditioned numerically but is unable to place multiple poles in the same location. Numerical stability becomes important when dealing with large problems (usually considered as problems having more than ten states). Function *place* returns a message under the label *ndigits* that gives an estimate of the number of digits of accuracy produced in the pole positions. The functions are called by the expressions:

```
k=acker(A,b,p)
k=place(A,b,p)
```

where `p` is a vector giving the desired pole positions.

The position of the poles of a system can be checked using the function

```
lambda=eig(A)
```

This function returns a vector, lambda, containing the eigenvalues of the square matrix A. The function can also be used to find the eigenvectors of a square matrix. The statement

```
[V,D] = eig(A)
```

produces a matrix $\boldsymbol{V}$ where columns are the right eigenvector of $\boldsymbol{A}$ and a diagonal matrix $\boldsymbol{D}$ with the respective eigenvalues.

Example 3.5-5. Give the computer solution of the pole placement problem solved in Example 3.5-1 and Example 3.5-2.

```
>>A=[0 1 0;0 -1 2;0 0 -4];
b=[0; 0; 2];
c=[1; 0; 0];
d=[0];
%input the desired pole polynomial
des=[16; 16; 6];
%compute the controllability matrix for A,b
CAb=[b A*b A*A*b]
CAb =
 0  0   4
 0  4 -20
 2 -8  32

CAbinv=inv(CAb)
CAbinv =
  1.0000  1.0000  0.5000
  1.2500  0.2500       0
  0.2500       0       0

%find the transfer function
[num,den]=ss2tf(A,b,c',d,1);
%form the a vector
a=[den(4); den(3); den(2)]
a =
  0
  4
  5

%form the phase variable A matrix
Ap=[0 1 0; 0 0 1; -a']
Ap =
  0  1  0
  0  0  1
  0 -4 -5
```

```
%form the phase variable b vector
bp=[0;0;1];
%form the controllability matrix for Ap,bp
CApbp=[bp Ap*bp Ap*Ap*bp]
CApbp =
  0  0   1
  0  1  -5
  1 -5  21

%let K=1;
K=1;
%Find k transpose
kT=(1/K)*(des'-a')*CApbp*CAbinv
kT =
 4.0000  2.7500  0.5000

%verify the pole positions
[V,D]=eig(A-b*kT)
V =
  0.4082  -0.1030 + 0.2050i  -0.1030 - 0.2050i
 -0.8165  -0.2039 - 0.6160i  -0.2039 + 0.6160i
  0.4082   0.7180 + 0.1041i   0.7180 - 0.1041i

D =
 -2.0000        0                   0
       0  -2.0000 + 2.0000i         0
       0        0             -2.0000 - 2.0000i

%find k using Ackerman's formula
%input desired poles
p=[-2;-2-2*j;-2+2*j]
p =
 -2.0000
 -2.0000 - 2.0000i
 -2.0000 + 2.0000i

k=acker(A,b,p)
k =
  4.0000  2.7500  0.5000

%find k using the place function
k=place(A,b,p)
place: ndigits= 16
k =
  4.0000  2.7500  0.5000
```

Exercises 3.5

3.5-1. You are given state-variable descriptions for two plants

$$\boldsymbol{A}_1 = \begin{bmatrix} -5 & 1 \\ -6 & 0 \end{bmatrix} \qquad \boldsymbol{b}_1 = \begin{bmatrix} 0 \\ 1 \end{bmatrix} \qquad \boldsymbol{c}_1 = \begin{bmatrix} 1 \\ 0 \end{bmatrix}$$

$$\boldsymbol{A}_2 = \begin{bmatrix} -1.2 & +0.4 \\ +0.4 & -1.8 \end{bmatrix} \qquad \boldsymbol{b}_2 = \begin{bmatrix} 1 \\ -2 \end{bmatrix} \qquad \boldsymbol{c}_2 = \begin{bmatrix} 1 \\ 1 \end{bmatrix}$$

Determine if each of these representations is controllable.

Answer: Plant 1 is controllable; Plant 2 is not.

3.5-2. For the plant A_1, b_1, c_1 given in Exercise 3.5-1, find k so that $u = k^T x$ places the closed-loop poles of the system at $s = -5$ and $s = -6$.
Answer:

$$k = \left(d^T - a^T\right) C_{A_p b_p} C_{Ab}^{-1} = \begin{bmatrix} -6 \\ 6 \end{bmatrix}$$

3.5-3. For the plant A_2, b_2, c_2 given in Exercise 3.5-1

(*a*) Find any uncontrollable mode and its associated left eigenvector.

(*b*) Compute $(sI - A_2)^{-1} b_2$ to see the pole-zero cancellation in the transfer function for any c vector.

(*c*) Define a new set of coordinates $x^* = Px$ with

$$P = \begin{bmatrix} 2 & 1 \\ -1 & 2 \end{bmatrix}$$

where x is the state vector associated with system A_2, b_2, c_2.

Draw the elementary block diagram associated with the x^* state vector representation. This diagram should make the uncontrollability clear.
Answers:
(*a*)

$$\lambda_i = -1 \text{ is controllable}$$

$$w_i = \begin{bmatrix} 2 \\ 1 \end{bmatrix}$$

(*b*)

$$(sI - A_2)^{-1} b_2 = \frac{\begin{bmatrix} s+1 \\ -2(s+1) \end{bmatrix}}{(s+2)(s+1)}$$

3.6 CONCLUSIONS

This section concludes the discussion of representation. In Chap. 2 the emphasis was on plant representation and description, whereas in this chapter we have concentrated solely on the representation and description of the closed-loop system. It is important once again to emphasize the fundamental difference between the plant and the system. The plant is the physical entity being controlled. The system is created through the use of feedback. The differential equations that describe the resulting system are different from those of the plant, or, alternatively, the closed-loop or system transfer function is different from the plant transfer function. The point of this chapter is to emphasize that this difference exists and to illustrate ways in which this difference can be expressed quantitatively.

As in Chap. 2, we have again stressed the dual approach to system representation through matrix methods and transfer functions. The H equivalent and G equivalent representations help unify the matrix and transfer function methods. Table 3.4-1 summarizes the relationships between these quantities and the state-variable equations, expressed in matrix notation.

A major focus of this chapter has been the development of how the closed-loop transfer function can be altered by output feedback or state-variable feedback. In the output feedback problem, it was found that the closed-loop zeros are made up of the plant zeros and zeros introduced by the controller. The poles can be placed arbitrarily. In the state feedback problem, it was found that the closed-loop zeros are the same as the plant zeros while the closed-loop poles can be arbitrarily placed if the state-variable representation is controllable. In either case, the closed-loop transfer function can be fine tuned using a command shaping prefilter.

At the beginning of this chapter the basic effects of feedback were pointed out. Feedback reduces the sensitivity of the closed-loop response to plant variations and increases the ability of the system to reject output disturbances. In addition, feedback allows control over bandwidth and the system transient response. We are now prepared to begin the analysis portion of the book. We will find that the transient response of a system is determined by its poles and zeros. By learning how pole and zero positions affect transient response we will learn how we should use the pole placement techniques developed in this chapter.

It should be noted here that if the objective of control system design were merely to make a system respond in a certain way to a particular class of inputs, then this book could end at the conclusion of the next chapter. In this chapter we have learned how to place poles arbitrarily; in the next chapter we learn how the pole positions need to be selected to create a desired transient response. However, despite the emphasis in many other texts on control theory, control of a system's transient response is not the only benefit realized in designing a control system. A control design should achieve all the benefits mentioned in Sec. 3.2 and it should achieve these benefits even in the face of the kinds of modeling uncertainty introduced in Sec. 2.7.

PROBLEMS

3.1. Find $G_c(s)$ so that Fig. P3.1-2 is equivalent to Fig. P3.1-1. That exercise should help you answer the following questions. Give the correct response, either *better*, or *worse*.

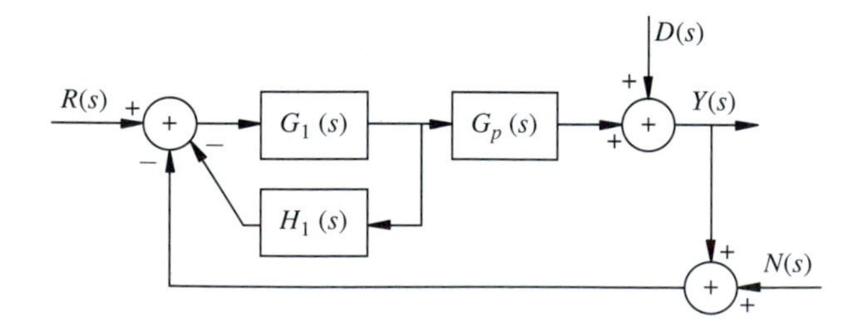

FIGURE P3.1-1
Equivalent representation of Problem 3.1.

(*a*) As $H_1(s)$ gets larger, the rejection of the disturbance $D(s)$ gets better or worse?

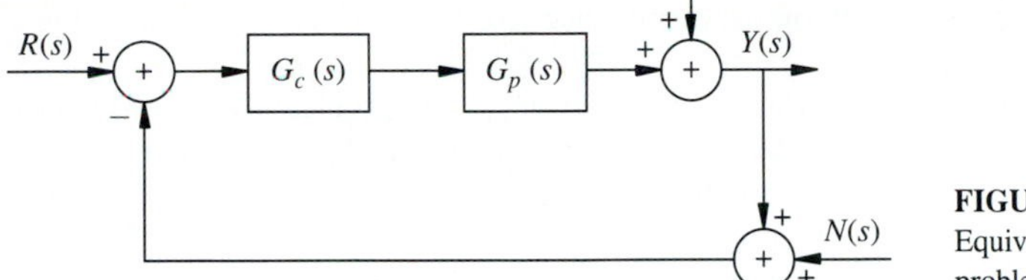

FIGURE P3.1-2
Equivalent representation of problem 3.1.

(*b*) Let $H_1(s)$ be small, then as $G_1(s)$ gets larger the elimination of the sensor noise gets better or worse?

(*c*) As $H_1(s)$ gets larger, the ability for $Y(s)$ to follow $R(s)$ precisely gets better or worse?

(*d*) Let $H_1(s)$ be small, then as $G_1(s)$ gets larger the sensitivity of the closed-loop transfer functions to changes in the plant gets better or worse?

3.2. Given the block diagram of Fig. P3.2, explain succinctly why the goal of reduced sensitivity to output disturbances D conflicts with the goal of attenuation of sensor noise N. How is the conflict usually resolved?

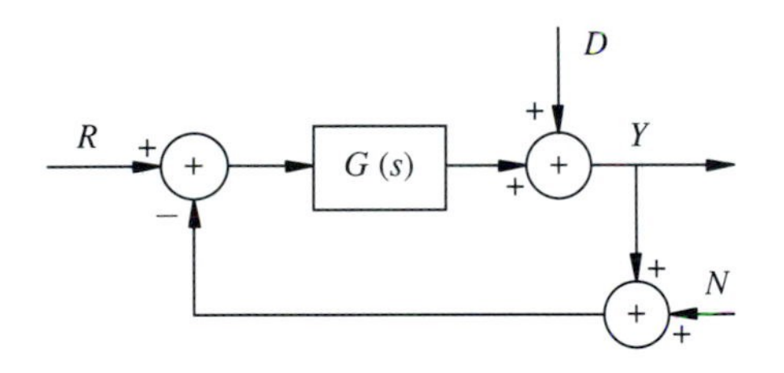

FIGURE P3.2
Problem 3.2.

3.3. Essay Question

"Good things happen when the loop gain is large."

Explain what good things happen and why. (You should have four good things in your list.) Explain any bad things that can happen. (You should have three bad things in your list.)

3.4. Very often the desired response of a control system is to have the output $y(t)$ match the input $r(t)$ as closely as possible, i.e., the desired closed-loop transfer function, $M_d(s)$, is approximately unity for all frequencies. This goal is achieved when a series compensator such as that of Fig. P3.4-1 is used and $G_c(s)$ is allowed to be very large.

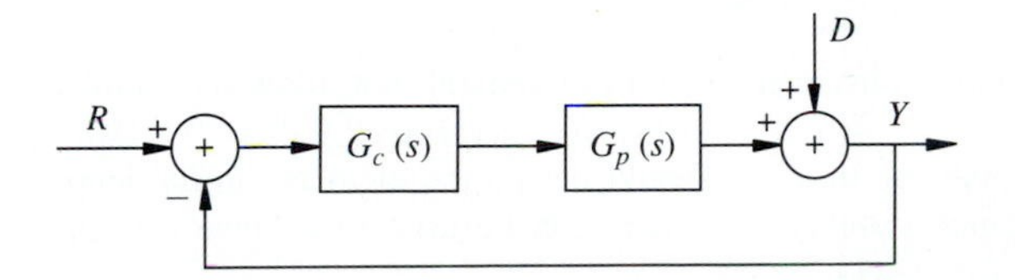

FIGURE P3.4-1
Problem 3.4.

(*a*) Find the closed-loop transfer function $M_C(s)$ for the series compensator configuration. What happens to $M_C(s)$ as $G_c(s)$ becomes large?

(*b*) Find the sensitivity of the closed-loop response to changes in the plant transfer function and find the response to output disturbances for the series compensator configuration when $G_c(s)$ is large.

If the desired closed-loop transfer function $M_d(s)$ is significantly different from unity we may still be able to use $G_c(s)$ to achieve the given $M_d(s)$. (We may be able to solve a Diophantine equation to provide the desired poles after canceling undesirable zeros.) However, we now show that the ability to reduce sensitivities may be lost.

(*c*) Set the closed-loop transfer function, $M_C(s)$ from (*a*) equal to a desired transfer function $M_d(s)$ and solve for the required $G_c(s)$.

(*d*) Keeping $G_c(s)$ fixed to the value found in (*c*), find the sensitivity of the closed-loop transfer function, $M_C(s)$, to changes in the plant and the response to output disturbances when $M_d(s)$ is very large or very small.

(*e*) Unless it is desired to make the closed-loop transfer function close to unity, we see that achieving a desired closed-loop transient response may conflict with the goals of sensitivity reduction and disturbance rejection. The problem can be solved by using a two-degree of freedom controller, i.e., by adding a prefilter as in Fig. P3.4-2.

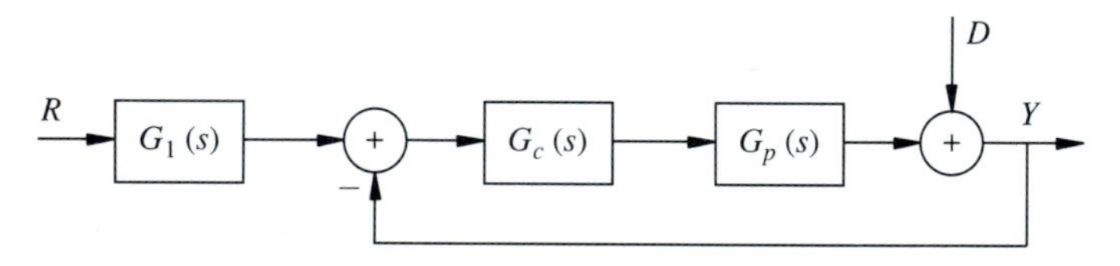

FIGURE P3.4-2
Two degree of freedom controller.

Now find the closed-loop transfer function $M_C(s)$ of Fig. P3.4-2. What happens as $G_c(s)$ becomes large with $G_1(s) = M_d(s)$? What are the sensitivities to plant perturbations and output disturbances in this system?

(*f*) If I choose to use an H configuration with a prefilter as in Fig. P3.4-3, how should I select $G_1(s)$ and $H(s)$ to achieve the same results as in (*e*).

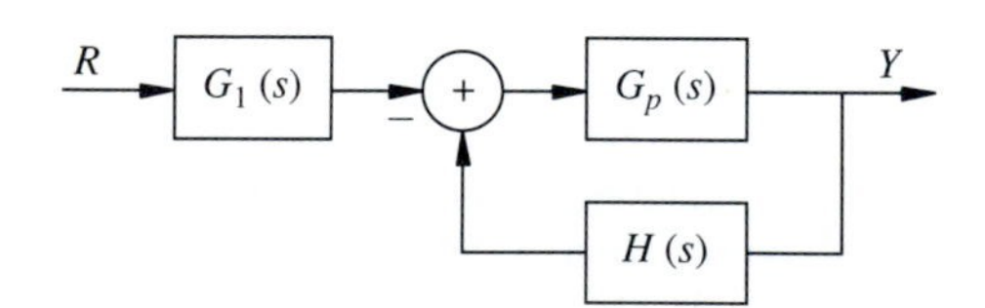

FIGURE P3.4-3
Two degree of freedom controller—H configuration.

A word of caution: Stability considerations will, in general, not allow the simple solutions given in (*e*) and (*f*).

This problem seems to indicate that one should design the elements in the loop to reduce sensitivity (and enhance stability). Prefilters can be used to achieve desired transient response.

Some recent research has supported the contention that this statement is universally true. This position is adopted in this book. Unlike most control texts, in which compensators are assigned primarily to achieve acceptable transient response, in this text we will design loop elements to achieve stability and reduce sensitivities and will assume that the resulting transient response can be fine tuned by the addition of a prefilter.

3.5. The effects of feedback on sensitivity reduction occur with large gains in the loop. If there are many loops, as is often the case with state-variable feedback, it makes sense that the elements with larger gains around them should have less sensitivity. For the system in Fig. P3.5, determine the sensitivity of the closed-loop transfer function $M_C(s)$ first with respect to $G_1(s)$ and then with respect to $G_2(s)$. Evaluate these expressions when $|G_1(s)| = |G_2(s)| = 1$. Determine the response of $Y(s)$ first to $D_1(s)$ then to $D_2(s)$.

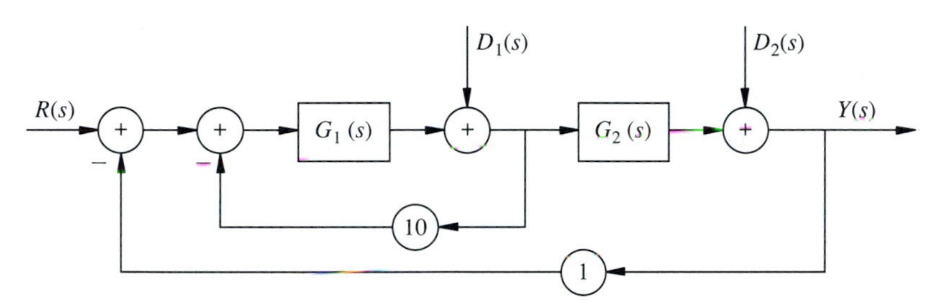

FIGURE P3.5
Problem 3.5.

3.6. With respect to the state variable feedback problem indicate the correct answer or answers to the following:

(a) The zeros of $H_{eq}(s)$ are
  (i) The poles of $G_p(s)$
  (ii) The zeros of $G_p(s)$
  (iii) The poles of $Y(s)/R(s)$
  (iv) The zeros of $Y(s)/R(s)$
  (v) None of the above

(b) The poles of $H_{eq}(s)$ are
  (i) The poles of $G_p(s)$
  (ii) The zeros of $G_p(s)$
  (iii) The poles of $Y(s)/R(s)$
  (iv) The zeros of $Y(s)/R(s)$
  (v) None of the above

(c) The zeros of $Y(s)/R(s)$ are
  (i) The zeros of $G_{eq}(s)$
  (ii) The poles of $G_{eq}(s)$
  (iii) The zeros of $G_p(s)$
  (iv) The poles of $G_p(s)$
  (v) None of the above

(d) By the proper selection of K and $\boldsymbol{k}$ it is possible to position arbitrarily
  (i) The poles of $G_p(s)$
  (ii) The zeros of $G_p(s)$
  (iii) The poles of $H_{eq}(s)$
  (iv) The zeros of $H_{eq}(s)$
  (v) The poles of $Y(s)/R(s)$
  (vi) The zeros of $Y(s)/R(s)$

3.7. A control system is described by

$$\boldsymbol{A} = \begin{bmatrix} -4.5 & -1 \\ -1.5 & -2 \end{bmatrix} \qquad \boldsymbol{c} = \begin{bmatrix} 1 \\ 0 \end{bmatrix}$$

$$\boldsymbol{b} = \begin{bmatrix} 1 \\ 1 \end{bmatrix} \qquad \boldsymbol{k} = \begin{bmatrix} 0.5 \\ 1 \end{bmatrix}$$

$$K = 10$$

Use matrix methods to find

(*a*) $G_P(s)$
(*b*) $Y(s)/R(s)$
(*c*) $G_{\text{eq}}(s)$
(*d*) $H_{\text{eq}}(s)$

3.8. Given the state-variable representation

$$\dot{\boldsymbol{x}} = \begin{bmatrix} 0 & 1 \\ -1 & -1 \end{bmatrix} \boldsymbol{x} + \begin{bmatrix} 0 \\ 1 \end{bmatrix} u \qquad y = [1 \quad 1]\boldsymbol{x}$$

$$u = r - \boldsymbol{k}^T \boldsymbol{x}$$

where

$$\boldsymbol{k} = \begin{bmatrix} 1 \\ 0.5 \end{bmatrix}$$

$$K = 1$$

(*a*) Draw the block diagram representation for the plant and for the system.
(*b*) Using matrix techniques, find
  (*i*) $\boldsymbol{\Phi}(s)$
  (*ii*) $\boldsymbol{\Phi}_k(s)$
  (*iii*) $H_{\text{eq}}(s)$
  (*iv*) $G_P(s)$
  (*v*) $Y(s)/R(s)$

3.9. In this problem the state-variable feedback problem is solved using properties of the H_{eq} configuration and the solution of the Diophantine equation for pole placement. The Diophantine equation for the closed-loop poles of a system in the H configuration is given in Eq. (3.2-19). If the H transfer function is really a non-realized H_{eq} transfer function arising from state-variable feedback then the denominator of H_{eq} is the same as the numerator of the plant, and the Diophantine equation becomes

$$N_P(s)\left(D_P(s) + N_{H_{\text{eq}}}(s)\right) = D_d(s)$$

If H_{eq} were realized there would always be closed-loop poles at the zeros of the plant. However, in state-variable feedback the pole-zero cancellations are not realized and neither are these poles. There are the same number of poles in the closed-loop system as there are in the plant and these poles are given by the simplified Diophantine equation

$$D_P(s) + N_{H_{\text{eq}}}(s) = D_d(s)$$

where D_p and D_d are monic nth order polynomials and $N_{H_{\text{eq}}}$ is an arbitrary polynomial of order $n-1$.

$$N_{H_{\text{eq}}}(s) = h_{n-1}s^{n-1} + h_{n-2}s^{n-2} + \cdots + h_1 s + h_0$$

(*a*) Given the state-variable representation

$$\dot{\boldsymbol{x}} = \begin{bmatrix} -2 & 1 \\ -1 & -1 \end{bmatrix} \boldsymbol{x} + \begin{bmatrix} 0 \\ 1 \end{bmatrix} u \qquad y = [1 \quad 1]\boldsymbol{x}$$

Find the denominator polynomial of the plant. (This is, of course, equal to the characteristic polynomial of $\boldsymbol{A}$.) Find the coefficients of $N_{H_{eq}}$ needed to solve the simplified Diophantine equation when the desired closed-loop poles are located at $s = -3$ and $s = -5$.

(*b*) Since $N_{H_{eq}}$ arises from state-variable feedback we know that

$$N_{H_{eq}}(s) = \boldsymbol{k}^T \operatorname{adj}(s\boldsymbol{I} - \boldsymbol{A})\boldsymbol{b}$$

Since the desired $N_{H_{eq}}$ is known from (*a*), use this equation to solve for the required values of $\boldsymbol{k}$.

3.10. (*CAD Problem*) The linearized state equation developed in Example 2.7-5 for a chemical reaction system is repeated here

$$\begin{bmatrix} \delta\dot{C}_a/\bar{C}_a \\ \delta\dot{T}/\bar{T} \\ \delta\dot{T}_j/\bar{T}_j \end{bmatrix} = \begin{bmatrix} -1.7 & -0.12 & 0 \\ 1.218 & 2.9 & 1.6 \\ 0 & 9.75 & -19.5 \end{bmatrix} = \begin{bmatrix} \delta C_A/\bar{C}_A \\ \delta T/\bar{T} \\ \delta T_j/\bar{T}_j \end{bmatrix} + \begin{bmatrix} 0 \\ 0 \\ -2.1 \end{bmatrix} \delta F_j/\bar{F}j$$

$$\delta T/\bar{T} = [0 \quad 1 \quad 0] \begin{bmatrix} \delta C_a/\bar{C}_A \\ \delta T/\bar{T} \\ \delta T_j/\bar{T}_j \end{bmatrix}$$

Assume that all the states can be measured and used for feedback. (In reality, the concentration would be difficult to measure and the temperature measurements would be produced with some time delay, but the assumption serves our academic purposes here.)

(*a*) Find the poles of the plant. Notice that there is a pole with a positive real part. In the next chapter we learn that such a pole is responsible for the poor response seen in Example 2.7-5. Such a pole is called *unstable*.

(*b*) Find the state-variable feedback vector $\boldsymbol{k}$ that places the poles at $s = -5$, $s = -10$, and $s = -10$. Since all three poles have negative real parts the closed-loop system is stable. The equation for the input u is given by

$$u(t) = \boldsymbol{k}^T \boldsymbol{x}(t) + r(t)$$

(*c*) Run a simulation of the closed-loop system with the same 1 percent initial offset in concentration as was done in Example 2.7-5. Now that the system is stable each state-variable returns to its equilibrium position.

(*d*) Notice that the plant input, $u(t)$ is F_J, the flow of fluid in the cooling jacket. The input to the closed-loop system is $r(t)$, the reference input. From the study in Sec. 3.2 we now might expect the output $y(t)$ to follow the reference input. Run a simulation with zero initial conditions and let $r(t)$ be a step of height 0.1. This reference input is indicating that the output, i.e., the temperature in the reaction vessel, should rise by ten per cent from its equilibrium value. What does the temperature do? What is the dc gain of the closed-loop system; i.e., what is the value of the closed-loop transfer function when $s = 0$?

In the next chapter we learn about the relationship between the placement of the poles of a system and the speed and shape of the transient response of a system. We also learn about the steady-state error in response to a unit step function and how this error relates to the loop gain of the system.

3.11. (*CAD Problem*) This problem is similar to Problem 3.10 but uses the transfer function and Diophantine equation method of pole-placement. It can be done without doing Problem 3.10 except the last comparison cannot be made.

(*a*) Find the transfer function of the chemical reaction plant given in Problem 3.10.

(*b*) Use the Diophantine equation approach to find a $G_c(s)$ that places a pole at $s = -5$, two poles at $s = -10$, and as many poles $s = -100$ as needed to create a causal controller with a proper transfer function.

(*c*) Run a simulation with zero initial conditions and let $r(t)$ be a step of height 0.1. This is the same conditions as in Problem 3.10*d*. How do the results compare? Find the value of the closed-loop transfer function when $s = 0$ for this problem.

3.12. It is sometimes useful to represent a feedback loop with a transfer function compensator as a closed-loop state-variable system. The composite state of such a system is composed of the state of the plant concatenated with the state of the differential equations associated with the controller's transfer function. We explore this situation in this problem.

(*a*) Since we are allowing controllers that are proper but not strictly proper, the first step in representing the closed-loop system is to find the state-variable representation for a proper but not strictly proper transfer function. Find the transfer function from $U(s)$ to $Y(s)$ associated with the state equation

$$\dot{\boldsymbol{x}} = \boldsymbol{A}\boldsymbol{x} + \boldsymbol{b}u$$

$$y = \boldsymbol{c}^T\boldsymbol{x} + du$$

A transfer function with the same order in both numerator and denominator may be turned into this state-variable form by long division. For example

$$G_c(s) = \frac{n_2 s^2 + n_1 s + n_0}{s^2 + d_1 s + d_0} = n_2 + \frac{(n_1 - n_2 d_1)s + (n_0 - n_2 d_0)}{s^2 + d_1 s + d_0}$$

By setting the d of the state equation given above equal to the n_2 of $G_c(s)$ and using a phase variable representation for the second term, give a state-variable representation of the form given above for the $G_c(s)$ given above.

(*b*) Now we are ready to represent the closed-loop system. Given the plant state-variable representation

$$\dot{\boldsymbol{x}}_{\boldsymbol{p}} = \boldsymbol{A}_{\boldsymbol{p}}\boldsymbol{x}_{\boldsymbol{p}} + \boldsymbol{b}_{\boldsymbol{p}}u$$

$$y = \boldsymbol{c}_{\boldsymbol{p}}^T\boldsymbol{x}_{\boldsymbol{p}}$$

and the controller state-variable representation

$$\dot{\boldsymbol{x}}_{\boldsymbol{c}} = \boldsymbol{A}_{\boldsymbol{c}}\boldsymbol{x}_{\boldsymbol{c}} - \boldsymbol{b}_{\boldsymbol{c}}y + \boldsymbol{b}_r r$$

$$u = \boldsymbol{c}_{\boldsymbol{c}}^T\boldsymbol{x}_{\boldsymbol{c}} - d_c y + d_r r$$

find a state-variable representation of the closed-loop system where the state-variables are given by $\boldsymbol{x}^T = [\boldsymbol{x}_{\boldsymbol{p}}{}^T \boldsymbol{x}_{\boldsymbol{c}}{}^T]$, r is the reference input and y is the output.

CHAPTER 4

TIME RESPONSE

4.1 INTRODUCTION

This is the first chapter concerned primarily with analysis. Chapters 2 and 3 dealt with methods of representing a plant and a closed-loop system. Now we are interested in how the signals associated with a system behave for a variety of inputs and initial conditions.

Although the reader is assumed to have a basic knowledge of Laplace transform methods, this chapter begins with a review of partial-fraction expansion methods. The algebraic and residue methods that are useful in finding inverse Laplace transforms are treated along with the closely related graphical approach. Graphical procedures are a powerful tool in control engineering. The accuracies obtained are usually compatible with the accuracies with which the given plant or system is described and additional insights may often be obtained using the graphical approach.

In the previous chapter, we learned how to arbitrarily place the poles of a closed-loop system. To decide where to place these poles, we must ascertain how pole position relates to the time response of the system.

The review of inverse Laplace transforms is followed by a discussion of first-order and second-order system responses. The specific interest is the response to a step-function input and the time domain specifications associated with the step response. The effect of additional poles and zeros on the nominal first-order or second-order responses is explored. The following section treats the response of the general nth-order system to general inputs with initial conditions on all the state-variables. The general case is treated from the matrix point of view, and alternative approaches related to the system block diagrams are also discussed.

The last section of this chapter deals with time domain methods of determining total system response. Again the general case is treated, this time using the state transition matrix and the convolution integral.

Although the use of digital computers has to some extent eliminated the need for the hand computation of time response, the material of this chapter still plays an important role in control theory, especially in the area of system design. Of particular importance is the ability to achieve a reasonably accurate knowledge of the qualitative nature of the response without extensive computation. This is the basic goal of this chapter.

4.2 PARTIAL-FRACTION EXPANSION METHODS

It has been assumed thus far that the reader has a working knowledge of Laplace transform methods. In fact, the whole underlying philosophy of this book is that dynamic systems, those described by differential equations, are easier to handle in the frequency or s domain than in the time domain.

In working in the complex frequency domain, two problems arise: first, the transformation of the given differential equations to the s domain and, second, the transformation of the problem solution back to the time domain. The two previous chapters described a variety of methods of representing both open-loop and closed-loop systems as sets of differential equations and, equivalently, through block diagrams or transfer functions. Here we consider in detail the inverse problem, that is, the problem of going from the complex s domain back to the time domain.

The basic assumption is that the function whose inverse Laplace transform is desired is given as a ratio of polynomials in s and that the denominator of this function appears in factored form. For the purpose of discussion, we assume a general function $F(s)$ given as

$$F(s) = \frac{N(s)}{D(s)} = \frac{N(s)}{(s+s_1)(s+s_2)\dots(s+s_n)} \tag{4.2-1}$$

Here the numerator of $F(s)$ is a polynomial in s of order less than n. In the factored form of $D(s)$, the s_i terms may be zero, real, or complex, and two or more values of s_i may be equal. To emphasize the different types of the roots of $D(s)$, Eq. (4.2-1) is made more specific and is written as

$$F(s) = \frac{N(s)}{D(s)} = \frac{N(s)}{(s+s_1)(s+s_2)^k(s+\alpha+j\beta)(s+\alpha-j\beta)} \tag{4.2-2a}$$

$$F(s) = \frac{N(s)}{D(s)} = \frac{N(s)}{(s+s_1)(s+s_2)^k\left[(s+\alpha)^2+\beta^2\right]} \tag{4.2-2b}$$

Here the s_1 and s_2 may be any real numbers, and one of these roots is assumed to be repeated k times. A single set of complex conjugate roots is indicated either separately, as in Eq. (4.2-2*a*), or as a product, as in Eq. (4.2-2*b*). Although Eq. (4.2-2) is less general than Eq. (4.2-1), the latter form is sufficient to illustrate the partial-fraction inversion method used here. The two forms of Eq. (4.2-2) indicate alternative approaches to the complex conjugate terms.

Let us assume initially that no complex conjugate terms are present, so that Eq. (4.2-2) may be further reduced to

$$F(s) = \frac{N(s)}{D(s)} = \frac{N(s)}{(s+s_1)(s+s_2)^k} \tag{4.2-3}$$

Again the order of $N(s)$ is assumed to be strictly less than that of $D(s)$. Partial-fraction expansion requires that Eq. (4.2-3) be expressed as

$$F(s) = \frac{N(s)}{D(s)} = \frac{R_{11}}{s+s_1} + \frac{R_{2k}}{(s+s_2)^k} + \frac{R_{2(k-1)}}{(s+s_2)^{k-1}} + \cdots + \frac{R_{21}}{s+s_2} \tag{4.2-4}$$

The problem is to find the numerater constants R_{ij}. Here the i subscript identifies the index of the root, and j indicates the power of the associated denominator term. Once the numerator constants are known, the inversion of the individual terms is a simple matter that may be carried out with the aid of Table 4.2-1 or Appendix B. The actual partial-fraction expansion may be accomplished by three methods: algebraic procedures, the method of residues, or graphical methods from a pole-zero plot of $F(s)$. These three approaches complement each other, and all are discussed in this section.

4.2.1 Algebraic Approach

The algebraic approach involves equating $N(s)/D(s)$ as given by Eq. (4.2-3) to the right-hand side of Eq. (4.2-4), forming a common denominator for the partial-fraction expansion, and finally equating the coefficients of equal powers of s in the result with those of $N(s)$. This procedure is illustrated in the following example.

Example 4.2-1. Let the closed-loop transfer function of a particular system be

$$\frac{Y(s)}{R(s)} = \frac{2.5(s+2)}{(s+5)(s+1)^2}$$

Let us assume that the input is a unit step, or $R(s) = 1/s$, so that $Y(s)$ becomes

$$Y(s) = \frac{2.5(s+2)}{s(s+5)(s+1)^2}$$

This $Y(s)$ corresponds to the $F(s)$ in the previous discussion, and $Y(s)$ can be expanded in partial-fraction form as

$$Y(s) = \frac{N(s)}{D(s)} = \frac{R_{11}}{s} + \frac{R_{21}}{s+5} + \frac{R_{32}}{(s+1)^2} + \frac{R_{31}}{s+1}$$

If the right-hand side of the above equation is written as a ratio of polynomials in s by use of the common denominator $[s(s+5)(s+1)^2]$, it becomes

$$Y(s) = \left[\frac{s^3(R_{11}+R_{21}+R_{31}) + s^2(7R_{11}+2R_{21}+R_{32}+6R_{31})}{s(s+5)(s+1)^2} + \frac{s(11R_{11}+R_{21}+5R_{32}+5R_{31})+5R_{11}}{s(s+5)(s+1)^2}\right]$$

TABLE 4.2-1
Table of Laplace transform pairs

	$F(s)$	$F(t)$
1.	1	Impulse, $\delta(t)$
2.	$\dfrac{1}{s}$	Step, $u(t)$
3.	$\dfrac{1}{s^2}$	Ramp, $tu(t)$
4.	$\dfrac{n!}{s^{n+1}}$	$t^n u(t)$
5.	$\dfrac{1}{s+\alpha}$	$e^{-\alpha t}u(t)$
6.	$\dfrac{1}{(s+\alpha)^2}$	$te^{-at}u(t)$
7.	$\dfrac{\beta}{s^2+\beta^2}$	$\sin\beta t u(t)$
8.	$\dfrac{s}{s^2+\beta^2}$	$\cos\beta t u(t)$
9.	$\dfrac{\beta}{(s+\alpha)^2+\beta^2}$	$e^{-\alpha t}\sin\beta t u(t)$
10.	$\dfrac{s+\alpha}{(s+\alpha)^2+\beta^2}$	$e^{-\alpha t}\cos\beta t u(t)$
11.	$\dfrac{s+\alpha_0}{(s+\alpha)^2+\beta^2}$	$\dfrac{1}{\beta}\left[(a_0-\alpha)^2+\beta^2\right]^{1/2}e^{-\alpha t}\sin(\beta t+\psi)u(t)$ where $\psi = \arctan\dfrac{\beta}{a_0-\alpha}$
12.	$\dfrac{\|R\|e^{j\phi_R}}{s+\alpha-j\beta}+\dfrac{\|R\|e^{-j\phi_R}}{s+\alpha+j\beta}$	$2\|R\|e^{-\alpha t}\cos(\beta t+\phi_R)u(t)$

Since the numerator of $Y(s)$ is just $2.5(s+2)$, four simultaneous linear equations result by equating the coefficients of equal powers of s in the two numerator expressions for $Y(s)$.

$$
\begin{aligned}
R_{11}+R_{21}+R_{31} &= 0 \\
7R_{11}+2R_{21}+R_{32}+6R_{31} &= 0 \\
11R_{11}+R_{21}+5R_{32}+5R_{31} &= 2.5 \\
5R_{11} &= 5
\end{aligned} \tag{4.2-5}
$$

These equations may be solved to yield

$$
R_{11} = 1 \qquad R_{21} = \tfrac{3}{32}
$$
$$
R_{32} = -\tfrac{5}{8} \qquad R_{31} = -\tfrac{35}{32}
$$

Therefore the partial-fraction form for $Y(s)$ becomes

$$Y(s) = \frac{1}{s} + \frac{\frac{3}{32}}{s+5} - \frac{\frac{5}{8}}{(s+1)^2} - \frac{\frac{35}{32}}{s+1}$$

and from the Table 4.2-1, $y(t)$ is found to be

$$y(t) = 1 + \tfrac{3}{32}e^{-5t} - \tfrac{5}{8}te^{-t} - \tfrac{35}{32}e^{-t}$$

The above example illustrates the conceptual simplicity of the algebraic approach, as well as the computational difficulties involved. To find all the unknown constants in the partial-fraction expansion, it is necessary to solve four simultaneous equations. In general it is necessary to solve n equations for the nth-order expression for $F(s)$.

4.2.2 Residue Method

The residue method avoids this difficulty but adds the difficulty of differentiation if multiple roots are involved. To illustrate this method, let us consider again Eqs. (4.2-3) and (4.2-4). The right-hand sides of these two equations are set equal, to yield

$$\frac{N(s)}{(s+s_1)(s+s_2)^k} = \frac{R_{11}}{s+s_1} + \frac{R_{2k}}{(s+s_2)^k} + \frac{R_{2(k-1)}}{(s+s_2)^{k-1}} + \cdots + \frac{R_{21}}{s+s_2} \tag{4.2-6}$$

If Eq. (4.2-6) is multiplied by $s + s_1$, all terms on the right, except R_{11}, become zero when s is set equal to $-s_1$, and R_{11} is evaluated by the resulting expression

$$R_{11} = \left.\frac{N(s)}{(s+s_2)^k}\right|_{s=-s_1}$$

A similar procedure may be used to evaluate R_{2k}. In this case, however, both sides of Eq. (4.2-6) must be multiplied by $(s+s_2)^k$ and then evaluated at $s = -s_2$. The result of this procedure is the expression

$$R_{2k} = \left.\frac{N(s)}{s+s_1}\right|_{s=-s_2}$$

Unfortunately, the only coefficient associated with the multiple pole at $s = -s_2$ that may be evaluated in this simple fashion is R_{2k}. The remaining coefficients R_{ij}, $j < k$ must be evaluated by differentiation. The constant $R_{2(k-1)}$ may be isolated from Eq. (4.2-6) by multiplying both sides by $(s+s_2)^k$, then differentiating with respect to s, and finally evaluating $s = -s_2$. This is done in two steps

$$\frac{N(s)}{s+s_1} = \frac{R_{11}(s+s_2)^k}{s+s_1} + R_{2k} + R_{2(k-1)}(s+s_2) + \cdots + R_{21}(s+s_2)^{k-1}$$

$$\left.\frac{d}{ds}\frac{N(s)}{s+s_1}\right|_{s=-s_2} = R_{2(k-1)}$$

In an entirely similar manner, the R_{ij} coefficients for a kth-order root can be established by successive differentiation to be

$$R_{ij} = \left\{ \frac{1}{(k-j)!} \frac{d^{k-j}}{ds^{k-j}} \left[(s+s_i)^k \frac{N(s)}{D(s)} \right] \right\} \Bigg|_{s=-s_i} \tag{4.2-7}$$

If k is equal to 1, that is, if the root is simple, this expression leads to the previous expression

$$R_{i1} = \frac{(s+s_i)N(s)}{D(s)} \Bigg|_{s=-s_i} \tag{4.2-8}$$

The general expression of Eq. (4.2-7) involves tedious differentiation if k is larger than 2. Fortunately multiple roots are rare in practice except possibly at $s = 0$. The coefficients R_{i1} are called *residues* in complex-variable theory; hence the name *residue method.*

To demonstrate the residue method, let us rework the problem of Example 4.2-1.

Example 4.2-2. Again we wish to express $Y(s)$ in the form

$$Y(s) = \frac{2.5(s+2)}{s(s+5)(s+1)^2} = \frac{R_{11}}{s} + \frac{R_{21}}{s+5} + \frac{R_{32}}{(s+1)^2} + \frac{R_{31}}{s+1}$$

The residues associated with the poles at $s = 0$ and $s = -5$ are easily determined from Eq. (4.2-8) as

$$R_{11} = \frac{2.5(s+2)}{(s+5)(s+1)^2} \Bigg|_{s=0} = \frac{5}{5} = 1 \tag{4.2-9}$$

$$R_{21} = \frac{2.5(s+2)}{s(s+1)^2} \Bigg|_{s=-5} = \frac{-7.5}{-5(16)} = \frac{3}{32} \tag{4.2-10}$$

Application of Eq. (4.2-7) is illustrated in the evaluation of both R_{32} and R_{31}. Since the root at $s = -1$ is repeated twice, the index $k = 2$ is used. The determination of R_{32} involves no differentiation, and substitution into Eq. (4.2-7) results in

$$R_{32} = \frac{2.5(s+2)}{s(s+5)} \Bigg|_{s=-1} = \frac{2.5}{-4} = -\frac{5}{8} \tag{4.2-11}$$

The evaluation of R_{31}, on the other hand, does require differentiation and is given by

$$R_{31} = \frac{1}{1!} \frac{d}{ds} \frac{2.5(s+2)}{s(s+5)} \Bigg|_{s=-1}$$

$$= \left[\frac{-2.5(s+2)(2s+5)}{s^2(s+5)^2} + \frac{2.5}{s(s+5)} \right] \Bigg|_{s=-1} = -\frac{35}{32}$$

These are the same values for the partial-fraction expansion coefficients determined in Example 4.2-1.

Although the differentiation involved in Example 4.2-2 is not difficult, it would have been considerably more complicated if k had been 3, rather than 2. By combining the algebraic and residue approaches, the disadvantages of both methods are largely overcome. Using the residue method, three of the four unknown constants are evaluated with ease. With three of the four constants in Eq. (4.2-5) known, the solution of the four simultaneous equations becomes trivial. Thus the two methods of partial-fraction expansion complement each other.

Thus far, attention has been focused upon cases involving repeated roots. Such cases are much more likely to appear as textbook examples and problems than to appear in practice, particularly in closed-loop systems. The reader may not appreciate the truth of this statement until the root locus method is discussed in Chap. 7, but it is nevertheless a fact. A much more common occurrence than repeated roots is the presence of complex conjugate roots. Here, to avoid unnecessary computational difficulties, we assume that complex conjugate roots are always simple, an assumption that the authors have never seen violated in practice.

To initiate the discussion of complex conjugate roots, let us return to Eqs. (4.2-2*a*) and (4.2-2*b*). We ignore the possibility of repeated roots by assuming that $k = 0$, so that Eqs. (4.2-2*a*) and (4.2-2*b*) become

$$F(s) = \frac{N(s)}{D(s)} = \frac{N(s)}{(s+s_1)(s+\alpha+j\beta)(s+\alpha-j\beta)} \tag{4.2-12a}$$

$$F(s) = \frac{N(s)}{D(s)} = \frac{N(s)}{(s+s_1)\left((s+\alpha)^2+\beta^2\right)} \tag{4.2-12b}$$

These two equations indicate the two methods of representing complex conjugate roots. If Eq. (4.2-12*a*) is rewritten and expanded into partial-fraction form, it becomes

$$F(s) = \frac{N(s)}{D(s)} = \frac{N(s)}{(s+s_1)(s+s_2)(s+s_3)} = \frac{R_{11}}{s+s_1} + \frac{R_{21}}{s+s_2} + \frac{R_{31}}{s+s_3} \tag{4.2-13}$$

where $s_2 = \alpha - j\beta$ and $s_3 = \alpha + j\beta$.

Here the fact that two of the roots are complex conjugates is not important, and the residues at each simple pole, whether complex or not, may be evaluated in the usual way. Thus the three residues are

$$R_{11} = \left.\frac{N(s)}{(s+s_2)(s+s_3)}\right|_{s=-s_1} = \left.\frac{N(s)}{(s+\alpha-j\beta)(s+\alpha+j\beta)}\right|_{s=-s_1}$$

$$R_{21} = \left.\frac{N(s)}{(s+s_1)(s+s_3)}\right|_{s=-s_2} = \left.\frac{N(s)}{(s+s_1)(s+\alpha+j\beta)}\right|_{s=-\alpha+j\beta}$$

$$R_{31} = \left.\frac{N(s)}{(s+s_1)(s+s_2)}\right|_{s=-s_3} = \left.\frac{N(s)}{(s+s_1)(s+\alpha-j\beta)}\right|_{s=-\alpha-j\beta}$$

It will be shown later in this section that R_{21} and R_{31} are always complex conjugates, so that Eq. (4.2-13) can be written

$$F(s) = \frac{N(s)}{D(s)} = \frac{R_{11}}{s+s_1} + \frac{R}{s+\alpha-j\beta} + \frac{\bar{R}}{s+\alpha+j\beta}$$

where the overbar indicates the complex conjugate. Once either R or $\bar{R}$ is found, the other is known.

When roots appear as complex conjugate pairs it is convenient to find the inverse transform of the pair taken together. The inverse transform of the last two terms in the expression for $F(s)$ can be derived starting with the inverse transform of each first-order term separately. Representing the complex number R in magnitude and phase form

$$R = |R|e^{j\phi_R}$$

the inverse transform of

$$G(s) = \frac{|R|e^{j\phi_R}}{s + \alpha - j\beta} + \frac{|R|e^{-j\phi_R}}{s + \alpha + j\beta}$$

can be written

$$\begin{aligned} g(t) &= |R|e^{j\phi_R}e^{-\alpha t}e^{j\beta t} + |R|e^{-j\phi_R}e^{-\alpha t}e^{-j\beta t} \\ &= 2|R|e^{-\alpha t}\frac{e^{j(\beta t+\phi_R)} + e^{-j(\beta t+\phi_R)}}{2} \\ &= 2|R|e^{-\alpha t}\cos(\beta t + \phi_R) \end{aligned}$$

It is important to remember when using the inverse transform just derived that the angle ϕ_R must come from the residue associated with the pole at $s = -\alpha + j\beta$ and not from the residue associated with the pole at $s = -\alpha - j\beta$.

The use of the residue method to obtain the partial-fraction expansion of an $F(s)$ involving complex conjugate roots is illustrated in the following example.

Example 4.2-3. Let the closed-loop transfer function $Y(s)/R(s)$ be given as

$$\frac{Y(s)}{R(s)} = \frac{16(s+5)}{(s+10)\left[(s+2)^2 + 2^2\right]} = \frac{16(s+5)}{(s+10)(s+2-j2)(s+2+j2)}$$

To find the impulse response of this system let $R(s) = 1$, and then $Y(s)$ is

$$\begin{aligned} Y(s) &= \frac{16(s+5)}{(s+10)(s+2-j2)(s+2+j2)} \\ &= \frac{R_{11}}{s+10} + \frac{R}{s+2-j2} + \frac{\bar{R}}{s+2+j2} \end{aligned}$$

R_{11} is easily found to be $-80/68$, and R is

$$R = 1.75e^{-j(70.4^\circ)} = 0.588 - j1.65$$

Therefore $\bar{R}$ is

$$\bar{R} = 1.75e^{+j(70.4^\circ)} = 0.588 + j1.65$$

and $Y(s)$ in partial-fraction form is

$$Y(s) = \frac{-\frac{80}{68}}{s+10} + \frac{1.75e^{-j(70.4^\circ)}}{s+2-j2} + \frac{1.75e^{+j(70.4^\circ)}}{s+2+j2} \tag{4.2-14}$$

(Note that, as is common in engineering, the exponential here is shown in degrees rather than in radians.)

The inverse function is found easily using entry 12 in Table 4.2-1.

$$y(t) = -1.176e^{-10t} + 3.50e^{-2t} \cos\left(2t - 70.4^\circ\right)$$

The algebraic approach to partial-fraction expansion involving complex conjugate roots is based on writing $F(s)$ as in Eq. (4.2-12*b*). In expanded form, Eq. (4.2-12*b*) is written

$$Y(s) = \frac{R_{11}}{s + s_1} + \frac{As + B}{(s + \alpha)^2 + \beta^2} \tag{4.2-15}$$

Using the algebraic method, it is most natural to find A and B of Eq. (4.2-15) directly and use entry 11 of Table 4.2-1 for the inverse transform. The algebraic approach to Example 4.2-3 is illustrated below.

Example 4.2-4. Let $Y(s)$ be written in the partial-fraction expansion form of Eq. (4.2-15) as

$$Y(s) = \frac{16(s+5)}{(s+10)\left[(s+2)^2 + 2^2\right]} = \frac{R_{11}}{s+10} + \frac{As + B}{(s+2)^2 + 2^2}$$

$$= \frac{s^2(R_{11} + A) + s(4R_{11} + 10A + B) + 8R_{11} + 10B}{(s+10)\left[(s+2)^2 + 2^2\right]}$$

Equating the numerator terms, the three equations that must be solved simultaneously for R_{11}, A, and B are

$$R_{11} + A = 0 \qquad 4R_{11} + 10A + B = 16 \qquad 8R_{11} + 10B = 80$$

Of course, R_{11} may be found more easily from the residue formula, yielding both A and B with little effort. The values of these unknown coefficients are

$$R_{11} = -1.175 \qquad A = 1.175 \qquad B = 8.94 = 1.175(7.60)$$

The resulting $Y(s)$ is given by

$$Y(s) = \frac{-1.175}{s+10} + \frac{1.175(s + 7.60)}{(s+2)^2 + 2^2}$$

The inverse transform may be written directly from Table 4.2-1 as

$$y(t) = -1.175e^{-10t} + \frac{1.175}{2}\left[(7.60 - 2)^2 + 2^2\right]^{1/2} e^{-2t} \sin(2t + \psi)$$

where

$$\psi = \arctan\left[2/(7.60 - 2)\right]$$

or

$$y(t) = -1.175e^{-10t} + 3.50e^{-2t} \sin\left(2t + 19.65^\circ\right)$$

This result is seen to be equal to the $y(t)$ found using the residue method by remembering the identity

$$\cos(x) = \sin(90^\circ + x)$$

Thus far in this section, we have worked two example problems, each by the residue method and by the algebraic method. In each case the resulting $y(t)$ is a sum of time functions rather than a plot of $y(t)$ versus time. In one sense, a plot of $y(t)$ versus time is a more satisfactory answer than a sum of time functions, since by looking at the plot we know the behavior of the output for all values of time. To obtain the same information from a series of time functions, each function must be evaluated at a number of different values of time and the results added and plotted. Often we accept the analytic expression for $y(t)$, even though the exact nature of the output for all time values can be determined only from the analytic expression after considerable computation.

In attempting to ascertain the nature of the output from its component parts, one automatically looks to see which terms have the largest coefficients, the longest time constants, etc. In short, one often makes a mental approximation of the output. This section concludes with a discussion of graphical methods of determining residues. The object is not necessarily to replace the analytic methods of calculating residues but rather to provide a method by which the approximate contribution of each pole to the total output may be gauged.

4.2.3 Graphical Approach

The key to the graphical approach to evaluating a complex expression is the replacement of algebraic factors by vectors in the complex variable s plane. A complex number s is represented by a vector from the origin to the point s with the correct real and imaginary parts. Associated with this vector is an angle measured counterclockwise from the positive real axis, which is the same angle as the phase angle of the complex number s. For example, to represent $s = -1 + j2$, a vector is drawn from the origin to the point $(-1, 2)$ in the s plane, as shown in Fig. 4.2-1a. The vector has

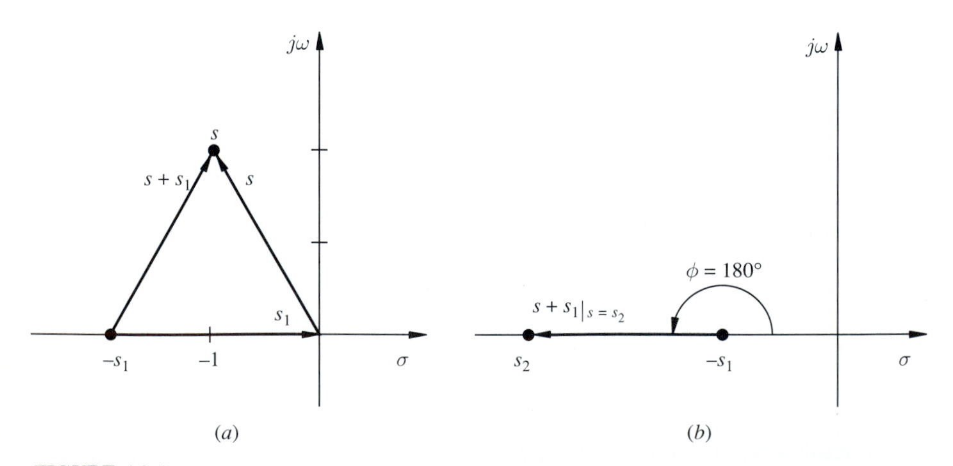

FIGURE 4.2-1
Graphical evaluation of residues. (a) Illustration of the distance $s + s_1$ for an arbitrary s; (b) the distance $s + s_1$ evaluated at $s = -s_2$.

length $\sqrt{5}$ and angle 116.6°, matching the magnitude and phase of the complex number. A second vector, shown in Fig. 4.2-1*a*, is a vector associated with the complex number s_1. This vector is shown as $-(-s_1)$. The vector $-s_1$ would be represented by a vector from the origin to the point $-s_1$. Changing the sign of a complex number changes its phase by 180° so that the direction of the complex number is rotated by 180°, reversing the direction of the arrow. The key utility of vector representations of complex numbers is that vectors add and subtract in the same way that complex numbers do. The complex number that results from the expression $s + s_1 = s - (-s_1)$ has the same magnitude and phase as a vector beginning at the beginning of s_1 and terminating at the end of s as shown in Fig. 4.2-1*a*.

In Fig. 4.2-1*b* the point s is located at $s = -s_2$. Thus the term $s + s_1$ evaluated at $s = -s_2$ has its magnitude equal to the length of the vector from s_1 to s_2 and its phase equal to the angle of this vector, in this case, 180°.

Of course, terms such as this one are needed in evaluating residues. Therefore, for simple poles the values of the residues, and hence the contribution to the total output, may be determined by measuring or estimating distances on the s plane.

To be specific, consider the s plane plot of Fig. 4.2-2*a*. There four poles are indicated at $s = -1, -2, -7$, and -12, and a zero at $s = -1.1$. The corresponding

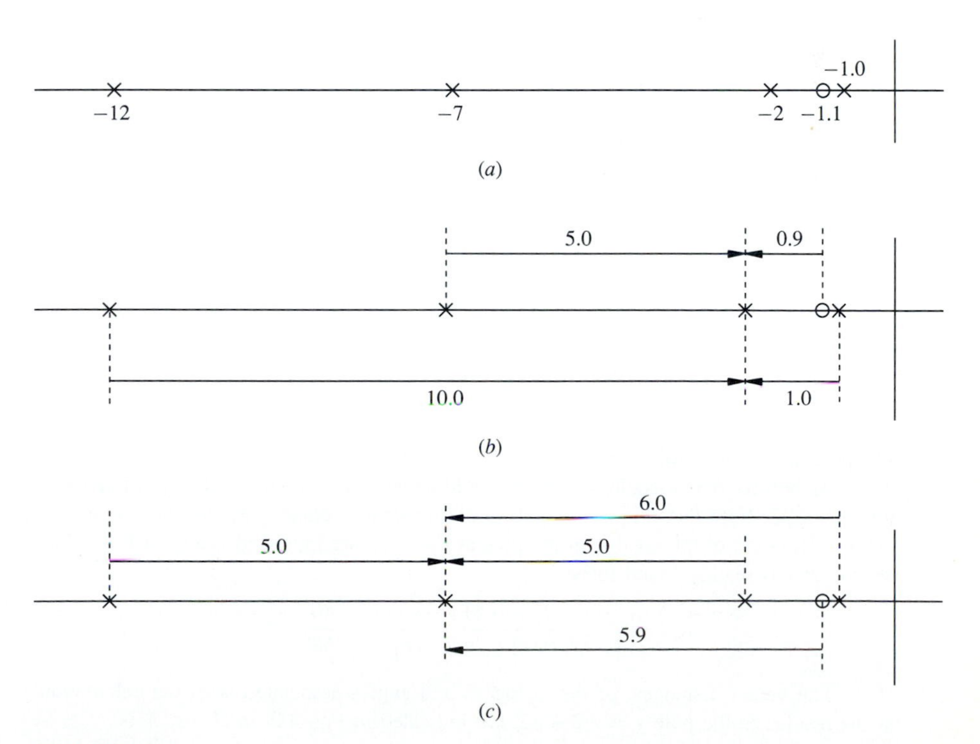

FIGURE 4.2-2
$F(s) = K(s + 1.1)/[(s + 1)(s + 2)(s + 7)(s + 12)]$. (*a*) Pole-zero plot; (*b*) residue at $s = -2$; (*c*) residue at $s = -7$.

$F(s)$ is

$$F(s) = \frac{K(s+1.1)}{(s+1)(s+2)(s+7)(s+12)}$$

If the exact inverse transformation is to be taken, the gain K must be known. However, to estimate only the relative contribution of each of the poles to the time response $f(t)$, the value of K is unimportant. The distances involved in calculating the residue at $s = -2$ are indicated in Fig. 4.2-2*b*, and the distances involved in calculating the residue at $s = -7$ are indicated in Fig. 4.2-2*c*. From Fig. 4.2-2*b*, the residue at $s = -2$ is

$$R_{21} = \frac{-0.9K}{(-1)(5)(10)} = \frac{0.9K}{50}$$

For a rough approximation, the distances to the pole and zero near 1 are almost the same, and they might be ignored. This approximation is even better in the evaluation of the residue at $s = -7$, which from Fig. 4.2-2*c* is given by

$$R_{31} = \frac{-5.9K}{(-6)(-5)(+5)} \approx \frac{-K}{25}$$

By comparing Fig. 4.2-2*b* and *c*, it is easily seen that the residue for the pole at $s = -2$ is smaller than that for the pole at $s = -7$. However, because the time constant is larger, 1/2 sec compared with 1/7 sec, the transient term associated with the residue at $s = -2$ contributes to the output for a longer time.

Because of the very short distance between the pole at -1 and the zero at -1.1, the residue of the pole at $s = -1$ is very small, as this distance appears in the numerator when evaluating residues. Therefore, even though this pole has the smallest time constant, it does not affect the output to a great extent because its residue is so small.

The use of graphical procedures to determine the complete system response is illustrated in the following example for a case involving complex conjugate poles.

Example 4.2-5. As an example, consider the partial-fraction expansion of $Y(s)$ used in Examples 4.2-3 and 4.2-4.

$$Y(s) = \frac{16(s+5)}{(s+10)\left[(s+2)^2+2^2\right]} = \frac{R_{11}}{s+10} + \frac{R}{s+2-j2} + \frac{\overline{R}}{s+2+j2}$$

The pole-zero plot of this function is shown in Fig. 4.2-3.

As before, R is associated with the pole located at $s = -\alpha + j\beta$, or, in this case, $s = -2 + j2$. Note that this is the pole in the upper half-plane. The distances involved in the calculation of the residue of the pole at $s = -10$ are indicated in Fig. 4.2-4*a*. This residue R_{11} is readily found to be

$$R_{11} = \frac{(16)(-5)}{\left(\sqrt{68}e^{-j\phi}\right)\left(\sqrt{68}e^{+j\phi}\right)} = -\frac{80}{68}$$

The vector distances, or the distances and angles associated with the calculation of the residue at the pole $s = -2 + j2$, are indicated in Fig. 4.2-4*b*. Thus, R is

$$R = \frac{\left(16\sqrt{13}e^{j33.7^\circ}\right)}{\left(4e^{j90^\circ}\right)\left(\sqrt{68}e^{j14.1^\circ}\right)} = 1.75e^{-j70.4^\circ} = 0.588 - j1.65$$

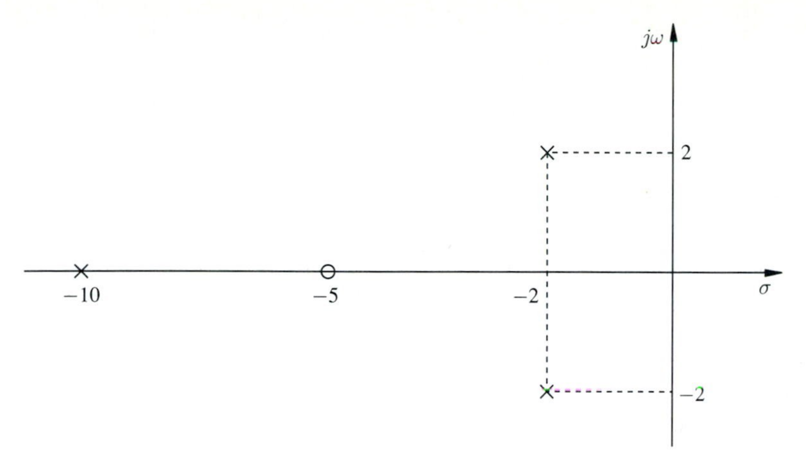

FIGURE 4.2-3
Pole-zero plot of $y(s) = 16(s+5)/(s+10)[(s+2)^2 + 2^2]$.

Consider now the evaluation of $\overline{R}$, the residue associated with the pole at $s = -2 - j2$. All the distances are the same as those in the calculation of R, and all the angles are equal but opposite in sign. As a consequence, the real part of each vector distance remains the same but the complex part is opposite, or the result is just the complex conjugate of R. Thus $Y(s)$ is completely known and is identical to that of Eq. (4.2-14) of Example 4.2-3.

On the basis of this short review of partial-fraction expansion methods, we are ready to consider the problem of finding the total system response. As a start, in the next section the time response of a dominantly first-order system is considered in detail.

4.2.4 CAD Notes

CAD programs like MATLAB are very useful in factoring polynomials and calculating residues. MATLAB has a form of representing a transfer function in addition to the state-space and transfer function forms, called the zero-pole form. The statement

```
[z,p,k]=tf2zp(num,den)
```

takes the transfer function determined by `num` and `den` and produces zeros and poles in the column vectors `z` and `p` respectively. The constant `k` gives the gain of the transfer function in zero-pole form as

$$G(s) = \frac{k\,(s-z_1)\,(s-z_2)\cdots(s-z_m)}{(s-p_1)\,(s-p_2)\cdots(s-p_n)}$$

The version of MATLAB available to the authors at the time of publication did not include a function for calculating the residues needed for a partial-fraction expansion

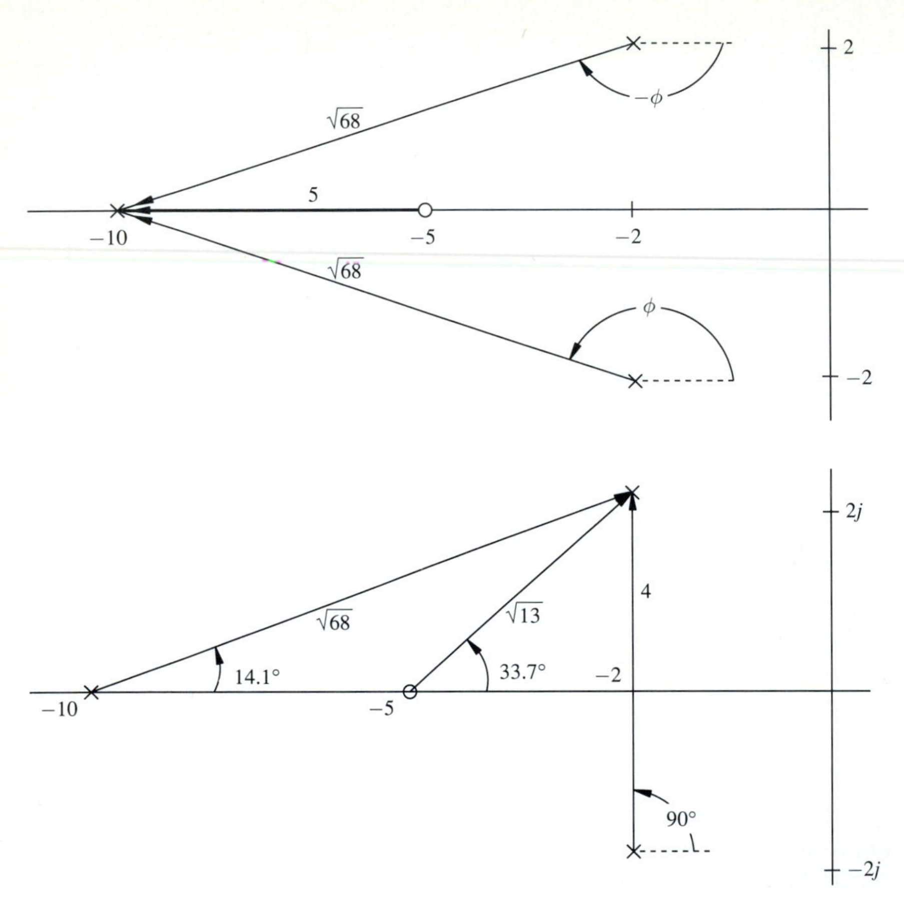

FIGURE 4.2-4
Distances and angles involved in residue evaluation. (*a*) Pole at $s = -10$; (*b*) Pole at $s = -2 - j2$.

although this function apparently has now been added. This circumstance provides an opportunity to show how functions can be added to MATLAB with simple routines. We add the function `residue1` to our Control Toolbox. It is similar in form to the function `residue` that was later included in the MATLAB Control Toolbox although it is less versatile. The function `residue1` finds the residues of a scalar transfer function in zero-pole form assuming that the transfer function has no repeated poles.

```
function [r,p]=residue1(num,den)
% This function returns in the vector r the partial
% fraction residues of a transfer function
% given by num and den
% The poles are given in the column vector p.
% The ith element of r is the residue for the ith
% pole as given in p
```

```
% THIS FUNCTION IS FOR STRICTLY PROPER TRANSFER FUNCTIONS
% THIS FUNCTION DOES NOT WORK IF THERE ARE REPEATED
% POLES IN p.
% This function is for scalar transfer functions only.
%
%
% change from transfer function to zero-pole form
[z,p,k]=tf2zp(num,den);
% Strip infinities and throw away.
z = z(abs(z) ~= inf);
% get the length of z
if isempty(z), nz=0;
else
        [nz,m] = size(z);
end
%get the length of p
[np,m] = size(p);
% compute the residue for each pole
for ip=1:np,
        r(ip)=k;
        for iz=1:nz,
                r(ip)=r(ip)*(p(ip)-z(iz));
        end
        for jp=1:np
                if jp~=ip,
                        r(ip)=r(ip)/(p(ip)-p(jp));
                end
        end
end
```

This function should be filed as `residue1.m` in the Control Toolbox directory.

Example 4.2-6. MATLAB is used to recompute the residues from Examples 4.2-3 and 4.2-5.

```
>>num=[16 80];
den1=[1 4 8];
den2=[1 10];
den=conv(den1,den2);
[z,p,k]=tf2zp(num,den)
z =
  -5
  ∞
  ∞
p =
 -10.0000
  -2.0000 + 2.0000i
  -2.0000 - 2.0000i
k =
  16
```

```
[r,p]=residuel(num,den)
r =

   -1.1765 - 0.0000i    0.5882 - 1.6471i    0.5882 + 1.6471i
p =
  -10.0000
   -2.0000 + 2.0000i
   -2.0000 - 2.0000i

rmag=abs(r)
rmag =
   1.1765   1.7489   1.7489

rang=angle(r)*180/pi
rang =
  -180.0000   -70.3462   70.3462
```

Exercises 4.2

4.2-1. Write an algebraic expression for each of the vectors in the s plane in Fig. 4.2-5.
Answers:
(*a*) s_0
(*b*) $s_0 + 3$
(*c*) $s_0 + 3 - 2j$

4.2-2. Draw a vector in the s plane representing each of the following expressions. The vector should have the same distance as the magnitude of the matching complex number and the same directional angle as the phase of the complex number.
(*a*) $s + 1$ evaluated at $s = 3j$
(*b*) $s - 3$ evaluated at $s = 0$
(*c*) $s + 2 + j2$ evaluated at $s = 1$
Answer:

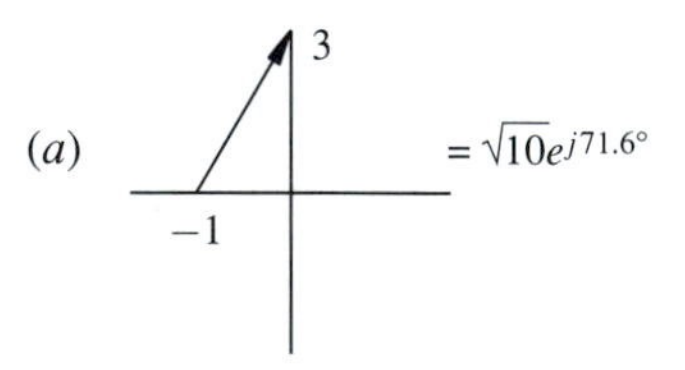

4.2-3. If $Y(s)$ is given as

$$Y(s) = \frac{3(s+4)}{(s+1)\left[(s+3)^2 + 3^2\right]}$$

find the partial-fraction expansion of $Y(s)$ by the algebraic approach, by residues, and through graphical means.
Answer:

$$Y(s) = \frac{\frac{9}{13}}{s+1} + \frac{\sqrt{5/26}e^{-j142.125^\circ}}{s+3-j3} + \frac{\sqrt{5/26}e^{+j142.125^\circ}}{s+3+j3}$$

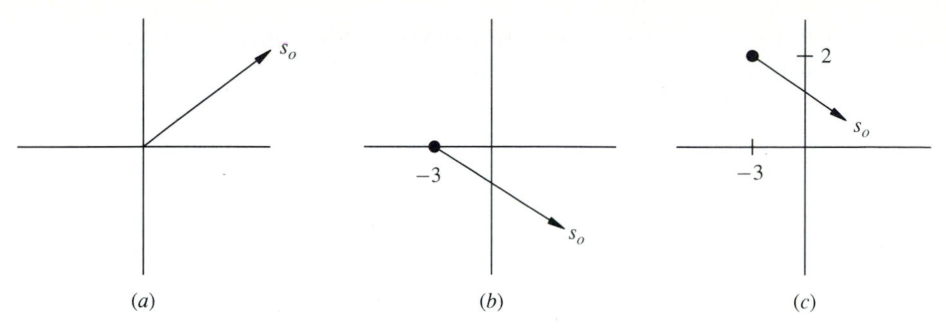

FIGURE 4.2-5
Exercise 4.2-1.

4.2-4. Find the step response of the system $G(s) = \dfrac{-10(s-1)}{s^2+2s+5}$
Use the graphical procedure for finding residues and use entry 12 of Table 4.2-1 to invert the transform.
Answer:

$$y(t) = 2 + 6.32e^{-t}\cos\left(2t + 108.5^\circ\right)$$

4.3 STEP-FUNCTION RESPONSE OF DOMINANTLY FIRST-ORDER SYSTEMS

In this section the time response of systems that can be well approximated by a first-order transfer function is discussed. We first discuss the step response of a true first-order system. A step command input is common in control theory as it is often desired to move the output of a system from one value to another. Also, by knowing the response to a step, insight into the ability of the system to respond to more general inputs can be obtained.

After establishing the properties of the step response of a true first-order system, the effect of an additional pole on the response of the system is discussed. We will see that, as long as the additional pole is placed significantly farther from the origin than the original pole, the response of the system is only slightly modified. This introduces the concept of a dominantly first-order system, i.e., a system whose transfer function has more than one pole but whose step response can be well approximated by a first-order system. Finally, we investigate the addition of a zero to a dominantly first-order system.

4.3.1 The First-order Step Response

The discussion here is based upon the first-order differential equation

$$\frac{dy(t)}{dt} + \frac{1}{\tau}y(t) = \frac{A}{\tau}r(t) \tag{4.3-1}$$

There are two parameters, τ and A, in this first-order system. The parameter τ is called the *time constant* of the system as it provides the information concerning the speed of the response of the system. The parameter A is called the *dc gain* of the system and establishes the final value the output approaches in response to a unit step.

The transfer function for the system of Eq. (4.3-1) is given by

$$\frac{Y(s)}{R(s)} = \frac{A/\tau}{s + 1/\tau} = \frac{A}{\tau s + 1} \tag{4.3-2}$$

There is a single pole located at $s = -1/\tau$. To find the step response, let $R(s) = 1/s$. Then $Y(s)$ is given by

$$Y(s) = \frac{A/\tau}{s(s + 1/\tau)}$$

By referring to Fig. 4.3-1 and using the graphical method for evaluating the residues, the partial-fraction expansion for $Y(s)$ is easily obtained.

$$Y(s) = \frac{\dfrac{A/\tau}{1/\tau}}{s} + \frac{\dfrac{A/\tau}{-1/\tau}}{s + 1/\tau} = \frac{A}{s} - \frac{A}{s + 1/\tau} \tag{4.3-3}$$

The step response $y(t)$ is given by

$$y(t) = A - Ae^{-t/\tau} \tag{4.3-4}$$

The first thing to notice is that if τ is negative, so that the pole is in the right half-plane, the magnitude of the step response grows exponentially with time. Such a system is unstable. If any pole is in the right half-plane the exponential growth associated with it eventually dominates all other responses. Stability considerations will be studied in Chap. 6.

A graph of $y(t)$ versus time, as given by Eq. (4.3-4), appears in Fig. 4.3-2. The time constant τ determines the speed with which the system responds. As time approaches infinity the response approaches the value of the transfer function of Eq. (4.3-2) evaluated at $s = 0$, in this case, A. This is called the *final value* of the response. The output is plotted on the ordinate as a percentage of the final value. Time is plotted on the abscissa as the number of time constants, i.e., it is given by t/τ. At time τ the response has reached 63.2 percent of its final value; at time 2τ, it has reached 86.5 percent. The values at three, four, and five time constants can be seen in Fig. 4.3-2. The value of five time constants is often used as a rule of thumb for the time when the effect of a pole is essentially complete.

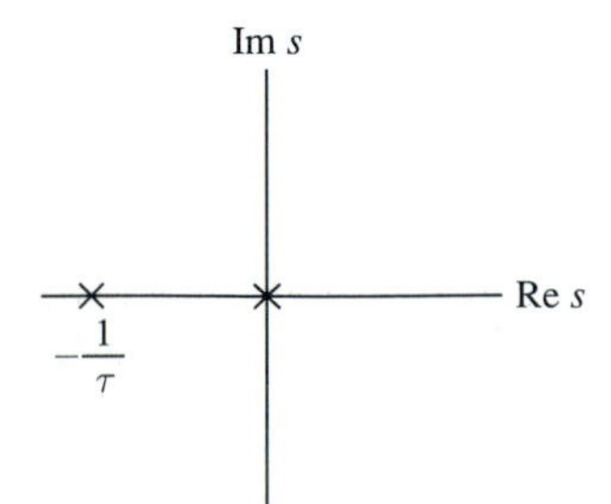

FIGURE 4.3-1
Plot of poles for first-order step response.

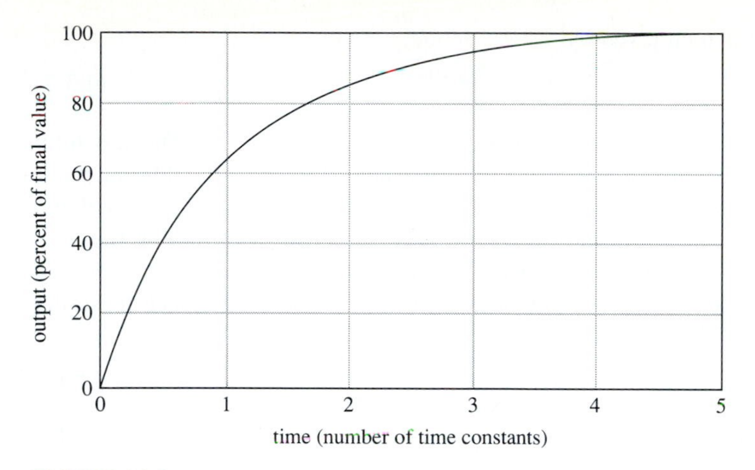

FIGURE 4.3-2
A first-order step response.

Note that as τ increases, the pole of the first-order system moves along the negative real axis toward the origin and the response of the system slows. In general, if all the poles of a system are in the left half-plane, the farther the poles of a system are from the origin, the faster the system responds.

4.3.2 The Effect of an Additional Pole

We now examine the step response of a system whose transfer function is given

$$\frac{Y(s)}{R(s)} = \frac{p}{(s+1)(s+p)} \tag{4.3-5}$$

The step response will be examined for different values of p. The transfer function of Eq. (4.3-5) has been chosen so that the step response of the system will approach the value of one for large values of t, independent of the value of p. Note that the transfer function of Eq. (4.3-5) evaluated at $s = 0$ equals one. Let $R(s) = 1/s$ and perform a partial-fraction expansion on the resulting step response transform $Y(s)$ to obtain

$$Y(s) = \frac{1}{s} - \frac{\dfrac{p}{p-1}}{s+1} + \frac{\dfrac{1}{p-1}}{s+p} \tag{4.3-6}$$

The step response $y(t)$ is given by

$$y(t) = 1 - \frac{p}{p-1}e^{-t} + \frac{1}{p-1}e^{-pt} \tag{4.3-7}$$

Let's examine the step response of Eq. (4.3-7) for the case $p = 10$. We will consider this response as the sum of two terms. The slow or dominant term is given by the first two terms of Eq. (4.3-7). The dominant response $y_d(t)$ in this case is

$$y_d(t) = 1 - \frac{10}{9}e^{-t} = 1 - R_1 e^{-t} \tag{4.3-8}$$

The fast term $y_f(t)$ is then given by

$$y_f(t) = \frac{1}{9}e^{-10t} \tag{4.3-9}$$

The two terms and the total response are plotted in Fig. 4.3-3. The fast term associated with the pole at $s = -10$ has a time constant of 0.1 sec. In addition, since this pole is far from the other poles, the residue associated with this pole is small and the value of $y_f(t)$ is relatively small for all values of t. The result is that the total response is close to the dominant response, especially for all time values greater than 0.5 seconds. The dominant response can be rewritten to appear as a typical first-order step response that is delayed in time.

$$y_d(t) = 1 - e^{\ln R_1}e^{-t} = 1 - e^{-(t-\ln R_1)} \tag{4.3-10}$$

The total response starts at zero and moves quickly toward the response that results from the dominant pole at $s = -1$. The overall effect of adding a second pole located far out the negative real axis from the position of the original pole is to modify the step response of the system by slightly delaying or retarding the original step response. Since the response remains very similar to the response for a single pole system the system is said to be *dominantly* first-order.

We can now examine how varying the parameter p in Eq. (4.3-5) varies the position of the second pole and affects the step response of the system. We first write the expression for the dominant and fast parts of the step response for $p > 1$.

$$y_d(t) = 1 - \frac{p}{p-1}e^{-t} = 1 - R_p e^{-t} = 1 - e^{-(t-\ln R_p)} \tag{4.3-11}$$

$$y_f(t) = \frac{1}{p-1}e^{-pt} \tag{4.3-12}$$

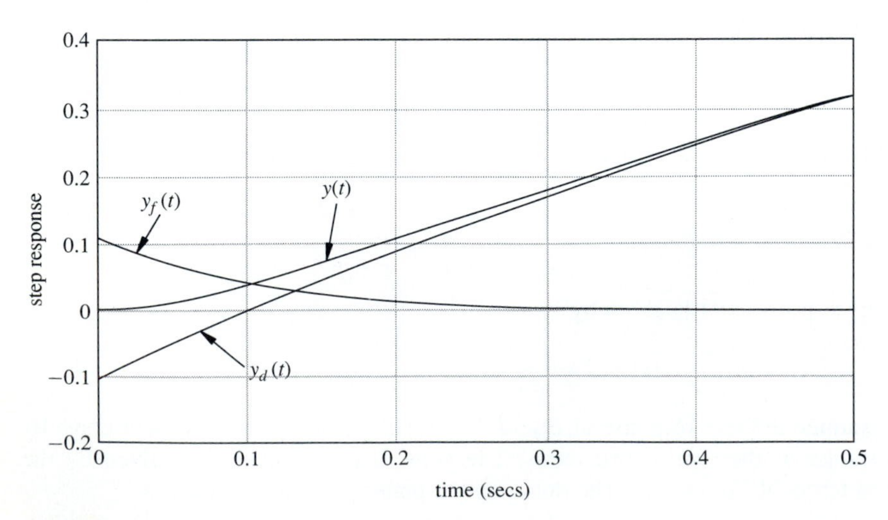

FIGURE 4.3-3
Single pole step response with an additional pole at $s = -10$.

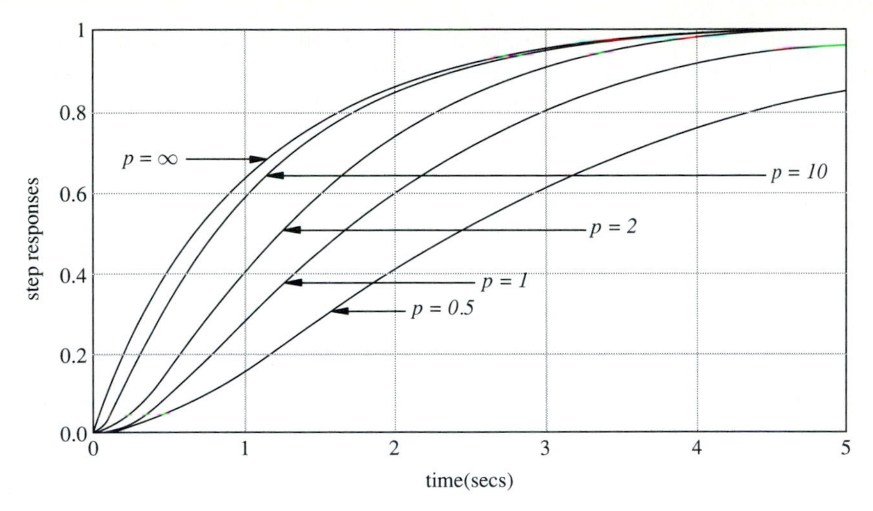

FIGURE 4.3-4
Single pole step response with additional pole at $-p$.

As p decreases from a large value toward one and the second pole moves along the negative real axis toward the dominant pole located at $s = -1$, the effect of the second pole on the step response is magnified. The residue associated with the fast response of Eq. (4.3-12) increases as the parameter p decreases. The time constant associated with the fast pole also increases so that as p decreases, the fast response is larger and takes longer to decay. More importantly, the dominant response of Eq. (4.3-11) is delayed longer as p decreases. The overall result is that the system's step response is continually retarded as a second pole moves in from negative infinity toward the dominant pole.

The step responses of the system of Eq. (4.3-5) are plotted in Fig. 4.3-4 for the values $p = \infty$, 10, 2, 1, and 0.5. As the pole moves in from negative infinity to negative ten to negative two, the response slows, as our analysis predicts. When the pole moves to negative one, the system's response further slows although now neither pole is dominant and our previous analysis is not entirely valid. With $p = 0.5$, this pole at $s = -0.5$ is now closer to the origin and becomes the dominant pole. The response looks more like the response of a first-order system with a time constant of 2 sec that is slowed further by the presence of the pole at negative one.

4.3.3 The Effect of a Zero on a Dominantly First-order System

Now that we have seen that an additional pole retards a dominantly first-order system as the pole moves along the negative real axis, we now examine the effect of moving a zero in along the negative real axis. The system to be studied is given by

$$\frac{Y(s)}{R(s)} = \frac{\frac{10}{z}(s + z)}{(s + 1)(s + 10)} \tag{4.3-13}$$

Again, the system is chosen so that the steady state response to a unit step is one. An additional pole is placed at $s = -10$ so that the transfer function remains strictly proper and the step response is continuous. We have seen that a pole this far out has little effect on a dominantly first-order system with the dominant pole at $s = -1$. Again, letting $R(s) = 1/s$ we take the partial-fraction expansion of the transform of the step response

$$Y(s) = \frac{1}{s} - \frac{\frac{10}{9}\frac{(z-1)}{z}}{s+1} + \frac{\frac{1}{9}\frac{(z-10)}{z}}{s+10} \tag{4.3-14}$$

When z is somewhat larger than one, the system has a fast pole at $s = -10$ and a dominant pole at $s = -1$. The total response moves quickly from zero at $t = 0$ to the dominant response, which is modified by the presence of the zero. The dominant response is given by

$$y_d(t) = 1 - \frac{10}{9}\frac{(z-1)}{z}e^{-t} = 1 - R_1 e^{-t} = 1 - e^{-(t - \ln R_1)} \tag{4.3-15}$$

If the zero is farther out on the negative real axis than the pole at $s = -10$, the zero has little effect and the dominant response is very close to the slightly delayed first-order response expected from having a pole at $s = -1$ and a pole at $s = -10$. As the zero moves in toward $s = -10$, it negates the effect of the delay caused by the pole at $s = -10$ until, when $z = 10$, the zero cancels the pole and a pure first-order response results.

From Eq. (4.3-15), we can see that for the case where $z < 10$, and the zero is inside the fast pole, the quantity R_1 is less than one so that the logarithm is negative and the dominant part of the response is advanced in time rather than delayed in time. Fig. 4.3-5 shows the dominant response, fast response, and total response for the case when $z = 5$. The total response moves quickly from zero to the dominant response and is faster than the response of a system with a single pole at $s = -1$ and no zero.

The speed of the response continues to increase as the zero moves along the negative real axis toward the pole at $s = -1$. When the zero gets near the pole at $s = -1$ the response of Eq. (4.3-15) speeds up to the point where it is faster than the previously fast part of the response. When z is near one, the response of Eq. (4.3-15) is close to a simple unit step but the critical rise time of the system is given by the remaining pole at $s = -10$. The time constant of 0.1 sec was previously much faster than the dominant response but is now slower than the immediate rise of the unit step response of Eq. (4.3-15). The effect of a zero near $s = -1$ can be seen in the transfer function of Eq. (4.3-13). The pole at $s = -1$ and the zero near $s = -1$ almost cancel so that the residue for the pole at $s = -1$ in the partial-fraction expansion of Eq. (4.3-14) is very small. Since the term associated with the pole at $s = -1$ almost vanishes, the system response is now dominated by the remaining term from the pole at $s = -10$. When $z = 1$ exactly, the step response is exactly the same as the step response of a system with a single pole at $s = -10$. Thus we see that poles that are close to the origin in the s plane and have large time constants

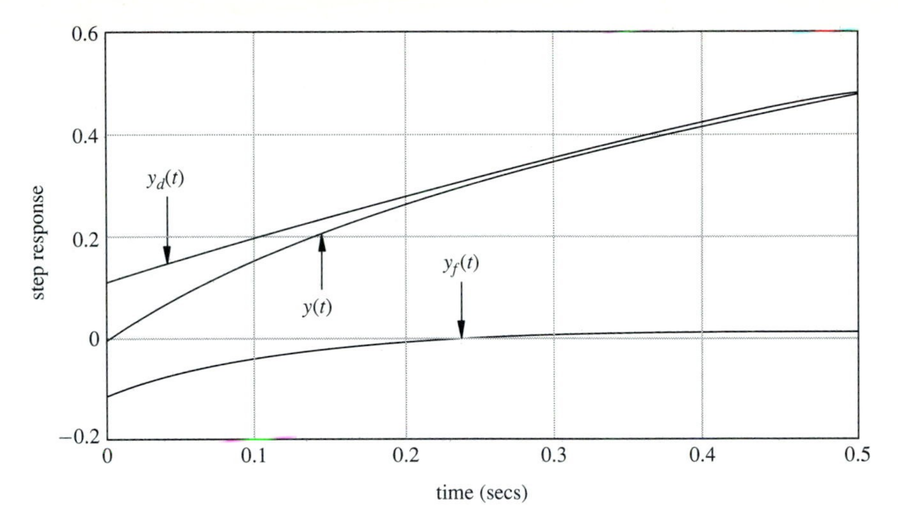

FIGURE 4.3-5
Dominantly first-order step response with zero at $s = -5$.

may not dominate a step response if there is a zero very close. The pole and zero almost cancel, leaving the pole too small a residue to make its response dominant even though it is slow.

When z is significantly less than one in Eq. (4.3-13), a new effect occurs. The zero is now closer to the origin than any pole and, in some sense, the presence of the zero dominates the response. The partial-fraction expansion of Eq. (4.3-14) for the step response of the system remains valid. The residue for the pole at $s = -1$ is again large enough for the effect of the first two terms of Eq. (4.3-14) to dominate. Since z is now less than one, the inverse transform of the first two terms should now be written in a slightly different form than Eq. (4.3-15).

$$y_d(t) = 1 + \frac{10}{9}\frac{(1-z)}{z}e^{-t} = 1 + R_1'e^{-t} \tag{4.3-16}$$

The dominant response now consists of a step with a decaying exponential added positively to the step. The dominant response begins at the value $1 + R_1'$ and decays with a time constant of 1 sec toward a value of one. The total response starts at zero and moves toward this dominant response with a time constant of 0.1 sec. Since the dominant response is greater than one for all time values, the total response will move from zero to a value greater than one before settling back toward one. Figure 4.3-6 demonstrates how the step response for $z = 0.5$ overshoots its final value and settles toward one. Figure 4.3-6 shows the dominant response, the fast response, and the total response of the system.

In Eq. (4.3-16) we see that as z approaches zero, so that the zero approaches the origin, the initial value for dominant response grows without bound. The overshoot becomes very large. The effect can be explained using Laplace transforms. A single

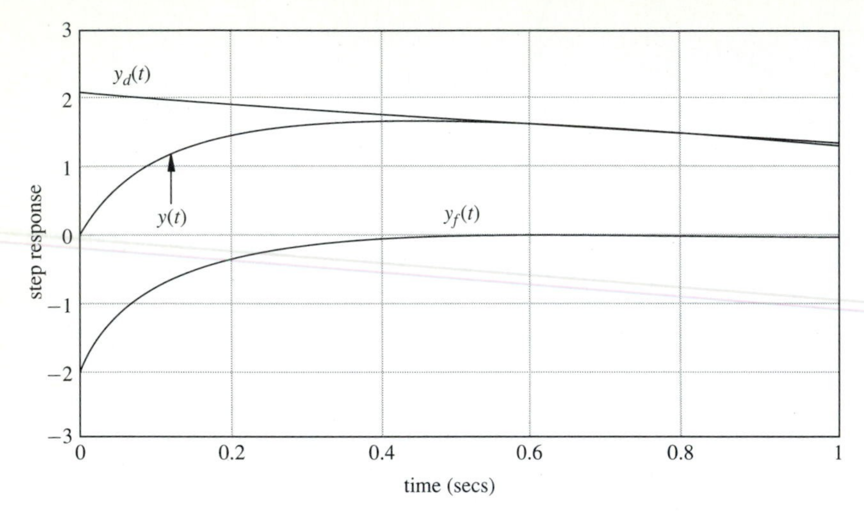

FIGURE 4.3-6
Dominantly first-order step response with zero at $s = -0.5$.

zero at the origin with no poles is the Laplace transform of a differentiator. The step response of a differentiator is an impulse. As the zero in Eq. (4.3-13) approaches the origin it becomes dominant, causing the step response to begin to resemble the step response of a differentiator and become highly spiked.

Figure 4.3-7 gives the step response of the system of Eq. (4.3-13) for the cases where $z = \infty$, 5, 1, 0.5, and 0.2. We can see the general effect that the speed of the response increases as the zero moves along the negative real axis from negative

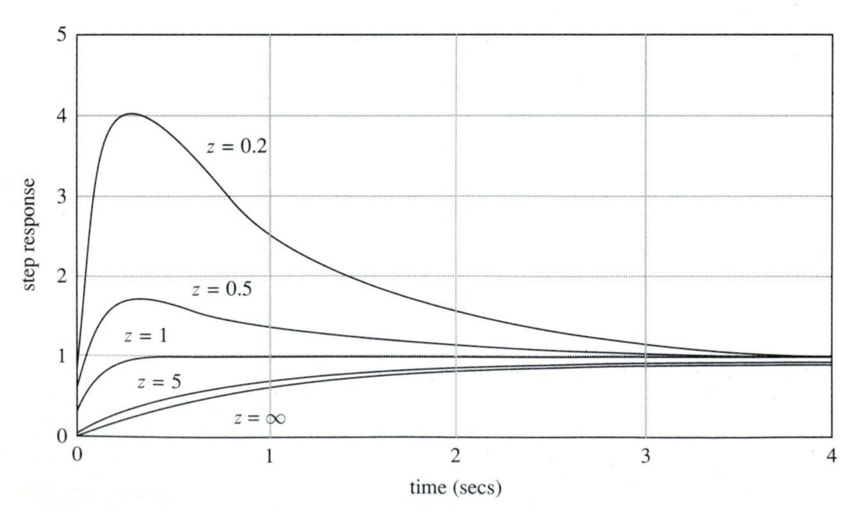

FIGURE 4.3-7
Dominantly first-order step response with zero at $s = -z$.

infinity. When the zero moves inside the pole closest to the origin, an overshoot occurs.

4.3.4 The Effect of a Right Half-plane Zero

We now examine the step response for the system given by Eq. (4.3-13) for the case when z is negative. A negative value for z corresponds to a zero in the right half-plane. Right half-plane zeros are usually referred to as *non-minimum phase zeros* for reasons that will be discussed in the next chapter.

When z is negative, Eq. (4.3-15) remains valid for the dominant response. Since R_1 is now greater than one, the dominant response is that of a delayed first-order system, as it was when a left half-plane pole was added. However, the delaying effect on the dominant response is more severe for a right half-plane zero than it is for the left half-plane pole. Figure 4.3-8 shows the dominant response, the fast response, and the total response for the case of $z = -10$ with the zero located at $s = 10$. We see that the dominant response is delayed enough so that the total response starts off in a negative direction before moving toward the dominant response. The "wrong way" step response as seen here is characteristic of systems with non-minimum phase zeros. From Eq. (4.3-15) we can see that as the zero moves in along the positive real axis from infinity the delay of the dominant response is lengthened, causing the dominant response to start from larger negative values. Fig. 4.3-9 shows the step responses of the system given by Eq. (4.3-13) with $z = -\infty$, -5, -0.5, and -0.2. As the zero moves toward the origin the wrong way response is greatly exaggerated.

It is interesting to examine the derivative of the step response of the system given by Eq. (4.3-13) evaluated at $t = 0$. This operation may be carried out using

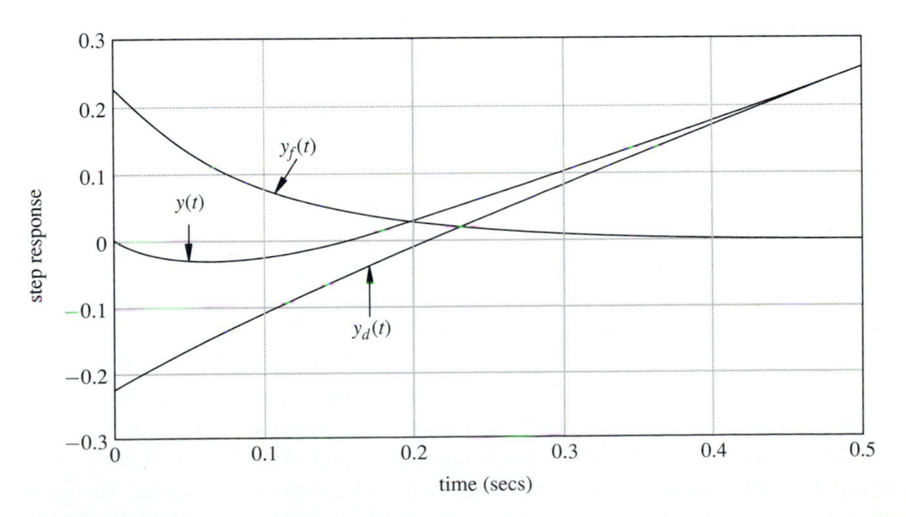

FIGURE 4.3-8
Dominantly first-order step response with zero at $s = +10$.

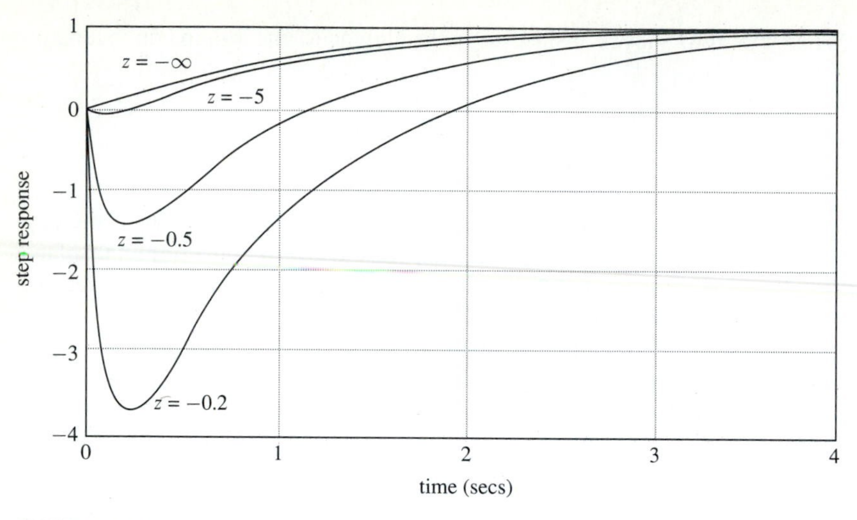

FIGURE 4.3-9
Dominantly first-order step response with zero at $s = +z$.

the initial value theorem of Laplace transforms. The Laplace transform of the step response is given by

$$Y(s) = \frac{\frac{10}{z}(s+z)}{s(s+1)(s+10)}$$

Since differentiation in the time domain corresponds to multiplication by s in the transform domain, the Laplace transform of the derivative of the step response is given as

$$\pounds\left\{\frac{dy(t)}{dt}\right\} = sY(s) = \frac{\frac{10}{z}(s+z)}{(s+1)(s+10)}$$

Using the initial value theorem of Laplace transforms we obtain

$$\left.\frac{dy(t)}{dt}\right|_{t=0} = \lim_{s\to\infty} s\pounds\left\{\frac{dy(t)}{dt}\right\} = \lim_{s\to\infty} \frac{s\frac{10}{z}(s+z)}{(s+1)(s+10)} = \frac{10}{z} \tag{4.3-17}$$

When z is positive, corresponding to a left half-plane zero, the derivative is also positive; when z is negative, corresponding to a right half-plane zero, the derivative is negative, indicating that the response starts off in the negative direction. As the magnitude of a positive z decreases, corresponding to a zero moving along the negative real axis toward the origin, the increasing magnitude of the derivative indicates, first, a more rapid response caused by the zero and, then, an increase in overshoot. For right half-plane zeros moving along the positive real axis toward the origin the increase in the magnitude of the derivative indicates an increase in the magnitude of the wrong way response and an increase in the delay before the response approaches its steady-state value.

4.3.5 Summary

We have seen that if there is a single pole that is significantly closer to the origin than other poles and zeros of a transfer function with all its poles in the left half-plane, the time constant of that pole closest to the origin dominates the response of the system. The dominance of the pole closest to the origin is removed if there is a zero very close to that pole; in this case, the pole next closest to the origin becomes dominant.

The response from the dominant pole is modified from a pure first-order system response by the presence of other poles and zeros. Additional poles delay or retard the response of the system while left half-plane zeros speed up the response. Right half-plane zeros cause the response to start off in the wrong direction before recovering. The effect increases as either a pole or zero moves in toward the origin. If there is a zero and a pole modifying the effect of another dominant pole and the modifying zero is closer to the origin than the modifying pole, the response from the dominant pole is modified more by the zero than the pole and the response is slightly advanced or sped up in time. If the modifying pole is closer to the origin than the modifying zero, the response is modified more by the pole than the zero and the response is slightly delayed or retarded.

In this section, we examined the problem where a single pole is dominant. It often occurs that the poles closest to the origin are a complex pole pair. In Sec. 4.4, we investigate the response of a second-order system. We will see that in the presence of additional poles and zeros the response of a second-order system is modified similarly to the way the response of a first-order system is modified.

4.3.6 CAD Notes

Step responses of systems are easily obtained from MATLAB by using the `lsim` function as explained in the CAD Notes of Sec. 2.7. The labels on the graphs of this section were altered after the graphs were produced by MATLAB.

Exercises 4.3

4.3-1. The graphs of Fig. 4.3-10 show the step responses of simple first-order transfer functions. Estimate each of the transfer functions.
Answers:

$$(a) \quad G(s) = \frac{4}{s+2}$$

$$(b) \quad G(s) = \frac{18}{s+6}$$

4.3-2. (*a*) Which of the following closed-loop transfer functions has the step response with the longest (slowest) rise time?

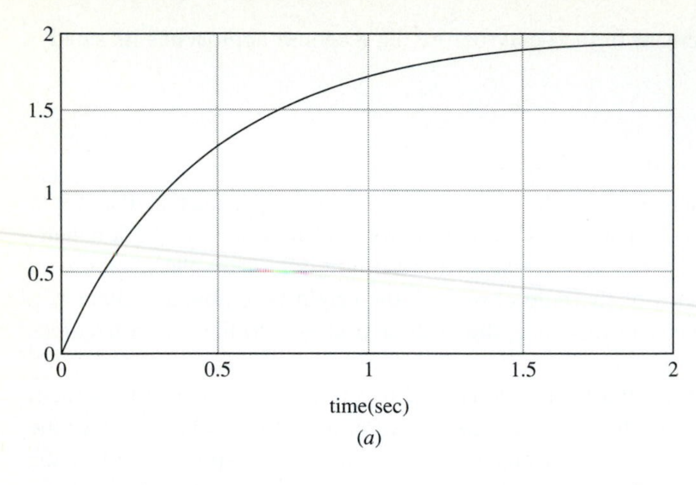

(*a*)

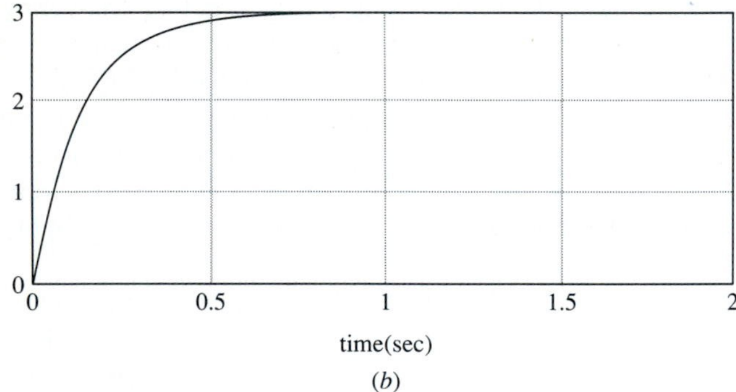

(*b*)

FIGURE 4.3-10
Exercise 4.3-1.

(*b*) Which, if any, has the greatest overshoot?

$$G_1(s) = \frac{0.5(s+10)}{(s+1)(s+5)} \qquad G_2(s)\frac{50(s+1)}{(s+10)(s+5)}$$

$$G_3(s) = \frac{10(s+5)}{(s+10)(s+5)} \qquad G_4(s)\frac{2.5(s+10)}{(s+5)(s+5)}$$

Answers: (*a*) $G_1(s)$ (*b*) $G_2(s)$

4.3-3. How long do you expect to wait for the following system to reach the value 0.32 in response to a unit step input?

$$G(s) = \frac{5(s+0.9)}{0.9(s+10)(s+1)}$$

Answer: ~0.1 sec

4.3-4. The dotted line curves of Fig. 4.3-11 are all step responses of the transfer function

$$G_1(s) = \frac{1}{\left(\frac{s}{5}+1\right)}$$

The solid line curves of Fig. 4.3-11 are all step responses of the transfer function

$$G(s) = \frac{\left(\dfrac{s}{a}+1\right)}{\left(\dfrac{s}{b}+1\right)\left(\dfrac{s}{5}+1\right)}$$

for different values of a and b. Possible values for a and b are 10, 5, 1, and -5. In each solid line curve either a or b equals 10. Determine the value of the other variable from the shape of the curve. Explain your choice.

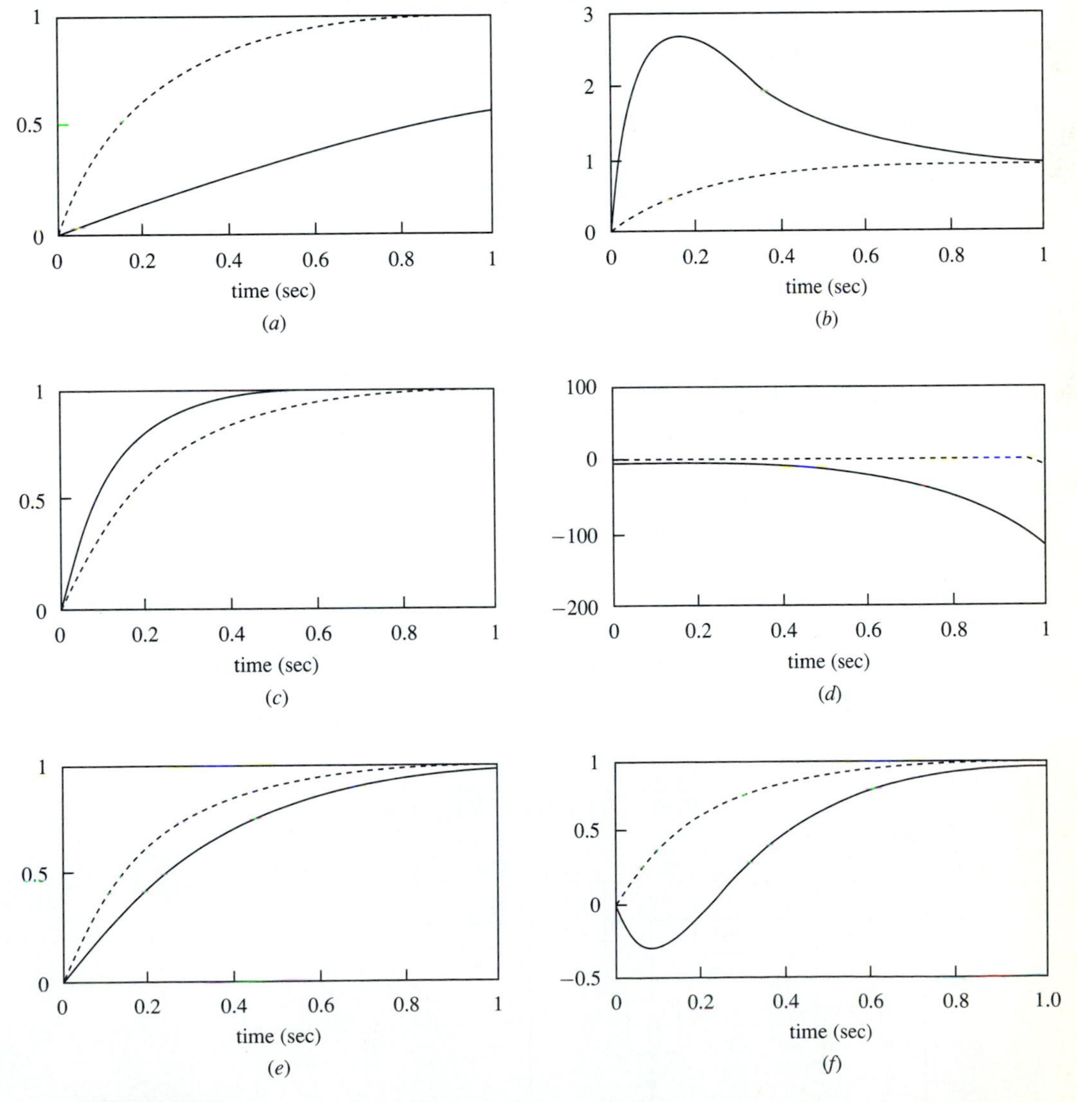

FIGURE 4.3-11
Exercise 4.3-4.

Answers: (*a*) $a = 10 \quad b = 1$ (*b*) $a = 1 \quad b = 10$
(*c*) $a = 5 \quad b = 10$ (*d*) $a = 10 \quad b = -5$
(*e*) $a = 10 \quad b = 5$ (*f*) $a = -5 \quad b = 10$

4.4 STEP FUNCTION RESPONSE OF DOMINANTLY SECOND-ORDER SYSTEMS

In this section the time response of a dominantly second-order system to a step function input is discussed in detail. The second-order system is important because it is easy to analyze and understand. For this reason closed-loop-system specifications are often expressed in comparison to behavior typical of the second-order case. In other words, the performance of a high-order system is related to a dominant set of second-order poles; that is, the high-order system is approximated by a second-order system. Thus a thorough understanding of the second-order case is important before the total time response for the general case is discussed.

The discussion here is based upon the second-order differential equation

$$\frac{d^2y(t)}{dt^2} + 2\zeta\omega_n \frac{dy(t)}{dt} + {\omega_n}^2 y(t) = {\omega_n}^2 r(t) \tag{4.4-1}$$

where $r(t)$ is a step-function input and the initial conditions are assumed to be zero. Here the variable ζ (the Greek letter *zeta*) is called the *damping ratio*, and ω_n (the Greek letter *omega*) is referred to as the *undamped natural frequency*. Equation (4.4-1) is purposely written in terms of these parameters to express the solution conveniently. In addition, a number of aspects of the solution are most easily written in terms of ζ and ω_n.

The physical origins of the second-order differential equation (4.4-1) are many and varied. For example, a series electrical circuit containing resistive, inductive, and capacitive elements is typical. A system consisting of a spring, mass, and viscous friction is a mechanical analog. Here attention is directed to the feedback system shown in Fig. 4.4-1. This system may be thought of as a simple positioning servomechanism in which the actuator is either a field-controlled or armature-controlled dc motor.

The closed-loop transfer function $Y(s)/R(s)$ associated with Fig. 4.4-1 is

$$\frac{Y(s)}{R(s)} = \frac{K}{s^2 + s(\lambda + Kk_2) + K} \tag{4.4-2}$$

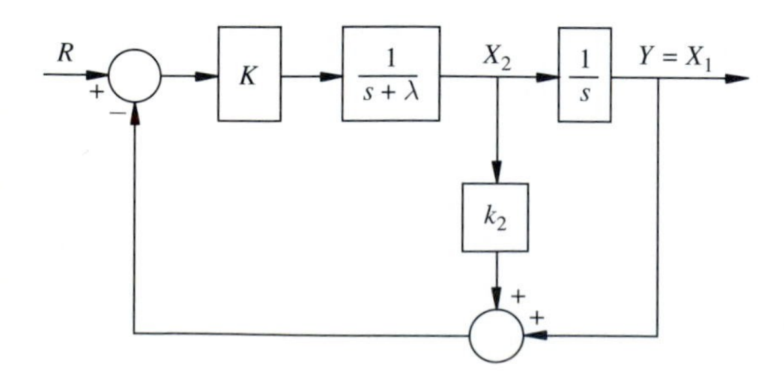

FIGURE 4.4-1
Second-order system.

The Laplace transform of Eq. (4.4-1), with zero initial conditions, on the other hand, is

$$\frac{Y(s)}{R(s)} = \frac{\omega_n{}^2}{s^2 + 2\zeta\omega_n s + \omega_n{}^2} \tag{4.4-3}$$

and Eqs. (4.4-2) and (4.4-3) are equal when

$$K = \omega_n{}^2 \qquad \lambda + Kk_2 = 2\zeta\omega_n \tag{4.4-4}$$

Therefore by proper selection of K and k_2 it is possible to cause ζ and ω_n to have any desired values. In other words, by varying K and k_2 it is possible to adjust ζ and ω_n and therefore the closed-loop response of the system of Fig. 4.4-1. We shall consider this process in more detail later, but first let us investigate the basic nature of the time response.

The closed-loop transfer function (4.4-3) has poles located at

$$s = -\zeta\omega_n \pm \sqrt{(\zeta^2 - 1)\,\omega_n{}^2}$$

The nature of the closed-loop response depends almost completely on the damping ratio. The value of ω_n simply adjusts the time scale, as we shall see later. If ζ is greater than 1, both closed-loop poles are real; if ζ is equal to 1, both closed-loop poles are equal to $-\zeta\omega_n$; and if ζ is less than 1, the closed-loop poles are complex conjugates, given as

$$s = -\zeta\omega_n \pm j\omega_n\sqrt{1 - \zeta^2} \tag{4.4-5}$$

The first two cases were covered in the previous section. We shall concentrate our discussion on the last case.

The location of the closed-loop poles with respect to the damping ratio and the undamped natural frequency is indicated in Fig. 4.4-2. To investigate the response for

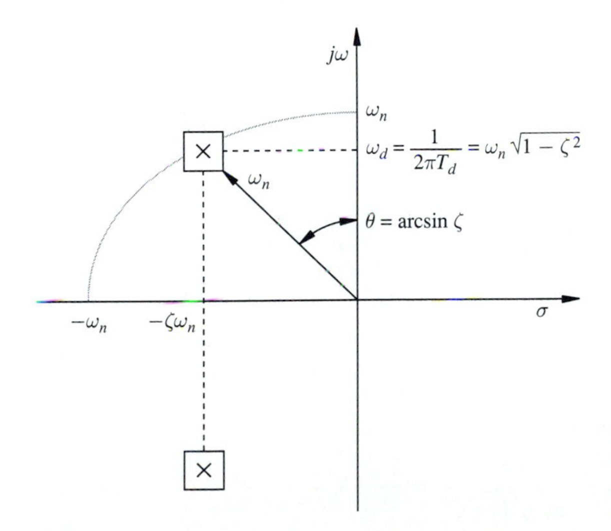

FIGURE 4.4-2
Location of the closed-loop poles with respect to ζ and ω_n.

the closed-loop system, it is necessary to determine the inverse Laplace transform of $Y(s)$ for some $R(s)$. In this chapter the input is chosen as a step since this is a typical and convenient test input. For $R(s) = 1/s$, $Y(s)$ is given by

$$Y(s) = \frac{{\omega_n}^2}{s\left[(s+\zeta\omega_n)^2 + \left(\omega_n\sqrt{1-\zeta^2}\right)^2\right]} \tag{4.4-6}$$

Equation (4.4-6) has been written in this form to emphasize the complex conjugate structure of the closed-loop system. To find $y(t)$ it is convenient to expand Eq. (4.4-6) into partial-fraction form and use the short table of transforms (Table 4.2-1). Thus $Y(s)$ is

$$Y(s) = \frac{1}{s} + \frac{\frac{-1}{2\sqrt{1-\zeta^2}}e^{j(\psi-90^\circ)}}{s+\zeta\omega_n - j\omega_n\sqrt{1-\zeta^2}} + \frac{\frac{-1}{2\sqrt{1-\zeta^2}}e^{j(-\psi+90^\circ)}}{s+\zeta\omega_n + j\omega_n\sqrt{1-\zeta^2}}$$

where
$$\psi = \arctan\frac{\sqrt{1-\zeta^2}}{\zeta} \tag{4.4-7}$$

With the aid of the short table of transform pairs, $y(t)$ is written directly as

$$\begin{aligned} y(t) &= 1 - \frac{1}{\sqrt{1-\zeta^2}}e^{-\zeta\omega_n t}\cos\left(\omega_n\sqrt{1-\zeta^2}t + \psi - 90^\circ\right) \\ &= 1 - \frac{1}{\sqrt{1-\zeta^2}}e^{-\zeta\omega_n t}\sin\left(\omega_n\sqrt{1-\zeta^2}t + \psi\right) \end{aligned} \tag{4.4-8}$$

Here the actual frequency of oscillation in radians per second is $\omega_n\sqrt{1-\zeta^2}$ and is known as the *damped frequency* ω_d. A typical oscillatory response is given in Fig. 4.4-3, with pertinent characteristics of the curve labeled.

The period of oscillation, T_d, associated with the damped frequency ω_d is

$$T_d = \frac{2\pi}{\omega_d} = \frac{2\pi}{\omega_n\sqrt{1-\zeta^2}} \tag{4.4-9}$$

The time to the maximum or peak value of the output is t_p. The maximum value of $y(t)$ may be found by taking the derivative of $y(t)$ with respect to t and equating the result to zero.

$$\frac{dy(t)}{dt} = \frac{\zeta\omega_n}{\sqrt{1-\zeta^2}}e^{-\zeta\omega_n t}\sin(\omega_d t + \psi) - \omega_n\sqrt{1-\zeta^2}e^{-\zeta\omega_n t}\cos(\omega_d t + \psi) = 0$$

This is zero when t is

$$t = \frac{n\pi}{\omega_n\sqrt{1-\zeta^2}} \qquad n = 0, 1, 2, \ldots \tag{4.4-10}$$

The peak value of $y(t)$ occurs when $n = 1$ or

$$t_p = \frac{\pi}{\omega_n\sqrt{1-\zeta^2}} = \frac{T_d}{2} \tag{4.4-11}$$

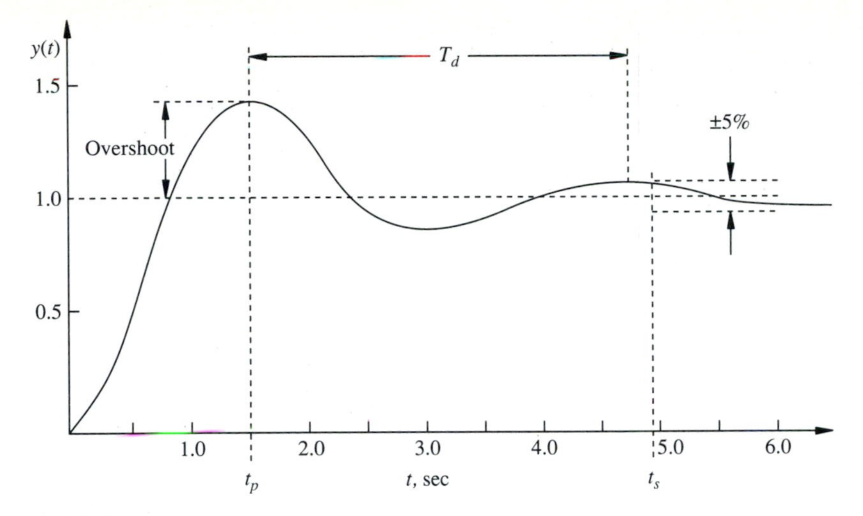

FIGURE 4.4-3
Typical underdamped ($0 < \zeta < 1$) second-order system response for a unit step input.

The percent overshoot, PO, is 100 times the peak value of $y(t)$ minus the step size, divided by the step size. This is a normalized quantity that is independent of the step size, although here we have assumed a unit step. The peak value of the response, $y(t)_{\text{max}}$, can be found by substituting t_p into the expression for $y(t)$. The result can be reduced to

$$y(t)_{\text{max}} = 1 + e^{-\left(\frac{\zeta\pi}{\sqrt{1-\zeta^2}}\right)} \tag{4.4-12}$$

Therefore the percent overshoot is simply

$$\text{PO} = 100e^{-\left(\frac{\zeta\pi}{\sqrt{1-\zeta^2}}\right)} \tag{4.4-13}$$

The overshoot for a range of values of ζ is shown in Fig. 4.4-4.

The settling time t_s is normally defined as the time required for the response to remain within 5 percent of its final value. The magnitude of the sinusoidal portion of Eq. (4.4-8) is always less than or equal to 1. For convenience, the settling time is often approximated as the time beyond which the envelope of the sinusoid is less than 0.05 times its original value,

$$e^{-\zeta\omega_n t_s} < 0.05$$

or when

$$\zeta\omega_n t_s > 3$$

Therefore, t_s can be given by

$$t_s = \frac{3}{\zeta\omega_n} \tag{4.4-14}$$

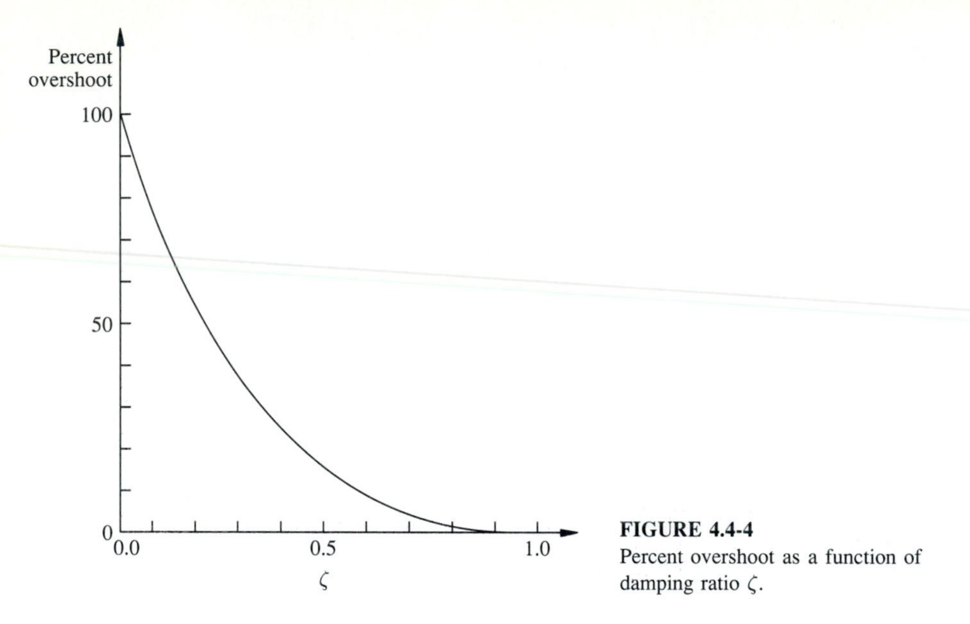

FIGURE 4.4-4
Percent overshoot as a function of damping ratio ζ.

Note that T_d, t_p, and t_s are all inversely proportional to ω_n. In other words, the quantities $\omega_n T_d$, $\omega_n t_p$, and $\omega_n t_s$ are independent of ω_n. The percent overshoot is completely independent of ω_n. If the time response $y(t)$ is plotted against the normalized variable $\omega_n t$, the response is solely a function of ζ. This justifies the comment made earlier that the nature of the response is almost completely determined by the value of ζ and that ω_n serves merely to affect the time scale of the response. To illustrate more clearly the influence of ζ and ω_n on the time response, let us consider three special cases: (1) vary ζ with ω_n constant, (2) vary ω_n with ζ constant, and (3) vary ζ and ω_n with the product $\zeta\omega_n$ constant. Each of these cases is discussed separately.

First we view the variation of ζ while holding ω_n constant. Figure 4.4-5*a* indicates the position of the closed-loop poles for a number of values of ζ. The locus of the closed-loop poles as ζ is varied from 0 to 1 is a semicircle of radius ω_n shown by the dashed line in Fig. 4.4-5*a*. The corresponding step responses are shown in Fig. 4.4-5*b*. Note that T_d, t_s, and t_p as well as the percent overshoot vary as ζ is changed.

An interesting case that is often used as a goal in practical closed-loop specifications as well as academic examples is $\zeta = 0.707$. A system with this damping ratio has the property that, for a given ω_n, it will have a fast convergence to the step value with very little overshoot. Thus it is a reasonable practical goal. It also has the property that the poles lie on a 45° angle from the origin. Thus, it possesses academic convenience.

Next we vary ω_n while keeping ζ constant. Figure 4.4-6*a* shows the locus of the closed-loop poles in this case; it is a straight line passing through the origin. The step responses for three values of ω_n are shown in Fig. 4.4-6*b*. In this case, we see

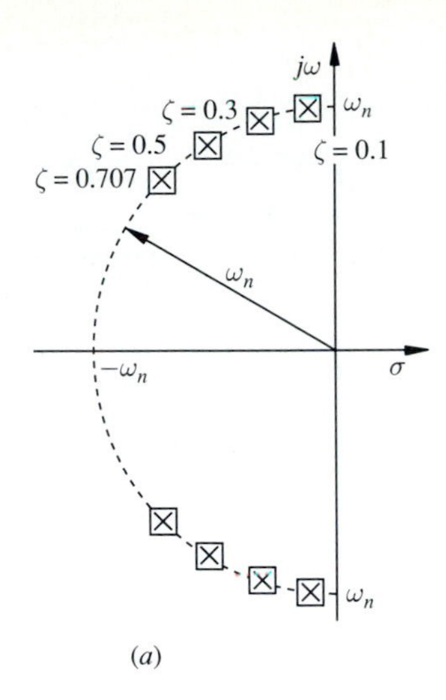

(a)

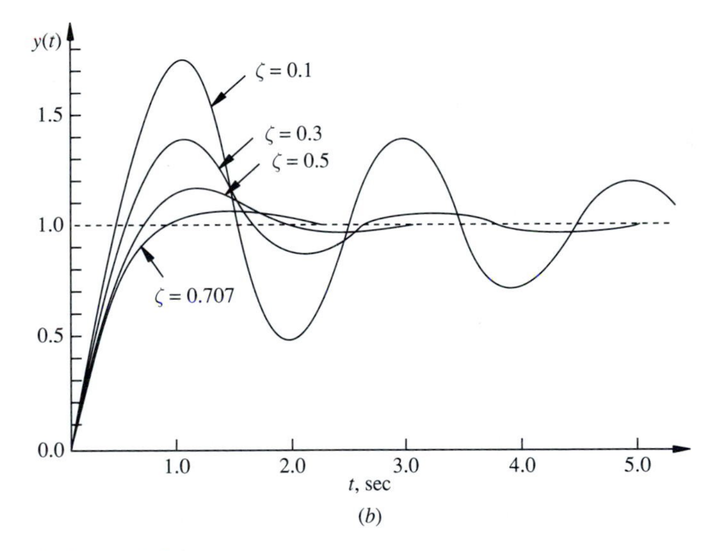

(b)

FIGURE 4.4-5
Variation of ζ with $\omega_n = 1.0$. (a) Plot of the closed-loop poles; (b) step response.

that the basic character of the response, in particular the overshoot, is constant since ζ is constant but that T_d, t_s, and t_p decrease as ω_n is increased. Indeed, changing ω_n is equivalent to changing the time scale.

When ζ and ω_n are both varied while the product $\zeta\omega_n$ is kept constant, the locus of the closed-loop poles becomes the vertical line $s = -\zeta\omega_n$ as shown in Fig. 4.4-7a. The step responses are shown in Fig. 4.4-7b. Note that, in this case, t_s remains constant although T_d, t_p, and the overshoot vary.

The quantities t_s, t_p, and PO can be expressed in terms of the second-order parameters ζ and ω_n. In addition, there are a number of other terms used to describe

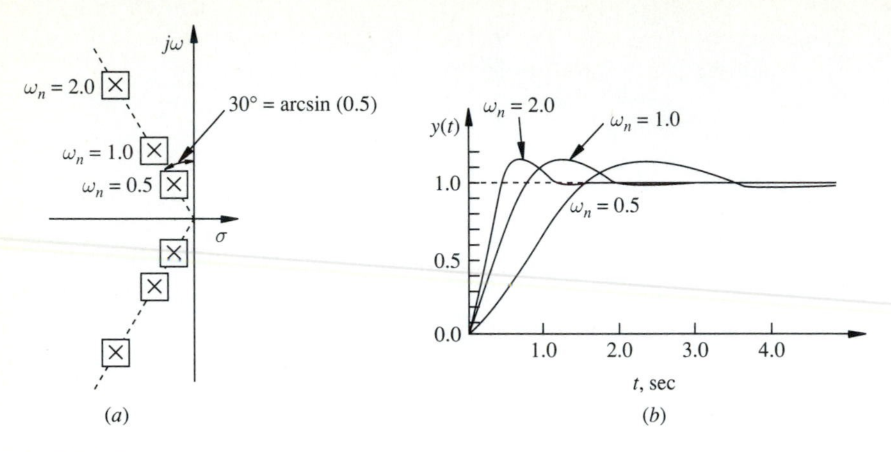

FIGURE 4.4-6
Variation of ω_n with $\zeta = 0.5$. (*a*) Plot of the closed-loop poles; (*b*) step response.

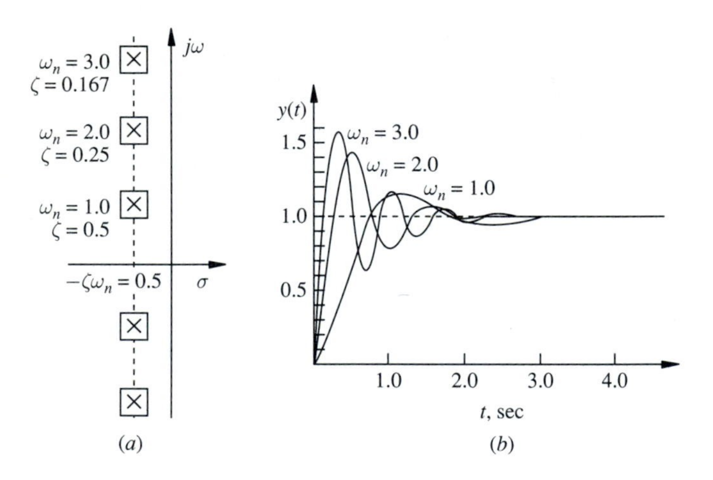

FIGURE 4.4-7
Variation of ζ and ω_n with $\zeta\omega_n = 0.5$. (*a*) Plot of the closed-loop poles; (*b*) step response.

the system response. These quantities are the rise time t_r and the delay time t_d. Rise time is defined as the time required by the system to rise from 10 percent to 90 percent of its final value.

Delay time is defined as the time required for the system to reach 50 percent of its final value. While the performance measures have been defined here in the context of a pure second-order system, these measures are used for more general control responses with roughly the same shape, i.e., dominantly second-order systems.

The discussion above has centered on the case in which $0 < \zeta < 1$, that is, the case of the underdamped system. If $\zeta = 1$, the step-function response is said to be critically damped and is of the form

$$y(t) = 1 - \omega_n t e^{-\omega_n t} - e^{-\omega_n t}$$

In the overdamped case, the damping ratio is greater than 1, and the response contains two exponential terms with different time constants. This case was covered in Sec. 4.3.

The effects that added poles and zeros have on dominantly second-order systems are similar to the effects that added poles and zeros have on dominantly first-order systems. From Equation (4.4-8) it is seen that a second-order step response consists of a dc component and a damped sinusoid. If Eq. (4.4-8) is rederived after including a zero in the transfer function in such a way that the dc gain remains the same, the magnitude of the residue associated with the sinusoid and the phase angle associated with sinusoid both change. With some fairly painful derivation it can be shown that the changes occur in a specific manner. As the zero moves in along the negative real axis toward the origin, the time to the first peak of the step response decreases monotonically while the percent overshoot increases monotonically. The most dramatic effect is created when the zero moves closer to the origin than the real part of the complex pole pair. At this point, the magnitudes of the residues associated with the complex poles increase dramatically and thus the size of the sinusoidal component in the dominant response increases dramatically. In general, an additional zero makes the system faster and more oscillatory as that zero moves in the negative real axis toward the origin.

Conversely, as an added pole moves in along the negative real axis toward the origin the time to the first peak of the step response increases and the percent overshoot decreases. Again, the most dramatic effect occurs as the pole moves inside the real part of the complex pole pair. The magnitudes of the residues associated with the complex poles decrease and the sinusoidal component in the dominantly second-order step response also decreases.

Figures 4.4-8 and 4.4-9 display the step responses of the systems given by

$$\frac{Y(s)}{R(s)} = \frac{\frac{\omega_n^2}{a}(s+a)}{s^2 + 2\zeta\omega_n s + \omega_n^2} \tag{4.4-15}$$

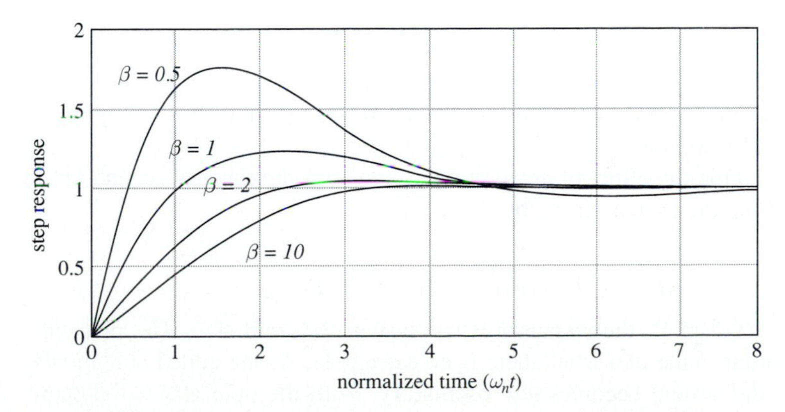

FIGURE 4.4-8
Second-order step responses ($\zeta = .7$), each with an additional zero.

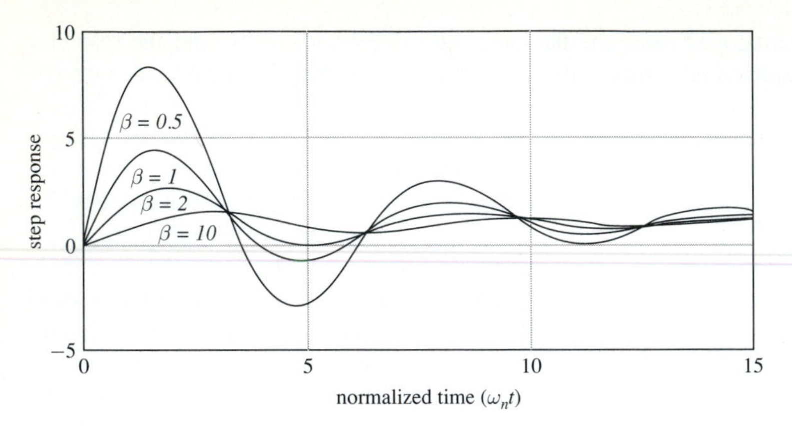

FIGURE 4.4-9
Second-order step responses ($\zeta = 0.2$), each with an additional zero.

The responses are given using a normalized time scale, $\omega_n t$, and various values of the zero locations a are used. The curves are parameterized by the zero position using the parameter

$$\beta = \frac{a}{\zeta \omega_n}$$

Thus, β positions the extra pole relative to the real part of the complex pole pair. Figure 4.4-8 uses a damping ratio of $\zeta = 0.7$ while Figure 4.4-9 uses a damping ratio of $\zeta = 0.2$. The plot with $\beta = 10$ is very similar to the plot when there is no zero. As the zero moves in along the negative real axis toward the origin, the sinusoidal component is increased creating a more oscillatory response with a shorter rise time but significantly more overshoot and a longer settling time. An added zero in the right half-plane also creates a more oscillatory system and, as in the dominant first-order system, the response starts off in the opposite direction to the step input. Again, the effect becomes more pronounced as the zero approaches the origin.

From the above discussion, it is clear that, if one is placing dominant second-order poles to achieve a particular shape of step response, one must be sure that all zeros are much farther from the origin than the dominant pole pair. If there are zeros that are too close, the dominant pole pair must be placed with greater damping to achieve the desired response.

Next we examine the effect of an additional pole on a dominantly second-order system. We examine the system given by

$$\frac{Y(s)}{R(s)} = \frac{p}{(s+p)(s^2 + 2(0.2)s + 1)}$$

Figure 4.4-10 displays the response of the system to a unit step. The plot with $p = 5$ is very similar to the plot when there is no extra pole. As the added pole moves in from infinity, the system becomes less oscillatory. With the pole at $s = -1$ there is some effect upon the pole pair at $\omega_n = 1$. The system exhibits only 35 percent overshoot when $p = 1$ instead of the 53 percent overshoot usually associated with

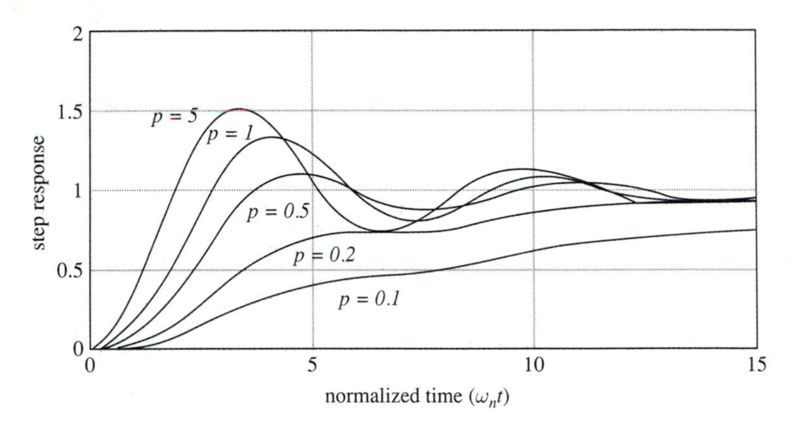

FIGURE 4.4-10
Second-order step responses ($\zeta = 0.2$), each with an additional pole at $s = -p$.

a damping ratio of $\zeta = 0.2$. By the time the added pole has moved in to the point $s = -0.2$, a value equal to the real part of the pole pair, the single pole has become dominant with its response slightly modified by the oscillatory poles.

Thus we see, as in the dominantly first-order system, the response of a dominantly second-order system is sped up by an additional zero and slowed by an additional pole. In the dominantly second-order system the added zero also has the important effect of increasing the amount of oscillation in the system while an added pole has the effect of decreasing the amount of oscillation. A right half-plane zero also causes a wrong way response. All effects become more pronounced as the added zero or pole approaches the origin.

Closed-loop poles and zeros in a control system may be manipulated to achieve a desired closed-loop transient response. Often, the final configuration consists of a dominant pair of poles with some additional modifying dynamics. Thus, the results of this section should enable the reader to approximately translate specifications on a closed-loop step response into a set of closed-loop pole and zero positions that are likely to produce an acceptable transient response. While performing analysis, we are taking steps toward design.

Exercises 4.4

4.4-1. Figure 4.4-11 shows the step response of two simple second-order systems with no zeros. The transfer functions can be parameterized as

$$G(s) = \frac{\omega^2}{s^2 + 2\omega\zeta s + \omega^2}$$

For each case, estimate the values of ζ and ω_n from the step responses.
Answers:

$$(a) \quad \zeta = 0.05 \qquad \omega = 4\pi$$

$$(b) \quad \zeta = 0.7 \qquad \omega = 2\pi$$

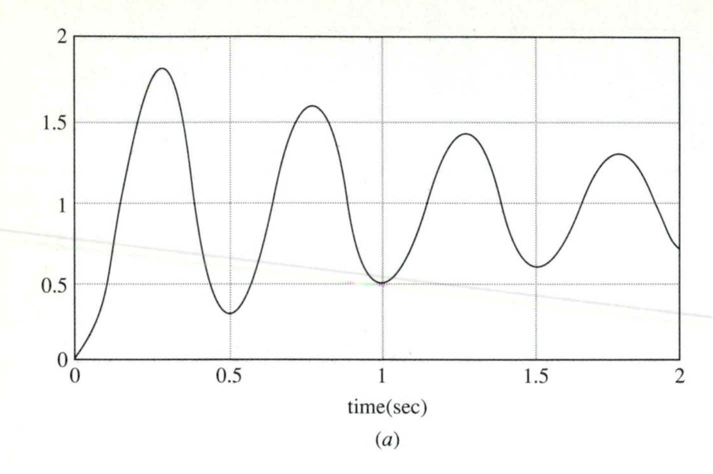

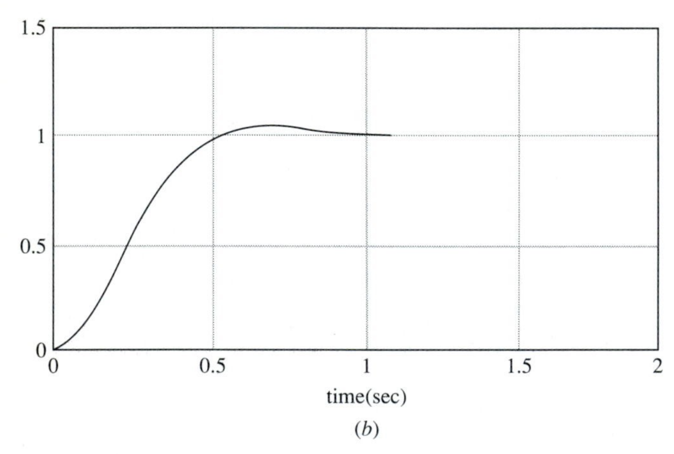

FIGURE 4.4-11
Exercise 4.4-1.

4.4-2. (*a*) Which of the following closed-loop transfer functions has the step response with the longest settling time?

(*b*) Which have step responses with no overshoot (the answer could be none, one or more than one)?

$$G_1(s) = \frac{0.1(s+10)}{s^2+2s+1} \qquad G_2(s) = \frac{0.4(s+10)}{s^2+2s+4}$$

$$G_3(s) = \frac{0.1(s+10)}{s^2+s+1} \qquad G_4(s) = \frac{0.9(s+10)}{s^2+3s+9}$$

Answers: (*a*) $G_3(s)$ (*b*) $G_1(s)$

4.4-3. (*a*) What is the expected time for the step response of the following system to settle to within 5 percent of its final value?

$$G(s) = \frac{100}{s^2+4s+100}$$

(*b*) The step response of which of the following two systems oscillates more?

$$G_1(s) = \frac{200(s+5)}{(s+10)(s^2+4s+100)} \qquad G_2(s) = \frac{50(s+10)}{(s+5)(s^2+4s+100)}$$

Answers: (*a*) 1.5 sec (*b*) $G_1(s)$

4.4-4. The dotted line curves of Fig. 4.4-12 are all step responses of the transfer function

$$G_2(s) = \frac{25}{s^2+5s+25}$$

The solid line curves of Fig. 4.4-12 are all step responses of the transfer function

$$G(s) = \frac{25\left(\frac{s}{a}+1\right)}{\left(s^2+5s+25\right)\left(\frac{s}{b}+1\right)}$$

for different values of a and b. Possible values for a and b are 10, 1, and −5. In each graph either a or b equals 10. Determine the value of the other variable from the shape of the graph. Explain your answer.

Answers: (*a*) $a = 1$ $b = 10$ (*b*) $a = 10$ $b = -5$
(*c*) $a = -5$ $b = 10$ (*d*) $a = 10$ $b = 1$

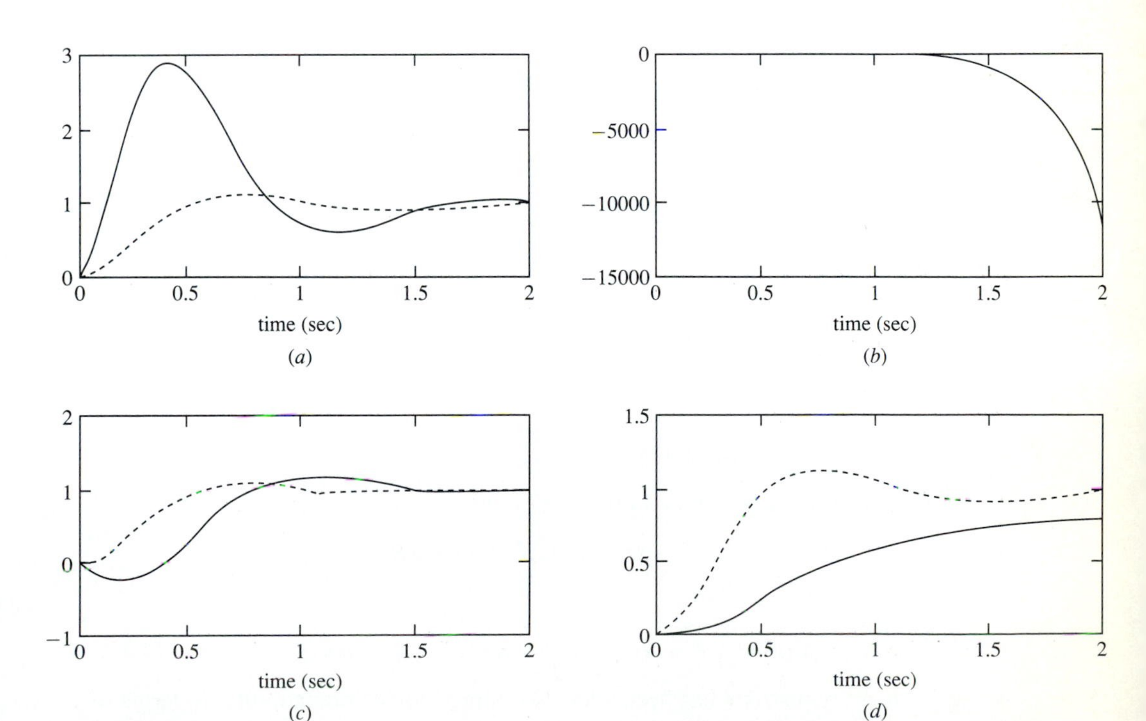

FIGURE 4.4-12
Exercise 4.4-4.

4.5 STATE-VARIABLES TIME RESPONSE

In the preceding section the output response was determined as a function of time for a dominantly second-order system subject to a unit step input and zero initial conditions. In the general case treated in this section, the order of the system is not restricted nor is the nature of the input. Initial conditions are allowed on any of the state-variables and, in addition to the output response, here we determine the behavior of all the state-variables as functions of time.

As noted in Chaps. 2 and 3, it is easiest to formulate a general problem in matrix notation. Assume that the plant to be controlled is described by Eq. ($\boldsymbol{Ab}$),

$$\dot{\boldsymbol{x}} = \boldsymbol{A}\boldsymbol{x} + \boldsymbol{b}u \tag{$\boldsymbol{Ab}$}$$

and that the control is achieved by state-variable feedback so that u is given by

$$u = K(r - \boldsymbol{k}^T\boldsymbol{x})$$

The closed-loop system is again

$$\dot{\boldsymbol{x}} = (\boldsymbol{A} - K\boldsymbol{b}\boldsymbol{k}^T)\boldsymbol{x} + K\boldsymbol{b}r = \boldsymbol{A}_k\boldsymbol{x} + K\boldsymbol{b}r \tag{$\boldsymbol{A}_k\boldsymbol{b}$}$$

where $\boldsymbol{A}_k$ is the describing matrix of the closed-loop system defined as before by

$$\boldsymbol{A}_k = \boldsymbol{A} - K\boldsymbol{b}\boldsymbol{k}^T$$

The output y is given by

$$y = \boldsymbol{c}^T\boldsymbol{x} \tag{$\boldsymbol{c}$}$$

Thus the two equations that concern us are Eqs. ($\boldsymbol{A}_k\boldsymbol{b}$) and ($\boldsymbol{c}$). These must be solved for $\boldsymbol{x}(t)$ and $y(t)$. The procedure that we use here is based on the Laplace transform. In the next section an alternative time domain solution will be considered. Since we are interested in the response to initial conditions we include the effects of the initial conditions in the Laplace transforms.

The Laplace transform of each side of Eq. ($\boldsymbol{A}_k\boldsymbol{b}$) is

$$\pounds[\dot{\boldsymbol{x}}(t)] = s\boldsymbol{X}(s) - \boldsymbol{x}(0)$$

and

$$\pounds[\boldsymbol{A}_k\boldsymbol{x} + K\boldsymbol{b}r] = \boldsymbol{A}_k\boldsymbol{X}(s) + K\boldsymbol{b}R(s)$$

so that the transformed version of Eq. ($\boldsymbol{A}_k\boldsymbol{b}$) is just

$$s\boldsymbol{X}(s) - \boldsymbol{x}(0) = \boldsymbol{A}_k\boldsymbol{X}(s) + K\boldsymbol{b}R(s)$$

or

$$\boldsymbol{X}(s) = (s\boldsymbol{I} - \boldsymbol{A}_k)^{-1}\boldsymbol{x}(0) + K(s\boldsymbol{I} - \boldsymbol{A}_k)^{-1}\boldsymbol{b}R(s) \tag{4.5-1}$$

where the identity matrix $\boldsymbol{I}$ has been added for dimensional compatibility. In terms of the resolvent matrix, Eq. (4.5-1) becomes

$$\boldsymbol{X}(s) = \boldsymbol{\Phi}_k(s)\boldsymbol{x}(0) + K\boldsymbol{\Phi}_k(s)\boldsymbol{b}R(s) \tag{4.5-2}$$

since

$$\Phi_k(s) = (s\boldsymbol{I} - \boldsymbol{A}_k)^{-1} \tag{4.5-3}$$

The total time response $\boldsymbol{x}(t)$ is therefore

$$\boldsymbol{x}(t) = \pounds^{-1}[\Phi_k(s)]\boldsymbol{x}(0) + K\pounds^{-1}[\Phi_k(s)\boldsymbol{b}R(s)] \tag{4.5-4}$$

Notice that the solution for $\boldsymbol{x}(t)$ is divided into two parts: one associated with initial conditions and one associated with the forcing function.

The apparent simplicity of the solution for $\boldsymbol{x}(t)$, that is, Eq. (4.5-4), is deceiving. To obtain the solution, one must first find the resolvent matrix of the closed-loop system, as discussed in Chap. 3. Then the indicated multiplications must be carried out, which is a simple job but one that produces, in the general case, $2n^2$ terms. The state-variables are then found as functions of time by taking the inverse Laplace transform of these $2n^2$ terms. Once $\boldsymbol{x}(t)$ is known, $y(t)$ is easily found.

As a second-order example, let us consider the system shown in Fig. 4.5-1. This system might be a dc positioning system with tachometer feedback to produce a voltage proportional to the derivative of the output. On the basis of the block diagram, the system equations are written as

$$\dot{x}_1 = x_2$$

$$\dot{x}_2 = -3x_2 + 25\left(r - \frac{3}{25}x_2 - x_1\right)$$

or

$$\dot{x}_1 = x_2$$

$$\dot{x}_2 = -25x_1 - 6x_2 + 25r$$

so that $\boldsymbol{A}_k$ and $K\boldsymbol{b}$ are

$$\boldsymbol{A}_k = \begin{bmatrix} 0 & 1 \\ -25 & -6 \end{bmatrix} \qquad \text{and} \qquad K\boldsymbol{b} = \begin{bmatrix} 0 \\ 25 \end{bmatrix}$$

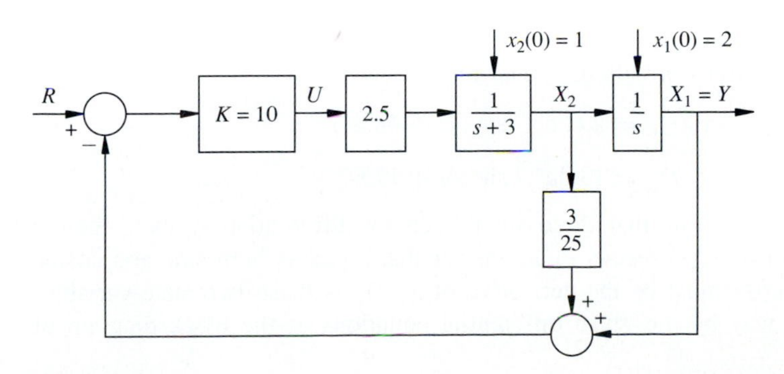

FIGURE 4.5-1
Second-order system.

Also

$$(s\boldsymbol{I} - \boldsymbol{A}_k) = \boldsymbol{\Phi}_k^{-1} = \begin{bmatrix} s & -1 \\ 25 & s+6 \end{bmatrix}$$

The resolvent matrix is therefore

$$\boldsymbol{\Phi}_k(s) = \frac{1}{(s+3)^2 + 4^2} \begin{bmatrix} s+6 & 1 \\ -25 & s \end{bmatrix}$$

Note particularly that the determinant of the system matrix $(s\boldsymbol{I} - \boldsymbol{A}_k)$ appears in the denominator of every element of $\boldsymbol{\Phi}_k(s)$. This determinant is a polynomial in s and, as previously mentioned, is commonly referred to as the characteristic polynomial of the closed-loop system. The *form* of the behavior of each state-variable is the same, since they have the same characteristic equation and hence the same pole locations.

Once $\boldsymbol{\Phi}_k(s)$ is known, we are ready to substitute into Eq. (4.5-2). In expanded form this equation is now

$$\boldsymbol{X}(s) = \frac{1}{(s+3)^2+4^2} \begin{bmatrix} s+6 & 1 \\ -25 & s \end{bmatrix} \begin{bmatrix} x_1(0) \\ x_2(0) \end{bmatrix} + \frac{1}{(s+3)^2+4^2} \begin{bmatrix} s+6 & 1 \\ -25 & s \end{bmatrix} \begin{bmatrix} 0 \\ 25 \end{bmatrix} R(s)$$

For the initial conditions of $\boldsymbol{x}(0) = [2 \quad 1]^T$, $\boldsymbol{X}(s)$ is given by

$$\boldsymbol{X}(s) = \frac{1}{(s+3)^2+4^2} \begin{bmatrix} 2(s+6) + 1 + 25R(s) \\ s - 50 + 25sR(s) \end{bmatrix}$$

$$= \frac{1}{(s+3)^2+4^2} \begin{bmatrix} 2s+13 \\ s-50 \end{bmatrix} + \frac{1}{(s+3)^2+4^2} \begin{bmatrix} 25 \\ 25s \end{bmatrix} R(s)$$

For the moment let us discuss the initial condition response in isolation by assuming that $R(s)$ is zero. In this case the expressions for $X_1(s)$ and $X_2(s)$ reduce to the following, where the subscript *ic* is used to indicate response to initial conditions only.

$$X_1(s)_{\text{ic}} = \frac{2(s+6.5)}{(s+3)^2+4^2}$$

$$X_2(s)_{\text{ic}} = \frac{s-50}{(s+3)^2+4^2}$$

After inverse Laplace transformation, $x_1(t)_{\text{ic}}$ and $x_2(t)_{\text{ic}}$ are

$$\boldsymbol{x}_1(t)_{\text{ic}} = 2.65e^{-3t} \sin(4t + 48.8^\circ)$$

$$\boldsymbol{x}_2(t)_{\text{ic}} = 13.3e^{-3t} \sin(4t + 175.7^\circ)$$

The reader is urged to show that these are related by differentiation, even though direct differentiation of $x_1(t)$ results in an answer that contains both sine and cosine terms. Of course, $x_2(t)$ must be the derivative of $x_1(t)$, as these two state-variables are defined in this way by the given differential equations or the block diagram of Fig. 4.5-1.

In this example the roots of the characteristic equation, that is, the closed-loop poles, are located at $s = -3 \pm j4$, and the damping ratio is $\zeta = 0.6$. Both x_1 and x_2

are plotted as functions of time in Fig. 4.5-2. An alternative state-space representation of the initial-condition response is given in Fig. 4.5-3. Here the coordinates are the state-variables, and time is a parameter along the curve and does not appear explicitly. Common points are labeled in Figs. 4.5-2 and 4.5-3 to aid in correlation of the two plots. Much of modern control theory is oriented toward state-variables and their behavior in state-space. Unfortunately it is difficult to draw a state-space of more than two dimensions.

The forced response is contributed by the last term of Eq. (4.5-4). If $R(s)$ is assumed to be a unit step function, the problem of finding the inverse transform is

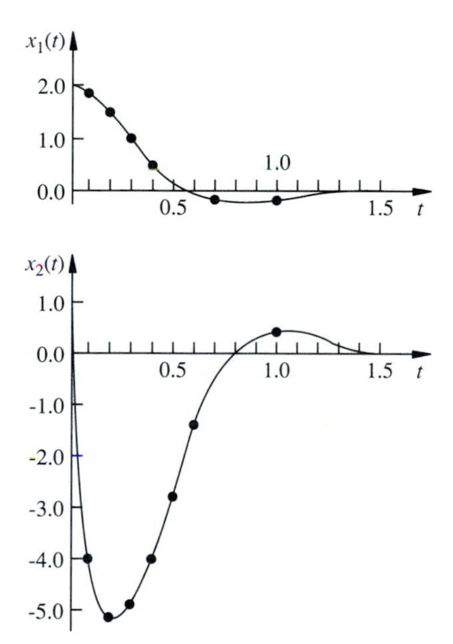

FIGURE 4.5-2
Time response of $x_1(t)$ and $x_2(t)$.

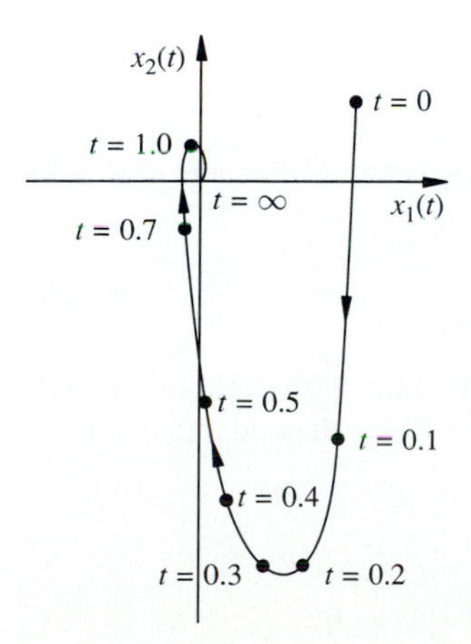

FIGURE 4.5-3
Alternative state-space representation of $x_1(t)$ and $x_2(t)$.

identical to that treated in Sec. 4.2, with the result that

$$x_1(t)_f = 1 - 1.25e^{-3t} \sin(4t + 53.2°)$$

$$x_2(t)_f = 6.25e^{-3t} \sin 4t$$

Here the subscript f indicates that only the forced response is considered. Although in this example we consider the input forcing function to be a unit step, it should be emphasized that the solution given in the frequency domain in Eq. (4.5-4) is valid for any $r(t)$ having a Laplace transform.

Since this system is linear, the principle of superposition applies, and the total response is found by simply adding the initial-condition and forced responses. Thus in this problem the total response is

$$\begin{aligned} x_1(t) &= 1 - 1.25e^{-3t} \sin(4t + 53.2°) + 2.65e^{-3t} \sin(4t + 48°) \\ &= 1 + 1.414e^{-3t} \sin(4t + 45°) \\ x_2(t) &= 6.25e^{-3t} \sin 4t + 13.3e^{-3t} \sin(4t + 175.7°) \\ &= 7.08e^{-3t} \sin(4t + 171.9°) \end{aligned} \tag{4.5-5}$$

Again it is important to emphasize that, even though $\boldsymbol{x}(t)$ may be completely determined in an analytic sense, we still have little idea of the actual behavior of the various state-variables unless we plot them. The preceding example serves to illustrate this point. If in that example we were to plot Eq. (4.5-5), the response curves for any initial conditions other than those specifically assumed at the beginning of the example would be unknown. This is always the case, and for this reason it is very common to assume that initial conditions are zero. This leads to thinking in terms of transfer functions, as a transfer function is defined for zero initial conditions. One danger of this approach is a tendency to concentrate on overall input-output relationships, rather than on the transfer functions that relate the individual state-variables.

In the above development, the response of the individual state-variables was carried out as separate scalar problems of inverse Laplace transforms. In other words, the residues for each $X_i(s)$ were computed separately. An alternative procedure is to determine the inverse transform in matrix notation. Although the latter approach probably does not provide a significant computational advantage, it makes the labor somewhat more systematic.

Let us suppose, for example, that $\boldsymbol{X}(s)$ has been determined from Eq. (4.5-2) and written in the form[1]

$$\boldsymbol{X}(s) = \frac{\boldsymbol{N}(s)}{D(s)} = \frac{\boldsymbol{N}(s)}{(s + s_1)(s + s_2)\cdots(s + S_n)} \tag{4.5-6}$$

Here $\boldsymbol{N}(s)$ is a vector of polynomials in s that are the numerator polynomials of each of the state-variables. The coefficients of the elements of $\boldsymbol{N}(s)$ depend on the $\boldsymbol{x}(0)$

[1] Only simple poles are represented here; the case of multiple poles follows directly.

and $R(s)$ used. In addition, $D(s)$ contains any poles that have been added by the forcing function, if one is present.

Now let us expand $\boldsymbol{X}(s)$ in the usual fashion as

$$\boldsymbol{X}(s) = \frac{\boldsymbol{r}_{11}}{s + s_1} + \frac{\boldsymbol{r}_{21}}{s + s_2} + \cdots + \frac{\boldsymbol{r}_{n1}}{s + s_n} \tag{4.5-7}$$

where the residue vectors are given by

$$\boldsymbol{r}_{i1} = \left. \frac{(s + s_1)\boldsymbol{N}(s)}{D(s)} \right|_{s=s_1} \tag{4.5-8}$$

Using partial-fraction expansion of Eq. (4.5-7), the time response $\boldsymbol{x}(t)$ may be written

$$\boldsymbol{x}(t) = \boldsymbol{r}_{11}e^{-s_1 t} + \boldsymbol{r}_{21}e^{-s_2 t} + \cdots + \boldsymbol{r}_{n1}e^{-s_n t} \tag{4.5-9}$$

The advantage of this procedure is that it groups the evaluation of the residues associated with each pole into one operation.

Example 4.5-1. To illustrate the use of the above procedure, let us determine again the total time response of the system shown in Fig. 4.5-1. Again we assume that $\boldsymbol{x}(0) = [2 \quad 1]^T$ and $R(s) = 1/s$ so that $\boldsymbol{X}(s)$ from Eq. (4.5-2) becomes

$$\boldsymbol{X}(s) = \frac{1}{s\left[(s+3)^2 + 4^2\right]} \begin{bmatrix} 2s^2 + 13s + 25 \\ s^2 - 25s \end{bmatrix} \tag{4.5-10}$$

Rather than treat the complex conjugate poles as separate complex terms, let us expand $\boldsymbol{X}(s)$ in the following form, in the manner of the development of Example 4.2-4:

$$\boldsymbol{X}(s) = \frac{\boldsymbol{r}_{11}}{s} + \frac{\boldsymbol{\alpha} s + \boldsymbol{\beta}}{(s+3)^2 + 4^2} \tag{4.5-11}$$

Here $\boldsymbol{r}_{11}$ is easily found to be

$$\boldsymbol{r}_{11} = \left. \frac{1}{(s+3)^2 + 4^2} \begin{bmatrix} 2s^2 + 13s + 25 \\ s^2 - 25s \end{bmatrix} \right|_{s=0} = \begin{bmatrix} 1 \\ 0 \end{bmatrix}$$

Therefore Eq. (4.5-11) becomes

$$\begin{aligned} \boldsymbol{X}(s) &= \frac{1}{s}\begin{bmatrix} 1 \\ 0 \end{bmatrix} + \frac{\boldsymbol{\alpha} s + \boldsymbol{\beta}}{(s+3)^2 + 4^2} \\ &= \frac{1}{s\left[(s+3)^2 + 4^2\right]} \begin{bmatrix} (1 + \alpha_1)s^2 + (6 + \beta_1)s + 25 \\ \alpha_2 s^2 + \beta_2 s \end{bmatrix} \end{aligned} \tag{4.5-12}$$

If the coefficients of equal powers of s in Eqs. (4.5-10) and (4.5-12) are equated, we find that $\boldsymbol{\alpha} = [1 \quad 7]^T$ and $\boldsymbol{\beta} = [1 \quad -25]^T$ so that the partial-fraction expansion of Eq. (4.5-11) becomes

$$\boldsymbol{X}(s) = \frac{1}{s}\begin{bmatrix} 1 \\ 0 \end{bmatrix} + \frac{1}{(s+3)^2 + 4^2}\left(s\begin{bmatrix} 1 \\ 1 \end{bmatrix} + \begin{bmatrix} 7 \\ -25 \end{bmatrix}\right)$$

Now, by using the short table of transform pairs, we find again that $\boldsymbol{x}(t)$ is

$$\boldsymbol{x}(t) = \begin{bmatrix} 1 \\ 0 \end{bmatrix} + \begin{bmatrix} 1.414e^{-3t}\sin(4t + 45^\circ) \\ 7.07e^{-3t}\sin(4t + 171.9^\circ) \end{bmatrix}$$

Before leaving this section, let us examine one additional way of looking at the problem of finding the total system response. The use of the resolvent matrix is a very systematic approach, and if one is interested in the response for all initial conditions and any input, that is probably the best way to approach the problem. This is particularly true if all the elements of the vector $\boldsymbol{b}$ are nonzero. Often, as mentioned above, initial conditions are assumed to be zero. Also, we have often observed that in cascade-type systems, that is, systems in which the transfer function blocks appear in series, only one element of $\boldsymbol{b}$ is nonzero, the nth element. Under these circumstances it may be easier just to find $x_1(t)$ and then find all the other state-variable time descriptions from $x_1(t)$. Example 4.5-2 illustrates the procedure.

Example 4.5-2. This example is concerned with the determination of the total time response of the system of Fig. 4.5-4. The initial conditions on all the state-variables are zero; the input is a step function at $t = 0$. The output y is just $x_1(t)$, and since $Y(s)/R(s)$ is

$$\frac{Y(s)}{R(s)} = \frac{KG_p(s)}{1 + KG_p(s)H_{eq}(s)}$$

$X_1(s)$ is easily found from a knowledge of $H_{eq}(s)$ and the plant transfer function. Here $H_{eq}(s)$ is

$$H_{eq}(s) = \frac{1}{80}\left(7s^2 + 38s + 80\right)$$

so that $Y(s)/R(s)$ becomes

$$\frac{Y(s)}{R(s)} = \frac{80}{s^3 + 14s^2 + 48s + 80} = \frac{80}{\left[(s+2)^2 + 2^2\right](s+10)}$$

For a step function input, $X_1(s)$ is

$$Y(s) = X_1(s) = \frac{80}{s\left[(s+2)^2 + 2^2\right](s+10)}$$

Therefore $x_1(t)$ is

$$x_1(t) = 1 - \frac{2}{17}e^{-10t} - 1.72e^{-2t}\sin\left(2t + 30.9^\circ\right)$$

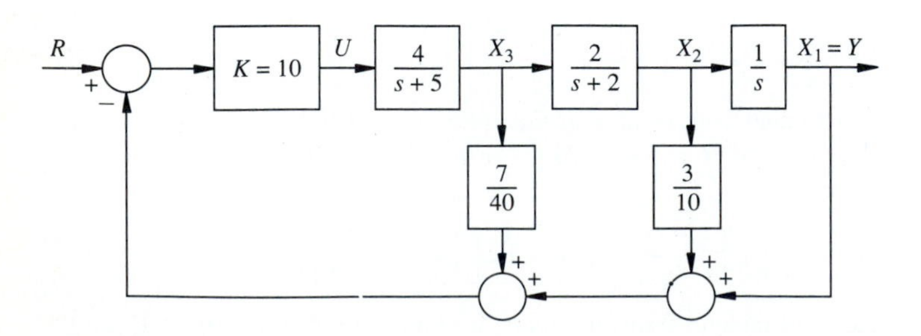

FIGURE 4.5-4
System of Example 4.5-2.

From the block diagram, or alternatively, from the defining differential equations, it is seen that

$$x_2(t) = \dot{x}_1(t)$$

Thus $x_2(t)$ may be found by just differentiating $x_1(t)$. Similarly $x_3(t)$ is just

$$x_3(t) = \frac{1}{2}\left[\frac{dx_2(t)}{dt} + 2x_2(t)\right]$$

and the complete solution can be computed in analytic form.

As yet we have made no mention of why anyone should desire to know the total system response. The answer is that one must be sure that the system under consideration will indeed behave as the analysis predicts. The analysis predicts the behavior of a linear system subject to a particular input. One must be sure that the linear model of the system is valid by verifying that none of the state-variables exceeds its range of linear operation. In addition, it may be necessary to verify that certain physical limitations or specifications of performance have not been violated. To illustrate the point, let us reexamine the system of Fig. 4.5-1. Again we think of this system as a positioning servomechanism with a dc motor for a power element and tachometer feedback. The dc motor is a velocity limited device. If one increases the input voltage to a dc motor, its output velocity increases proportionately until magnetic field saturation occurs. The output velocity then remains essentially constant regardless of the voltage applied.

Assume in the initial discussion of the system of Fig. 4.5-1 that the largest input that could be expected is a unit step input. The problem is worked for a unit step input, and the velocity $x_2(t)$ for this input is found to be

$$x_2(t) = 6.25e^{-3t} \sin 4t$$

Suppose we rework this problem with the same plant but change the forward gain and the feedback coefficient so that $Y(s)/R(s)$ is now

$$\frac{Y(s)}{R(s)} = \frac{100}{(s+6)^2 + 8^2}$$

The new system requires a K of 40 and a feedback coefficient k_2 of 0.09. The pole locations for the closed-loop system are indicated in Fig. 4.5-5, where they are compared with the original closed-loop poles. The damping ratio has been kept the same, although the poles are now located twice as far from the origin of the s plane.

To find the response of both state-variables to a unit step input, let us proceed as in Example 4.5-2. Again $Y(s)/R(s)$ is equal to $X_1(s)/R(s)$, or

$$X_1(s) = \frac{100}{s\left[(s+6)^2 + 8^2\right]} = \frac{1}{s} - \frac{s+12}{(s+6)^2 + 8^2}$$

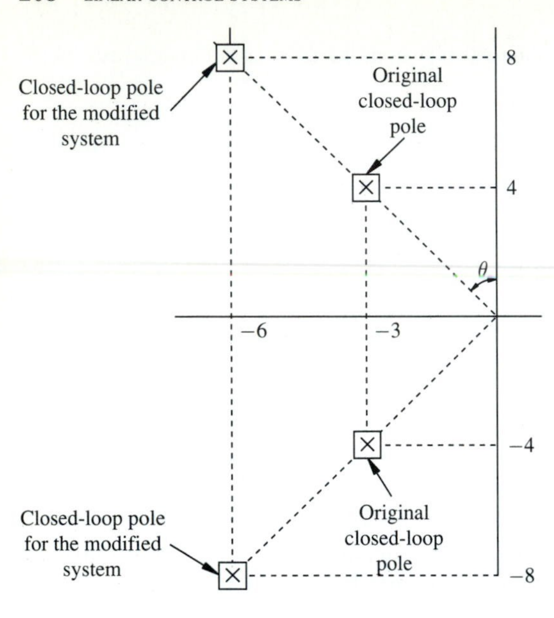

FIGURE 4.5-5
Comparison of closed-loop pole locations.

and $x_1(t)$ is

$$x_1(t) = 1 - \frac{5}{4}e^{-6t}\sin(8t + 53.2^\circ)$$

Here the time constant is shorter, and the frequency of oscillation is higher. The system responds faster and hence might be considered better.

The velocity is easily found, since

$$x_2(t) = \frac{dx_1(t)}{dt}$$

then

$$x_2(t) = 12.5e^{-6t}\sin 8t$$

The maximum velocity is twice the maximum velocity found previously. Assume that the system had originally been designed so that a unit step input required the maximum effort from the power element; that is, the step input drove the motor at its maximum velocity. Then, by altering the gain and feedback coefficient, it only *appears* that the resulting system is faster and better. For the same unit step input, the velocity of the modified system saturates, and we have no idea whether this system is better or worse than the previous one, because the linear model is no longer adequate. For steps of magnitude 1/2, the system is clearly faster, but it was initially postulated that the steps of unit magnitude could be expected.

As yet we have discussed only the analysis of given control system configurations. Ultimately we shall be concerned with synthesis; however, the last step of any synthesis procedure is always one of analysis.

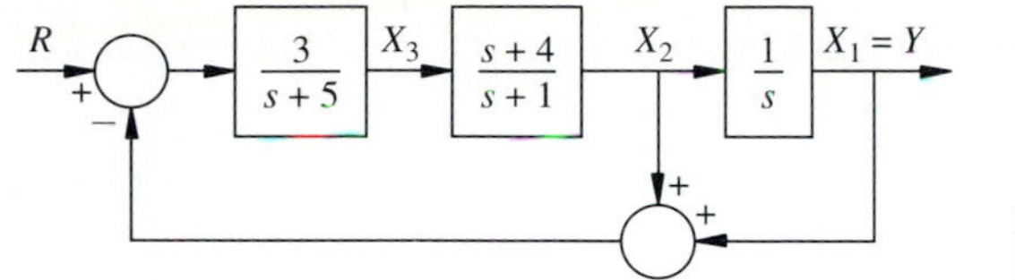

FIGURE 4.5-6
Exercise 4.5-2.

Exercises 4.5

4.5-1. Consider the system described by the two first-order differential equations

$$\dot{x}_1 = x_2$$

$$\dot{x}_2 = -3x_1 - 4x_2$$

Find the response of this system to the initial conditions $x_1(0) = 1$, $x_2(0) = 1$.
Answer:

$$x(t) = \begin{bmatrix} -e^{-3t} + 2e^{-t} \\ 3e^{-3t} - 2e^{-t} \end{bmatrix}$$

4.5-2. Consider the block diagram shown in Fig. 4.5-6. Determine the resolvent matrix for this closed-loop system and find $x_1(t)$ and $x_2(t)$ if $r(t)$ is a unit step and $\boldsymbol{x}(0) = 0$.
Answer:

$$x_1(t) = 1 - 1.8e^{-t} + 0.75e^{-2t} + 0.05e^{-6t}$$

$$x_2(t) = 1.8e^{-t} - 1.5e^{-2t} + 0.3e^{-6t}$$

4.5-3. Repeat Exercise 4.5-2 but use the closed-loop transfer function $Y(s)/R(s)$ to find $X_1(s) = Y(s)$. From $x_1(t)$ find $x_2(t)$ by using the state-variable equations.

4.6 TIME-DOMAIN METHODS

The object of this section is to present an alternative time-domain technique for determining the total system response $\boldsymbol{x}(t)$. The method makes use of the inverse Laplace transform of the resolvent matrix, which is referred to as the *state transition matrix* $\phi_k(t)$, and requires the use of the convolution integral. The evaluation of the convolution integral proves to be difficult, and there is no shortcut approach analogous to partial-fraction expansion techniques. Hence this section also serves to justify our almost exclusive preference for Laplace transform methods as the most convenient mathematical medium for the discussion of analysis and synthesis procedures in automatic control. The next chapter on frequency response will serve as further justification for our concentration on s plane methods.

As an introduction, let us return to the solution of Eq. $(\boldsymbol{A}_k\boldsymbol{b})$, given as Eq. (4.5-4), which is repeated.

$$\boldsymbol{x}(t) = \pounds^{-1}[\boldsymbol{\Phi}_k(s)]\boldsymbol{x}(0) + K\pounds^{-1}[\boldsymbol{\Phi}_k(s)\boldsymbol{b}R(s)] \tag{4.6-1}$$

To find the contribution to $x(t)$ due to the forcing function $r(t)$, it is necessary to find the inverse Laplace transform

$$\boldsymbol{x}_f(t) = K\pounds^{-1}\left[\boldsymbol{\Phi}_{\boldsymbol{k}}(s)\boldsymbol{b}R(s)\right] \tag{4.6-2}$$

This inverse Laplace transformation involves the inverse of the product of two Laplace transforms, $R(s)$ and $\boldsymbol{\Phi}_{\boldsymbol{k}}(s)$. We know that $x_f(t)$ requires the convolution of time signals since multiplication in the Laplace transform domain corresponds to convolution in the time domain.

The state transition matrix $\phi_{\boldsymbol{k}}(t)$ of the closed-loop system is defined as the inverse Laplace transform of the resolvent matrix, or

$$\phi_{\boldsymbol{k}}(t) = \pounds^{-1}\left[\boldsymbol{\Phi}_{\boldsymbol{k}}(s)\right]$$

Then we may take the inverse transform required by Eq. (4.6-1) by replacing the multiplication in the s domain by the convolution integral. The result, still in matrix notation, is then

$$\boldsymbol{x}(t) = \phi_{\boldsymbol{k}}(t)\boldsymbol{x}(0) + K\int_0^t \phi_{\boldsymbol{k}}(t-\tau)\boldsymbol{b}r(\tau)d\tau \tag{4.6-3}$$

In expanded form the equation for the ith state-variable is just

$$\begin{aligned} x_i(t) &= x_1(0)\phi_{\boldsymbol{k}(i1)}(t) + x_2(0)\phi_{\boldsymbol{k}(i2)}(t) + \cdots + x_n(0)\phi_{\boldsymbol{k}(in)}(t) \\ &\quad + K\int_0^t \phi_{\boldsymbol{k}(i1)}(t-\tau)b_1 r(\tau)d\tau + \cdots + K\int_0^t \phi_{\boldsymbol{k}(in)}(t-\tau)b_n r(\tau)d\tau \end{aligned}$$

The convolution integrals here are the time-domain equivalents of the expression noted in Eq. (4.6-2). It is evident that in general one must evaluate n^2 convolution integrals to find the total system response, just as one must take $2n^2$ inverse Laplace transforms in Eq. (4.6-1). The very practical questions then arises: Which is easier, finding inverse transforms or evaluating convolution integrals?

To answer this question, let us take a particular case in which a transfer function, $W(s)$, is second-order, or

$$W(s) = \frac{1}{(s+\alpha)^2 + \beta^2}$$

or

$$w(t) = \frac{1}{\beta}e^{-\alpha t}\sin\beta t$$

and $w(t-\tau)$ is

$$w(t-\tau) = \frac{1}{\beta}e^{-\alpha(t-\tau)}\sin\beta(t-\tau)$$

For zero initial conditions and an arbitrary input $r(t)$, the output is

$$y(t) = \int_0^t r(\tau)\left[\frac{1}{\beta}e^{-\alpha(t-\tau)}\sin\beta(t-\tau)\right]d\tau$$

Only in the case when $r(t)$ is a step function or an exponential is this integral easy to evaluate. In that case evaluation is done by looking up the integral in a set of standard integral tables. If $r(t)$ were allowed to be t, or a sinusoidal driving force, the usual tables are no longer adequate. However, the alternative approach through partial-fraction expansion of the transformed quantity would be no more difficult than problems already solved in this chapter.

In defense of the time domain approach through the use of convolution, it should be mentioned that graphical and digital computer methods are available to evaluate the convolution integral. These computer methods are particularly suited to cases in which the input or weighting functions are not known analytically. Of course, before these methods may be used, it is necessary to know the weighting function. If an analytical description of the system is known, then $w(t)$ may be determined analytically. In addition, $w(t)$ may be determined experimentally in various ways even if a mathematical model for the system is not completely known.

The most obvious method of determining $w(t)$ is to apply an impulse to the system and measure the output. Of course, it is impossible to apply an exact impulse, but a pulse of very short duration is often an acceptable substitute. A second method for finding $w(t)$ is to differentiate, usually numerically, the step response of the system. The step response can normally be obtained directly and without difficulty. The weighting function can also be obtained by measuring the response of the system to sinusoidal inputs at several different frequencies. This frequency response approach will be discussed in detail in the next chapter.

Exercise 4.6

4.6-1. Consider Exercise 4.5-1 with the addition of a forcing term $r(t)$. The system equations become

$$\dot{x}_1 = x_2$$

$$\dot{x}_2 = -3x_1 - 4x_2 + 2r(t)$$

If the initial conditions are again $x_1(0) = 1$ and $x_2(0) = 1$, use time domain methods to find $x_1(t)$ and $x_2(t)$ if $r(t)$ equals a step function, $u(t)$

Answer:

$$x_1(t) = \frac{2}{3} - \frac{2}{3}e^{-3t} + e^{-t}$$

$$x_2(t) = 2e^{-3t} - e^{-t}$$

4.7 STEADY-STATE ERRORS TO SIMPLE INPUTS

We have spent some time examining the response of systems to unit step inputs. Step responses are important because a linear time-invariant system can be completely characterized by its unit step response. Also, the response of a control system to a step input is of concern in its own right. Of particular interest is the steady-state error in response to a step in either the reference input or a disturbance. If you command a telescope positioning system to point in a new direction, you are clearly interested in how closely the final position of the telescope matches the commanded position. You are also interested in resetting the desired position after the onset of a constant disturbance. One of the major advantages of feedback control systems is the ability to track reference inputs precisely and to maintain this precision in the face of disturbances. In addition, precision tracking can be achieved with very low sensitivity to changes or modeling errors in the plant.

The tool we use to investigate the steady-state operation is the final value theorem of Laplace transforms. The theorem states that if a signal $e(t)$ tends toward a constant value, the steady-state value can be computed by the equation

$$e_{ss} = \lim_{t\to\infty} e(t) = \lim_{s\to 0} sE(s) \tag{4.7-1}$$

We will deal with the G configuration repeated in Fig. 4.7-1. Remember that $G(s)$ is the product of the plant and the series compensator transfer functions. Notice that in the G configuration, the error between the reference input, R, and the output, Y, appears as a signal, E. It is easy to write the relationship between the error and the reference input or disturbance.

$$E(s) = \frac{1}{1+G(s)}(R(s) - D(s)) \tag{4.7-2}$$

Except for a sign change, the error (not the plant output) responds identically to reference inputs and disturbances. If the error is driven to zero in response to a reference input to the plant, then the plant output is driven to match the reference input. If the error is driven to zero in response to a disturbance, then the effect of the disturbance is removed from the plant output. We will use the response of the error to the reference input to demonstrate the points, but you should remember that a system that follows certain signals well when they are input as reference inputs also eliminates those same signals well when they arrive as disturbances. From Eq. (4.7-1) we can see that the basic method of keeping the error small is to make $G(s)$ large. Next, we will formalize this concept in the context of a few simple inputs. In the remainer of this section we assume that $G(s)$ has no zeros at the origin so that there are no pole-zero cancellations at the origin.

4.7.1 Step Inputs

Let $r(t)$ be a step of height A so that

$$R(s) = \frac{A}{s} \tag{4.7-3}$$

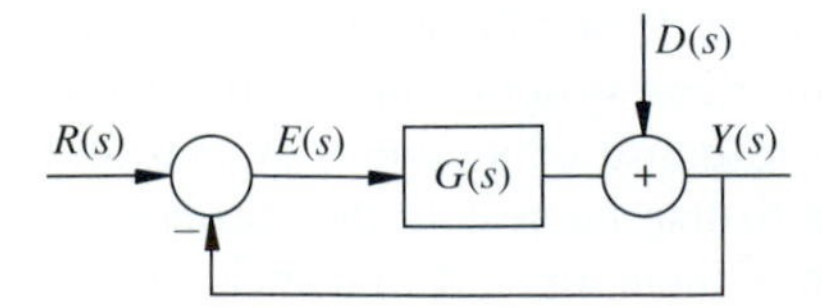

FIGURE 4.7-1
The G configuration.

Then

$$e_{ss} = \lim_{s\to 0} s \frac{1}{1+G(s)} \frac{A}{s} = \frac{A}{1 + \lim_{s\to 0} G(s)} \tag{4.7-4}$$

Clearly, we would like the limit of $G(s)$ as s goes to zero to be as large as possible. Indeed, if $G(s)$ has one or more poles at $s = 0$ then $G(s)$ goes to infinity as s goes to zero and there will be a zero steady-state error in response to a step reference input or a constant disturbance. Notice that as long as $G(s)$ has one or more poles at the origin and the closed-loop system is stable, the zero steady-state error property remains in force despite large modeling errors or changes in the plant.

The number of integrators, i.e., the number of poles at the origin of the loop gain transfer function, defines the system's *type number*. For a type zero system, there is no integrator and the quantity $G(0)$ is finite. A constant called the *position error constant*, K_p, is used to indicate the size of $G(0)$.

$$K_p = G(0) \tag{4.7-5}$$

The constant K_p has significance only in reference to step inputs. A step input has the physical interpretation as a change in reference position for a positioning system and thus the name position error constant. For a type zero system the steady-state error in response to a step input of height A is

$$e_{ss} = \frac{A}{1+G(0)} = \frac{A}{1+K_p} \tag{4.7-6}$$

Note that a large position error constant corresponds to a small steady-state error.

The error constant K_p can be expressed as the steady-state output, y_{ss}, divided by the steady-state error, e_{ss}.

$$\frac{y_{ss}}{e_{ss}} = \frac{r_{ss} - e_{ss}}{e_{ss}} = \frac{A - \dfrac{A}{1+K_p}}{\dfrac{A}{1+K_p}} = K_p \tag{4.7-7}$$

Remember that for a system of type one or higher the steady-state error for step inputs is zero. In this case, K_p is said to be infinite.

4.7.2 Ramp Input

Now consider the steady-state error in response to a ramp input

$$R(s) = \frac{A}{s^2} \tag{4.7-8}$$

$$e_{ss} = \lim_{s\to 0} s \frac{1}{1+G(s)} \frac{1}{s^2} = \frac{A}{\lim_{s\to 0} sG(s)}$$

Since ramp input corresponds to a change in velocity in a position control system, we call the error constant associated with ramp inputs the *velocity error constant*, K_v.

$$K_v = \lim_{s\to 0} sG(s) \tag{4.7-9}$$

If $G(s)$ is a type zero system with no poles at the origin then K_v is zero and the steady-state error is infinite, that is, the error grows with time. If you have a type zero system and you wish to be able to follow a ramp within some degree of accuracy, you must add at least one integrator to your controller.

If $G(s)$ is a type two or higher system the steady-state error for a ramp input or disturbance is zero and K_v can be said to be infinite. A type one system will have a finite, non zero K_v and the steady-state error is given by

$$e_{ss} = \frac{A}{K_v} \tag{4.7-10}$$

For a type one system the steady-state error to ramp input is constant. This situation is demonstrated in Fig. 4.7-2. The derivative of the output $y(t)$ approaches the derivative of the input $r(t)$.

$$\left.\frac{dy(t)}{dt}\right|_{ss} = \left.\frac{dr(t)}{dt}\right|_{ss} = A \tag{4.7-11}$$

Using this relationship, K_v can be expressed as

$$K_v = \frac{A}{e_{ss}} = \frac{\left.\frac{dy(t)}{dt}\right|_{ss}}{e_{ss}} \tag{4.7-12}$$

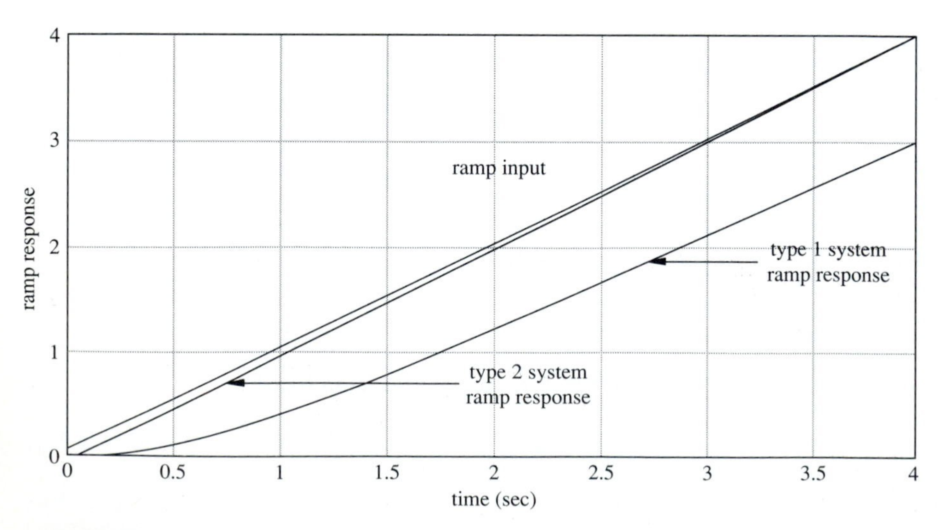

FIGURE 4.7-2
Ramp responses of type 1 and type 2 systems.

Notice again that a large K_v indicates a small steady-state error. Also, notice that K_v is defined only with respect to a ramp input.

4.7.3 Parabolic Input

Let $R(s)$ be given as

$$R(s) = \frac{A}{s^3} \tag{4.7-13}$$

This is a parabolic input in the time domain. The associated error constant, K_a, is called the *acceleration error constant* in reference to a positioning system.

For a parabolic input

$$e_{ss} = \lim_{s \to 0} s \frac{1}{1 + G(s)} \frac{A}{s^3} = \frac{A}{\lim_{s \to 0} s^2 G(s)} \tag{4.7-14}$$

The acceleration error constant is given by

$$K_a = \lim_{s \to 0} s^2 G(s) \tag{4.7-15}$$

Type zero and type one systems have $K_a = 0$ and an infinitely growing error to a parabolic input. A type three or higher system has an infinite K_a and zero steady-state error. A type two system has a finite K_a and a finite steady-state error with a parabolic point.

The acceleration error constant can be expressed as

$$K_a = \frac{\left. \dfrac{dy^2}{dt^2} \right|_{ss}}{e_{ss}} \tag{4.7-16}$$

4.7.4 Summary

The table summarizes the results of this section for steady-state errors.

Input	**Step** A/s	**Ramp** A/s^2	**Parabola** A/s^3
Error Constant	$K_p = \lim_{s \to 0} G(s)$	$K_v = \lim_{s \to 0} sG(s)$	$K_a = \lim_{s \to 0} s^2 G(s)$
System type		**Steady-state error**	
0	$\dfrac{A}{1+K_p}$	∞	∞
1	0	$\dfrac{A}{K_v}$	∞
2	0	0	$\dfrac{A}{K_a}$

We have seen that steady-state errors due to low frequency inputs and disturbances such as steps, ramps, and parabolas can be eliminated by adding integrators to the loop gain transfer function. Such integral action raises the low frequency loop gain and can accomplish the steady-state error reduction even in the face of large modeling errors in the plant

Exercises 4.7

4.7-1. Integrators are often added to a control system to reject constant disturbances in steady-state operation. For the system given in Fig. 4.7-1, find the position error constant, and find the steady-state error when the input is zero, the disturbance is a step of height 2 and $G(s)$ is given as

$(a)\ G(s) = \dfrac{10(s+1)}{(s+2)(s+5)}$

$(b)\ G(s) = \dfrac{10(s+1)}{s(s+2)(s+5)}$

Answers: $(a)\quad K_p = 1 \quad e_{ss} = -1 \qquad (b)\quad K_p = \infty \quad e_{ss} = 0$

4.7-2. The effect of integral action occurs even in the face of gross modeling errors as long as the system remains stable. In the control system of Fig. 4.7-1, assume that $G(s)$ consists of a plant $G_p(s)$ in series with a compensator $G_C(s)$.

$$G(s) = G_C(s)G_p(s)$$

Let $G_C(s)$ include an integrator

$$G_C(s) = \frac{1}{s}G'_C(s)$$

(*a*) Show that if the closed-loop system is stable, the response to a step input results in zero steady-state error no matter what $G_p(s)$ and $G'_C(s)$ are.

(*b*) What property should $G_C(s)$ have if a ramp input is to be followed with zero steady-state error whenever the closed-loop system is stable?

Answer:

$$(b)\quad G_C(s) = \frac{G'_C(s)}{s^2},\ G_C(s) \text{ is a type 2 system.}$$

4.7-3. (*a*) Given that $G(s) = \frac{1}{s+1}$ find the response of the closed-loop system of Fig. 4.7-1 to a unit ramp input.

(*b*) Find the value of $y(t)$, $t = 1$, 10, and 100.

(*c*) Find the steady-state error.

(*d*) Sketch the input and output.

Answers:

$(a)\ y(t) = \left(\frac{1}{2}t - \frac{1}{4} + \frac{1}{4}e^{-2t}\right)\boldsymbol{u}(t)$

$(b)\ y(1) = 0.28,\quad y(10) = 4.75,\quad y(100) = 49.75$

$(c)\ e_{ss} = \infty$

4.8 CONCLUSIONS

This chapter is the first one that has been concerned with analysis, and in particular, with the determination of the time response of a closed-loop control system. The assumptions made at the beginning of the chapter are worth repeating. It is assumed that the system is completely specified. That is, not only is the plant assumed to be known, but the gains and feedback coefficients are assumed to be specified as well.

Actually, the understanding of the relationship between a system's pole-zero configuration and its step response completes the basic information needed to use the pole placement technique of control system design. The objective of this design technique is to achieve a specified system transient response, usually given as an envelope of acceptable step response characteristics. Using the knowledge obtained by the general analysis of this chapter a target set of desired closed-loop poles can be hypothesized. (All poles should be chosen to lie in the left half-plane for stability.) The pole placement techniques developed in Chap. 3 can be used to find the controller which produces the desired closed-loop pole positions. If all goes well the resulting response possesses acceptable characteristics. Note, however, that this design technique provides no systematic way of ensuring the other major benefits of control that are discussed in Sec. 3.2. Problems 4.11 and 4.12 explore pole placement designs.

As we proceeded through this chapter it was pointed out that it is often difficult to obtain the time response for any input or initial conditions exactly. In many cases we might be quite content with much less information, assuming that less information would be easier to obtain. That is the point of view we shall adopt in the next two chapters. In fact, in the next chapter, one section begins with the assumption that the plant to be controlled exists physically but not as a transfer function written on a piece of paper. Before control can be attempted, the plant must first be identified. Once the plant is known, its stability must be examined before the system can be given any inputs or initial conditions. This is commonly done in the frequency domain, the topic of the following chapter.

PROBLEMS

4.1. The pole-zero plot of a given transfer function $Y(s)/R(s)$ is shown in Fig. P4.1. If the input $r(t)$ is an impulse, the output $y(t)$ is of the form

$$y(t) = A + Be^{-\alpha t} + Ce^{-\gamma t}\cos(\omega t + \phi)$$

Let $A = 1$. What are the values of B, C, α, γ, φ and ω?

4.2.

$$Y(s) = \frac{4(s+8)}{s\left[(s+4)^2+16\right]} = \frac{1}{s} + \frac{B}{s+4-j4} + \frac{\overline{B}}{s+4+j4}$$

Find B (magnitude and phase) graphically.

4.3. If a closed-loop transfer function is

$$M_C(s) = \frac{\frac{15}{1.1}(s+1.1)}{(s+1)\left[(s+1)^2+2^2\right]}$$

find the unit step response of the closed-loop system. Tell how this response compares with the unit step response associated with

$$M_C(s) = \frac{15}{\left[(s+1)^2+2^2\right]}$$

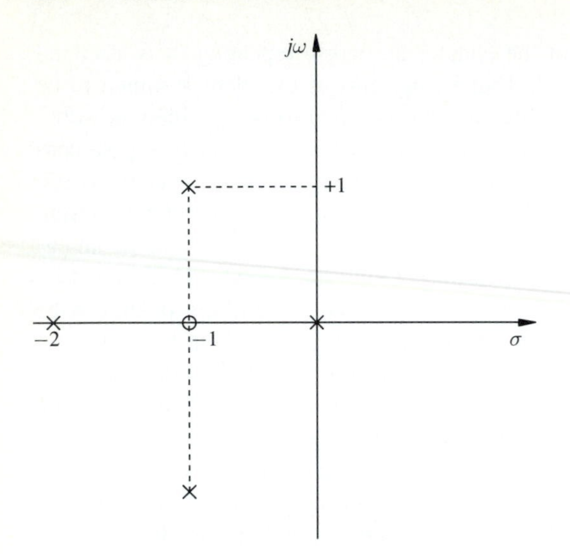

FIGURE P4.1
Problem 4.1.

4.4. Answer the following questions with a short (one line) explanation as well as the answer. The questions refer to the pole-zero plots of system transfer functions in Fig. P4.4. For this problem, $\omega_1 > \omega_0$.

(*a*) The step response of which system, (*a*) or (*e*), has a shorter settling time?

(*b*) The step response of which system, (*a*) or (*f*), has a greater overshoot?

(*c*) The step response of which system, (*a*) or (*b*), arrives at its maximum overshoot sooner?

(*d*) The step response of which system, (*a*) or (*d*), is more oscillatory?

(*e*) What characteristic of the step response of the system (*g*) is different from all the other systems, (*a*) through (*f*)?

4.5. (*a*) For the two systems below, which one will have the greater percent overshoot and why?

$$(i) \quad G_a(s) = \frac{1}{s^2 + 1.4s + 1}$$

$$(ii) \quad G_b(s) = \frac{4}{s^2 + 4s + 4}$$

(*b*) For the two systems below, which one will have a longer setting time and why?

$$(i) \quad G_a(s) = \frac{(s+5)(2)}{(s+10)\left(s^2 + 1.4s + 1\right)}$$

$$(ii) \quad G_b(s) = \frac{1}{s^2 + 1.4s + 1}$$

(*c*) If

$$G(s) = \frac{\frac{10}{z}(s+z)}{(s+1)(s+10)}$$

what can you say about z given that the step response overshoots 1?

(*d*) With $G(s)$ as in (*c*) what can you say about z given that the step response starts off negative?

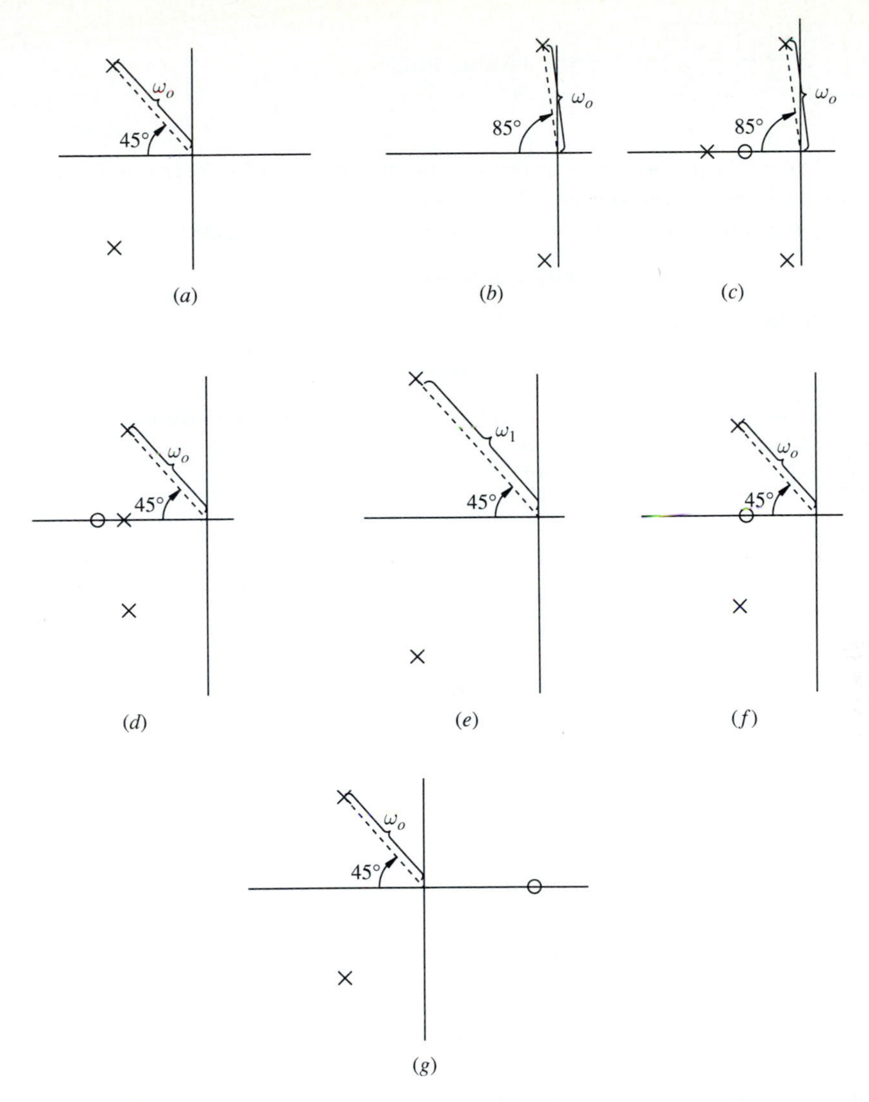

FIGURE P4.4
Pole-zero plots for Problem 4.4.

4.6. The resolvent matrix $\mathbf{\Phi}(s)$ for a given plant is

$$\mathbf{\Phi}(s) = \begin{bmatrix} 1/s & 1/s(s+3) & 10/[s(s+3)(s+10)] \\ 0 & 1/(s+3) & 10/[(s+3)(s+10)] \\ 0 & 0 & 1/(s+10) \end{bmatrix}$$

and $$\boldsymbol{b} = [0 \quad 0 \quad 5]^T$$

Find $x_2(t)$ if $u(t)$ is a step input and $\boldsymbol{x}(0) = [1 \quad 2 \quad 3]^T$.

4.7. Find the total time response of the plant shown in Fig. P4.7 if $U(s) = 1/s$ and $\boldsymbol{x}(0) = 0$.
(*a*) Use the resolvent matrix to find the answer.

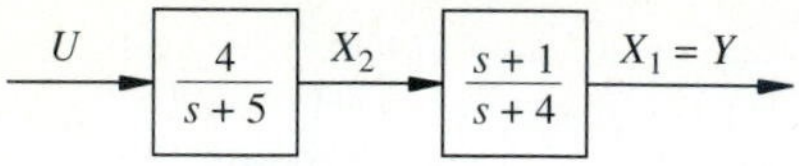

FIGURE P4.7
Problem 4.7.

(*b*) Find the answer by using the plant transfer function $G_p(s)$ to find $Y(s) = X_1(s)$ and then applying the state-variable relations to find $x_2(t)$.

4.8. If the input to a plant is zero, the total time response may be represented as

$$\boldsymbol{x}(t) = \phi(t)\boldsymbol{x}(0)$$

For $t = 0$, this result becomes

$$\boldsymbol{x}(0) = \phi(0)\boldsymbol{x}(0)$$

and we see that $\phi(0)$ must equal the identity matrix $\boldsymbol{I}$ if this equation is satisfied for every $\boldsymbol{x}(0)$. This fact provides a simple check on the validity of $\phi(t)$. In addition, by using the initial-condition theorem, we may write

$$\phi(t)|_{t=0} = \lim_{s\to 0}[s\boldsymbol{\Phi}(s)] = \boldsymbol{I}$$

and use this expression as a check on $\boldsymbol{\Phi}(s)$. Show that the above expressions are satisfied for $\phi(t)$ and $\boldsymbol{\Phi}(s)$ associated with the plant

$$\dot{\boldsymbol{x}} = \begin{bmatrix} 0 & 1 \\ -2 & -3 \end{bmatrix}\boldsymbol{x} + \begin{bmatrix} 0 \\ 1 \end{bmatrix}u$$

4.9. Given that when there is no input, the following relations hold for all $\boldsymbol{x}(0)$

$$\boldsymbol{x}(t) = \phi(t)\boldsymbol{x}(0)$$

$$\dot{\boldsymbol{x}}(t) = \boldsymbol{A}\boldsymbol{x}(t)$$

show that the state transition matrix satisfies the homogeneous plant equation, i.e.,

$$\dot{\phi}(t) = A\phi(t)$$

Verify this expression for the $\phi(t)$ found in Prob. 4.8.

4.10. A very important electrical circuit in many applications is called a *phase locked loop*. This circuit enables one subsystem to lock on to the timing of another subsystem. It is important in computer systems and other systems where digital signals are transmitted across a printed circuit board, across a backplane, or across a country. A receiving subsystem needs to know at what time in a waveform it needs to sample the waveform and decode the information. A phase locked loop can be analyzed essentially as a feedback control circuit.

The components of a phase locked loop are displayed schematically in Fig. P4.10*a*. An incoming clock signal may be a square wave of a certain frequency and phase offset. The job of the phase locked loop is to create an identical square wave of the same frequency and phase offset as the incoming clock signal. The phase discriminator is an electronic device that produces a voltage proportional to the differences in phase of its two input signals. The voltage-controlled oscillator is an electronic device that produces a local clock signal (a square wave) with a frequency (the instantaneous derivative of its phase) proportional to its input voltage. The combination of the voltage-controlled oscillator and the phase discriminator takes an input voltage that commands the frequency of the voltage controlled oscillator and produces an output voltage that measures the phase difference between the incoming signal and the local clock signal. Because of the change from a command in frequency to an output of phase, the combination can be modeled as an integrator and a subtracter as shown in Fig. P4.10*b*. Let the filter be represented by $F(s)$. This must be designed. *Please note:* The symbol ϕ in this problem bears no relationship to the ϕ in Problems 4.8 and 4.9. It is coincidence that the same symbol happens to be traditionally used in these two different settings.

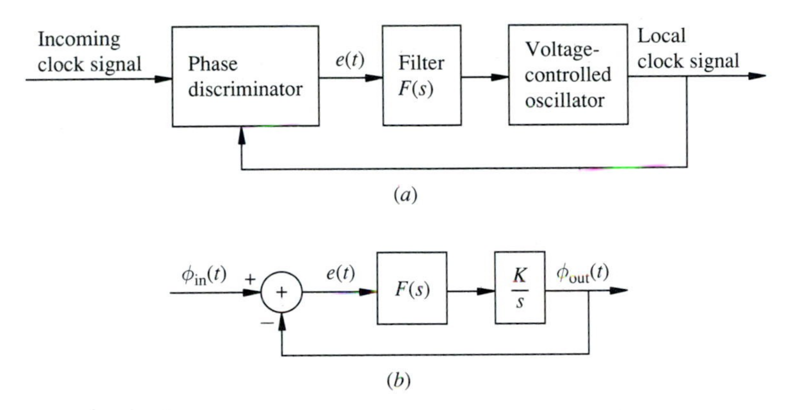

FIGURE P4.10
Problem 4.10 on phase locked loops. (*a*) Schematic diagram; (*b*) block diagram.

(*a*) The major objective of a phase locked loop is to have the error between the phases of the local and incoming clock go to zero in the steady-state when there is a step change in the frequency of the incoming clock. (This corresponds to a ramp in the phase of the incoming clock.) What property must $F(s)$ have to achieve this objective?

(*b*) Let $K = 2$. Use a transfer function pole placement design to achieve a closed-loop transfer function so that $\phi_{\text{out}}(t)$ can follow step change in $\phi_{\text{in}}(t)$ with a response comparable to a dominantly first-order system with a time constant of 0.2 sec?

(*c*) Find the response of the system designed in (*b*) to a unit ramp input in the incoming phase. What is the steady-state error? Does the $F(s)$ designed in (*b*) meet the requirement given in (*a*)?

(*d*) Let

$$F(s) = \frac{n_1 s + n_0}{s}$$

Now try to pick n_1 and n_0 so that the system meets the requirements of both (*a*) and (*b*).

(*e*) This problem can be solved in a more systematic fashion. You know that the factor $1/s$ must be part of the controller. Start by putting the $1/s$ factor in your controller. However, to proceed, consider this factor as part of the plant. Now, design a pole placement transfer function controller to meet the specifications given in (*b*). This systematic process is often referred to as augmenting the plant dynamics with integrators. The only problem with this process is that you often obtain a higher-order controller than is necessary to meet the design criteria. Did you find that true in this case?

4.11. (*CAD Problem*) In this problem a controller is designed to smooth the phugoid mode of the Boeing 747 jumbo jet introduced in Sec. 2.A. The state equation for the longitudinal dynamics of the plane are given in Eq. (2.A-9), which is repeated here.

$$\frac{d}{dt}\begin{bmatrix} \Delta V \\ \Delta\alpha \\ q \\ \Delta\theta \end{bmatrix} = \begin{bmatrix} -0.0188 & 11.5959 & 0 & -32.2 \\ -0.0007 & -0.5357 & 1 & 0 \\ 0.000048 & -0.4944 & -0.4935 & 0 \\ 0 & 0 & 1 & 0 \end{bmatrix}\begin{bmatrix} \Delta V \\ \Delta\alpha \\ q \\ \Delta\theta \end{bmatrix}$$
$$+\begin{bmatrix} 0 & 1 \\ 0 & 0 \\ -0.5632 & 0 \\ 0 & 0 \end{bmatrix}\begin{bmatrix} \Delta\delta_e \\ \Delta\delta_T \end{bmatrix} \tag{2.A-9}$$

The elevator command $\Delta\delta_e$ is used as the control input, so that $\boldsymbol{b} = [0 \quad 0 \quad -0.5632 \quad 0]^T$ and the pitch angle $\Delta\theta$ is used as the plant output, so that $\boldsymbol{c}^T = [0 \quad 0 \quad 0 \quad 1]$. Let $\Theta(s)$ and $\Delta_e(s)$ be the transforms of $\Delta\theta(t)$ and $\Delta\delta_e(t)$, respectively.

(*a*) Find the transfer function,

$$G_p(s) = \frac{\Theta(s)}{\Delta_e(s)}$$

Find the zero-pole representation of this transfer function. From the zero-pole plot use the ideas of this chapter to describe the step response of this transfer function. Explain the reasoning in your description. Use a CAD tool to get the step response. (You may run into an example of numerical problems using CAD tools in this problem. If you use `ss2tf` to get the transfer function and follow it directly with `tf2zp` to get the zero-pole representation, you will get some very strange zeros and an apparently miscalculated gain. The problem is that the numerator polynomial is computed as a third-order polynomial with an extremely small (10^{-16}) leading coefficient. This coefficient should be set to zero before proceeding.)

(*b*) Design a realizable series compensator in the G configuration using polynomial pole placement. The only measurable state is the pitch angle $\Delta\theta$ and the control variable is the elevator deflection $\Delta\delta_e$. Select a dominant pole pair that, if it were the only dynamics, would result in a step response that has a 5 percent overshoot and time to the first peak of 22 sec. Place all the other poles about a decade further out from the origin than the dominant poles. Find the poles and zeros of the resulting closed-loop system. Predict the step response of the closed-loop system from the poles and zeros. Use a CAD tool to plot the actual closed-loop step response. Is this step response near the desired step response of the dominant poles? If it is not, why not?

(*c*) The reason that undesirable results may have been attained in (*b*) is that pole placement algorithms do nothing about the closed-loop zeros. The closed-loop zeros in the G configuration come from the union of the plant zeros and the controller zeros. In this design method there is nothing we can do about the controller zeros since they are determined by selecting the closed-loop poles and are unknown until the algorithm calculates the controller zeros. However, we can eliminate the effect of the plant zeros from the closed-loop transfer function since these zeros are known before we start the design and they can be canceled by some of the closed-loop poles we select.

Design another realizable polynomial pole placement controller in the G configuration. Use the same dominant poles as in (*b*), but this time choose two closed-loop poles to cancel the two plant zeros. Choose the remainder of the closed-loop poles about a decade further out from the origin than the dominant poles. Find the poles and zeros of the resulting closed-loop system. Predict the step response of the closed-loop system from the poles and zeros. Use a CAD package to plot the actual closed loop step response. Is this response better than that of (*b*)? Why or why not?

4.12. (*CAD Problem*) This problem requires having solved Problem 3.12 and Problem 4.11. In this problem we examine how the controller designed in Problem 4.11 smoothes out a climbing maneuver in the Boeing 747 jet.

Find a state-variable representation for the controller which was designed in Problem 4.11(*c*). The controller should have the form

$$\dot{\mathbf{x}} = \mathbf{A}_C \mathbf{x}_C - \mathbf{b}_C \Delta\theta$$

$$\Delta\delta_e = \mathbf{c}_C^T \mathbf{x}_C - d_C \Delta\theta$$

Consider this extended form of the state-variable equations for the jet

$$\frac{d}{dt}\begin{bmatrix} \Delta V \\ \Delta\alpha \\ q \\ \Delta\theta \\ \Delta h/10V_o \end{bmatrix} = \begin{bmatrix} -0.0188 & 11.5959 & 0 & -32.2 & 0 \\ -0.0007 & -0.5357 & 1 & 0 & 0 \\ 0.000048 & -0.4944 & -0.4935 & 0 & 0 \\ 0 & 0 & 1 & 0 & 0 \\ 0 & -0.1 & 0 & 0.1 & 0 \end{bmatrix}\begin{bmatrix} \Delta V \\ \Delta\alpha \\ q \\ \Delta\theta \\ \Delta h/10V_o \end{bmatrix} + \begin{bmatrix} 0 & 1 \\ 0 & 0 \\ -0.5632 & 0 \\ 0 & 0 \\ 0 & 0 \end{bmatrix}\begin{bmatrix} \Delta\delta_e \\ \Delta\delta_T \end{bmatrix}$$

Combine the equation for the extended plant with the equation for the controller to arrive at a singe extended state-variable equation with $\Delta\delta_T$ as the input and $\Delta h/10V_o$ as the output. Compare the response to a step change in throttle position to the response of the open-loop plant as given in Fig. 2.A-4. Explain the differences and the reasons for the differences.

Plot $\Delta\theta(t)$, the input to the controller and $\Delta\delta_e$, the output of the controller, on the same plot so that you can see the relationship between the two. Might a simple controller where $\Delta\delta_e$ is proportional to $\Delta\theta(t)$ work acceptably on this system?

CHAPTER
5

FREQUENCY RESPONSE

5.1 INTRODUCTION

This is the second chapter concerned with analysis and is intended to complement and extend the results of Chap. 4. The reader may be somewhat surprised that any further discussion of analysis is necessary, as we have given the complete solution for any initial conditions and forcing functions for either an open-loop or a closed-loop system. Thus the complete time response for all the state-variables may be calculated, and the behavior of the system for all values of time is known.

The complete solution for a single-input, single-output closed-loop system was given as Eq. (4.5-7) and is repeated here for reference.

$$\boldsymbol{x}(t) = \phi_k(t)\boldsymbol{x}(0) + K\int_0^t \phi_k(t-\tau)\boldsymbol{b}r(\tau)\,d\tau \tag{5.1-1}$$

Equation (5.1-1) is a matrix equation, and although the time solution is completely indicated, the actual determination of the solution is tedious. Even if this equation is solved so that the response $\boldsymbol{x}(t)$ is known analytically, to plot $\boldsymbol{x}(t)$ versus time it is necessary to proceed in a point-by-point fashion. This is even more work.

Although Eq. (5.1-1) is a general solution for any input, the discussion in Chap. 4 was restricted mainly to step inputs. Another important class of inputs that was not mentioned in the preceding chapter is the sinusoidal function. This chapter is devoted entirely to a discussion of the response to sinusoidal inputs. The discussion is in terms of a general transfer function, $W(s)$. Particular transfer functions that are of interest are the transfer functions of the plant, $G_p(s)$, the closed-loop system, $Y(s)/R(s)$, and the internal transfer functions relating the various state-variables with the input $U(s)$ and with themselves. That is, we are interested in $X_i(s)/U(s)$ and $X_i(s)/X_j(s)$, $i \neq j$. The response of the transfer functions to a sinusoidal input is particularly important if no plant model is known and one is seeking to determine a plant model by measurement of its transfer function.

Section 5.2 is devoted to a discussion of the frequency response function $W(j\omega)$. Of particular interest are its magnitude and phase angle. Section 5.3 discusses methods of sketching the magnitude of this function. Specifically, it is important to develop methods of plotting the frequency response function that are less tedious than the point-by-point method required to plot $\boldsymbol{x}(t)$. Section 5.3 uses straight line approximations to make the plot of the magnitude of the frequency response function almost trivial to accomplish. The plots that are developed plot the logarithm of the magnitude versus the logarithm of the frequency and the phase versus the logarithm of the frequency. Such plots are called *Bode plots* after a pioneer of control theory, Hendrik Bode.

Section 5.4 is a discussion of the phase plot; in particular, a method which makes it possible to determine the phase angle directly from the magnitude plot for certain systems is presented. A straight line approximation for the Bode phase plot is developed.

Sections 5.3 and 5.4 fall into the category of "how to do it" sections; that is, they explain the mechanics of handling the frequency response function but give little indication as to why the frequency response function is important. Its importance is emphasized in Sec. 5.5 on plant identification. The assumption of the preceding sections is that the transfer function is known, and we are asked to represent the frequency function associated with the given transfer function. In Sec. 5.5 the inverse problem is discussed; that is, we have a plot of a frequency response function, and it is desired to find the transfer function. This problem occurs in practice when a plant whose transfer function is unknown is to be controlled. As a first step in the design, the plant must be identified; that is, its transfer function must be determined so that methods of control can be calculated.

5.2 FREQUENCY RESPONSE FUNCTION

Let us consider the time response of a rather general linear system driven by the sinusoidal time function $r(t) = A\sin\omega t$, as pictured in Fig. 5.2-1. The output is $y(t)$, and the transfer function of the system is $W(s)$, where $W(s)$ is assumed to be a ratio of polynomials in s. The output $y(t)$ contains, among other things, a sinusoidal component of the same frequency as the input of the form $AR(\omega)\sin[\omega t + \phi(\omega)]$. Here $R(\omega)$ is the ratio of the magnitude of the sinusoidal component of the output to the input, and the phase angle $\phi(\omega)$ is the phase difference between the input and the output. Both R and ϕ have been designated as functions of ω since, in general, they

FIGURE 5.2-1
General linear system with a sinusoidal input.

vary as the input frequency is varied. The object of this section is to show that

$$R(\omega) = |W(s)|\big|_{s=j\omega} = |W(j\omega)| \tag{5.2-1a}$$

$$\phi(\omega) = \text{phase angle of } W(s)\big|_{s=j\omega} = \arg\, W(j\omega) \tag{5.2-1b}$$

This means that the magnitude and phase of the output for a sinusoidal input may be found by simply determining the magnitude and phase of $W(j\omega)$. The complex function $W(j\omega)$ is referred to as the *frequency response function.*

To demonstrate the truth of Eq. (5.2-1), let us examine the Laplace transform of $y(t)$ for the indicated input. Since the Laplace transform of $A \sin \omega t$ is $A\omega/\left(s^2+\omega^2\right)$, $Y(s)$ is

$$Y(s) = W(s)R(s) = \frac{A\omega W(s)}{s^2+\omega^2} = \frac{A\omega W(s)}{(s-j\omega)(s+j\omega)} \tag{5.2-2}$$

If a partial-fraction expansion of $Y(s)$ is made, using any of the approaches of Sec. 4.2, we obtain

$$Y(s) = \underbrace{\frac{R_{11}}{s-j\omega} + \frac{R_{21}}{s+j\omega}}_{\substack{\text{Particular solution} \\ \text{(steady-state)}}} + \underbrace{\text{other terms}}_{\substack{\text{Complementary} \\ \text{solution (transient)}}} \tag{5.2-3}$$

The "other terms" arise from the poles of $W(s)$. In mathematical terminology, the first two terms on the right side of Eq. (5.2-3) are due to the sinusoidal forcing function and are therefore related to the particular solution; the "other terms" comprise the unforced or complementary solution.

One often refers to the particular solution as the steady-state solution and the complementary solution as the transient solution. Although these designations are often appropriate, there are situations in control system applications where such labeling is inappropriate. In some plants, for example, which are referred to as unstable, it is possible for the complementary solution to grow without bound as time increases. In such cases the use of the term transient response to describe the complementary solution is meaningless.

For the majority of systems, however, the complementary solution is, in fact, transient in nature and the particular solution becomes the usual steady-state response. In these cases, the frequency response function may be given a particularly useful physical meaning; we shall discuss this interpretation later.

Independent of whether the complementary solution is transient or not, our present interest is only in the particular solution. We therefore ignore the complementary portion of the solution. To find the sinusoidal portion of $y(t)$, which we designate as $y_s(t)$, it is necessary to find R_{11} and R_{21}. This is easily done by the method of residues, so that

$$R_{11} = \left.\frac{A\omega W(s)}{s+j\omega}\right|_{s=+j\omega} = \frac{AW(+j\omega)}{2j}$$

$$R_{21} = \left.\frac{+A\omega W(s)}{s-j\omega}\right|_{s=-j\omega} = \frac{-AW(-j\omega)}{2j}$$

The two residues are, of course, complex conjugates of each other. Letting $R(\omega)$ and $\phi(\omega)$ be defined as in Eq. (5.2-1), R_{11} can be represented in polar form.

$$R_{11} = \frac{AR(\omega)}{2} e^{j(\phi(\omega)-90^\circ)} \tag{5.2-4}$$

Taking the inverse Laplace transform, $y_s(t)$ is found.

$$y_s(t) = AR(\omega)\cos\left[\omega t + \phi(\omega) - 90^\circ\right] = AR(\omega)\sin\left[\omega t + \phi(\omega)\right] \tag{5.2-5}$$

Hence we have established that the frequency response function may be obtained from the transfer function by letting $s = j\omega$.

This is an important result. Given the transfer function, the frequency response function follows immediately. On the other hand, if the frequency response function is known, as the result of experiment, for instance, then the transfer function may be found by replacing $j\omega$ by s. That is, the frequency response function and the transfer function are directly related to one another. If one is known, the other is also known. Of course, if the transfer function is known, the response to any input can be found; hence we may make the following statement: *The frequency response function of a linear time-invariant system uniquely determines the time response of the system to any known input.*

Let's examine the validity of the converse of this statement. In Chap. 4 one method discussed for finding the weighting function of a system was the use of an impulse input. Then the Laplace transform of the weighting function is the transfer function of the system in question. An alternative approach is to use a step input and differentiate the resulting output to find the impulse response. For the system of Fig. 5.2-1, $w(t)$, the weighting function, is

$$w(t) = \pounds^{-1}\left[\frac{Y(s)}{R(s)}\right]$$

and this is true regardless of the input. It is often convenient to use a step and an impulse as test inputs to determine $w(t)$. With $w(t)$ known, $W(s)$ may be found and hence the frequency response function $W(j\omega)$ is found. On the basis of these observations, we note the following property: *The frequency response function of a linear time-invariant system is uniquely determined by its time response to an impulse or step input.*

The combination of these two statements establishes the complete interdependence of the time and frequency response methods. The importance of this fact is that it allows one to translate time domain specifications, such as rise time and overshoot, into frequency domain specifications. One may often, therefore, carry out a complete design in the frequency domain where the procedures are usually simpler and more systematic.

There was some discussion of the interdependence of the time and frequency domains in Chap. 4 with regard to second-order systems in particular. In addition, the reader is undoubtedly already familiar with the relationship of such properties as bandwidth, a frequency domain concept, and rise time, a time domain characteristic.

In this and the following chapters, this interplay of the time and frequency domains will be further developed to give the reader an appreciation for the usefulness of this duality of the time and frequency domains.

At the present, however, our interest is simply in the meaningful representation of the frequency response information. In particular, we are interested in graphical means of representation, since this allows us to examine the frequency response data most easily. Initially, the amplitude and phase characteristics are plotted separately; in the next chapter, the representation of both items on a single polar plot with frequency as a parameter will be considered.

To introduce the concept of graphical representation of the frequency response information, let us consider the simple electrical network shown in Fig. 5.2-2. In discussing this example, we illustrate the problems associated with a usual graphical representation.

The transfer function for this network is easily found to be

$$\frac{Y(s)}{U(s)} = G_p(s) = \frac{1}{1 + R_1C_1s}$$

and in this simple example $W(s)$ is a plant transfer function, $G_p(s)$. The frequency response function for this network may be obtained by setting $s = j\omega$ so that

$$G_p(j\omega) = \frac{1}{1 + jR_1C_1\omega}$$

The magnitude and phase of the frequency response function are therefore

$$R(\omega) = |G_p(j\omega)| = \frac{1}{\sqrt{1 + (\omega R_1C_1)^2}}$$

$$\phi(\omega) = \arg G_p(j\omega) = -\tan^{-1}\omega R_1C_1$$

A plot of these two functions versus frequency is shown in Fig. 5.2-3.

In addition to the asymptotic behavior of the plots for very low and very high frequencies, the value of the magnitude and phase functions can be determined *by inspection* at only one finite value of frequency, namely, $\omega = 1/R_1C_1$. Therefore, to obtain reasonably accurate plots, it is necessary to substitute various values of ω and carry out the computations in detail. Even though in this case the computations are not difficult and usually five to ten appropriately selected points are sufficient, the effort involved is not trivial.

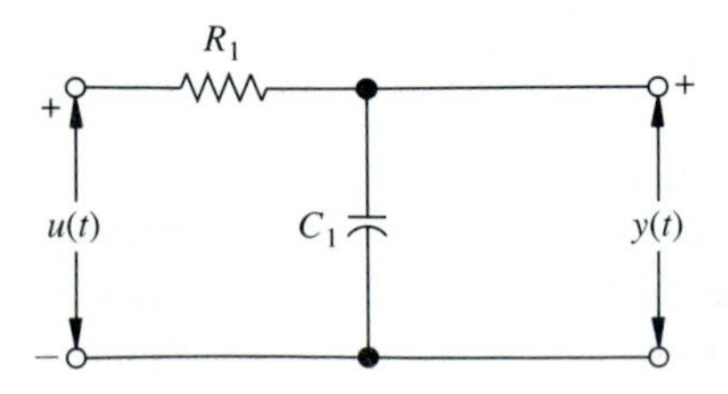

FIGURE 5.2-2
Simple electrical network.

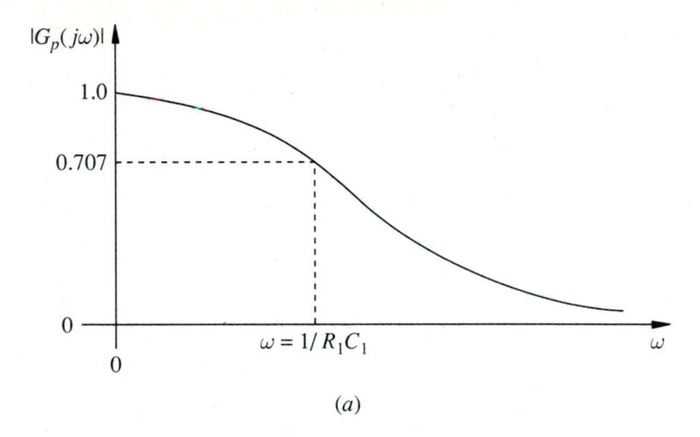

(a)

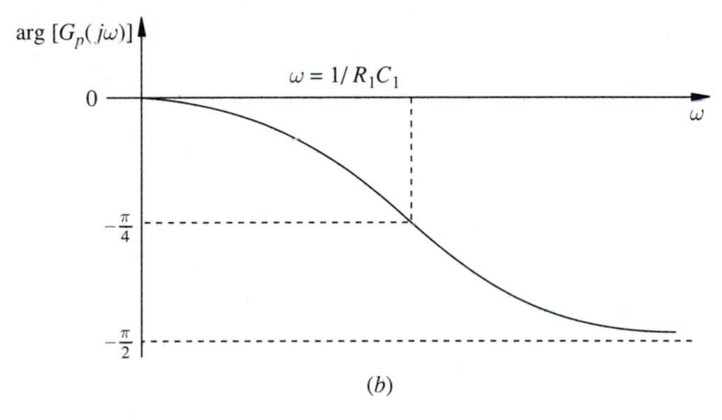

(b)

FIGURE 5.2-3
Frequency response function for the network of Fig. 5.2-2. (a) Magnitude; (b) phase.

The situation becomes much worse when systems are cascaded. Consider, for example, the problem of plotting the magnitude and phase of the frequency response function for the network shown in Fig. 5.2-4. This network is formed by cascading a second RC network with the RC network shown in Fig. 5.2-2.

If we assume that the impedance levels of the two networks are such that no loading occurs, for example if R_2 is very large or an impedance matching buffer network not shown in Fig. 5.2-2 is placed between the two networks, the transfer

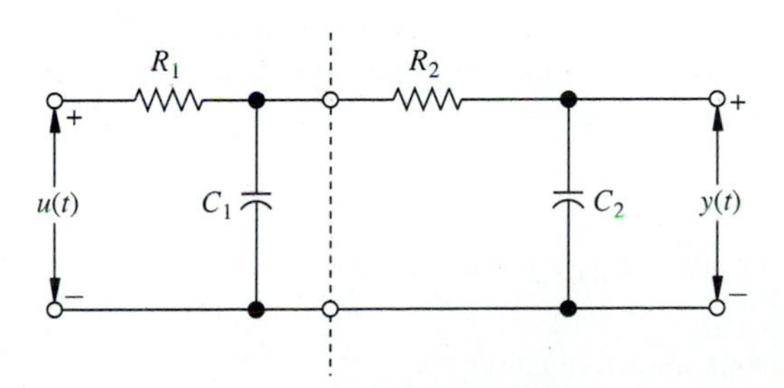

FIGURE 5.2-4
Cascade network.

function of the cascade network is

$$G_p(s) = \frac{1}{(1 + sR_1C_1)(1 + sR_2C_2)}$$

and the frequency response function is therefore

$$G_p(j\omega) = \frac{1}{(1 + j\omega R_1C_1)(1 + j\omega R_2C_2)}$$

Even with the no-loading assumption to simplify the task, plotting the magnitude and phase of $G_p(s)$ is not easy. Although the asymptotic character of the two plots may still be easily determined, there is no longer even a single frequency besides $\omega = 0$ for which the value of the magnitude or phase may be determined by inspection. In addition, since the character of the cascade frequency response function may not be as obvious as its simple components, more points may be necessary to achieve an accurate representation.

Even if the magnitude plots of the two networks were known exactly, the job of finding the cascade plot would not be trivial, since point-by-point multiplication would be necessary. Since the cascading of various component blocks is commonplace in control system design and analysis, the above direct method is obviously inadequate. The approach is cumbersome for handling simple networks, and its inability to facilitate cascading of component systems makes the technique completely unacceptable.

Hence we are led to seek alternative procedures for plotting the magnitude and phase of the frequency response function. Such a procedure should have one or more of the following characteristics. First, the procedure should allow a reasonably accurate plot to be made quickly. Second, composite systems should be handled with relative ease. Third, it should be possible to make the plots as accurate as needed in a given frequency range or ranges by a reasonably small amount of additional work.

Although it may appear that achieving the above requirements is too much to hope for, in the following sections we shall develop procedures that more or less meet all these requirements.

Of course, there are now generally available a variety of computer programs for plotting the magnitude and phase of frequency responses from transfer function data. The availability of these programs does not eliminate the need to master the techniques of approximating these plots. The simple techniques developed here allow a designer to see how certain elements of a transfer function affect a system's frequency response. This permits insight into how the frequency response can be altered by the addition of designed elements in a system's transfer function. This insight is required to perform design work. In the next section we begin by considering the representation of the magnitude plot by means of a straight line approximation.

Exercises 5.2

5.2-1. Given $G(s) = \frac{6}{s+2}$ find the steady-state response to $u(t) = 4\sin 4t$.
Answer:

$$y_{ss}(t) = 5.36\sin(4t - 63.4^\circ)$$

5.2-2. Given $G(s) = \frac{6}{s+2}$ find the complete response to $u(t) = 4\sin 4t$ when the initial conditions are zero.

Answer:

$$y(t) = 5.36\sin(4t - 63.4^\circ) + 4.8e^{-2t}$$

5.3 BODE MAGNITUDE PLOT—STRAIGHT LINE APPROXIMATION

In the preceding section we showed that the ratio of the magnitude of the sinusoidal component of the output to the magnitude of the sinusoidal input is simply the magnitude of the frequency response function, or $|W(j\omega)|$. In addition, the phase angle of the sinusoidal component of the output relative to the phase angle of the sinusoidal input is just the argument or angle of $W(j\omega)$. Thus for any particular frequency it is possible to plot $y_s(t)$, since the amplitude, frequency, and phase angle of the response are known. However, because the plot of sinusoidal functions is so familiar, this is usually not done; rather, the amplitude and phase angle of $W(j\omega)$ are plotted for all frequencies. These curves supply all the information necessary to draw $y_s(t)$ for any frequency if such a plot is ever required.

In this chapter the amplitude and phase diagrams, known as the Bode plots, are considered separately; the polar plot, or Nyquist plot, is introduced in this chapter and discussed more fully in the following chapter. In this section we concentrate on the methods by which the magnitude of the frequency response function may be sketched with a minimum of difficulty. The phase plot will be discussed in Sec. 5.4. The procedure developed in this section involves the approximation of the magnitude plot as a sequence of straight lines. To develop the method, we consider a rather simple example transfer function of the form

$$W(s) = \frac{K(s + \omega_1)}{s(s + \omega_2)} \tag{5.3-1}$$

This specific example is used to derive the procedure, since it illustrates the principal features of the approach without extensive notational problems. The extension of the approach to the general situation is relatively direct. In addition, several numerical examples are used to demonstrate further the application of the technique.

At the outset, notice that $W(s)$ in Eq. (5.3-1) has been factored and written in pole-zero form. That is, the coefficient of s in each factor is 1, as the poles and zeros are each written as $s + \omega_i$. Let us rewrite Eq. (5.3-1) in an alternative form, so that the constant term in each factor is 1. This is referred to as the *time constant form*, since each of the quantities $1/\omega_1$ and $1/\omega_2$ that multiply the s terms is the time constant of the first-order real pole. Although it is not necessary to write $W(s)$ in this form, the results obtained below are more convenient if this time constant form is used. In time constant form Eq. (5.3-1) becomes

$$W(s) = \frac{K\omega_1}{\omega_2}\frac{1 + s/\omega_1}{s(1 + s/\omega_2)}$$

The frequency response function in time constant form is then

$$W(j\omega) = \frac{K\omega_1}{\omega_2} \frac{1 + j\omega/\omega_1}{j\omega\,(1 + j\omega/\omega_2)}$$

The magnitude of the frequency response function is

$$|W(j\omega)| = \left|\frac{K\omega_1}{\omega_2}\right| \frac{|1 + j\omega/\omega_1|}{|j\omega||1 + j\omega/\omega_2|} \tag{5.3-2}$$

using the fact that the magnitude of a product of complex numbers is the product of the magnitudes.

Evaluating $|W(j\omega)|$ for various values of ω is difficult because we must multiply (or divide) the various components of the magnitude expressions of Eq. (5.3-2) for each value of ω. Note that the same problem occurs when system components are cascaded. Since multiplication is one of the problems, let us consider the logarithm of $|W(j\omega)|$ so that the multiplication involved is replaced by the simpler job of adding logarithms. If this were the only advantage gained by this change, it would hardly be worthwhile. However, several other advantageous features are also associated with this logarithmic approach.

If we now take the logarithm of the magnitude of $W(j\omega)$, we obtain

$$\log|W(j\omega)| = \log\left|\frac{K\omega_1}{\omega_2}\right| + \log\left|1 + \frac{j\omega}{\omega_1}\right| - \log|j\omega| - \log\left|1 + \frac{j\omega}{\omega_2}\right| \tag{5.3-3}$$

Let us examine in detail one term of Eq. (5.3-3), the term $\log|1 + j\omega/\omega_1|$. In particular, let us examine the asymptotic behavior for both large and small ω. For small ω, the magnitude of the quantity $j\omega/\omega_1$ is small relative to 1, and since addition of these terms resembles vector addition at right angles, a reasonable approximation is that

$$\log\left|1 + \frac{j\omega}{\omega_1}\right| \approx \log 1 = 0 \text{ for } \omega \ll \omega_1 \tag{5.3-4}$$

Rather than plot this quantity against ω, as done in Sec. 5.2, we assume that the quantity is plotted with $\log\omega$ as the horizontal variable rather than just ω. The slope of the asymptotic approximation expressed in Eq. (5.3-4) is

$$\frac{d(0)}{d(\log\omega)} = 0 \tag{5.3-5}$$

Similarly for large ω, 1 is small compared with $j\omega/\omega_1$, and again the components add analagous to vectors at right angles so that

$$\log\left|1 + \frac{j\omega}{\omega_1}\right| \approx \log\left|\frac{j\omega}{\omega_1}\right| \quad \text{for } \omega \gg \omega_1$$

Since we are concerned only with positive values of frequency, and multiplication by j doesn't affect the magnitude, this result becomes

$$\log\left|1 + \frac{j\omega}{\omega_1}\right| \approx \log\frac{\omega}{|\omega_1|} = \log\omega - \log|\omega_1| \text{ for } \omega \gg \omega_1 \tag{5.3-6}$$

The slope of this asymptotic approximation for large ω when plotted against $\log \omega$ is

$$\frac{d}{d(\log \omega)}(\log \omega - \log |\omega_1|) = +1$$

This constant slope is the main advantage of using $\log \omega$ rather than ω as the x coordinate on Bode plots. Constant slopes correspond of course to straight lines on the plot.

The exact logarithmic plot of $|1 + j\omega/\omega_1|$ is plotted versus $\log \omega$ as the dashed line in Fig. 5.3-1. The straight line asymptotic approximations for large and small ω are also indicated in that figure. Note that the two straight line approximations that form the high and low frequency asymptotes for the exact plot intersect at $\omega = |\omega_1|$. For this reason, ω_1 is often referred to as a *break frequency*, since it is the point at which the corner, or break, of the straight line approximation occurs. From Fig. 5.3-1, we observe that, whenever $\omega < |\omega_1|/10$ or $\omega > 10|\omega_1|$, the exact and approximate plots are almost identical.

For most control applications, the straight line asymptotic approximations are often of sufficient accuracy. However, if a more accurate approximation is needed, an exact calculation for one or two points near the break point can be made. Note, in particular, that for $\omega = |\omega_1|$ the exact value of $|1 + j\omega/\omega_1|$ is $\sqrt{2}$. If correction of the straight line approximation is necessary, using this point is often adequate.

In practical applications, it is often more convenient to plot the actual values of $|1 + j\omega/\omega_1|$ and ω on log-log paper, as in Fig. 5.3-2a, rather than the logarithms of these quantities on linear paper, as in Fig. 5.3-1. In this way, one may work directly with the quantities of interest and still gain the advantages of the logarithmic plot. Notice that $\omega = 3$ is about halfway between $\omega = 1$ and $\omega = 10$ because $\log 1 = 0$, $\log 10 = 1$ and $\log 3 \approx 0.5$.

Note that in this case the reference for the vertical scale becomes the $|1 + j\omega/\omega_1| = 1$ line, and the low frequency asymptote is coincident with this line. The slope of the high frequency asymptote is indicated here by the value +1 on the

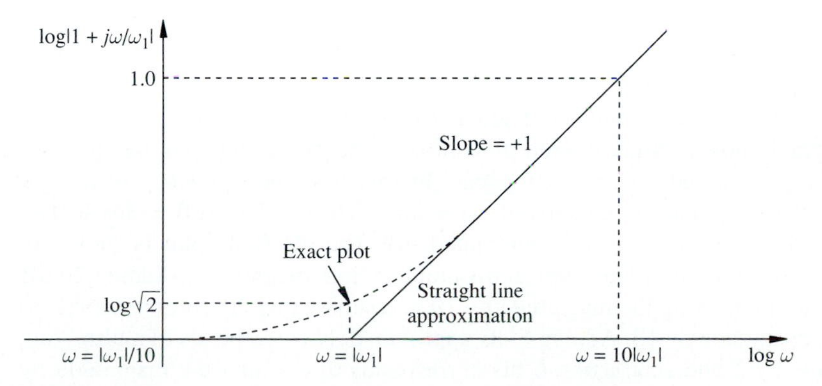

FIGURE 5.3-1
Bode plot of $|1 + j\omega/\omega_1|$.

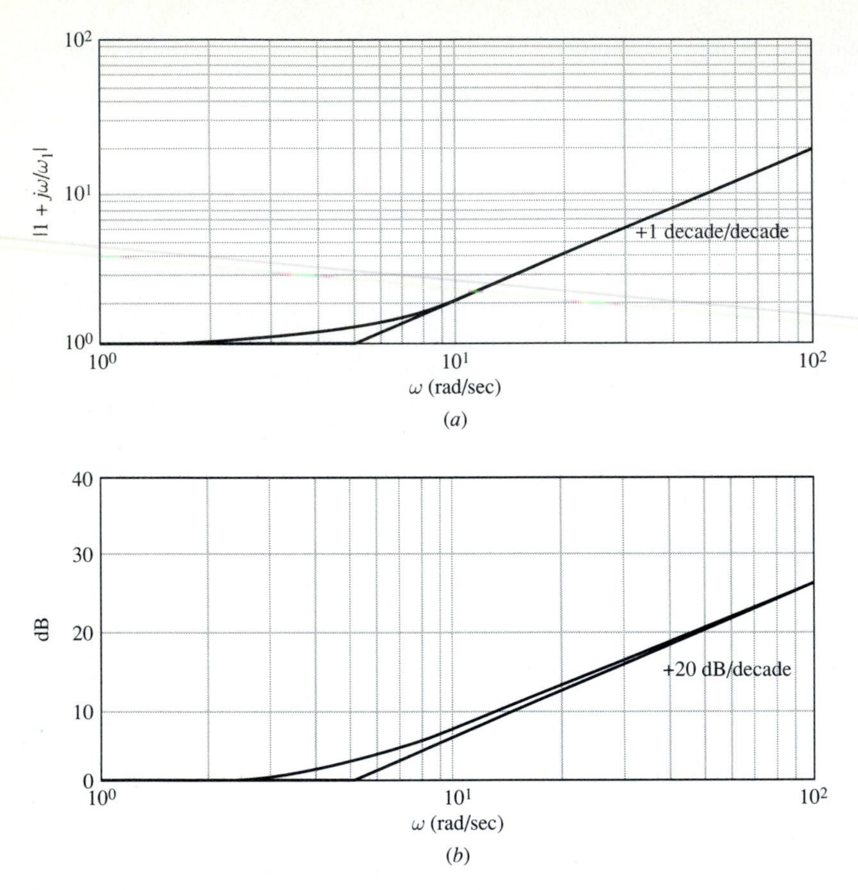

FIGURE 5.3-2
Bode magnitude plot of $|1+j\omega/\omega_1|$. (*a*) Log-log plot of $|1+j\omega/\omega_1|$; (*b*) dB versus $\log \omega$ plot of $|1+j\omega/\omega_1|$.

high frequency asymptote. On log-log paper, a slope of $\pm k$ implies that the magnitude changes by $\pm k$ decades as the frequency changes by one decade. For example, in Fig. 5.3-2*a* the magnitude of $1 + j\omega/\omega_1$ changes from 1 to 10 as ω increases from 5 to 50, thereby indicating a $+1$ slope.

There is a third commonly used graph for displaying Bode magnitude plots. This is the graph most commonly used by control engineers. In this plot the quantity $20 \log |W(j\omega)|$ is plotted on semi-log paper. In this way the frequency ω is read directly but the magnitude is thought of on a log scale relative to $0 = \log 1$. The units on the ordinate are decibels, abbreviated dB. The dB Bode plot is shown in Fig. 5.3-2*b*. Table 5.3-1 translates typical dB numbers into magnitudes. Adding 20 dB corresponds to multiplying the magnitude by 10 and subtracting 20 dB corresponds to dividing the magnitude by 10. Adding 6 dB approximately corresponds to multiplying the magnitude by 2 and subtracting 6 dB corresponds to dividing the magnitude by 2. Using the table one can see that the level 36 dB equals 20 dB + 10 dB + 6 dB and corresponds to a magnitude of $10 \times 3 \times 2 = 60$. A slope of $+1$ on the log-log plot

TABLE 5.3-1
Values of dB and magnitudes

$\mathrm{dB} = 20 \log \lvert W(j\omega) \rvert$	$\lvert W(j\omega) \rvert$
-20	0.1
-10	$0.316 \approx 0.1/3$
-6	0.5
0	1
6	2
10	$3.16 \approx 3$
20	10

corresponds to the effect of one zero and also to a slope of +20 dB per decade on the dB versus $\log \omega$ plot.

If we examine the entire expression for $\log |W(j\omega)|$ as given by Eq. (5.3-3), we see that the last term is identical in form to the term $\log |1 + j\omega/\omega_1|$ except that it is preceded by a minus sign. Hence, at its break frequency of ω_2, the plot of this term breaks downward with a slope of -20 dB per decade. Otherwise, it is the same as the $|1 + j\omega/\omega_1|$ term.

The other two terms of Eq. (5.3-3), namely, the $\log |K\omega_1/\omega_2|$ and the $\log |j\omega|$ terms, have even simpler representations on the logarithmic plot. The constant term, since it is frequency independent, describes a horizontal line passing through $\log |W(j\omega)| = \log |K\omega_1/\omega_2|$. If the plot is made on log-log paper, the line coincides with the $|W(j\omega)| = |K\omega_1/\omega_2|$ line. If the plot is made on the dB versus $\log \omega$ plot the horizontal line is at $20 \log |K\omega_1/\omega_2|$.

To treat the $j\omega$ term, observe that $\log |j\omega| = log\omega$ and the exact expression for the slope is

$$\frac{d(\log \omega)}{d(\log \omega)} = +1$$

Thus the plot of $\log |j\omega|$ versus $j\omega$ is a straight line with slope of +1 passing through the origin on the logarithmic plot. On log-log paper, this means that the line passes through the point $|W(j\omega)| = 1$, $\omega = 1$. On a dB versus $\log \omega$ plot the line has a slope of 20 dB per decade and passes through the point 0 dB at $\omega = 1$. Note that no approximation is involved in this case and that the exact plot is a single straight line. In Eq. (5.3-3), this term appears with a negative sign so that the slope becomes -20 dB per decade rather than +20 dB per decade.

Let us plot each of the four constituent terms of the $\log |W(j\omega)|$ as given in Eq. (5.3-3) on a single dB versus $\log \omega$ plot, as in Fig. 5.3-3*a*. To make the discussion more specific, numerical values have been chosen for K, ω_1, and ω_2 so that $W(j\omega)$ becomes

$$W(j\omega) = \frac{10(1 + j\omega/5)}{(j\omega)(1 + j\omega/50)}$$

In other words, $K = 100$, $\omega_1 = 5$, and $\omega_2 = 50$.

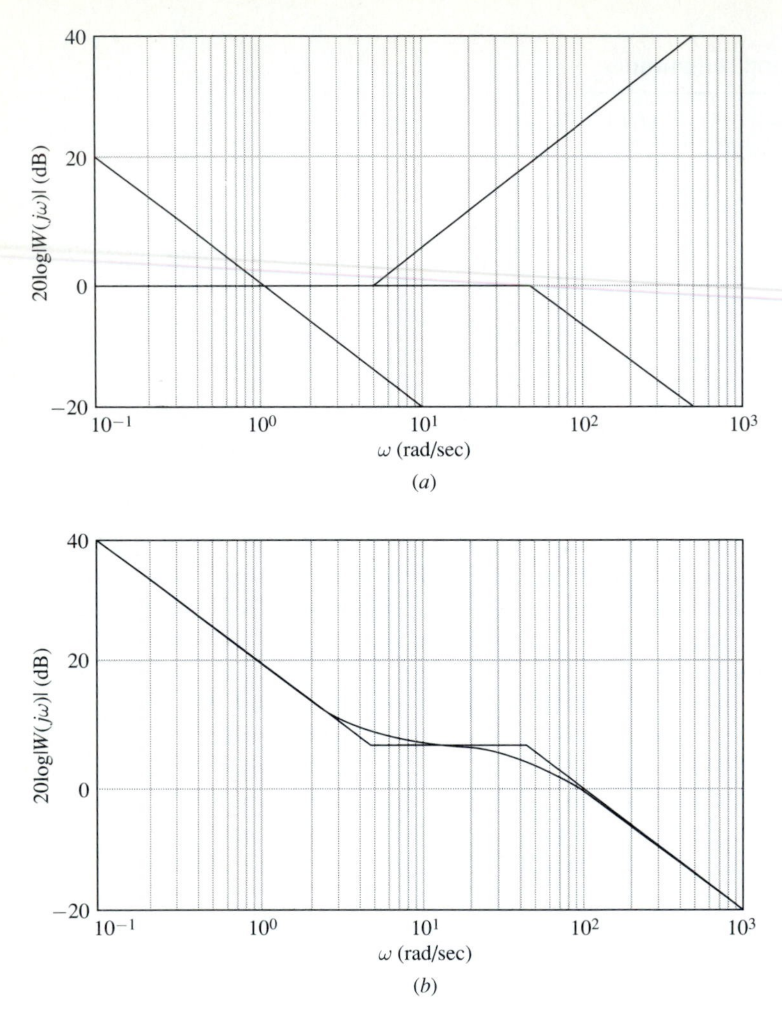

FIGURE 5.3-3
Bode magnitude plot of $|W(j\omega)|$. (*a*) Constituent terms of $|W(j\omega)|$; (*b*) resultant plot of $|W(j\omega)|$.

All that remains to be done is to add the contributions of the four terms in order to find the magnitude plot of $W(j\omega)$, since addition of decibels is equivalent to multiplication of magnitudes. The resulting asymptotic plot of $20\log|W(j\omega)|$ is shown in Fig. 5.3-3*b*, along with the exact plot for comparison. Once again, note that the approximation is reasonably accurate.

It must be remembered that when adding the effects of the basic terms we wish to add the decibels. On the dB plot, we must add using units of the distance from the 0 dB axis.

Even the job of finding the composite plot of $|W(j\omega)|$ can be considerably simplified, since each of the four terms involves only straight lines of integer slopes. The $|W(j\omega)|$ plot must be similarly composed. Hence the plot of $|W(j\omega)|$ is determined by fixing a single point on the plot by adding the effects of the four terms and then computing the slope of each straight line segment by adding the slopes of the four terms.

Since the slope of the composite plot of $|W(j\omega)|$ can change only at the break points of the constituent terms, it is a very simple job to find the entire plot. For example, for the frequency range $5 < \omega < 50$, the slope of the $|W(j\omega)|$ is equal to $-20 + 20 = 0$.

The reason for selecting the time constant form for $W(j\omega)$ rather than the original form of Eq. (5.3-1) should now be clear. If the latter representation had been selected, the low frequency asymptote would have been of the form $20 \log |\omega_1|$. Therefore, rather than being a line coincident with the 0 dB line, the asymptote would have been a line coincident with a line at $20 \log |W(j\omega)| = 20 \log |\omega_1|$. In this case, each of the factors of this form would have had a different low frequency asymptote, making the addition of their effects more difficult. The time constant form is used whenever we sketch Bode diagrams.

Example 5.3-1. As another illustration of the use of the asymptotic magnitude diagram, let us determine the magnitude plot for the transfer function

$$\frac{U(s)}{E(s)} = \frac{0.1\,(s+10)^2}{s(s+100)}$$

This is the transfer function of the so-called lead-lag compensator that is often put in series with the plant to modify the closed-loop frequency response and to ensure stability. The use of such series compensators will be discussed in Chap. 8.

After putting $U(s)/E(s)$ into time constant form, we have

$$\frac{U(s)}{E(s)} = \frac{0.1\,(1+s/10)^2}{s(1+s/100)}$$

Taking $20 \log |U(j\omega)/E(j\omega)|$ we get

$$20 \log \left| \frac{U(j\omega)}{E(j\omega)} \right| = 20 \log 0.1 + 40 \log \left| 1 + \frac{j\omega}{10} \right| - 20 \log |j\omega| - 20 \log \left| 1 + \frac{j\omega}{100} \right|$$

The only difference between the form of this expression and the form of Eq. (5.3-3) multiplied by 20 is that a factor of 40 rather than 20 precedes the $1 + j\omega/10$ term. This coefficient is twice as large because this term is repeated; that is, it is a double zero. This means that the slope of the high frequency asymptote is +40 rather than +20 so that beyond the break frequency $\omega = 10$ the plot rises by 40 dB of magnitude for every decade of frequency.

With this slight modification, the constituent terms of $|U(j\omega)/E(j\omega)|$ may now be placed on the Bode magnitude plot, as shown in Fig. 5.3-4*a*. The Bode magnitude composite plot of $|U(j\omega)/E(j\omega)|$ is obtained as before and shown in Fig. 5.3-4*b*. Again, the exact plot of the magnitude is also shown for comparison.

With some practice, the reader should be able to draw the asymptote plot for $|W(j\omega)|$ directly from the time constant form for $W(s)$ or $W(j\omega)$ without expanding the equation into $20 \log |W(j\omega)|$ or showing component terms on the diagram. It is

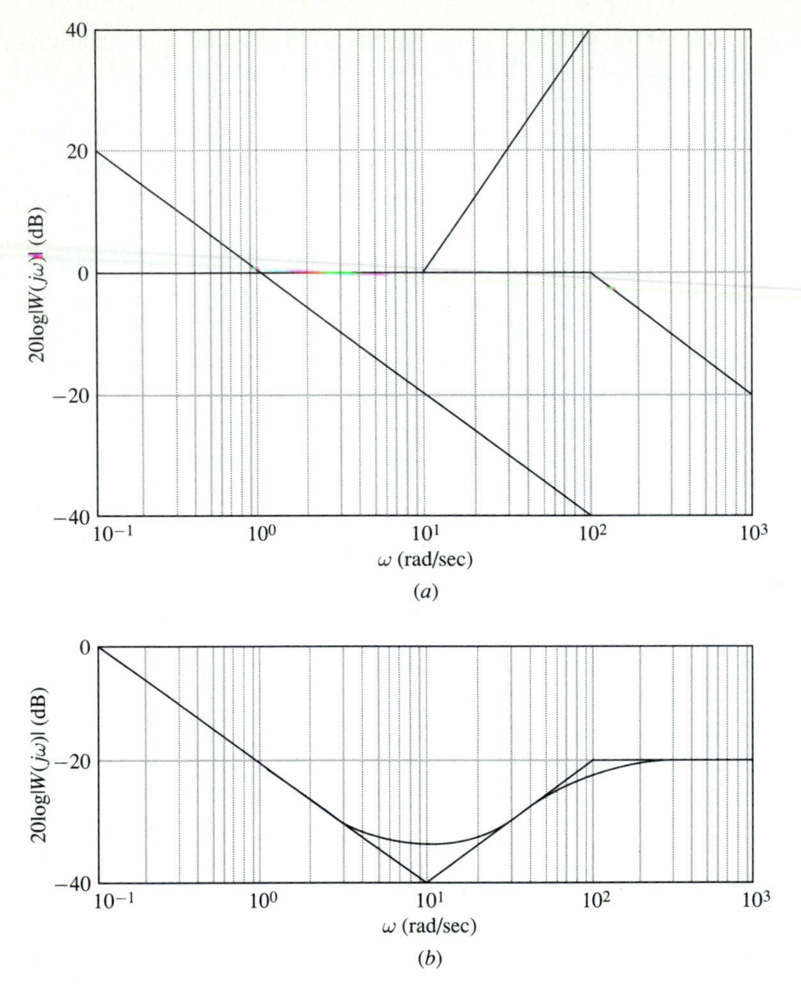

FIGURE 5.3-4
Bode plot for Example 5.3-1. (*a*) Constituent parts of Bode plot; (*b*) Composite Bode plot.

suggested that the reader work several of the exercises at the end of this section until such a facility is gained, since it is valuable in later work. Once such an ability is achieved, the reader should be able to determine the asymptotic plot for the $|W(j\omega)|$ with very little effort. Perhaps, even more importantly, the student should be able to sketch the general nature of the plot rapidly.

If the transfer function $W(s)$ contains complex conjugate poles or zeros, the approach must be somewhat altered. To illustrate the treatment of such terms, let us suppose that $W(s)$ is actually a plant transfer function consisting of only a pair of complex conjugate poles so that

$$W(s) = G_p(s) = \frac{\omega_n^2}{s^2 + 2\zeta\omega_n s + \omega_n^2} = \frac{1}{1 + 2\zeta\,(s/\omega_n) + (s/\omega_n)^2}$$

Again the use of the time constant form is emphasized. The magnitude of $G_p(j\omega)$ is now

$$\left|G_p(j\omega)\right| = \frac{1}{\left|(1-\omega^2/\omega_n^2)+2j\zeta\omega/\omega_n\right|} \tag{5.3-7}$$

For $\omega \ll \omega_n$ the unity term in the real part dominates and $20\log|G_p(j\omega)|$ is therefore

$$20\log\left|G_p(j\omega)\right| \approx -20\log 1 = 0 \qquad \text{for } \omega \ll \omega_n$$

On the other hand, for $\omega \gg \omega_n$, the quadratic term of the real part becomes dominant, and $20\log|G_p(j\omega)|$ becomes

$$20\log\left|G_p(j\omega)\right| \approx -20\log\left|\left(\frac{\omega}{\omega_n}\right)^2\right| = -40\log\omega + 40\log|\omega_n|$$

Taking the derivative with respect to $\log\omega$ yields

$$\frac{d\,(-40\log\omega + 40\log|\omega_n|)}{d(\log\omega)} = -40$$

Hence the high frequency asymptote has a slope of -40 dB per decade.

From the above discussion, we see that the pair of complex conjugate poles is identical to two real poles located at $s = \omega_n$ as far as the asymptotic plot is concerned. In other words, the asymptotic, straight line approximation treats the situation as if $\zeta = 1$. A normalized asymptotic plot is shown in Fig. 5.3-5 where the double break occurs as $\omega/\omega_n = 1$.

Although the value of the damping ratio ζ does not influence the straight line approximation, it does have an effect on the exact plot, as shown in Fig. 5.3-5. As ζ approaches zero, the exact curve begins to rise above the asymptotic plot. To determine the exact value of ζ for which this effect first occurs, we solve for the frequency at which the maximum occurs with the expression

$$\left.\frac{d\left|G_p(j\omega)\right|}{d\omega}\right|_{\omega=\omega_p} = 0$$

From this equation we find

$$\omega_p = \omega_n\sqrt{1-2\zeta^2} \tag{5.3-8}$$

The frequency at which the maximum p magnitude occurs is often called the *peak frequency,* $\boldsymbol{\omega_p}$.Therefore there is a maximum in the magnitude plot because there is a real value for ω_p only if $\zeta < \sqrt{2}/2$. Otherwise, the exact plot always lies below the asymptotic plot and no maximum occurs.

If a maximum does occur, that is, if $\zeta < \sqrt{2}/2$, the maximum value of $|G_p(j\omega)|$ may be obtained by substituting ω_p for ω in Eq. (5.3-7), with the result that

$$\left|G_p\left(j\omega_p\right)\right| = \frac{1}{2\zeta\sqrt{1-\zeta^2}} \qquad \text{for } \zeta \le 0.707 \tag{5.3-9}$$

Lest the reader become confused, it should be noted that the step response of the system has an overshoot whenever $\zeta < 1$. However, the magnitude portion of the

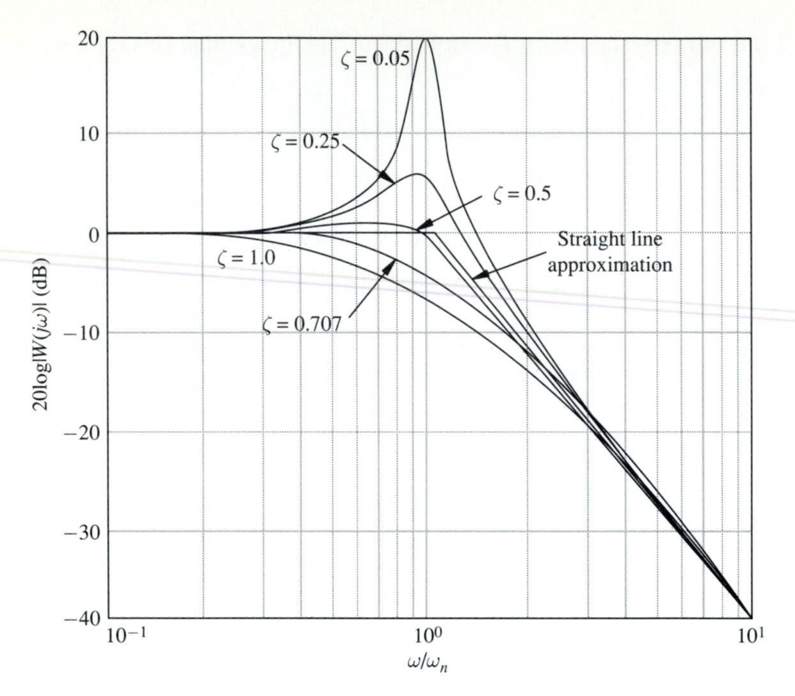

FIGURE 5.3-5
Asymptotic and exact plots of the magnitude of

$$\frac{1}{1 + 2\zeta(s/\omega_n) + (s/\omega_n)^2}$$

frequency response function does not display a maximum peak until $\zeta < \sqrt{2}/2$. Note also that the value of $|G_p(j\omega)|$ for $\omega = \omega_n$ is given by

$$\left|G_p\,(j\omega_n)\right| = \frac{1}{2\zeta} \tag{5.3-10}$$

From Fig. 5.3-5 we observe that for $0.5 < \zeta < 1$ the exact plot and the asymptotic approximation are actually closer than in the case of real roots. If ζ becomes larger than 1, the plant has two real poles and the previous procedures can be used.

If there is a complex pair of zeros the contribution to the Bode magnitude plot is similar to that for a pair of poles except that the sign is reversed. The straight line approximation for a complex zero pair is the same as the straight line approximation for two real zeros located at $s = -\omega_n$. If ζ is less than 0.3 there is a trough of depth 2ζ very near ω_n.

In addition to the use of the asymptotic approximations for determining the magnitude plot, straight line approximations may also be used to form approximate analytic expressions for $|W(j\omega)|$. These analytic expressions can often be used to augment the graphical procedure and to determine more precise values than may be obtained graphically.

Example 5.3-2. Consider, for example, the problem of determining the frequency for which a loop transfer function

$$W(s) = KG_p(s)H(s) = \frac{5}{s(1+s)}$$

has a magnitude of unity. The frequency for which a loop transfer function has a magnitude of unity is called the *crossover frequency* of the loop gain. The asymptotic plot of $|KG_p(j\omega)H(j\omega)|$ is shown in Fig. 5.3-6. From this figure we see that the crossover frequency ω_c is between 1 and 10, but without a more exact graphical plot a precise value cannot be determined. If we use the asymptotic approximations, then for $\omega > 1$, $|KG_p(j\omega)H(j\omega)|$ becomes

$$\left|KG_p(j\omega)H(j\omega)\right| = \frac{5}{\omega^2}$$

Therefore ω_c, which is given by $|KG_p(j\omega)H(j\omega)| = 1$, becomes

$$\omega_c \approx \sqrt{5} = 2.24$$

As a final example let us examine the case when $W(j\omega)$ represents the transfer function of a plant with a lightly damped complex pole pair.

Example 5.3-3. Assume that the transfer function of a particular closed-loop system is given as

$$\frac{Y(s)}{U(s)} = \frac{100(s+1)}{\left(s^2+4s+100\right)(s+.1)}$$

In time constant form, this becomes

$$\frac{Y(s)}{U(s)} = \frac{10(s+1)}{\left((s/10)^2+2(0.2)(s/10)+1\right)(s/0.1+1)} \tag{5.3-11}$$

The denominator term here contains a set of complex conjugate poles, with $\omega_n^2 = 100$, or $\omega = 10$. Thus a double break downward occurs at $\omega = 10$, owing to the denominator

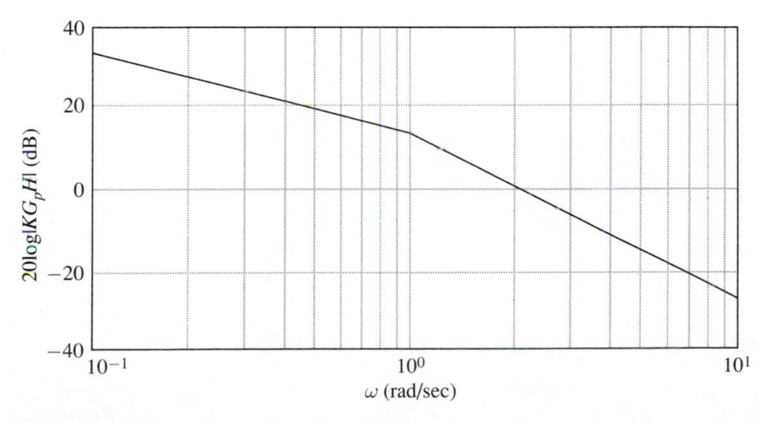

FIGURE 5.3-6
Bode magnitude plot of the loop gain for Example 5.3-2.

term. There is also a single break downward from the pole located at $\omega = 0.1$. There is a zero causing a single break upward at $\omega = 1$.

For small frequencies, as ω goes to zero, the magnitude of $Y(j\omega)/U(j\omega)$ is 10, and the slope at that point is zero. With this knowledge and the knowledge of the break points, the plot of the straight line approximation of the magnitude of the frequency response function is made directly without bothering to write the frequency response function. In other words, one works directly from the transfer function of Eq. (5.3-11). In addition, it seems unneccesary to indicate the constituent elements that make up the final magnitude plot. One can simply track the slope of the magnitude plot from break point to break point. From the left side of the plot until $\omega = 0.1$ the slope is zero. From $\omega = 0.1$ until $\omega = 1$ the slope is -20 dB per decade on the dB plot. From $\omega = 1$ until $\omega = 10$ the slope is $-20 + 20 = 0$. From $\omega = 10$ until the right side of the plot the slope is -40 dB per decade. Consequently the straight line approximation of the magnitude of the frequency response function is drawn directly from Eq. (5.3-11) as shown in Fig. 5.3-7.

When there is a lightly damped pole pair present the straight line approximation is inaccurate near the break point of the pole pair. The plot must be modified by establishing a point or two of the response near the break point and sketching the response. In that way the various differences in damping can be accounted for as they are in the plots of Fig. 5.3-5.

In this example the damping is $\zeta = 0.2$. Using Eq. (5.3-8), the contribution of the pole pair is found to peak at $\omega/\omega_n = 0.96$. Using Eq. (5.3-9), the magnitude at this value is 2.55 times the break point value. This point can be placed on the Bode plot and the hump in the plot sketched in. (In a dB plot, the plot lies $20 \log 2.55 = 8.1$ dB above the straight line approximation at this point.) Notice that, with ζ as small as it is here, the point occurs very near ω_n and the value is very close to $1/(2\zeta)$. These simplified values are usually good enough when $\zeta < 0.3$. Remember, when $\zeta > 0.5$ no serious inaccuracies are present in the straight line approximation.

In this section, we have considered the problem of determining the magnitude plot for the frequency response function. By making use of asymptotic approximations

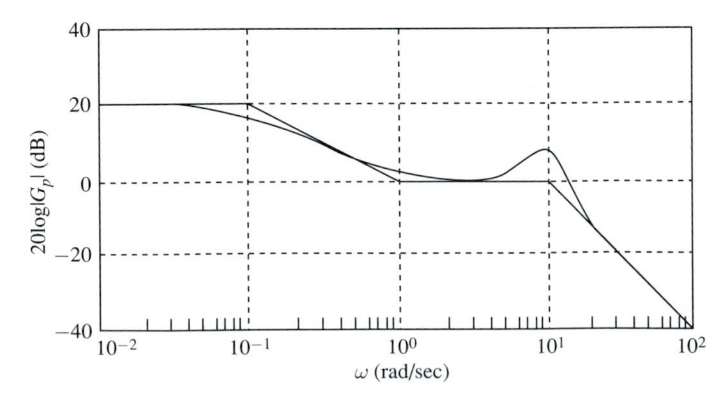

FIGURE 5.3-7
Bode magnitude plot for Example 5.3-3.

and a dB vs. log ω plot, we are able to obtain a very satisfactory solution to this problem. A similar solution for the phase plot is investigated in Sec. 5.4.

5.3.1 CAD Notes

The computer is very helpful in producing graphical information like Bode plots. The function

```
[mag,phase]=bode(num,den,w)
```

returns in the vectors `mag` and *`phase`* the magnitude and phase of the complex numbers which result when the transfer function given by *`num`* and *`den`* is evaluated at the frequencies given in the vector *`w`*. A set of equally spaced frequencies on the log plot can be achieved by using the function

```
w=logspace(p1,p2,n)
```

This function returns `n` points equally spaced on a log plot between 10^{p1} and 10^{p2}.

As an illustration the base plot for Fig. 5.3-5 can be obtained with the following macro.

```
>>%set up a for loop with different dampings
for z=[.05 .25 .5 .707 1]
        %enter the transfer function
        num=[1];
        den=[1 2*z 1];
        %set up the samples of omega needed
        w = logspace(-1,1,200);
        %get the magnitude and phase information
        [mag,phase] = bode(num,den,w);
        %convert the magnitude to dB
        mag1 = 20*log10(mag);
        %plot
        semilogx(w,mag1), xlabel('w/wn'),
               ylabel('20log|W(jw)| (dB)'), grid
        %set up to plot the other plots on the same graph
        hold on
end
%set up the points to be joined in a
%straight line approximation
x=[.1 1 10];
y=[0 0 -40];
%plot the straight line approximation
semilogx(x,y)
%turn hold off so that next plot starts anew
hold off
```

The labeling of the plots that is produced by MATLAB is not always what one might wish. The graph of Fig. 5.3-5 has been relabeled.

Exercises 5.3

5.3-1. Draw the asymptotic, straight line magnitude plots for each of the following transfer functions. In addition, determine at least one point of the exact plot near the break points to help sketch the plot. If you have access to a CAD package, have the computer plot the exact plots for comparison.

(*a*) $G(s) = \dfrac{1000}{(s+1)(s+10)(s+100)}$ (*b*) $G(s) = \dfrac{1000}{s(s+10)(s+100)}$

(*c*) $G(s) = \dfrac{900(s+10)}{s(s+3)(s+300)}$ (*d*) $G(s) = \dfrac{1000(s+1)}{s^2(s+10)}$

5.3-2. Repeat the instructions of Exercise 5.3-1 with the following transfer functions.

(*a*) $\dfrac{(100)^2}{\left(s^2+20s+(100)^2\right)}$ (*b*) $\dfrac{(100)^2(s+10)}{s\left(s^2+2s+(100)^2\right)}$

(*c*) $\dfrac{(100)^2\left(s^2+s+100\right)}{s^2(s+100)^2}$ (*d*) $\dfrac{\frac{10}{9}\left(s^2+6s+9\right)}{(s+1)(s+100)^2}$

5.3-3. Repeat the instructions of Exercise 5.3-1 with the following transfer functions.

(*a*) $\dfrac{-1000}{(s+1)(s+300)}$ (*b*) $\dfrac{s-1}{s(s+10)}$

(*c*) $\dfrac{s+3}{s^2(s-10)}$ (*d*) $\dfrac{1000\left(s^2-s+100\right)}{s\left(s^2-2s+(100)^2\right)}$

5.4 BODE PHASE PLOT—STRAIGHT LINE APPROXIMATION

In this and the following section, attention is concentrated upon the second component of the frequency response function, namely, the phase. At any one frequency the phase angle may be determined by finding the phase angle of $W(j\omega)$. However, if we wish to know the plot of the phase angle for all frequencies, this calculation has to be repeated many times. Methods of avoiding this detailed calculation are discussed here. More specifically, we examine a class of transfer functions that are referred to as minimum phase. A *minimum phase* system has all its poles and zeros in the left half-plane. For this class of systems it is possible to demonstrate that the magnitude and phase characteristics are not independent but are, in fact, closely related. In particular, Hendrick Bode[1] showed that the phase shift, arg $W(j\omega)$, at a frequency ω_1 is related to $|W(j\omega)|$ by the integral expression

$$\arg W(j\omega_1) = \frac{1}{\pi}\int_{-\infty}^{\infty} \frac{d\log|W(j\omega)|}{d\log\omega} \ln\coth\left|\frac{u}{2}\right| du \quad \text{radians} \tag{5.4-1}$$

[1] Hendrick Bode, *Network Analysis and Feedback Amplifier Design*, D. Van Nostrand Company, Inc., 1945.

where $u = \ln(\omega/\omega_1)$. Note the appearance of the logarithms of the magnitude and frequency, once again pointing out that logarithmic representation is a natural one.

We do not attempt to derive Eq. (5.4-1) since this would involve mathematics beyond the intended scope of this book. In addition, no attempt is made to use Eq. (5.4-1) by making a substitution for $|W(j\omega)|$ and carrying out the indicated integration. In fact, the work involved in such a process exceeds by far the work involved in a direct calculation of the phase shift. However, this does not mean that Eq. (5.4-1) is not of interest; on the contrary, the expression is useful in our study of the phase plot.

An examination of Eq. (5.4-1) reveals that phase shift is obtained by integrating the product of the slope of the log-log magnitude plot and a weighting factor $\ln\coth|u/2|$, much like the convolution procedure of Chap. 4. Hence we obtain the valuable clue that *the phase shift is related to the slope of the log-log magnitude plot.* (The slope of the log-log magnitude plot is obtained from the slope of the dB versus $\log\omega$ plot by dividing by 20.) In fact, we can show by the use of Eq. (5.4-1) that, if the slope of the log-log magnitude plot is constant at a value of α for all frequencies, the phase shift is given by

$$\arg W(j\omega) = \alpha\frac{\pi}{2} \text{ radians } = \alpha(90) \text{ degrees} \tag{5.4-2}$$

Although theoretically Eq. (5.4-1) indicates that the phase shift at any frequency is dependent on the slope of the log-log magnitude plot for all frequencies, an investigation of the function $\ln\coth|u/2|$, as shown in Fig. 5.4-1, indicates that in practice we need only consider frequencies for *one decade on either side of the frequency of interest.* In other words, if the slope of the log-log magnitude plot has constant value of α for the decade immediately preceding and following the frequency of interest, the phase shift is closely approximated by Eq. (5.4-2).

If the slope is not constant over this two-decade range, we may still use Eq. (5.4-2) by calculating an average slope to use in place of α. To compute this average slope, we could use the $\ln\coth|u/2|$ function shown in Fig. 5.4-1. This, however, is

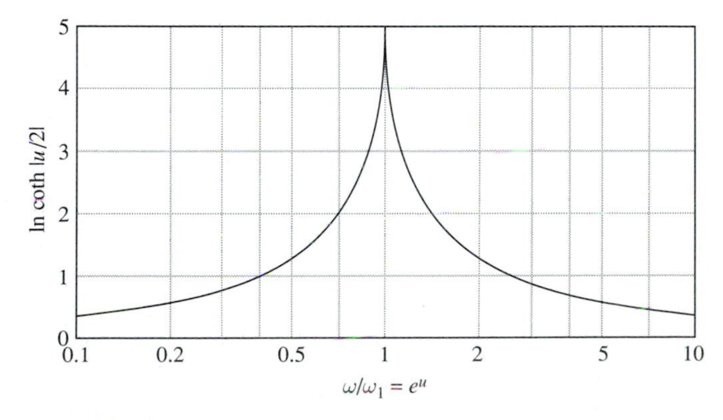

FIGURE 5.4-1
The function $\ln\coth|u/2|$.

not the procedure recommended, since the work involved would defeat the purpose of using an approximation, namely, to obtain answers quickly and easily. Rather, it is recommended that the average slope be obtained by simple linear averaging, with bias given to the frequency range near the frequency of interest. This should be done without making any calculations but by simply "eyeballing" the magnitude plot. Such a procedure provides good qualitative answers and often quite accurate quantitative information. If a more exact plot is needed, one can evaluate arg $W(j\omega)$ for particular values of ω that are of special interest. Often, computer programs are available for accurate phase plots.

Example 5.4-1. To illustrate the use of this procedure, let us consider the determination of the phase shift plot of $W(s) = Y(s)/U(s)$, where

$$W(s) = \frac{Y(s)}{U(s)} = \frac{5(1+s)(1+s/10)}{(1+s/0.1)(1+s/100)}$$

This transfer function is similar to that of the lead-lag compensator discussed in Example 5.3-1. The straight-line magnitude plot for this transfer function is shown in Fig. 5.4-2*a*.

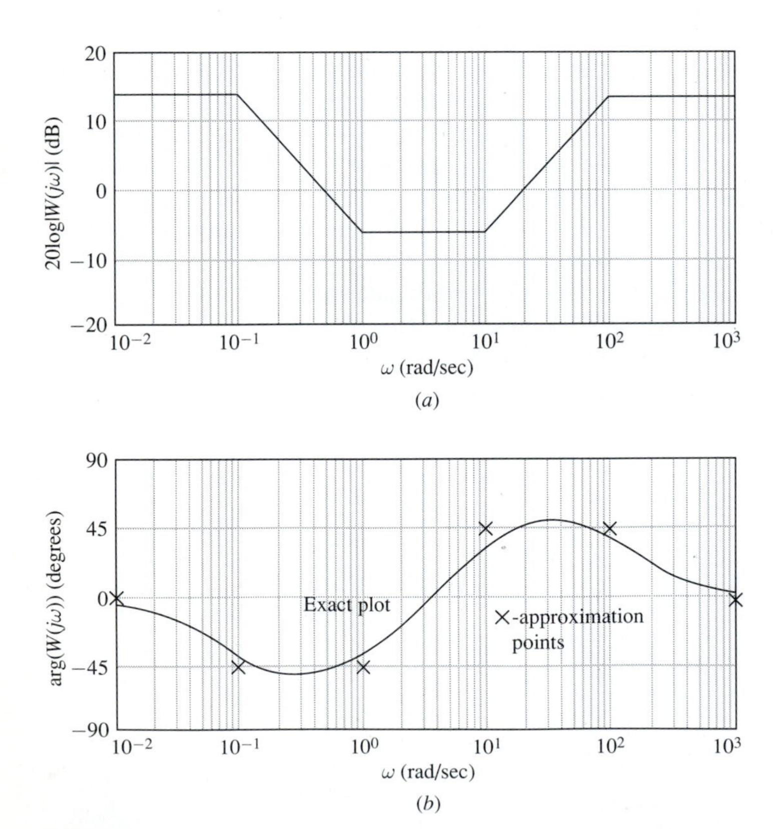

FIGURE 5.4-2
Bode plots for Example 5.4-1. (*a*) Magnitude; (*b*) phase.

The poles and zeros of this transfer function are all in the left half-plane and our minimum phase method applies.

Let us determine the phase shift for $\omega = 0.01$. If we examine the magnitude plot for one decade above and below this frequency, we see that the slope has a constant value of zero. Hence

$$\arg W(j0.01) \approx 0(90^\circ) = 0^\circ$$

For $\omega = 0.1$ the average slope is -10 dB per decade, which corresponds to a slope of $-1/2$ on the log-log plot, and therefore the phase shift becomes

$$\arg W(j0.1) \approx \left(-\frac{1}{2}\right) 90^\circ = 45^\circ$$

The phase shift for $\omega = 1.0$, 10, 100, and 1,000 is shown as Fig. 5.4-2*b*. In addition, the exact phase plot is shown for comparison. For this example, the approximation is good and gives a good deal of qualitative information concerning the character of the phase plot.

The approximation procedure discussed here works best if the break points of the magnitude plot are separated by at least a decade. The most accurate information is obtained for frequencies that are either at break points or well removed from any break point. This explains the success of the procedure for Example 5.4-1. However, even if these features are not present, a qualitative picture of the phase plot can be obtained quickly and easily by using this procedure.

5.4.1 Straight Line Approximation

There is a second method to sketch the phase curve approximately. A reasonably accurate straight line approximation to the phase plot is now explained. First note that the phase of any transfer function $W(j\omega)$ at each frequency ω arises from the sum of the individual contributions from poles and zeros of the transfer function. Also, the phase contribution of a pole is merely the negative of the phase contribution of a zero at the same location. If we establish straight line approximations for real and complex conjugate zeros, straight line approximations for real and complex conjugate poles follow immediately. A straight line approximation for an entire transfer function can be achieved by adding the plots of the contributions from the individual components as was done on the log magnitude plot of the previous section.

Following the derivation of the log magnitude approximation from the previous sections, we examine the frequency response function

$$W(j\omega) = \frac{K(j\omega + \omega_1)}{j\omega(j\omega + \omega_2)} = \frac{K\omega_1}{\omega_2}\frac{1 + j\omega/\omega_1}{j\omega(1 + j\omega/\omega_2)}$$

The phase function is given by

$$\arg(W(j\omega)) = \arg\left(\frac{K\omega_1}{\omega_2}\right) + \arg\left(1 + \frac{j\omega}{\omega_1}\right) - \arg(j\omega) - \arg\left(1 + \frac{j\omega}{\omega_2}\right)$$

The log magnitude plot of the term

$$\arg\left(1 + \frac{j\omega}{\omega_1}\right)$$

is given in Fig. 5.3-1. We can use an approximation similar to that used to obtain the straight line approximation to the magnitude plot over two intervals

$$\arg\left(1 + \frac{j\omega}{\omega_1}\right) \approx 0^\circ \qquad \omega < \omega_1/10$$

$$\arg\left(1 + \frac{j\omega}{\omega_1}\right) \approx 90^\circ \qquad \omega > 10\omega_1$$

We complete the approximation by connecting the points at $\omega_1/10$ and $10\omega_1$ by a straight line on the arg $(1 + j\omega/\omega_1)$ versus $\log \omega$ graph. The approximation and the actual plot of arg $(1 + j\omega/\omega_1)$ are shown in Fig. 5.4-3. Fortunately, the two match fairly well. The straight line approximation to an entire transfer function is achieved by adding the straight line contribution from each pole or zero factor, similar to the procedure for achieving the straight line magnitude approximation. This procedure is illustrated in the following example.

Example 5.4-2. Let

$$W(j\omega) = \frac{10(1 + j\omega/5)}{j\omega(1 + j\omega/10)}$$

$$\arg(W(j\omega)) = \arg(10) + \arg(1 + j\omega/5) - \arg(1 + j\omega/10) - \arg(j\omega)$$

The term arg(10) contributes zero phase. (If the gain were negative, the term arg(−10) would contribute 180° phase at all frequencies, but this is not the case.) The term $-\arg(j\omega)$ contributes -90° phase at all frequencies. The straight line approximations

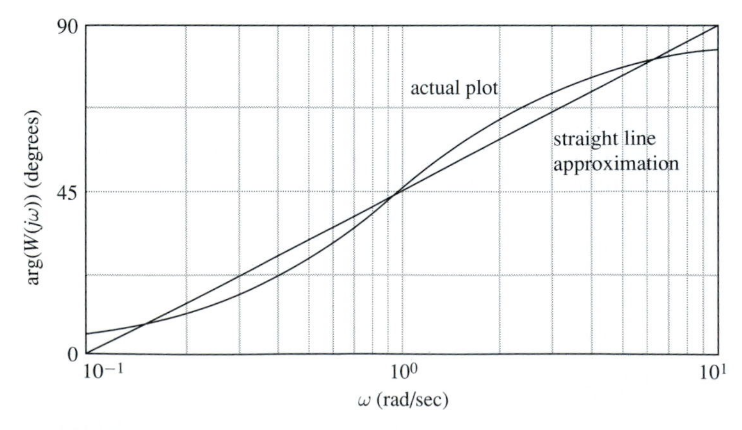

FIGURE 5.4-3
Straight line approximation for arg $\left(1 + \frac{j\omega}{\omega_1}\right)$.

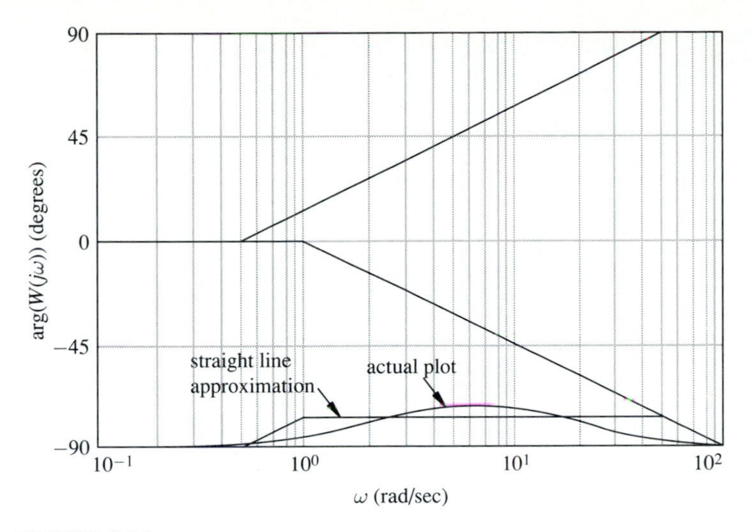

FIGURE 5.4-4

Plot of arg $W(j\omega)$; $W(j\omega) = \dfrac{10\left(1+\frac{j\omega}{5}\right)}{j\omega\left(1+\frac{j\omega}{10}\right)}$.

for the remaining two terms appear in Fig. 5.4-4 along with the total straight line approximation of the phase of $W(s)$ and an actual plot of the phase of $W(s)$.

The straight line approximation for the phase plot of a transfer function is derived similarly to the straight line approximation of the log-magnitude plot and it shares the property that the straight line approximation for a complex pole pair is given by the straight line approximation of a pair of real poles located at $-\omega_n$. The approximation is

$$\arg\left(1+2\zeta\left(\frac{j\omega}{\omega_n}\right)+\left(\frac{j\omega}{\omega_n}\right)^2\right) \approx 0^\circ \qquad \text{for } \omega < 0.1\omega_n$$

and

$$\arg\left(1+2\zeta\left(\frac{j\omega}{\omega_n}\right)+\left(\frac{j\omega}{\omega_n}\right)^2\right) \approx 180^\circ \qquad \text{for } \omega > 10\omega_n$$

A straight line is drawn between the values at $0.1\omega_n$ and $10\omega_n$ on the $\log\omega$ plot. Of course, the approximation becomes poorer and should be modified as ζ decreases from one to zero. The graph of the straight line approximation and the actual plots of $\arg W(j\omega)$ for

$$W(j\omega) = \frac{1}{1+2\zeta\frac{j\omega}{\omega_n}+\left(\frac{j\omega}{\omega_n}\right)^2}$$

for various values of ζ are given in Fig. 5.4-5. Notice here that the approximation becomes exact at $\omega = \omega_n$ independent of the damping ratio ζ. When ζ is small

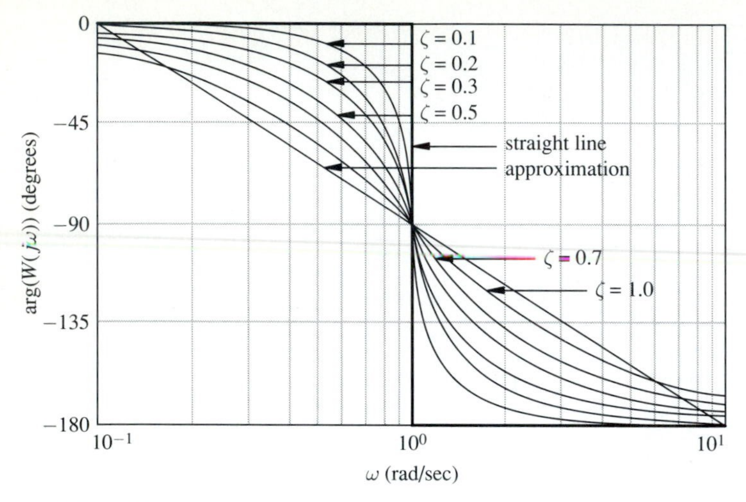

FIGURE 5.4-5
Plot of arg $W(j\omega)$; $W(j\omega) = \dfrac{1}{1 + 2\zeta\frac{j\omega}{\omega_n} + \left(\frac{j\omega}{\omega_n}\right)^2}$.

enough ($\zeta < 0.25$) the phase plot drops so quickly that it is better to approximate the phase curve with a discontinuity at ω_n. The appropriate approximation has the lightly damped pole pair contributing zero phase shift until ω_n and then dropping immediately to $-180°$ and contributing a constant $-180°$ for the rest of the frequency range. This approximation is also shown in Fig. 5.4-5. Of course, a minimum phase complex pair of zeros produces a contribution which is opposite in sign to the contribution of a minimum phase complex pole pair.

5.4.2 Minimum-Phase Condition

As we indicated, the relationship between the magnitude and phase plots given by Eq. (5.4-1) is valid only for minimum phase systems. To understand the necessity of this restriction, let us consider the frequency response of the three systems

$$
\begin{aligned}
W_1(s) &= \frac{10(s+1)}{s+10} = \frac{(1+s/1)}{1+s/10} \\
W_2(s) &= \frac{10(s-1)}{s+10} = \frac{-(1-s/1)}{1+s/10} \\
W_3(s) &= \frac{10(s+1)}{s-10} = \frac{-(1+s/1)}{1-s/10}
\end{aligned}
\tag{5.4-3}
$$

Here all three transfer functions have been expressed in both time constant and pole-zero form. The magnitude and phase plots for these three systems are shown in Fig. 5.4-6. Since the magnitude plots of the three systems are identical, only one is shown. Although the three systems have the same magnitude plot, they have radically

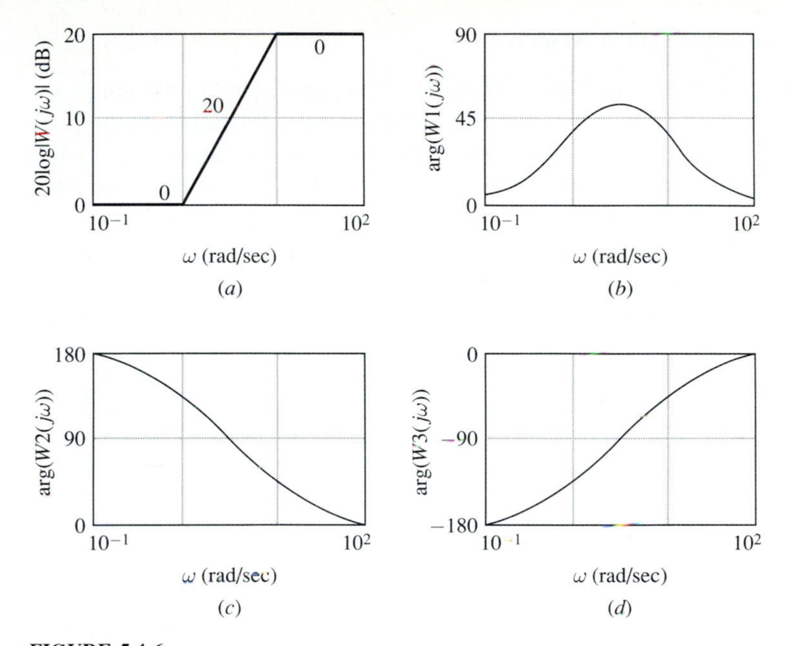

FIGURE 5.4-6
Magnitude and phase plots for minimum phase and non-minimum phase systems. (*a*) Magnitude; (*b*) arg $W_1(j\omega)$; (*c*) arg $W_2(j\omega)$; (*d*) arg $W_3(j\omega)$.

different phase characteristics. This difference between the phase plots for the three cases is due to the location of a pole or zero in the right half of the s plane. Only for $W_1(s)$ is it possible to infer the phase angle diagram from the amplitude diagram using Eq. (5.4-1), because only $W_1(s)$ is minimum phase.

Systems such as $W_1(s)$ that have no poles or zeros in the right half of the s plane are referred to as minimum phase systems, because they have the minimum amount of negative phase shift or phase lag for a given magnitude plot. For this simple example, we have established that the Bode relation is valid only if the system is minimum phase, that is, if it has no poles or zeros in the right half-plane. We have shown this condition to be true in only one specific example, but Bode[1] has shown that Eq. (5.4-1) is valid for any minimum phase transfer function. Thus, in the general case, the phase angle diagram may be determined uniquely from the amplitude diagram using Eq. (5.4-1) if the corresponding transfer function has no poles or zeros in the right half of the s plane.

Special note should also be made of the fact that a system with a negative value for the gain K is non-minimum phase. Obviously the addition of a negative sign does not change the magnitude plot but does alter the phase characteristic by adding 180° of phase lag at all frequencies. This point can be overlooked if one is not careful.

---

[1] *Ibid.*

5.4.3 Non-minimum Phase Systems

Fortunately, the phase plot is still readily attainable if a system is not minimum phase. One approach is to return to the pole-zero form of the transfer function and the pole-zero plot on the s plane. For this reason, the three transfer functions of Eqs. (5.4-3) are written in pole-zero form as well as the time constant form associated with Bode diagrams.

To discuss the graphical approach on the s plane, it is necessary to recall the graphical procedures used in Chap. 4. In Fig. 5.4-7, a pole is located at $s = -\alpha$, and an arbitrary point s on the plane has been chosen. A vector starting at $s = -\alpha$ and ending at the point s represents the length and angle of the vector $s + \alpha$. In plotting the frequency response, values of s along the $j\omega$ axis are of interest.

To illustrate the phase angle calculation for a non-minimum phase system, consider $W_2(s)$ as given in Eqs. (5.4-3). The pole-zero plot for this transfer function is given in Fig. 5.4-8*a*. Suppose we are interested in the phase angle for all positive frequencies, starting at $\omega = 0$. For $\omega = 0$, the components of the transfer function

$$W_2(s) = \frac{10(s-1)}{s+10}$$

are shown in Fig. 5.4-8*b*. The angle associated with the vector originating on the zero at $s = +1$ is 180°. The angle associated with the pole is 0°, so that at $s = 0$ the phase angle of $W_2(s)$ is just

$$\arg W_2(j) = 180^\circ - 0^\circ = 180^\circ$$

Figure 5.4-8*c* illustrates the calculation of the phase angle at $s = j1$. Again the argument of $W_2(s)$ is just $\theta_1 - \theta_2$ where the angles are $\theta_1 = 135^\circ$ and $\theta_2 \approx 6^\circ$, or

$$\arg W_2(j1) = \theta_1 - \theta_2 = 135^\circ - 6^\circ = 129^\circ$$

At $s = j10$, θ_1 is approximately 90°, and θ_2 is 45°. As the frequency approaches infinity, both θ_1 and θ_2 approach 90° so that in the limit the argument of $W_2(j\omega)$ approaches zero. The phase plot has already been given as Fig. 5.4-6*c*.

If the transfer function $W_3(s)$ is considered, as in Fig. 5.4-9, note that, at zero frequency, the phase angle is -180°, since the phase shift is due to the angle associated

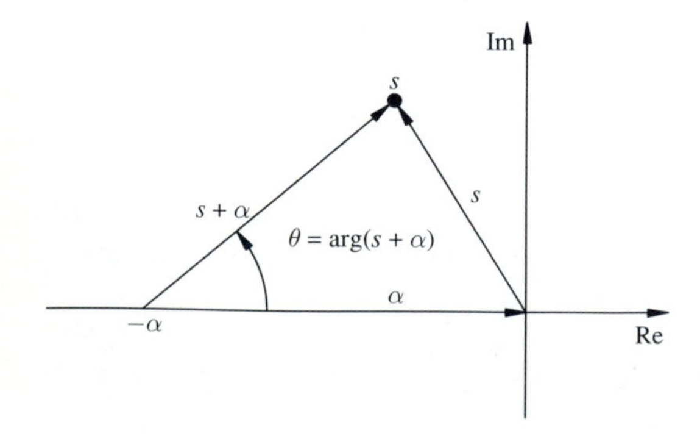

FIGURE 5.4-7
Method of evaluating the magnitude and the phase angle of the vector $s + \alpha$.

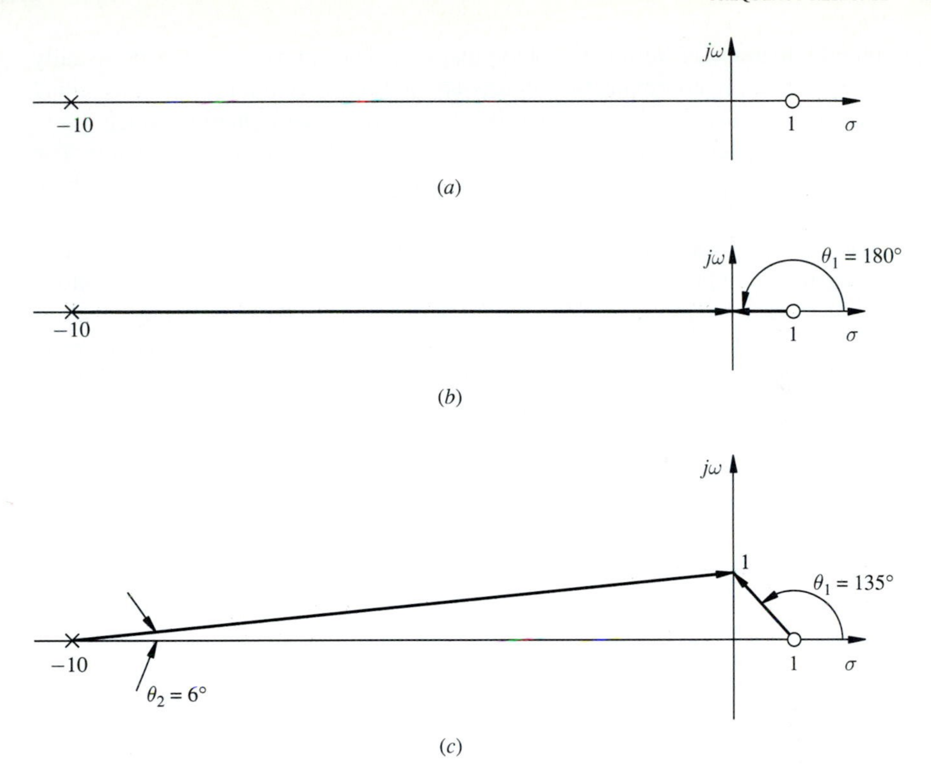

FIGURE 5.4-8
Steps in calculating the phase angle of $W_2(j\omega)$ in Eq. (5.4-3).

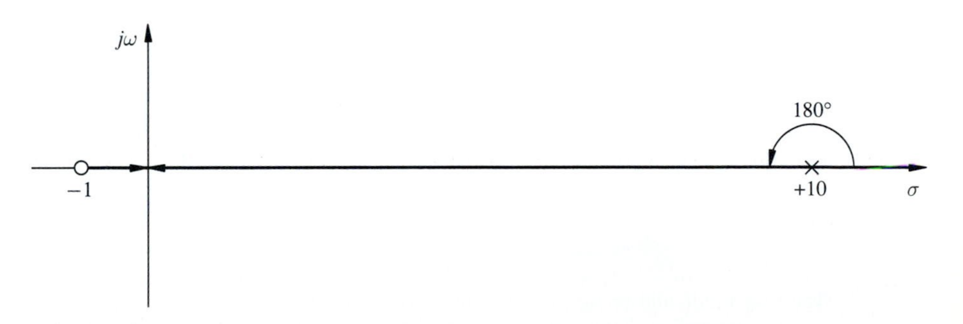

FIGURE 5.4-9
Calculation of the phase angle of $W_3(j0)$.

with the pole. As ω passes through 1, 10, and on to infinity, the phase angle diagram of Fig. 5.4-6*d* is determined.

The approach described here is applicable not only to non-minimum phase transfer functions but to any transfer function. Also, if the gain is taken into account and the vector distances measured, this approach is readily suited to the calculation of the amplitude diagram as well as the phase diagram. The approach, however, usually provides only a crude approximation unless applied with attention to detail. Actual phase diagrams are usually drawn using the straight line approximation which is extended to non-minimum phase plants below, or they can be drawn by computer. The graphical approach here is useful for understanding the general shape of the phase curve.

The phase plot for a non-minimum phase transfer function can be determined by applying the straight line approximation to a *pseudo* minimum phase magnitude plot. The method is based on the observation that a right half-plane zero contributes the same amount of phase shift but opposite in sign from a left half-plane zero. Thus a right half-plane zero affects the phase plot the same way that a left half-plane pole at the mirror image position does. Likewise, a right half-plane pole affects the phase plot the same way that a left half-plane zero does. The truth of these statements becomes apparent with consideration of the graphical approach described above.

The formation of this pseudo minimum phase transfer function may be easily accomplished. One must only replace every pole in the right half-plane at $s = +\alpha$ by a left half-plane zero at $s = -\alpha$ and every right half-plane zero by a left half-plane pole. The resulting transfer function is then minimum phase, and its phase characteristic is identical to the phase characteristic of the original system. The transfer function should be put into time constant form *before* interchanging poles and zeros to form the pseudo minimum phase transfer function. The straight line approximation can then be applied to this pseudo minimum phase system to obtain the same phase plot as the original non-minimum phase transfer function.

Example 5.4-3. As an illustration of this procedure, consider the phase representation of the non-minimum phase transfer function $W_2(s)$ discussed above.

$$W_2(s) = \frac{-(1 - s/1)}{1 + s/10}$$

The magnitude plot is given in Fig. 5.4-6*a*. Applying the above procedure, we form the pseudo minimum phase system by replacing the zero at $s = +1$ by a pole at $s = -1$, so that

$$W_{\text{pseudo}}(s) = \frac{-1}{(1 + s)(1 + s/10)}$$

The magnitude and phase plots for this system are shown in Fig. 5.4-10. A comparison of Figs. 5.4-6*c* and 5.4-10*b* reveals that the phase plot of $W_2(s)$ is the same as the phase plot of $W_{\text{pseudo}}(s)$, as predicted. Note that 180° has been added to the phase plot to account for the negative constant in the time constant form.

Once familiarity with the use of this method has been gained, the reader may see that it is no longer necessary to find the pseudo transfer function or its magnitude plot. Instead, one may simply determine the slopes of the pseudo magnitude plot mentally and perhaps write them on the actual magnitude plot as shown in Fig. 5.4-11. The phase shift plot is then obtained by using the pseudo slopes rather than the actual slopes. If using the straight line approximation procedure the sign of the phase of each contributing right half-plane zero or pole is reversed.

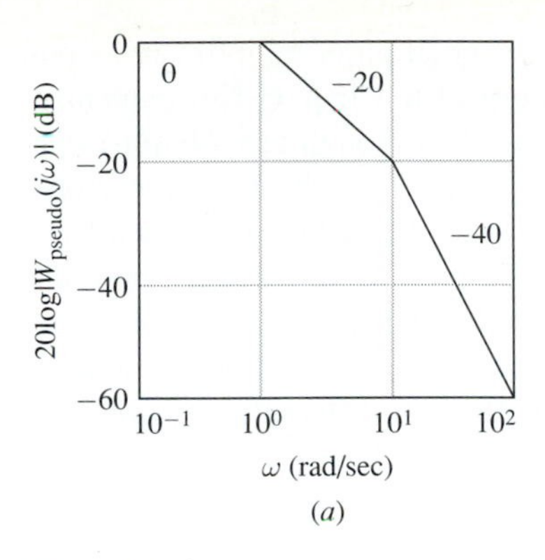

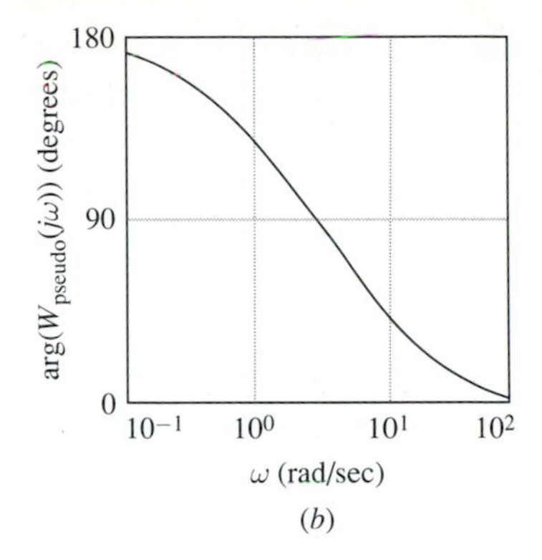

FIGURE 5.4-10
Pseudo minimum phase system. (*a*) Magnitude; (*b*) phase.

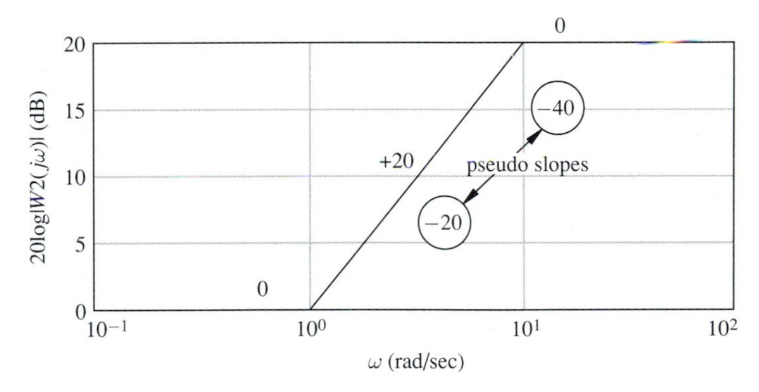

FIGURE 5.4-11
Magnitude plot for $W_2(j\omega)$ with pseudo slopes shown.

By making use of this pseudo minimum phase procedure, it is now possible to approximate the phase plot of any system. Although this procedure is adequate for an initial investigation of the qualitative nature of the phase characteristic, it often becomes necessary to refine the approximation to obtain more quantitative information. At this point a CAD package is usually used.

5.4.4 Nyquist Polar Plots

There is a second way of representing the information contained on the Bode plots that finds great utility in control theory. Instead of plotting the magnitude and the phase versus the frequency on separate plots, a polar plot of the complex number that gives the frequency response at each frequency can be made. The points that result on the polar plot can be joined to produce a curve with frequency as a parameter that references the points on the polar plot to the appropriate points on the Bode plots.

The polar plots are usually made without taking logarithms. There are no simple shortcuts in producing the polar plot. Usually the best way of producing the polar plot is to read the data and the trends off the Bode plots. The polar plots are intimately connected with the Nyquist stability theorem which is derived in Chap. 6. Consequently, the polar plots are usually referred to as Nyquist plots even though true Nyquist diagrams contain additional information. In most cases a crude sketch of the polar plot is sufficient to obtain the utility from the plot. Detailed information is needed only near the point $s = -1$. Since logarithms are not used in creating the Nyquist plot, the scale of the plot is often too large to see the plot clearly near the $s = -1$ point. For this reason, Nyquist plots are often sketched out of scale. When a computer is used and out of scale plots are not produced, two or more plots on different scales may be necessary.

Example 5.4-4. In this example, sketches are made of the Bode plots and the Nyquist or polar plot of the transfer function

$$W(s) = \frac{100(-s+10)}{(s+0.1)^2(s^2+0.1s+100)} = \frac{1000\left(\dfrac{-s}{10}+1\right)}{\left(\dfrac{s}{0.1}+1\right)^2\left(\dfrac{s^2}{100}+\dfrac{0.01s}{10}+1\right)}$$

The straight line approximation of the Bode magnitude plot is shown in Fig. 5.4-12*a*. The peak associated with the underdamped pole pair has been plotted and the plot has been sketched in around that point. The straight line approximation of the Bode phase plot is shown in Fig. 5.4-12*b*. The damping of the pole pair at $\omega_n = 10$ is small enough that a vertical line is used in the phase approximation for this pole pair.

A sketch of the Nyquist or polar plot is made in Fig. 5.4-13. Different parts of the plot are out of scale with each other but the general shape of the plot can be seen as can more detailed behavior near the point $s = -1$. The information for the sketch is obtained by examining the behavior of the Bode plots as the frequency increases.

At low frequencies the Bode plots show a large magnitude and small phase shift. The corresponding points on the polar plot are far from the origin slightly below the positive real axis. As the frequency increases the magnitude decreases while the phase decreases, i.e., becomes more negative. This corresponds to the polar plot spiraling in towards the origin in a counterclockwise fashion. The point at $\omega = 0.1$ where the phase passes -90° has been marked. The phase continues to decrease past -180° before the magnitude plot crosses a magnitude of 1 or 0 dB. The point at $\omega = 1$ where the phase crosses -180° has been marked. Notice that since the magnitude is greater than one, this point is outside the point $s = -1$. The complex pole pair produces a spike in the magnitude response and a sharp 180° phase shift in the phase plot. This corresponds to a loop in the polar plot. The point $\omega = 10$ where the magnitude reaches a peak has been marked on the polar plot.

Clearly the scaling on this plot is warped. The magnitude at the top of the loop created by the pole pair is around 10 while the magnitude near $\omega = 0$ is 1000. Nyquist plots are often sketched like this since the important information in this plot comes from the general shape of the plot and some detailed behavior around $s = -1$. Why this is true is shown in Chap. 6.

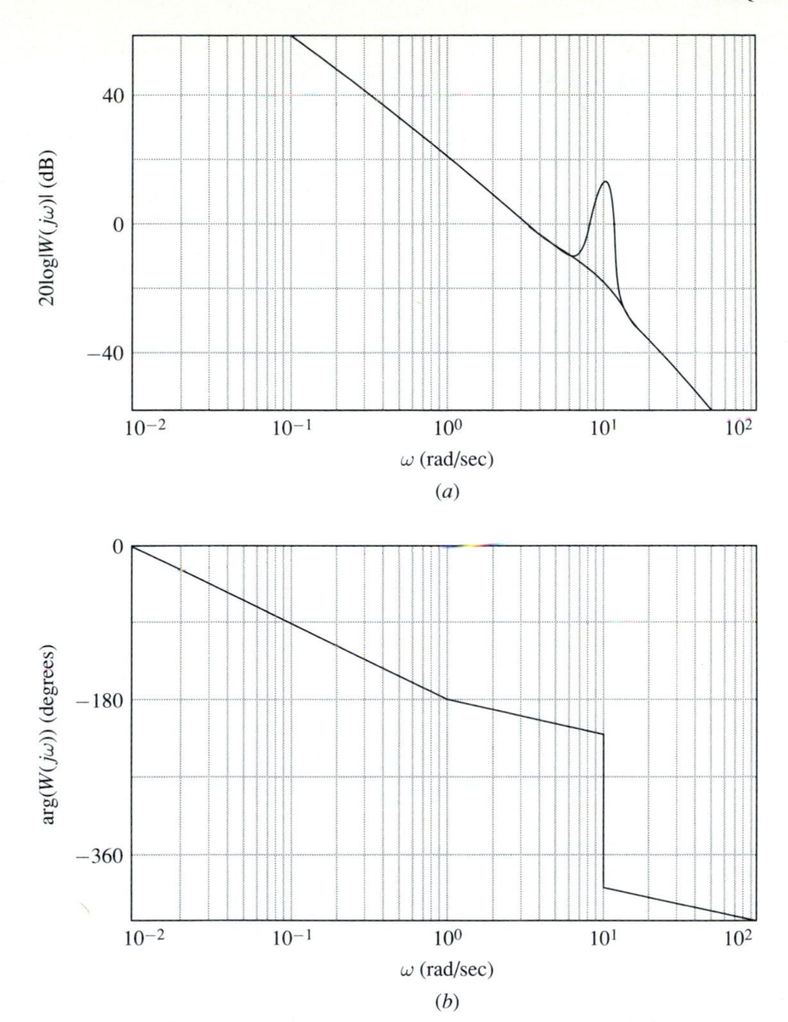

FIGURE 5.4-12
Bode sketches for Example 5.4-4. (*a*) Magnitude (*b*) phase.

5.4.5 CAD Notes

In the following example the Bode and Nyquist plots sketched in Example 5.4-4 above are accurately plotted using MATLAB. As mentioned in the CAD Notes for Sec. 5.3, the function *bode* is used to obtain an evaluation of the magnitude and phase of a transfer function. The function

```
[re,im]=nyquist(num,den,w)
```

evaluates the transfer function determined by *num* and *den* at each point in the vector *w* and returns the real part of the complex frequency response in the vector *re* and

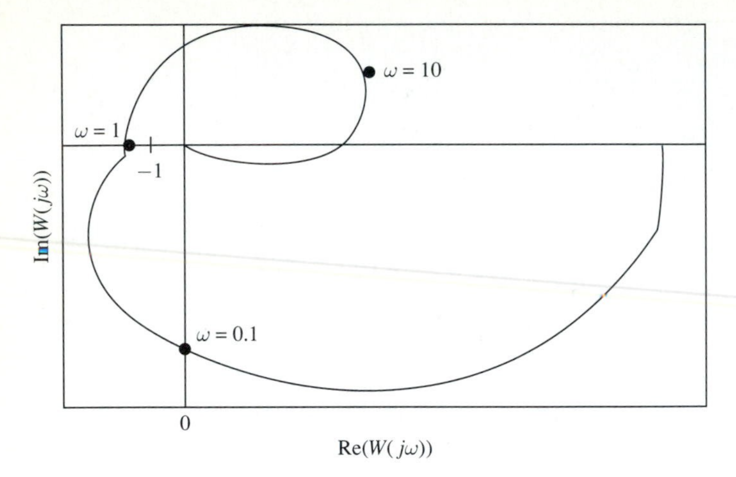

FIGURE 5.4-13
Nyquist sketch for Example 5.4-4.

the imaginary part in the vector `im`. These vectors can then be plotted with `im` as the ordinate and `re` as the abscissa to obtain the polar or Nyquist plot.

Example 5.4-5. The MATLAB program creates the Bode and Nyquist plots for the transfer function given in Example 5.4-4. The Bode plots are given in Figs. 5.4-14. To achieve enough detail in the plot to follow accurately the points around the underdamped pole pair, the frequency vector `w` is built in three parts with different densities of sample points. It should be noted on Fig. 5.4-14 that, in producing the scaling and grid marks, the plot program of MATLAB prefers to use round numbers based on factors of 10. This means that even though the axis limits are set in the magnitude plot, the grid markings appear at -50 and 50 rather than at multiples of 20 as would be preferred by a control engineer. Also, the phase plot is marked off in multiples of 100 rather than the preferable multiples of 90. Most of the plots in this book have been altered in another program to better fit the control setting.[1]

The polar plots are displayed in Fig. 5.4-15. In Fig. 5.4-15*a*, the entire polar plot is shown. On this scale the general shape of the plot can be seen but the loop near the origin is invisible. Usually, it is easy enough to ascertain the general shape of the polar plot from the Bode plot so that the accuracy provided by the computer is not needed and this plot can be safely sketched. In Fig. 5.4-15*b* the axis limits have been set so that the behavior of the polar plot near the point $s = -1$ can be observed. It is here that the computer accuracy is useful in showing the correct relationship between the plot and the point $s = -1$.

It should be noted that unless there is a sufficiently dense grid of points near an underdamped pole pair, strange looking plots may result. These strange plots occur

[1]For example, to obtain a phase plot, the phase data are multiplied by 100/90 before plotting. The markings on the resulting plot are then changed in another graphical program: 100 is changed to 90, 200 is changed to 180, etc. The result is an accurate plot with the desired end points and grid marks.

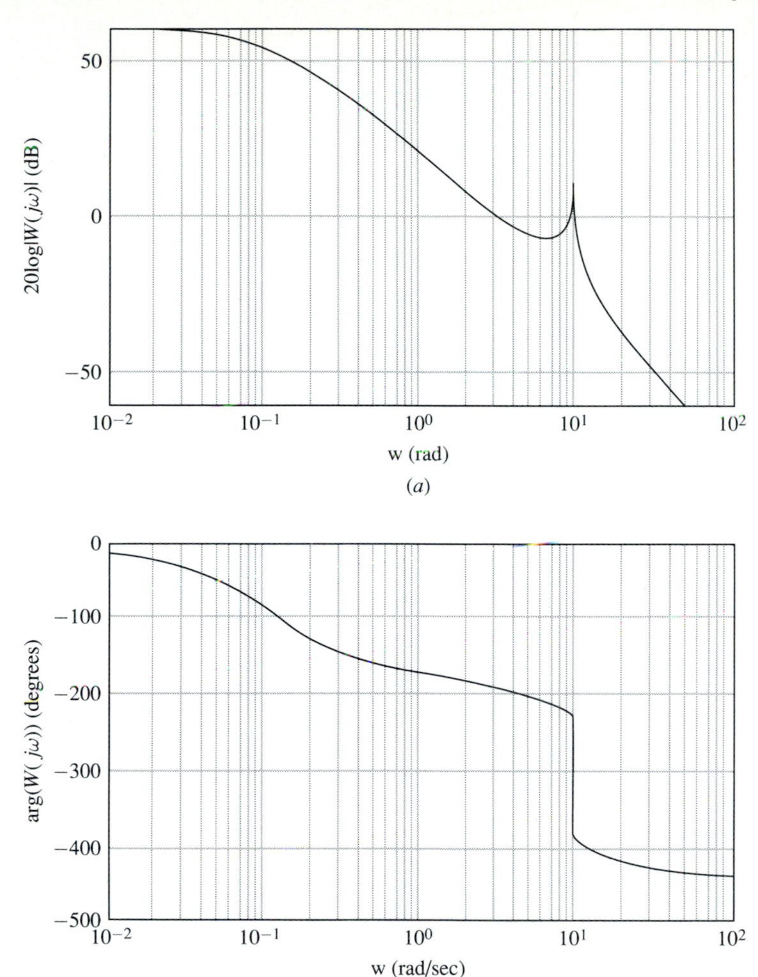

FIGURE 5.4-14
Bode plot for Example 5.4-5. (*a*) Magnitude; (*b*) phase.

because the frequency response is changing very rapidly with frequency. The computer simply connects the responses of the frequency points it is given. Often a control engineer needs to have a good idea what the plot should look like generally in order to properly interpret the specific information available on the plot.

```
% set up for the transfer function
num=[-100 1000];
den1=[1 .2 100];
den2=[1 .2 .01];
den=conv(den1,den2);
%create a frequency vector with extra points near
%the underdamped pole pair at w=10
```

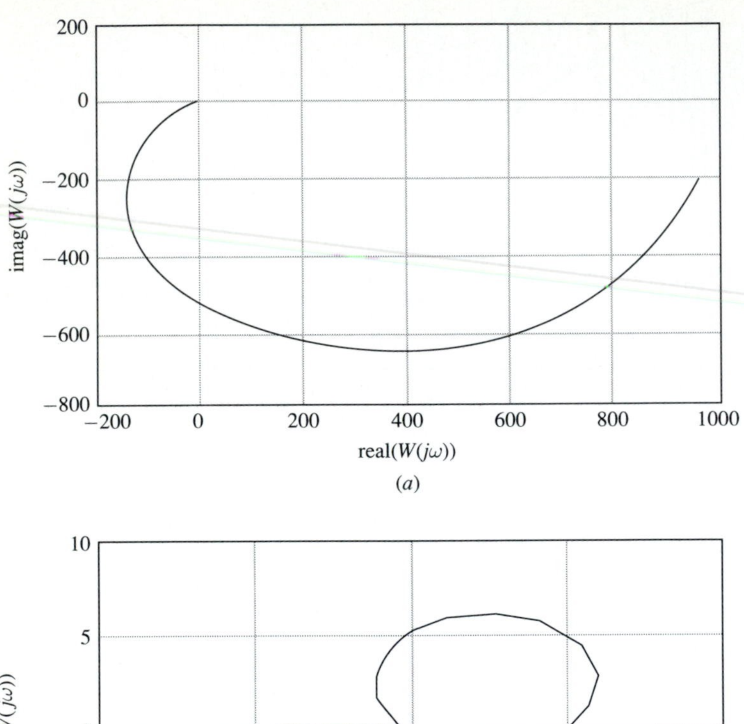

FIGURE 5.4-15
Nyquist plots for Example 5.4-5. (*a*) Large scale; (*b*) Small scale.

```
w1=logspace(-2,.5,75);
w2=logspace(.5,1.5,900);
w3=logspace(1.5,2,25);
w=[w1 w2 w3];
%get the magnitude and phase vectors
[mag,phase] = bode(num,den,w);
%convert the magnitude vector to dB
mag1=20*log10(mag);
%define the axis limits
a=[-2 2 -60 60];
axis(a);
```

```
%plot the magnitude plot
semilogx(w,mag1),xlabel('w (rad)'),
            ylabel('20log|W(jw)| (dB)'), grid
%return the axis to automatic scaling for the next plot
axis;
pause
%plot the phase plot
semilogx(w,phase), xlabel('w (Radians/s)'),
            ylabel('arg(W(jw))(degrees)'), grid
pause
%create the real part and the imaginary part vectors
[re,im]=nyquist(num,den,w);
%plot the whole polar plot to obtain the general shape
plot(re,im), xlabel('real(W(jw))'),ylabel('imag(W(jw))'),grid
pause
%set the axis for an expanded plot
a=[-10 10 -10 10];
axis(a);
%plot the expanded polar plot near the origin
plot(re,im), xlabel('real(W(jw))'),ylabel('imag(W(jw))'),grid
%hold this plot and add the point s=-1
hold on
plot(-1,0,'+')
%restore hold to off for the next usage
hold off
%return the axis to automatic scaling for the next usage
axis;
```

Exercises 5.4

5.4-1. Find the approximate phase plot for the following transfer functions by applying the straight line approximation. Sketch in an approximation to the actual phase plot around the straight line approximation. Don't plot more than one or two points for each pole or zero. If you have access to a CAD package, check your sketches with the plots of the package. (The transfer functions are the same as in Exercise 5.3-1.)

(a) $G(s) = \dfrac{1000}{(s+1)(s+10)(s+100)}$ (b) $G(s) = \dfrac{1000}{s(s+10)(s+100)}$

(c) $G(s) = \dfrac{900(s+10)}{s(s+3)(s+300)}$ (d) $G(s) = \dfrac{1000(s+1)}{s^2(s+10)}$

5.4-2. Repeat the instructions of Exercise 5.4-1 for the following transfer functions, which are the same transfer functions as in Exercise 5.3-2.

(a) $G(s) = \dfrac{(100)^2}{\left(s^2+20s+(100)^2\right)}$ (b) $G(s) = \dfrac{(100)^2(s+10)}{s\left(s^2+2s+(100)^2\right)}$

(c) $G(s) = \dfrac{(100)^2\,(s+s+100)}{s^2(s+100)^2}$ (d) $G(s) = \dfrac{\frac{10}{9}\left(s^2+6s+9\right)}{(s+1)(s+100)^2}$

5.4-3. Repeat the instructions of Exercise 5.4-1 for the following transfer functions, which are the same transfer functions as in Exercise 5.3-3.

$$(a)\quad G(s) = \frac{-1000}{(s+1)(s+300)} \qquad (b)\quad G(s) = \frac{s-1}{s(s+10)}$$

$$(c)\quad G(s) = \frac{s+3}{s^2(s-10)} \qquad (d)\quad G(s) = \frac{1000\left(s^2 - s + 100\right)}{s\left(s^2 - 2s + (100)^2\right)}$$

5.4-4. For the transfer functions referred to below, sketch the polar plot of the frequency response showing both the general nature of the response and the behavior of the plot near the point $s = -1$.

(*a*) The transfer function given in Exercise 5.4-1*a*
(*b*) The transfer function given in Exercise 5.4-1*b*
(*c*) The transfer function given in Exercise 5.4-2*b*
(*d*) The transfer function given in Exercise 5.4-1*b*

5.5 PLANT IDENTIFICATION

The early parts of this chapter developed the frequency response function $W(j\omega)$ and related this function to the steady-state response of the transfer function $W(s)$ to a sinusoidal input. Subsequent sections discussed the various means by which this frequency response function can be easily sketched, both in magnitude and phase. These sections are purely manipulative, and the reader may wonder what the ultimate utility of the frequency response function is. The frequency response function plays an important part in plant identification, closed-loop system specification, and system synthesis. This section discusses plant identification.

Throughout this chapter we have treated the frequency response function $W(j\omega)$ as though it were related to any transfer function $W(s)$; that is, $W(s)$ was not assumed to be related exclusively to either an open-loop plant or a closed-loop system. Since this section is concerned with plant identification, here we specifically restrict $W(s)$ to be associated with the plant to be controlled. Thus $W(s)$ may be the plant transfer function itself

$$W(s) = G_p(s) = \frac{Y(s)}{U(s)}$$

or $W(s)$ may relate any one of the state-variables to either the input or to another state-variable.

In the discussion no distinction is made between the three cases. In fact, the restriction that $W(s)$ be associated with the open-loop plant is in itself somewhat artificial. What is said in this section still applies to any general transfer function. The discussion here is directed toward the open-loop plant transfer function to emphasize the practical and important problem of plant identification that is readily solved through the use of frequency response methods.

Plant identification is the determination of the plant transfer function from experimental measurements. This problem often arises in actual control situations. A given, unalterable plant is to be controlled by an automatic control system. The transfer function of this plant may be completely unknown or the form of the transfer function may be known but not the numerical values. The problem is to identify the important state-variables and the transfer functions relating these state-variables so that the plant may be controlled in an intelligent manner.

Three factors make the asymptotic amplitude and phase diagrams ideally suited to the solution of this problem. First, it is fairly easy to approximate an experimentally determined magnitude plot by means of a straight line magnitude plot. Since the plot may have only integer multiples of 20 dB per decade for its slope, the trial and error fitting of the straight line approximation is considerably simplified.

Second, the form of the transfer function may be read, almost by inspection, from the straight line approximation. This feature may be traced to the fact that the break points of the straight line approximations correspond to the pole and the zero locations of the transfer function. Only a slight difficulty arises with regard to complex conjugate poles or zeros. This problem will be discussed in more detail later.

Third, joint consideration of the experimentally determined phase and magnitude plots uniquely specifies the transfer function. From the magnitude plot alone, the form of the transfer function is specified; that is, the break frequencies are all known. However, it is not known whether the system is minimum phase or not, and thus some ambiguity remains unless the phase plot is considered. This is an important point because, although many plants are minimum phase, such an assumption could prove disastrous if not true.

The use of the frequency response function and the asymptotic approximation of this frequency response function is illustrated in the following example.

Example 5.5-1. Assume that the form of a transfer function is known to be

$$W(s) = \frac{Y(s)}{U(s)} = \frac{K(s+\omega_1)}{s+\omega_2} = \frac{K(\omega_1/\omega_2)(1+s/\omega_1)}{1+s/\omega_2} \tag{5.5-1}$$

In this simple example we have assumed that the form of the transfer function has been determined from a knowledge of the physical structure of the actual system in question. The problem is to find K, ω_1, and ω_2.

A typical set of experimental data points for this transfer function is shown in Fig. 5.5-1 where both the amplitude and phase information are plotted. From the phase plot Fig. 5.5-1*b* it is seen that the phase angle initially starts at zero and that it also ends at zero. This can be true only if the system is minimum phase. Once the minimum phase nature of the transfer function has been established, the frequency response function corresponding to Eq. (5.5-1) may still take two different forms, depending on the relative magnitudes of ω_1 and ω_2. This excludes the highly unlikely situation that ω_1 is equal to ω_2, in which case the frequency response is flat at all frequencies. If $\omega_1 < \omega_2$, the zero will "break" before the pole, and the frequency response function will have the general form of Fig. 5.5-2*a*. On the other hand, if $\omega_1 > \omega_2$ the pole will "break" first, and the general shape of the magnitude plot will be that of Fig. 5.5-2*b*.

A comparison of Figs. 5.5-1*a* and 5.5-2 easily reveals that the frequency response must be of the form of Fig. 5.5-2*a*, or $\omega_1 < \omega_2$. Once this decision has been reached, all that remains to be done is to approximate the experimental data by horizontal (slope = 0) high and low frequency asymptotes and an interconnecting line segment with a slope equal to +20 dB per decade. This has been done in Fig. 5.5-3. From this figure we may

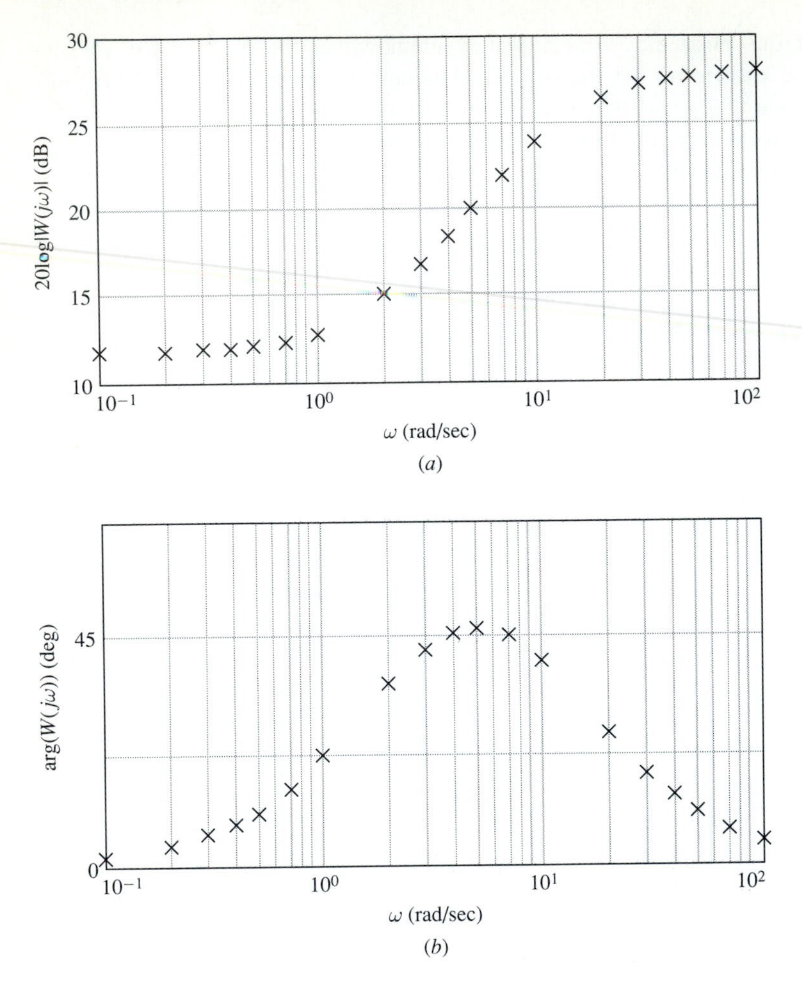

FIGURE 5.5-1
Experimental data points for Example 5.5-1. (*a*) Magnitude; (*b*) phase.

easily read off the values of ω_1, ω_2, and K as $\omega_1 = 2.0$ rad/sec, $\omega_2 = 13.0$ rad/sec, and $K = 4$. Therefore the transfer function $Y(s)/U(s)$ is

$$\frac{Y(s)}{U(s)} = \frac{4(1 + s/2)}{1 + s/13} = \frac{26(s + 2)}{s + 13} \tag{5.5-2}$$

Although this problem is now completely solved, it is well to point out that the phase plot of Fig. 5.5-1*b* might also have been used to aid in writing Eq. (5.5-2). Note that the maximum value of the phase angle occurs at $\omega = 5$. The center of the line of slope +20 dB per decade on Fig. 5.5-3 must also pass through this point, as it does in that figure.

Exactly this same procedure may be used in the general case when the plant is *n*th-order. Of course, in the general case the number of possible forms that the

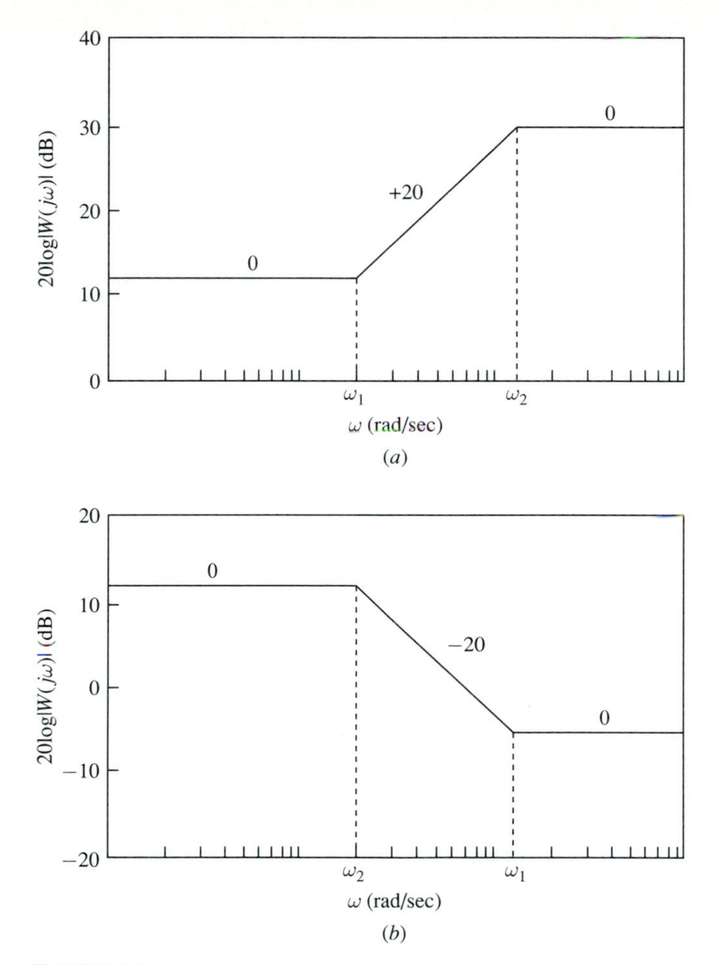

FIGURE 5.5-2
Possible forms for the magnitude plot for $Y(s)/U(s) = K(1+s/\omega_1)/(1+s/\omega_2)$. (*a*) $\omega_1 < \omega_2$; (*b*) $\omega_2 < \omega_1$.

magnitude plot may take is much greater than the two found in Example 5.5-1. Because of this fact, it is generally not practical to examine an exhaustive list of all possible forms to select the proper one. Rather, it is necessary to fit the experimental data with possible forms for the magnitude plots by trial and error. Once again this procedure is considerably simplified by the fact that the plot may have only integer multiples of 20 dB per decade for the slope.

If, in addition to the input-output data, one is able to make measurements of the internal state variables, it is possible to use this information to determine the state-variable representation of the plant in terms of the real, physical variables. This is done by finding each of the internal transfer functions and then writing the corresponding differential equations. These equations may then be combined to form the state-variable representation. In addition, these internal transfer functions serve

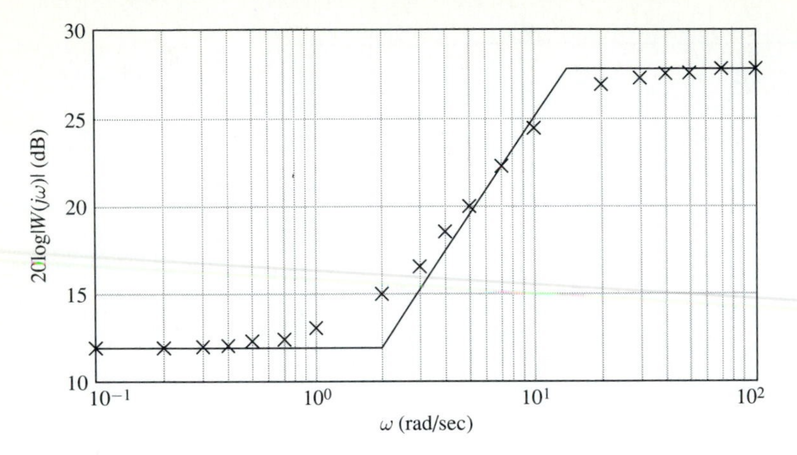

FIGURE 5.5-3
Approximation of experimental data points by straight lines.

as consistency checks on each other as well as on the input-output transfer function. Note also that, since these internal transfer functions are of lower order than the input-output transfer function, their determination is simpler since there are fewer possible forms to consider. This is an important practical consideration.

If it is not possible to measure all the internal variables, it is necessary to determine some of the state-variable equations from a knowledge of the physical origin of the plant or by simply making mathematical definitions. This situation of inaccessible states often occurs if the plant transfer function contains one or more sets of complex conjugate poles.

In addition, complex conjugate poles present a more difficult identification problem since the straight line approximation is not adequate for a complete identification. This inadequacy is due to the fact that the approximation is not affected by the value of the damping ratio ζ. This difficulty is overcome through the use of Eq. (5.3-10), which relates the magnitude of the transfer function at $\omega = \omega_n$ to the damping ratio ζ.

As in Sec. 5.3, let us assume that the plant is a typical second-order system so that

$$G_p(s) = \frac{\omega_n^2}{s^2 + 2\zeta\omega_n s + \omega_n^2} = \frac{1}{1 + 2\zeta s/\omega_n + s^2/\omega_n^2} \tag{5.5-3}$$

The magnitude of $G_p(j\omega)$ is then

$$\left|G_p(j\omega)\right| = \frac{1}{\left|(1 - \omega^2/\omega_n^2) + 2j\zeta(\omega/\omega_n)\right|}$$

and when ω equals ω_n, this magnitude becomes

$$\left|G_p(j\omega_n)\right| = \frac{1}{2\zeta} \tag{5.5-4}$$

The frequency ω_n is readily determined by the intercept of the two asymptotes with slopes 0 and -40 dB per decade, as in Fig. 5.3-5. Or the frequency ω_n may be determined from the phase plot of Fig. 5.4-5. Here it is seen that, regardless of the damping ratio, the additional phase shift cause by the complex pole pair is $-90°$ when ω equals ω_n. The damping ratio may then be determined by the use of Eq. (5.5-4) and the measured value of the amplitude at the frequency ω_n. The following example illustrates the determination of a plant transfer function involving complex conjugate poles.

Example 5.5-2. The transfer function of a completely unknown plant is to be determined by frequency response measurements. The amplitude and phase data that result from frequency response experiments are plotted in Fig. 5.5-4. As in Example 5.5-1, the amplitude information is plotted on a dB versus $\log \omega$ plot. The phase information is

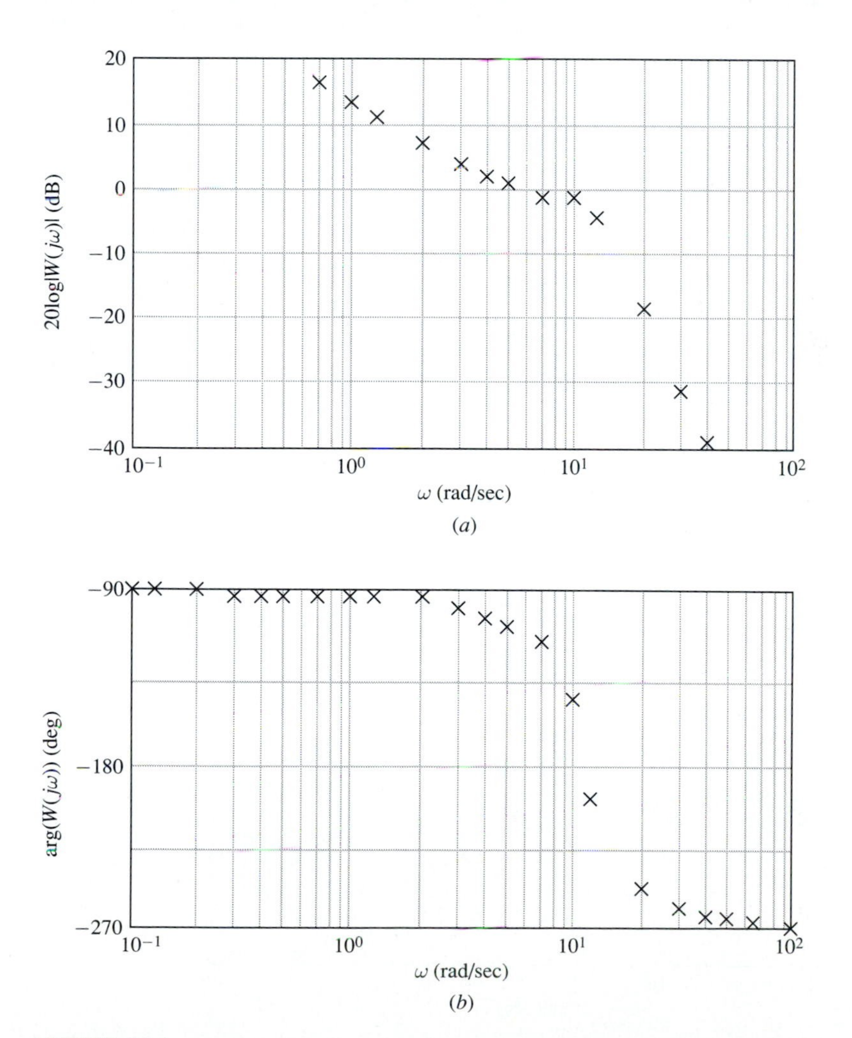

FIGURE 5.5-4
Plot of the data for the unknown plant of Example 5.5-2. (*a*) Magnitude; (*b*) phase.

plotted on semi-log paper. From the data of Fig. 5.5-4 we must determine the unknown plant transfer function.

In the amplitude diagram, the initial slope is -20 dB per decade, and the final slope is -60 dB per decade. In the phase diagram, the initial phase lag is $-90°$, whereas the final phase lag is $-270°$. Because of the initial -20 dB per decade slope, with the corresponding $-90°$ phase shift, the plant must contain an integrator. Because of the rise in the amplitude diagram, the plant contains either zeros or complex conjugate poles. Because of the final -60 dB per decade slope, and the corresponding $270°$ phase lag, we may hypothesize a minimum phase plant with a pole-zero excess of 3. Thus from a rather casual examination of the amplitude and phase plots, we already have considerable information about the plant transfer function.

Since it is clear that the plant contains an integrator, let us remove the frequency response of the integrator from the amplitude and phase plots. Removal of the $90°$ phase lag associated with the integrator is easily accomplished, and the resulting phase plot appears in Fig. 5.5-5. Removal of the effect of the integrator in the amplitude diagram is somewhat more difficult, as it must be done in a point-by-point fashion. It is required that $-20 \log \omega$ be subtracted from the amplitude plot of Fig. 5.5-4*a*. This is just the reverse of the process used in early sections of this chapter to construct a composite transfer function from the straight line approximations that made up its component parts. The subtraction is accomplished in Fig. 5.5-6. In the resultant figure it is seen that the dc gain is 5. If this gain is also removed (division becomes subtraction when dealing with decibels), the resulting amplitude diagram is as shown in Fig. 5.5-7. If this amplitude diagram is compared with those of Fig. 5.3-5, it is apparent that complex conjugate poles exist. We still must find the undamped natural frequency ω_n and the damping ratio ζ to determine the transfer function completely.

From the phase plot of Fig. 5.5-5 it is seen that the $-90°$ phase shift occurs near $\omega = 12$. On the amplitude diagram the extension of the low and high frequency asymptotes also meet at $\omega = 12$. From either diagram we may thus conclude that ω_n is 12. To determine ζ, we examine the value of $|G_p(j\omega_n)|$ and use Eq. (5.5-4) to determine

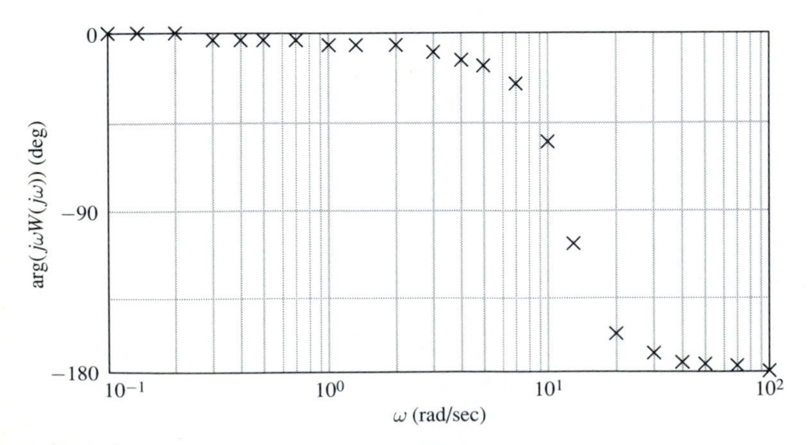

FIGURE 5.5-5
Phase plot after the integrator's phase shift has been removed.

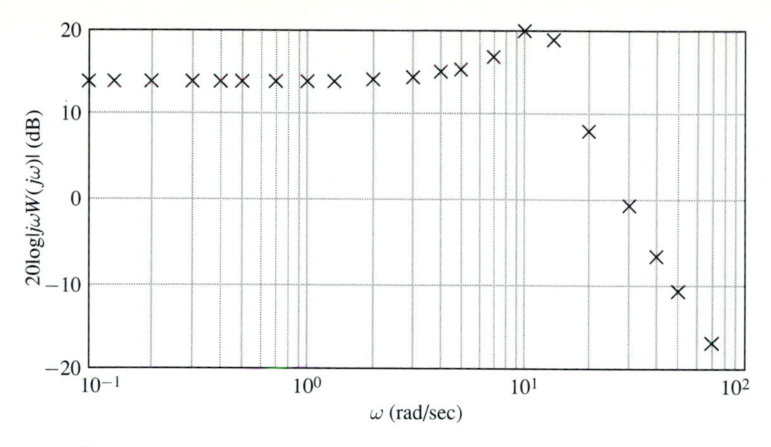

FIGURE 5.5-6
Magnitude response after the effect of the integrator has been removed.

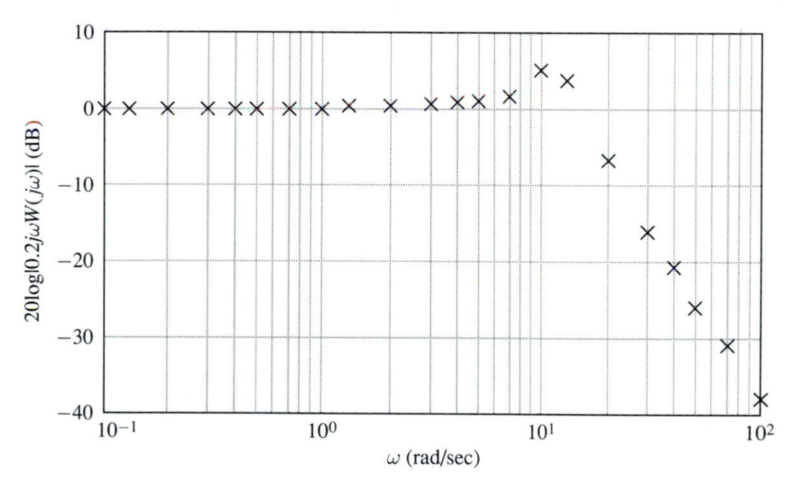

FIGURE 5.5-7
Magnitude response after the effect of the integrator and the dc gain of 5 have been removed.

ζ. From Fig. 5.5-7, it is seen that $|G_p(j\omega_n)|$ is 2 so that $\zeta = 0.25$. The plant transfer function is now known to be

$$\left|G_p(s)\right| = \frac{5}{s\left[1 + (2\zeta/\omega_n)\,s + s^2/\omega_n^2\right]} = \frac{5}{s(1 + 0.5s/12 + s^2/144)}$$

or, in pole-zero form,

$$\left|G_p(s)\right| = \frac{720}{s(s^2 + 6s + 144)}$$

It is important to check the phase plot as well as the magnitude plot to correctly identify a non-minimum phase system as shown in the following example.

Example 5.5-3. Assume that the magnitude plot of a transfer function is given as in Fig. 5.5-1 but the phase is given by Fig. 5.5-8. A straight line fit to the magnitude curve as shown in Fig. 5.5-3 shows a break point at $\omega = 2$. With a slope on the magnitude plot changing from zero to +20 dB per decade. For a minimum phase system, we would expect a phase shift of slightly less than $+45^\circ$ around $\omega = 2$ since the average slope here is slightly less than +10 dB per decade. However, from the plot of Fig. 5.5-8 we see that the phase has decreased to closer to -45°. The change in the magnitude curve to an increasing slope along with the decreasing of the phase shift indicates a zero in the right half-plane. Thus we consider the magnitude plot to have a pseudo slope of -20 dB per decade between $\omega = 2$ and $\omega = 5$. We also identify the zero at $\omega = 2$ in time constant form as (1-s/2).

The decrease of slope in the magnitude plot at $\omega = 15$ would change the psuedo slope from -20 dB per decade to -40 dB per decade if it were due to a left half-plane pole. This agrees with the high frequency phase approaching -180° in Fig. 5.5-8 and we identify a left half-plane pole in time constant form as $(1 + s/15)^{-1}$. The low frequency gain is 4 and zero low frequency phase shift tells us that we should associate a positive sign with the gain in time constant form. The final transfer function is identified.

$$G(s) = \frac{4(1 - s/2)}{1 + s/15} = \frac{-30(s - 2)}{s + 15}$$

An increase of the actual magnitude slope combined with a decrease of the pseudo slope as indicated by the phase curve indicates a right half-plane zero. A decrease in actual slope combined with an increase in the pseudo slope indicates a right half-plane pole. An increase in both the actual slope and the pseudo slope indicates a left half-plane zero while a decrease in both the actual slope and pseudo slope indicates a left half-plane pole.

There are a few practical considerations in identifying a plant by its frequency response. If it is possible, find an intermediate variable in the plant where a measurement can be made. It is often easier to break the plant into two pieces and

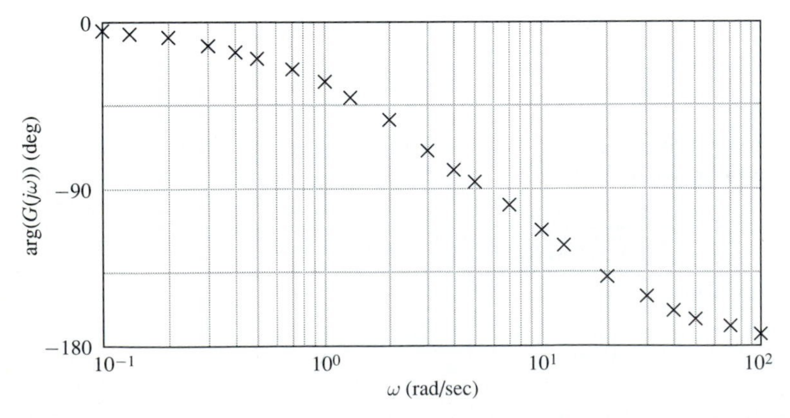

FIGURE 5.5-8
Phase plot for Example 5.5-3.

identify each piece separately. Thus in the field-controlled motor of Fig. 2.2-10 it may be easier to excite the input voltage $e(t)$ with a sinusoid and measure the current $i_f(t)$ and the output velocity $\dot{\theta}_0(t)$. The transfer function from $e(t)$ to $i_f(t)$ can then be identified. The transfer function from $i_f(t)$ to $\dot{\theta}_0(t)$ can be identified with the same data since in steady state $i_f(t)$ will consist of a sinusoid of the same frequency as $e(t)$. Finally, $\theta_0(t)$ is obviously just the integral of $\dot{\theta}_0(t)$ and the whole model is identified. The advantage of this method is that a clearer picture of each part of the transfer function is more easily obtained without the presence of the other pieces.

The last practical issue is how to identify unstable systems. Clearly a straightforward attempt to measure the frequency response by sinusoidal excitation of the plant will not work. An unstable plant will not reach a steady-state value. The magnitude of the output just continues to increase. The frequency response of the system is not well defined. However, the system function of the plant is well defined and can be identified.

First, a closed-loop controller which stabilizes the plant must somehow be designed, probably using a trial and error method. How this can be done will be better understood after studying Chaps. 6 and 7. Information similar to frequency response information can then be attained by introducing sinusoids into the loop through the reference input. A steady state sinusoid results at the signal points $u(t)$ at the input to the plant and $y(t)$ at the plant output. If the magnitude and phase relationship between $y(t)$ and $u(t)$ is recorded, the transfer function of the unstable plant can be identified using the methods of this section. Remember that the unstable pole(s) will cause a decrease in the actual slope of the magnitude while creating an increase in the pseudo slope as seen in the phase plot. Identification of an unstable system is explained in Problem 5.8.

Exercises 5.5

5.5-1. Find the transfer function for each of the three magnitude plots shown in Fig. 5.5-9. Assume that the plants are minimum phase and stable.
Answers:

$$(a) \qquad G_p(s) = \frac{100(s+10)}{s(s+100)}$$

$$(b) \qquad G_p(s) = \frac{1000s}{(s+100)(s^2+.01s+1)}$$

$$(c) \qquad G_p(s) = \frac{1000\left(s^2+0.015s+9\right)}{s^2(s+30)^2}$$

5.5-2. Find the transfer function for each of the two Bode plots shown in Fig. 5.5-10 and Fig. 5.5-11. The plants are *not necessarily* minimum phase or stable.
Answers:

$$(a) \qquad G_p(s) = \frac{3(s-100)}{(s-1)(s+10)}$$

$$(b) \qquad G_p(s) = \frac{100(s+1)}{s^2(s^2-5s+100)}$$

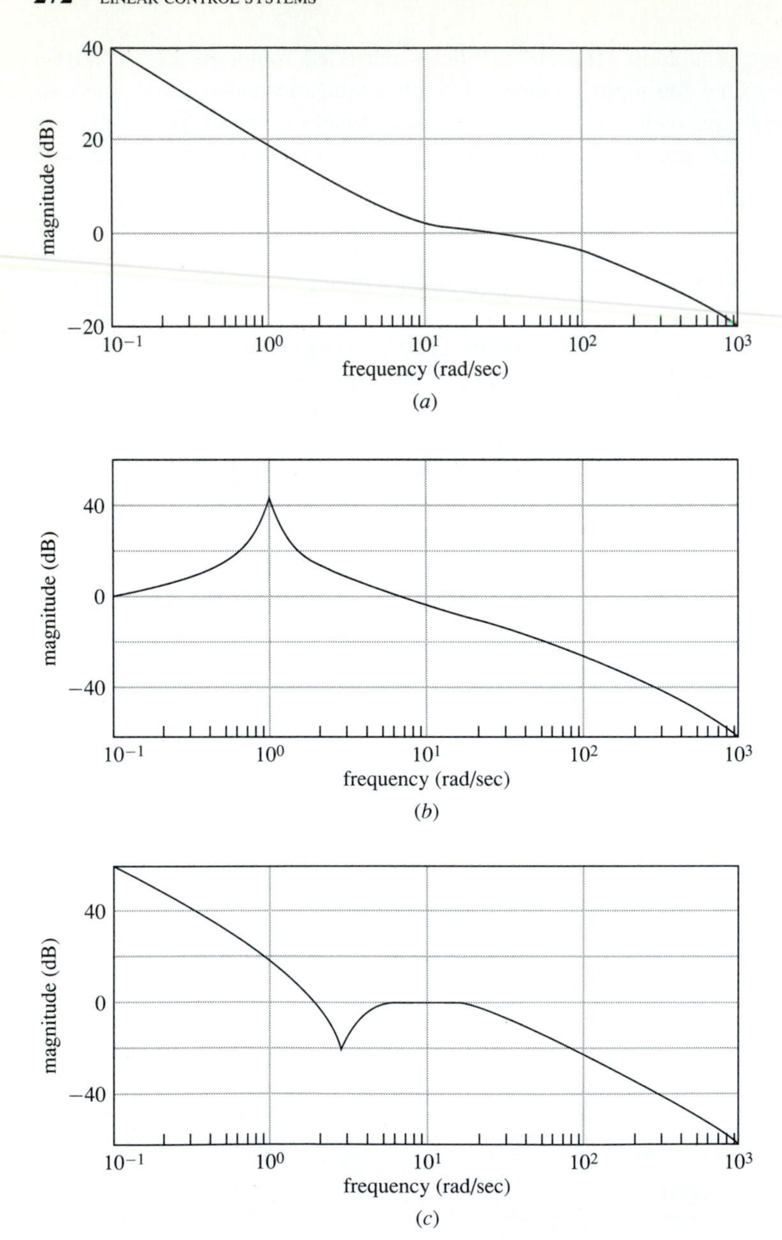

FIGURE 5.5-9
Exercise 5.5-1.

5.6 ACCOUNTING FOR UNCERTAINTY IN MODELING

In the previous section, we learned how to identify a system from its frequency response. Imagine that you are a manufacturer of position control systems consisting

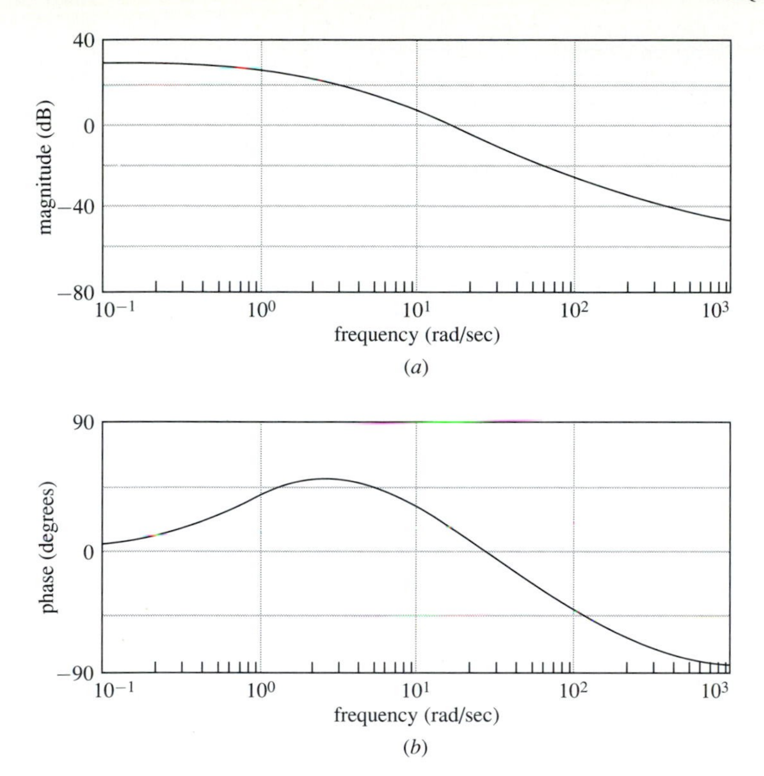

FIGURE 5.5-10
Bode plots for Exercise 5.5-2*a*. (*a*) Magnitude; (*b*) phase.

of a dc motor with a shaft and gears leading to a platform. After manufacturing a number of platform systems, you have technicians measure the frequency response of each system to see how well it matches the model your engineers derived for the system. The following patterns may arise.

Assuming you have a reasonably precise manufacturing operation, the frequency response of the different systems may match the model fairly well at low frequencies. There may be some discrepancies due to slightly different parameters among systems but the discrepancies should be within an acceptably small level.

As the frequency increases, you may find that the frequency responses start to deviate from the model and even start to deviate from each other. Phase measurements of different systems at individual frequencies may vary by 180°. At higher frequencies the technician is likely to be unable to get the system to settle down enough to give you an accurate measurement of the frequency response. It may be possible to obtain only a range for the magnitude of the response and may not be possible to obtain reliable phase information at all. The reasons for these problems in obtaining high frequency measurements are unmodeled dynamics.

The physical nature of unmodeled dynamics was discussed in Sec. 2.7. In this section we discuss a method of extracting the important and obtainable features of

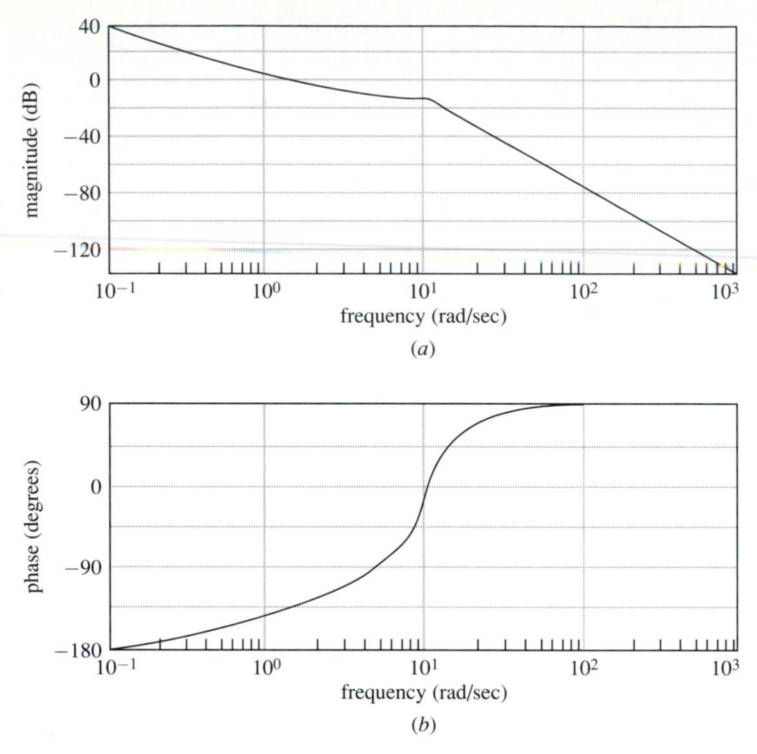

FIGURE 5.5-11
Bode plots for Exercise 5.5-2*b*. (*a*) Magnitude; (*b*) phase.

these modeling uncertainties. The goal is to produce a simple model that is reasonably accurate for all the systems that have been and will be manufactured. Thus we need to model a class of systems, not just one system. One way to accomplish this is to think about a nominal plant model in conjunction with a perturbation model. The range of frequency responses of a class of plants may be displayed graphically on Bode plots as in Figs. 5.6-1*a* and 5.6-1*b*. Fig. 5.6-1*a* shows the usual characteristics of increasing magnitude uncertainty with increasing frequency and Fig. 5.6-1*b* shows total phase uncertainty above some frequency.

A second way of viewing the uncertainty of the frequency response at each frequency is by using the polar plot of Fig. 5.6-1*c*. The area of the drawing around the point labeled ω_1 represents the situation where a range of magnitude responses and a range of phase responses are observed. The resulting uncertainty region is a wedge in the polar plane. The actual frequency response lies somewhere in this wedge. It may be possible by closer observation to find a different irregular shape for the uncertainty region. This is represented by the area in Fig.5.6-1*c* surrounding ω_2. To simplify the mathematical model in either case the uncertainty region is enclosed in a circle. The center of the circle is used as the nominal plant model's frequency response and the radius of the circle is the maximum magnitude of an additive perturbation.

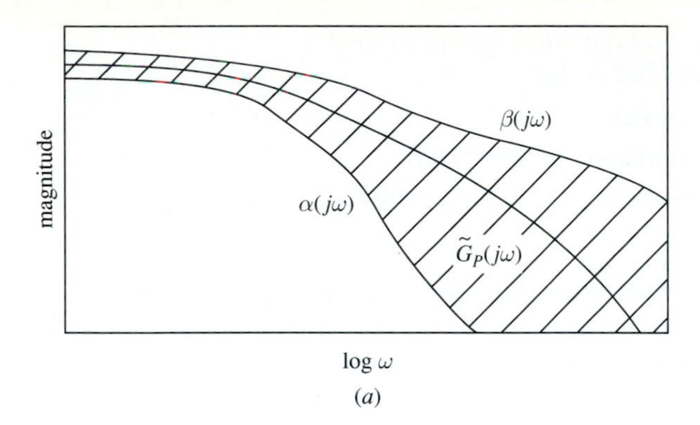

(a)

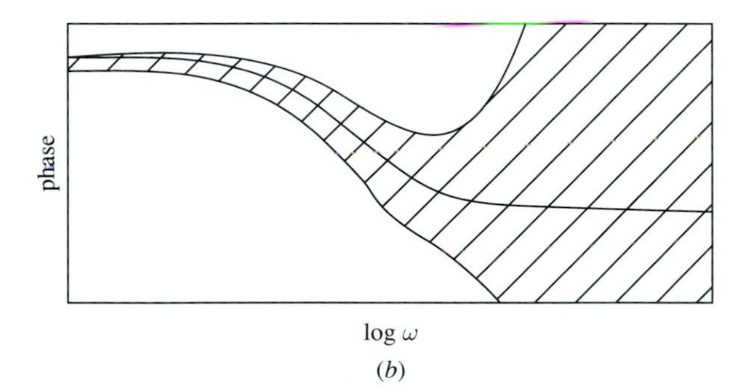

(b)

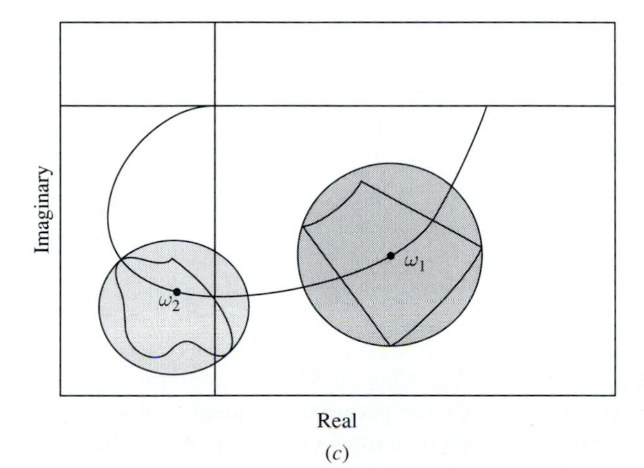

(c)

FIGURE 5.6-1
Frequency plot uncertainty regions; (a) Bode magnitude plot, (b) Bode phase plot, (c) Nyquist plot

This uncertainty model can be described mathematically as follows.

$$\tilde{G}_p(s) \approx G_p(s) + L_a(s) \tag{5.6-1a}$$

$$|L_a(j\omega)| < l_a(j\omega) \tag{5.6-1b}$$

where $G_p(s)$ is the nominal model, i.e., the basic plant model used for the controller design

$\tilde{G}_p(s)$ is the perturbed model

$L_a(s)$ is an additive perturbation to the model

$l_a(j\omega)$ is a bound on the magnitude of the additive perturbation

Equations (5.6-1a) and (5.6-1b) actually specify a large class of plants. Each different perturbation $L_a(s)$ defines a different perturbed plant. To encompass a class of plants we allow a class of perturbations as defined by Eq. (5.6-1b). In general, control system design is carried out on the nominal plant model but techniques are available to assure that a controller designed for the nominal plant remains acceptable for any of the possible perturbed plants defined by Eqs. (5.6-1a) and (5.6-1b).

Referring back to Fig. 5.6-1c, it can be seen that the figure matches the description if the frequency response points marked on the figure as ω_1 and ω_2 are points on the nominal plant response function, $G_p(j\omega)$, while the radii of the shaded circles are given by $l_a(j\omega_1)$ and $l_a(j\omega_2)$. Then all the points within the shaded circle around ω_1 are the possible frequency responses around ω_1 for the set of plants given by $\tilde{G}_p(j\omega_1)$. The same is true at ω_2.

The connection between Eq. (5.6-1) and Fig. 5.6-1c works both ways. The model of Eq. (5.6-1) may have been developed by other means and physical insights. In this case Fig. 5.6-1c is the proper way to interpret the model. Conversely, a model such as Eq. (5.6-1) may be derived using data represented as is done in Fig. 5.6-1c. In the latter case care must be taken that the number of plants observed and the conditions under which the observations are made are varied enough to encompass all plant responses which can reasonably be expected while operating the plant.

With a little manipulation it is also possible to write the expression for the perturbed plant as a multiplicative perturbation.

$$\tilde{G}_p(s) = G_p(s)\,(1 + L_m(s)) \tag{5.6-2a}$$

$$|L_m(j\omega)| < l_m(j\omega) \tag{5.6-2b}$$

There are a few things to note about Eq. (5.6-2). The perturbation $L_m(s)$ is referred to as a multiplicative perturbation since the perturbation quantity $(1 + L_m(s))$ multiplies the plant in a series connection. The quantity $(1 + L_m(s))$ rather than just $L_m(s)$ is used as the multiplier so that a small perturbation corresponds to $\tilde{G}_p(s) \approx G_p(s)$.

Equating $\tilde{G}_p(s)$ in Eq. (5.6-1) and Eq. (5.6-2) shows us that the multiplicative perturbation is a relative perturbation having the interpretation of a percentage change in the plant model.

$$L_m(s) = \frac{L_a(s)}{G_p(s)} \tag{5.6-3}$$

The relationship between Eq. (5.6-2) and the magnitude plot of Fig. 5.6-1a is straightforward. From the magnitude plot in Fig. 5.6-1a it can be seen that

$$\alpha(j\omega) < \left|\tilde{G}_p(j\omega)\right| < \beta(j\omega) \tag{5.6-4}$$

After $G_p(s)$ is chosen, $l_m(j\omega)$, the bound on the perturbation, must be set large enough to encompass both magnitude and phase uncertainty. This can be accomplished for magnitude uncertainty by defining

$$\left|G_p(j\omega)\right| = \frac{\alpha(j\omega) + \beta(j\omega)}{2} \tag{5.6-5}$$

and

$$\left|G_p(j\omega)l_m j\omega)\right| = \frac{\beta(j\omega) - \alpha(j\omega)}{2} \tag{5.6-6}$$

Equations (5.6-2), (5.6-5), and (5.6-6) describe all the $\tilde{G}(s)$ outlined in the magnitude plot of Figure 5.6-1a.

The relationship of the phase plot of Fig. 5.6-1b and Eq. (5.6-2) is slightly more subtle. When $|l_m(j\omega)| > 1$ the phase of $\tilde{G}_p(j\omega)$ is completely uncertain, i.e., given any angle there exists an $L_m(j\omega)$ with magnitude slightly larger than one, such that $\arg(\tilde{G}_p(j\omega))$ matches the given angle. Let's show this. From Eq. (5.6-2) we can write

$$\begin{aligned}\arg(\tilde{G}_p(j\omega)) &= \arg\left(G_p(j\omega)\,(1 + L_m(j\omega))\right) \\ &= \arg\left(G_p(j\omega) + G_p(j\omega)L_m(j\omega)\right)\end{aligned} \tag{5.6-7}$$

which shows that $\arg(\tilde{G}_p(j\omega))$ is equal to the argument of the sum of two complex numbers. Since the phase of $L_m(j\omega)$ is arbitrary the phase of $G_p(j\omega)L_m(j\omega)$ is arbitrary. Since $L_m(j\omega)$ has magnitude greater than one, $G_p(j\omega)L_m(j\omega)$ has magnitude greater than the magnitude of $G_p(j\omega)$. All the possible complex numbers that can be values of $\tilde{G}_p(j\omega)$ are shown in Fig. 5.6-2. The possible values of $\tilde{G}_p(j\omega)$ are given

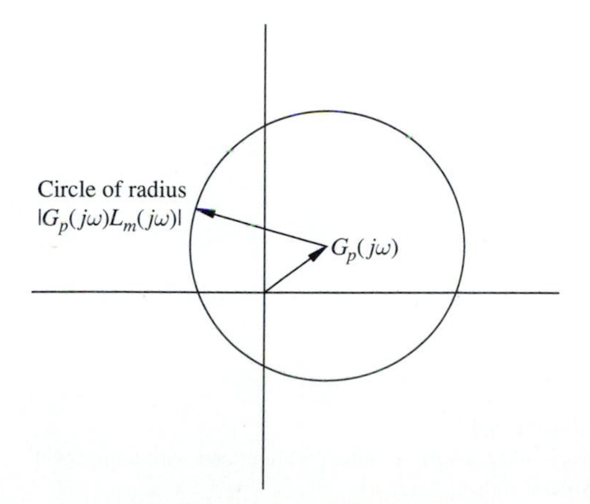

FIGURE 5.6-2
Possible complex numbers for $\arg\left(G_p(j\omega) + G_p(j\omega)L_m(j\omega)\right)$

by a circle in the complex plane centered at $G_p(j\omega)$ with radius $|G_p(j\omega)L_m(j\omega)|$, which is greater than $|G_p(j\omega)|$.

We now look at typical shapes of $l_m(j\omega)$ that may be expected in practice. We refer back to Sec. 2.7 where typical unmodeled dynamics were discussed in the context of a dc motor positioning system. We will discuss two commonly occurring modeling problems from that section and investigate how these problems are captured by Eqs. (5.6-1) and (5.6-2).

5.6.1 Unmodeled Dynamics

As discussed in Section 2.7 there are always dynamics that are part of the plant but are excluded from the nominal plant model. These dynamics are excluded for various reasons: the nominal model must remain simple enough to be computationally tractable, the unmodeled dynamics may be so highly sensitive to parameters that their models are inaccurate for most of the systems built, or it is just impossible to model all dynamics. In a mechanical system we can expect poorly damped bending modes to be typical of dynamics that are too difficult to model accurately.

Example 5.6-1. Refer to Fig. 2.7-2 from Example 2.7-2. A nominal rigid body positioning system model is displayed in Figure 2.7-2*a* and a model including the first bending mode is displayed in Fig. 2.7-2*b*. The forms of the plant models are repeated in Fig. 5.6-3.

Let the nominal plant be given by

$$G_p(s) = G_1(s)G_2(s) \tag{5.6-8}$$

The perturbed plant is given by

$$\begin{aligned}\tilde{G}_p(s) &= G_1(s)\,[G_2(s) - L(s)] \\ &= G_1(s)G_2(s) - G_1(s)L(s) \\ &= G_1(s)G_2(s)\left[1 - \frac{L(s)}{G_2(s)}\right]\end{aligned} \tag{5.6-9}$$

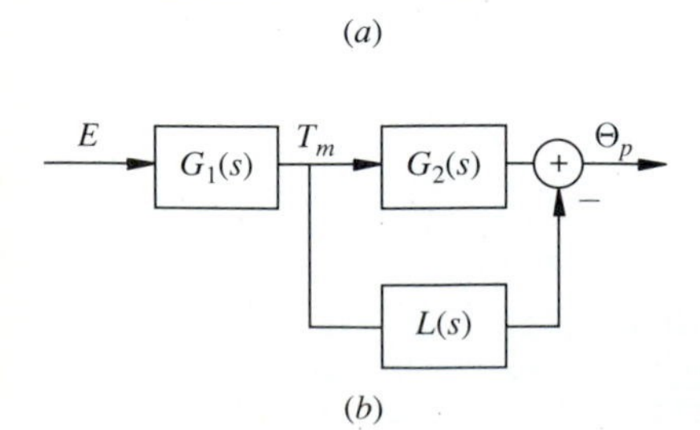

FIGURE 5.6-3
Block diagrams for a system without and with unmodeled dynamics.

From Eq. (5.6-9) the additive and multiplicative perturbations can respectively be defined as

$$L_a(s) = -G_1(s)L(s) \tag{5.6-10}$$

$$L_m(s) = \frac{L(s)}{G_2(s)} \tag{5.6-11}$$

Equations (5.6-10) and (5.6-11) satisfy Eq. (5.6-3). In our example, $L(s)$ is given by

$$L(s) = \frac{1/J}{s^2 + (\beta/J)s + K_s/J} \tag{5.6-12}$$

The parameters β and K_s are difficult to determine and will vary from one realization of the system to another. Also there is in reality a series of bending modes with higher-order modes having higher frequencies. The form of Eq. (5.6-12) is that of a complex pole pair. Assume that the pole pair is poorly damped so that $L(j\omega)$ has a small value of around $1/K_s$ at low frequencies, peaks to a value of $\left(\beta\sqrt{K_s/J}\right)^{-1}$ at $\omega = \sqrt{K_s/J}$ and then falls off. The magnitude of $L(j\omega)$ is plotted in Fig. 5.6-4 for $J = 1, \beta = 0.15$ and $K_s = 9,\ 100$, and 900. In general we can expect to know a minimum value of K_s but little else about it.

To account for the possibility of any subset of these bending modes appearing in the actual plant, we can bound in magnitude all of the curves in Figure 5.6-4.

For this example an appropriate bound is given by the magnitude of the following transfer function.

$$l'(s) = 4\frac{(s+1)^2(s^2+3s+9)}{(s+3)^3(s^2+0.6s+9)} \tag{5.6-13}$$

The magnitude of $l'(j\omega)$ and the possible perturbations are plotted in Fig. 5.6-4.

Now there is a bound on $|L(j\omega)|$

$$|L(j\omega)| < \left|l'(j\omega)\right| \tag{5.6-14}$$

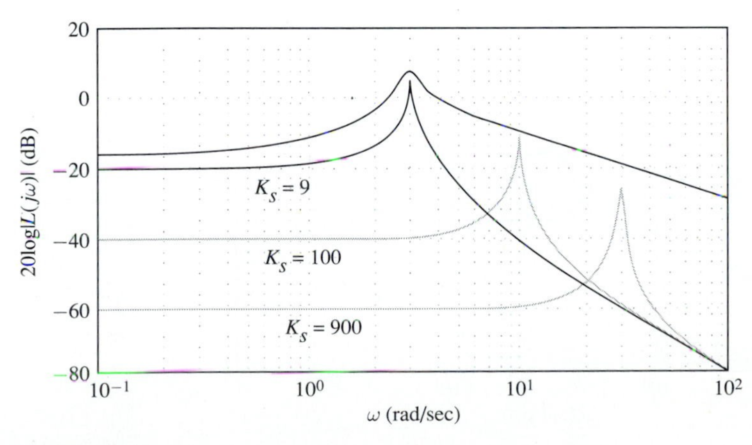

FIGURE 5.6-4
Bending modes and bounding $|l(j\omega)|$.

Combining Eqs. (5.6-11) and (5.6-14) produces the desired bound on the possible multiplicative perturbations

$$|L_m(j\omega)| < l_m(j\omega) \tag{5.6-15}$$

where

$$l_m(j\omega) = \frac{|l'(j\omega)|}{|G_2(j\omega)|} \tag{5.6-16}$$

The quantity $l_m(j\omega)$ is plotted in Fig. 5.6-5 for a typical system using the following $G_2(j\omega)$.

$$G_2(j\omega) = \frac{1}{s(s+0.1)}$$

While the bounds on additive perturbations may level off or decrease at high frequencies, the bounds on multiplicative perturbations tend to increase as frequency increases. The reasons for this increase can be seen from Eq. (5.6-3). The plant divides the additive perturbation to form the multiplicative perturbation. Since plant magnitudes roll off (decrease in magnitude with increasing frequency), if the additive perturbation levels off, the multiplicative perturbation increases with increasing frequency.

5.6.2 Time Delays

There are always some time delays in the actuation of control commands. In process control systems there may be long time delays involved with material flow. In mechanical and aerospace systems, non-oscillatory unmodeled dynamics such as those associated with the movement of actuators can be thought of as delays in command actuation. As was displayed in Sec. 2.7, the primary effect of certain nonlinearities can be summarized as a time delay. The model of a time delay does not yield a rational transfer function. Also, we will see in later chapters that time delays produce inherent

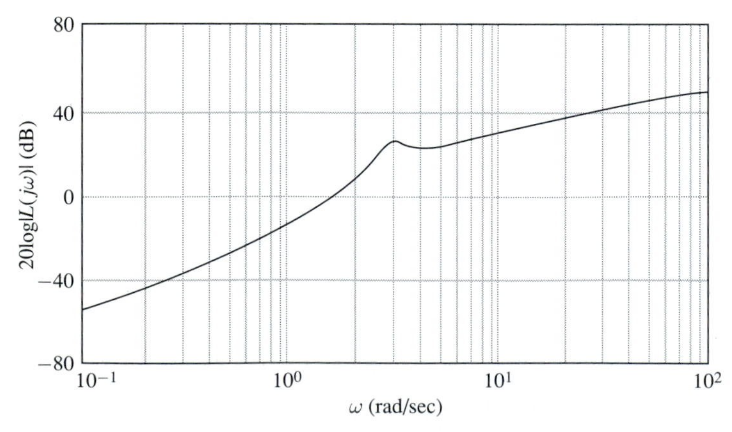

FIGURE 5.6-5
The bound $l_m(j\omega)$.

limitations to control systems that are not easily overcome. If there is a time delay in the system, that system cannot react to a command input in a time frame any faster than the length of the delay. Time delays are often left out of the nominal plant model and thus become part of the possible perturbations to the plant model.

Consider the system of Fig. 5.6-6 where there is an actuation delay connected with a plant.

The impulse response for a delay of T seconds is

$$h(t) = \delta(t - T) \tag{5.6-17}$$

The transfer function is given by the transform of the impulse response. We denote the transform of the delay by $L(s)$

$$L(s) = e^{-sT} \tag{5.6-18}$$

Notice that $L(s)$ is not a rational function, i.e., a ratio of polynomials in s. There are no finite poles or zeros.

The multiplicative perturbation can be derived from the perturbed plant

$$\tilde{G}_s(s) = e^{-sT} G_p(s) = (1 + L_m(s))\, G_p(s) \tag{5.6-19}$$

$$L_m(s) = e^{-sT} - 1 \tag{5.6-20}$$

The magnitude of $L_m(s)$ is plotted in Fig. 5.6-7 for the values $T = 2$ and $T = 4$. An upper bound for the magnitude of the multiplicative perturbations associated with

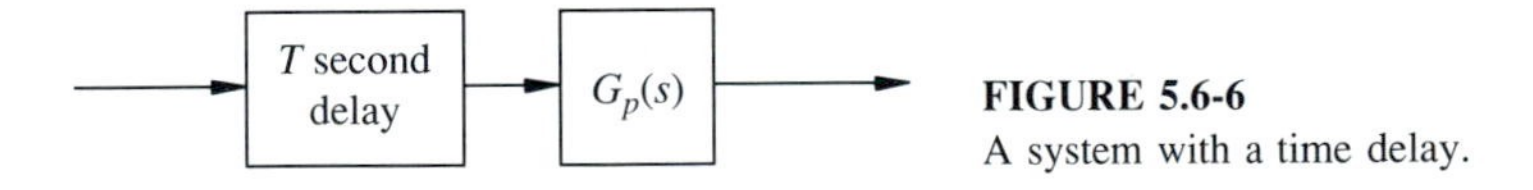

FIGURE 5.6-6
A system with a time delay.

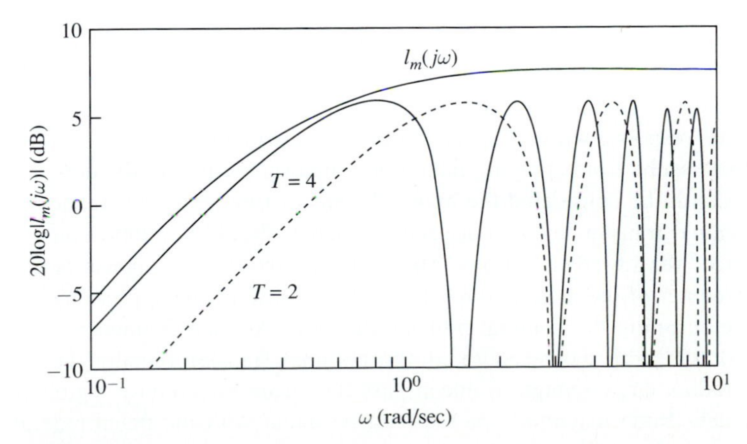

FIGURE 5.6-7
Magnitude of $l_m(j\omega)$ for a time delay perturbation.

time delays of less than $T_{\max}$ sec is

$$l_m(s) = \left| \frac{2.5s}{s + \frac{2}{T_{\max}}} \right| \tag{5.6-21}$$

Of interest is the frequency at which the magnitude of $l_m(j\omega)$ becomes greater than one. A direct calculation indicates that $l_m(j\omega)$ becomes greater than one when $\omega \approx 1/T_{\max}$.

In summary, Eqs. (5.6-1) and (5.6-2) provide a mechanism to allow the control designer to work with a simple nominal model while accounting for the remainder of the dynamics in a less specific, less detailed manner. This structure allows the designer to ascertain whether or not a controller design is able to function adequately for a whole class of plants. It is important to realize, however, that the ultimate goal of the controller is to work well on the physical system that is to be controlled. This is different from the controller working well on any of the mathematical models within the class defined by Eqs. (5.6-1) or (5.6-2).

The following procedure is given as a sensible method of moving from a set of data resulting from a number of identification experiments to plant models of the form given in Eqs. (5.6-1) and (5.6-2). First, it should be realized that the data coming from an identification experiment is never as clean as the "experimental" plots that appear in Sec. 5.5. Most systems are not purely linear and do not yield a pure sinusoidal output when excited by a sinusoid. Also, identification experiments ought to be made over the entire range of the expected operating environment of the system including changes of temperature and system wear. Lastly, in some applications, a large number of systems may need to be tested in the identification process since each realization of the plant design may vary somewhat from the others and a controller should work acceptably on each realization.

When a reasonably complete set of identification experiments is performed the response of the plant at each frequency is likely to fill a region on the polar or Nyquist plot as shown in Fig. 5.6-1*c*. At each frequency, the corresponding region on the plot should be encompassed by a circle of as small a radius as possible. The center of this circle can then be taken as the nominal experimental response at that frequency. The entire nominal experimental frequency response can be plotted in this way. The method of approximating this response with a transfer function is developed in Sec. 5.5. The transfer function that results from applying that approximation method to the nominal experimental data should be considered the nominal transfer function. The frequency response of the nominal transfer function at each frequency should be plotted on the original polar plots of the raw data regions. The nominal frequency response is not likely to match the center of the original circles exactly because of the approximation processs used in determining the nominal transfer function. At each frequency, new circles should be drawn with the response of the nominal transfer function at the center and with a radius large enough to encompass the entire uncertainty region at that frequency. A new Bode magnitude plot should be made with the magnitude at each frequency equal to the radius of the circle at that frequency. The method of Sec. 5.5 should now be used to find a transfer function $l_a(s)$ whose magnitude doesn't just approximate this magnitude plot but overbounds it. The transfer function $l_a(s)$ is the bound on the additive perturbation. One may need to increase the magnitude of this bound further to capture reasonably likely effects that, for some reason, are not present in the data of the identification experiments. The multiplicative perturbation bound can be computed using Eq. (5.6-3).

Equations (5.6-1) and (5.6-2) were developed to account for as much of the discrepancy between the physical system and mathematical model as is mathematically tractable. However, the final implementation of a controller requires a leap of faith on the part of the control designer. The control designer must be convinced that the final design will indeed work on the physical system. This confidence comes from experience in using the design tools provided in this book and in developing an understanding of the real limitations of mathematical models.

We stop momentarily to warn the student that the subject of how best to account for modeling inaccuracies is a subject of current research in control theory at the time this book is being printed. Approaches to this problem that differ from those presented in this section are being studied. In particular, work is being done on the effect of perturbations in the parameters of the plant transfer function. (These kinds of perturbations are often called structured perturbations.) We choose to emphasize perturbations to the frequency response of the plant because we feel that the additive or multiplicative perturbations (often called unstructured perturbations) lead clearly to important design principles. The reader may consult the references for more information.

Exercise 5.6-1

5.6-1. A plant is modeled by

$$G_p(s) = \frac{3}{s+1}$$

but in reality the plant contains unmodeled dynamics and its transfer function is

$$\tilde{G}_p(s) = \frac{15(s+5)}{(s+1)(s^2+3s+25)}$$

Find the additive perturbation $L_a(s)$ and the multiplicative perturbation $L_m(s)$ for this case.

Answers:

$$L_m(s) = \frac{-s(s-2)}{s^2+3s+25} \qquad L_a(s) = \frac{-3s(s-2)}{(s+1)(s^2+3s+25)}$$

5.6-2. A real plant transfer function is the product of a modeled portion and an unmodeled portion

$$\tilde{G}_p(s) = G_p(s)G_u(s) \qquad G_u(s) = \frac{A}{s^2+s+A}$$

(*a*) Briefly explain how $G_u(s)$ affects the phase of $\tilde{G}_p(s)$.

(*b*) Sketch the Bode magnitude plot of $L_m(s) = G_u(s) - 1$ for A=1, 10, 100, and 1000 on the semi-log plot using the dB scale. Sketch in an approximation to $20\log|l_m(j\omega)|$ so that $|l_m(j\omega)| > |L_m(j\omega)|$ for all ω and all A. Find a transfer function for $l_m(s)$.

Answer: (*b*) $|l_m(j\omega)| = 4/\omega$

5.7 CONCLUSIONS

In this chapter the frequency response function has been introduced, and several approximate methods of representing the frequency response have been described. Initially the discussion considered the general transfer function $W(s)$ and no attempt was made to associate $W(s)$ with either the open-loop or closed-loop system. In Sec. 5.5, $W(s)$ was restricted to the open-loop plant, and methods were described by which the transfer function $W(s)$ could be identified. We will find in the next chapter that much information concerning the frequency response of the closed-loop transfer function can be obtained from the frequency response of the open-loop transfer function.

The results of Chaps. 4 and 5 can be combined. In Chap. 5, we learned how to go from a frequency response plot to a transfer function description. In Chap. 4 we learned how to go from a transfer function to a step response. Thus we can tell qualitatively how a system will react to a unit step input by examining its frequency response. (See Problem 5.9.) This helps make frequency response methods useful tools in controller design. The usefulness of these tools is expanded by the ability of frequency response data to capture information on plant uncertainty.

Another of the important uses of the frequency response information is in answering the question of system stability, one of the basic questions in any control problem. Stability is the minimum requirement that any system design must satisfy. The frequency response provides a useful graphical tool for analyzing system stability. This subject is discussed in the next chapter.

PROBLEMS

5.1. The closed-loop transfer function for a particular positioning servomechanism has been experimentally determined to be

$$\frac{Y(s)}{R(s)} = \frac{125(s+4)}{\left[\left((s+3)^2+4^2\right)(s+5)\right]}$$

From the frequency-response function, determine the analytic expression for $y(t)$ for the values of ω of 1, 5, 10, and 100 if the input is $r(t) = 4\sin\omega t$. Assume steady-state conditions.

5.2. Plot the approximate Bode plots, i.e., the approximate magnitude and phase plots, for the given transfer functions.

$$(a)\quad W(s) = \frac{90000(s+1)}{s(s+3)(s+10)(s+30)}$$

$$(b)\quad W(s) = \frac{100(s+3)}{s(s+1)^2(1+s/30)}$$

5.3. Plot the approximate Bode plots, i.e., the approximate magnitude and phase plots, for the following two transfer functions involving complex poles and zeros.

$$(a)\quad W(s) = \frac{11.1(s^2+0.1s+9)}{s(1+s/0.1)(1+s/10)}$$

$$(b)\quad W(s) = \frac{125(s+1)}{\left[(s+3)^3+4^2\right](s+10)}$$

5.4. Plot the approximate Bode plots, i.e., the approximate magnitude and phase plots, for the following transfer functions.

$$(a) \quad W(s) = \frac{-1000s + 30000}{s^2 - s}$$

$$(b) \quad W(s) = \frac{200(s+5)}{s^2\left(s^2 - 0.5s + 100\right)}$$

5.5. For the transfer functions listed below, sketch the polar plot of the frequency response showing both the general nature of the response and the behavior of the plot near the point $s = -1$.

(*a*) The transfer function given in Problem 5.2(*a*)
(*b*) The transfer function given in Problem 5.3(*a*)
(*c*) The transfer function given in Problem 5.4(*b*)

5.6. The form of an unknown plant transfer function is indicated in Fig. P5.6, and the various parameter values are to be found. This is done by making frequency response measurements, where

$$u(t) = 7 \sin \omega t$$

Amplitude and phase data are taken between u and the available states, and from one available state to the other, that is, between u and x_3, u and x_1, and x_3 and x_1, and the data are recorded in Table P5.6. Use the data from u to x_3 and x_3 to x_1 to determine the unknown plant parameters. Then use only input-output data, u to x_1, and repeat the problem. Compare the difficulty involved with the two approaches.

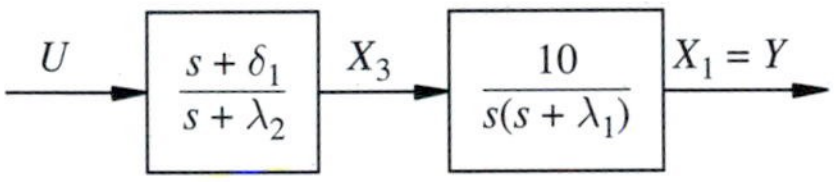

$$U \rightarrow \boxed{\frac{s+\delta_1}{s+\lambda_2}} \xrightarrow{X_3} \boxed{\frac{10}{s(s+\lambda_1)}} \xrightarrow{X_1 = Y}$$

FIGURE P5.6
Problem 5.6.

TABLE P5.6
Data for Problem 5.6

ω	$\left\lvert\frac{X_1(j\omega)}{U(j\omega)}\right\rvert$	$\arg \frac{X_1(j\omega)}{U(j\omega)}$, deg	$\left\lvert\frac{X_3(j\omega)}{U(j\omega)}\right\rvert$	$\arg \frac{X_3(j\omega)}{U(j\omega)}$, deg	$\left\lvert\frac{X_1(j\omega)}{X_3(j\omega)}\right\rvert$	$\arg \frac{X_1(j\omega)}{X_3(j\omega)}$, deg
0.01	6.70	−84.6	0.020	5.6	333	−90.2
0.02	3.41	−79.3	0.021	11.0	167	−90.4
0.04	1.80	−69.5	0.021	21.2	83.7	−90.8
0.10	0.942	−48.0	0.028	43.8	33.3	−91.9
0.20	0.743	−32.7	0.045	61.1	16.6	−93.8
0.40	0.679	−26.2	0.082	71.3	8.30	−97.6
1.00	0.623	−35.6	0.197	73.0	3.16	−108.4
2.00	0.516	−58.6	0.371	65.4	1.39	−123.6
4.00	0.314	−93.0	0.623	50.0	0.504	−143.0
10.0	0.0857	−137.3	0.894	26.0	0.0958	−163.3
20.0	0.0241	−157.7	0.970	13.8	0.0248	−171.4
40.0	0.062	−168.7	0.992	7.0	0.00629	−175.7
100	0.0010	−175.5	0.999	2.8	0.00100	−178.3

5.7. (*Nichols charts*). For the plant transfer function

$$G_p(s) = \frac{10}{s(1+s)(1+s/10)}$$

draw the Bode magnitude and phase plots for the frequency range $0.1 < \omega < 10$. Now plot the frequency response information on a single rectangular plot, using the dB scale on the vertical axis and the phase angle in degrees on the horizontal axis, with ω as a parameter along the plot. This type of frequency response plot is known as a *Nichols chart* and is an alternative method of representing the frequency response information. Note that semi-log paper may be used to plot magnitude versus phase angle directly.

5.8. A plant that is known to be unstable is to be identified. Through trial and error a constant gain stabilizing series compensator is found. The block diagram of the closed-loop system is shown in Fig. P5.8-1. The value $K = 100$ stabilizes the system. Various sinusoids of unity height were input as R. These reference inputs have the form $r(t) = \cos \omega t$. The magnitudes and phases of the steady-state sinusoids that result in the signals U and Y are recorded on the plots given in Fig. P5.8-2.

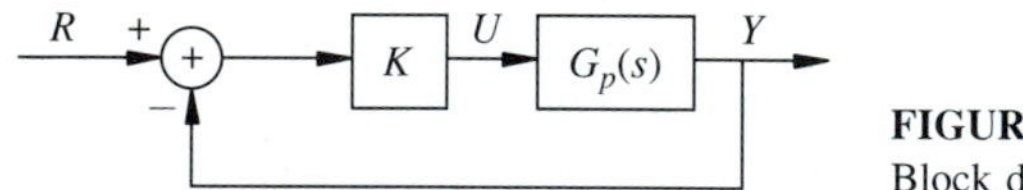

FIGURE P5.8-1
Block diagram for Problem 5.8.

(*a*) Use the identification method given in Sec. 5.5 to find the transfer function of the closed-loop system, $M_C(s)$. Then use this information and the fact that $K = 100$ to find $G_p(s)$.

(*b*) Identify $G_p(s)$ directly by using the ratio of the measured magnitudes and the difference of the measured phases of $Y(s)$ and $U(s)$ and applying the identification method of Sec. 5.5

(*c*) Use the ideas of Sec. 3.2 to explain why one of the above identifications of $G_p(s)$ is more reliable than the other.

5.9. In Chap. 4 you learned how to look at the zero-pole form of a transfer function and determine an approximation to the step response of the system. In Sec. 5.5, you learned how to look at the frequency response of a system and determine an approximation to its zero-pole form. In this problem the two lessons are placed back to back. Look at the frequency response plots of Fig. P5.9 and describe the step response of the underlying transfer functions by first establishing the zero-pole forms of the systems. (Note: The phase of (*b*) matches the phase of (*a*) for the first decade. It then falls sharply.) From working these problems certain patterns emerge that relate features of frequency responses to features of step responses. The feature demonstrated in each problem is given below:

(*a*) The bandwidth of the frequency response determines the speed of the step response.
(*b*) A peak in the magnitude response corresponds to an oscillation in the step response.
(*c*) A large phase lag at low frequency while the magnitude is not falling too rapidly indicates a delay in the step response.

5.10. A known field-controlled dc motor similar to the one modeled in Example 2.2-2 has been fitted with a flexible coupling in the shaft. Frequency domain measurements have been

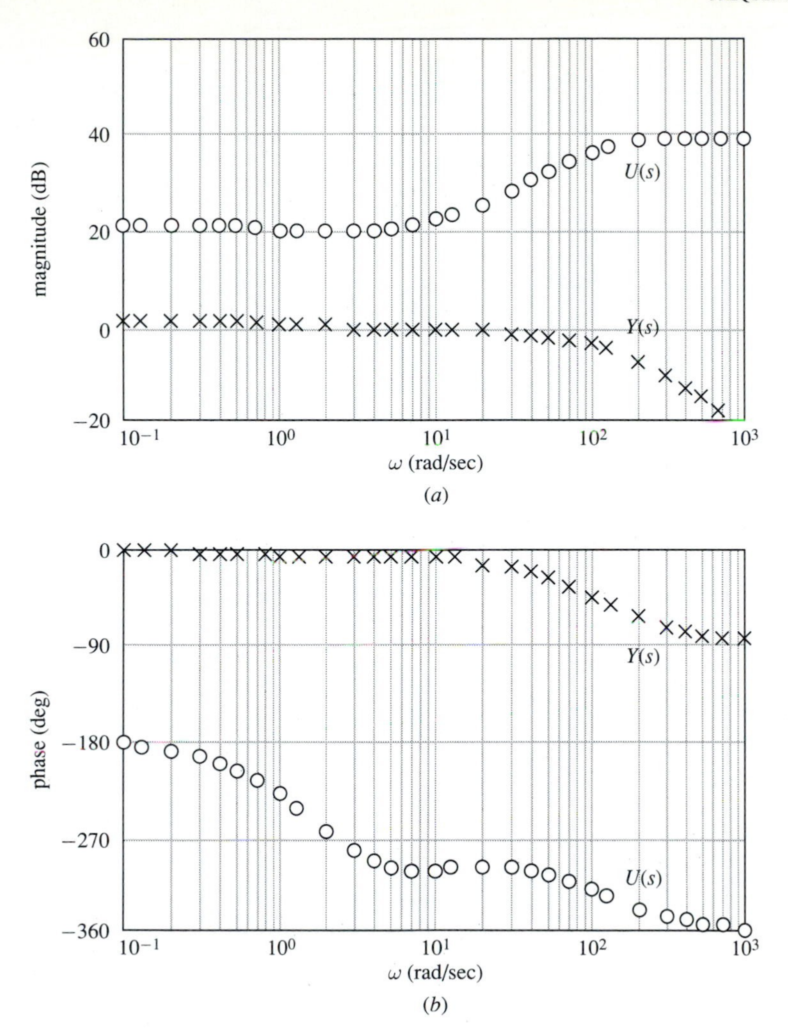

FIGURE P5.8-2
Bode plots for Problem 5.8. (*a*) Magnitude; (*b*) phase.

taken on this motor using the input voltage as the input and the angular velocity of the motor as the output. The results are given in Table P5.10. At each frequency a range of magnitudes and a range of phases were taken to reflect that the actual response is only approximately sinusoidal and all measurements are difficult to make precisely. Use the procedure given at the end of Sec. 5.6 to produce a nominal transfer function model and a multiplicative uncertainty bound for the data of Table P5.10.

5.11. The Bode diagram approach is particularly useful in producing series compensators. Suppose that

$$G_p(s) = \frac{100}{s+1}$$

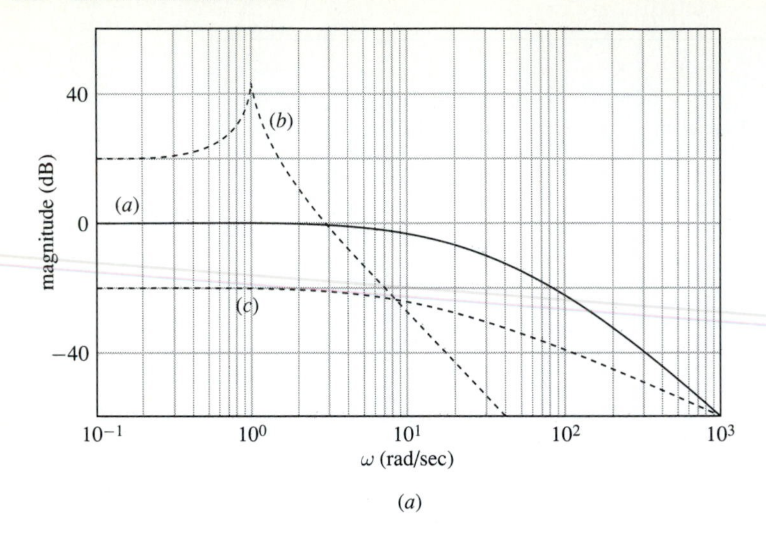

(a)

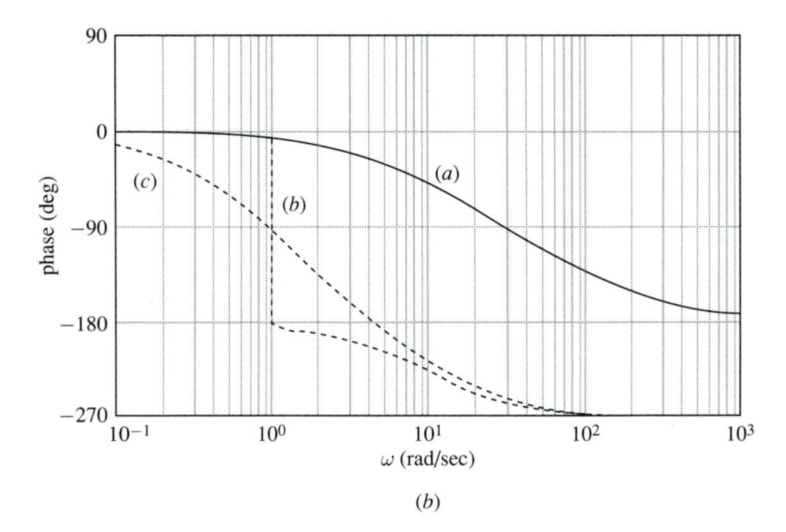

(b)

FIGURE P5.9
Bode plots for Problem 5.9. (*a*) Magnitude; (*b*) phase.

Sketch the straight line approximation to the Bode magnitude plot. Now a series compensator $G_c(s)$ is to be designed so that the loop gain

$$G(s) = G_c(s)G_p(s)$$

stays near 100 for $\omega < 0.3$ but also has the property that the magnitude is less than 0.1 for $\omega > 100$. Use the property that

$$\log|G(s)| = \log|G_c(s)| + \log|G_p(s)|$$

and the Bode plot to find an acceptable $G_c(s)$.

TABLE P5.10
Data for Problem 5.10

ω (rad/sec)	Magnitude	Phase(deg)
2π (.03)	6.6 to 6.7	$-3°$ to $-5°$
2π (.1)	6.3 to 6.5	$-10°$ to $-11°$
2π (.3)	5.4 to 5.6	$-23°$ to $-25°$
2π (1)	2.6 to 2.65	$-50°$ to $-60°$
2π (3)	1.12 to 1.15	$-50°$ to $-70°$
2π (10)	0.8 to 0.85	$-55°$ to $-75°$
2π (30)	0.32 to 0.5	$-70°$ to $-113°$
2π (100)	0.06 to 0.15	$-90°$ to $-130°$
2π (300)	0.037 to 0.120	$-80°$ to $-214°$
2π (1000)	0.01 to 0.08	$-214°$ to $-326°$

5.12. The multiplicative form of perturbations is also useful for series compensator design. Often plant mismodeling can be described as a multiplicative or additive perturbation on $G_p(s)$.

$$\begin{aligned}\tilde{G}_p(s) &= G_p(s) + L_{pa}(s)\\ &= G_p(s)\left(1 + L_{pm}(s)\right)\end{aligned}$$

However we are usually interested in analyzing the effect of the perturbations on the loop gain of a system with a known series compensator

$$\tilde{G}(s) = \tilde{G}_p(s)G_c(s)$$

We write $\tilde{G}(s)$ in perturbed form as

$$\begin{aligned}\tilde{G}(s) &= G(s) + L_a(s)\\ &= G(s)\,(1 + L_m(s))\end{aligned}$$

where $G(s) = G_p(s)G_c(s)$

(*a*) Find $L_a(s)$ first in terms of $L_{pa}(s)$, then in terms of $L_{pm}(s)$

(*b*) Find $L_m(s)$ first in terms of $L_{pa}(s)$, then in terms of $L_{pm}(s)$

Notice that dealing with all loop gain perturbations as multiplicative perturbations is easy because the perturbations to the loop gain do not change when the compensator changes.

5.13. (*CAD Problem*) An area in system theory closely related to communication system design and control design is called filter design. The objective in filter design is to find a transfer function whose frequency response most closely resembles a desired frequency response. The problem is closely related to the transfer function identification problem. In filter design certain prototype desired filters such as a low pass filter have become the focal point of many results. These filters are designed to produce a very sharp transition from near unity gain to small gain on the magnitude plot. The phase characteristics are often viewed as an interesting but less significant factor. In this problem we investigate the frequency response of three standard forms of low pass filters. A third-order (three pole) filter is examined in each case. The prototypes can all be found for higher-order filters and these higher-order filters result in sharper transition regions. In each part, first sketch the straight line approximation of the Bode plots, making the usual alterations for complex pole pairs. Then get precise plots using CAD tools.

(*a*) Butterworth filters:

$$G_B(s) = \frac{1000}{\left(s^2 + 10s + 100\right)(s + 10)}$$

A Butterworth filter transfer function is easy to find. To find an nth order Butterworth filter place $2n$ poles equally spaced around a circle of radius ω_c, where ω_c is the desired cutoff frequency. If n is odd a pole should lie on the negative real axis; if n is even the negative real axis should lie equidistant between two poles. The resulting frequency magnitude plot is monotone decreasing with the complex pair holding up the magnitude plot as much as possible just before ω_c. The phase plot is also monotone.

(*b*) Chebyshev filters:

$$G_C(s) = \frac{252}{\left(s^2 + 9s + 84\right)(s + 3)}$$

The Chebyshev filter achieves a tighter transition from the passband to the stopband than the Butterworth filter by allowing ripples in the passband. The passband magnitude must fall off from unity at some point. The Chebyshev filter spreads the error more evenly throughout the passband and achieves a tighter transition region by doing so. The poles in a Chebyshev filter are actually on an ellipse in the complex plane so that the more highly damped poles are at lower frequency than the less damped poles. The magnitude plot droops between the poles and is held up at the poles. The phase plot is more severely curved in the passband than the Butterworth filter.

(*c*) Elliptical filters:

$$G_E(s) = \frac{252/2153\left(s^2 + 2153\right)}{\left(s^2 + 9s + 84\right)(s + 3)}$$

The elliptical filter uses ripples in the passband as the Chebyshev filter does and it also introduces ripples in the stopband. The poles are usually close to the Chebyshev poles. (They are not always exactly the same as they are here.) Zeros are placed on the imaginary axis in the stopband to pull the magnitude response down more quickly. The tradeoff in doing this is that the final slope of the magnitude plot is only −20 dB per decade. The magnitude plot gets to a certain level of attenuation (in this case, 60 dB) more quickly but rises to that level again at a higher frequency before rolling off more slowly.

CHAPTER

6

STABILITY

6.1 INTRODUCTION

This chapter is devoted to one of the most basic requirements of any control system: stability. Independent of other performance specifications that the closed-loop system is to meet, it must always exhibit the characteristics that we shall define for stability. For this reason, the study of system stability forms an important part of both analysis and design of control systems.

In a sense, the concept of this chapter is the opposite of that of Chap. 4. There we sought the exact and complete behavior of a system due to a specific input for all time values. Here, on the other hand, our interest is in the asymptotic behavior of a system for all inputs as time approaches infinity.

The study of stability of linear systems deals with three basic questions. The first question, the absolute-stability problem, is qualitative in nature and seeks a simple "yes" or "no" statement concerning the system stability. The second question, the relative stability problem, is quantitative in nature and is associated with the problem of determining how stable a system is. The third question examines the qualities of robustness, trying to determine how much a plant can be perturbed and still remain stable.

Although the second and third questions are more difficult to answer, the answers are by far the more valuable. In effect, relative stability and robustness information provide a bridge between the minimal information of absolute stability and the complete information provided by the total time response, with labor also falling somewhere between the two extremes. That is, the use of the relative stability concept is an attempt to characterize the behavior of a system by means of one or more relative stability measures. Of course, relative stability is only one of the important performance measures that must be satisfied to ensure satisfactory system performance. The robustness question needs to be answered so that a designer can be confident that a controller designed for a nominal plant model works as expected for an actual plant. The actual plant's response should be within a typical perturbation of the modeled plant response, as discussed in Sec. 5.6.

In Sec. 6.2 several definitions of stability are presented and discussed in order to make the concept of stability more precise. An algebraic method of determining whether the characteristic polynomial of a closed-loop system contains poles on the $j\omega$ axis or in the right half-plane is discussed in Sec. 6.3. This is the Routh-Hurwitz approach. A quite different approach to the stability problem, through the use of the frequency response function and the Nyquist criterion, is discussed in Sec. 6.4. The Nyquist criterion for stability is an extremely powerful tool in the analysis and design of control systems. It received its name from Harry Nyquist, the inventor of this method of analyzing control systems. In Secs. 6.5 to 6.7, questions of relative stability are addressed. A closed-loop system that oscillates has poles closer to the imaginary axis than a system that doesn't oscillate. An oscillatory system is said to possess less relative stability than a non-oscillatory one. In Sec. 6.5 we discover how to obtain qualitative information about the response of a closed-loop system from the Nyquist plot of the loop gain when using a series compensator. A second notion of relative stability is that a system with much relative stability should remain stable in the face of modeling errors. Such a system is said to have a significant amount of robustness. The question of robustness in stability is examined in Sec. 6.6. This concept is extended to robustness in performance in Sec. 6.7. A system with a large amount of performance robustness can maintain desired performance levels in the face of uncertainty in the plant models.

6.2 DEFINITIONS OF STABILITY

In the preceding chapter, the concept of stability was used in two places in a rather loose sense to describe two different properties. In Sec. 5.2, a stable system was described as a system in which the complementary solution decays to zero as time approaches infinity. In the discussion of the Bode relation in Sec. 5.4, on the other hand, we defined a stable system as one in which all the poles are in the left half of the s plane.

This dual use of the concept of stability is acceptable for two reasons. First, each of these interpretations of stability has some intuitive appeal, and second, as we shall see later, these two concepts of stability are equivalent. Hence there is no lack of consistency in our use of them.

It may appear that, if stability has an intuitive meaning, there is no need for this section. On the contrary, just the opposite is the case. Concepts having intuitive meaning usually escape analytic treatment until precise mathematical definitions are imposed. At the same time, care must be taken in making the definition to ensure that the intuitive appeal is not lost.

As the title of this section indicates, there are not one but several definitions of stability. In particular, there are three definitions that are most widely used within our setting of linear time-invariant systems. It is the normal procedure to select one of these definitions as *the* definition and then prove that the other two follow directly. Rather than follow this procedure and elevate one of them above the others, a slightly different approach is taken here. All three definitions are offered as definitions of equivalent importance, and we then undertake to establish that the three are equivalent. The three equivalent definitions are:

1. A linear time-invariant system is stable if its output remains bounded for every bounded input.
2. A linear time-invariant system is stable if its weighting function is absolutely integrable over the infinite range, that is, if $\int_0^\infty |w(t)|\,dt$ is finite.
3. A linear time-invariant system is stable if all the poles of the closed-loop transfer function Y(s)/R(s) lie in the left half of the s plane.

The reader will note that the first two definitions deal with the time domain, whereas the third is a frequency domain condition. Although the third definition proves to be the most useful in our work, the first definition is more closely related to observable phenomena. The second definition serves as a bridge between the other two. Hence, the three definitions tend to complement each other by providing alternative ways of viewing the same problem.

Let us begin by showing the equivalence of the first and second definitions; we then demonstrate that the second and third definitions are equivalent.

We consider first the problem of showing that, if the weighting function is absolutely integrable, every bounded input yields a bounded output. By using the convolution integral, we may write the output $y(t)$ to an input $r(t)$ as

$$y(t) = \int_0^\infty w(\tau)r(t-\tau)d\tau \tag{6.2-1}$$

If we take the absolute value of both sides of Eq. (6.2-1), we obtain

$$|y(t)| = \left|\int_0^\infty w(\tau)r(t-\tau)d\tau\right| \le \int_0^\infty |w(\tau)|\,|r(t-\tau)|\,d\tau \tag{6.2-2}$$

Since the input is assumed to be bounded, there must be a positive number M_1 such that $|r(t)| \le M_1$ for all t, so that Eq. (6.2-2) becomes

$$|y(t)| \le M_1 \int_0^\infty |w(\tau)|\,d\tau$$

If the weighting function is absolutely integrable, there must exist a positive number M_2 such that

$$\int_0^\infty |w(\tau)|\,d\tau \le M_2$$

Therefore the magnitude of the output is bounded by $M_1 M_2$, or

$$|y(t)| \leq M_1 M_2$$

which is the desired result since $|y(t)|$ is thus bounded for all t.

So far, we have shown that, if the second definition of stability is satisfied, the first one is also satisfied. To establish the complete equivalence of these two definitions, we must now show that if the first definition is satisfied then the second is. That is, we must show that, if every bounded input yields a bounded output, the weighting function is absolutely integrable.

An alternative but entirely equivalent approach is to show that, if the second definition is not satisfied, the first also fails. In other words, we must prove that if the weighting function is *not* absolutely integrable then there exists at least one bounded input that yields an unbounded output. This is the approach that is employed.

We consider once again Eq. (6.2-1). For any t, let us construct an input whose value as a function of τ is given by the relation

$$r(t-\tau) = \begin{cases} +1 & \text{if } w(\tau) \geq 0 \text{ and } t > \tau \\ -1 & \text{if } w(\tau) < 0 \text{ and } t > \tau \\ 0 & \text{if } t < \tau \end{cases}$$

This input is bounded by ± 1, and for this $r(t-\tau)$ the product $w(\tau)r(t-\tau)$ becomes

$$w(\tau)r(t-\tau) = |w(\tau)|\,\mu(t-\tau)$$

where $\mu(t-\tau)$ is a unit step function. If this change is made in the integrand of Eq. (6.2-1), that equation becomes

$$y(t) = \int_0^{\infty} |w(\tau)|\,\mu(t-\tau)d\tau$$

Since the unit step is zero for $t-\tau < 0$, the range of τ in the upper limit of integration may be made t, or

$$y(t) = \int_0^{\infty} |w(\tau)|\,\mu(t-\tau)d\tau = \int_0^{t} |w(\tau)|\,d\tau \qquad (6.2\text{-}3)$$

Now let us use the assumption that the magnitude of $w(\tau)$ is not absolutely integrable. If the weighting function is not absolutely integrable, it must be possible to make the right-hand side of Eq. (6.2-3) larger than any given number by selecting t large enough. Since $y(t)$ is equal to the right-hand side of Eq. (6.2-3), then of course $y(t)$ also can not be bounded. Therefore, although the input is bounded, the assumption that the weighting function is not absolutely integrable allows the output to become unbounded, thereby establishing the complete equivalence of the first two definitions.

It is interesting to compare the second definition with our previous requirement that the impulse response or weighting function of the system must decay to zero as time increases to infinity. Obviously the weighting function must decay to zero if it is to be absolutely integrable. But consider the function $w(t) = 1/(t+1)$. Although this function approaches zero as t approaches infinity, it is not absolutely integrable. Hence it appears that the second definition is actually more restrictive than our previous requirement. However, for the class of linear systems in which we are interested, i.e., those whose transfer functions are ratios of polynomials in s, this distinction becomes nonexistent. This is shown as part of the next development, which demonstrates the equivalence of the second and third definitions of stability.

Let us begin by showing that the condition of the third definition is met if the second definition is satisfied. In other words, we wish to demonstrate that all the poles of the system transfer function are in the left half-plane if the weighting function is absolutely integrable over the infinite time range. To do this, we make use of the Laplace transform definition to write $W(s)$ as

$$W(s) = \int_0^\infty w(t)e^{-st}\,dt \tag{6.2-4}$$

Taking the magnitude of the complex functions on both sides of Eq. (6.2-4), we obtain

$$|W(s)| = \left|\int_0^\infty w(t)e^{-st}\,dt\right| \le \int_0^\infty |w(t)|\left|e^{-st}\right|dt$$

For all values of s outside the left half-plane, that is, those values of s with $\text{Re}(s) \ge 0$, the magnitude of the function e^{-st} is less than or equal to unity so that

$$|W(s)| \le \int_0^\infty |w(t)|\,dt \qquad \text{for Re}(s) \ge 0 \tag{6.2-5}$$

Since the weighting function is absolutely integrable, there exists a positive number M such that

$$\int_0^\infty |w(t)|\,dt \le M$$

Equation (6.2-5) therefore becomes

$$|W(s)| \le M \qquad \text{for Re}(s) \ge 0 \tag{6.2-6}$$

indicating that $W(s)$ is bounded for all values of s outside the left half-plane. If this is true, all poles of $W(s)$ must lie in the left half-plane. If a pole of $W(s)$ were to lie on the $j\omega$ axis or in the right half of the s plane, then there is a value of s with a real part greater than or equal to zero for which the denominator of $W(s)$ is zero. This would mean $W(s)$ would not be bounded at that value of s, a contradiction of Eq. (6.2-6). Hence the poles of $W(s)$ must lie in the left half of the s plane, which is the desired result.

Next we must show that the conditions of the second definition are satisfied if the third definition holds; that is, the weighting function is absolutely integrable if all the poles of $W(s)$ are in the left half-plane. To establish this property, we need only to recall that the weighting function is composed of a finite sum of terms whose general form is

$$w_i(t) = \alpha t^r e^{\lambda_i t} \tag{6.2-7}$$

Here λ_i is a pole of $W(s)$ that may be repeated. It is easy to show that functions of the form described by Eq. (6.2-7) are always absolutely integrable over the infinite time range if the $\text{Re}(\lambda_i) < 0$, i.e., if the poles of $W(s)$ are in the left half-plane. This is true because the exponential decreases faster than t^r grows for larger values of t. Therefore we have established that the weighting function $W(t)$ is absolutely integrable if the poles of the transfer function $W(s)$ are all in the left half-plane.

In the light of the above discussion, the requirement that the weighting function decay to zero is seen to be equivalent to the requirement of absolute integrability. Because of the exponential nature of each term, the rate of decay to zero is sufficiently rapid so that the function is always absolutely integrable. A possible fourth definition of stability is therefore

4. A linear time-invariant system is stable if its weighting function decays to zero as time approaches infinity, that is,

$$\lim_{t\to\infty} w(t) = 0$$

The equivalence of this definition to the third definition assumes that $W(s)$ is a ratio of polynomials, and that the terms that make up $w(t)$ are of the form of Eq. (6.2-7).

Of the four definitions of stability given above, all but the third definition deal with the time domain behavior of the system or its weighting function. The third definition, on the other hand, deals with the location of the closed-loop poles, a frequency domain feature of the system. It is for exactly this reason that the third definition is so useful in our study of closed-loop-system stability. In fact, our entire study of stability is based on an examination of the closed-loop pole locations.

Since the eigenvalues of the matrix $\boldsymbol{A_k}$ are identical to the closed-loop poles, the third definition can be equivalently phrased for the state-variable representation

3a. A linear time-invariant system is stable if all the eigenvalues of the matrix $\boldsymbol{A_k}$ lie in the left half of the s plane.

Although the transfer function approach is, in general, more convenient, there are situations in which the state-variable approach is preferable. The latter situation occurs, for example, in cases where the plant is initially described in state-variable form. Also, it is computationally preferable to compute the eigenvalues of a matrix rather than search for the roots of a polynomial. To retain flexibility, both the transfer function and state-variable representations are discussed in the following sections.

In discussing the topic of stability, the question of the meaning of instability cannot be avoided. A system is unstable if it is not stable; hence a system is unstable if any one of the equivalent stability definitions is not satisfied. Specifically, a system is unstable if its output does not remain bounded for all bounded inputs, *or* if the weighting function is not absolutely integrable, *or* if the poles of the closed-loop transfer function lie on the $j\omega$ axis or in the right half of the s plane, *or* if the eigenvalues of $\boldsymbol{A_k}$ have a zero or positive real part.

Some authors choose to make a distinction between systems that have poles on the $j\omega$ axis but not in the right half of the s plane, as compared with systems that have poles with the real part of s greater than zero. If the poles of the system are in the right half-plane, the output becomes unbounded for any nonzero input. On the other hand, if the poles are on the $j\omega$ axis, the output becomes unbounded only for special inputs. If the system in question contains an integrator, i.e., a pole at the origin of the s plane, one input that causes the output to become unbounded is just a constant input. For example, a dc motor is a physical device which contains an integrator. For a constant-voltage input the output angle θ continues to increase without bound. Of course, this is the type of behavior normally associated with a dc motor; hence one does not intuitively feel that such a device should be classified as unstable. For this reason systems with poles on the $j\omega$ axis are often classified as *marginally stable*. "Marginally unstable" may actually be a better description.

In this book we make no distinction between systems that have poles only on the $j\omega$ axis and those that have poles in the right half of the s plane. All are classified as unstable, and all violate *each* of the equivalent stability definitions given at the beginning of this section.

6.3 THE ROUTH-HURWITZ CRITERION

One obvious and direct method for determining the stability of a closed-loop system is to factor the denominator of the closed-loop transfer function $Y(s)/R(s)$. Once all the roots are known, it is immediately obvious whether all the poles lie in the left half-plane. The actual factoring of the denominator of $Y(s)/R(s)$, however, may be very difficult or even impossible if literal coefficients are present. In addition, this complete factoring procedure provides far more information than is actually needed to answer the stability question.

We do not need to know the exact location of each pole, which we obtain by factoring, but only whether each pole is in the left half-plane. In fact, it is necessary to know only if there are any poles that are not in the left half-plane.

Fortunately, there are procedures that enable one to obtain this information with much less work than complete factoring requires. As an added bonus, these procedures are also useful for system design. The remainder of this chapter presents two of the more common and more useful procedures of obtaining stability information.

In this section we consider an approach known as the *Routh-Hurwitz criterion.*[1] This approach is based on algebraic tests on the coefficients of the denominator polynomial of $Y(s)/R(s)$. In the following sections the Nyquist criterion is discussed. The Nyquist criterion employs a graphical procedure based on the frequency response functions of Chap. 5. Because of the sharp differences in their treatment of the stability problem, these two approaches complement each other well.

Before beginning the development of the Routh-Hurwitz criterion, let us review briefly the various procedures for finding the denominator polynomial of $Y(s)/R(s)$. We consider first the transfer function procedures for determining the closed-loop

---

[1]Routh and Hurwitz independently developed equivalent criteria for stability nearly a century ago. The form used here is due to Routh but the criterion usually bears both names.

transfer function $Y(s)/R(s)$. Using the H and G configurations, shown in Fig. 6.3-1, the closed-loop transfer function is given by

$$\frac{Y(s)}{R(s)} = \frac{KG_p(s)}{1 + KG_p(s)H(s)} = \frac{G(s)}{1 + G(s)} \tag{6.3-1}$$

where $G_p(s)$, $H(s)$, and $G(s)$ are in the form of a ratio of polynomials in s.

$$G_p(s) = \frac{K_p N_p(s)}{D_p(s)} \tag{6.3-2}$$

$$H(s) = \frac{K_H N_H(s)}{D_H(s)} \tag{6.3-3}$$

$$G(s) = \frac{K_G N_G(s)}{D_G(s)} \tag{6.3-4}$$

$N_p(s)$ is assumed to be of the form $N_p(s) = (s + \delta_1)(s + \delta_2) \cdots (s + \delta_m)$, and K_p is the plant gain. $N_H(s)$ and $N_G(s)$ are of the same form as $N_p(s)$.

Let us now find expressions for $Y(s)/R(s)$ in terms of the polynomial numerators and denominators defined in Eqs. (6.3-2) to (6.3-4). By writing $Y(s)/R(s)$ in terms of $H(s)$ we obtain

$$\begin{aligned} \frac{Y(s)}{R(s)} &= \frac{KK_pN_p(s)/D_p(s)}{1 + \left[KK_pN_p(s)/D_p(s)\right]\left[K_HN_H(s)/D_H(s)\right]} \\ &= \frac{KK_pN_p(s)D_H(s)}{D_H(s)D_p(s) + KK_pK_HN_H(s)N_p(s)} = \frac{KK_pN_p(s)D_H(s)}{D_k(s)} \end{aligned}$$

Therefore the denominator of the closed-loop transfer function $Y(s)/R(s)$ is given by

$$D_k(s) = D_H(s)D_p(s) + KK_pK_HN_p(s)N_H(s) \tag{6.3-5}$$

Using the $G(s)$ representation, we find that the closed-loop transfer function is

$$\frac{Y(s)}{R(s)} = \frac{K_GN_G(s)/D_G(s)}{1 + K_GN_G(s)/D_G(s)} = \frac{K_GN_G(s)}{D_G(s) + K_GN_G(s)} = \frac{K_GN_G(s)}{D_k(s)}$$

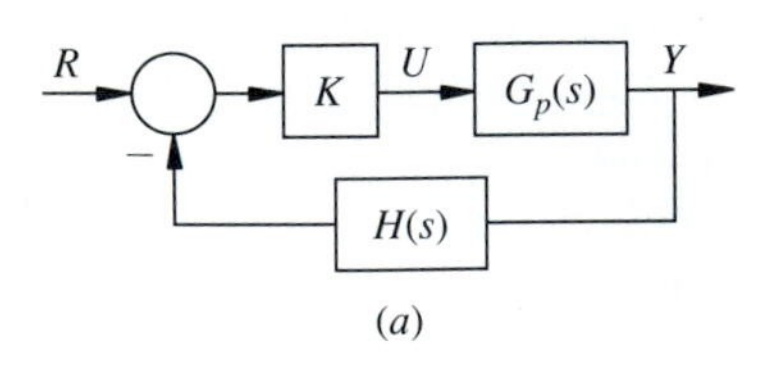

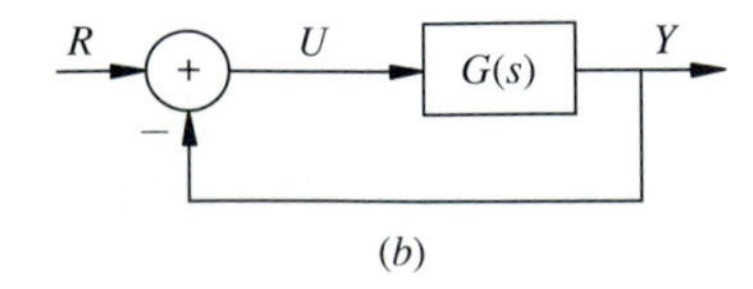

FIGURE 6.3-1
The H and G configurations.

Here we find that $D_k(s)$ is equal to the sum of the denominator polynomial of $G(s)$ and K_G times the numerator polynomial of $G(s)$, or

$$D_k(s) = D_G(s) + K_G N_G(s) \tag{6.3-6}$$

If we choose to make use of the state-variable representation, we can use the expression for $Y(s)/R(s)$ from Chap. 3

$$\frac{Y(s)}{R(s)} = K\mathbf{c}^T(s\mathbf{I} - \mathbf{A}_k)^{-1}\mathbf{b} = K\mathbf{c}^T(s\mathbf{I} - \mathbf{A} + K\mathbf{b}\mathbf{k}^T)^{-1}\mathbf{b}$$

The denominator polynomial is therefore equal to the characteristic polynomial of the matrix $\mathbf{A}_k$, or

$$D_k(s) = \det(s\mathbf{I} - \mathbf{A}_k) = \det(s\mathbf{I} - \mathbf{A} + K\mathbf{b}\mathbf{k}^T) \tag{6.3-7}$$

The three procedures discussed above are only three of many possible ways to find the denominator of $Y(s)/R(s)$ and are usually the most convenient. Using Eq. (6.3-5), (6.3-6), (6.3-7), or some other method, one may find the denominator polynomial of the closed-loop transfer function in unfactored form, as

$$D_k(s) = s^n + d_{n-1}s^{n-1} + \cdots + d_1 s + d_0 \tag{6.3-8}$$

The problem is then to determine, without factoring $D_k(s)$, whether all its roots lie in the left half-plane. The Routh-Hurwitz criterion provides a simple method for solving this problem. The procedure consists of an initial screening that provides a necessary condition for stability plus the application of the complete Routh-Hurwitz test to establish a necessary and sufficient condition for stability.

6.3.1 Initial Screening Test

The initial screening test consists of a trivial examination of the coefficients of the polynomial to ensure the following:

1. All the d_i coefficients are present.
2. All the d_i coefficients are positive.

For a polynomial to have all its roots in the left half-plane, it is necessary but not sufficient that it pass the initial screening test. In other words, if the characteristic polynomial fails to meet either or both of these conditions, one may immediately conclude that the polynomial has at least one root that does not lie in the left half-plane, and hence the closed-loop system is unstable.

On the other hand, if the polynomial satisfies both conditions, no conclusion about stability may be reached. Consider, for example, the third-order polynomial

$$D_k(s) = s^3 + 0.5s^2 + 3.5s + 4$$

Although this polynomial passes the initial screening test, it has two roots in the right half-plane.

The two test conditions are a direct consequence of the algebraic property that

$$d_i = (-1)^{n-i} \sum (\text{product of the roots taken } n - i \text{ at a time})$$
$$i = 0, 1, \ldots, n-1 \qquad (6.3\text{-}9)$$

For example,

$$(s + \delta_1)(s + \delta_2)(s + \delta_3) = s^3 + (\delta_1 + \delta_2 + \delta_3)s + (\delta_1\delta_2 + \delta_1\delta_3 + \delta_2\delta_3)s + \delta_2\delta_2\delta_3.$$

If all the roots are in the left half-plane, each of the coefficients must be nonzero and positive.

Because it may be checked easily, the initial screening test plays a valuable role in stability analysis. However, for a large percentage of cases, the polynomial satisfies both conditions. In these situations the Routh-Hurwitz criterion discussed below may be used to determine the stability of the system.

6.3.2 Routh-Hurwitz Criterion

The first step in the application of the Routh-Hurwitz criterion is the formation of an array of numbers based on the coefficients of the denominator polynomial. The array is called the Routh array and it takes the form

s^n	$\delta_{01} = 1$	$\delta_{02} = d_{n-2}$	$\delta_{03} = d_{n-4}$	$\cdots$
s^{n-1}	$\delta_{11} = d_{n-1}$	$\delta_{12} = d_{n-3}$	$\delta_{13} = d_{n-5}$	$\cdots$
s^{n-2}	δ_{21}	δ_{22}	δ_{23}	$\cdots$
s^{n-3}	δ_{31}	δ_{32}	δ_{33}	$\cdots$
$\vdots$	$\vdots$	$\vdots$	$\vdots$	$\vdots$
s^1	$\delta_{n-1,1}$	0	0	$\cdots$
s^0	δ_{n1}			
	Pivot column			

where

$$\delta_{ij} = \frac{\delta_{i-1,1}\delta_{i-2,j+1} - \delta_{i-2,1}\delta_{i-1,j+1}}{\delta_{i-1,1}} \qquad i = 2, 3, \ldots, n \qquad j = 1, 2, \ldots \qquad (6.3\text{-}10)$$

The first column of the Routh array is called the *pivot column*. The first two rows of the array are formed by simply writing the coefficients of $D_k(s)$ in the order of decreasing powers of s alternately in the first and second row. The remaining rows of the array are then consecutively formed using Eq. (6.3-10). Note that the computation of the remaining elements always involves the two elements in the pivot column of the two preceding rows. The pivot column is listed in the prototype Routh array shown above.

Although the procedure Eq. (6.3-10) looks somewhat complicated, once one has mastered the procedure, one can compute the elements of the array quite easily. The procedure used is similar to that employed in computing determinants, although the reader should note that the signs are reversed. The following schematic illustrates the calculation of the δ_{22} element.

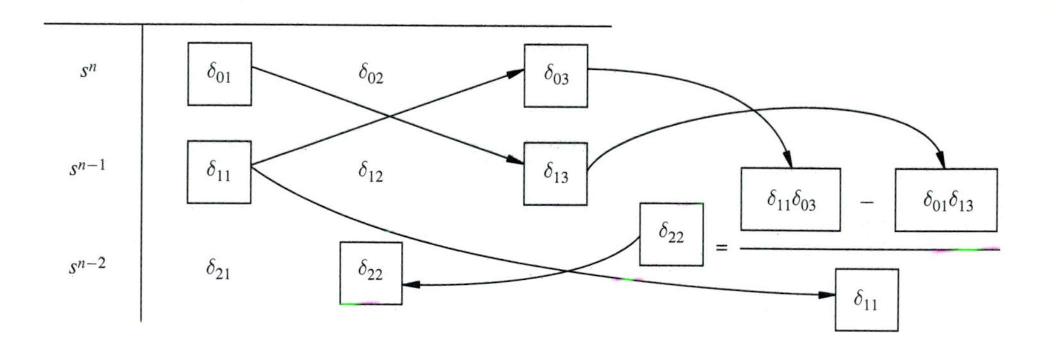

The rows of the Routh array are labeled with descending powers of s mainly to provide a rough check on the completion of the array. The number of nonzero elements in any row cannot be greater than one plus one-half the power of s associated with that row. For example, in the s^6 row, there cannot be more than $1+(6/2) = 4$ nonzero elements. In the s^1 and s^0 rows there cannot be more than one nonzero element.

The Routh-Hurwitz criterion may now be stated

The characteristic polynomial $D_k(s)$ has roots on the $j\omega$ axis or in the right half of the s plane if there are any zeros or sign changes in the pivot column of the Routh array.

Of course, if the characteristic polynomial has roots on the $j\omega$ axis or in the right half-plane, then the system from which this polynomial originated is unstable.

Example 6.3-1. Consider, for example, the polynomial

$$D_k(s) = s^3 - 2s^2 - 5s + 6$$

Since the polynomial does not pass the initial screening test, it is obvious that it must have at least one root that is not in the left half-plane. Hence there is no need to use the more elaborate Routh-Hurwitz criterion.

Example 6.3-2. As a second example, let us consider the general third-order polynomial given by

$$D_k(s) = s^3 + d_3 s^2 + d_2 s + d_1$$

To satisfy the initial screening condition, it is necessary that d_3, d_2, and d_1 be greater than zero. The first two rows of the Routh array are written immediately, with the remaining elements indicated in terms of δ_{ij}.

s^3	1	d_2
s^2	d_3	d_1
s^1	δ_{21}	0
s^0	δ_{31}	0

The δ_{21} and δ_{31} elements of the array are found from Eq. (6.3-10) to be

$$\delta_{21} = \frac{(\delta_{11})(\delta_{02}) - (\delta_{01})(\delta_{12})}{\delta_{11}} = \frac{d_3 d_2 - 1(d_1)}{d_3}$$

$$\delta_{31} = \frac{\delta_{21} d_1}{\delta_{21}} = d_1$$

and the complete array is finally

s^3	1	d_2
s^2	d_3	d_1
s^1	$(d_3 d_2 - d_1)/d_3$	
s^0	d_1	

There are no roots of the given polynomial $D_k(s)$ in the right half of the s plane or on the $j\omega$ axis as long as the elements of the pivot column do not change sign. This is true if

$$\begin{aligned} d_3 &> 0 \\ d_3 d_2 - d_1 &> 0 \\ d_1 &> 0 \end{aligned} \tag{6.3-11}$$

Note that from the Routh-Hurwitz criterion alone it does not appear, at first glance, that the requirement $d_2 > 0$ is necessary. However, if d_3 and d_1 are both greater than zero, then d_2 must also be greater than zero for

$$d_3 d_2 - d_1 > 0$$

to be true. Hence the initial screening and the Routh-Hurwitz criteria are in complete agreement.

For the general second-order polynomial, it is easy to show that the initial screening condition is both necessary and sufficient. Therefore a second-order polynomial has both its roots in the left half of the s plane if all the coefficients are nonzero and of the same sign.

Example 6.3-3. Let us consider the development of the Routh array for the polynomial

$$D_k(s) = s^4 + 10s^3 + s^2 + 15s + 3$$

The Routh array is

s^4	1	1	3
s^3	10	15	
s^2	δ_{21}	δ_{22}	
s^1	δ_{31}		
s^0	δ_{41}		

Here δ_{21} is

$$\delta_{21} = \frac{(10)(1) - 1(15)}{10} = -0.5$$

Since δ_{21} has a negative sign, whereas $\delta_{11} = 10$ is positive, a sign change has occurred in the pivot column, and there is no need to proceed. The system for which $D_k(s)$ of this example is the characteristic polynomial is unstable.

In the formation of the Routh array, the following property often proves to be convenient:

Any row of the Routh array may be multiplied or divided by any positive constant without changing the sign of any element in the array.

The use of this property can simplify the development of the array by making the numbers more convenient. One approach is to reduce the numbers in every row by a common factor, if one exists. If this is done for the array of Example 6.3-3, the second row is divided by 5. That array then becomes

s^4	1	1	3
s^3	2	3	
s^2	δ_{21}	δ_{22}	
s^1	δ_{31}		
s^0	δ_{41}		

Now the δ_{ij}'s may have values that are different from the corresponding elements in Example 6.3-3. This is unimportant since only the sign of the elements in the pivot column indicates stability or instability and these signs do not change.

Another approach that makes use of the property that any row may be divided by a positive constant is to force all the numbers in the pivot column to be one by dividing every element of the row by the number in the pivot column.

Example 6.3-4. Consider, for example, the formation of the Routh array for the polynomial

$$D_k(s) = s^5 + s^4 + 2s^3 + 2s^2 + 3s + 15$$

The partial Routh array is

s^5	1	2	3
s^4	1	2	15
s^3	0	δ_{22}	

Again there is no need to proceed further, as δ_{21} is zero, and, according to the Routh-Hurwitz criterion, either a zero or a sign change in the pivot column is sufficient to indicate that all the roots are not in the left half-plane.

On the basis of Examples 6.3-3 and 6.3-4, it may appear that δ_{21} always changes sign if the poles are outside the left half-plane. This is often, but not always, the case, as illustrated by the next example.

Example 6.3-5. The characteristic polynomial to be tested in this example is

$$D_k(s) = s^4 + 2s^3 + 11s^2 + 18s + 18$$

The Routh array for this polynomial is

s^4	1	11	18
s^3	2	18	0
s^2	2	18	
s^1	0		
s^0	δ_{41}		

Since δ_{31} is zero, $D_k(s)$ has at least one root on the $j\omega$ axis or the right half s plane.

The most important uses of the Routh-Hurwitz criterion are not in those cases for which $D_k(s)$ is completely specified, but rather in the stability analysis of closed-loop systems. The roots of completely specified $D_k(s)$ polynomials can be found by computer using eigenvalue routines on the phase variable system matrix associated with the pole polynomial. The utility of the Routh-Hurwitz criteria is best seen in cases where the $D_k(s)$ polynomial retains some literal coefficients. The following two examples illustrate various applications of the Routh-Hurwitz criterion to the study of system stability in which parameter adjustments are available.

Example 6.3-6. As our first example of this type, let us consider the closed-loop system shown in Fig. 6.3-2. The closed-loop transfer function for this system is

$$\frac{Y(s)}{R(s)} = \frac{K}{s(s+2)(s+3)+K} = \frac{K}{s^3+5s^2+6s+K}$$

so that $D_k(s)$ is

$$D_k(s) = s^3 + 5s^2 + 6s + K$$

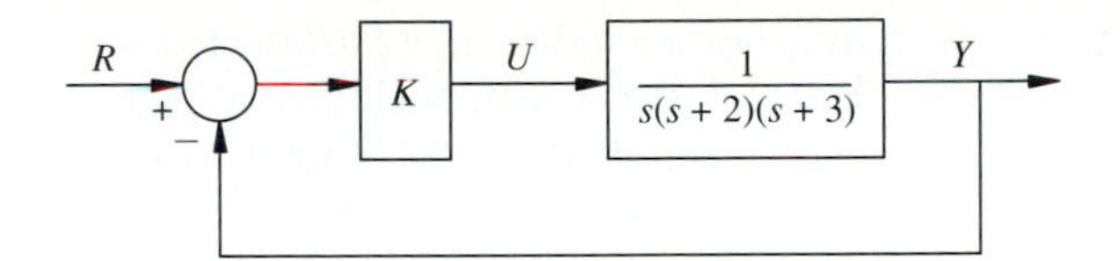

FIGURE 6.3-2
Closed-loop system for Example 6.3-6.

The Routh array for this polynomial is

s^3	1	6
s^2	5	K
s^1	$(30 - K)/5$	
s^0	K	

In order that there be no sign changes in the pivot column, it is necessary that

$$K > 0 \text{ and } 30 - K > 0$$

or

$$0 < K < 30$$

Therefore the system shown in Fig. 6.3-2 is stable if the above condition on K is satisfied. Note that the above result could have been obtained directly by making use of the Routh-Hurwitz conditions given in Example 6.3-2, namely, Eq. (6.3-11).

In the situation where a parameter, such as the gain K above, has a maximum value beyond which instability occurs, it is possible to define a stability margin for that parameter. In effect, a stability margin describes the amount by which a given value of the parameter may be increased just to achieve marginal stability. As a mathematical expression, this becomes

$$\text{Stability margin} = \frac{\text{maximum stable value}}{\text{actual value}}$$

In Example 6.3-6, we may speak of the stability margin of the gain K, or more commonly just the *gain margin*, GM, which becomes

$$\text{GM} = \frac{\text{maximum stable gain}}{\text{actual gain}} = \frac{30}{K}$$

If the gain $K = 15$, then GM = 2, indicating that the gain may be increased by a factor of 2 before instability occurs.

The gain margin is one of the simplest relative-stability indices or figures of merit for a closed-loop system. Unfortunately, it is also one of the least useful. First, the concept of a gain margin does not apply to all systems, since in some systems there is no maximum stable gain. In addition, even in the case of systems to which it applies, the gain margin does not give uniformly reliable information. For example, two systems that have the same gain margin may have entirely different behaviors.

Before proceeding to the Nyquist diagram, it should be noted that not all the information which can be extracted from the Routh array has been presented here.

Methods are available for determining how many poles are in the right half-plane and how many are on the $j\omega$ axis. The number of roots of $D_k(s)$ with positive real parts is equal to the number of sign changes of the coefficients of the pivot column of the Routh array. The number of poles on the $j\omega$ axis may be determined by a slightly more complicated procedure from the rows having all zero elements. The point of view adopted here is that, from the standpoint of stability, it is unimportant how many poles are outside the left half-plane. A much more important question than how "unstable" is a system is the question of relative stability. The questions of stability, relative stability and robustness are adequately handled by the Nyquist diagram.

Exercises 6.3

6.3-1. Each of the polynomials shown below represents the characteristic polynomial of some closed-loop system. Determine as much of the Routh array as necessary to make a definite statement concerning the stability or instability of the system.

(*a*) $s^8 + 7s^7 + 4s^6 - s^5 + 7s^4 + 2s^2 + s + 3$
(*b*) $s^4 + 9s^3 + 13s^2 + 54s + 40$
(*c*) $s^4 + 6s^3 + 13s^2 + 30s + 40$
(*d*) $s^3 + 4s^2 + 8s + 40$

Answers:

(*a*) Unstable (*b*) Stable
(*c*) Unstable (*d*) Unstable

6.3-2. Find the values of K for which the system shown in Fig. 6.3-3 is stable when $G(s)$ is given as below.

(*a*) $G(s) = \dfrac{9K}{s^3 + 3s^2 + 9s}$

(*b*) $G(s) = \dfrac{K(s+2)(s+4)}{(s+5)(s+1)(s+3)}$

(*c*) $G(s) = \dfrac{K(s^2 + 15s + 45)}{s^3 + s^2 + 10s}$

(*d*) $G(s) = \dfrac{K}{s^4 + 6s^3 + 13s^2 + 12s + 4}$

Answers:

(*a*) $0 < K < 3$
(*b*) $K > -15/8$
(*c*) $K > 0$
(*d*) $-4 < K < 18$

6.3-3. Find the values of K and T for which the system shown in Fig. 6.3-3 is stable when $G(s)$ is given by

$$G(s) = \frac{9K(s+T)^2}{s^3}$$

Answer: $K > 0$ $2K > T > 0$

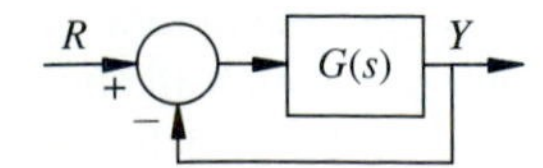

FIGURE 6.3-3
Exercise 6.3-3.

6.4 THE NYQUIST CRITERION

The Routh-Hurwitz procedure discussed in the preceding section is analytic in nature and deals directly with the closed-loop characteristic polynomial $D_k(s)$. The Nyquist criterion presented here, on the other hand, is essentially a graphical method and deals with the loop gain transfer function. The graphical character of the Nyquist criterion is probably one of its most appealing features.

Since the key quantity in the Nyquist criterion is the loop gain transfer function, we develop the criterion in the context of the G configuration. In the G configuration, the loop gain transfer function is given simply by $G(s)$. The closed-loop transfer function is given by

$$\frac{Y(s)}{R(s)} = \frac{G(s)}{1 + G(s)} = \frac{K_G N_G(s)/D_G(s)}{1 + K_G N_G(s)/D_G(s)} = \frac{K_G N_G(s)}{D_G(s) + K_G N_G(s)} = \frac{K_G N_G(s)}{D_k(s)}$$

where $D_k(s)$ is the denominator of the closed-loop transfer function. We see that the closed-loop poles are equal to the zeros of the function

$$1 + G(s) = 1 + \frac{K_G N_G(s)}{D_G(s)} = \frac{D_G(s) + K_G N_G(s)}{D_G(s)} \tag{6.4-1}$$

Of course, the numerator of Eq. (6.4-1) is just the closed-loop denominator polynomial $D_k(s)$ from Eq. (6.3-6), so that

$$1 + G(s) = \frac{D_k(s)}{D_G(s)} \tag{6.4-2}$$

In other words, we can determine the stability of the closed-loop system by locating the zeros of $1 + G(s)$. This is not a new result, but the reader's attention is drawn to this fact, since it is of prime importance in the following development.

For the moment, let us assume that $1 + G(s)$ is known in factored form so that we have

$$1 + G(s) = \frac{(s + \lambda_{k1})(s + \lambda_{k2}) \cdots (s + \lambda_{kn})}{(s + \lambda_1)(s + \lambda_2) \cdots (s + \lambda_n)} \tag{6.4-3}$$

Obviously, if $1 + G(s)$ were known in factored form, there would be no need for the use of the Nyquist criterion, since we could simply observe whether any of the zeros of $1 + G(s)$, poles of $Y(s)/R(s)$, lie in the right half of the s plane. In fact, the primary reason for using either the Routh-Hurwitz or Nyquist criterion is to avoid this factoring. Although it is convenient to think of $1 + G(s)$ in factored form at this time, no actual use is made of that form.

Let us suppose that the pole-zero plot of $1 + G(s)$ takes the form shown in Fig. 6.4-1a. Consider next an arbitrary closed contour, such as that labeled Γ in Fig. 6.4-1a, which encloses one and only one zero of $1 + G(s)$ and none of the poles. Associated with each point on this contour is a value of the complex function $1 + G(s)$. The value of $1 + G(s)$ for any value of s on Γ may be found analytically by substituting the appropriate complex value of s into the function. Alternatively the value may be found graphically. The magnitude value of the function may be

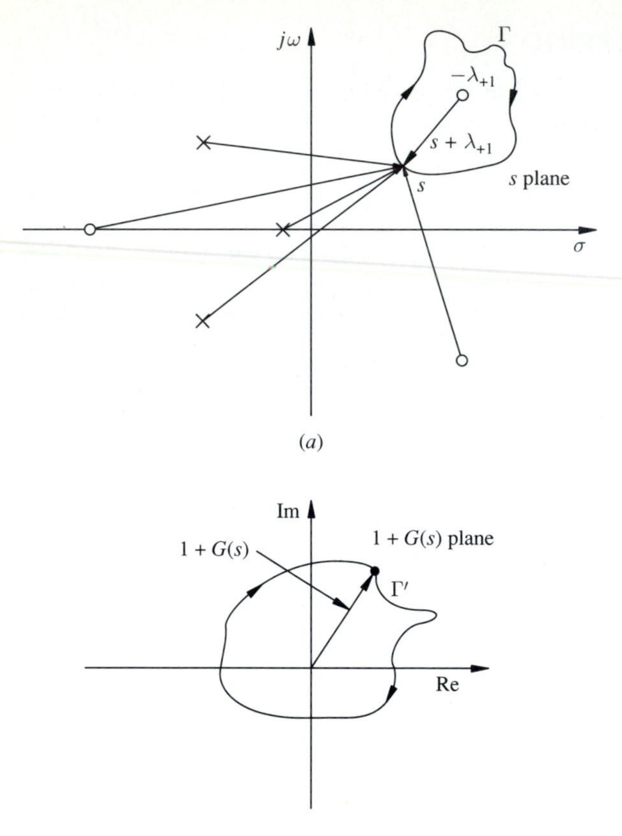

FIGURE 6.4-1
(a) Pole-zero plot of $1 + G(s)$ in the s plane: (b) plot of the Γ' contour in the $1 + G(s)$ plane.

found as was done in Chap. 4 by multiplying the distances from s on Γ to the zeros and dividing by the distances to the poles. The phase value of the function may be found by subtracting the phase angles associated with the poles from the phase angles associated with the zeros.

If the complex value of $1+G(s)$ associated with every point on the contour Γ is plotted, another closed contour Γ' is created in the complex $1+G(s)$ plane, as shown in Fig. 6.4-1b. The function $1+G(s)$ is said to map the contour Γ in the s plane into the Γ' contour in the $1+G(s)$ plane. What we wish to demonstrate is that, if a zero is enclosed by the contour Γ, as in Fig. 6.4-1a, the contour Γ' encircles the origin of the $1+G(s)$ plane in the same sense that the Γ contour encircles the zero in the s plane. In the s plane, the zero is encircled in the clockwise direction; hence we must show that the origin of the $1+G(s)$ plane is also encircled in the clockwise direction. This result is known as the *Principle of the Argument.*

The key to the Principle of the Argument rests in considering the value of the function $1+G(s)$ at any point s as simply a complex number. This complex number has a magnitude and a phase angle. Since the contour Γ in the s plane does not pass through a zero, the magnitude is never zero. Now we consider the phase angle by rewriting Eq. (6.4-3) in polar form

$$1 + G(s) = \frac{|s + \lambda_{k1}| \underline{/ \arg(s + \lambda_{k1})} \cdots |s + \lambda_{kn}| \underline{/ \arg(s + \lambda_{kn})}}{|s + \lambda_1| \underline{/ \arg(s + \lambda_1)} \cdots |s + \lambda_n| \underline{/ \arg(s + \lambda_n)}}$$

$$= \left[\frac{|s + \lambda_{k1}| \cdots |s + \lambda_{kn}|}{|s + \lambda_1| \cdots |s + \lambda_n|}\right] \tag{6.4-4}$$

$$\times \left[\underline{/ \arg(s + \lambda_{k1}) + \cdots + \arg(s + \lambda_{kn}) - \arg(s + \lambda_1) - \cdots - \arg(s + \lambda_n)}\right]$$

We assume that the zero encircled by Γ is at $s = -\lambda_{k1}$. Then the phase angle associated with this zero changes by a full $-360°$ as the contour Γ is traversed clockwise in the s plane. Since the argument or angle of $1 + G(s)$ includes the angle of this zero, the argument of $1 + G(s)$ also changes by $-360°$. As seen from Fig. 6.4-1a, the angles associated with the remaining poles and zeros make no net change as the contour Γ is traversed. For any fixed value of s, the vector associated with each of these other poles and zeros has a particular angle associated with it. Once the contour has been traversed back to the starting point, these angles return to their original value; they have not been altered by plus or minus 360° simply because these poles and zeros are not enclosed by Γ.

In a similar fashion, we could show that, if the Γ contour were to encircle two zeros of $1 + G(s)$ in the clockwise direction on the s plane, the Γ' contour would encircle the origin of the $1+G(s)$ plane twice in the clockwise direction. On the other hand, if the Γ contour were to encircle only one pole and no zero of $1 + G(s)$ in the *clockwise* direction, then the contour Γ' would encircle the origin of the $1+G(s)$ plane once in the *counterclockwise* direction. This change in direction comes about because angles associated with poles are accompanied by negative signs in the evaluation of $1 + G(s)$, as indicated by Eq. (6.4-4). In general, the following conclusion can be drawn.

The net number of clockwise encirclements by Γ' of the origin in the $1+G(s)$ plane is equal to the difference between the number of zeros n_z and the number of poles n_p of $1 + G(s)$ encircled in the clockwise direction by Γ.

This result means that the difference between the number of zeros and the number of poles enclosed by *any* closed contour Γ may be determined simply by counting the net number of clockwise encirclements of the origin of the $1+G(s)$ plane by Γ'. For example, if we find that Γ' encircles the origin three times in the clockwise direction and once in the counterclockwise direction, then $n_z - n_p$ must be equal to $3 - 1 = 2$. Therefore, in the s plane, Γ must encircle two zeros and no poles, three zeros and one pole, or any other combination such that $n_z - n_p$ is equal to 2.

For stability analysis, the problem is to determine the number of zeros of $1 + G(s)$, i.e., the number of poles of $Y(s)/R(s)$, that lie in the right half of the s plane. Accordingly, the contour Γ is chosen as the entire $j\omega$ axis and an infinite semicircle enclosing the right half-plane as shown in Fig. 6.4-2a. This contour is known as the *Nyquist D-contour* as it resembles the capital letter D.

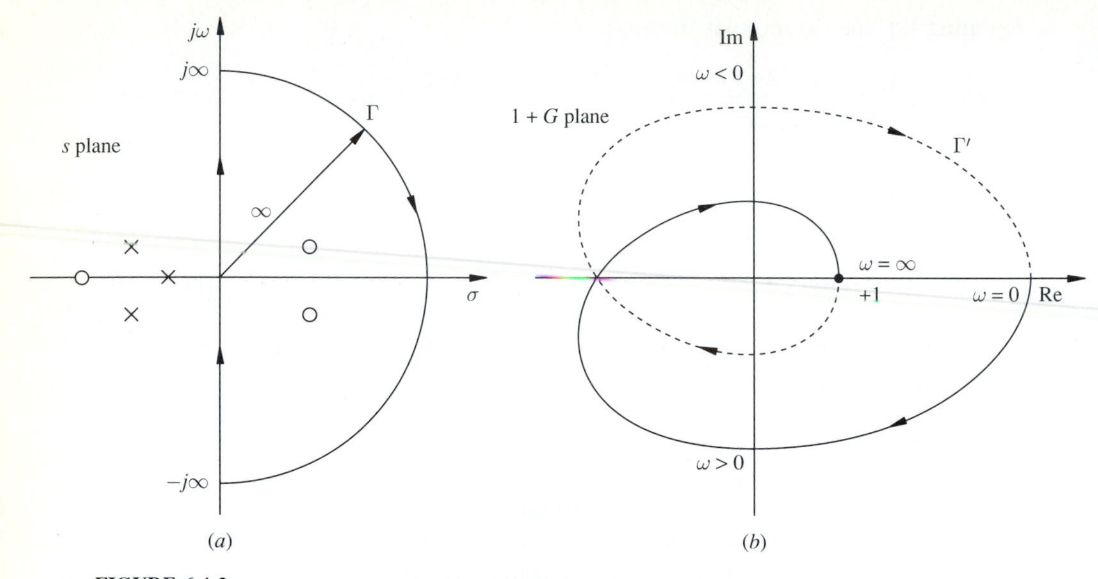

FIGURE 6.4-2
(a) Γ contour in the s plane—the Nyquist D-contour; (b) Γ' contour in the $1 + G$ plane.

To avoid any problems in plotting the values of $1 + G(s)$ along the infinite semicircle, let us assume that

$$\lim_{|s| \to \infty} G(s) = 0$$

This assumption is justified since, in general, the loop gain transfer function $G(s)$ is strictly proper. With this assumption, the entire infinite semicircle portion of Γ maps into the single point $s = +1 + j0$ on the $1 + G(s)$ plane.

The mapping of Γ therefore involves simply plotting the complex values of $1 + G(s)$ for $s = j\omega$ as ω varies from $-\infty$ to $+\infty$. For $\omega \geq 0$, Γ' is nothing more than the polar plot of the frequency response of the function $1 + G(s)$ as discussed in Chap. 5. The values of $1 + G(j\omega)$ for negative values of ω are the mirror image of the values of $1 + G(j\omega)$ for positive values of ω reflected about the real axis. The Γ' contour may therefore be found by plotting the frequency response $1 + G(s)$ for positive ω and then reflecting this plot about the real axis to find the plot for negative ω. The Γ' plot is always symmetrical about the real axis of the $1 + G$ plane. Care must be taken to establish the direction in which the Γ' plot is traced as the D-contour moves up the $j\omega$ axis, around the infinite semicircle and back up the $j\omega$ axis from $-\infty$ towards 0.

From the Γ' contour in the $1 + G(s)$ plane, as shown in Fig. 6.4-2b, the number of zeros of $1 + G(s)$ in the right half of the s plane may be determined by the following procedure. The net number of clockwise encirclements of the origin by Γ' is equal to the number of zeros minus the number of poles of $1 + G(s)$ in the right half of the s plane. Note that we must know the number of poles of $1 + G(s)$ in the right half-plane if we are to be able to ascertain the exact number of zeros in the right half-plane and therefore determine stability. This requirement usually poses no problem since the poles of $1 + G(s)$ correspond to the poles of the loop gain transfer function. In Eq. (6.4-2) the denominator of $1 + G(s)$ is just $D_G(s)$, which is usually described in factored form. Hence the number of zeros of $1 + G(s)$ or the number of poles of $Y(s)/R(s)$ in the right half-plane may be found by determining the net number of clockwise encirclements of the origin by Γ' and then adding the number of poles of the loop gain located in the right half s plane.

At this point the reader may revolt. Our plan for finding the number of poles of $Y(s)/R(s)$ in the right half of the s plane involves counting encirclements in the $1 + G(s)$ plane and observing the number of loop gain poles in the right half of the s plane. Yet we were forced to start with the assumption that all the poles and zeros of $1 + G(s)$ are known, so that the Nyquist contour can be mapped by the function of $1 + G(s)$. Admittedly we know the poles of this function, because they are the poles of the loop gain, but we do not know the zeros; in fact, we are simply trying to find how many of these zeros lie in the right half of the s plane.

What we do know are the poles and zeros of the loop gain transfer function $G(s)$. Of course, this function differs from $1 + G(s)$ only by unity. Any contour that is chosen in the s plane and mapped through the function $G(s)$ has exactly the same shape as if the contour were mapped through the function $1 + G(s)$ except that it is displaced by one unit. Figure 6.4-3 is typical of such a situation. In this diagram the -1 point of the $G(s)$ plane is equivalent to the origin of the $1 + G(s)$ plane. If we now map the boundary of the right half of the s plane through the mapping function $G(s)$, which we often know in pole-zero form, information concerning the zeros of $1 + G(s)$ may be obtained by counting the encirclements of the -1 point.

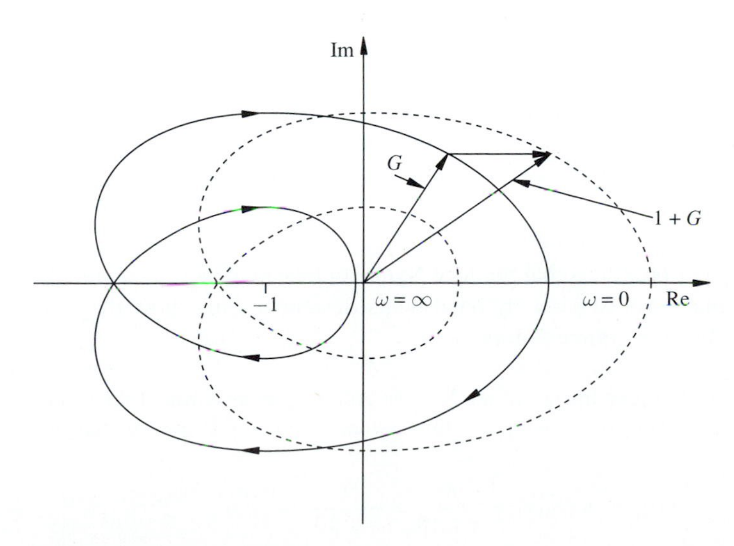

FIGURE 6.4-3
Comparison of the $G(s)$ and $1 + G(s)$ plots.

The important point is that, by plotting the open-loop frequency response information, we may reach stability conclusions regarding the closed-loop system.

As mentioned previously, contour Γ of Fig.6.4-2*a* is referred to as the Nyquist D-contour. The map of the Nyquist D-contour through $G(s)$ is called the *Nyquist diagram* of $G(s)$. There are three parts to the Nyquist diagram. The first part is the polar plot of the frequency response of $G(s)$ from $\omega = 0$ to $\omega = \infty$. The second part is the mapping of the infinite semicircle around the right half-plane. If $G(s)$ is strictly proper, this part maps entirely into the origin of the $G(s)$ plane. The third part is the polar plot of the negative frequencies, $\omega = -\infty$ to $\omega = 0$. The map of these frequencies forms a mirror image in the $G(s)$ plane about the real axis from the first part.

In terms of the Nyquist diagram of $G(s)$, the Nyquist stability criterion may be stated:

The closed-loop system is stable if and only if the net number of clockwise encirclements of the point $s = -1 + j0$ by the Nyquist diagram of $G(s)$ plus the number of poles of $G(s)$ in the right half-plane is zero.

Notice that while the *net* number of clockwise encirclements is counted in the first part of the Nyquist criterion only the number of right half-plane *poles* of $G(s)$ is counted in the second part. Right half-plane zeros of $G(s)$ are not part of the formula in determining stability using the Nyquist criterion.

Because the Nyquist diagram involves the loop gain transfer function $G(s)$, a good approximation of the magnitude and phase of the frequency response plot can be obtained by using the Bode diagram straight line approximations for the magnitude and phase. The Nyquist plot can then be obtained by transferring the magnitude and phase information to a polar plot. If a more accurate plot is needed, the exact magnitude and phase may be determined for a few values of ω in the range of interest. However, in most cases, the approximate plot is accurate enough for practical problems.

An alternative procedure for obtaining the Nyquist diagram is to plot accurately the poles and zeros of $G(s)$ and obtain the magnitude and phase by graphical means. In either of these methods, the fact that $G(s)$ is known in factored form is important. Even if $G(s)$ is not known in factored form, the frequency response plot can still be obtained by simply substituting the values $s = j\omega$ into $G(s)$ or by measuring frequency response on the actual system.

Of course, computer programs that produce Nyquist plots are generally available. Again, the ability to plot Nyquist plots by hand helps designers know how they can affect such plots by adjusting compensators.

Example 6.4-1. To illustrate the use of the Nyquist criterion, let us consider the simple first-order system shown in Fig. 6.4-4*a*. For this system the loop gain transfer function takes the form

$$G(s) = KG_p(s) = \frac{K}{s+10} = \frac{50}{s+10}$$

The magnitude and phase plots of the frequency response of $KG_p(s)$ are shown. From these plots the Nyquist diagram for $KG_p(s)$ may be easily plotted, as shown in

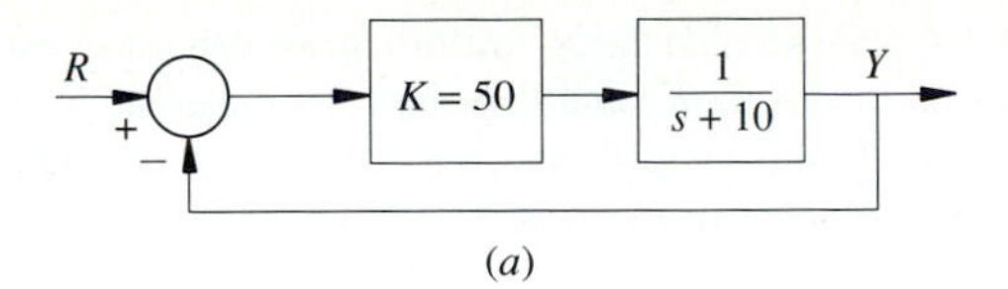

(a)

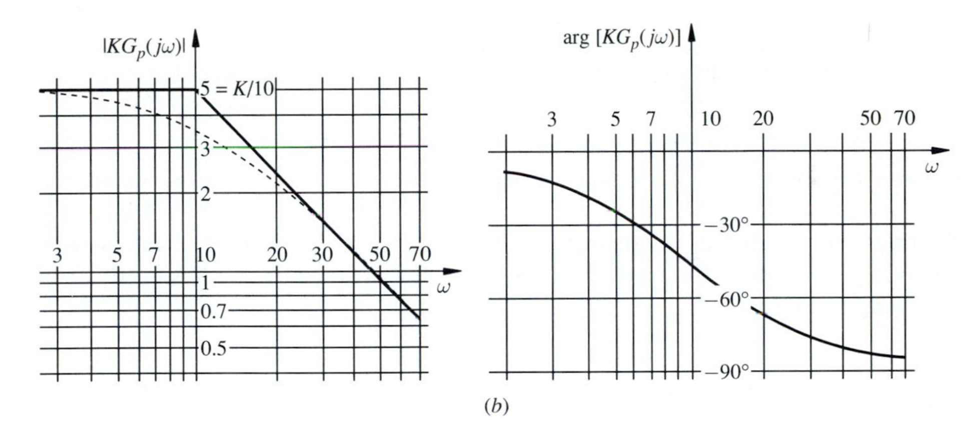

(b)

FIGURE 6.4-4
Simple first-order example. (a) Block diagram; (b) magnitude and phase plots.

Fig. 6.4-5. For example, the point associated with $\omega = 10$ rad/sec is found to have a magnitude of $K/\left(10\sqrt{2}\right)$ and a phase angle of $-45°$. The point at $\omega = -10$ rad/sec is just the mirror image of the value at $\omega = 10$ rad/sec.

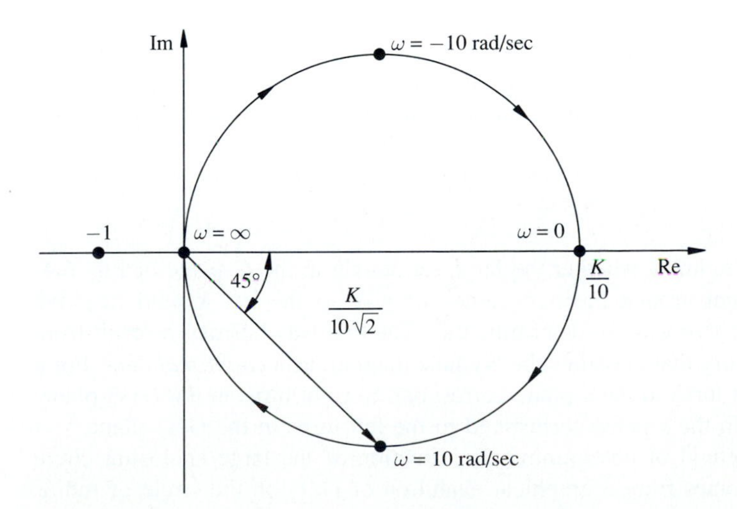

FIGURE 6.4-5
Nyquist diagram for Example 6.4-1.

From Fig. 6.4-5 we see that the Nyquist diagram can never encircle the $s = -1 + j0$ point for any positive value of K, and therefore the closed-loop system is stable for all positive values of K. In this simple example, it is easy to see that this result is correct since the closed-loop transfer function is given by

$$\frac{Y(s)}{R(s)} = \frac{K}{s + 10 + K}$$

For all positive values of K, the pole of $Y(s)/R(s)$ is in the left half-plane.

In the above example, $G(s)$ remains finite along the entire Nyquist contour. This is not always the case even though we have assumed that $G(s)$ approaches zero as $|s|$ approaches infinity. If a pole of $G(s)$ occurs on the $j\omega$ axis, as often happens at the origin because of an integrator in the plant, a slight modification of the Nyquist contour is necessary. The method of handling the modification is illustrated in the following example.

Example 6.4-2. Consider a system whose loop transfer function is given by

$$G(s) = \frac{(2K/7)\left[(s + 3/2)^2 + \left(\sqrt{5/2}\right)^2\right]}{s(s + 2)(s + 3)}$$

The pole-zero plot of $G(s)$ is shown in Fig. 6.4-6a. Since a pole occurs on the standard Nyquist contour at the origin, it is not clear how this problem should be handled. As a beginning, let us plot the Nyquist diagram for $\omega = +\epsilon$ to $\omega = -\epsilon$, including the infinite semicircle; when this is done, the small area around the origin is avoided. The resulting plot is shown as the solid line in Fig. 6.4-6b with corresponding points labeled.

From Fig. 6.4-6b we cannot determine whether the system is stable until the Nyquist diagram is completed by joining the points at $\omega = -\epsilon$ and $\omega = +\epsilon$. To join these points, let us use a semicircle of radius ϵ to the right of the origin, as shown in Fig. 6.4-6a. Now $G(s)$ is finite at all points on the contour in the s plane, and the mapping to the G plane can be completed as shown by the dashed line in Fig. 6.4-6b. The small semicircle used to avoid the origin in the s plane maps into a large semicircle in the G plane.

It is important to know whether the large semicircle in the G-plane of Fig. 6.4-6b swings to the right around positive values of s or to the left around negative values of s. There are two ways to determine this. The first way borrows a result from complex variable theory that classifies the Nyquist diagram as a *conformal map*. For a conformal map, right turns in the s plane correspond to right turns in the $G(s)$-plane. Likewise, left turns in the s plane correspond to the left turns in the $G(s)$-plane.

The second method of determining the direction of the large enclosing circle on the $G(s)$-plane comes from a graphical evalution of $G(s)$ on the circle of radius ϵ in the s plane. The magnitude is very large here due to the proximity of the pole. The phase at $s = -\epsilon$ is slightly larger than $+90^\circ$ as seen from the solid line of the

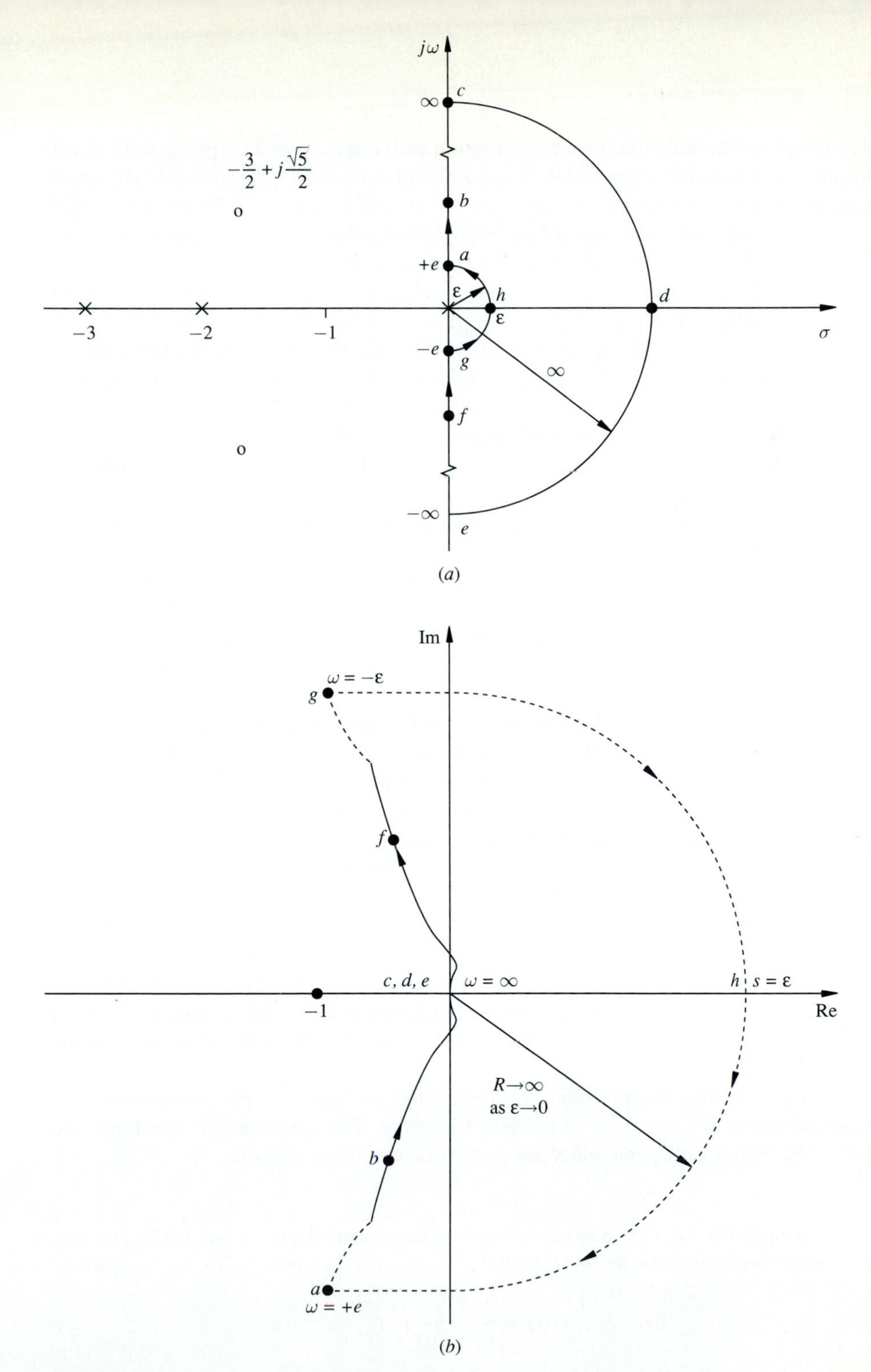

FIGURE 6.4-6
Example 6.4-2. (*a*) Pole-zero plot; (*b*) Nyquist diagram.

Nyquist plot. The phase contribution from all poles and zeros except the pole at the origin does not change appreciably as the circle of the radius ϵ is traversed. The angle from the pole at the origin changes from -90° through 0° to $+90^\circ$. Since angles from poles contribute in a negative manner, the contribution from the pole goes from $+90^\circ$ through 0° to -90°. Thus, as the semicircle of radius ϵ is traversed in the s plane a semicircle moving in a clockwise direction through about 180° is traversed in the $G(s)$-plane. The semicircle is traced in the clockwise direction as the angle associated with $G(s)$ becomes more negative. Notice that this is consistent with the conformal mapping rule that matches right turns of 90° at the top and bottom of both circles.

To ensure that no right half-plane zeros of $1 + G(s)$ can escape discovery by lying in the semicircular indentation of radius ϵ in the s plane, ϵ is made arbitrarily small, with the result that the radius of the large semicircle in the G-plane approaches infinity. As ϵ approaches zero, the shape of the Nyquist diagram remains unchanged, and we see that there are no encirclements of the $s = -1 + j0$ point. Since there are no poles of $G(s)$ in the right half-plane, the system is stable. In addition, since changing the value of K can never cause the Nyquist diagram to encircle the -1 point, the closed-loop system must be stable for all positive values of K. This result agrees with the result obtained for this system by the use of the Routh-Hurwitz approach.

We could just as well close the contour with a semicircle of radius ϵ into the left half-plane. If we did this the D-contour would have encircled the pole at the origin and this pole would have been counted as a right half-plane pole of $G(s)$. In addition, by applying either the conformal mapping with left turns or the graphical evaluation, we would have closed the contour in the $G(s)$-plane by encircling the negative real axis. There would have been 1 counterclockwise encirclement (-1 clockwise encirclement) of the -1 point. The Nyquist criterion would have indicated that -1 counterclockwise encirclement plus 1 right half-plane pole of $G(s)$ yields zero closed-loop right half-plane poles. The result that the closed-loop system is stable for all positive values of K remains unchanged, as it must. The two approaches are equally valid although philosphically the left turn contour, which places the pole on the $j\omega$ axis in the right half-plane, is more in keeping with our definition of poles on the $j\omega$ axis being unstable.

In each of the two preceding examples, the system was open-loop stable; that is, all the poles of $G(s)$ were in the left half-plane. The next example illustrates the use of the Nyquist criterion when the system is open-loop unstable.

Example 6.4-3. This example is based on the system shown in Fig. 6.4-7. The loop gain transfer function for this system is

$$G(s) = \frac{K(s+1)}{(s-1)(s+2)}$$

Let's use the Bode diagrams of magnitude and phase to help with plotting the Nyquist diagram. The magnitude and phase plots are shown in Fig. 6.4-8a. If these plots

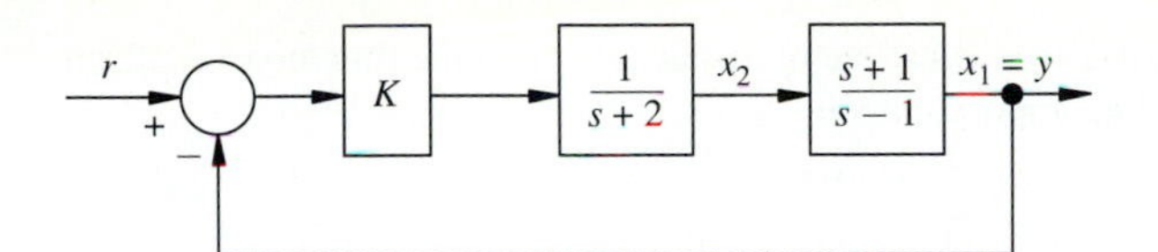

FIGURE 6.4-7
Example 6.4-3.

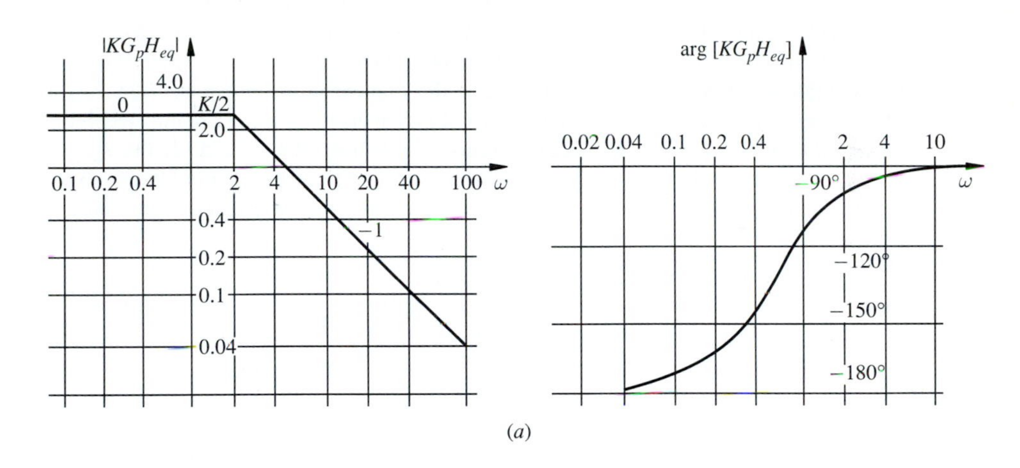

(*a*)

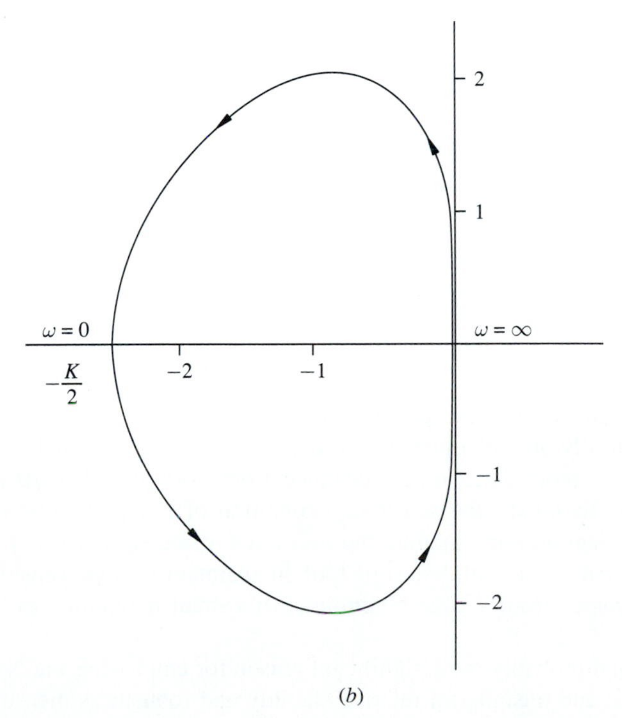

(*b*)

FIGURE 6.4-8
Example 6.4-3. (*a*) Magnitude and phase plots; (*b*) Nyquist diagram.

look strange, the reader should recall that we must put $G(s)$ into the time constant form before we use the straight line approximation.

$$G(s) = \frac{-K/2(1+s)}{(1-s)(1+s/2)}$$

In addition, it must be remembered that the unstable pole at $s = +1$ requires the use of the pseudo minimum phase approach of Sec. 5.4 to obtain the phase plot.

The Nyquist diagram for this system is shown in Fig. 6.4-8*b* for the case where $K > 2$. Note that the exact shape of the plot is not very important since the only information we wish to obtain at this time is the number of encirclements of the $s = -1 + j0$ point. It is easy to see that the Nyquist diagram encircles the -1 point once in the *counterclockwise* direction if $K > 2$ and has no encirclements if $K < 2$. Since this system has one right half-plane pole in $G(s)$, it is necessary that there be one counterclockwise encirclement if the system is to be stable. Therefore, this system is stable if and only if $K > 2$.

Since it is possible to determine stability of all closed-loop systems by means of the Routh-Hurwitz procedure, the reader may justifiably question why we introduce the Nyquist criterion. Assuredly, the Nyquist criterion is not easier to use than the Routh-Hurwitz procedure. The answer to this question is threefold.

First, because the Nyquist diagram is just a frequency response plot of the loop gain transfer function, it may be determined experimentally. In this way, one can determine the stability properties of a closed-loop system without any analytical knowledge of the system transfer function. In fact, as we shall see later, it may even be possible to carry out a system design based on this information.

Second, the Nyquist procedure is graphical and therefore, by its very nature, appealing to engineers. The old cliché that "one picture is worth a thousand words" is very applicable to the engineering profession. An excellent example of the advantage gained by the use of a graphical representation is the sensitivity condition developed in Sec. 3.2. There it was shown that, to reduce the sensitivity of the closed-loop transfer function to variations in the plant and to reduce the effects of disturbances on the output, the following requirement is imposed on the return difference function.

$$|1 + G(j\omega)| \geq 1 \qquad \text{for all } \omega \tag{6.4-5}$$

This requirement takes the particularly appealing form of a circle criterion if interpreted graphically on a Nyquist diagram, as shown in Fig. 6.4-9. There the quantity $|1 + G(j\omega)|$ is represented as the distance from the $s = -1 + j0$ point to the Nyquist diagram. To satisfy the sensitivity condition of Eq. (6.4-5), it is necessary that the Nyquist diagram not penetrate the unit circle centered at $s = -1 + j0$. Thus the Nyquist diagram is a useful design tool in maintaining low sensitivity as the loop gain itself passes from a large magnitude, to a small magnitude as it inevitably must.

The third and probably most significant reason for employing the Nyquist criterion is that simple and meaningful relative stability and robustness measurements are based on the Nyquist diagram. These measurements are the topics of the next three sections.

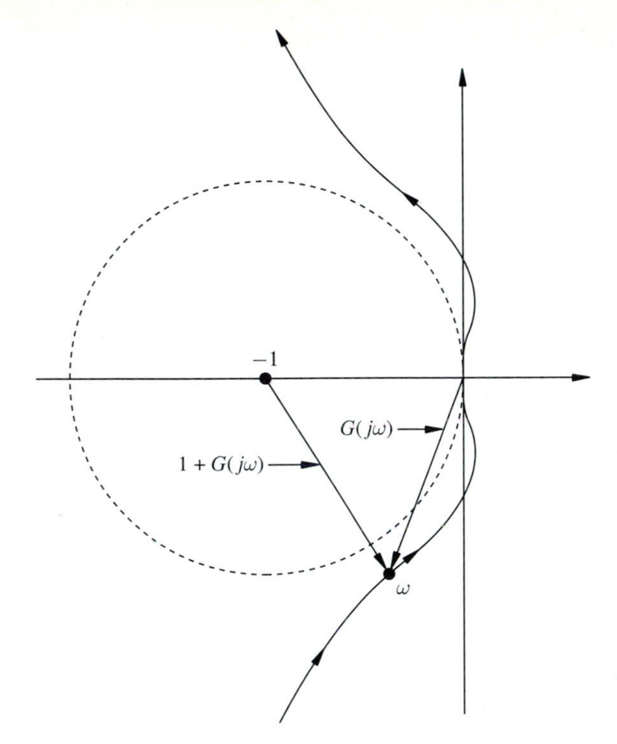

FIGURE 6.4-9
Interpretation of the sensitivity condition.

Exercises 6.4

6.4-1. Draw the Nyquist diagram corresponding to the loop gain polynomial

$$G(s) = \frac{K}{s(s+1)^2}$$

assuming $K = 1$.

The following questions refer to the closed-loop system

$$\frac{G(s)}{1+G(s)}$$

Answer "stable" or "unstable" to each question based on your plot and explain why.

(a) When $K = 0.5$ the closed-loop system is __________.
(b) When $K = 5$ the closed-loop system is __________.
(c) When $K = 50$ the closed-loop system is __________.

Answers:
(a) Stable
(b) Unstable
(c) Unstable

6.4-2. The Nyquist diagram of the following loop gain polynomial is given in Figure 6.4-10 for some gain K

$$G(s) = \frac{K(s+T_1)^2}{(s+T_2)(s+T_3)(s+T_4)(s+T_5)^2}$$

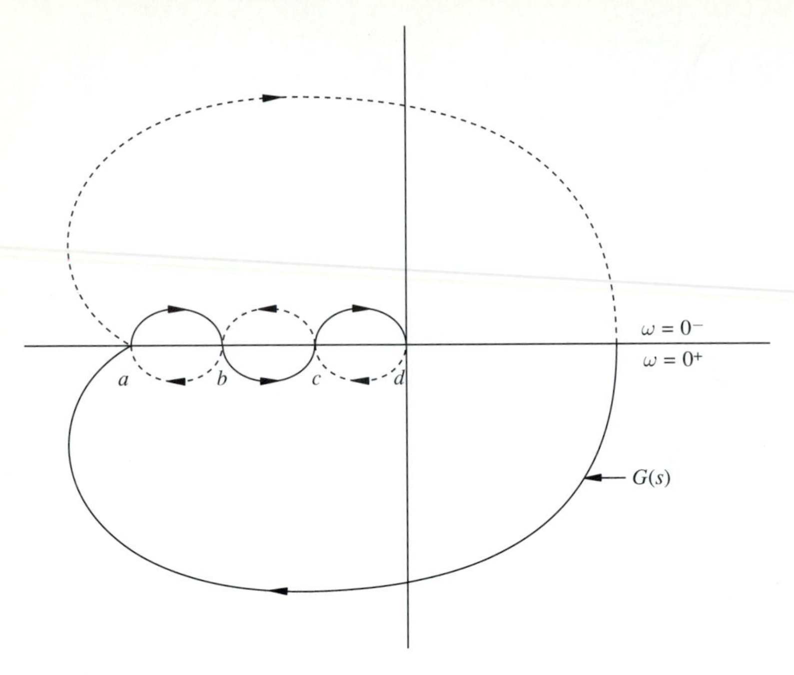

FIGURE 6.4-10
Exercise 6.4-2.

The diagram assumes $T_5 > T_1 > (T_2, T_3, \text{ and } T_4) > 0$. Answer the following questions:

(*a*) There are ________ right half-plane loop gain poles.
Fill out the following chart given the ranges of locations of the -1 point.

	Range where -1 point is found	Number of clockwise encirclements of -1	Closed-loop system is stable or unstable?
(*b*)	To the left of *a*		
(*c*)	Between *a* and *b*		
(*d*)	Between *b* and *c*		
(*e*)	Between *c* and *d*		

Answers: (*a*) 0 (*b*) 0, stable
(*c*) 2, unstable (*d*) 0, stable
(*e*) 2, unstable

6.4-3. The Nyquist diagram of the following loop gain polynomial is given in Figure 6.4-11 for some gain K. Assume $T_1 > 0$ and $T_2 > 0$.

$$G(s) = \frac{K(s + T_2)}{s(s - T_1)}$$

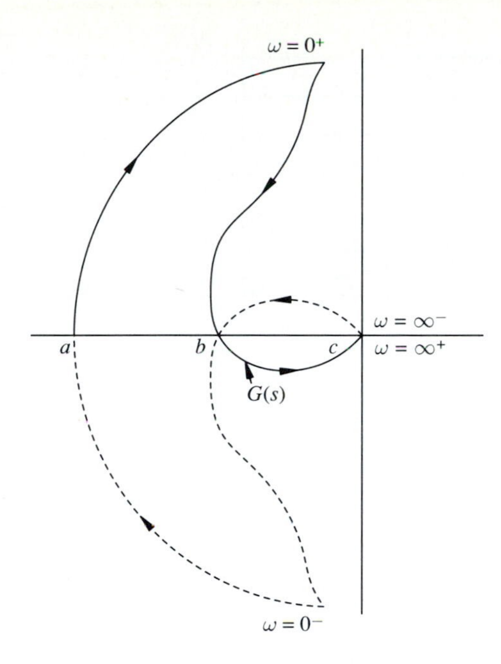

FIGURE 6.4-11
Exercise 6.4-3

Answer the following questions:

(*a*) There are __________ right half-plane loop gain poles.

Fill out the following chart given the ranges of locations of the -1 point.

	Range where -1 point is found	Number of clockwise encirclements of -1	Closed-loop system is stable or unstable?
(*b*)	Between *a* and *b*		
(*c*)	Between *b* and *c*		

Answers: (*a*) 1 (*b*) 1, unstable
(*c*) -1, stable

6.4-4. The open-loop system

$$G(s) = \frac{K(s+10)^2}{100s^2(s+1)}$$

has the Bode plot given in Fig. 6.4-12 when $K = 1$.

Sketch the complete Nyquist diagram. If the system is placed in the G configuration use the Bode and Nyquist plots to determine the values of K for a stable system.

Answer: $K > 400$

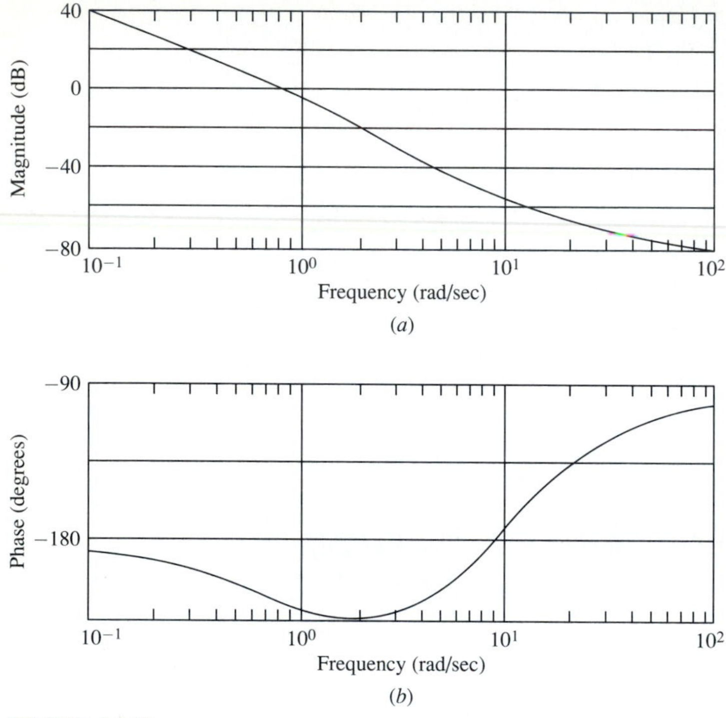

FIGURE 6.4-12
Exercise 6.4-4.

6.5 CLOSED-LOOP RESPONSE AND NYQUIST DIAGRAMS

Previously in this chapter, we have learned methods to determine if a closed-loop system is stable or not. However, there are many possible designs which result in closed-loop systems that are stable but have highly oscillatory and thus unsatisfactory responses to inputs and disturbances. Systems that are oscillatory are often said to be relatively less stable than systems that are more highly damped.

In this section, we start with an open-loop system in the G configuration and learn how the closed-loop frequency response of a stable closed-loop system can be determined from the same Nyquist plot used to determine stability. From the closed-loop frequency response the closed-loop pole-zero configuration can be determined using the methods of Sec. 5.5. Once the closed-loop system is identified, the closed-loop step response can be predicted using the methods of Chap. 4.

Instead of following the procedure outlined above to find the exact closed-loop step response, we consider only general characteristics of the step response by approximating the given closed-loop frequency response with a closed-loop frequency response of a second-order system. This will enable us to summarize general information about the closed-loop step response for most systems in terms of the closed-loop system's natural frequency and damping ratio. We are then able to predict the general shape of the closed-loop step response from a few key parameters on the Nyquist plot of the loop gain.

We must start with a system in the G configuration. The key for extracting information about the closed-loop system is to determine the frequency response function of the closed-loop system, often referred to as the M curve. The M curve is, of course, a function of frequency and may be determined analytically as

$$M(j\omega) = \frac{Y(j\omega)}{R(j\omega)} = \frac{G(j\omega)}{1 + G(j\omega)} \tag{6.5-1}$$

Figure 6.5-1 illustrates how the value of $M(j\omega_1)$ may be determined directly from the Nyquist diagram of $G(j\omega)$ at one particular frequency, ω_1. In this figure the vectors -1 and $G(j\omega_1)$ are indicated, as is the vector $(G(j\omega_1) - (-1)) = 1 + G(j\omega_1)$. The length of the vector $G(j\omega_1)$ divided by the length of $1 + G(j\omega_1)$ is thus the value of the magnitude $M(j\omega_1)$. The argument of $M(j\omega_1)$ is determined by subtracting the angle associated with the $1 + G(j\omega_1)$ vector from that of $G(j\omega_1)$. The complete M curve may be found by repeating this procedure over the range of frequencies of interest. From the completed M curve, the bandwidth of the closed-loop system may be read by inspection. Since most closed-loop frequency response functions look like low pass systems the closed-loop bandwidth (BW) may usually be defined as that frequency at which the frequency response function falls below 0.707 times its low frequency value.

For the magnitude portion of the $M(j\omega)$ plot, the point-by-point procedure illustrated above may be considerably simplified by plotting contours of constant $|M(j\omega)|$ on the Nyquist plot of $G(s)$. The magnitude plot of $M(j\omega)$ can then be read directly from the Nyquist diagram of $G(s)$. Fortunately, these contours of constant $|M(j\omega)|$ have a particularly simple form. For $|M(j\omega)| = M$, the contour is simply a circle centered at the point $-M^2/(M^2 - 1) + j0$ with a radius of $|M/(M^2 - 1)|$. This property is easily demonstrated.

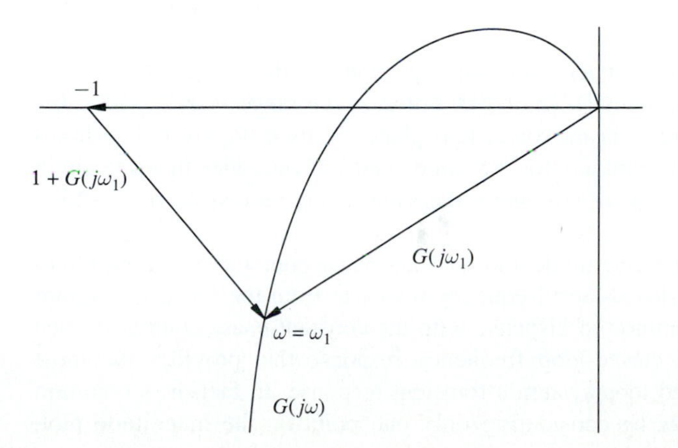

FIGURE 6.5-1
Graphical determination of $M(j\omega)$.

Let the complex function $G(j\omega)$ be written as

$$G(j\omega) = X(\omega) + jY(\omega) \tag{6.5-2}$$

where $X(\omega)$ and $Y(\omega)$ are real functions of ω and are, respectively, the real and imaginary parts of $G(j\omega)$. In this notation $M(j\omega)$ becomes

$$M(j\omega) = \frac{X(\omega) + jY(\omega)}{[1 + X(\omega)] + jY(\omega)}$$

Therefore the magnitude of $M(j\omega)$ is given by

$$|M(j\omega)| = M = \frac{|X(\omega) + jY(\omega)|}{|[1 + X(\omega)] + jY(\omega)|} = \frac{\sqrt{X^2 + Y^2}}{\sqrt{(1 + X)^2 + Y^2}}$$

so that

$$M^2 = \frac{X^2 + Y^2}{(1 + X)^2 + Y^2} \tag{6.5-3}$$

where the dependence on ω has been dropped from the notation.

After a number of algebraic steps, this last result may be written as

$$\left(X + \frac{M^2}{M^2 - 1}\right)^2 + Y^2 = \left(\frac{M}{M^2 - 1}\right)^2 \tag{6.5-4}$$

This expression is the equation for a circle of radius $|M/(M^2 - 1)|$ centered at $X = -M^2/(M^2 - 1)$ and $Y = 0$ as predicted. These circles are referred to as constant M-circles or simply M-circles.

If these constant M-circles are plotted together with the Nyquist diagram of $G(s)$, as shown in Fig. 6.5-2 for the system $G(s) = \frac{42}{s(s+2)(s+15)}$, the values of $|M(j\omega)|$ may be read directly from the plot. Note that the $M = 1$ circle degenerates to the straight line $X = -0.5$. For $M < 1$, the constant M-circles lie to the right of this line, whereas, for $M > 1$, they lie to the left. In addition, the $M = 0$ circle is the point $0 + j0$, and $M = \infty$ corresponds to the point $-1.0 + j0$.

In an entirely similar fashion, the contours of constant $\arg(M(j\omega))$ can be found. Surprisingly, these contours turn out to be segments of circles. In this case, however, the circles are centered on the line $X = -0.5$. The contour of the $\arg(M(j\omega)) = \beta$ for $0 < \beta < 180°$ is the upper half-plane portion of the circle centered at $-0.5 + j1/(2\tan\beta)$ with a radius $|1/(2\sin\beta)|$. For β in the range $-180° < \beta < 0°$, the portions of the same circles in the lower half-plane are used. Figure 6.5-3 shows the plot of the constant phase contours for the same values of β. Notice that one circle represents $\beta = 45°$ above the real axis while the same circle represents $\beta = -135°$ below the real axis.

By using these constant magnitude and constant phase contours, it is possible to read directly the complete closed-loop frequency response from the Nyquist diagram of $G(s)$. In practice it is common to dispense with the constant phase contours, since it is the magnitude of the closed-loop frequency response that provides the most information about the closed-loop system's transient response. In fact, it is common to simplify the labor further by considering only one point on the magnitude plot,

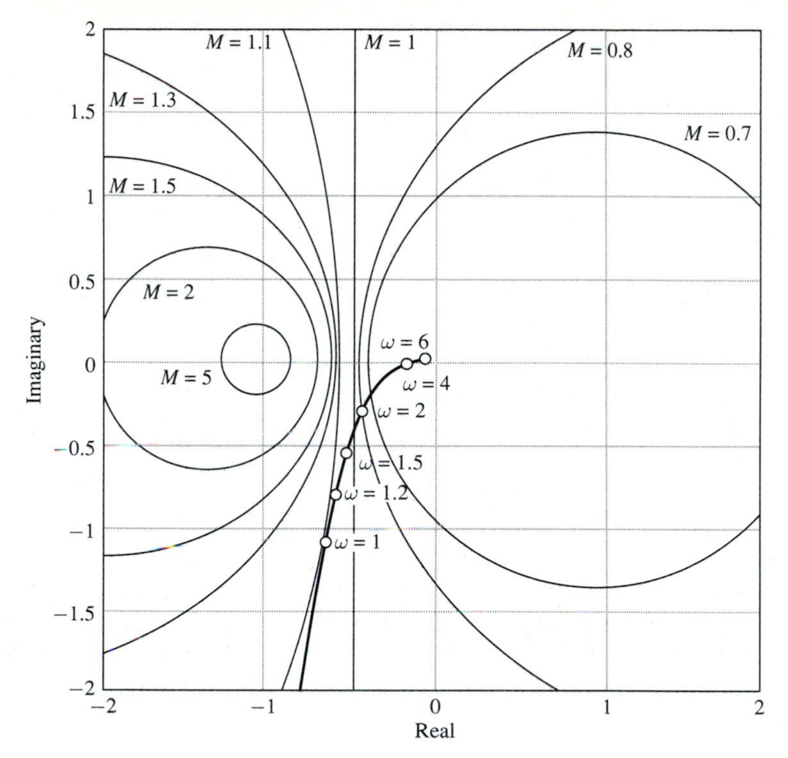

FIGURE 6.5-2
Constant *M* contours.

namely, the point at which M is maximum. This point of peak magnitude is referred to as M_p, and the frequency at which the peak occurs is ω_p. The point M_p may be easily found by considering the contours of larger and larger values of M until the contour is found that is just tangent to the plot of $G(s)$. The value associated with this contour is then M_p, and the frequency at which the M_p contour and $G(s)$ touch is ω_p. In the plot of $G(s)$ shown in Fig. 6.5-2, for example, the value of M_p is 1.1 at the frequency $\omega_p \approx 1.1$ rad/sec.

One of the primary reasons for determining M_p and ω_p, in addition to the obvious saving of labor as compared with the determination of the complete frequency response, is the close correlation of these quantities with the behavior of the closed-loop system. In particular, for the simple second-order closed-loop system

$$\frac{Y(s)}{R(s)} = \frac{\omega_n^2}{s^2 + 2s\zeta\omega_n + \omega_n^2} \tag{6.5-5}$$

the values of M_p and ω_p completely characterize the system. In other words, for this second-order system, M_p and ω_p specify ζ and ω_n, the only parameters of the system. To demonstrate this property, we need only recall Eqs. (5.3-8) and (5.3-9), which relate the maximum point of the frequency response of Eq. (6.5-5) to the values of ζ and ω_n. In our present terminology for a second-order system, these equations

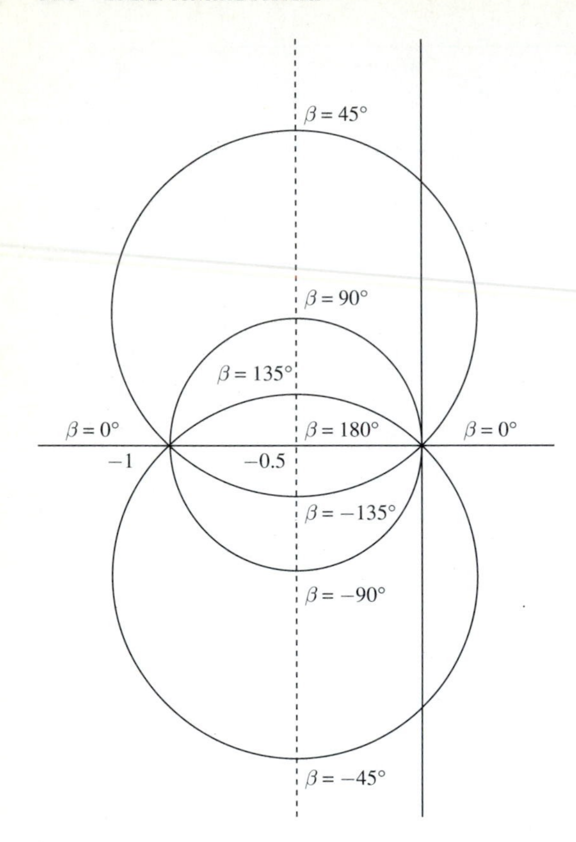

FIGURE 6.5-3
Constant-phase contours.

become

$$\omega_p = \omega_n \sqrt{1 - 2\zeta^2} \tag{6.5-6}$$

$$M_p = \frac{1}{2\zeta\sqrt{1 - \zeta_2}} \quad \text{for} \quad \zeta \leq 0.707 \tag{6.5-7}$$

From these equations one may determine ζ and ω_n if M_p and ω_p are known, and vice versa. Figure 6.5-4 graphically displays the relationships between M_p and ω_p and ζ and ω_n. Once ζ and ω_n are known, the results of Sec. 4.4 may be used to determine the time behavior of this second-order system.

Not all systems are of a simple second-order form. However, it is common practice to assume that the behavior of many high-order systems is closely related to that of a dominantly second-order system with the same M_p and ω_p. This is, in fact, one of the primary reasons for the extensive consideration given to second-order systems in Sec. 4.4.

Two other measures of the qualitative nature of the closed-loop response that may be determined from the Nyquist diagram of $G(s)$ are the phase margin and crossover frequency. The *crossover frequency* ω_c is the positive value of ω for which the magnitude of $G(j\omega)$ is equal to unity, that is,

$$|G(j\omega_c)| = 1 \tag{6.5-8}$$

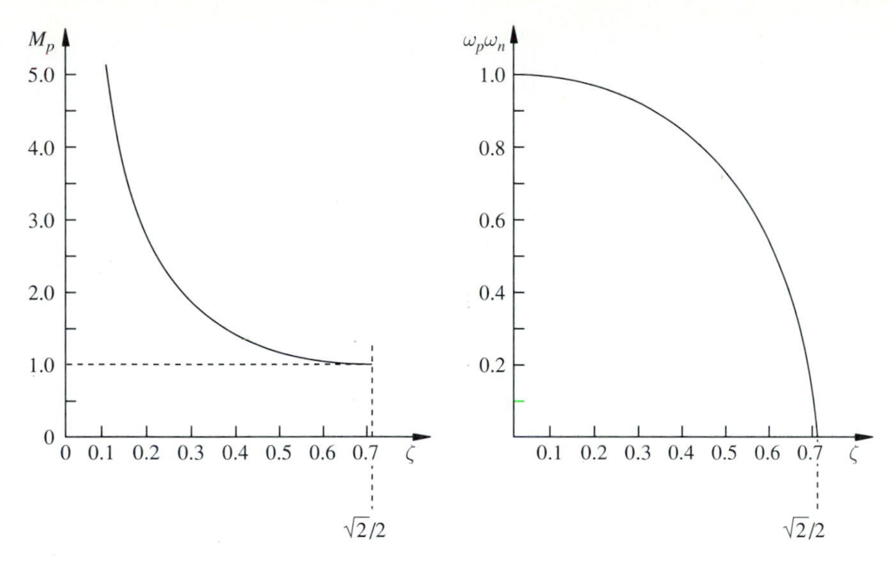

FIGURE 6.5-4
Plots of M_p and ω_p/ω_n versus ζ for a simple second-order system.

The *phase margin* ϕ_m is defined as the difference between the argument of $G(j\omega_c)$ (evaluated at the crossover frequency) and $-180°$. In other words, if we define β_c as

$$\beta_c = \arg(G(j\omega_c)) \tag{6.5-9}$$

the phase margin is given by

$$\phi_m = \beta_c - (-180°) = 180° + \beta_c \tag{6.5-10}$$

While it is possible for a complicated system to possess more than one crossover frequency, most systems are designed to possess just one. The phase margin takes on a particularly simple and graphic meaning in the Nyquist diagram of $G(s)$. Consider, for example, the Nyquist diagram shown in Fig. 6.5-5. In that diagram, we see that the phase margin is simply the angle between the negative real axis and the vector $G(j\omega_c)$. The vector $G(j\omega_c)$ may be found by intersecting the $G(s)$ locus with the unit circle, as shown in Fig. 6.5-5. The frequency associated with the point of intersection is ω_c.

It is possible to determine ϕ_c and ω_c with greater accuracy directly from the Bode plots of the magnitude and phase of $G(s)$. The appropriate value of ω for which the magnitude crosses unity is ω_c. The phase margin is then determined by inspection from the phase plot by noting the difference between the phase shift at ω_c and $-180°$. Consider, for example, the magnitude and phase plots shown in Fig. 6.5-6 for the $G(s)$ function of Fig. 6.5-2. In time constant form this transfer function is

$$G(s) = \frac{1.4}{s(1+s/2)(1+s/15)}$$

From this figure we see that $\omega_c = 1.4$ and $\phi_m = 60°$.

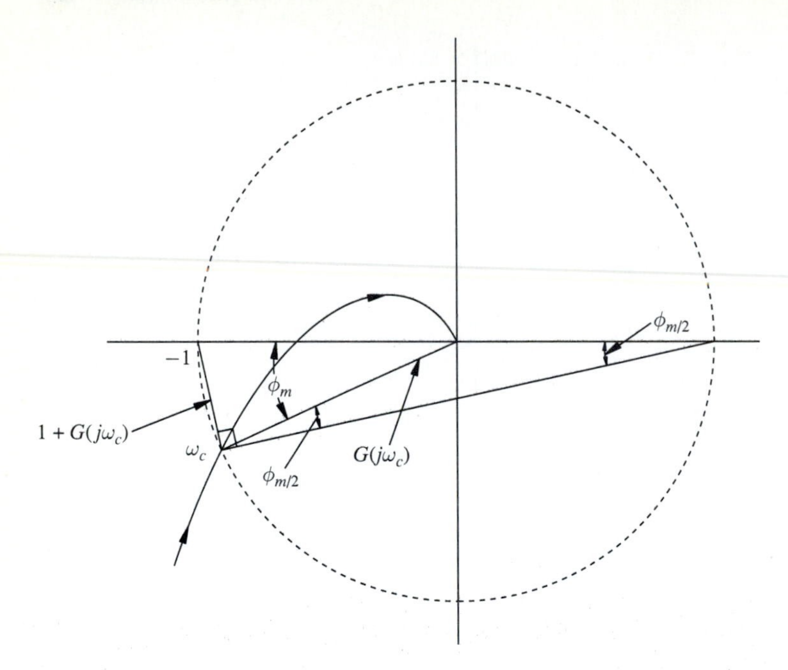

FIGURE 6.5-5
Definition of phase margin.

The magnitude of the closed-loop frequency response at ω_c can be derived from ϕ_m. We shall call this value M_c. Often the closest point to the -1 point on a Nyquist plot occurs at a frequency close to ω_c. This means that M_c is often a good approximation to M_p. The geometric construction in Fig. 6.5-5 shows a right triangle with a hypotenuse of 2, one side of length $|1 + G(j\omega_c)|$, and the opposite angle of $\phi_m/2$ where ϕ_m is the phase margin.[1] From this construction, we see

$$\sin(\phi_m/2) = \frac{|1 + G(j\omega_c)|}{2} \tag{6.5-11}$$

Since, at $\omega = \omega_c$,

$$|G(j\omega_c)| = 1 \tag{6.5-12}$$

$$M_c = \frac{|G(j\omega_c)|}{|1 + G(j\omega_c)|} = \frac{1}{2\sin(\phi_m/2)} \tag{6.5-13}$$

Thus, an oscillatory characteristic in the closed-loop time response can be identified by a large peak in the closed-loop frequency response, which, in turn, can be identified

[1] The triangle is known to be a right triangle because it is a triangle determined by the endpoints of the diameter of the circle and a third point on the circle. The opposite angle is $\phi/2$ because it is one of the equal angles of an isosceles triangle whose third angle is $180° - \phi_m$.

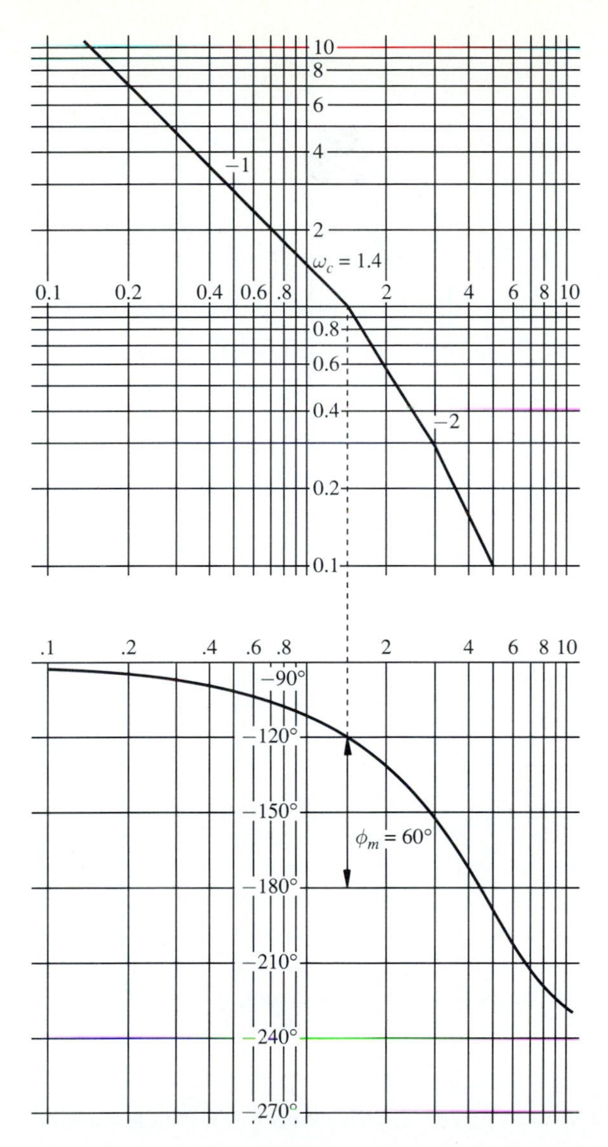

FIGURE 6.5-6
Magnitude and phase plots of $G(s)$.

by a small phase margin and the corresponding large value of M_c. Unfortunately the correlation between the closed-loop time response and phase margin is somewhat poorer than the correlation between closed-loop time response and M_p. This lower reliability of the phase margin measure is a direct consequence of the fact that ϕ_m is determined by considering only one point, ω_c on the G plot, whereas M_p is found by examining the entire plot to find the maximum. Consider, for example, the two Nyquist diagrams shown in Fig. 6.5-7. The phase margin for these two diagrams is identical; however, it is obvious that the closed-loop step response resulting from closing the loop around the loop gain of Fig. 6.5-7b is far more oscillatory and underdamped than

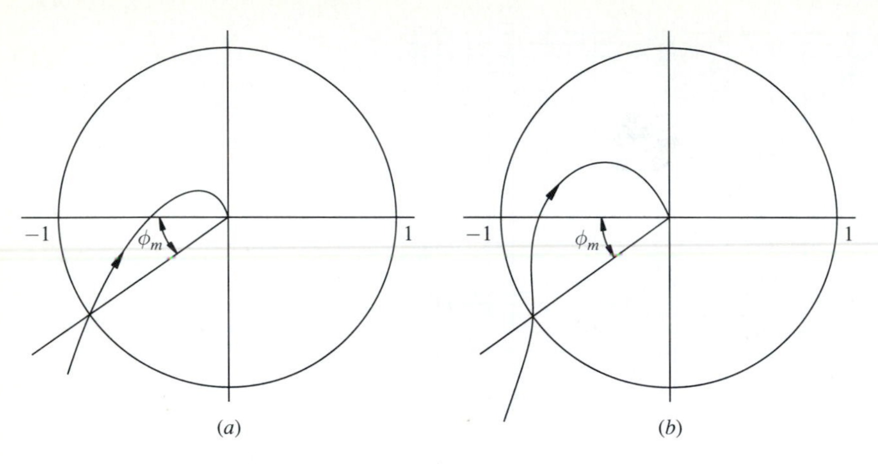

FIGURE 6.5-7
Two systems with the same phase margin but different M_p.

the closed-loop step response resulting from closing the loop around the loop gain of Fig. 6.5-7*a*.

In other words, the relative ease of determining ϕ_m as compared with M_p has been obtained only by sacrificing some of the reliability of M_p. Fortunately, systems such as that of Fig. 6.5-7*b* are fairly rare, and the phase margin provides a simple and effective means of estimating the closed-loop response from the $G(j\omega)$ plot. We make considerable use of the phase margin in Chap. 8 on design.

A system such as that shown in Fig. 6.5-7*b* could be identified as a system having a fairly large M_p by checking another parameter, the gain margin. The gain margin is easily determined from the Nyquist plot of the system. The gain margin is the ratio of the maximum possible gain for stability to the actual system gain. If a plot of $G(s)$ for $s = j\omega$ intercepts the negative real axis at a point $-a$ between the origin and the critical -1 point, then the gain margin GM is simply

$$GM = \frac{1}{a}$$

If a gain greater than or equal to $1/a$ were placed in series with $G(s)$, the closed-loop system would be unstable.

Example 6.5-1. Consider

$$G(s) = \frac{10}{s(s+2)(s+3)}$$

The portion of the Nyquist diagram for $\omega \geq 0$ for this loop transfer function is shown in Fig. 6.5-8. The frequency response function crosses the negative real axis at approximately 0.35, and the gain margin is approximately

$$GM \approx \frac{1}{0.35} \approx 2.86$$

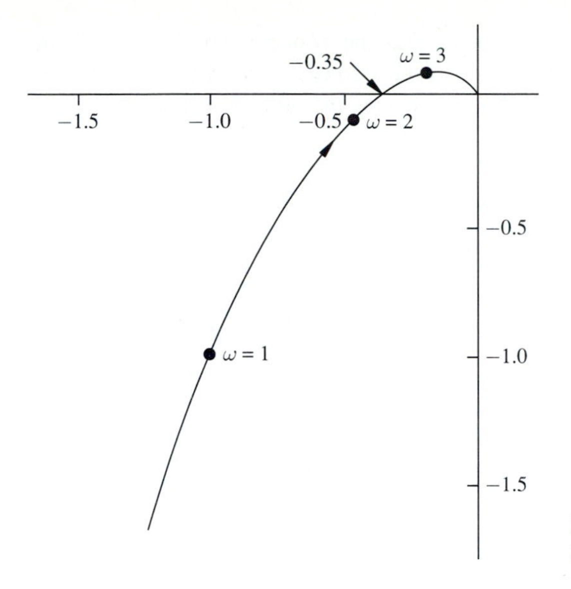

FIGURE 6.5-8
Example 6.5-1.

The gain margin can be determined more accurately by using the Bode plot if a system has been determined to be closed-loop stable. From the Bode plot an exact reading can be made of a, the magnitude of $G(j\omega)$ evaluated at the frequency where the phase is 180°.

While the gain margin does not provide very complete information about the response of the closed-loop system, a small gain margin indicates a Nyquist plot which approaches the −1 point closely at the frequency where the phase shift is 180°. Such a system will have a large M_p and an oscillatory closed-loop time response independent of the phase margin of the system.

While a system may have a large phase margin and a large gain margin and still get close enough to the critical point to create a large M_p, such a phenomenon can occur only in high-order loop gains. However, one should never forget to check any results obtained by using phase margin and gain margin as indicators of the closed-loop step response, lest an atypical system slip by. A visual check to see if the Nyquist plot approaches the critical −1 point too closely should be sufficient to determine if the resulting closed-loop system may be too oscillatory. In Sec. 6.6 we find that a Nyquist plot that approaches the −1 point too closely also indicates a system that may become unstable in the presence of modeling errors and uncertainty.

Equipped with the concepts that give rise to the M-circles a designer can develop a reasonably accurate impression of the nature of the closed-loop transient response by examining the loop gain Bode plots. The chain of reasoning is as follows: From the loop gain Bode plots the shape of the Nyquist plot of the loop gain can be envisioned. From the shape of the loop gain Nyquist plot, the shape of the Bode magnitude plot of the closed-loop system can be envisioned using the concepts of this section. Certain important points are evaluated by returning to the loop gain Bode plots. From the shape of the Bode magnitude plot of the closed-loop system, the dominant poles of the closed-loop transfer function are identified using the concepts of Chap. 5. From the knowledge of the dominant poles the shape of the step response of the closed-loop system is determined. The following example illustrates this chain of thought.

Example 6.5-2. Consider the loop gain transfer function

$$G(s) = \frac{80}{s(s+1)(s+10)}$$

The Bode plots for this loop gain are given in Fig. 6.5-9.

From the Bode plots the Nyquist plot can be envisioned. The Nyquist plot begins far down the negative imaginary axis since the Bode plot has large magnitude and -90° phase at low frequency. It swings to the left as the phase lag increases, and then spirals clockwise towards the origin, cutting the negative real axis and approaching the origin from the direction of the positive imaginary axis, i.e., from the direction associated with -270° phase. From the Bode plot it is determined that the Nyquist plot does not encircle the -1 point since the Bode plot shows that the magnitude crosses unity (0 dB) before the phase crosses -180°.

From the Bode plot it can be seen that the Nyquist plot passes very close to the -1 point near the crossover frequency. In this case $\omega_p \approx \omega_c$ and the phase margin is a key parameter to establish the value of M_p, the peak in the closed-loop frequency magnitude plot. The crossover frequency is read from the Bode magnitude plot as $\omega_c = 2.5$ rad/sec and the phase margin is read from the Bode phase plot as $\phi_m = 6^\circ$.

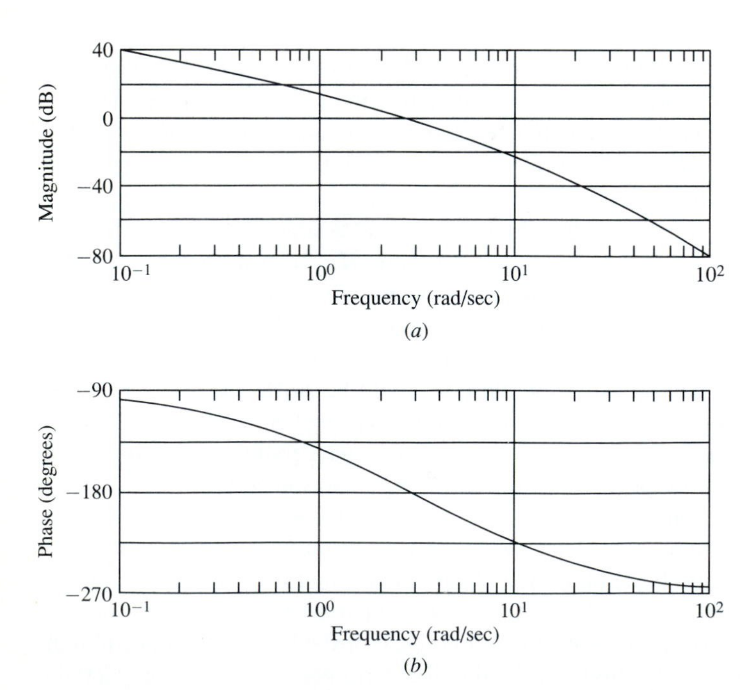

FIGURE 6.5-9
Bode plots of loop gain. (*a*) Magnitude, (*b*) phase.

Our visualization of the Nyquist plot is confirmed by the diagram of the actual Nyquist plot shown in Fig. 6.5-10.

The magnitude of the closed-loop frequency response for this system can be envisioned using the techniques learned in this section. At low frequencies $G(s)$ is very large, the distance from the origin to the Nyquist plot is very nearly the same as the distance from the -1 point to the Nyquist plot and the closed-loop frequency response has magnitude near one. As the Nyquist plot of the loop gain approaches -1, the magnitude of the closed-loop frequency response function increases to a peak. At higher frequencies the loop gain becomes small and the closed-loop frequency response decreases with the loop gain since the distance from -1 point to the loop gain Nyquist plot approaches unity. Thus, the closed-loop frequency response starts near 0 dB, peaks as the loop gain approaches -1 and then falls off.

The key point occurs when the loop gain approaches the -1 point and the closed-loop frequency response peaks. The closest approach to the -1 point occurs at a frequency very close to the crossover frequency, which has been established as $\omega_c = 2.5$ rad/sec. The height of the peak can be established using the phase margin, which has been established as $\phi_m = 6°$, and using Eq. (6.5-13). The height of the peak should be very close to $(2\sin(\phi_m/2))^{-1} = 9.5 = 19.6$ dB.

Our visualization of the magnitude of the closed-loop frequency response is confirmed by the actual plot shown in Fig. 6.5-11.

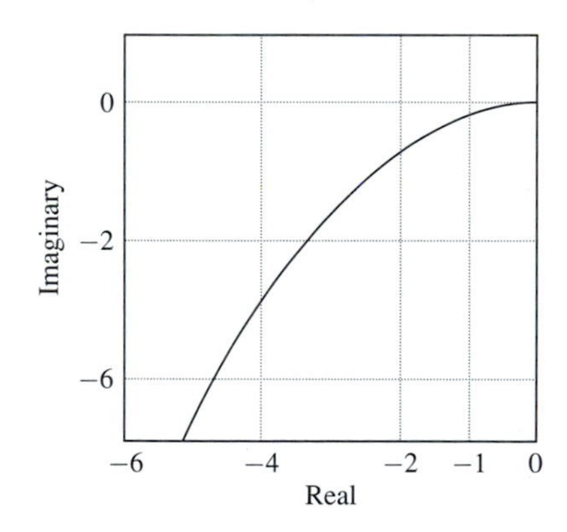

FIGURE 6.5-10
Nyquist plot of loop gain.

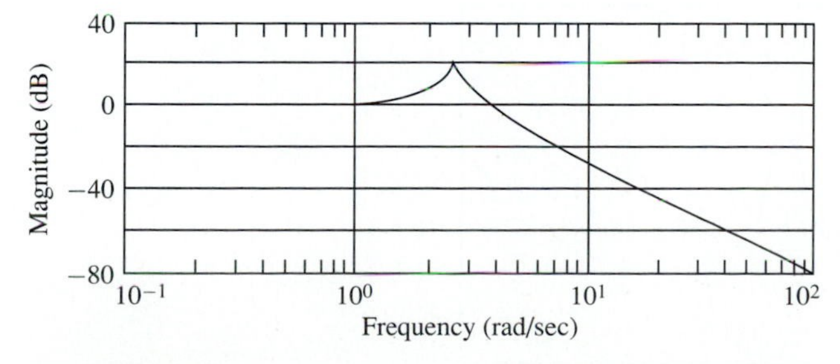

FIGURE 6.5-11
Magnitude plot of closed-loop frequency response.

From the visualization of the closed-loop frequency response function and the information about the peak of the frequency response, it is possible to use the ideas on identification from Chap. 5 to identify the dominant closed-loop poles. The frequency of the peak identifies the natural frequency of a pair of complex poles and the height of the peak identifies the damping ratio. More precisely, from Eq. (5.3-9)

$$M_p = \frac{1}{2\zeta\sqrt{1-\zeta^2}} \approx \frac{1}{2\zeta} \text{ for } \zeta \text{ small}$$

and from Eq.(5.3-8)

$$\omega_p = \omega_n\sqrt{1-2\zeta^2} \approx \omega_n \text{ for } \zeta \text{ small}$$

Using the approximations for ω_p and M_p obtained from the loop gain crossover frequency and phase margin, the following values are obtained: $\zeta \approx 1/(2M_p) \approx 0.05$ and $\omega_n \approx \omega_p \approx \omega_c \approx 2.5$ rad/sec.

If the Nyquist plot of a loop gain had not passed too closely to the -1 point the closed-loop frequency response would not have exhibited a sharp peak. In such a case, the dominant poles are well damped or real. The distance of these dominant poles from the origin can be identified by the system's bandwidth, which is given by the frequency at which the closed-loop frequency response begins to decrease. From the M-circle concept it can be seen that the frequency at which the closed-loop frequency response starts to decrease is well approximated by the crossover frequency of the loop gain.

Having established the position of the dominant closed-loop poles, it is easy to use the information of Chap. 4 to describe the closed-loop step response. The step response has an overshoot given by Eq. (4.4-13).

$$PO = 100e^{-\left(\frac{\zeta\pi}{\sqrt{1-\zeta^2}}\right)} \approx 85\%$$

The period of the oscillation is given by Eq. (4.4-9).

$$T_d = \frac{2\pi}{\omega_d} = \frac{2\pi}{\omega_n\sqrt{1-\zeta^2}} \approx 2.5 \text{ sec}$$

The first peak in the step response occurs at a time equal to half of the period of oscillation, or about 1.25 sec. The envisioned step response is confirmed in the plot of the actual closed-loop step response shown in Fig. 6.5-12.

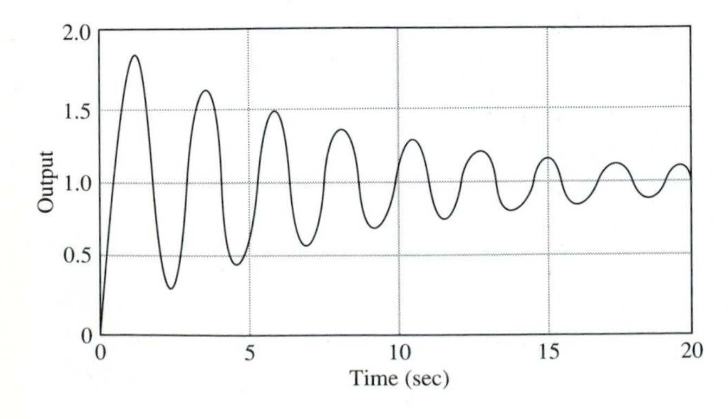

FIGURE 6.5-12
Closed-loop step response.

The method of the previous example may seem like it creates a long way to go in order to get an approximation to the closed-loop step response. Indeed it is much simpler to calculate the closed-loop transfer function directly from the loop gain transfer function. The importance of the logic in the example is not to create a computational method; the importance lies in the insight that is achieved in predicting problems with the closed-loop transient response by examining the Bode plot of the loop gain. The essence of the insight can be summarized in a few sentences: Assume that the Nyquist plot of the loop gain indicates a stable closed-loop system. *If the Nyquist plot of the loop gain approaches the* -1 *point too closely, the closed-loop system's transient response characteristics are oscillatory. The speed of the transient response of the closed-loop system is usually indicated by the loop gain crossover frequency.* Detailed information about the loop gain Nyquist plot is available in the loop gain Bode plots. In particular, the crossover frequency and the phase margin can be read from the Bode plots.

Any information that can be wrenched out of the Bode plots of the loop gain is critically important for two reasons. First, the Bode plots are a natural place to judge the properties of the feedback loop. When the magnitude of the loop gain is large, positive feedback properties such as good disturbance rejection and good sensitivity reduction are obtained. When the magnitude of the loop gain is small these properties are not enhanced. The work of this section completes the missing information about transient response that can be read from the loop gain Bode plots. Second, it is the Bode plot that we will be able to manipulate directly using the series compensation techniques of Chap. 8. It is important to be able to establish the qualities of the Bode plots that produce positive qualities in a control system because only then can the Bode plots be manipulated to attain the desired qualities.

We have now established that a good control system results when the Bode plots of the loop gain indicate a large loop gain for as large a frequency range as possible and a large phase margin, indicating that the Nyquist plot of the loop gain avoids proximity to the -1 point when the loop gain rolls off from large values to small values. In the next section we study how the presence of modeling inaccuracies creates a limitation on the range of frequencies over which the loop gain can be made large.

6.5.1 CAD Notes

CAD programs can be used to add M-circles or β-circles to Nyquist plots. The following MATLAB program is used to generate the Nyquist plot with M-circles presented in Fig. 6.5-2. You might try to write a similar program to draw β-circles.

```
[%set up the spacing in frequency
w=logspace(-4,1,100);
%enter the transfer function;
ng=[0 0 0 42];dg=[1 17 30 0];
```

```
%compute the real and imaginary parts
[rg,ig]=nyquist(ng,dg,w);
%set the axis limits
v=[-2 2 -2 2];
axis(v);
%force the graph area to be square
axis('square')
%plot the Nyquist plot;
plot(rg,ig),xlabel('real'),ylabel('imaginary'),grid;
%hold the graph so the M-circles can be added
hold on;
%replot some Nyquist points to mark off the plot
w1=[1 1.2 1.5 2 4 6];
[rg1,ig1]=nyquist(ng,dg,w1);
plot(rg1,ig1,'o')
%plot the degenerate M-circle at Re(s)=-0.5
x1=[-.5 -.5];
y1=[v(3) v(4)];
plot(x1,y1)
%run a for loop with the desired values of m
%DO NOT INCLUDE M=1
for m=[.7 .8 1.1 1.3 1.5 2.0 5.0]
       %compute the radius r and the center xc
       %of the M-circle
       msq=m*m;
       r=(m/(abs(msq-1)));
       xc=msq/(msq-1);
       %initialize an index for the vector of points
       %in an M-circle
       ind=1;
       %run a for loop parameterized by the angle
       %around the circle
       for q1=0:.1:2*pi+.1
              %for the circles which close to the left,
              %start at -pi
              if(m>1)q=q1-pi;
              else q=q1;
              end;
              %compute the x and y coordinates of the circle
              x(ind)=r*cos(q)-xc;
              y(ind)=r*sin(q);
              ind=ind+1;
       end;
       %plot an individual M-circle
       plot(x,y)
end;
%return the hold and axis to their usual status
hold off;
axis('normal')
```

Exercises 6.5

6.5-1. The question refers to a system in the G configuration with no loop gain right half-plane poles. The loop gain polar plot is shown in Fig. 6.5-13. The values at the point labeled $\times$ are

$$\omega_o = 10$$

$$|G(j\omega_o)| = 1$$

$$\text{phase } (G(j\omega_o)) = -185^\circ$$

Sketch the closed-loop frequency response.

(*a*) Label the value of the dc gain, the size of the peak amplitude and the frequency of the peak.

(*b*) What is the approximate percent overshoot and natural frequency of the step response of the closed loop system?

Answers: (*a*) $M(0) = 0.91$

(*b*) $M_p = 11.5 = 21\ dB, \omega_p = 10$

(*c*) $PO = 87\% \quad \omega_n = 10$

6.5-2. We wish a closed-loop system to have dominant poles with a damping ratio of at least $\zeta = 0.2$. In two steps calculate the minimum phase margin that should be allowed in the loop gain.

Answer: 23°

6.5-3. In Fig. 6.5-14 are Nyquist plots (positive imaginary axis image only) of the loop gains of two systems in the G configuration. Assume that each loop gain has no right half-plane poles or zeros except at the origin where each loop gain has one pole. Answer the following questions with a short (one or two line) explanation.

(*a*) The closed-loop system corresponding to which Nyquist plot is more oscillatory and why?

(*b*) Which system has a larger gain margin and why?

(*c*) Sketch *roughly* the shape of the closed-loop frequency response (Bode magnitude) of each system on the same plot by reading the closed-loop frequency response from the open-loop Nyquist plot.

Answers: (*a*) B (*b*) B

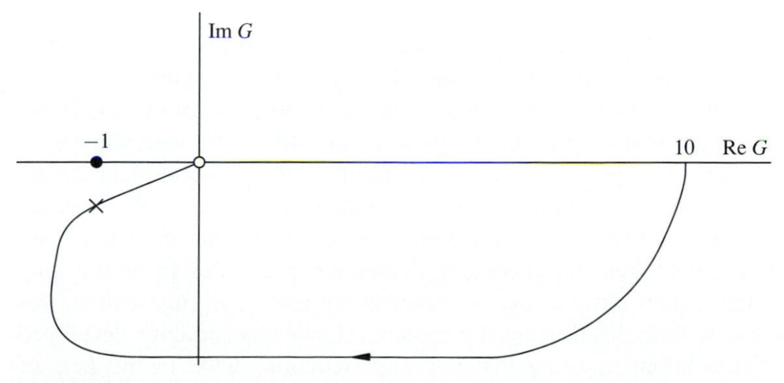

FIGURE 6.5-13
Exercise 6.5-1.

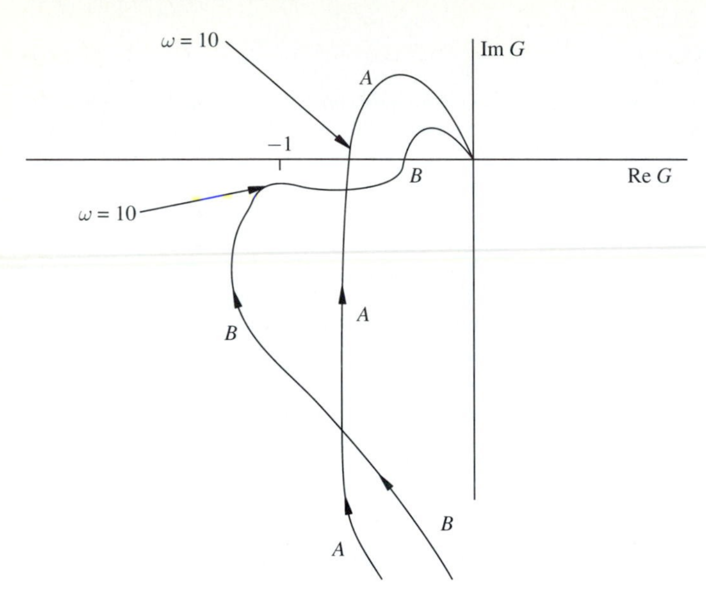

FIGURE 6.5-14
Exercise 6.5-3.

6.6 ROBUST STABILITY

In the previous section one notion of relative stability was discussed. In the view of that section, one system is considered less stable than another if its time response is more oscillatory and less highly damped. The relative stability decreases as the Nyquist plot of the loop gain transfer function approaches the -1 point. Besides an oscillatory time response, a system whose Nyquist plot passes close to the -1 point suffers from another problem. If such a system is even slightly mismodeled, the Nyquist plot can be easily perturbed in such a way that the number of encirclements of the -1 point changes without changing the number of right half-plane poles in the loop gain transfer function. The Nyquist theorem indicates that the number of right half-plane poles in the perturbed closed-loop system changes, so that, if the nominal closed-loop system is stable, the perturbed closed-loop system is unstable.

It is important that a control system that is designed to be stable and perform well when used in conjunction with a nominal plant model still works well when used in conjunction with an actual physical plant. The output of the physical plant can be expected to behave in a manner similar to but not exactly the same as the nominal model. A controller design that works well with a large set of plant models is said to be *robust*. It is apparent from the preceding discussion that a design with a loop gain Nyquist plot that passes close to the -1 point is not robust. In this section, this notion of robustness is formalized using the models of plant uncertainty developed in Sec. 5.7. First, conditions assuring that a design remains *stable* in the face of plant perturbations is developed. Clearly, maintaining stability for expected modeling errors is an absolute requirement. In addition, it is desirable to maintain adequate performance in the face of modeling errors. After the question of stability robustness is resolved in this section the question of robust performance is addressed in Sec. 6.5.

In Sec. 5.7, we developed techniques for describing a set of possible plant models using a nominal plant model and a perturbation transfer function with a known magnitude bound. Consider first the description using an additive perturbation as described in Sec. 5.7

$$\tilde{G}_p(s) = G_p(s) + L_a(s) \tag{6.6-1a}$$

where $L_a(s)$ is itself a stable transfer function containing no right half-plane poles. A bound $l_a(j\omega)$ on the magnitude of $L_a(j\omega)$ is known, i.e.,

$$|L_a(j\omega)| < l_a(j\omega) \tag{6.6-1b}$$

but otherwise $L_a(s)$ is a completely unknown transfer function. We are interested in the stability of the G configuration system of Fig. 6.6-1.

Let $G(s) = G_C(s)G_p(s)$ be the nominal loop gain, and $\tilde{G}(s) = G_C(s)\tilde{G}_p(s)$ be the perturbed loop gain.

Assume that the nominal design is stable, that is, that the closed-loop system is stable when $L_a(s) = 0$. The robust stability question is formulated as follows: What conditions must be placed on $G(s)$ so that the configuration of Fig. 6.6-1 remains stable for all $\tilde{G}_p(s)$ satisfying Eq. (6.6-1)?

The robust stability question can be answered using the same tool that was used to answer the nominal stability and relative stability questions—the Nyquist diagram. Consider Fig. 6.6-2, which contains a typical Nyquist plot of $G(s)$ and $\tilde{G}(s)$. From the definitions above we can see that

$$\tilde{G}(s) = G_C(s)\left(G_p(s) + L_a(s)\right) = G(s) + G_C(s)L_a(s) \tag{6.6-2}$$

The plot of $\tilde{G}(j\omega)$ can be obtained from the plot of $G(j\omega)$ at each value of ω by adding a vector corresponding to $G_C(j\omega)L_a(j\omega)$ to $G(j\omega)$. Indeed, if each possible $L_a(j\omega)$ satisfying Eq. (6.6-1b) is used in turn, the set of all possible $\tilde{G}(j\omega_o)$ at the frequency ω_o is given by the interior of a circle centered at $G(j\omega_o)$ with radius $G_C(j\omega_o)l_a(j\omega_o)$. This circle is shown on Fig. 6.6-2 assuming that the $L_a(j\omega)$ chosen for display has the maximum magnitude. The key observation for the desired result is that each $L_a(s)$ is assumed to be stable itself, so that the number of right half-plane poles of the loop gain $\tilde{G}(s)$ is unchanged from the number of right half-plane poles of the nominal loop gain $G(s)$. Therefore, the closed-loop stability assumed for the configuration in Fig. 6.6-1 with $L_a(s) = 0$ is maintained for nonzero $L_a(s)$ if and only if the number of encirclements of the -1 point is unchanged in going from $G(j\omega)$ to

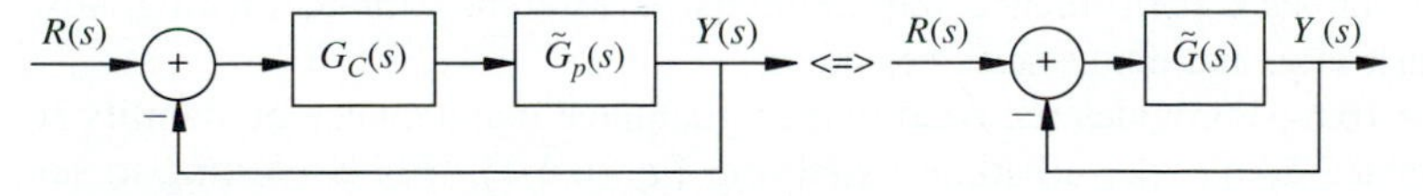

FIGURE 6.6-1
The perturbed G configuration.

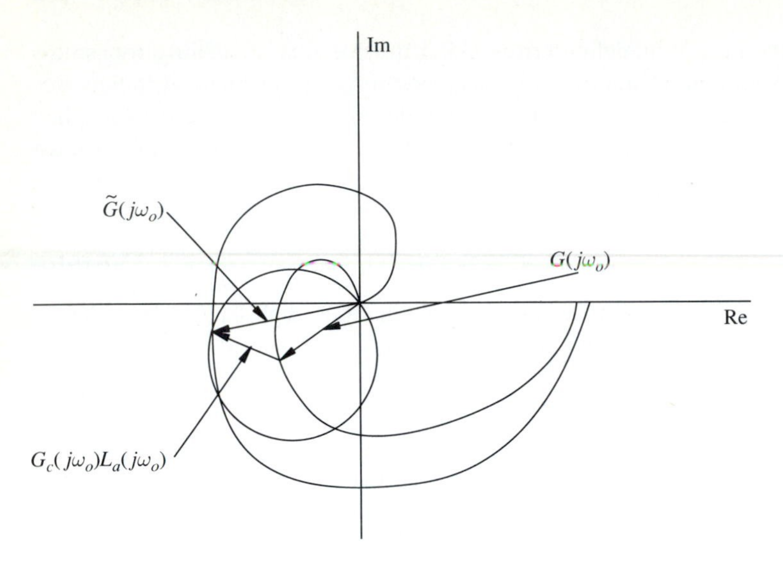

FIGURE 6.6-2
Nyquist plots of $\tilde{G}(s)$ and $G(s)$.

$\tilde{G}(j\omega)$. One way to assure that the number of encirclements remains unchanged is to assure that, at every single frequency, $l_a(j\omega)$ is small enough so that $\tilde{G}(j\omega)$ cannot reach the -1 point.

At each ω, the distance between $G(j\omega)$ and the -1 point is given by

$$|G(j\omega) - (-1)| = |1 + G(j\omega)|$$

Therefore, if at every ω we have the condition

$$|G_C(j\omega)|\, l_a(j\omega) < |1 + G(j\omega)|$$

or

$$l_a(j\omega) < \frac{|1 + G(j\omega)|}{|G_C(j\omega)|} = \left|G_C^{-1}(j\omega) + G_p(j\omega)\right| \tag{6.6-3}$$

then the perturbed system $\tilde{G}(j\omega)$ is stable since the perturbation in the Nyquist plot cannot change the number of encirclements.

From Eq. (6.6-3) we see that, as expected, a measure of the robustness of a control system is given by the distance the Nyquist plot of the loop gain transfer function maintains from the -1 point. This distance can be expressed as the magnitude of the return difference transfer function, a quantity that we already have seen should be kept large for such performance requirements as reference input tracking, low parameter sensitivity, and disturbance rejection.

Equation (6.6-3) provides the condition to guarantee maintenance of stability in the face of stable additive perturbations satisfying Eq. (6.6-1). It is interesting to see what can be said if Eq. (6.6-3) is violated. If Eq. (6.6-3) is violated the closed-loop system remains stable for many of the possible $L_a(s)$ satisfying Eq. (6.6-1); however, there is no longer any guarantee that stability is maintained for every perturbation satisfying Eq. (6.6-1). Indeed, if $l_a(j\omega)$ is continuous in ω and if, for some ω_o and $\epsilon > 0$,

$$l_a\,(j\omega_o) \geq \frac{|1 + G\,(j\omega_o)|}{|G_C\,(j\omega_o)|} + \epsilon \tag{6.6-4}$$

then there exists a perturbation $L_a(j\omega)$ satisfying Eq. (6.6-1) that causes instability. The destabilizing perturbation is constructed by choosing $L_a(s)$ stable so that

$$|L_a\,(j\omega_o)| = \frac{|1 + G\,(j\omega_o)|}{|G_C\,(j\omega_o)|} + \epsilon/2$$

and $$\arg\,(L_a\,(j\omega_o)) = \arg\,(1 + G\,(j\omega_o)) - \arg G_C\,(j\omega_o) + 180^\circ.$$

The ensuing Nyquist plot of $\tilde{G}(j\omega)$ has a different number of encirclements of the -1 point than the Nyquist plot of $G(j\omega)$ and the closed-loop stability is lost.

As discussed in Sec. 5.7 it is often easier to express a class of possible plant models using a multiplicative perturbation rather than an additive perturbation. This is particularly true for stability robustness conditions since the condition of Eq. (6.6-3) has the compensator appearing separately from the loop gain. The robust stability condition for a multiplicative perturbation is a function of only the loop gain. Assume a set of possible plant models is

$$\tilde{G}_p(s) = G_p(s)\,(1 + L_m(s)) \tag{6.6-5}$$

where $L_m(s)$ is itself a stable transfer function containing no right half-plane poles. A bound $l_m(j\omega)$ on the magnitude of $L_m(j\omega)$ is known, i.e.,

$$|L_m(j\omega)| < l_m(j\omega) \tag{6.6-6}$$

Then Eqs. (6.6-5) and (6.6-6) are equivalent to Eqs. (6.6-1a) and (6.6-1b) with the identification

$$L_a(s) = G_p(s)L_m(s) \tag{6.6-7}$$

However, the equivalent expression for the loop gain is

$$\tilde{G}(s) = G_C(s)G_p(S)\,(1 + L_m(s)) = G(s)\,(1 + L_m(s)) = G(s) + G(s)L_m(s) \tag{6.6-8}$$

The Nyquist plot is perturbed from $G(s)$ to $\tilde{G}(s)$ by the addition of $G(s)L_m(s)$. The condition for robustness in the multiplicative perturbation setting is obtained by observing again that if, at each frequency, the distance that the Nyquist plot of $G(j\omega)$ can be perturbed is less than the distance to the -1 point, a nominally stable closed-loop system is guaranteed to remain stable. Thus, if the closed loop system of Fig. 6.6-1 is stable for $L_m(s) = 0$, if each $L_m(s)$ is stable, and if for all ω

$$|l_m(j\omega)G(j\omega)| < |1 + G(j\omega)| \tag{6.6-9}$$

then the closed-loop system of Fig. 6.6-1 remains stable. Equation (6.6-9) can be manipulated into the equivalent forms

$$l_m(j\omega) < \frac{|1 + G(j\omega)|}{|G(j\omega)|} \tag{6.6-10}$$

$$\left| \frac{G(j\omega)}{1 + G(j\omega)} \right| < \frac{1}{l_m(j\omega)} \tag{6.6-11}$$

$$l_m(j\omega) \left| \frac{G(j\omega)}{1 + G(j\omega)} \right| < 1 \tag{6.6-12}$$

As with the robustness condition for the additive perturbation formulation, something can be said if Eq. (6.6-11) is violated. If Eq. (6.6-11) is violated, then there is a multiplicative perturbation satisfying Eq. (6.6-6) that causes the perturbed closed-loop system of Fig. 6.6-1 to be unstable.

The robustness condition of Eq. (6.6-11) usually poses a constraint on the allowable bandwidth of a control system as can be seen by the following argument. Recall from Sec. 5.7 that usually a multiplicative perturbation is small at low frequencies and grows to be larger than unity at higher frequencies. Let ω_1 be the frequency where $l_m(j\omega_1) = 2$. Assume that for $\omega > \omega_1$, $l_m^{-1}(j\omega) < \frac{1}{2}$. Now notice that the left hand side of Eq. (6.6-11) is simply the nominal closed-loop transfer function. It is usually desirable to keep the closed-loop transfer function close to unity for as large a range of frequencies as possible. However, the constraint of Eq. (6.6-11) dictates that the magnitude of the nominal closed-loop transfer function be less than $\frac{1}{2}$ and the nominal loop gain be less than 1 and for all $\omega > \omega_1$. Thus the maximum bandwidth of the system is limited to less than ω_1 if the controller is to result in a stable closed-loop system for all possible models as given by Eqs. (6.6-5) and (6.6-6).

Example 6.6-1. Let a collection of possible plant models be given by Eqs. (6.6-5) and (6.6-6) with

$$G_p(s) = \frac{1}{s} \tag{6.6-13}$$

$$l_m(s) = \frac{1}{10}|(s + 1)| \tag{6.6-14}$$

A series compensator given by

$$G_C(s) = a, \ a \text{ constant} \tag{6.6-15}$$

makes

$$G(s) = \frac{a}{s} \tag{6.6-16}$$

By straight calculation, the closed-loop transfer function is

$$\frac{G(s)}{1 + G(s)} = \frac{a}{s + a} \tag{6.6-17}$$

which includes a single closed-loop pole at $s = -a$. The closed-loop bandwidth covers $\omega = 0$ to $\omega = a$.

Figure 6.6-3 plots are derived from Eq. (6.6-14) and the closed-loop frequency response for various values of a. From the figure and a few calculations we can see that the robustness condition of Eq. (6.6-11) is satisfied for all $a < 10$. Thus, the maximum bandwidth is roughly equal to the frequency where $l_m^{-1}(j\omega) = 1$.

Now, let's take $a = 11$ so that the robustness condition is violated. We show that there is a multiplicative perturbation satisfying the magnitude bound given by Eq. (6.6-14) which produces an unstable closed-loop system.

To construct a destabilizing perturbation first choose a frequency ω_o where the robustness constraint of Eq. (6.6-11) is violated. In this example, we can choose $\omega_o = 100$ since

$$\left|\frac{G(j100)}{1+G(j100)}\right| = \left|\frac{11}{j100+11}\right| = \frac{11}{\sqrt{10121}} \geq \frac{1}{l_m(j100)} = \frac{10}{\sqrt{10001}}$$

The concept is to select $l_m(j\omega_o)$ so that the equivalent additive perturbation given by $l_m(j\omega_o)G(j\omega_o)$ moves the Nyquist plot of $G(j\omega)$ to and past the -1 point at the frequency ω_o. Thus we select the magnitude of $L_m(j\omega_o)$ to be slightly larger than $|1+G(j100)|/|G(j100)|$ while remaining smaller than $l_m(j\omega_o)$ and we select the phase of the $L_m(j\omega_o)$ to point the vector $L_m(j\omega_o)G_m(j\omega_o)$ in the Nyquist plane from $G(j\omega_o)$ towards -1. This can be achieved using the relationship

$$\arg(L_m(j\omega_o)) = \arg(1 + G(j\omega_o)) - 180^\circ - \arg(G(j\omega_o))$$

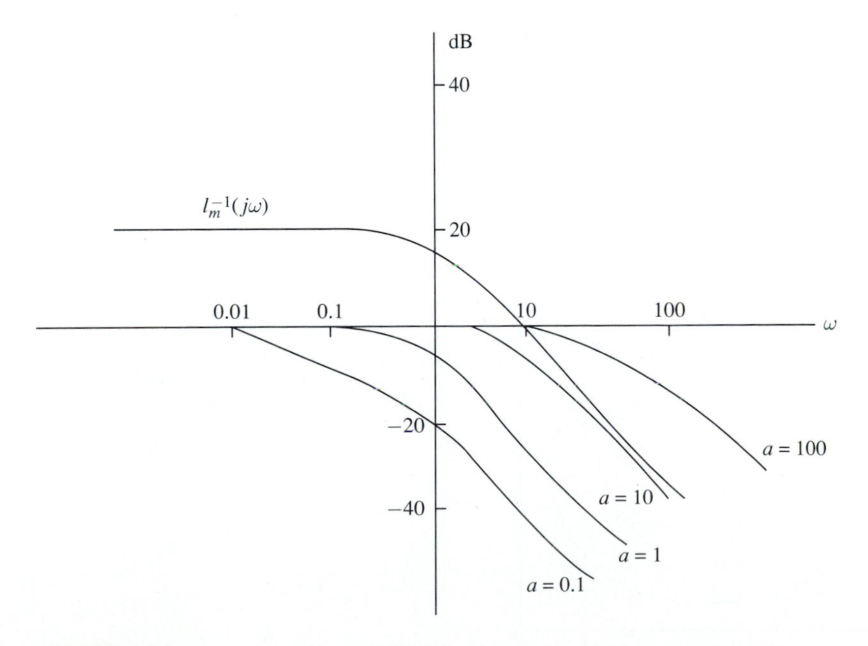

FIGURE 6.6-3
Bode magnitude plots of Eq. (6.6-14) and Eq. (6.6-17) for various a.

Performing these calculations the phase of the destabilizing $L_m(j\omega_o)$ for this example is -96° while 9.5 is a destabilizing magnitude. Once we know what $L_m(j\omega_o)$ should equal we only need to fit a transfer function that also satisfies the upper bound of Eq. (6.6-6) to that point. For example, one destabilizing $L_m(s)$ is given by

$$L_m(s) = \frac{0.95(s+1)}{(0.001s+10)} \frac{(-0.0103s+1)^2}{(0.0103s+1)^2}$$

First note that $L_m(s)$ is stable and that

$$|L_m(s)| < \frac{0.95|j\omega+1|}{10} < l_m(j\omega)$$

Then we can draw the Nyquist plot for the loop gain $G(s)(1+L_m(s))$ as in Fig. 6.6-4 and see that the resulting closed-loop system is unstable.

We have seen in this section that the stability of the nominal system can be guaranteed for a system with a stable multiplicative perturbation if Eq. (6.6-12) is satisfied. Such a system is said to possess robust stability with respect to the perturbation in question. Multiplicative perturbations tend to get large at high frequencies. Equation (6.6-12) requires that the nominal loop gain is made small at high frequencies.

Exercises 6.6

6.6-1. A plant is modeled by

$$G_p(s) = \frac{3}{s+1}$$

but in reality the plant contains unmodeled dynamics. Its perturbed transfer function is

$$\tilde{G}_p(s) = \frac{15(s+5)}{(s+1)\left(s^2+3s+25\right)}$$

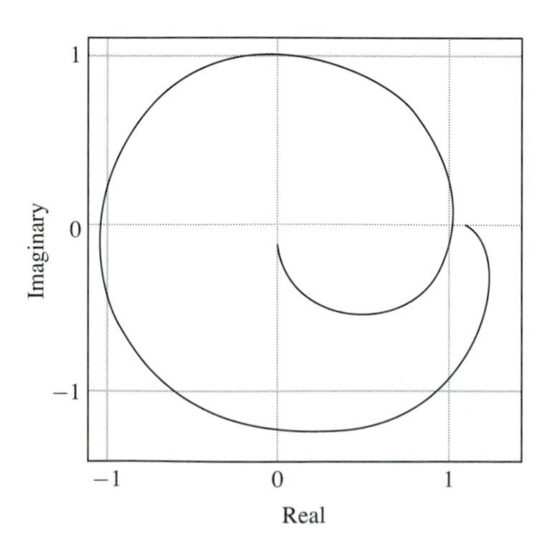

FIGURE 6.6-4
Perturbed loop gain for Example 6.6-1.

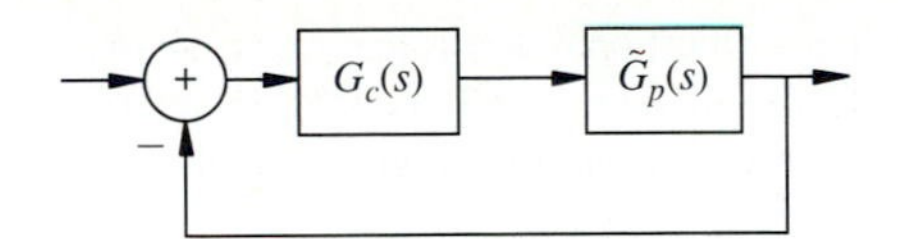

FIGURE 6.6-5
Exercise 6.6-1.

(a) Find the additive perturbation $L_a(s)$ and the multiplicative perturbation $L_m(s)$ for this case. Sketch the Bode magnitude plots of $L_a(j\omega)$ and $L_m(j\omega)$. Consider the closed-loop system of Figure 6.6-5. Let $G_C(s) = K$, a constant.

(b) For $K = 1$, check to see if the design that is clearly stable for the nominal model is stable for the actual plant by sketching the Bode magnitude plot of

$$\frac{\left|1 + G_C(j\omega)G_p(j\omega)\right|}{|G_C(j\omega)|}$$

on the same plot as $L_a(j\omega)$. Also, sketch the Bode magnitude plot of

$$\frac{G_C(j\omega)G_p(j\omega)}{1 + G_C(j\omega)G_p(j\omega)}$$

on the same plot as $|L_m^{-1}(j\omega)|$.

(c) Repeat (b) for $K = 8$.

Answers:

(a) $L_a(s) = \dfrac{\frac{6}{25}s(1 - 0.5s)}{(s+1)\left(1 + \frac{0.6}{5}s + \left(\frac{s}{5}\right)^2\right)}$ $\qquad L_m(s) = \dfrac{2}{25}\dfrac{s(1 - 0.5s)}{\left(\left(\frac{s}{5}\right)^2 + \frac{0.6}{5}s + 1\right)}$

(b) Robustly stable

(c) Not robustly stable

Remember, the magnitude tests in (b) and (c) are sufficient for stability, but not necessary. If a system passes the magnitude test, then the perturbed system is stable. However, if the system fails the magnitude test the perturbed system may be stable or unstable. (Option: To see this analyze (b) and (c) with $K = 2$. Check the robust stability test and check stability directly.)

6.6-2. The magnitude of the possible multiplicative perturbations that are likely to occur between a plant $\tilde{G}_p(s)$ and a model $G_p(s)$ can be bounded by $|L_m(s)| < l_m(s)$, where $l_m(s)$ is shown in Fig. 6.6-6.

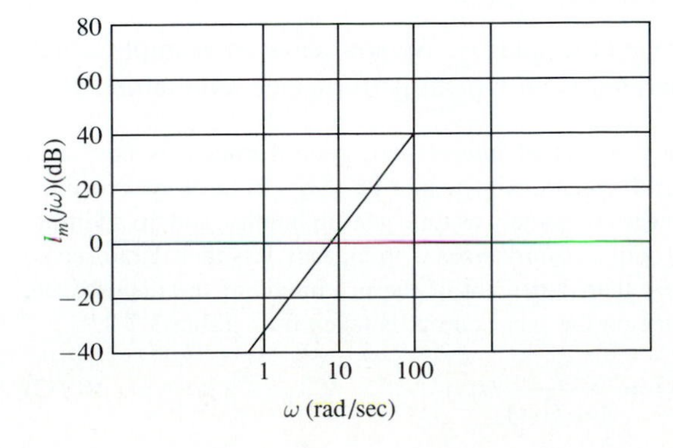

FIGURE 6.6-6
Exercise 6.6-2.

The model is given by $G_p(s) = \frac{1}{s+1}$. If a proportional controller $G_c(s) = K$ is used the nominal closed-loop system is stable for all $K > -1$. For what values of $K > 0$ can stability be guaranteed in the presence of all multiplicative perturbations obeying the above bound? Explain your answer briefly.
Answer: $K < 1$

6.7 PERFORMANCE AND ROBUSTNESS

In Sec. 6.6, we developed the constraint of Eq. (6.6-11), which characterizes a collection of loop gain transfer functions for which a perturbed closed-loop system remains stable. If such a constraint is satisfied and the actual plant's behavior is adequately described by one of the models in the collection of models given by Eqs. (6.6-5) and (6.6-6), the implementation of the control system produces a stable closed-loop response. However, even though that response is stable it may produce unacceptable oscillatory responses. We need some way to assure that the performance aspects required for the system and met by the nominal design are also met when the controller is implemented on the actual plant. We cannot assure 100 percent performance on the actual plant since even the collection of plants described by Eq. (6.6-5) and (6.6-6) cannot perfectly model the actual plant. We can arrive at a constraint that will assure that some performance measure is achieved for any plant that is a member of the collection of plants described by Eqs. (6.6-5) and (6.6-6).

We must first decide on a performance measure that suitably describes such diverse control system performance requirements as possessing desirable transient responses to command inputs, quickly and completely rejecting certain disturbances, and providing a small sensitivity in response to small parameter changes. One way to abstract these goals into a single performance specification is to recall that, in Sec. 3.2, we came to the realization that all these performance objectives can be met if we can make the return difference transfer function large enough. While the return difference cannot capture every aspect of performance that might be specified, a controller with a large return difference over a broad frequency band can be expected to perform well in most common tests of control system performance. With that in mind we can write down an interesting general performance requirement for a control system.

We can say that a control system in the G configuration performs adequately if

$$|1 + \tilde{G}(j\omega)| > p(j\omega) \tag{6.7-1}$$

where $p(j\omega)$ is some specified function of frequency. We now show an example which demonstrates how $p(j\omega)$ may be derived from typical performance requirements.

Example 6.7-1. One performance aspect of interest in a control system is how well disturbances are rejected. A typical specification would be that a control system must reject all constant output disturbances completely as time goes to infinity, and, in addition, it must attenuate the effect of all output disturbances of frequency less than 1 rad/sec so that the output is disturbed by less than 1 percent of the magnitude of the disturbance. The effect of an output disturbance on the plant output is taken from Table 3.2-2.

$$Y(s) = \frac{1}{1 + \tilde{G}(s)} D(s) \tag{6.7-2}$$

where the class of perturbed plants $\tilde{G}(s)$ is used to indicate the desire that the system reject disturbances when used with any of the possible plant models. The disturbance rejection requirement above can be translated to the form of Eq. (6.7-1) by requiring

$$p(j\omega) > 100 \text{ for } \omega < 1 \tag{6.7-3}$$

and

$$\lim_{\omega \to 0} p(j\omega) = \infty \tag{6.7-4}$$

If Eqs. (6.7-1) and (6.7-3) are satisfied, then

$$|Y| \le \frac{1}{|p(j\omega)|}|D| \le 0.01|D| \qquad \text{for } \omega < 1$$

A second common specification involves the closed-loop transfer function's sensitivity to changes in the plant transfer function. A specification may read that for all input frequencies less than 10 rad/sec the closed loop transfer function's magnitude response should not differ from a nominal design by more than 1 percent of the change in the plant model. Again, from Table 3.2-2 it is seen that

$$S_{\tilde{G}_p}^{M} = \frac{1}{1 + \tilde{G}(s)} \tag{6.7-5}$$

The sensitivity requirement is met if

$$p(j\omega) > 100 \text{ for } \omega < 10 \tag{6.7-6}$$

Specifications on the transient response of the system translate into specifications on $p(j\omega)$ less directly than do specifications on disturbance rejection and sensitivity reduction. To translate typical step response information into a specification on $p(j\omega)$, two intermediate steps are used. First, the transient response specifications are translated into desired dominant closed-loop pole positions. Then, the desired closed-loop pole positions are translated into a desired closed-loop frequency response. Finally, the closed-loop frequency response is translated into a specification on the return difference function or, equivalently, $p(j\omega)$.

Suppose that there is a requirement to produce a closed-loop step response which has less than 20 percent overshoot and a 5 percent settling time of less than 5.5 sec. From Eqs. (4.4-13) and (4.4-14), it is seen that these specifications are satisfied by a closed-loop transfer function whose dominant behavior is characterized by a single pole pair with damping ratio $\zeta = 0.47$ and natural frequency $\omega_n = 1.2$. From Sec. 5.3 it is known that the closed-loop frequency response for such a transfer function has magnitude very close to unity for all frequencies less than 1.2 rad/sec and rolls off at higher frequencies. From Eq. (5.3-9) the peak of the magnitude plot is computed to be less than 1.2.

In the G configuration the magnitude of the closed-loop frequency response $|\tilde{M}_c(j\omega)|$ is related to the magnitude of the loop gain $|\tilde{G}(j\omega)|$ and the magnitude of the return difference $|1 + \tilde{G}(j\omega)|$ by the expression

$$\left|\tilde{M}_c(j\omega)\right| = \frac{\left|\tilde{G}(j\omega)\right|}{\left|1 + \tilde{G}(j\omega)\right|} \tag{6.7-7}$$

If $|1 + \tilde{G}(j\omega)|$ is much greater than one, then $|\tilde{G}(j\omega)|$ is much greater than one and $|\tilde{M}_c(j\omega)|$ is very close to one. This logic can be quantified using the inequalities

$$\left|1 + \tilde{G}(j\omega)\right| - 1 < \left|\tilde{G}(j\omega)\right| < \left|1 + \tilde{G}(j\omega)\right| + 1 \tag{6.7-8}$$

Dividing Inequalities (6.7-8) by $|1 + \tilde{G}(j\omega)|$ and using Eq. (6.7-7) we can arrive at an expression relating the magnitude of the closed-loop frequency response to the magnitude of the return difference.

$$1 - \frac{1}{\left|1 + \tilde{G}(j\omega)\right|} < \left|\tilde{M}_c(j\omega)\right| < 1 + \frac{1}{\left|1 + \tilde{G}(j\omega)\right|} \tag{6.7-9}$$

If the magnitude of the return difference is greater than 5 for $\omega < 1.2$, the magnitude of the closed-loop frequency response remains between 0.8 and 1.2 for those frequencies. The inequalities given by (6.7-9) provide only a rough guideline to what is needed to achieve a certain closed-loop response. To meet the step response specifications, not only must the magnitude of the closed-loop response be kept near one for the appropriate frequencies ($\omega < 1.2$ in this case) but also peaks in the magnitude of the closed-loop frequency response at higher frequencies must be avoided. This requires that the return difference be kept from being too small at any frequency.

The guidelines needed to produce an adequate closed-loop frequency response become clearer by looking at Fig. 6.7-1. This figure shows a polar frequency plot complete with M-circles for a typical loop gain transfer function. Recall from Section 6.5 that, from the M-circles, the magnitude of the closed-loop frequency response can be read off at any frequency. The plot of Fig. 6.7-1 would be an acceptable loop gain for this

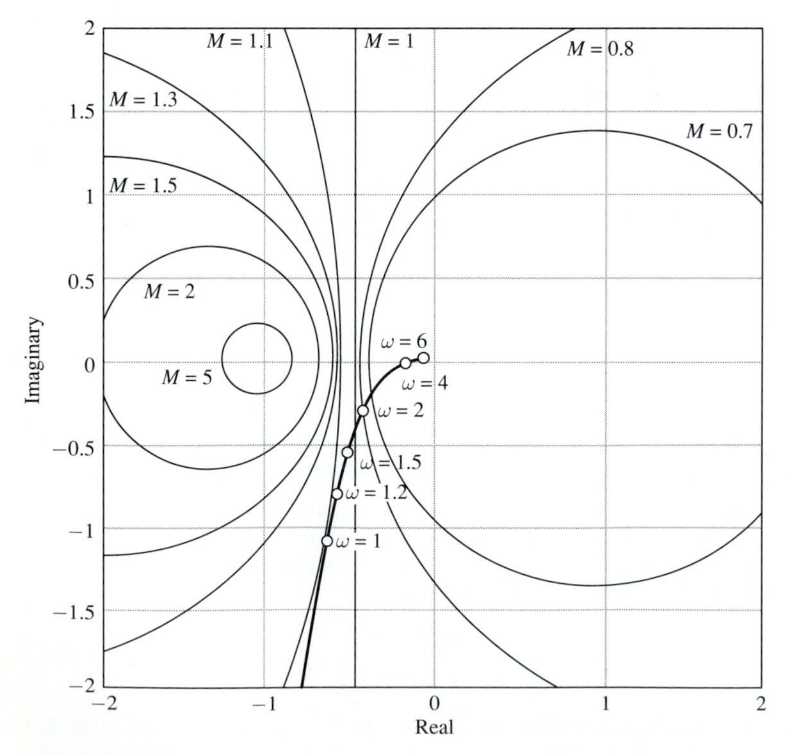

FIGURE 6.7-1
Performance specifications for transient response illustrated with M-circles.

example since the peak M-circle reached is $M = 1.1$ near $\omega = 1.2$ and the plot moves through M-circles of smaller values for higher frequencies.

In this typical example, the guidelines of Eq. (6.7-9) are conservative in that it is not required to have the return difference greater than 5 for all $\omega \leq 1.2$. The guidelines are somewhat lacking as mentioned above in that they don't guarantee that a peak in the closed-loop frequency won't occur at a higher frequency, causing oscillations.

In spite of the shortcomings in precise guarantees, it is still useful to provide a guideline for adequate transient response of the closed-loop system by bounding the return difference from below. For many control systems, the loop gain transfer function decreases monotonically producing a smooth polar frequency plot similar to Fig. 6.7-1. For these systems the range of frequencies where the magnitude of the return difference stays greater than one provides a good estimate of the bandwidth of the controller. The greater the frequency range over which the magnitude of the return difference stays larger than one, the wider is the bandwidth of the system and the faster can the system respond to step inputs. In addition, the magnitude of the return difference should be kept as large as possible over all frequencies to guard against high frequency oscillations.

In the example, we have shown that the concept of providing a performance measure for a control system by bounding the magnitude of the return difference as in Eq. (6.7-1) works well for specifications involving disturbance rejection and sensitivity reduction. The bound of Eq. (6.7-1) also provides a guideline for transient response specifications. This lack of precision for transient response specification is acceptable because the transient response of a robustly stable control loop that responds almost fast enough can usually be modified to precisely meet specifications using a prefilter on the command input.

Equation (6.7-1) can be rewritten as

$$\left| \frac{1}{1 + \tilde{G}(j\omega)} \right| < \frac{1}{p(j\omega)} \tag{6.7-10}$$

or

$$p(j\omega) \left| \frac{1}{1 + \tilde{G}(j\omega)} \right| < 1 \tag{6.7-11}$$

The function $\tilde{S}(j\omega) = (1 + \tilde{G}(j\omega))^{-1}$ is called the *sensitivity function* of the perturbed control system since the response of the control system is relatively insensitive to parameter changes and disturbances if this function is small.

Notice that the performance requirement of Eq. (6.7-1) is written as a function of $\tilde{G}(j\omega)$. It is desirable to design a controller based upon the nominal model of the plant $G(j\omega)$ that meets the performance requirement of Eq. (6.7-1) for any plant $\tilde{G}(j\omega)$ in the collection given by the multiplicative perturbation model of Eqs. (6.6-5) and (6.6-6).

To assure that the sensitivity function of the perturbed control system $\tilde{S}(j\omega)$ is small for any allowable perturbation $L_m(j\omega)$ it is useful to find how large $\tilde{S}(j\omega)$ can become when facing the most damaging perturbation allowable. The goal is to find the maximum value of $\tilde{S}(j\omega)$ as $L_m(j\omega)$ is allowed to vary over all allowable perturbations.

$$\max_{|L_m(j\omega)|<l_m(j\omega)} \left|\tilde{S}(j\omega)\right|$$

$$= \max_{|L_m(j\omega)|<l_m(j\omega)} \frac{1}{\left|1+\tilde{G}(j\omega)\right|}$$

$$= \max_{|L_m(j\omega)|<l_m(j\omega)} \frac{1}{|1+G(j\omega)+G(j\omega)L_m(j\omega)|}$$

$$= \frac{1}{\min\limits_{|L_m(j\omega)|<l_m(j\omega)} |1+G(j\omega)+G(j\omega)L_m(j\omega)|} \tag{6.7-12}$$

Clearly, minimizing the denominator maximizes the expression of Eq. (6.7-12) where various definitions have been substituted. The minimization is accomplished by realizing that the worst case $L_m(j\omega)$ has maximal magnitude and has phase aligned to subtract this maximal magnitude away from the first two terms.

$$\min_{|L_m(j\omega)|<l_m(j\omega)} |1+G(j\omega)+G(j\omega)L_m(j\omega)|$$

$$= |1+G(j\omega)| - |G(j\omega)l_m(j\omega)| \tag{6.7-13}$$

So

$$\max_{|L_m(j\omega)|<l_m(j\omega)} \left|\tilde{S}(j\omega)\right| = \frac{1}{|1+G(j\omega)| - |G(j\omega)|l_m(j\omega)}$$

$$= \frac{1}{|1+G(j\omega)|}\left(\frac{1}{1-\left|\dfrac{G(j\omega)}{1+G(j\omega)}\right| l_m(j\omega)}\right) \tag{6.7-14}$$

The nominal *complementary sensitivity function*, $T(j\omega)$ is defined as

$$T(j\omega) = \frac{G(j\omega)}{1+G(j\omega)} = 1 - \frac{1}{1+G(j\omega)} = 1 - S(j\omega) \tag{6.7-15}$$

The complementary sensitivity function equals one minus the sensitivity function. (The complementary sensitivity function is also equal to the closed-loop response function for the G configuration.) A final expression for the sensitivity under the worst case perturbation in terms of the nominal sensitivity and complementary sensitivity functions can be written as

$$\max_{|L_m(j\omega)|<l_m(j\omega)} \left|\tilde{S}(j\omega)\right| = |S(j\omega)|\left(\frac{1}{1-|T(j\omega)|l_m(j\omega)}\right) \tag{6.7-16}$$

If the worst case perturbed sensitivity function remains less than $p^{-1}(j\omega)$ then the sensitivity functions for all allowable perturbations are less than $p^{-1}(j\omega)$. The robust performance condition of Eq. (6.7-10) is satisfied if

$$|S(j\omega)|\left(\frac{1}{1-|T(j\omega)|l_m(j\omega)}\right)<\frac{1}{p(j\omega)} \tag{6.7-17}$$

Equation (6.7-17) can be rewritten as

$$|S(j\omega)|p(j\omega)+|T(j\omega)|l_m(j\omega)<1 \tag{6.7-18a}$$

or, more explicitly,

$$\left|\frac{1}{1+G(j\omega)}\right|p(j\omega)+\left|\frac{G(j\omega)}{1+G(j\omega)}\right|l_m(j\omega)<1 \tag{6.7-18b}$$

Equation (6.7-18) provides some interesting information. First note that keeping the second term less than one matches Eq. (6.6-12) and guarantees stability robustness. Keeping the first term less than one matches Eq. (6.7-10) with $G(j\omega)$ replacing $\tilde{G}(j\omega)$ and this provides for acceptable nominal performance, i.e., acceptable performance would result if the plant actually responded like $G_p(j\omega)$. When there is mismodeling as represented by $l_m(j\omega)$ the nominal design must exceed the performance specification by enough of a margin to account for modeling error. Alternatively, the nominal design must not only allow enough stability margin to ensure stability but must allow a greater margin to maintain performance.

It is perhaps even more interesting to view Eq. (6.7-18) as a weighted tradeoff between two terms. The first term contains the magnitude of the nominal sensitivity function, weighted by the performance requirement, which is large at frequencies where good performance is required. The second term contains the magnitude of the complementary sensitivity function weighted by the bound on the modeling error, which is large at frequencies where the plant is not well modeled.

Since by Eq. (6.7-15) the sensitivity function and the complementary sensitivity function sum to one they cannot both be small at the same frequency. Thus, by using Eq. (6.7-18) it can be seen that good control system performance can be maintained only at frequencies where the plant is well modeled. The modeling error quantified by $l_m(j\omega)$ is usually large at high frequencies. The complementary sensitivity function is then required to be small at high frequencies. A small complementary sensitivity function means a sensitivity function very near one, dictating that the achievable performance function $p(j\omega)$ be somewhat less than one at high frequencies. The resulting implication that the magnitude of the sensitivity function cannot be kept smaller than one for all frequencies means poor performance at some frequencies in the areas of disturbance rejection, sensitivity reduction and reference input tracking. Luckily, in most situations, a large performance bound is required only for low frequency reference inputs and disturbances. Similar logic then dictates that $l_m(j\omega)$, the modeling error, be small at low frequencies. If the control designers are asked to produce strong performance results at frequencies where the modeling error is large, they must reply that they can not. Either the performance requirements must be relaxed at those frequencies or a more accurate model must be obtained at those frequencies. Much of control design is involved in judiciously squeezing the tradeoff between the two conflicting requirements of good performance and robustness to modeling errors.

6.7.1 CAD Notes

It is easy to compute the sensitivity function and the complementary sensitivity function. The denominator of each function is equal to the sum of the denominator and the numerator of G. Care must be taken when adding polynomials in MATLAB. Since MATLAB is really just adding vectors it requires them to be the same length. Leading zeros must be included in the lower degree polynomial before adding. The numerator of the sensitivity function is the denominator of G while the numerator of the complementary sensitivity function is the numerator of G.

Example 6.7-2. Assume that $G(s)$ is given by

$$G(s) = \frac{10}{s^2}$$

and

$$l_m(s) = 0.01s + 0.1$$

The following MATLAB macro computes the numerator and denominator of the sensitivity function (`snum` and `sden`) and the numerator and denominator of the complementary sensitivity function (`tnum` and `tden`). It then computes the maximum perturbed sensitivity function as defined by Eq. (6.7-16) and finds the maximum of this function and the frequency at which the maximum occurs.

```
%enter the transfer function so that the num and den are equal
%length
gnum=[0 0 1];
gden=[1 10 0];
%compute the den of s and t
sden=gnum+gden;
tden=sden;
%compute the num of s
snum=gden;
%compute the num of t
tnum=gnum;
%enter the transfer function of the perturbation
lmnum=[.01 .1];
lmden=[1];
%compute the magnitudes of t,s,and lm
w=logspace(-1,2,100);
[tmag,tphase]=bode(tnum,tden,w);
[smag,sphase]=bode(snum,sden,w);
[lmmag,lmphase]=bode(lmnum,lmden,w);
%compute and plot max perturbed s
for ind=1:1:100
    maxsp(ind)=smag(ind)*(1/(1-tmag(ind)*lmmag(ind)));
end
```

```
%find maximum value of this vector and the frequency where
%max occurs
[maxelsp,maxind]=max(maxsp);
maxelsp
wmaxsp=w(maxind)
maxelsp =
     1.0152
wmaxsp =
     1
```

Exercises 6.7

6.7-1. You are given a nominal plant model $G(s)$ and a multiplicative perturbation $|L_m(j\omega)| < l_m(j\omega)$ such that the real plant is $\tilde{G} = G(1 + L_m)$. You are also given that, at $\omega = 10$, the following relations hold:

$$\left|\frac{G(j10)}{1 + G(j10)}\right| = 0.09$$

$$|1 + G(j10)| = 1$$

$$l_m(j10) = 10$$

What is the minimum $|1 + \tilde{G}|$ at $\omega = 10$?

Answer: 0.1

6.7-2. Show that one cannot guarantee robust performance if for some frequency ω_{bad} the bound on the multiplicative perturbation $l_m(j\omega_{\text{bad}})$ is greater than one and, at the same frequency, $p(j\omega_{\text{bad}})$ is greater than one, where $p(j\omega)$ is the performance requirement bound as in the expression

$$\left|1 + \tilde{G}(j\omega)\right| > p(j\omega)$$

6.8 CONCLUSIONS

To summarize effectively the contents of this chapter, it is necessary to view the material in light of all that is said in the analysis portion of this book. Chapter 4 discusses the means of obtaining the time solution for any input or initial conditions. As far as system response is concerned, this is the most complete information that one might seek. In fact, it is often more information than can be effectively used, particularly in the early stages of system design or evaluation.

We now seek ways of obtaining a less complete answer to the question of system response, ideally at the cost of less energy and more insight. The most basic response question that can be asked is whether the closed-loop system is stable or not. That question can be answered in a yes–no fashion by the use of the Routh-Hurwitz criterion. If a system proves to be unstable, this often ends the analysis, as unstable operation is usually unacceptable from a physical as well as a mathematical point of view. The system must somehow be modified to make it stable before any more penetrating questions regarding system response can be answered.

If it is assumed that the answer to the stability question is "yes" and that the system in question is actually stable, the next level of questioning regards relative stability. Relative stability measures are easily discussed in the frequency domain, and for this reason the frequency response is introduced in Chap. 5. In this chapter we use the frequency response function expressed in the Nyquist diagram as an alternative to the Routh-Hurwitz criterion to answer the absolute stability question. More importantly, a method is now available for discussing relative stability and robustness.

As is seen in Chap. 8 the combination of the Bode diagrams of Chap. 5 and the Nyquist plot of this chapter form extremely powerful analysis and design tools. The Bode diagram gives immediate information about the size of the loop gain at various frequencies. If the closed loop system is stable the results of Sec. 3.2 let us use this information to evaluate the system's response to disturbances, parameter variations, and sensor noise. The Bode diagram is also used to aid in drawing the Nyquist plot. From the Nyquist plot the stability of the closed-loop system is established. In addition, the Nyquist plot provides information about the closed-loop frequency response and therefore, through the results of Chap. 5 and Chap. 4, the closed-loop transient response to step inputs. In addition, the combination of the Nyquist and Bode plots provide information about the robustness of the system to modeling errors. In Chap. 8, these analysis tools are used to provide insight into controller design.

In using the frequency response function, the complex variable $s = \sigma + j\omega$ is allowed to be a function only of ω; that is, the frequency response function is formed from the transfer function by setting $s = j\omega$, with $\sigma = 0$. On the s plane, we make use only of the $j\omega$ axis and, in forming the Nyquist diagram, the infinite semicircle bounding the right half s plane. One way in which more information regarding system behavior might be obtained is to make use of the entire s plane. This is the procedure to be used in the next chapter on the root locus method.

PROBLEMS

6.1. Use the Routh-Hurwitz criterion to determine for what values of K the closed-loop systems corresponding to the following open-loop transfer functions are stable.

$$(a)\ G(s) = \frac{100K}{(s+1)(s+10)(s+100)}$$

$$(b)\ G(s) = \frac{100K}{(s-1)\left(s^2+14s+100\right)}$$

$$(c)\ G(s) = \frac{K\left(s^2+s+100\right)}{(s+1)^3}$$

$$(d)\ G(s) = \frac{K(s+3)}{(s-1)\left(s^2+3s+10\right)(s+10)}$$

6.2. Given the following loop gain transfer functions arising from systems in the G configuration consider the resulting closed-loop systems. Use the Nyquist criterion to determine the values of $K > 0$ for which the closed-loop systems are stable. For each system that is stable when $K = 1$ determine the gain margin and the phase margin for $K = 1$.

$$(a)\ G(s) = \frac{10{,}000K}{s(s+10)(s+100)}$$

$$(b)\ G(s) = \frac{(10)^6 K(s+10)}{(s+1)^2\left(s^2+20s+(100)^2\right)}$$

$$(c)\ G(s) = \frac{10K(s+1)}{s^2(s+10)}$$

$$(d)\ G(s) = \frac{0.1K(s+10)}{s^2(s+1)}$$

$$(e)\ G(s) = \frac{1000K(s+1)(s+10)}{s^3(s+100)}$$

6.3. Determine M_p, ω_p, and ω_c for the Nyquist diagram of $G(s)$ given in Fig. P6.3. Also determine the phase margin using the size of the sensitivity function at the crossover frequency.

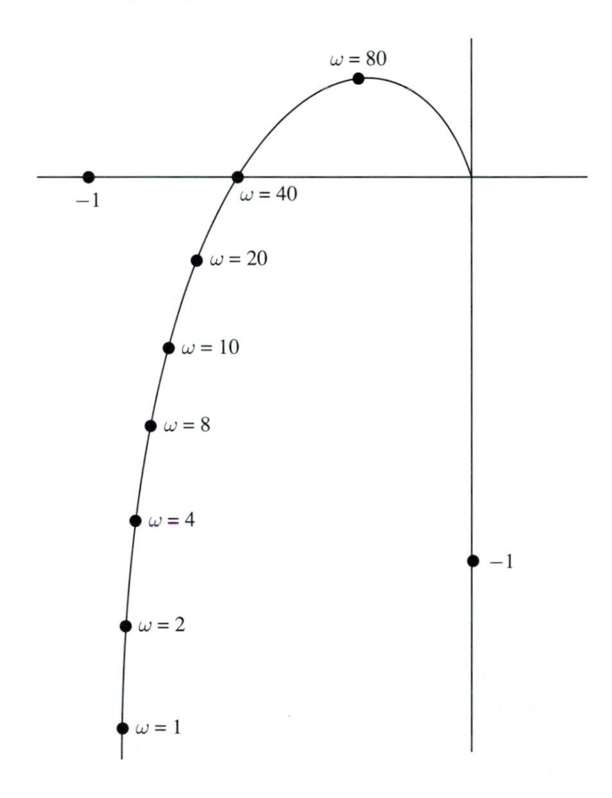

FIGURE P6.3
Problem 6.3.

6.4. Determining closed-loop characteristics from the loop gain frequency response.

Consider the following loop gain transfer function that arises from a system in the G configuration

$$G(s) = \frac{-3(s-10)}{(s+1)(s+3)}$$

In this problem we go through a series of steps using sketches and approximations to determine graphically the closed-loop step response from the loop gain transfer function. When you've finished this problem you may want to repeat the steps more precisely using a CAD program if you have one available.

1. Sketch the Bode plot of $G(s)$.
2. Sketch the Nyquist plot of this system's loop gain. Verify that the closed-loop system is stable. (You may have to more accurately compute the values of a few points in this and other parts.)
3. From the loop gain Nyquist plot sketch the magnitude plot of the closed-loop frequency response using the M-circle concepts. (Don't forget to check the dc gain.)
4. Make a rough sketch of the closed-loop frequency response phase plot.
5. From the results of 3 and 4, identify the closed-loop system's transfer function.
6. From the result of 5, describe the step response of the closed-loop system.

6.5. The straight line approximation of the Bode magnitude plots of two loop gain transfer functions appear in Fig. P6.5. Both systems are open-loop stable except for poles at the origin and minimum phase and are in the G configuration. Neither system has any complex conjugate poles. Both closed-loop systems are stable.

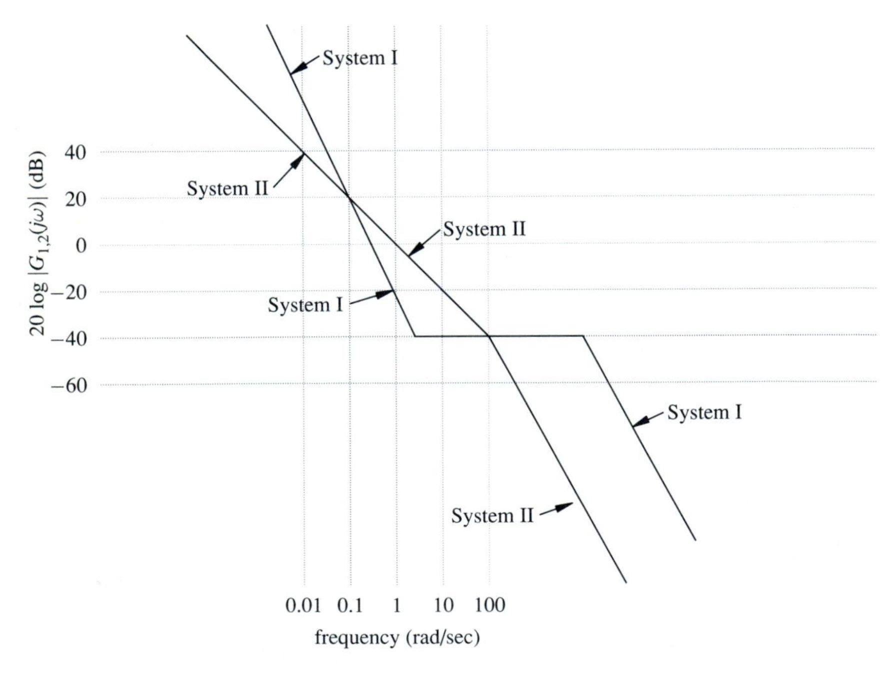

FIGURE P6.5
Problem 6.5.

(*a*) Which closed-loop system has a larger overshoot and why?
(*b*) Which closed-loop system has a longer settling time and why?
(*c*) In steady-state which system rejects a sinusoidal output disturbances at $\omega = 0.01$ better and why?
(*d*) Which system is more susceptible to instability in the presence of unmodeled dynamics and why?

6.6. A plant transfer function is the product of a modeled portion and an unmodeled portion,

$$\tilde{G}(s) = G(s)G_u(s) \qquad G_u(s) = \frac{A}{s^2 + s + A}$$

(*a*) Briefly explain how $G_u(s)$ affects the phase of $\tilde{G}(s)$.

(*b*) Sketch the Bode magnitude plot of $L_m(s) = G_u(s) - 1$ for $A = 1, 10, 100$, and 1000 all on the same plot. Sketch in and determine a function $20 \log |l_m(s)|$ so that $|l_m(j\omega)| > |L_M(j\omega)|$ for all ω and all A.

6.7. In Fig. P6.7 is a Bode plot for a nominal plant transfer function $G_p(s)$. A unity compensator, i.e., $G_c(s) = 1$, is placed in series with this plant and the loop is closed.

(*a*) Sketch a magnitude plot of the closed-loop frequency response. Label an estimate of the height and the location in frequency of any peaks.

(*b*) Describe the response of the closed-loop system to a unit step input.

(*c*) Remember that the loop gain arises from a plant with a unity compensator so that $G(s) = G_p(s)$. Assume that there is an additive perturbation on the plant so that $\tilde{G}(s) = G_p(s) + L_a(s)$ with $|L_a(j\omega)| < 0.05$ for all ω. What happens to the step response if the worst case perturbation is present?

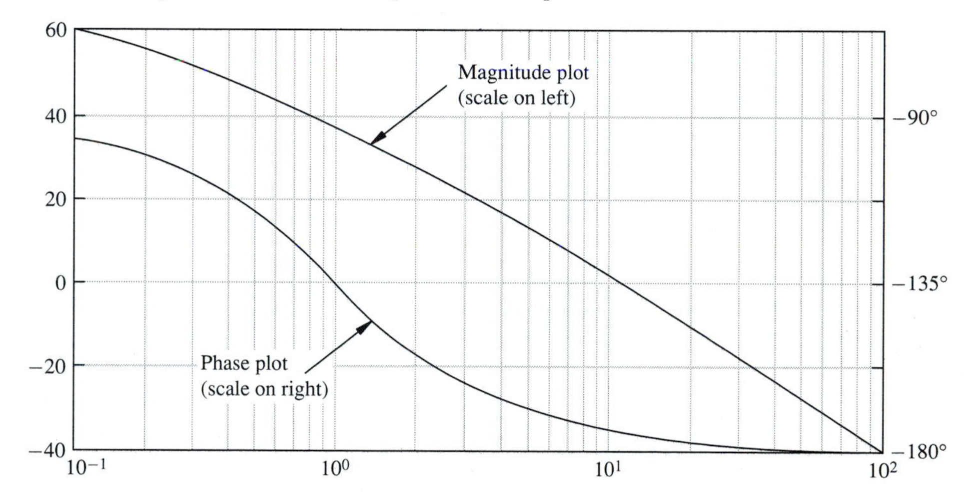

FIGURE P6.7
Problem 6.7.

6.8. Consider a system in the H configuration as defined in Fig. 3.2-9*a*. From Table 3.2-2 we see that such a system reacts the same as a system in the G configuration with regard to sensitivity to plant perturbations, output disturbance rejection, and sensor noise reduction if the loop gain for the G configuration, $G(s) = G_c(s)G_p(s)$, is replaced by the loop gain for the H configuration, $KG_p(s)H(s)$. Thus it is natural to define the sensitivity function and the complementary sensitivity function for the H configuration as

$$S_H(j\omega) = \frac{1}{\big(1 + KG_p(j\omega)H(j\omega)\big)}$$

$$T_H(j\omega) = 1 - S_H(j\omega) = \frac{KG_p(j\omega)H(j\omega)}{1 + KG_p(j\omega)H(j\omega))}$$

(a) If $\tilde{G}_p(s) = G_p(s)\,(1 + L_m(s))$
with $|L_m(j\omega)| \leq l_m(j\omega)$
prove that Eq. (6.7-16) holds for the H configuration.

(b) Show the relationship between the complementary sensitivity function and the closed-loop transfer function for the H configuration.

(c) Explain how the concept and use of M-circles changes for a system in the H configuration.

6.9. (This problem can be done without CAD assistance but it is much easier and just as useful if CAD tools are used.) A system is in the G configuration with the loop gain given by

$$G(s) = \frac{3(s+1)}{s^2(s+3)}$$

so that

$$1 + G(s) = \frac{s^3 + 3s^2 + 3s + 1}{s^2(s+3)} = \frac{(s+1)^3}{s^2(s+3)}$$

The set of perturbed plants are described as in Eqs. (6.6-5) and (6.6-6) with

$$l_m(s) = 0.5(s + 0.01)$$

(a) Is this design robustly stable for this set of plants?

(b) If it is not robustly stable find a destabilizing perturbation.
If it is robustly stable find the maximum value that is achieved by the *perturbed* sensitivity function $\tilde{S}(j\omega)$.

6.10. (*CAD Problem*) (This problem requires the solution of Problem 3.11 as a prerequisite.) In this problem the pole placement controller designed in Problem 3.11 is analyzed.

(a) For the pole placement controller that was found in Problem 3.11 plot the Bode and Nyquist plots of the loop gain. Sketch the magnitude and phase of the closed-loop frequency response from the Nyquist plot of the loop gain. Find the closed-loop transfer function. Also plot the sensitivity and complementary sensitivity functions for this loop gain. Comment on how well this system might behave with respect to transient response, sensitivity to plant mismodeling, rejection of disturbances, and robustness to unmodeled dynamics.
What do these results say about using the mechanized design of pole placement blindly?

(b) It is clearly a poor decision to place the closest pole to the origin at $s = -5$ when the plant contains a zero at $s = -1.7$. Why?
Now solve the pole placement problem again placing a closed-loop pole to cancel the zero at $s = -1.7$, a pole at $s = -5$, a pole at $s = -10$ and two poles at $s = -100$. Repeat the analysis of (a) for this control system.

CHAPTER 7

THE ROOT LOCUS METHOD

7.1 INTRODUCTION

In the discussion of the frequency response methods in the two preceding chapters, it was assumed that the complex Laplace variable s was replaced by an imaginary variable $j\omega$. In this chapter we consider the entire complex plane rather than restrict attention to that line in the complex plane where $s = j\omega$. Once again, however, we wish to obtain information about the closed-loop system by examining the loop gain transfer function. In this case the entire closed-loop transfer function is to be obtained in factored form. From this information the closed-loop time and frequency responses to any input may be determined. The root locus method studied here is also a useful synthesis tool.

The root locus method is a graphical means of determining the poles of the closed-loop transfer function $Y(s)/R(s)$. Conceptually, the procedure may be different from any the reader has previously encountered. The root locus is a picture of the behavior of the closed-loop poles as one of the system parameters, usually the gain K, is varied. If the closed-loop poles are found for all values of the parameter, then the closed-loop poles for the specific value of interest are also known.

In Chap. 4 root locus diagrams were drawn for a second-order system, though at that time they were not so called. The reader may recall that we considered the behavior of the closed-loop poles as ω_n, ζ and the product $\zeta\omega_n$ were varied. Let us repeat that development here but in slightly different terms. We consider the closed-loop system of Fig. 7.1-1*a* and the corresponding pole-zero plot of the loop gain transfer function shown in Fig. 7.1-1*b*.

The closed-loop transfer function of this system is

$$\frac{Y(s)}{R(s)} = \frac{K}{s^2 + \lambda s + K} = \frac{\omega_n^2}{s^2 + 2\zeta\omega_n s + \omega_n^2}$$

and the two closed-loop poles are given by

$$s = -\frac{\lambda}{2} \pm \sqrt{\frac{\lambda^2}{4} - K}$$

As K is varied from zero to infinity, the positions of the closed-loop poles change. If this change is plotted on the same s plane as that on which the open-loop poles are displayed, Fig. 7.1-2 results.

In Chap. 4 we also allowed ω_n^2 to be a constant and we varied ζ. The result was Fig. 4.3-5*a*, repeated here as Fig. 7.1-3. In this case the parameter that is varying is not the gain K but the damping coefficient ζ. However, the resulting picture once again displays the behavior of the closed-loop poles as this parameter is allowed to vary. Note that, in both Figs. 7.1-2 and 7.1-3, many problems are solved, depending upon the value of K or ζ of interest. If this general solution for a variety of values of K or ζ, as given pictorially by the diagrams, is known, it is easy to find the particular closed-loop poles for any given value of K or ζ.

In Figs. 7.1-2 and 7.1-3 the path of the closed-loop poles as a function of K or ζ was determined by actually solving for the closed-loop poles. This is possible because the system is second-order, and we could use the quadratic formula. If the system is of higher order, this convenience is lost. Hence the object of the early sections of

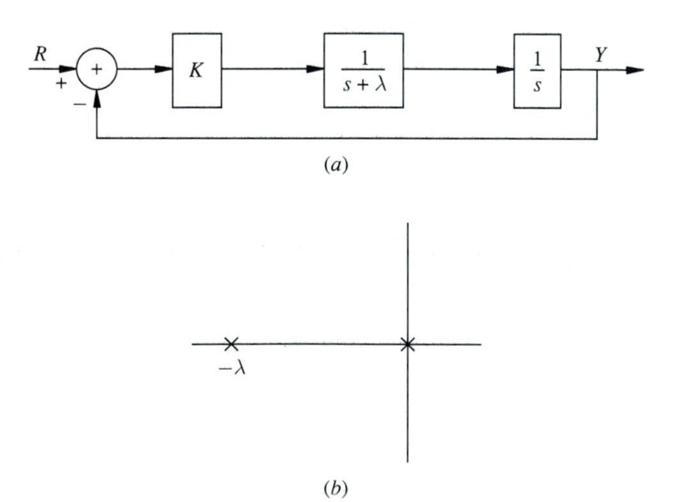

FIGURE 7.1-1
Simple second-order system. (*a*) Block diagram; (*b*) pole-zero plot of the loop gain transfer function.

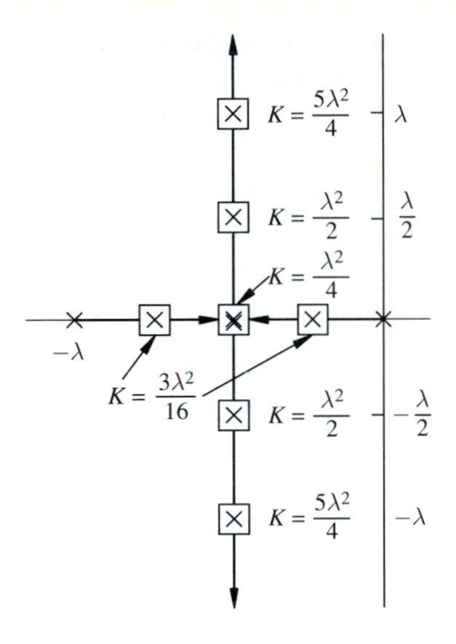

FIGURE 7.1-2
Variation of closed-loop poles as a function of K.

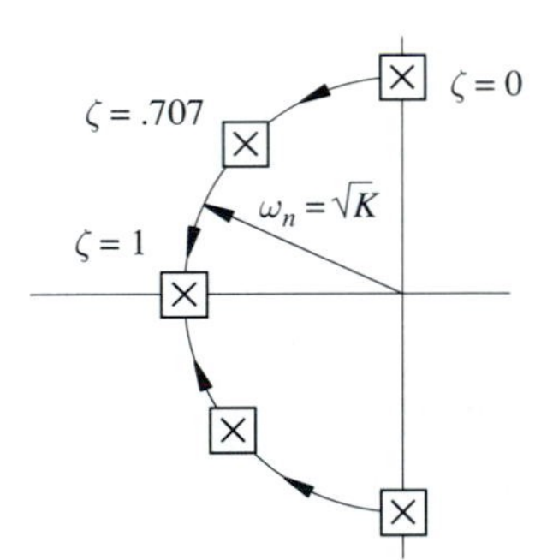

FIGURE 7.1-3
Variation of closed-loop poles as function of ζ.

this chapter is to enable the drawing of the root locus diagram without first having to solve for the locations of the closed-loop poles. In fact, that is just what we are trying to avoid: solving analytically for the closed-loop poles locations.

Sections 7.2 and 7.3 are devoted entirely to construction rules for root locus diagrams. Section 7.4 contains additional examples and continues the discussion of information that may be obtained from the completed root locus diagram. These three sections assume that the gain K is the parameter that is varying. Section 7.5 deals with the determination of the closed-loop response from the root locus diagram. In Sec. 7.6 methods for plotting the root locus versus any system parameter are discussed, particularly with regard to sensitivity.

7.2 THE ROOT LOCUS METHOD

As noted above, the root locus method provides a graphical means of determining the poles of the system transfer function $Y(s)/R(s)$, usually as a function of K. The approach may appear at first to be somewhat devious, like going from Chicago to

New York by proceeding west around the world. This is partially true, but the reader will find that, once facility with graphical construction procedures is achieved, the root locus technique proves to be a powerful tool for analysis and synthesis.

Initially, we consider only changes in the closed-loop poles as K is varied and make use of the H configuration shown in Fig. 7.2-1. The procedure for a G configuration controller follows directly from the procedure using the H configuration.

The closed-loop transfer function is given by

$$\frac{Y(s)}{R(s)} = \frac{KG_p(s)}{1 + KG_p(s)H(s)} \tag{7.2-1}$$

It is a simple matter to express the closed-loop transfer function in terms of $KG_p(s)$ and $H(s)$, but, as pointed out in the discussion of stability in Chap. 6, the poles of $Y(s)/R(s)$ are not known. Not only are the exact closed-loop pole locations unknown, but we may not even know if the closed-loop system is stable, that is, if the closed-loop poles lie in the left half of the s plane.

The necessity of factoring to determine the closed-loop poles is clearly illustrated by writing the closed-loop transfer function as a ratio of polynomials in s

$$\frac{Y(s)}{R(s)} = \frac{K\left[\dfrac{K_pN_p(s)}{D_p(s)}\right]}{1 + K\left[\dfrac{K_pK_hN_p(s)N_h(s)}{D_p(s)D_h(s)}\right]} \tag{7.2-2}$$

or

$$\frac{Y(s)}{R(s)} = \frac{KK_pN_p(s)D_h(s)}{D_h(s)D_p(s) + KK_pK_hN_h(s)N_p(s)} = \frac{KK_pN_p(s)D_h(s)}{D_k(s)} \tag{7.2-3}$$

Even if $G_p(s)$ and $H(s)$ are known in factored form, as they usually are, in order to find the closed-loop poles, the direct approach is to factor the denominator polynomial $D_k(s)$. This direct factoring approach usually requires computer assistance if the order of the system is higher than 2.

The root locus method provides a simple graphical method of factoring that gives considerable insight into the nature of the closed-loop system. Let us consider, once again, the closed-loop transfer function of Eq. (7.2-1). The closed-loop poles occur at the values of s for which the denominator of $Y(s)/R(s)$ is zero. Hence let us set the denominator equal to zero.

$$1 + KG_p(s)H(s) = 0 \tag{7.2-4}$$

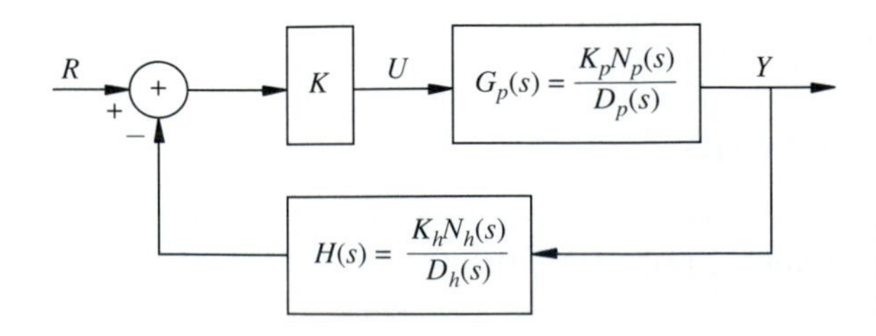

FIGURE 7.2-1
Basic block diagram used in the root locus method.

Again the loop gain transfer function plays a key role. As the gain K plays a key role we define a modification of the usual loop gain so that K can appear explicitly. We initially consider only positive values of K. Let $G_{L/K}(s) = G_p(s)H(s)$. The transfer function $G_{L/K}(s)$ is just the loop gain divided by K so that the loop gain equals

$$G_L(s) = KG_{L/K}(s) = KG_p(s)H(s)$$

The denominator polynomial of the modified loop gain $G_{L/K}(s)$ is the same as the loop gain denominator polynomial, and the pole-zero structure of the loop gain and $G_{L/K}(s)$ are identical. From Eq. (7.2-4), the closed-loop poles are given by the values of s for which the modified loop gain equals $-1/K$, where minus one is viewed as a complex number, i.e., the closed-loop poles are those values of s where

$$G_{L/K}(s) = -1/K = 1/K e^{j(180^\circ \pm N360^\circ)} \tag{7.2-5}$$

In Eq. (7.2-5) the term of $\pm N360^\circ$ recognizes the fact that angles are the same when any multiple of 360° is added or subtracted.

The key idea of the root locus is to write Eq. (7.2-5) as two equations separately satisfying the magnitude and phase angle requirements of Eq. (7.2-5).

$$\left|G_{L/K}(s)\right| = 1/K \tag{7.2-6}$$

$$\arg\left(G_{L/K}(s)\right) = 180^\circ \pm N360^\circ \tag{7.2-7}$$

For each value of s the transfer function $G_{L/K}(s)$ yields a complex number that can be broken into a magnitude and phase. The root locus diagram is a plot of all the values of s that satisfy the phase angle criterion of Eq. (7.2-7), independent of the magnitude criterion of Eq. (7.2-6). Once the phase angle condition is met, one may select K to satisfy the magnitude condition.[1] In other words, the root locus is the locus of the closed-loop poles for all positive values of K from 0 to ∞. This range for K obviously includes any specific positive value of K. The closed-loop poles for a specific K may be found by determining the points on the root locus that satisfy the magnitude criterion for the given K. Note that in this procedure we need only examine points on the root locus, not every point of the s plane, because the closed-loop poles for all positive K are by definition on the root locus. The advantage of this two-step procedure is that the root locus, that is, the points that satisfy the phase angle criterion, may be determined with relative ease. In addition, the location of the closed-loop poles for all positive values of gain have been found, as well as the manner in which the poles move with gain variations.

The construction of the locus of points that satisfy the phase angle criterion is accomplished directly on the complex s plane by the consideration of $G_{L/K}(s)$ as a complex number. We begin by placing the poles and zeros of $G_{L/K}(s)$ on the s plane.

---

[1] See Rule 5 given below.

The modified loop gain $G_{L/K}(s)$ may be written as the ratio of a numerator and a denominator polynomial

$$G_{L/K}(s) = \frac{N_{L/K}(s)}{D_{L/K}(s)} \tag{7.2-8}$$

If there are common factors in the numerator and denominator of Eq. (7.2-8), these must not be canceled. The reason for this restriction will become clear later in the discussion.

Once the poles and the zeros of $G_{L/K}(s)$ have been placed on the s plane, the determination of the root locus involves the determination of all values of s that satisfy the angle criterion of Eq. (7.2-7). One can find out if a given point satisfies the angle criterion by graphically determining the argument of $G_{L/K}(s)$ from the s plane on which the poles and zeros of $G_{L/K}(s)$ have been placed.

Consider, for example, the pole-zero plot for $G_{L/K}(s)$ shown in Fig. 7.2-2. At the point $s = -1 + j1$, the phase of $G_{L/K}(s)$ can be evaluated.

$$\arg\left(G_{L/K}(s)\right)\big|_{s=-1+j1} = -135^\circ - 26.5^\circ + 18.4^\circ = -143.1^\circ$$

This point does not lie on the root locus. On the other hand, the point $s = -2 + j0$ does lie on the root locus since

$$\arg\left(G_{L/K}(s)\right)\big|_{s=-2+j0} = -180^\circ - 0^\circ + 0^\circ = 180^\circ - 360^\circ$$

It is clear that, if we proceed in a random fashion to examine individual points on the s plane, the determination of the root locus would be a lengthy process. Fortunately

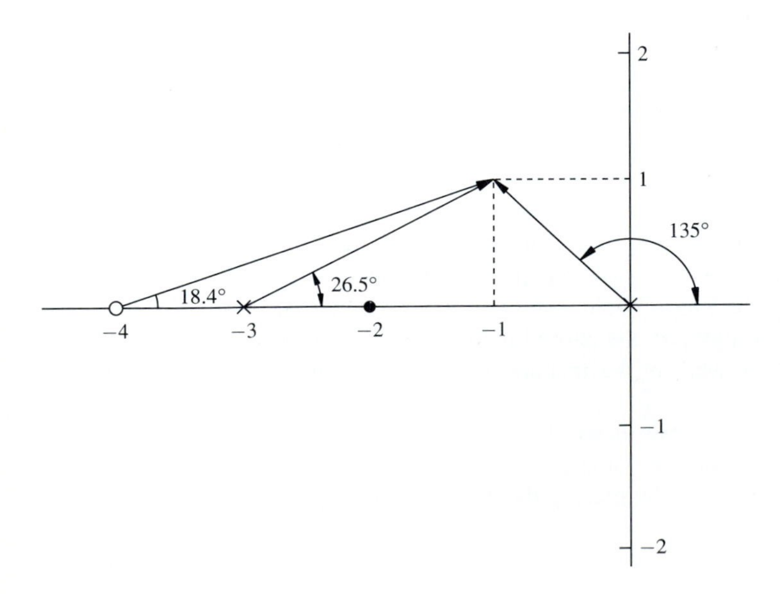

FIGURE 7.2-2
Pole-zero plot of $G_{L/K}(s)$ used to determine the root locus.

there are a number of simple rules that make sketching the root locus a simple task. In the remainder of this section we consider the simplest rules by discussing only situations in which the branches of the root locus remain on the real axis. These rules are applicable in the general case, however, and additional construction procedures are given in the following section.

RULE 1 NUMBER OF BRANCHES. There is one branch of the root locus for every closed-loop pole of $Y(s)/R(s)$, and the total number of branches is equal to the number of poles of the transfer function $G_{L/K}(s)$. It is assumed that $G_{L/K}(s)$ has more poles than zeros.

A consideration of Eq. (7.2-3) easily establishes the veracity of the rule. The denominator of the closed-loop transfer function $D_k(s)$ is the same order as the denominator of the loop gain since the numerator of the loop gain is of lower order than the denominator of the loop gain and cannot cancel the highest-order term in the denominator $D_k(s)$.

RULE 2 STARTING POINTS ($K = 0$). The branches of the root locus start at the poles of the loop gain transfer function $G_{L/K}(s)$.

The truth of this statement is easily seen from Eq. (7.2-6). As K approaches zero, $|G_{L/K}(s)|$ approaches infinity. The quantity $|G_{L/K}(s)|$ approaches infinity only as s approaches the poles of the loop gain transfer function. Thus the loop gain poles and the closed-loop poles are identical as K goes to zero.

An alternative method of establishing this rule is to consider Eq. (7.2-3) again. Since $D_k(s) = D_{L/K}(s) + KN_{L/K}(s)$, the zeros of $D_k(s)$, that is, the poles of $Y(s)/R(s)$, approach the zeros of $D_{L/K}(s)$, that is, the poles of $G_{L/K}(s)$, as K approaches zero. Obviously for $K = 0$, $D_k(s) = D_{L/K}(s)$ and the loop gain poles and the closed-loop poles are identical.

RULE 3 END POINTS ($K = \infty$). The branches of the root locus end at values of s for which the loop gain transfer function $G_{L/K}(s)$ is zero. If the loop gain has p poles and z zeros, z branches of the root locus end on the z zeros of the loop gain and p-z branches go to infinity.

The argument here is similar to that used in the discussion of Rule 2. Again look at Eq. (7.2-6). If K is now going to infinity, $G_{L/K}(s)$ must go to zero. Thus the closed-loop poles are at the zeros of the loop gain when K is infinite. We shall see in the examples of this section that a zero might lie at infinity, that is, if $G_{L/K}(s)$ has more poles than zeros, the value of $G_{L/K}(s)$ goes to zero as s goes to infinity. In the following section we shall be concerned with the question "How does it get to infinity?" For the present it is sufficient to point out that the branches of the root locus terminate at the open-loop finite zeros or go to infinity.

RULE 4 BEHAVIOR ALONG THE REAL AXIS. A point on the real axis is a point on the root locus if the sum of the number of poles and zeros lying to the right of the point is odd.

This statement is easily verified from the phase angle criterion of Eq. (7.2-7). Consider an arbitrary point on the real axis, such as the point $s = -a$ in Fig. 7.2-3. At this point the phase angle of the loop gain transfer function $G_{L/K}(s)$ is the sum of all the phase angles associated with the poles and zeros of $G_{L/K}(s)$. These phase angles are indicated in Fig. 7.2-3. The contribution to the phase angle from a set of complex conjugate poles or zeros is zero, since the two angles associated with the complex conjugates are equal and are of opposite signs. The phase angle contribution of each real pole or zero that lies to the left of the point $s = -a$ is zero. The contribution of each pole or zero which lies to the right of the point is $\pm 180°$. In Fig. 7.2-3 there are two poles and one zero that lie to the right of $s = -a$, so that

$$\arg(G_{L/K}(-a)) = 2(-180°) + 180° = -180°$$

The phase angle criterion is satisfied; hence the point $s = -a$ must be a point on the root locus.

RULE 5 GAIN DETERMINATION. The gain at an arbitrary point s_1 on the root locus is derived from Eq. (7.2-6) as

$$|K| = \frac{1}{|G_{L/K}(s)|}\bigg|_{s=s_1} = \frac{|D_{L/K}(s_1)|}{|N_{L/K}(s_1)|} \tag{7.2-9}$$

This rule follows directly from the magnitude criterion of Eq. (7.2-6), which can be rewritten

$$|K|\ |G_{L/K}(s)| = 1$$

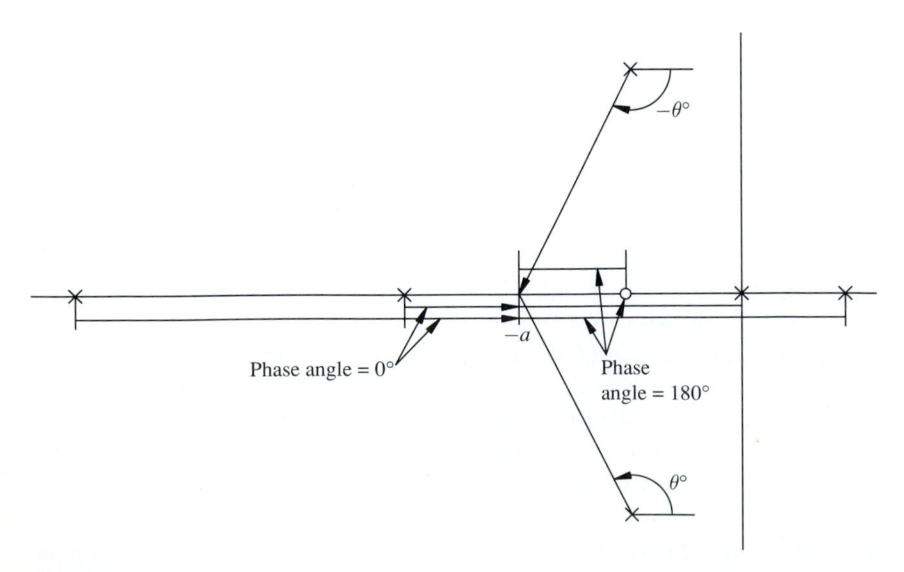

FIGURE 7.2-3
Illustration of Rule 4.

Then, for $s = s_1$, K is just

$$|K| = \left. \frac{1}{|G_{L/K}(s)|} \right|_{s=s_1}$$

The gain K at any point may be determined by simple substitution into $1/|G_{L/K}(s)|$. Usually this is done graphically, in much the same way that the residues are determined in Chap. 4. In applying the graphical method, it is important to place the transfer function in pole-zero form before starting the graphical evaluation. In this case, the gain associated with a point $s = s_1$ is proportional to the product of the vector distances from the poles of $G_{L/K}(s)$ to s_1 divided by the product of the vector distances from the zeros of $G_{L/K}(s)$ to s_1. The constant of proportionality is $K_{L/K}$, the gain of $G_{L/K}(s)$ after $G_{L/K}(s)$ is placed into pole-zero form. In other words, K is given by

$$|K| \Big|_{s=s1} = \frac{1}{K_{L/K}} \left. \frac{\prod[\text{distance to poles of } G_{L/K}(s)]}{\prod[\text{distance to zeros of } G_{L/K}(s)]} \right|_{s=s1} \tag{7.2-10}$$

These simple rules are applied to the construction of a root locus diagram in the following example.

Example 7.2-1. This example is based upon the system of Fig. 7.2-4. The problem is to determine the root locus diagram for this closed-loop configuration. Of particular interest is the nature of $Y(s)/R(s)$ for $K = 10$.

For this simple problem,

$$G_p(s) = \frac{1}{s(s+6)}$$

and

$$H(s) = \frac{1}{2}(s+2)$$

so that the loop gain transfer function is

$$G_{L/K}(s) = \frac{\frac{1}{2}(s+2)}{s(s+6)}$$

where $K_{L/K} = 1/2$. The first step is to mark on the s plane the loop gain pole and zero locations, as shown in Fig. 7.2-5*a*. The next step is to examine the real axis to determine regions in which the branches of the root locus might lie. Points on the positive real axis have no poles or zeros to the right of them. Hence no point on the positive real axis is a point on the root locus. Said in another way, there is no value of $K > 0$ for which a

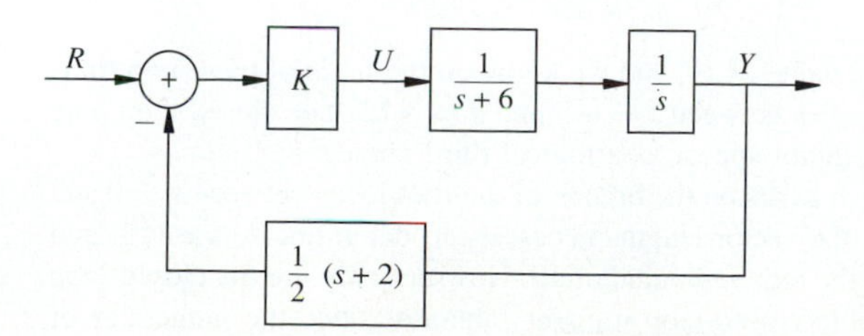

FIGURE 7.2-4
Closed-loop system for Example 7.2-1.

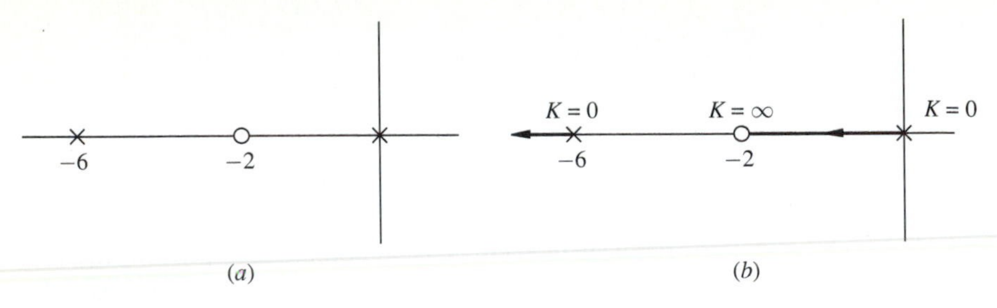

FIGURE 7.2.5
Development of the root locus for Example 7.2-1. (*a*) Pole-zero plot; (*b*) root locus.

closed-loop pole lies on the positive real axis. However, for points between $s = -2$ and $s = 0$, one pole lies to the right, namely, the pole at $s = 0$. Hence every point on the line from 0 to -2 must lie on the root locus.

For points on the real axis between $s = -6$ and $s = -2$, one pole and one zero lie to the right. Since the sum of the number of poles and zeros to the right of any point on a line between -2 and -6 is 2, the root locus does not coincide with any point on this line segment. For any real value of s less than -6, the sum of poles and zeros to the right is 3; hence this is also a branch of the root locus. The final root locus diagram is given in Fig. 7.2-5*b*. The arrows on each of the branches indicate the direction of increasing K. Note that one branch of the root locus is going to infinity along the negative real axis. It is easily verified that $s = -\infty$ is a zero of $G_{L/K}(s)$ since, at $s = -\infty$, $|G_{L/K}(s)|$ is zero, so that the two branches of the root locus do terminate on zeros of $G_{L/K}(s)$.

Now let us turn our attention to finding the closed-loop poles for $K = 10$. There is a point on *each* branch of the root locus corresponding to the gain of 10, and we must find these two points, as they determine the closed-loop pole locations. Suppose that we start with the leftmost branch of the root locus, where $-\infty < s < -6$. Our approach here is to guess an arbitrary point, to determine the gain at that point, and proceed to the correct answer. As an initial guess, we examine the closed-loop pole at $s = -8$. The gain at $s = -8$ is computed from the vector lengths indicated in Fig. 7.2-6*a*. This gain is

$$K = \frac{1}{\frac{1}{2}} \frac{(8)(2)}{6} = 5.3$$

This is below the desired gain of 10; hence we must move along the locus in the direction of higher gain. As a second try, let us examine the closed-loop pole at $s = -12$. The vector lengths involved in the calculation of K, indicated in Fig. 7.2-6*b*, lead to the result

$$K = \frac{1}{\frac{1}{2}} \frac{(6)(12)}{10} = 14.4$$

This value is above the desired value of 10, and we know on the basis of these two trials alone that the closed-loop pole lies between $s = -8$ and $s = -12$. The closed-loop pole is actually at $s = -10$, which might appear as a logical third choice.

One closed-loop pole also exists on the branch of the root locus between $s = 0$ and $s = -2$. At the point $s = -1$, the vector lengths necessary to determine K are 1, 1, and 5, so that, at this point, K has the required value of 10. Now not only are the closed-loop poles known but also the whole closed-loop transfer function, since the numerator of

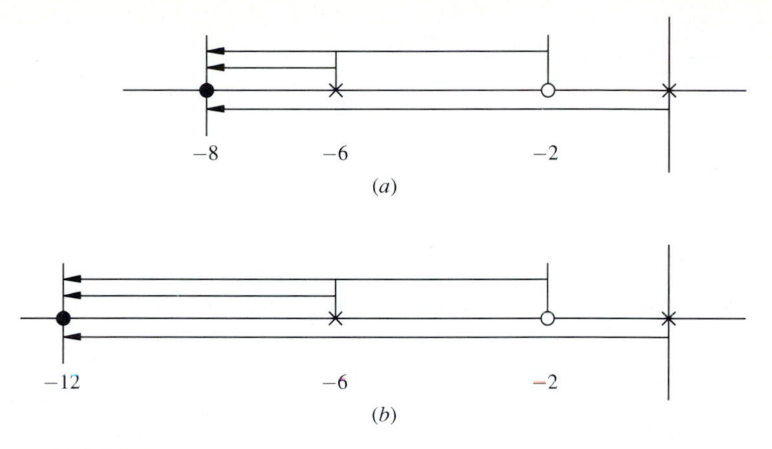

FIGURE 7.2-6
Vector lengths involved in the determination of K. (a) Point at $s = -8$; (b) point at $s = -12$.

$Y(s)/R(s)$ is the product of numerator of $G_P(s)$ and the denominator of $H(s)$. In this simple problem, $N_P(s)$ and $D_h(s)$ are just 1. By substituting into Eq. (7.2-3) it is seen that $Y(s)/R(s)$ is

$$\frac{Y(s)}{R(s)} = \frac{KN_p(s)D_h(s)}{D_k(s)} = \frac{K}{(s+1)(s+10)} = \frac{10}{(s+1)(s+10)} \tag{7.2-11}$$

Because this problem is second-order, clearly it would have been easier simply to factor the denominator to find the closed-loop poles. The point here is to illustrate the use of the simple rules for drawing the root locus and to show in a graphical way the additional information regarding system performance that is readily gained from the root locus method. Figure 7.2-7 is a redrawing of Fig. 7.2-5b with the addition of calculated gain values along the two branches of the root locus. From the very few values of gain indicated, it is clear that for a closed-loop gain of 10, the value of interest in this problem, the dominant time constant becomes −1. As gain is changed from 10 to infinity, this

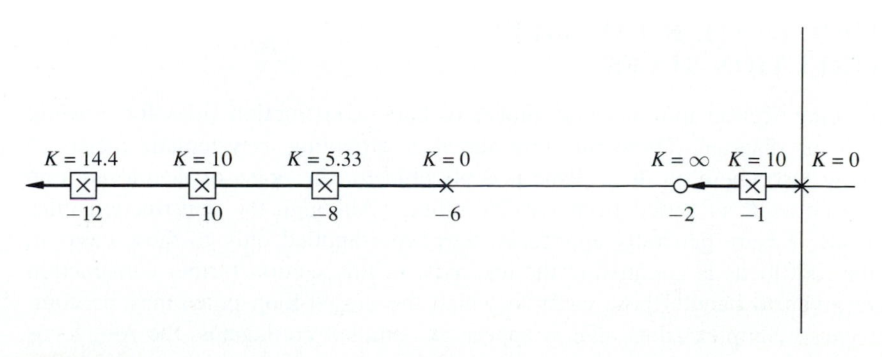

FIGURE 7.2-7
Root local for Example 7.2-1.

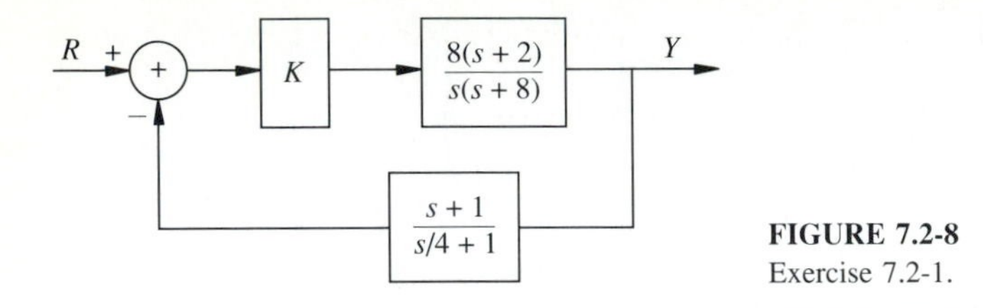

FIGURE 7.2-8
Exercise 7.2-1.

closed-loop pole moves only from $s = -1$ to $s = -2$. On the other hand, the closed-loop pole to the left of the point $s = -6$ moves toward infinity as the gain is increased.

Two other points are worth mentioning with respect to Example 7.2-1. In evaluating K from the measurement of vector lengths, no attention is given to the phase angles of the vectors involved since it is already known that any point along the root locus satisfies the angle condition. This is, in fact, the way in which the root locus is drawn, to satisfy the angle condition.

The second point concerns conclusions regarding the closed-loop system performance. As discussed in Chap. 4, it is possible to determine graphically the contribution to the output made by each pole. This essentially amounts to finding the residue, and we developed in Chap. 4 a method of finding the residues, based on measured lengths and angles of the system pole-zero plot. The root locus is used to find the closed-loop poles but *not* the closed-loop zeros; these are already known. In Example 7.2-1, for instance, one might be tempted to conclude that, as the gain is increased, the closed-loop pole that is approaching the zero at $s = -2$ also has a decreasing residue because of the proximity of the zero. This is not true, because this zero does not appear in the closed-loop system. More will be said on this point in Sec. 7.5.

Exercises 7.2

7.2-1. Find the root locus for the system in Fig. 7.2-8. Find the value of K that places the dominant closed-loop pole at $s = -1/2$.
Answer: $K = 35/64$

7.3 ADDITIONAL ROOT LOCUS CONSTRUCTION RULES

The preceding section introduces a number of basic construction rules for drawing the root locus diagram. These rules are aimed at providing a systematic means of determining just where on the s plane it is possible for the poles of the closed-loop system to lie as K is varied from zero to infinity. Although the construction rules cited in Sec. 7.2 are generally applicable, they were applied only to those cases in which the root locus is confined to the real axis. In this section further construction rules are given to handle those cases in which the closed-loop poles may be complex. Because complex roots always appear as complex conjugates, the root locus diagram is always symmetric about the real axis and hence we can state a governing rule.

RULE 6 SYMMETRY OF LOCUS. The root locus diagram is always symmetric with respect to the real axis.

If the loop gain transfer function $G_L(s)$ has complex conjugate poles, then branches of the root locus must lie off the real axis, since they start at these complex conjugate poles. However, complex conjugate poles in the closed-loop system may also arise even if the loop gain transfer function has only real poles. If this occurs, it is necessary for branches of the root locus to leave the real axis and enter the upper and lower portions of the s plane. We begin the discussion by considering the case of the breakaway from or reentry to the real axis.

RULE 7 BREAKAWAY OR REENTRY POINTS ON THE REAL AXIS. A point of breakaway from the real axis occurs at a relative maximum of the gain. Two branches of the root locus return to the real axis at a point of relative minimum gain.

Consider the situation in which every point on the real axis between two poles satisfies the angle condition. The simplest case in which this occurs is when the loop gain transfer function has only two real poles, as in Fig. 7.3-1*a*. In this simple case, it is known from Sec. 7.2 that there are two branches of the root locus and that each of these branches starts at one of the poles, as indicated in Fig. 7.3-1*b*. As the gain increases, these two separate branches of the root locus approach each other until they finally meet. Since the branches of the root locus must terminate on zeros, they cannot remain on the real axis, as no zeros exist there. Hence they must break away from

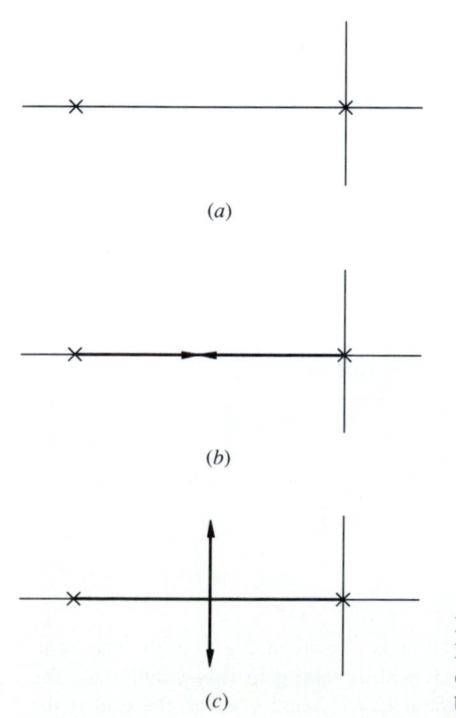

FIGURE 7.3-1
Illustration of breakaway. (*a*) Pole-zero diagram; (*b*) branches move together; (*c*) branches meet and break away from the real axis.

the real axis to reach the zeros, wherever they exist, as indicated in Fig. 7.3-1*c*. Our interest here is in finding the point of breakaway.

The point of breakaway is the point of highest gain on the line segment joining the two poles. This point of highest gain can be found in several ways. One way is simply to determine the gain at a number of points along the line segment. The point of highest gain is then the point of breakaway. The point may also be found by taking the derivative with respect to s of the expression for K given by Eq. (7.2-9) and setting it equal to zero. The solution with real values of s represents a relative maximum, a relative minimum or an inflection point of K on the real axis.

Before doing an example, let us consider the analogous situation in which the line connecting two real zeros satisfies the angle condition. For infinite gain, a branch of the root locus terminates on each of these zeros. Consequently, two branches of the root locus must meet somewhere between the two zeros, and the place of the meeting must be a point of minimum gain. The methods of finding this point of minimum gain are identical to those used to find the point of maximum gain associated with the breakaway point.

Once the point of breakaway or reentry has been established, the next question to be answered is: What angle do these branches of the root locus make with the real axis?

RULE 8 BREAKAWAY OR REENTRY ANGLE. The incoming and outgoing lines of the root locus are separated by an angle of $180°/\alpha$ at a point of breakaway or reentry, where α is the number of branches intersecting at the point.

For α is equal to 2, which is really the most practical case, the angle separating branches of the root loci is 90°.

A simple example involving breakaway points is the case of two simple poles, illustrated in Fig. 7.3-1. This case was also discussed in the introductory section of this chapter, and the final root locus diagram is given in Fig. 7.1-2. A simple root locus involving both a breakaway point and a reentry point involves two poles and one zero. This is the case chosen for Example 7.3-1.

Example 7.3-1. In this example we assume that the control system to be investigated is shown in Fig. 7.3-2*a*, so that

$$G_P(s) = \frac{4}{s+2} \qquad H(s) = \frac{\frac{1}{3}(s+3)}{s}$$

and the modified loop gain transfer function is

$$G_{L/K}(s) = G_p(s)H(s) = \frac{4}{3}\frac{s+3}{s(s+2)}$$

with

$$K_{L/K} = 4/3$$

The pole-zero plot of the loop transfer function is shown in Fig. 7.3-2*b*. The complete root locus diagram is given in Fig. 7.3-3, with several gain values indicated. The breakaway point is chosen as that point between $s = -2$ and 0 where the gain is the

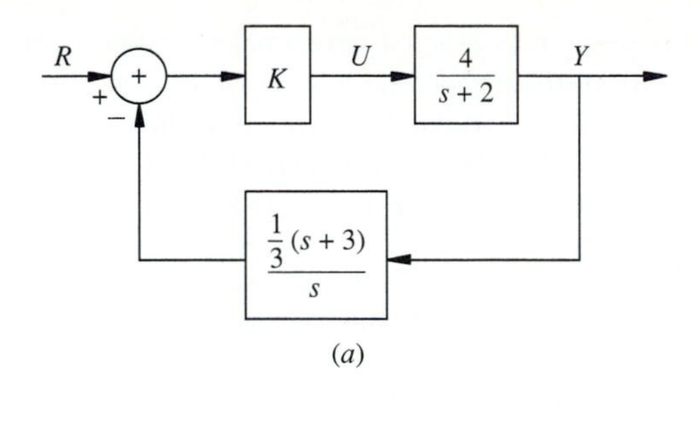

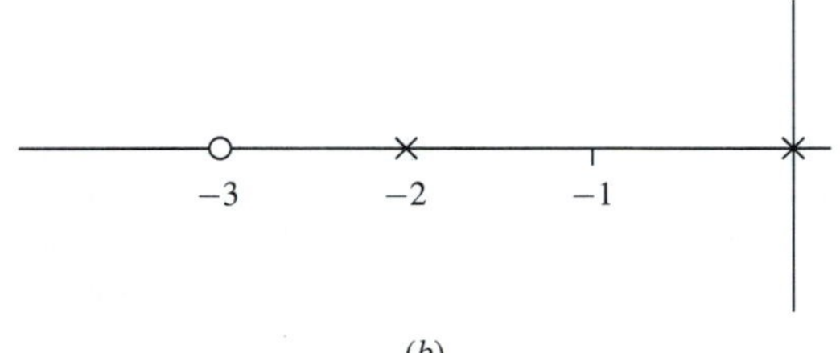

FIGURE 7.3-2
Example 7.3-1. (*a*) System; (*b*) pole-zero plot.

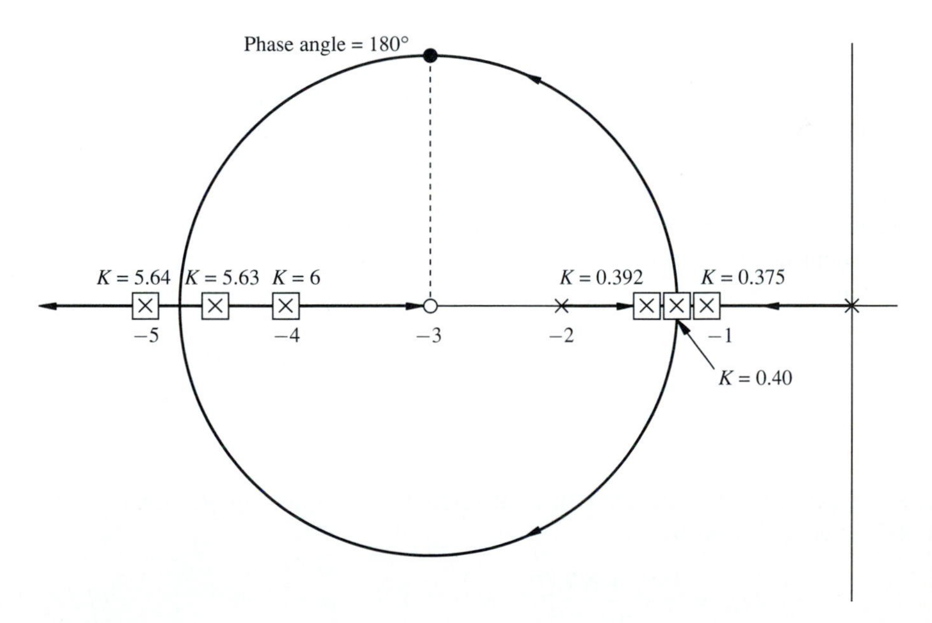

FIGURE 7.3-3
Root locus for the system of Fig. 7.3-2.

highest. The reentry point is chosen to the left of $s = -3$ at a point midway between two values of very nearly the same gain. Clearly the minimum value of gain must exist between these two values, if gain is to increase as one branch of the root locus moves from the reentry point to the real zero at $s = -3$ and the other moves to minus infinity.

To complete the root locus it is necessary to examine a few points in the upper half-plane. (Recall that the root locus is symmetrical about the real axis so that the lower half-plane need not be considered.) This is done by finding the point on the line $s = -3$ at which the phase angle equals $180° \pm N360°$. This point is also indicated in Fig. 7.3-3. By Rule 8, the lines of the root locus must be $90°$ apart at the breakaway and reentry points. Hence the angles made by the root loci with the real axis must be $\pm 90°$, as shown in Fig. 7.3-3. The remainder of the root locus is made up of the best smooth curve that passes through the point on the $s = -3$ line and is perpendicular to the real axis at the point of breakaway or reentry. In this case the curve is a semicircle, but in more complicated cases there may be no simple analytic expression for a given segment of the locus.

It is instructive to examine the effect of moving the zero's location. If the zero is moved to minus infinity, the system has only two poles, and the root locus diagram is that of Fig. 7.3-1. That is, the radius of the semicircle has gone from its present value to infinity, and the breakaway point is midway between the poles. If the zero is moved toward the pole at $s = -2$, the radius of the semicircle shrinks from its present value until, when the zero is at $s = -2$, the radius of the semicircle is zero. As the zero goes to $s = -2$ the breakaway point always moves toward the left. That is, the proximity of the zero to the pole influences the breakaway point. The closer the zero is to the left-most pole, the closer the breakaway point is to that pole.

One more comment is in order with regard to Rule 7. Note that the statement of the rule uses the terms *relative maximum* and *relative minimum.* In Example 7.3-1, on each segment of the root locus there was only one maximum or minimum so that the absolute and relative maximum or minimum are the same. However, it is entirely possible that there may be more than one breakaway or reentry point on any line segment of the real axis that satisfies the angle condition. Cases in which this happens are discussed in the following section, when more construction rules are available.

For large values of the gain K some branches of the root locus approach the zeros at infinity along straight lines called *asymptotes*. The following rule addresses this situation.

RULE 9 ASYMPTOTIC BEHAVIOR FOR LARGE *K*. The asymptotes intersect the real axis at angles given by

$$\phi_{asy} = \frac{180° + N360°}{p - z} \qquad N = 0, 1, \ldots, p - z - 1 \tag{7.3-1}$$

The single point of intersection on the real axis, called the origin of the asymptotes, OA, is given by

$$OA = \frac{\sum(\text{pole locations}) - \sum(\text{zero locations})}{p - z} \tag{7.3-2}$$

In Eqs. (7.3-1) and (7.3-2) the letter p designates the number of poles of the modified loop gain transfer function $G_{L/K}(s)$ and the letter z indicates the number of finite zeros of this same transfer function.[1] The quantity $p - z$ is referred to as the *pole-zero excess* of $G_{L/K}(s)$. It is also known as the *relative degree* of $G_{L/K}(s)$.

Here we are interested in cases where $p - z \geq 2$. In all the examples of Sec. 7.2, $p - z$ was 1, and in that case the existence of a branch along the negative real axis out to infinity at angle of 180° is apparent from Rule 4 concerning real axis behavior.

It is relatively easy to verify Eq. (7.3-1) using the condition

$$KG_{L/K}(s) = -1$$

As K becomes large, this equation can be satisfied only if $G_{L/K}(s)$ becomes small. Since this loop transfer function has more poles than zeros, $G_{L/K}(s)$ becomes small as $|s|$ is increased. For large values of s, the quantity $KG_{L/K}(s)$ behaves approximately as

$$KG_{L/K}(s) \approx \frac{KK_{L/K}}{s^{p-z}} = -1$$

or

$$s^{p-z} = -KK_{L/K} = KK_{L/K}e^{j(180^\circ + N360^\circ)}$$

Clearly the magnitude of s approaches infinity as K goes to infinity. Since the angle associated with s^{p-z} at any point is simply $(p - z)$ times the angle associated with s at that point, the angles of the values of the complex variable s that satisfy the above equation are

$$\arg(s) = \frac{180^\circ + N360^\circ}{p - z} \qquad N = 0, 1, \ldots, p - z - 1$$

For $p - z$ equal to 2, the angles of the asymptotes are just $\pm 90^\circ$, and for $p - z$ equal to 3, the angles of the asymptotes are $\pm 60^\circ$ and 180°. In every case the asymptotes are always separated by equal angles of $360^\circ/(p - z)$.

It is more difficult to establish the truth of Eq. (7.3-2). Rather than resort to merely quoting a theorem from the field of linear algebra, let us present a heuristic argument. Recall that we are concerned with large values of K and hence large values of s. From a point near infinity along one of the asymptotes the original collection of poles and zeros is somewhat obscure, because the distance separating the individual poles and zeros is very small compared with the distance from any one of them to the point near infinity. As a consequence the individual character of each pole and zero is lost to an observer far away, so that the original pole-zero plot looks just like an isolated system of $p - z$ poles. The analogy often used is that of positive and negative charges, analogous to the poles and zeros of $G_{L/K}(s)$. From far away a group of p-positive and z-negative charges appears as just a charge of magnitude $p - z$. The location of the single charge of magnitude $p - z$ is the centroid of the individual charges that make up the composite charge. Equation (7.3-2) is just the expression for

[1] In the plant transfer function $G_p(s)$ we have consistently used n to designate the number of poles and m the number of finite zeros.

the centroid of the poles and zeros of the open-loop transfer function, which, when viewed from a distance that is large compared with their separation, appears as $p - z$ poles located at their centroid.

The heuristic argument can be made somewhat more precise by realizing that the main idea of the asymptotic approach is that the loop gain transfer function can be approximated with a collection of $p - z$ poles located at the point OA.

$$G_{L/K}(s) \approx \frac{K_{L/K}}{(s + OA)^{p-z}} \quad \text{for } |s| \text{ large} \tag{7.3-3}$$

We will work to find OA but first we note that Rules 8 and 9 follow directly from the fact that the root locus of the right hand side of the Eq. (7.3-3) is a set straight lines eminating from the point OA at angles given by Rule 8.

We work on the two sides of Eq. (7.3-3) separately and bring them together. We manipulate the left hand side after expressing the transfer function in pole-zero form displaying its p poles and z zeros.

$$\begin{aligned} G_{L/K}(s) &= \frac{K_{L/K}(s + z_1)(s + z_2)\cdots(s + z_z)}{(s + p_1)(s + p_2)\cdots(s + p_p)} \\ &= \frac{K_{L/K}}{\left(\dfrac{(s + p_1)(s + p_2)\cdots(s + p_p)}{(s + z_1)(s + z_2)\cdots(s + z_z)}\right)} \\ &= \frac{K_{L/K}}{\left(\dfrac{s^p + \left(\sum_{i=1}^{p} p_i\right) s^{p-1} + \cdots}{(s^z) + \left(\sum_{i=1}^{z} z_i\right) s^{z-1} + \cdots}\right)} \end{aligned}$$

where the $p_i, i = 1, \ldots, p$ are p pole positions and $z_i, i = 1, \ldots, z$ are the z zero positions.

Carrying out the long division in the denominator

$$\begin{array}{r|l} & s^{p-z} + \left(\sum_{i=1}^{p} p_i - \sum_{i=1}^{z} z_i\right) s^{p-z-1} + \cdots \\ \hline s^z + \left(\sum_{i=1}^{z} z_i\right) s^{z-1} + \cdots & s^p + \left(\sum_{i=1}^{p} p_i\right) s^{p-1} + \cdots \\ & s^p + \left(\sum_{i=1}^{z} z_i\right) s^{p-1} + \cdots \\ \hline & \left(\sum_{i=1}^{p} p_i - \sum_{i=1}^{z} z_i\right) s^{p-z-1} + \cdots \end{array}$$

we obtain

$$G_{L/K}(s) \approx \frac{K_{L/K}}{s^{p-z} + \left(\sum_{i=1}^{p} p_i - \sum_{i=1}^{z} z_i\right) s^{p-z-1} + \cdots} \tag{7.3-4}$$

Working the right side of Eq. (7.3-3) we get

$$\frac{K_{L/k}}{(s + OA)^{p-z}} = \frac{K_{L/K}}{s^{p-z} + (p - z)OAs^{p-z-1} + \cdots} \tag{7.3-5}$$

For large $|s|$ the high-order terms of the denominator dominate. By equating the high-order terms of the denominators Eqs. (7.3-4) and (7.3-5) we see that Eq. (7.3-3) is accurate for large $|s|$ and

$$OA = \frac{\sum_{i=1}^{p} p_i - \sum_{i=1}^{z} z_i}{p - z}$$

The root locus Rules 8 and 9 then follow.

The application of the rule for the asymptotic behavior of the branches of the root locus for large K is illustrated in the following example of a constant gain feedback control system.

Example 7.3-2. This example is concerned with the system of Fig. 7.3-4, where the loop gain transfer function is given by

$$G_{L/K}(s) = \frac{6}{s(s+1)(s+2)}$$

Here the only parameter free for adjustment is K. A natural question concerns the behavior of this closed-loop system as K is varied. Of particular interest is the question of stability for large K. We know from the Bode method of Chap. 5, for example, that the final slope of the Bode magnitude plot is -60 dB/decade, with an associated phase shift approaching $-270°$. The Nyquist plot approaches the origin from an angle of $-270°$ for large ω, or from the positive imaginary axis of the polar plane. Thus there is danger of encirclement of the -1 point. Hence, for sufficiently high gain, the system can be unstable. The root locus diagram should give the same stability information, plus additional information on the location of the closed-loop poles for any value of gain.

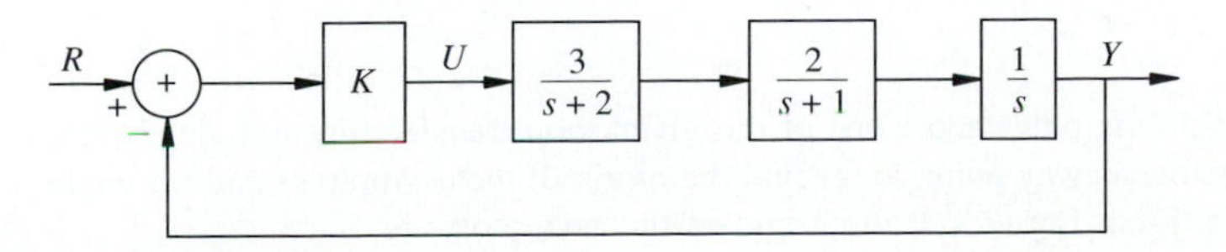

FIGURE 7.3-4
System for Example 7.3-2.

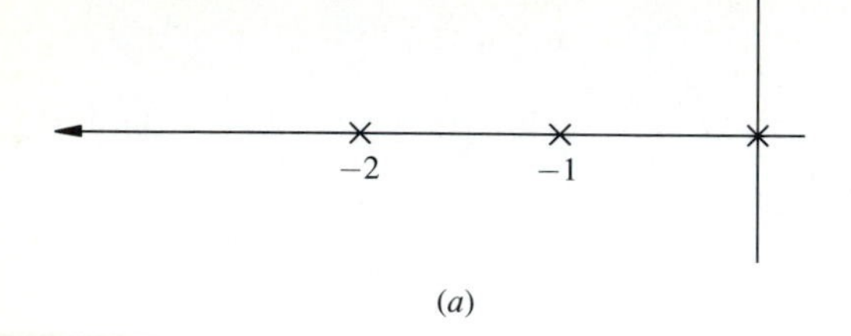

(a)

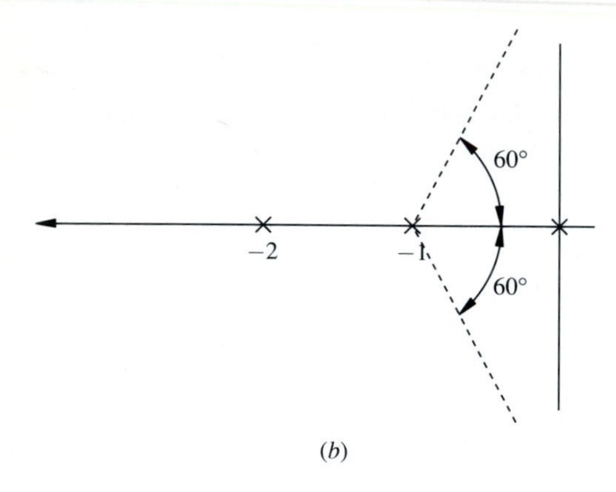

(b)

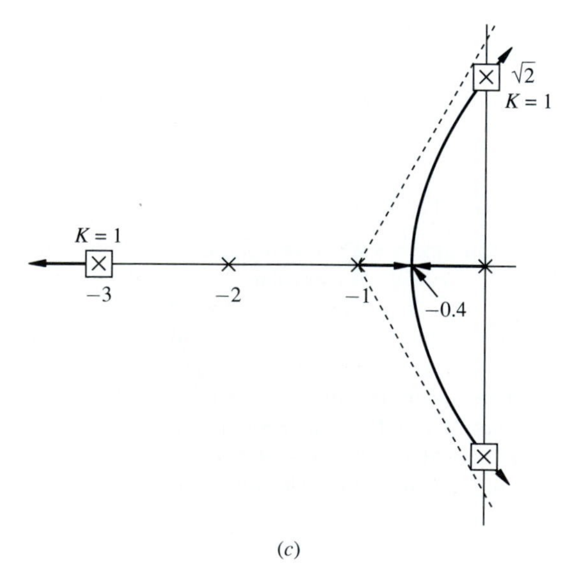

(c)

FIGURE 7.3-5
Example 7.3-2. (*a*) Pole-zero plot; (*b*) asymptotes; (*c*) final root locus.

In Fig. 7.3-5*a* the poles and zeros of the given loop transfer function are shown. Before finding the breakaway point, let us find the origin of the asymptotes and the angle of the asymptotes. From Eq. (7.3-2) the origin of the asymptotes is

$$OA = \frac{(-0 - 1 - 2) - 0}{3 - 0} = -1.0$$

and from Eq. (7.3-1) the angles of the asymptotes are

$$\phi_{asy} = 60^\circ, 180^\circ, 300^\circ$$

These asymptotes are indicated in Fig. 7.3-5*b*, and the branches of the root locus along the real axis have been completed. The breakaway point from the real axis is found, as discussed under Rule 7, to be approximately $s = -0.4$. To complete the locus, it would be helpful to know the point at which the branches of the root locus cross into the right half-plane. This is the topic of the next construction rule, and there it is shown how to determine that the crossover point is at $s = \pm j\sqrt{2}$. The measured gain at this point is $K = 1$. With this information, the final root locus diagram is given in Fig. 7.3-5*c*.

The reader may wonder at the use of the word *final* in the preceding sentence. Clearly we are guessing at the exact root locus diagram, since we have only approximately established the breakaway point and the point at which the root locus crosses into the right half s plane. The branches of the root locus that exist in the upper and lower half-planes are drawn on the basis of these two points and the asymptotes and hence are not exact. However, the degree of accuracy of Fig. 7.3-5*c*, for example, is usually enough to serve as the basis for the selection of a desired gain K. Any more accuracy than this would be warranted only if one were sure that the loop gain pole positions were exactly as indicated, at $s = 0$, -1, and -2. Because of the complex nature of many control systems, it is often difficult to establish the plant pole positions to within more than 5 or 10 percent. Of course, more accurate root locus plots can be obtained using computer programs.

Before leaving this example, note that, in this case, the breakaway point is to the right of the midpoint between the two poles at 0 and -1. In Example 7.3-1 the breakaway point was to the left of the midpoint. The difference is that, in Example 7.3-1, the additional singular point on the s plane was a zero rather than a pole, as it is in this case. It was pointed out with respect to the earlier example that, as the zero approached the left pole, the breakaway point also approached the left pole. In this case, if the pole at $s = -2$ is moved closer to the pole at $s = -1$, the breakaway point moves farther and farther to the right. The point here is that the breakaway point between two poles on the real axis is influenced by the presence of the other poles and zeros in the s plane. Often one can guess at which side of the midpoint the breakaway occurs. This is particularly true if there is another singularity on the real axis that is closer to either of the poles in question than other poles or zeros in the s plane. This is something of a rule of thumb but, with sufficient experience in drawing root locus diagrams, the breakaway or reentry point can often be estimated with enough accuracy so that no calculation is needed.

RULE 10 IMAGINARY-AXIS CROSSING. The branches of the root locus cross the imaginary axis at a point where the phase shift is 180°.

This rule is nothing more than a statement of the phase angle criterion, here with respect to the imaginary axis. To find the point on the imaginary axis where the phase shift is 180°, one must start at an arbitrary point and measure the total phase angle contribution from each of the poles and zeros of the loop gain transfer function. If the phase angle at the initially assumed point is not correct, that is, not 180°, the phase angle must be measured at another point on the imaginary axis. This second point may also not have the necessary phase angle, but the two points are usually sufficient to establish a trend so that the third guess is often all that is needed to find the point to within the degree of accuracy required.

Since the poles crossing the imaginary axis changes the number of unstable poles, the value of K for which the root locus crosses the imaginary axis can be found from among the values of K that zero an element in the first column of the Routh array for the problem.

In a problem such as Example 7.3-2, the point at which the asymptotes cross the imaginary axis is often a good starting point. Consider once again Fig. 7.3-5*b*. There the asymptote is at 60° and crosses the imaginary axis at $s = j\sqrt{3}$. The calculation of the phase angle at that point is indicated in Fig. 7.3-6. The phase angles that are shown are all positive, but since these phase angles are associated with poles and hence are in the denominator $G_{L/K}(s)$, these angles are actually negative in determining the phase angle of $G_{L/K}(s)$. Thus the phase angle at the point $s = j\sqrt{3}$ is $-191°$.

Consider the phase angle at the point $s = j\epsilon$. This angle is very nearly $-90°$, because the angle contributions from the poles at $s = -1$ and -2 are nearly zero, and the contribution from the pole at the origin is $-90°$. Thus the actual point of crossover into the right half-plane must lie somewhere between the crossing of the asymptote and the origin of the s plane. Clearly it must be nearer the crossover of the asymptotes, since the phase angle there is in error only by 11°. At $s = j\sqrt{2}$, the phase angle is measured to be

$$\phi = -(90° + 55° + 35°) = -180°$$

so that this establishes the crossover point.

RULE 11 SUM OF THE CLOSED-LOOP POLES. If the pole-zero excess of the modified loop gain transfer function $G_{L/K}(s)$ is $p - z \geq 2$, the sum of the closed-loop poles remains constant, independent of K, and is equal to the sum of the poles of the loop gain transfer function $G_{L/K}(s)$.

To demonstrate the truth of this statement, let us return to Eq. (7.2-3), which is repeated here in a slightly modified form.

$$\frac{Y(s)}{R(s)} = \frac{KK_pN_p(s)D_h(s)}{D_h(s)D_p(s) + KK_pK_hN_h(s)N_p(s)} = \frac{KK_pN_p(s)D_h(s)}{D_{L/K}(s) + KK_{L/K}N_{L/K}(s)} \tag{7.3-6}$$

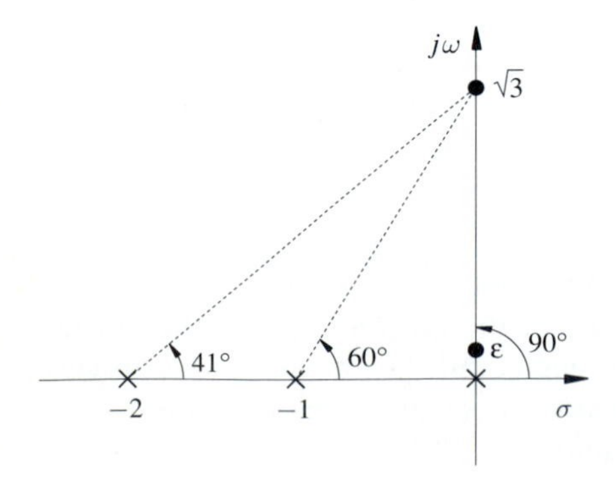

FIGURE 7.3-6
Calculation of phase shift at $s = +j\sqrt{3}$.

The denominator of $Y(s)/R(s)$ is the characteristic equation, which is assumed to be of order p so that the denominator of Eq. (7.2-6) may be rewritten as a polynomial in s with the highest power p. This is done as

$$\frac{Y(s)}{R(s)} = \frac{KK_pN_p(s)D_h(s)}{s^p + d_{p-1}s^{p-1} + \cdots + d_1 s + d_0} \tag{7.3-7}$$

If the denominator of $G_{L/K}(s)$ is of order 2 or higher than the numerator, from Eq. (7.2-3) it is seen in the transition from Eq. (7.3-6) to Eq. (7.3-7) that d_{p-1} is not a function of K. In factored form Eq. (7.3-7) becomes

$$\frac{Y(s)}{R(s)} = \frac{KK_pN_p(s)D_h(s)}{(s+\lambda_{k1})(s+\lambda_{k2})\ldots(s+\lambda_{kn})} \tag{7.3-8}$$

or by expanding this factored expression

$$\frac{Y(s)}{R(s)} = \frac{KK_pN_p(s)D_n(s)}{s^p + (\lambda_{k1}+\lambda_{k2}+\cdots+\lambda_{kp})s^{p-1} + \cdots + \lambda_{k1}\lambda_{k1}\ldots\lambda_{kp}}$$

Therefore

$$d_{p-1} = \lambda_{k1} + \lambda_{k2} + \cdots + \lambda_{kp} \tag{7.3-9}$$

where $-\lambda_{ki}$ is the ith closed-loop pole position. Since d_{p-1} is a constant, the sum of the closed-loop poles is also a constant and independent of K. When K is zero, the closed-loop poles correspond to the loop gain poles so that this constant is just the sum of the poles of the modified loop gain transfer function $G_{L/K}(s)$.

Example 7.3-2 serves to illustrate the point. In that example p is 3 and there are no zeros, so that Rule 11 applies. The sum of the loop gain poles is -3, so that the sum of the closed-loop poles is also -3. Thus, when a closed-loop pole exists at the point $s = -3$, there are two pole locations on the $j\omega$ axis for the same K. The gain K at $s = -3$ is just

$$K = \tfrac{1}{6}(1)(2)(3) = 1$$

Hence this must be the gain at which the two complex conjugate roots cross into the right half-plane, the maximum gain for stability. The gain corresponding to any other set of complex conjugate roots may be found in a similar way, as illustrated in the following example.

Example 7.3-3. In this example the full loop gain $G_L(s)$ is assumed to be

$$G_L(s) = \frac{K(s+1)}{s(s-1)(s+6)}$$

This example is chosen to illustrate further the use of Rules 10 and 11 and to demonstrate that the location of a pole in the right half-plane has no effect upon the construction rules. The root locus for this system is shown in Fig. 7.3-7. A breakaway point exists in the right half-plane between $s = 0$ and $s = +1$. Because of the zero at $s = -1$, it is felt that the breakaway point should be to the left of the midpoint between the two poles, or to the left of 0.5. The gains at $s = 0.5$ and 0.4 are 1.08 and 1.09, respectively, and the gain at $s = 0.45$ is 1.11, or very nearly the maximum. Along the imaginary axis the phase shift is 180° at approximately $s = j1.2$. Rather than measure the value of K at

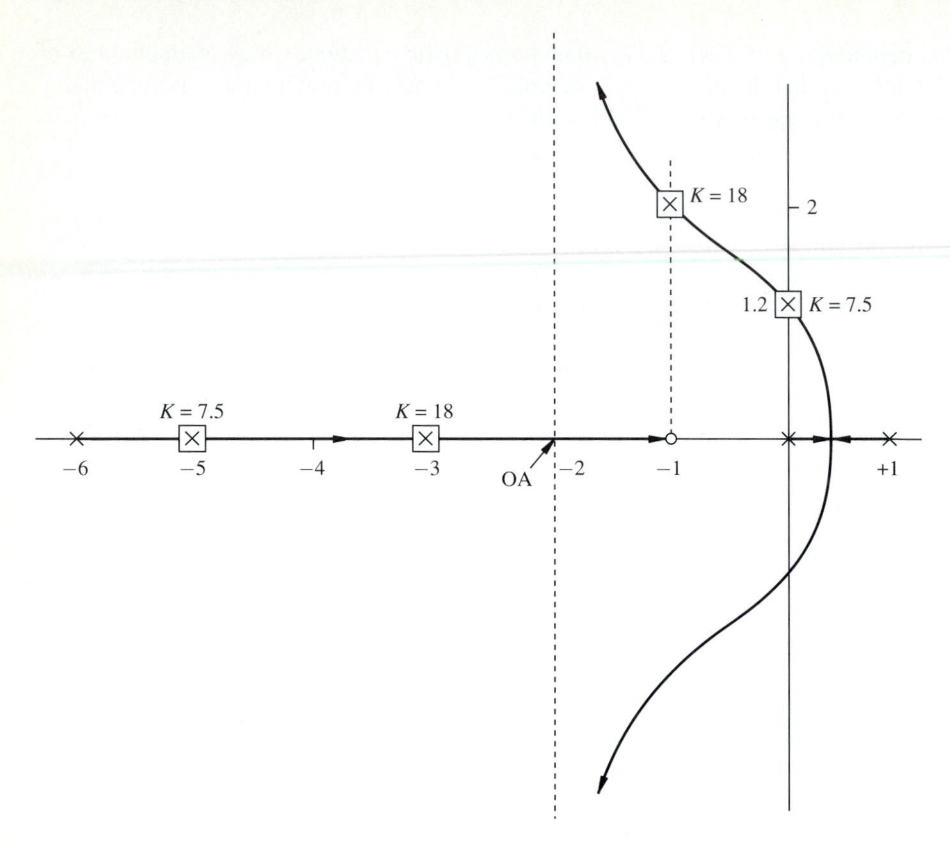

FIGURE 7.3-7
Example 7.3-3.

this point, let us apply Rule 11. Rule 11 can be applied because the relative degree of the modified loop gain is 2. Here the sum of the loop gain poles is -5, so that when the complex conjugate poles are on the $j\omega$ axis, the remaining closed-loop pole is at $s = -5$. The gain K at this point is readily found to be (6) (5) (1)/4 or 7.5. Thus, in this system, if the gain K ever falls below 7.5, the system is unstable.

To complete the locus in this example the asymptotes are drawn and it is considered necessary to find one additional point. This is done by finding the point along the line $s = -1$ at which the phase angle criterion is satisfied. The condition is met at $s \approx -1+j2$, and the corresponding gain is 18, and the real pole is at $s \approx -3$.

The examples thus far in this chapter have dealt exclusively with those cases in which both the loop gain poles and zeros have all been real. The last construction rule deals with the angles made by the branches of the root locus as they either leave a complex pole or arrive at a complex zero of the loop gain.

RULE 12 ANGLES OF DEPARTURE AND ARRIVAL. The angle of departure from a complex conjugate pole or the angle of arrival at a complex conjugate zero is determined by satisfying the angle criterion at a point very close to the pole or zero in question.

To make the meaning of this rule clear, let us consider the pole-zero plot shown in Fig. 7.3-8. Let us determine the departure angle associated with the pole at $s = -1 + j2$. Suppose we consider a point on a circle of small radius ϵ as shown in the expanded plot. The angle at the point for the pole at $s = -1 + j2$ is indicated as ϕ_d. The angles from this point to the other two poles are essentially the same as the angles from the pole at $s = -1 + j2$ to the other two poles since the radius of the circle is small compared with the other lengths involved. If we assume now that the point on the circle is also on the root locus, the angle criterion must be satisfied at this point so that

$$-(90^\circ + 126^\circ + \phi_d) = 180^\circ \pm N360^\circ$$

or

$$\phi_d = -36^\circ \text{ or } 324^\circ$$

Near the pole at $s = -1 + j2$, the root locus must depart from the pole at an angle of -36°. When this information is used, the root locus takes the form shown in Fig. 7.3-9. Because the lower half of the root locus diagram is just a mirror image about the real axis, the lower half of the diagram requires no further work.

The following example illustrates the use of Rule 12 in computing the arrival angle at a complex zero.

Example 7.3-4. Complex conjugate zeros are often introduced through the use of state-variable feedback. The block diagram of Fig. 7.3-10 illustrates such a case. Here $KG_P(s)H_{eq}(s)$ is

$$KG_p(s)H_{eq}(s) = \frac{(3K/4)\left[(s+2.89)^2 + 2.23^2\right]}{s(s+2)(s+8)}$$

The root locus for this system is shown in Fig. 7.3-11. Also included in this figure are the angles necessary for the calculation of the angle of arrival of the branch of the root

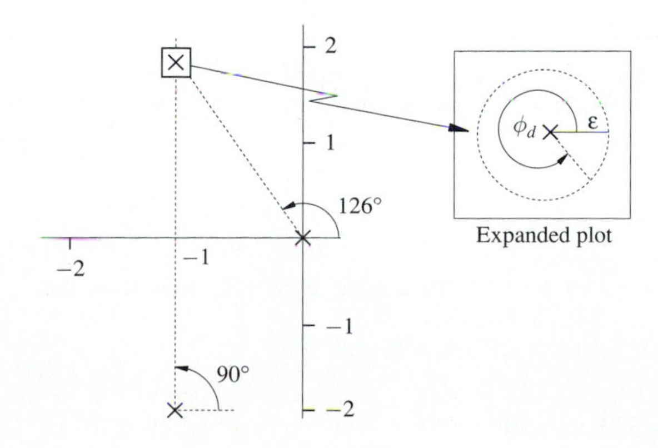

FIGURE 7.3-8
Computation of departure angle.

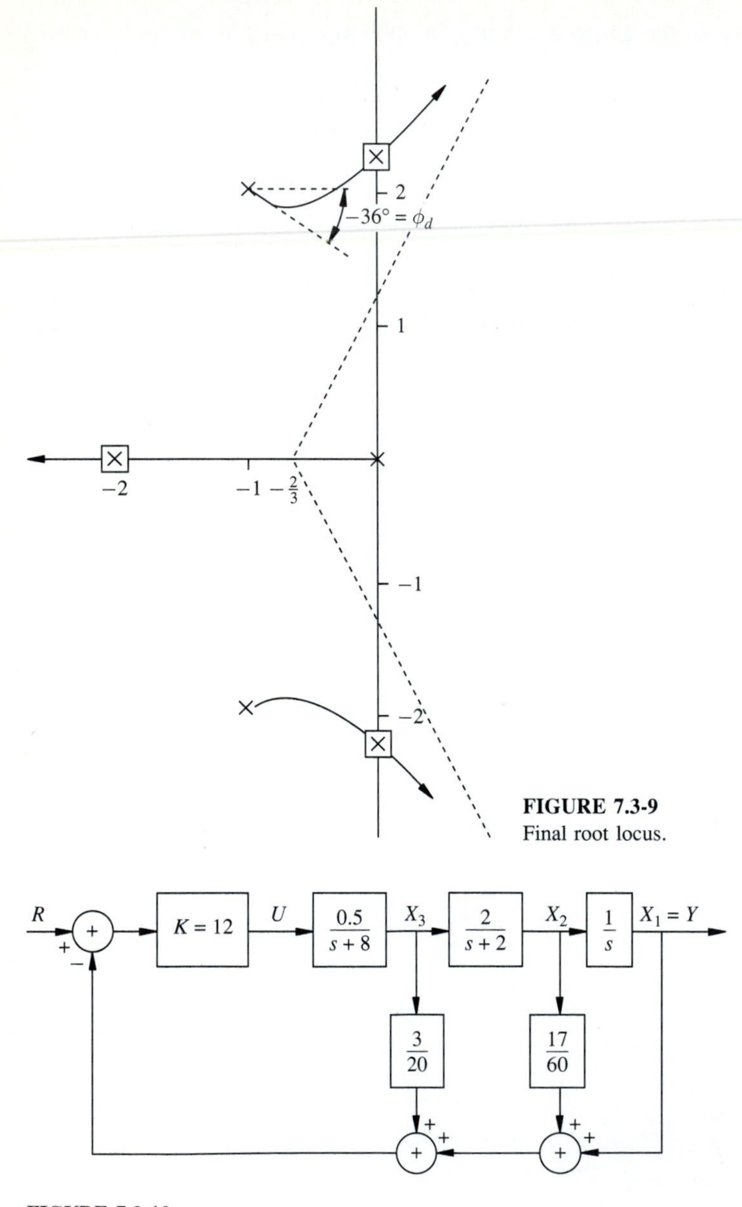

FIGURE 7.3-9
Final root locus.

FIGURE 7.3-10
Block diagram for Example 7.3-4.

locus at the zero located at $s = -2.89 + j2.23$. This angle ϕ_a is calculated from the formula

$$(90^\circ + \phi_a) - (143^\circ + 114^\circ + 23^\circ) = 180^\circ \pm N360^\circ$$

so that ϕ_a is

$$\phi_a = 370^\circ \pm N360^\circ = 10^\circ$$

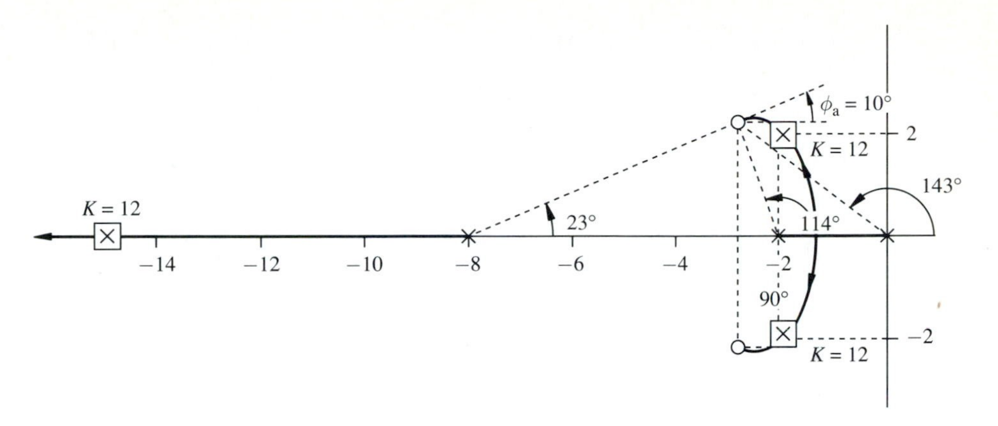

FIGURE 7.3-11
Completed root locus for Example 7.3-4.

In this example $p - z$ is 1, so that Rule 11 does not apply. To find the points corresponding to $K = 12$ the gain must be calculated along each branch of the root locus. These gain points are indicated in the figure. Because the angle of the asymptote is 180°, it is not shown.

Both this section and the preceding one have dealt with the construction rules that make the drawing of the root locus diagram a relatively simple matter. Occasionally there is a need for a few additional points, to ensure that the resulting root locus is an adequate approximation of the behavior of the closed-loop poles. In these cases one may always check the phase angle at any point on the s plane to see if it is on the root locus. This was done in Example 7.3-3 in a systematic fashion by examining the phase shift along a vertical line. Any other systematic approach is satisfactory, although one usually chooses to check the phase angle along a straight line through which it is certain the root locus passes.

The digital computer is also a useful tool for developing a root locus diagram. If a computer factors the characteristic equation or, equivalently, finds the eigenvalues of the system matrix $\boldsymbol{A}_k$ for a range of values of K, this information may be displayed in graphical form as a root locus diagram. However, even if a digital computer is available to plot the root locus, it is often helpful to be able to sketch the general shape of the root locus quickly by hand and then use the digital computer for refinements if needed. This computer-aided procedure is especially helpful in design problems.

7.3.1 CAD Notes

The function `rlocus` in MATLAB takes a numerator polynomial, a denominator polynomial, and a vector of gains K as arguments. It produces a matrix. Each row of the matrix contains the closed-loop poles for the corresponding value of K. These roots can then be plotted to produce a root locus. The following MATLAB statements produce the same root locus as Fig. 7.3-7. The resulting plot is shown in Fig. 7.3-12.

```
% enter transfer function
num=[1 1];
den=[1 5 -6 0];
% set up k vector of gains
k=0:.1:50;
% compute roots at each gain
r=rlocus(num,den,k);
% set up axis limits
v=[-9 1 -5 5];
axis(v)
% keep real and imag axes equal length
axis('square')
% plot points, not lines
% program doesn't know how to associate poles as k changes
% plot program plots each point in the matrix as a
% complex number
plot(r,'.'),xlabel('real'),ylabel('imaginary')
% place hold on to add open loop zeros and poles
hold on
% find open loop poles
r0=roots1(den);
% find real and imag part of poles
x=real(r0);
y=imag(r0);
% plot open loop poles with 'x'
plot(x,y,'x')
% plot open loop zeros with 'o'
r1=roots1(num);
x=real(r1);
y=imag(r1);
plot(x,y,'o')
% retrieve roots when k=48 from r and plot with stars
r2=r(181,:);
x=real(r2);
y=imag(r2);
plot(x,y,'*')
% set up and plot real axis
x=[-15,5];
y=[0 0];
plot(x,y,'--')
% set up and plot imag axis
x=[0 0];
y=[-15 15];
plot(x,y,'--')
% turn hold off
hold off
% restore axis to normal state and auto scale
axis('normal');
```

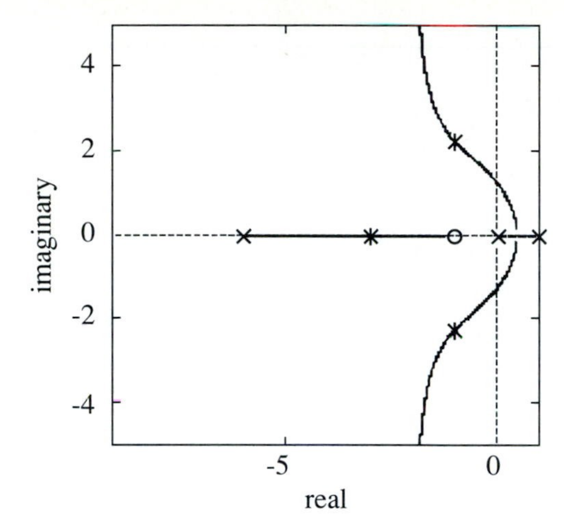

FIGURE 7.3-12
CAD generated root locus.

Exercises 7.3

7.3-1. Sketch the root locus diagram corresponding to the Fig. 7.3-13 for positive values of K when

$$G_p(s) = \frac{s+1}{s(s-1)(s+20)}$$

For this system, answer the following questions.

(*a*) The number of excess poles is ________.

(*b*) The origin of the asymptotes is ________.

(*c*) The angles that the asymptotes make with the real axis are (list all) ________.

(*d*) The locus crosses the $j\omega$ axis when K is (if no cross-over write "none") ________.
Table 7.3-1 lists, for selected points on the real axis, the corresponding value of K. Use this table to find the breakaway and reentry points for your plot in (*a*).

(*e*) breakaway points: ________.

(*f*) reentry points: ________.

Answers: (*a*) 2 (*b*) -9
(*c*) $90°, -90°$ (*d*) 21.1
(*e*) 0.45 (*f*) -2.75

7.3-2. Using root locus rules determine whether each of the following loop gain transfer functions produces a stable closed-loop system for all $K > 0$? Explain how you arrived at your answer. (You need not draw the root loci if you can determine the answer from the rules more directly. Do not use the Routh-Hurwitz test.)

(*a*) $G_1(s) = \dfrac{K(s+10)(s+2)}{s^2(s+1)}$ (*b*) $G_2(s) = \dfrac{K}{(s+1)(s+10)(s+100)}$

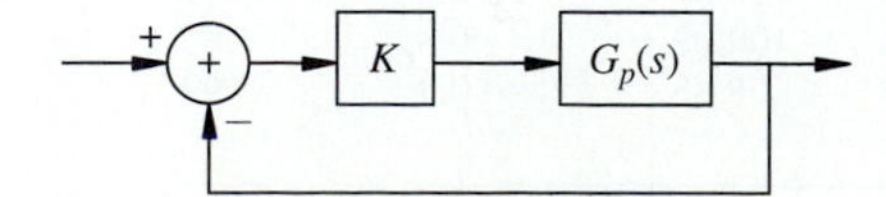

FIGURE 7.3-13
Exercise 7.3-1.

TABLE 7.3-1

Exercise 7.3-1

s	K	s	K	s	K	s	K
−20.0	0.00	−14.7	89.28	−9.4	123.37	−4.2	107.83
−19.9	2.20	−14.6	90.43	−9.3	123.49	−4.1	107.25
−19.8	4.38	−14.5	91.56	−9.2	123.59	−4.0	106.67
−19.7	6.54	−14.4	92.68	−9.1	123.68	−3.9	106.09
−19.6	8.68	−14.3	93.77	−9.0	123.75	−3.8	105.53
−19.5	10.80	−14.2	94.84	−8.9	123.80	−3.7	104.98
−19.4	12.91	−14.1	95.89	−8.8	123.83	−3.6	104.46
−19.3	14.99	−14.0	96.92	−8.7	123.85	−3.5	103.95
−19.2	17.05	−13.9	97.94	−8.6	123.84	−3.4	103.47
−19.1	19.09	−13.8	98.93	−8.5	123.82	−3.3	103.03
−19.0	21.11	−13.7	99.90	−8.4	123.78	−3.2	102.63
−18.9	23.11	−13.6	100.86	−8.3	123.72	−3.1	102.29
−18.8	25.09	−13.5	101.79	−8.2	123.64	−3.0	102.00
−18.7	27.06	−13.4	102.70	−8.1	123.54	−2.9	101.79
−18.6	29.00	−13.3	103.60	−8.0	123.43	−2.8	101.67
−18.5	30.92	−13.2	104.47	−7.9	123.30	−2.7	101.66
−18.4	32.82	−13.1	105.33	−7.8	123.15	−2.6	101.79
−18.3	34.71	−13.0	106.17	−7.7	122.98	−2.5	102.08
−18.2	36.57	−12.9	106.98	−7.6	122.80	−2.4	102.58
−18.1	38.41	−12.8	107.78	−7.5	122.60	−2.3	103.34
−18.0	40.24	−12.7	108.56	−7.4	122.38	−2.2	104.43
−17.9	42.04	−12.6	109.32	−7.3	122.14	−2.1	105.94
−17.8	43.82	−12.5	110.05	−7.2	121.89	−2.0	108.00
−17.7	45.59	−12.4	110.77	−7.1	121.62	−1.9	110.81
−17.6	47.33	−12.3	111.47	−7.0	121.33	−1.8	114.66
−17.5	49.05	−12.2	112.15	−6.9	121.03	−1.7	120.00
−17.4	50.76	−12.1	112.81	−6.8	120.71	−1.6	127.57
−17.3	52.44	−12.0	113.45	−6.7	120.38	−1.5	138.75
−17.2	54.11	−11.9	114.08	−6.6	120.03	−1.4	156.24
−17.1	55.75	−11.8	114.68	−6.5	119.66	−1.3	186.38
−17.0	57.38	−11.7	115.26	−6.4	119.28	−1.2	248.16
−16.9	58.98	−11.6	115.82	−6.3	118.88	−1.1	436.59
−16.8	60.57	−11.5	116.37	−6.2	118.47	−1.0	∞
−16.7	62.13	−11.4	116.89	−6.1	118.04	−0.9	−326.61
−16.6	63.68	−11.3	117.40	−6.0	117.60	−0.8	−138.24
−16.5	65.20	−11.2	117.89	−5.9	117.15	−0.7	−76.56
−16.4	66.71	−11.1	118.35	−5.8	116.68	−0.6	−46.56
−16.3	68.19	−11.0	118.80	−5.7	116.20	−0.5	−29.25
−16.2	69.66	−10.9	119.23	−5.6	115.70	−0.4	−18.29
−16.1	71.11	−10.8	119.64	−5.5	115.19	−0.3	−10.98
−16.0	72.53	−10.7	120.03	−5.4	114.68	−0.2	−5.94
−15.9	63.94	−10.6	120.40	−5.3	114.15	−0.1	−2.43
−15.8	75.33	−10.5	120.75	−5.2	113.61	0.0	0.00
−15.7	76.70	−10.4	121.08	−5.1	113.06	0.1	1.64
−15.6	78.04	−10.3	121.40	−5.0	112.50	0.2	2.69
−15.5	79.37	−10.2	121.69	−4.9	111.93	0.3	3.28
−15.4	80.68	−10.1	121.97	−4.8	111.36	0.4	3.50
−15.3	81.97	−10.0	122.22	−4.7	110.78	0.5	3.42
−15.2	83.24	−9.9	122.46	−4.6	110.20	0.6	3.09
−15.1	84.49	−9.8	122.68	−4.5	109.61	0.7	2.56
−15.0	85.71	−9.7	122.88	−4.4	109.02	0.8	1.85
−14.9	86.92	−9.6	123.06	−4.3	108.43	0.9	0.99
−14.8	88.11	−9.5	123.22				

(*c*) $G_3(s) = \dfrac{K(s+10)}{(s+1)(s-10)(s+100)}$ (*d*) $G_4(s) = \dfrac{K(s+0.1)^2}{s(s+3)(s^2+s+100)}$

Answers: (*a*) No (*b*) No (*c*) No (*d*) Yes

7.3-3. Plot the root locus for the loop gain transfer function given below. Determine the origin of the asymptotes, breakaway point, arrival angles, and $j\omega$ axis intersection. Find the value of K such that there is a closed-loop pole at $s = -4$.

$$G_L(s) = \frac{K\left[(s+2)^2 + 2^2\right]}{s(s+2)^2(s+6)}$$

Answers: Breakaway points at $s \approx -0.75$ and -4.25; $OA = -3$; $\phi_a = 71°$; no $j\omega$ axis intersection; for a pole at $s = -4$, $K = 4$.

7.4 ADDITIONAL EXAMPLES AND ROOT LOCUS RULES FOR NEGATIVE *K*

This section consists mainly of examples typical of situations that commonly occur in control work; they serve to illustrate further the construction rules of Secs. 7.2 and 7.3, which are summarized for easy reference in Table 7.4-1.

As noted in the introduction to this chapter, the root locus diagram is a plot of the locus of the roots of the closed-loop system as the gain K is allowed to vary from zero to infinity. From viewing the entire diagram, we appreciate how gain changes affect the closed-loop pole locations. For a specific value of gain we are able to state with reasonable accuracy the locations of the closed-loop poles. The results derived from the root locus diagram are, on one hand, very general, and on the other, very specific. The general results, concerning the overall behavior of the closed-loop poles as the gain is varied, are more important for synthesis when the pattern of system behavior is to be established. The specific results concerning the location of the closed-loop poles for one given value of gain are more important in analysis. Once the specific closed-loop pole locations are known, the analysis of the particular system may be carried out to any degree of completeness and accuracy desired. As one's facility with the root locus method develops, so does the appreciation for both the general information and the specific facts that may be acquired from the root locus diagram.

The first example of this section is meant to clarify the statement of Rule 7. Note the use of the term *relative* maximum or minimum. In the examples considered thus far, there has been only one maximum or minimum along each real axis segment of the root locus. This is not always the case, however, as the following example illustrates.

Example 7.4-1. Here it is assumed that the loop gain transfer function is

$$G_L(s) = \frac{K(s+2)}{s(s+1)(s+19)}$$

and the complete root locus diagram is desired. The pole-zero plot is shown in Fig. 7.4-1. Since the pole-zero excess is 2, the angles of the asymptotes are $\pm 90°$, and the origin of the asymptotes is

$$OA = \frac{(-19-1)-(-2)}{2} = -9$$

TABLE 7.4-1
Summary of root locus construction rules

Rule 1: Number of branches	There is one branch for each pole of the loop gain transfer function $G_L(s)$.
Rule 2: Starting points ($K = 0$)	The branches of the root locus start at the poles of $G_L(s)$.
Rule 3: End points ($K = \infty$)	The branches of the root locus end at the finite zeros of $G_L(s)$ or go to infinity.
Rule 4: Behavior along the real axis	The root locus exists on the real axis at every point for which the sum of the number of poles and zeros lying to the right of the point is odd.
Rule 5: Gain determination	At a point s_1 on the root locus, the gain is given by $$\lvert K\rvert = \left\lvert \frac{1}{G_{L/K}(s)} \right\rvert_{s=s_1}$$
Rule 6: Symmetry of locus	The root locus is always symmetric with respect to the real axis.
Rule 7: Breakaway or reentry points	The root locus breaks away from the real axis at a point of relative maximum gain and returns to the real axis at a point of relative minimum gain.
Rule 8: Breakaway or reentry angles	At points of breakaway or reentry, the incoming and outgoing lines of the root locus are separated by an angle of $180°/\alpha$, where α is the number of branches that intersect.
Rule 9: Asymptotic behavior for large K	The asymptote angles are given by $$\phi_{ary} = \frac{180° + k360°}{p - z} k = 0, 1, 2, \ldots, p - z - 1$$ and the origin of the asymptotes is $$OA = \frac{\sum(\text{pole locations}) - \sum(\text{zero locations})}{p - z}$$ where $p - z$ is the pole-zero excess.
Rule 10: Imaginary axis crossing	The branches of the root locus cross the imaginary axis at points where the phase shift is $180° + N360°$. Imaginary axis poles can often be found using Routh-Hurwitz techniques.
Rule 11: Sum of the closed-loop poles	If $p - z \geq 2$, the sum of the closed-loop poles is a constant.
Rule 12: Angles of departure and arrival	The angles of departure and arrival at complex conjugate poles and zeros are determined by satisfying the angle criterion near the pole or zero in question.

A breakaway point must exist between the poles at $s = -1$ and 0. Gain calculations at $s = -0.5$, -0.6, and -0.7 yield the following values for K: 3.08, 3.15, and 2.96, respectively. Thus the breakaway point is approximately $s = -0.6$.

One might suspect at first glance that this is the only breakaway point along the negative real axis. A branch of the root locus exists between -19 and -2, but since this

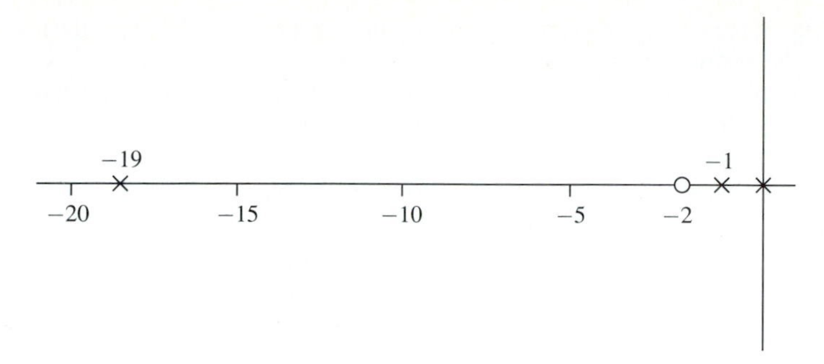

FIGURE 7.4.1
Pole-zero plot for Example 7.4-1.

branch exists between a pole and a zero, no breakaway need occur. In fact, there is one point of breakaway and one point of reentry, as shown in Fig. 7.4-2. One is led to this conclusion in two ways.

Once the breakaway point at $s = -0.6$ has been established, a first guess is that these two branches seek the asymptotes. Because these asymptotes are fairly far removed from the breakaway point, it is probably desirable to establish another point on the branches of the root locus that are approaching the asymptote. For example, one might select the line $s = j2$ and check the phase angle at various points along that line, looking for a point with a phase angle of 180°. Several such points are indicated in Fig. 7.4-2, and in the region near the breakaway point the phase angle is far from 180°. Not until the asymptote is almost reached does the phase angle become the desired value. The distance between these two points might indicate that the locus does not proceed directly from the breakaway point to the asymptote.

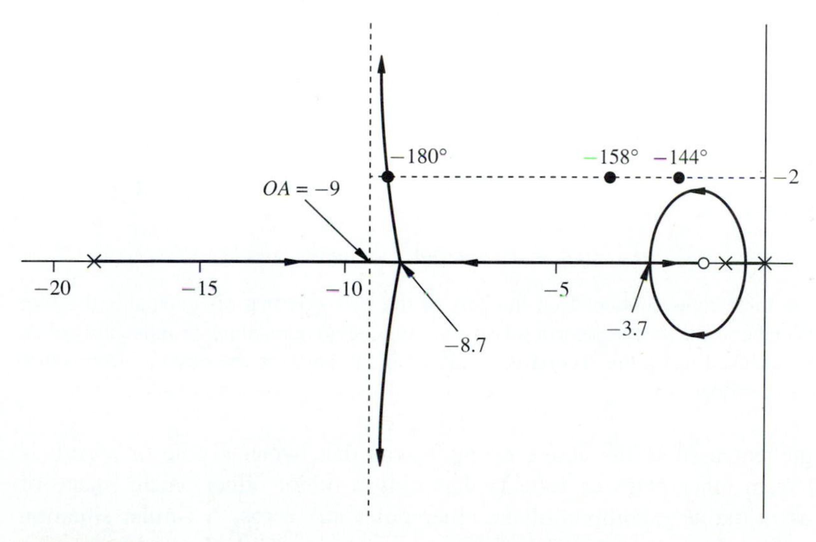

FIGURE 7.4-2
Root locus for Example 7.4-1.

The necessity of breakaway and reentry points becomes apparent when gain values are marked along the segment of the negative real axis from $s = -2$ to $s = -19$. At $s = -10$ the gain is 101, whereas at $s = -6$ the gain is only 97.5. If the branch of the root locus were proceeding directly from the pole at $s = -19$ to the zero at $s = -2$, this decrease in gain could not occur. A bit of probing soon establishes that a point of relative maximum gain exists at $s \approx -8.7$, and a point of relative minimum gain exists at $s \approx -3.7$. Since only two branches of the root locus are involved in each case, the breakaway and reentry angles must be $\pm 90°$. In this simple case the root locus can be completed without checking any further points. Since the breakaway point at $s = -8.7$ is near the origin of the asymptotes, and since the point at $s = -8.9 + j2$ near the asymptote has a phase angle of 180°, the leftmost branch of the root locus is sketched by connecting the two points. Because the pole at $s = -19$ is so far removed from the other poles and zeros, its effect on the shape of the root locus near the other poles and zeros is only minor. The breakaway-reentry combination near the zero was encountered previously in Example 7.3-1, and the resulting locus was a semicircle. Hence a semicircle was assumed to be a good approximation to the locus, as indicated in Fig. 7.4-2, which is the required root locus diagram.

The local minimum and maximum for this problem can also be found using calculus. An expression for K equivalent to Eq. (7.2-5) for this example is

$$K = \frac{-s^3 - 20s^2 - 19s}{s + 2}$$

Taking the derivative of K with respect to s yields

$$\frac{dK}{ds} = \frac{-(s + 2)(3s^2 + 40s + 19) + (s^3 + 20s^2 + 19)}{(s + 2)^2}$$

Setting the numerator equal to zero is equivalent to setting the derivative equal to zero except at the point $K = -2$. Thus the local extrema of K as a function of s are defined by the roots of the equation

$$-2s^3 - 26s^2 - 80s - 38 = 0$$

or equivalently

$$s^3 + 13s^2 + 40s + 19 = 0$$

The roots may be found by searching. (Many calculators are equipped to perform this function.) The roots are

$$s = -0.6, s = -3.8 \text{ and } s = -8.7$$

There are three roots contained on the part of the real axis that also contains the root locus. Whether these roots represent relative minima, relative maxima, or inflection points could be checked using the second derivative but the answers are obvious from other root locus considerations.

An argument used in the above example was that, when a pole or a zero is far removed from other poles or zeros, it has only a minor effect on the shape of the root locus in the neighborhood of the other poles and zeros. A similar situation often occurs with pole-zero cancellation. Since we are dealing with physical quantities, perfect cancellation is possible only on paper. However, if cancellation is attempted, a pole-zero pair results with a separation that is small compared with the distances to other poles and zeros. The next example considers pole-zero cancellation on the real axis.

Example 7.4-2. Let us consider the closed-loop system shown in Fig. 7.4-3 for which the open-loop transfer function is

$$G_L(s) = \frac{Kk_2(s + 1/k_2)}{s(s+1)(s+8)}$$

where k_2 and hence the zero location of $H(s)$ can be adjusted. It is desired to cancel the pole at $s = -1$ with this zero so that k_2 is selected as $k_2 = 1$. Note that in this case a zero of $H(s)$ is identical to a pole of $G_p(s)$ and it may appear that these terms should be canceled. The pole-zero combination should not be canceled, however, as indicated in the initial discussion of Sec. 7.2. Zeros of $H(s)$ must never be canceled with the poles of $G_p(s)$. To see why this is true, let us examine the root locus for $k_2 = 1$ as shown in Fig. 7.4-4.

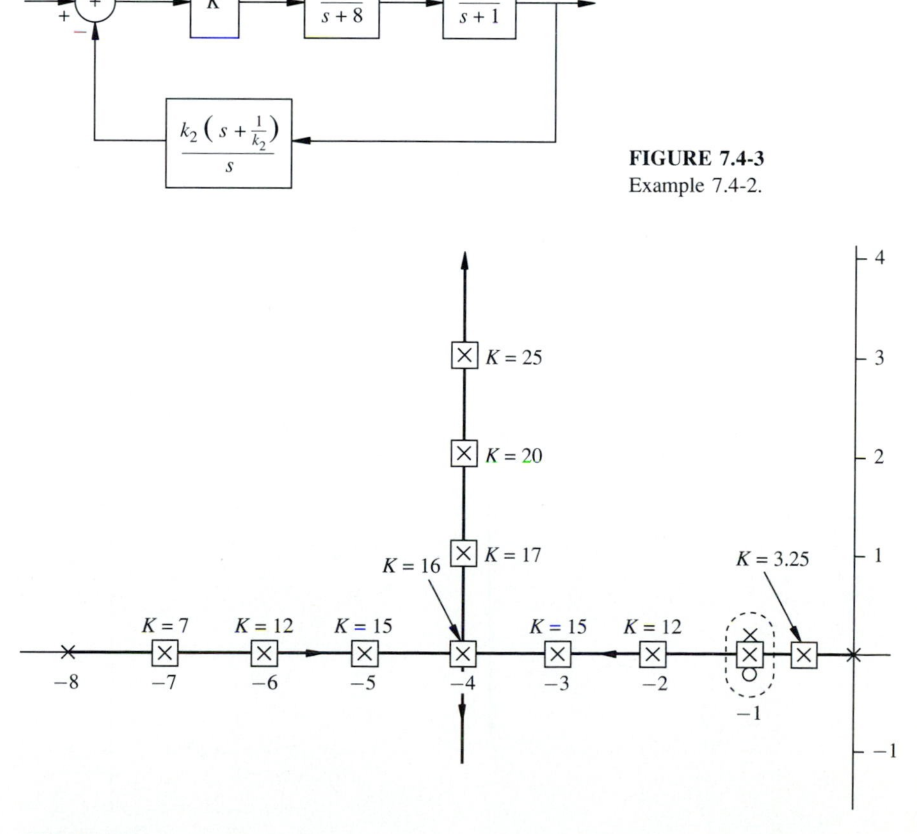

FIGURE 7.4-3
Example 7.4-2.

FIGURE 7.4-4
Root locus for Example 7.4-2, with $k_2 = 1$.

Note that at $s = -1$ we have indicated the location of a pole, a zero, and a closed-loop pole. These should all be located at the point $s = -1$, but the loop gain pole and zero are shown off the real axis to emphasize their distinct character. In every case of perfect cancellation, a closed-loop pole *always* exists at the point of cancellation. That is, a branch of the root locus is located between the pole and the zero that happen to be at the same point. In this case the branch takes the form of a single point. If the original loop gain pole-zero pair had been canceled, this branch of the locus would have been suppressed and the closed-loop pole at $s = -1$ overlooked. This pole is not canceled in the closed-loop transfer function.

The breakaway point is at $s = -4$, and the rest of the root locus diagram is identical to the result if the pole-zero pair at $s = -1$ had been removed entirely from the drawing. Angles and distances from the canceling pole and zero are, of course, equal, and they offset each other since one term appears in the numerator and the other in the denominator of the modified loop gain transfer function. With respect to the rest of the root locus diagram, the pole and zero that cancel might just as well not be present. However, they are included so that the closed-loop pole between them is not overlooked.

If k_2 is too small, say $k_2 = 1/1.2$, the zero is at $s = -1.2$ and the root locus of Fig. 7.4-5 results. Note that the origin of asymptotes is now at $s = -3.9$ rather than at $s = -4$ and that the breakaway point is slightly affected. On this diagram there are

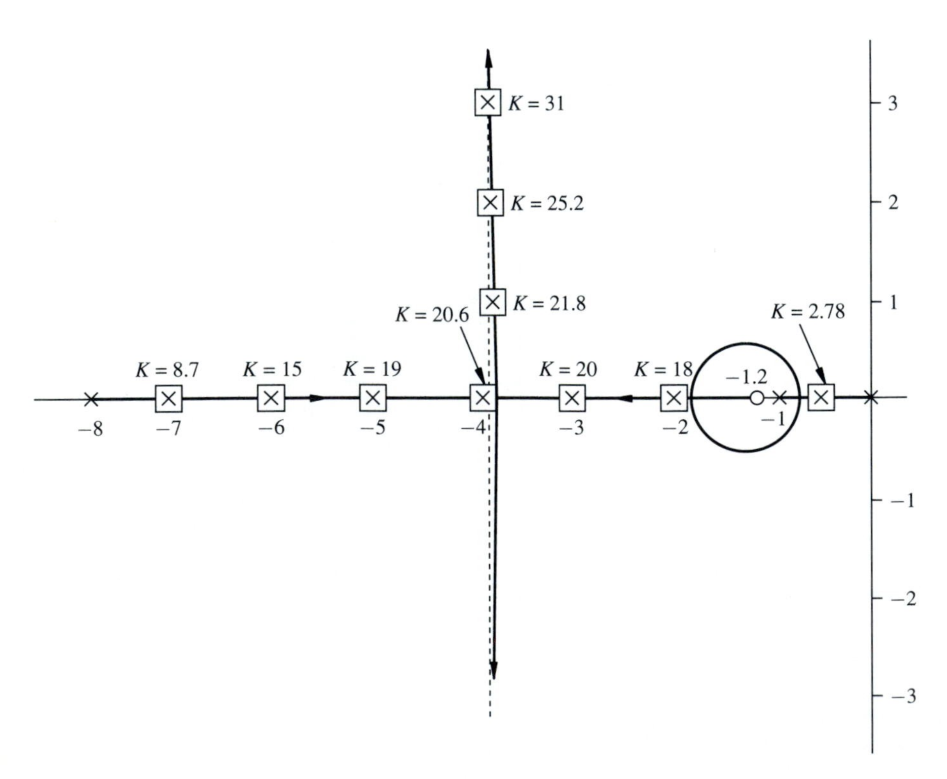

FIGURE 7.4-5
Root locus for Example 7.4-2, with $k_2 = 1/1.2$.

clearly three branches to the root locus and, of course, three closed-loop poles result, as in Fig. 7.4-4.

The main difference here is that, near the point $s = -1$, it is possible for complex conjugate roots to exist. Note one striking similarity, however. For gains beyond approximately 20, the complex conjugate roots with real part equal to -3.9 are very close to the same values that occurred when the cancellation was perfect except for the gain change due to the fact that k_2 is now equal to $1/1.2$. Also, the one remaining closed-loop pole is very close to the zero on the real axis, as before.

Figure 7.4-6 is the root locus diagram for the case of $k_2 = 1/0.8$, so that the zero is at $s = -0.8$. Again the three closed-loop poles are quite evident. This time the origin of the asymptotes is at $s = -4.1$, and there are no complex conjugate roots near $s = -1$. Again, though, for gains beyond 20, the complex conjugate poles with real parts equal to -4.1 are very close to those shown in Fig. 7.4-4 for perfect cancellation, and the real pole is very close to the real zero.

Two conclusions can be drawn from the above example. Often in synthesis work, pole-zero cancellation is used, and as far as the paper-and-pencil analysis is concerned, the resulting root locus is simple to draw. That is, it is simpler to draw than if cancellation were not perfect. If in the actual system the cancellation is not

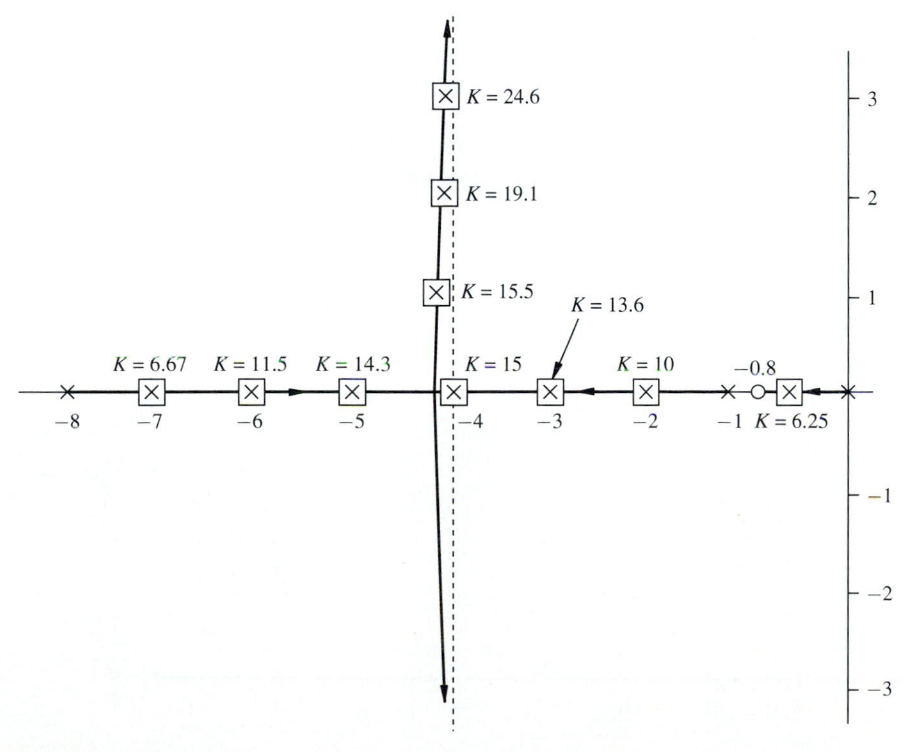

FIGURE 7.4-6
Root locus for Example 7.4-2, with $k_2 = 1/0.8$.

perfect, this is not of great consequence, particularly if the gain is high. The closed-loop poles in the rest of the drawing are largely unaffected, and a closed-loop pole near the canceling zero seeks that zero.

The other conclusion to be drawn is that cancellation of a right half-plane pole and zero is utterly impractical. A closed-loop pole always exists in the right half-plane, causing the system to be unstable. A problem at the end of this chapter considers the case of pole-zero cancellation for complex poles.

Example 7.4-3. Here the loop gain transfer function is given as

$$G_L(s) = \frac{K\left[(s+1.5)^2 + 1^2\right]}{s^2(s+0.5)(s+8)(s+9)}$$

A branch of the locus exists on the real axis for $s < -9$ and between $s = -0.5$ and $s = -8$. The maximum value of gain along the latter segment of the real axis occurs as $s = -2.4$, and this is the point of breakaway. The origin of the asymptotes is $(-17.5+3)/3 = -4.83$, and the angles of the asymptotes are $\pm 60^\circ$ and 180°. Because the breakaway point is near the origin of the asymptotes, one might expect these branches of the root locus to follow the asymptotes to infinity. If this were the case, the two branches starting at $s = 0$ would then terminate at the two finite complex conjugate zeros. However, the angle of arrival at the uppermost complex zero is 175°, and by determining one more point along the line $s = -2$ where the phase angle is satisfied it is evident that this portion of the locus is as indicated in Fig. 7.4-7.

The only remaining branch of the root locus to be determined starts at the origin and approaches infinity along the asymptotes. Points of 180° phase shift are found along the imaginary axis at $s = j2.4$ and $s = j3.5$. The completed root locus is shown in Fig. 7.4-7. Note that the system is stable only for a relatively narrow range of gain from

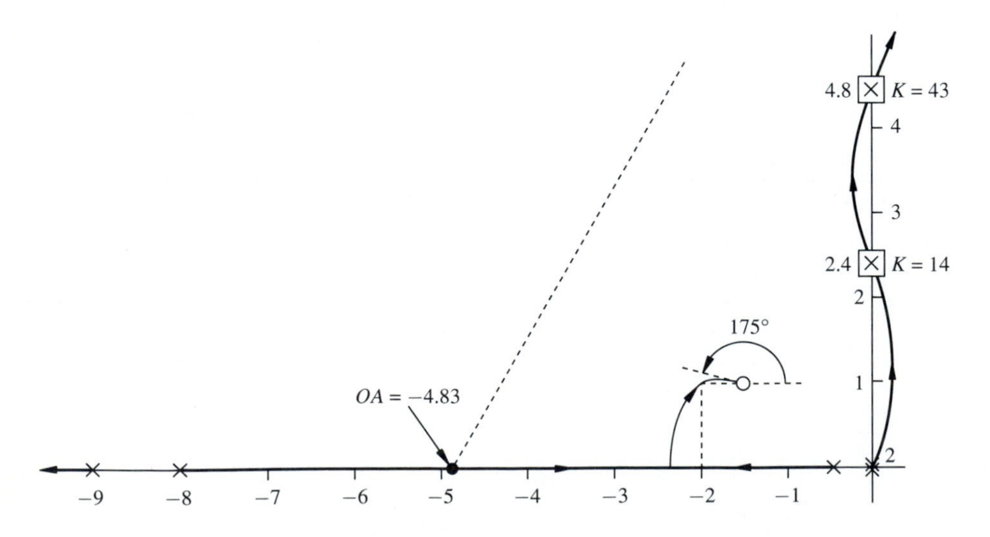

FIGURE 7.4-7
Completed root locus for Example 7.4-3.

$K = 142$ to $K = 433$. One can check this by performing a Routh-Hurwitz stability test on this system.

Systems like the one in Example 7.4-3 that are unstable for both high and low values of gain are said to be *conditionally stable*. Such systems are undesirable in a practical sense, because during the initial startup process the gain may be low enough to cause instability.

Example 7.4-4. The loop gain transfer function for a particular positioning system is

$$G_L(s) = \frac{0.0833K\left[(s+3.25)^2 + 1.21^2\right]}{s(s+1)(s+2.1)}$$

The complete root locus diagram is required, as well as the location of the closed-loop poles for a gain $K = 48$. The breakaway point is found at approximately $s = -0.5$, as the proximity of the pole at $s = -2$ is negated by the presence of the two complex zeros. The angle of arrival at the upper complex zero is 187°, and this is somewhat unexpected. The two branches of the root locus that break away from the real axis near $s = -0.5$ must negotiate a rather long path if they are to terminate on the complex zeros at an angle of 187°. In the final root locus a pole in the upper half-plane takes such an indirect path to the zero so that the root locus is somewhat more difficult to find than in the preceding examples. The construction rules give only limited help. It is necessary to find points along lines such as $s = j0.5$, $j1.0$, $j2.0$, $j3.0$, and $j4.0$ that satisfy the angle criterion to determine the shape of the complex branches. By connecting these points the locus of Fig. 7.4-8 results. The closed-loop poles at $K = 48$ are indicated by boxes. For an example such as this where the rules for determining the root locus do not offer significant help, it is useful to have a CAD program plot the root locus.

Up until this point the root locus rules were developed under the assumption that $K > 0$. Often situations arise where a root locus analysis is needed for $K < 0$.

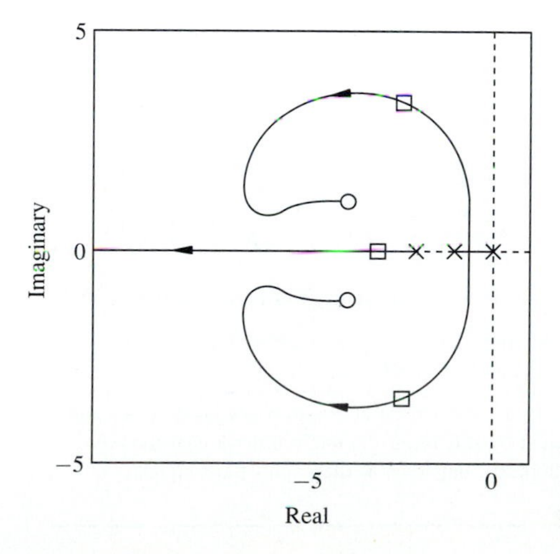

FIGURE 7.4-8
Root locus for Example 7.4-4. (Closed-loop poles for $K = 48$ indicated by boxes.)

These situations tend to occur when a right half-plane zero is present in the loop gain transfer function. For example, the loop gain

$$G_L(s) = \frac{K\left(2 - \frac{s}{2}\right)}{\left(1 + \frac{s}{4}\right)\left(1 + \frac{s}{7}\right)} = \frac{-14K(s-4)}{(s+4)(s+7)}$$

has positive dc gain but a negative constant when the transfer function is placed in pole-zero form. Some new root locus rules need to be formulated for the negative gain root locus. The difference between the negative gain root locus and the positive gain root locus is that, for negative K, the angle condition associated with Eq. (7.2-5) becomes

$$\arg\left(G_{L/K}(s)\right) = 0^\circ \pm N360^\circ$$

Only a few rules need to be changed. The negative root locus rules that are different from the positive root locus rules are given in Table 7.4-2. The rules can be derived using the same principles used to develop the original rules, taking into account that the angle condition now specifies points with an evaluation angle of $0^\circ \pm N360^\circ$ as points on the root locus rather than points with an evaluation angle of $180^\circ \pm N360^\circ$.

Example 7.4-5. The root locus for $G_L(s)$ given below is sketched in Fig. 7.4-9.

$$G_L(s) = \frac{K\left(1 - \frac{s}{2}\right)}{\left(\frac{s}{4} + 1\right)(s+1)\left((s+2)^2 + (2)^2\right)} = \frac{-2K(s-2)}{(s+4)(s+1)\left((s+2)^2 + (2)^2\right)}$$

The rules for negative K are used since gain constant for the loop gain when placed in pole-zero form is negative. The positive real axis for values greater than $s = 2$ has no poles or zeros lying to the right and is part of the root locus. The part of the real axis between $s = -1$ and $s = -4$ has one pole and one zero to the right and is part

TABLE 7.4.2
Root locus rules for negative *K*

Rule 4N: Behavior along the real axis	The root locus exists on the real axis at every point for which the sum of the number of poles and zeros lying to the right of the point is even.
Rule 9N: Asymptotic behavior for large K	The asymptote angles are given by $\phi_{asy} = \frac{0^\circ + N360^\circ}{p - z} N = 0, 1, \ldots, p - z - 1$ The origin of the asymptotes is the same as it is for positive K.
Rule 10N: Imaginary axis crossing	The branches of the root locus cross the imaginary axis at points where the phase shift is $0^\circ \pm N360^\circ$
Rule 12N: Angles of departure and arrival	The angles of departure and arrival at complex conjugate poles and zeros are determined by satisfying the angle criteria near the pole and zero in question. For negative K the angles must equal $0^\circ \pm N360^\circ$

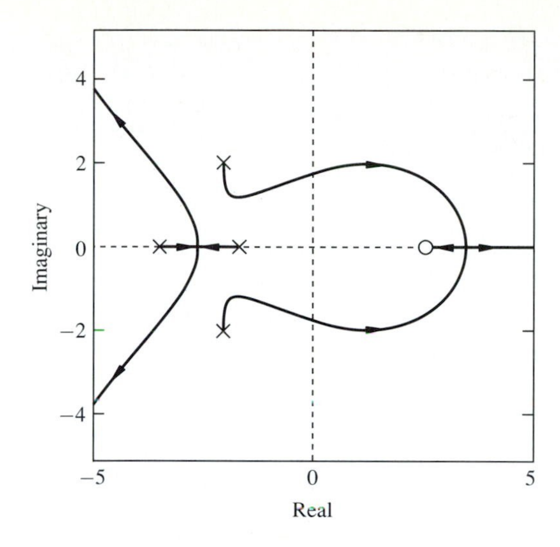

FIGURE 7.4-9
Root locus for Example 7.4-5.

of the root locus. There are three more poles than zeros so that there are asymptotes at $0°$, $+120°$, and $-120°$. The angle of departure ϕ_d for the complex pole with positive imaginary part is computed as

$$153° - 116° - 90° - 14° - \phi_A = 0° + N360°$$

$$\phi_d = -98°$$

A root crosses the imaginary axis near $s = j2$. At $s = j2$ the angle is

$$135° - 63° - 26° - 63° - 0° = -17°$$

which is near zero. The actual crossing point is slightly below $s = j2$. The remainder of the root locus rules are the same as in the case of positive K. Note how the right half-plane zero pulls two poles into the right half-plane creating an unstable closed-loop system for large K. The root locus rule stating that all poles end at zeros indicates that all systems with right half-plane zeros become unstable closed-loop systems for sufficiently large gain.

This concludes the discussion of drawing root locus diagrams. The following section is devoted to interpreting the diagram once it is completed.

Exercises 7.4

7.4-1. For each of the loop gain transfer functions below, sketch the root locus diagram with respect to the positive parameter K. Include on your plots all pertinent information, e.g., angles of departure or arrival, breakaway or reentry points from the real axis, asymptotes and origin of asymptotes, and the value of K where the locus crosses the $j\omega$ axis.

(*a*) $KG_p(s)H(s) = \dfrac{K(s+6)(s+3)}{s(s+1)(s+4)}$

(*b*) $KG_p(s)H(s) = \dfrac{K(s-5)(s-3)}{s(s+1)(s+5)}$

(*c*) $KG_p(s)H(s) = \dfrac{K(s+2)}{(s+1)(s-1)(s^2+14s+100)}$

(*d*) $KG_p(s)H(s) = \dfrac{K(s^2+6.4s+16)}{(s+10)(s+3)(s^2+5s+25)}$

7.4-2. Repeat Exercise 7.4-1 for negative K.

7.5 THE CLOSED-LOOP RESPONSE PLANE

In Chap. 4, we discussed the determination of transient response from an s plane diagram of poles and zeros. In particular, the graphical approach to the determination of both step response and impulse response is emphasized. In Chap. 5, the means of determining frequency response from a pole-zero diagram were examined. In each case the pole-zero diagram in question is assumed to represent a transfer function. This transfer function may be the plant transfer function, the loop gain transfer function, or the closed-loop transfer function.

Here we are concerned with any information about the behavior of the closed-loop system that may be gained from the pole-zero diagram of the closed-loop transfer function. We have noted that, if $G_p(s)$, $G_C(s)$ or $H(s)$ are given in factored form, the zeros of $Y(s)/R(s)$ are already known. The three preceding sections of this chapter have been concerned with the root locus method for determining the closed-loop pole locations. Once the closed-loop pole locations are known, the pole-zero information is complete.

To interpret results for a specific value of gain, it is necessary to transfer information from the s plane, on which the root locus is drawn, to a *closed-loop response plane* that contains only pole-zero information concerning the closed-loop system. From this closed-loop response plane we may find the transient and frequency responses of the closed-loop system. To illustrate the need for the use of the closed-loop response plane more clearly, let us consider the nature of the loop gain transfer function $KG_p(s)H(s)$ and the closed-loop transfer function $Y(s)/R(s)$ for the H configuration. The loop gain transfer function $KG_p(s)H(s)$ is given by

$$KG_p(s)H(s) = KK_pK_h\frac{N_h(s)N_p(s)}{D_p(s)D_h(s)}$$

In other words, the zeros of the loop gain transfer function are the zeros of $H(s)$ and the zeros of the plant. The zeros of the closed-loop transfer function, on the other hand, are the zeros of the plant and the poles of $H(s)$, so that $Y(s)/R(s)$ is

$$\frac{Y(s)}{R(s)} = \frac{KK_pN_p(s)D_h(s)}{D_k(s)}$$

Since the root locus is based on the pole-zero plot of the loop gain, the closed-loop response cannot be obtained directly from the root locus diagram, since the zeros of the closed-loop system are different from the zeros of the loop gain. Therefore it is necessary to transfer the closed-loop pole locations obtained from the root locus diagram to another s plane on which the zeros the closed-loop system have been placed. Only then may the closed-loop behavior be determined.

Example 7.5-1. As an example of the use of the closed-loop response plane, let us consider the state-variable feedback system pictured in Fig. 7.5-1*a*. This system is equivalent to that of Fig. 7.5-1*b*. The H_{eq} configuration naturally arises in the analysis of state-variable feedback systems. For this example

The loop gain transfer function is equal to $KG_p(s)H_{eq}(s)$.

$$G_L(s) = \frac{K(s+1)}{s(s+1)}$$

The zero in $H_{eq}(s)$ should not be canceled by the pole of $G_p(s)$, lest the closed-loop pole at this point be missed. The s plane on which the root locus is drawn is shown in Fig. 7.5-2*a*. Because of the location of the open-loop pole and zero at the same point, the root locus may be drawn without regard to either of these two singularities. In measuring angles or distances, the pole and zero produce equal and opposite effects. Thus the completed root locus is given in Fig. 7.5-2*b*, with the location of a second closed-loop pole shown at the point $s = -K$. One is now ready to transfer the closed-loop poles determined from the root locus diagram to the closed-loop response plane to determine the response of the closed-loop system. The closed-loop response plane showing the poles and zeros of the closed-loop system is given in Fig. 7.5-2*c*. (In this simple problem $G_p(s)$ has no zeros and $H_{eq}(s)$ has no poles so that no zeros occur in the closed-loop transfer function.)

It is a simple matter to verify analytically that the indicated closed-loop poles are indeed the correct ones. From the given $G_p(s)$ and $H_{eq}(s)$, the closed-loop transfer function is found to be

$$\frac{Y(s)}{R(s)} = \frac{K}{s^2 + s(K+1) + K} = \frac{K}{(s+1)(s+K)} \tag{7.5-1}$$

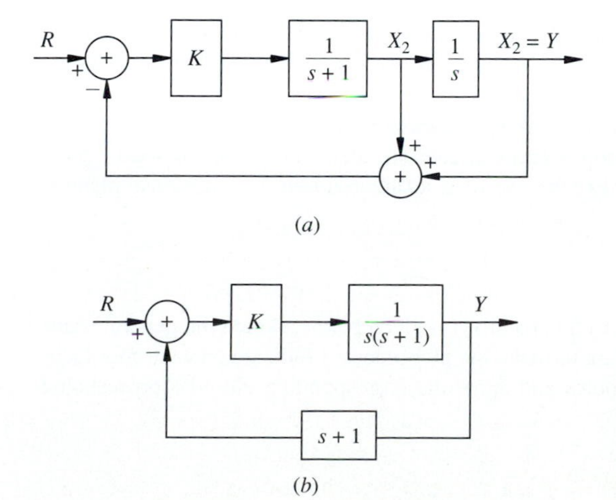

FIGURE 7.5-1
System for Example 7.5-1.

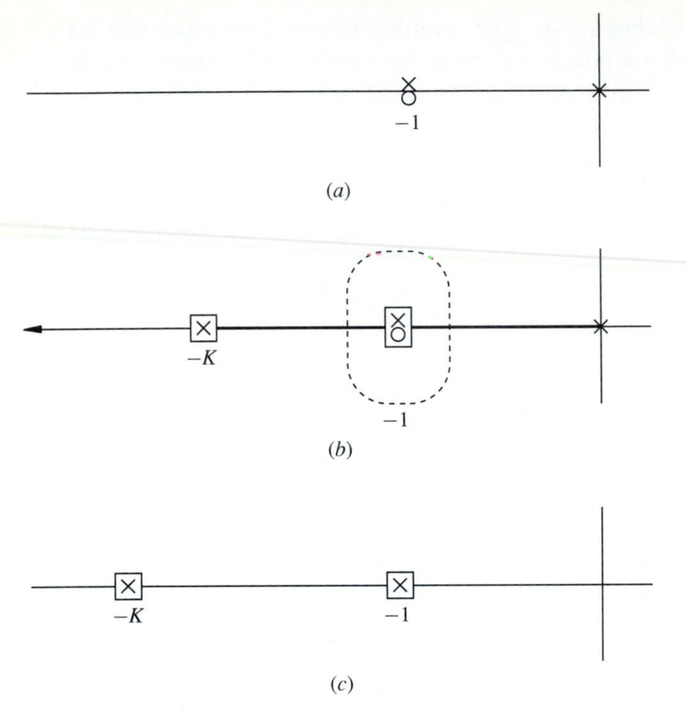

FIGURE 7.5-2
Example 7.5-1. (*a*) Pole-zero plot; (*b*) root locus; (*c*) closed-loop response plane.

If the root locus had been plotted for $G_p(s)H_{\text{eq}}(s)$ and the forbidden cancellation made, then it would appear that the closed-loop system had only one pole at $s = -K$ as given by

$$\frac{Y(s)}{R(s)} = \frac{K}{s+K} \tag{7.5-2}$$

This would, of course, have been a serious error. When K is large the step responses of the closed-loop transfer functions given by Eq. (7.5-1) and Eq. (7.5-2) are extremely different.

Notice that, for a system in the series compensator or G configuration, the closed-loop zeros used in the closed-loop response plane are identical to the loop gain zeros used in the root locus. In this case, the use of a separate closed-loop response plane is unnecessary.

Exercises 7.5

7.5-1. For each of the combinations of $G_p(s)$ and $H(s)$ given below, sketch the general shape of the root locus. (Do not attempt to make the plot precise.) Indicate on each root locus a possible set of closed-loop poles and show the corresponding closed-loop response plane for $Y(s)/R(s)$.

(*a*) $G_p(s) = \dfrac{1}{s(s+2)}$ $\qquad H(s) = \dfrac{1}{s+4}$

(*b*) $G_p(s) = \dfrac{2(s+3)}{s(s+2)} \qquad H(s)\dfrac{1}{(s+3)(s+4)}$

(*c*) $G_p(s) = \dfrac{s+1}{s(s+2)} \qquad H(s) = \dfrac{(s+3)}{(s+1)(s+4)}$

7.5-2. The plant and H_{eq} feedback transfer functions for several systems are given below. All these systems have the same root locus. Draw the root locus and indicate a possible set of closed-loop poles. Then draw the closed-loop response plane for each system, assuming that these pole locations are correct.

(*a*) $G_p(s) = \dfrac{1}{s(s+1)(s+4)} \qquad H(s) = s^2 + 2s + 2$

(*b*) $G_p(s) = \dfrac{s+1}{s(s+1)(s+4)} \qquad H(s) = \dfrac{s^2+2s+2}{s+1}$

(*c*) $G_p(s) = \dfrac{s+3}{s(s+1)(s+2)} \qquad H(s) = \dfrac{s^2+2s+2}{s+3}$

7.6 THE ROOT LOCUS USING PARAMETERS OTHER THAN *K*

As mentioned repeatedly throughout this chapter, the root locus provides a graphical method of finding the locus of the roots of the characteristic equation of the closed-loop system as a function of the gain parameter K. Drawing the root locus is accomplished by setting the denominator of $Y(s)/R(s)$ equal to zero

$$1 + KG_{L/K}(s) = 0 \tag{7.6-1}$$

or, in alternative form,

$$KG_{L/K}(s) = -1 = 1e^{j180^\circ} \tag{7.6-2}$$

This alternative form gives rise to the two criteria that must be satisfied to locate the actual closed-loop poles. The satisfaction of the phase angle criterion gives the locus of roots for any finite value of gain from zero to infinity. Satisfaction of the magnitude criterion locates the closed-loop poles for a specific value of gain.

In this section we wish to consider Eq. (7.6-2) in a more general context. The form of this equation is particularly interesting. Note that its left-hand side represents a system parameter times a transfer function. In Eq. (7.6-2) the system parameter is the controller gain K and the transfer function is $G_{L/K}(s)$. A more general form of Eq. (7.6-2) is

$$\alpha W(s) = -1 = 1e^{j180^\circ} \tag{7.6-3}$$

where α is any system parameter and $W(s)$ is any transfer function that is independent of α.

In this section the root locus method is applied to equations of the form of Eq. (7.6-3) to determine the closed-loop pole variations with respect to variations of system parameters other than the gain K.

In the conventional root locus, the system parameter that is allowed to vary is the gain K. From an examination of the completed root locus with various values of K indicated on the root locus diagram, it is possible to obtain very specific information on the behavior of the closed-loop poles as K is allowed to vary. It is tacitly assumed that the other plant parameters are fixed and not free to be set as K is. Now, we are interested in the movement of the closed-loop poles as some other plant parameter changes. The primary concern here is sensitivity: How much do the closed-loop poles vary if one parameter of the actual plant deviates from the model? Previously we found that, in general, sensitivity to plant perturbations decreases as the loop gain increases. Here we are concerned with detailed changes in the closed-loop pole positions for changes in a specific plant parameter. We are interested in much more detailed information.

The procedure is fairly simple. Rearrange the characteristic equation into the form of Eq. (7.6-3), with the parameter in question appearing as a pseudo gain term α. Then draw the root locus, marking on it the extremes expected in the system parameter.

The procedure is illustrated by the following example. In this example, root loci are drawn for the case in which a plant pole is allowed to vary and then for the case in which a feedback coefficient is allowed to vary. These two particular system parameters are chosen specifically to emphasize a point made in Chap. 2, namely, that feedback moves sensitivity problems from the plant, over which one may have no control, to the feedback paths, over which one may have complete control.

Example 7.6-1. This example is related to Example 7.3-4 and is described by the block diagram of Fig. 7.6-1. In the earlier example, the pole at $s = -a$ was actually at $s = -2$, and the feedback coefficient k_2 was 17/60. Under these conditions the closed-loop transfer function is

$$\frac{Y(s)}{R(s)} = \frac{120}{\left[(s+2)^2 + 2^2\right](s+15)} \tag{7.6-4}$$

The question we wish to ask now is how do the poles of the closed-loop system change if either the pole at -2 or the feedback coefficient k_2 is allowed to take on different values.

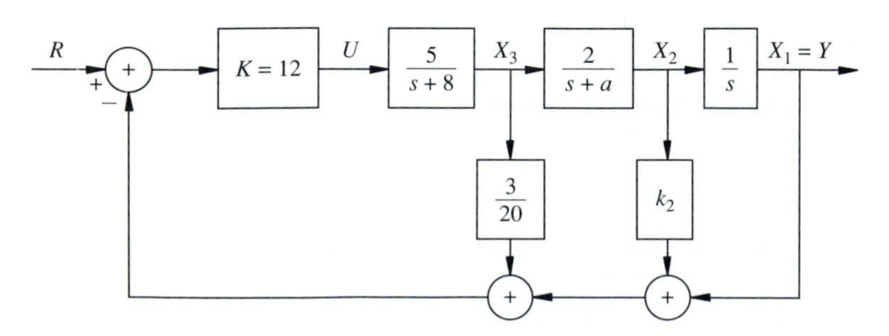

FIGURE 7.6-1
Block diagram of the system.

Let us consider first the case of the pole position. On the basis of information concerning the original plant model fixed plant, it was decided that the pole of the fixed plant was located at $s = -2$. Suppose that the pole location is different because of imprecise measurement or wrong initial assumptions concerning the fixed plant. We wish to find out how the closed-loop system is affected.

Assume that the feedback gain is set with $k_2 = 17/60$. Let us proceed by finding the characteristic equation in terms of the system parameter a and then by reshaping this equation into the form of Eq. (7.6-3). A simple procedure is to involve the variable parameter in the determination of the characteristic equation as few times as possible. Toward this end, the block diagram of Fig. 7.6-1 is reduced as indicated in Fig. 7.6-2.

Note that this is neither a G_{eq} nor an H_{eq} reduction but just a convenient way of involving the parameter a in the characteristic equation as little as possible. The characteristic equation is now

$$1 + \frac{120}{s(s+a)(s+17)}\left[\frac{17}{60}\left(s + \frac{60}{17}\right)\right] = 0$$

or

$$\frac{34(s+3.53)}{s(s+a)(s+17)} = -1 \tag{7.6-5}$$

If both sides of Eq. (7.6-5) are multiplied by the denominator of the left-hand side, the result is

$$34s + 120 = -\left(s^3 + (a+17)s^2 + 17as\right)$$

Isolating the factors of a yields

$$-a\left(s^2 + 17s\right) = s^3 + 17s^2 + 34s + 120$$

or

$$a\frac{s^2 + 17s}{s^3 + 17s^2 + 34s + 120} = -1 \tag{7.6-6}$$

This is the form of Eq. (7.6-3). Unfortunately the denominator is no longer in factored form. Hence before we can plot the root locus versus a, the denominator must be factored. The factored version of the denominator is

$$s^3 + 17s^2 + 34s + 120 \approx (s + 15.3)\left[(s+0.85)^2 + 2.7^2\right]$$

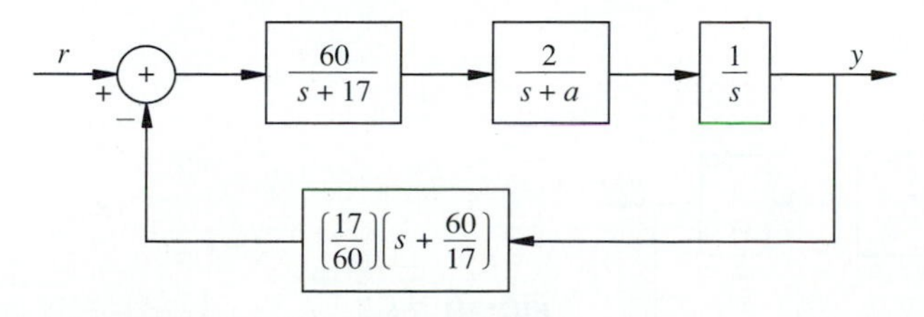

FIGURE 7.6-2
Reduction of the system to a more convenient form.

The root locus to be plotted versus a is based upon the equation

$$a\frac{s(s+17)}{(s+15.3)\left[(s+0.85)^2+2.7^2\right]} = -1 \qquad (7.6\text{-}7)$$

The root locus is shown in Fig. 7.6-3. It is known in advance from Example 7.3-4 and Eq. (7.6-4) that when $a = +2$, the closed-loop poles are at $s = -2 \pm j2$ and $s = -15$. Note that a portion of the locus is of little consequence and hence no effort is made to determine that part of the locus with any degree of accuracy. The dotted portions of the locus are said to be of little consequence because in that region the values of the parameter a are vastly different from the design value. It is seen that, regardless of how large a is allowed to become, the system is always stable. However, of more immediate importance is the location of the closed-loop poles if a is different from the anticipated value by 50 or 100 percent. It is seen from Fig. 7.6-3 that the pole farthest to the left on the s plane is relatively insensitive to the value of a. For values of a less than the design value, the damping ratio is less, and for $a = 3.6$, both roots become real. For very large values of a, the closed-loop response is dominated by the pole that is approaching the zero at the origin in the root locus plane, and the exact locations of the other closed-loop poles are unimportant.

To examine the effects of changes in k_2, it is convenient to redraw the block diagram of Fig. 7.6-1 as Fig. 7.6-4. The purpose is exactly as before, to involve the

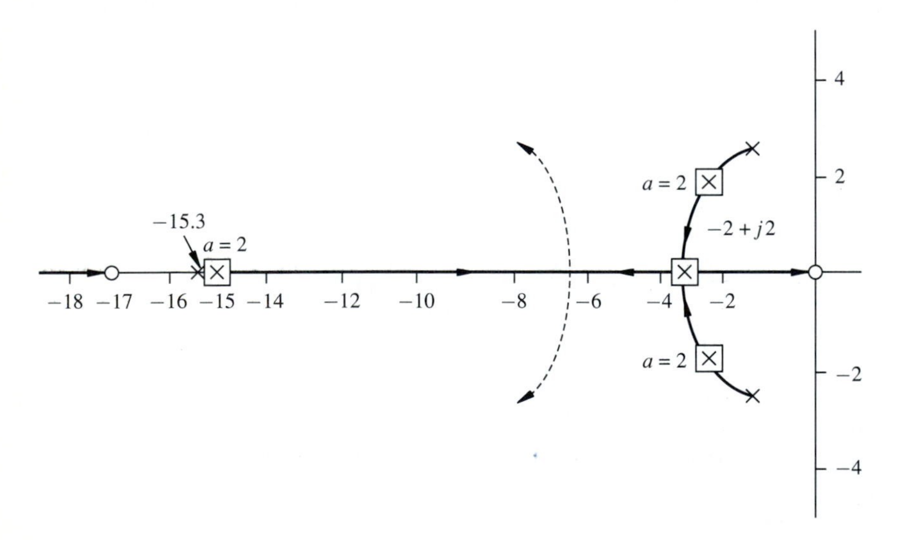

FIGURE 7.6-3
Root locus for Eq. (7.6-11).

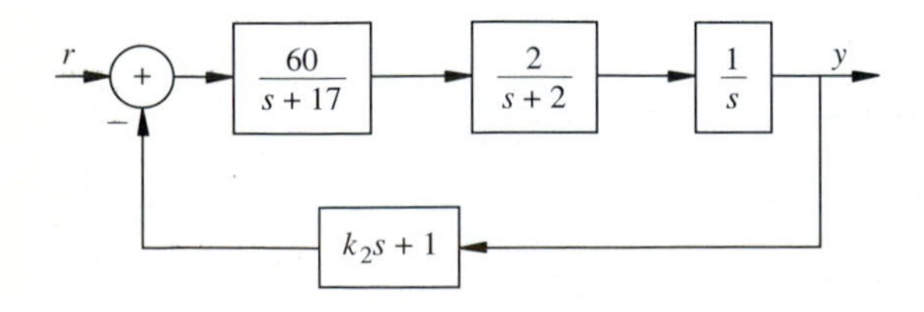

FIGURE 7.6-4
Block diagram of Fig. 7.6-3 redrawn in a more convenient form.

parameter k_2 as few times as possible in the characteristic equation. The characteristic equation for Fig. 7.6-4 is

$$\frac{120(k_2 s + 1)}{s(s+2)(s+17)} = -1 \tag{7.6-8}$$

This may be put into the form of Eq. (7.6-3), much as was done above. The resulting equation is

$$k_2 \frac{120s}{s^3 + 19s^2 + 34s + 120} = -1$$

Again it is seen that the denominator must be factored. The final form of the characteristic equation is

$$k_2 \frac{120s}{(s+17.45)\left[(s+0.775)^2 + 2.54^2\right]} = -1 \tag{7.6-9}$$

The root locus is sketched in Fig. 7.6-5. Perhaps the most important feature of system performance with regard to the feedback coefficient k_2 is the stability of the system when k_2 is zero.

What happens to the system if the feedback path involving k_2 is opened? In that case, the system is still stable, but the damping ratio decreases to $\zeta = \sin 17°$, or 0.29. If y represents the output position of a positioning system, then, from Fig. 7.6-1, it can be seen that k_2 scales the feedback of the velocity of the positioning mechanism. The parameter k_2 is called the *velocity feedback constant*. The effects of increasing the velocity feedback are to increase damping and produce a more sluggish system response. As the velocity feedback is increased by increasing k_2, the system once again becomes dominated by the closed-loop pole near the origin, and the remainder of the locus is unimportant.

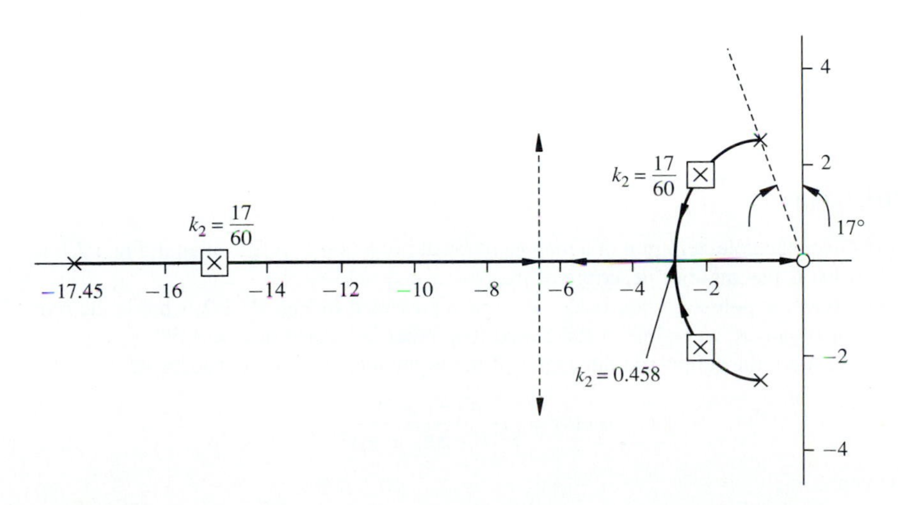

FIGURE 7.6-5
Root locus corresponding to Eq. (7.6-9).

Exercises 7.6

7.6-1. Poorly damped oscillatory modes in a plant cause closed-loop stability problems. Let $G_p(s)$ consist of a second-order system with a high frequency oscillatory mode.

$$G_p(s) = \frac{100}{(s+1)(s^2 + 20\zeta s + 100)}$$

Let $G_c(s) = 1.04$. Find the root locus versus ζ. Above what level of damping is the closed-loop system stable.

Useful fact: $s^3 + s^2 + 100s + 204 = (s+2)(s^2 - s + 102)$

7.6-2. The root locus can often be used to determine the sensitivity to some poorly known parameter in the plant model. Let

$$G_p(s) = \frac{-s + \dfrac{2}{T}}{s(s+1)\left(s + \dfrac{2}{T}\right)}$$

Such a plant model may arise from a process control system with a delay of T sec in the system. Suppose a control $G_c(s) = 1$ is used in the G configuration. Find the root locus as T goes from zero to infinity ($2/T$ goes from infinity to zero).

7.7 CONCLUSIONS

The root locus method of this chapter serves as an ideal end to our discussion of analysis methods. Not only is the root locus approach important as an analysis tool in its own right but, in addition, it may be used to supplement many of the procedures discussed in the three preceding chapters on analysis. Time or frequency responses may be determined through the use of the closed-loop response plane. The stability of the closed-loop system may be investigated by examining the possibility of the locus crossing the $j\omega$ axis. In addition, the graphical information dealing with the movement of the poles for parameter variations is of great assistance in sensing the general behavior of the closed-loop system.

PROBLEMS

7.1. (*a*) Given the pole-zero plot of a loop gain transfer function as represented in Fig. P7.1-1, what is the angle of departure of the pole at $s = -1 - j$?

(*b*) Given the pole-zero plot of $G_{L/K}(s)$ as represented in Fig. P7.1-2, what is the root locus gain K when one of the closed-loop poles is located at $s = -10$?

(*c*) What are the origin and the angles of the asymptotes for the root locus of

$$G_L(s) = \frac{K(-s+4)}{s^2(s+5)(-s+3)}$$

7.2. Given:

$$G_{L/K}(s) = \frac{(s+10)^2}{100(s+3)(s+2)^2}$$

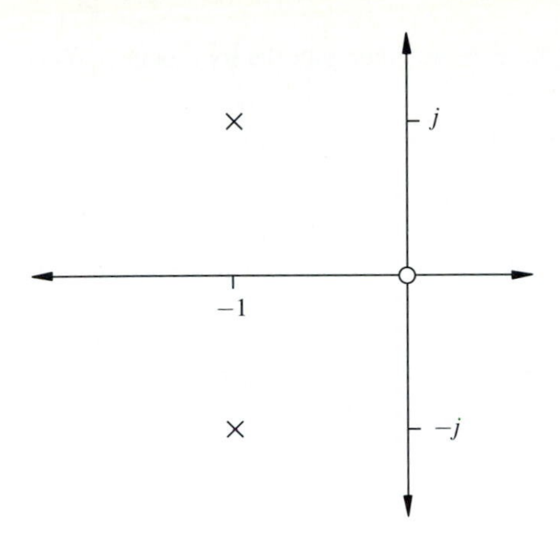

FIGURE P7.1-1
Problem 7.1(a).

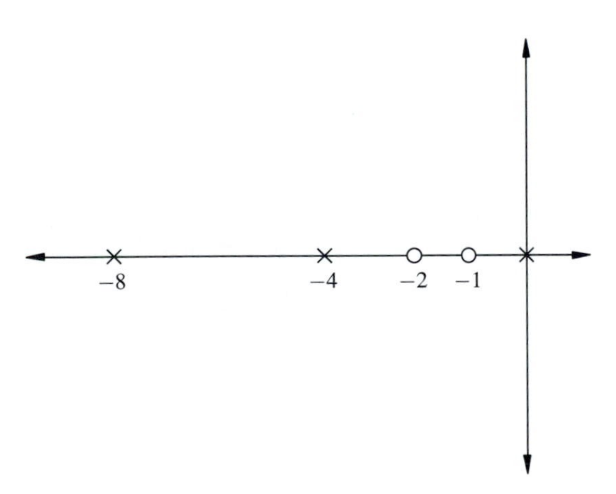

FIGURE P7.1-2
Problem 7.1(b).

Sketch the root locus of the closed-loop poles. Label on the sketch where applicable the values of the origin of the asymptotes, the asymptotes' angles of arrival and departure, and the $j\omega$ axis intersections. For what values of K is the closed-loop system stable?

7.3. Repeat Problem 7.2 for the two systems

$$G_{A:L/K}(s) = \frac{s+2}{s(s+1)(s+5)^2}$$

$$G_{B:L/K}(s) = \frac{(s+2)(s+3)}{s(s+1)(s+5)^2}$$

What benefit, if any, is gained by the extra zero in $G_{B:L/K}(s)$?

7.4. Plot the root locus diagram for the closed-loop poles of a system in the H configuration where

$$G_p(s) = \frac{1}{s(s+2)(s+4)} \qquad \text{and} \qquad H(s) = 1$$

Include the usual information on the plot. On the same graph plot the root locus if $H(s)$ is changed to

$$H(s) = \frac{(s+2)}{(s+8)}$$

A controller of the form $\frac{s+a}{s+b}$ with $a < b$ is called a *lead compensator*. From the root locus diagrams what advantage of lead compensation over simple constant feedback can you observe?

7.5. A lead compensator, introduced in Problem 7.4, can be used to stabilize unstable systems. We investigate this problem using the root locus. All root locus rules remain the same, independent of the side of the imaginary axis on which the poles and zeros lie. A plant is given by

$$G_P(s) = \frac{1}{s(s-1)}$$

The lead compensator used in series to stabilize the closed-loop system is defined by the equation

$$G_C(s) = \frac{0.1K(s+1)}{\left(\frac{s}{10}+1\right)}$$

Sketch the root locus of the closed-loop poles. Supply the usual information about the root locus. Find the gain K for which there is a closed-loop pole at $s = -2$. (If there is never a pole at $s = -2$, state that fact.) Give the closed-loop pole s for the system using the K that was just found.

7.6. Excess poles, which are often left out as unmodeled dynamics, can limit the loop gain of a control system.

(*a*) Let a plant of a positioning system be represented by an ideal plant model

$$G_{p1}(s) = \frac{1}{s(s+5)}$$

Assume that a constant controller $G_c(s) = K$ is used. Such a controller is called a proportional compensator. Sketch the root locus of the closed-loop poles and indicate the range of K for which the system is stable.

(*b*) Now assume that a flexible coupling in the positioning system is modeled more closely so that new plant model becomes

$$G_{p2}(s) = \frac{100}{s(s+5)(s^2+2s+100)}$$

Sketch this root locus and find the range of K for which the system is stable.

7.7. In determining how to use root locus ideas to design controllers, the following rule of thumb can be used. *Zeros pull roots towards themselves; poles push roots away from themselves.*

To see this behavior, sketch the following root loci.

(*a*) $G_{L/K}(s) = \dfrac{10}{s\left((s+8)^2+(2)^2\right)}$

(*b*) $G_{L/K}(s) = \dfrac{s+10}{s\left((s+8)^2+(2)^2\right)}$

(c) $G_{L/K}(s) = \dfrac{100}{(s+10)s\left((s+8)^2+(2)^2\right)}$

Notice how even the pole near the origin is affected by the additional pole or zero. Explain how the results confirm or deny the rule of thumb.

7.8. You wish to position a cart on a length of track. The cart is to move to one position on the track to be filled and other position to be emptied. A model for the motion of the cart is

$$F = m\ddot{x}$$

where x is the cart's position. For this problem the force will be considered as

$$F = u - \mu\dot{x}$$

where u is the force on the cart that you can control via a motor and $\mu\dot{x}$ is a counterforce arising from aerodynamic drag and friction in moving parts. The complete set of equations becomes

$$u - \mu\dot{x} = m\ddot{x}$$

$$m\ddot{x} + \mu\dot{x} = u$$

$$\left(ms^2 + \mu s\right) X(s) = U(s)$$

$$X(s) = \frac{U(s)}{ms^2 + \mu s}$$

Nominal values for m and μ are measured to be $m = 1$, $\mu = 1$.

We wish to explore the possibilities of using a cascade lead compensator (System I) or a velocity feedback compensator (System II) shown in Fig. P7.8.

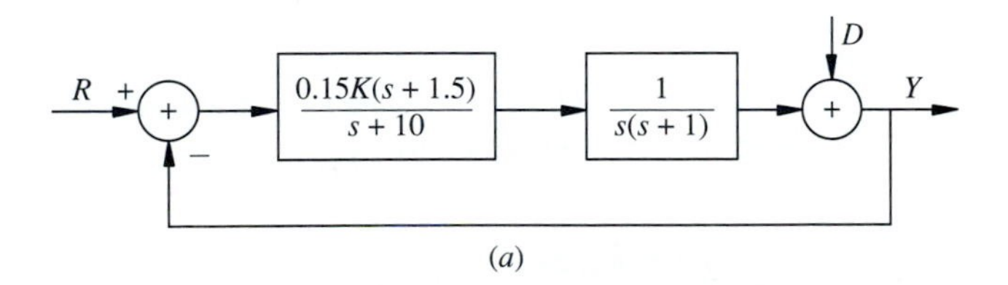

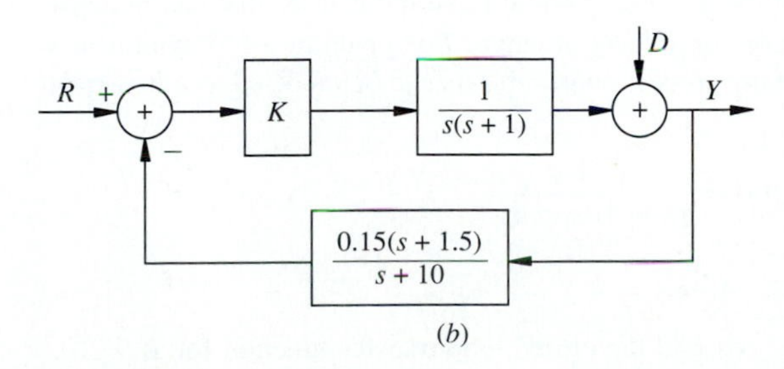

FIGURE P7.8
Problem 7.8. (a) System I; (b) System II.

(*a*) The root locus of the closed-loop system is the same for each system. Sketch the root locus indicating the asymptotes' positions and angles, and the breakaway and reentry points on the real axis.

The data in Table P7.8 should be helpful. Let

$$|W(s)| = \frac{|s||s+1||s+10|}{|s+1.5|}$$

TABLE P7.8

Problem 7.8

s	0	−0.2	−0.4	−0.6	−0.8	−1.0	−1.2	−1.4	−1.6	−1.8	2.0	−2.2	−2.4
$\|W(s)\|$	0	1.2	2.1	2.5	2.1	0	7.0	48.2	80.6	39.4	32.0	29.4	28.4
s	−2.6	−2.8	−3.0	−3.2	−3.4	−3.6	−3.8	−4.0	−4.2	−4.4	−4.6	−4.8	−5.0
$\|W(s)\|$	29.0	27.9	28.0	28.2	28.3	28.5	28.7	28.8	28.9	28.9	28.8	28.7	28.6

(*b*) A good-closed loop response for System I places the closed-loop poles at about $s = -4.65 + j4.65$, $s = -4.65 - j4.65$, $s = -1.7$. However, not nearly as fast a response can be attained for System II. Explain why this is true. A reasonable design for System II places the closed-loop poles at $s = -2.6$, $s = -3.0$, $s = -5.4$.

(*c*) Find the value of K that yields the poles suggested in (*b*) for each system. (Use graphical techniques to the accuracy those techniques imply.)

(*d*) Using the appropriate value of K for each system from (*c*), describe the unit step response of each system.

(*e*) Briefly explain the relative merits of each system design (with its appropriate K) in rejecting the disturbance D.

(*f*) The mass m of the cart will be somewhat different when the cart is full compared to when it is empty. Considering System I, use the value of K found in (*c*) and plot the root locus as a function of m.

(*g*) If the car is filled with liquid, the sloshing of the liquid will affect the dynamics of the cart. This effect might be modeled as additional dynamics cascaded with the original plant. If the "sloshing" dynamics are represented by

$$\frac{18}{(s+3+j3)(s+3-j3)}$$

how will this affect the root locus of our System I and the gain K that can be used?

7.9. The first order Padé approximation for a delay of length T is given by $\frac{-s+a}{s+a}$ with $a = \frac{2}{T}$ (This is derived in Sec. 8.10). Many process control plants can be modeled as a first-order system with a delay. Let

$$G_p(s) = \frac{-s+a}{(s+3)(s+a)}$$

and

$$G_c(s) = K$$

(*a*) When $a = 0$ (the delay $T = \infty$) find the closed-loop transfer function for $K = 10$.

(*b*) Plot the root locus with K as the variable parameter when $a = 10$. (Don't forget to place your system in pole-zero form before starting the root locus.)

(*c*) Now with $K = 10$, plot the root locus with a as a parameter. What happens as a decreases (the delay T increases)? Remember to interpret the system in the closed-loop response plane before answering.

7.10. (*CAD Problem*) It is often difficult to associate the proper poles with the proper branches on the root locus. However, it is rarely important to do so.

(*a*) Consider the plot of the root locus from Example 7.4-5 and Fig. 7.4-9. In that example

$$G_{L1}(s) = \frac{-2K(s-2)}{(s+4)(s+1)\left((s+2)^2+4\right)}$$

Now consider

$$G_{L2}(s) = \frac{-3K(s-2)}{(s+4)(s+1)\left((s+2)^2+8\right)}$$

Use a CAD program to plot each of the two root loci and notice the difference in shape. However, notice also that for most values of K the closed-loop poles of the two loop gains are very close.

(*b*) Repeat (*a*) using the following transfer function, which is similar to that used in Example 7.4-4.

$$G_{Li}(s) = \frac{0.04166aK\left[(s+3.25)^2+(1.21)^2\right]}{s(s+1)(s+a)}$$

For $G_{L1}(s)$, let $a = 2.1$. For $G_{L2}(s)$, let $a = 1.9$.

7.A APPENDIX TO CHAPTER 7—BYPASSING THE ROOT LOCUS

With the advent of readily accessible CAD programs for control systems design the utility of learning two of the classical topics in control theory has become a matter of debate. It is now questioned whether the Routh-Hurwitz stability criterion and the rules for sketching root locus plots need to be learned in depth. The root locus rules often involve a fair amount of calculation themselves and do not necessarily produce accurate sketches simply. It is felt by some that what insight is available in the root locus can be achieved learning a few of the root locus rules and examining a number of computer-generated root locus plots. A major reason for learning the Routh-Hurwitz stability criterion is to help in determining where the roots on the root locus cross the imaginary axis.

Others feel that both root locus and the Routh-Hurwitz methods are integral parts of the syllabus of classical control theory. They feel that these topics are part of the language of the control practitioner and that new control engineers must understand this language.

We choose not to enter into the debate but, in this appendix to the chapter on root locus techniques, we provide a prescription to address the root locus in much less depth than is available in the chapter body. We like the insight provided by root locus techniques and refer to root locus plots in the following chapter on design. We note, however, that the point of view stressed in this book is that the key properties of control design revolve around the sensitivity behavior of the loop gain rather than the system's nominal transient response. Because of this key, design methods involving the loop gain's Bode plots are given top priority while root locus ideas are used only for supplemental understanding.

What follows is a prescription for an abbreviated treatment of root locus diagrams. This approach bypasses the details of sketching root locus diagrams but provides enough information about the root locus plots to allow understanding of the material in the following chapter on design.

The reader should first read Sec 7.1 and the beginning of Sec. 7.2 to the place where the rules for drawing the root locus begin to be enumerated. This provides an introduction to the concept of the root locus diagram.

Next, we provide Table 7.A-1, an abbreviated table of root locus construction rules. These rules are stated here without explanation but the rules are numbered as they are in the text where their origin is explained.

Consider how the rules indicate the form of root loci in the following examples.

Example 7.A-1. Let the loop gain transfer function be given by

$$G_L(s) = \frac{2K}{s(s+1)(s+19)}$$

TABLE 7.A-1
An abbreviated summary of root-locus construction rules

Rule 1: Number of branches	There is one branch for each pole of the loop gain transfer function $G_L(s)$.
Rule 2: Starting points ($K = 0$)	The branches of the root locus start at the poles of $G_L(s)$.
Rule 3: End points ($K = \infty$)	The branches of the root locus end at the finite zeros of $G_L(s)$ or go to infinity.
Rule 4: Behavior along the real axis	The root locus exists on the real axis at every point for which the sum of the number of poles and zeros lying to the right of the point is odd.
Rule 6: Symmetry of locus	The root locus is always symmetric with respect to the real axis.
Rule 9: Asymptotic behavior for large K	The asymptote angles are given by $\phi_{asy} = \frac{180^\circ + N360^\circ}{p-z} \quad N = 0, 1, 2, \ldots, p-z-1$ and the origin of the asymptotes is $OA = \frac{\sum(\text{pole locations}) - \sum(\text{zero locations})}{p-z}$ where $p - z$ is the pole-zero excess.

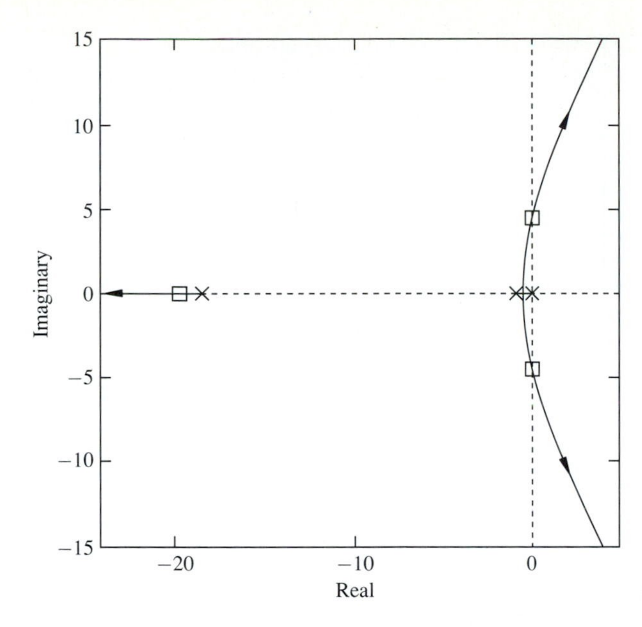

FIGURE 7.A-1
Example 7.A-1.

The root locus plot is shown in Fig. 7.A-1. The poles corresponding to $K = 250$ are marked by boxes. The key rule for this plot is the rule that the asymptotes make angles of $60°, -60°$ and $180°$ with the positive real axis. The origin of the asymptotes is $s = -20/3 = -6.67$. Notice that, where the pole-zero excess is three or greater, the asymptotes reach into the right half-plane and for large enough gain the system is unstable.

This root locus shows a very typical situation as the closed-loop poles are related to the closed-loop step response. As the gain is initially increased the dominant pole moves out along the negative real axis and the system's step response becomes faster. As the gain becomes larger a complex conjugate dominant pole pair emerges and an oscillatory step response results. As the gain is further increased the damping on the complex pole pair decreases until finally the system becomes unstable. The reader should relate this behavior back to what happens on the Nyquist diagram for this loop gain and the techniques of Sec. 6.5 relating the open-loop Nyquist plot with the closed-loop frequency response and the closed-loop step response. (See Example 6.5-2.)

Example 7.A-2. Consider adding a left half-plane zero to the transfer function of Example 7.A-1 while leaving the dc gain unchanged. Consider the transfer function

$$G_L(s) = \frac{K(s+2)}{s(s+1)(s+19)}$$

The root locus for this loop gain appears in Fig. 7.A-2. The closed-loop poles corresponding to the gain $K = 250$ are marked with boxes. An inset in the upper left corner provides an expanded view of the box near the origin.

This example demonstrates a root locus concept that loop gain zeros pull closed-loop poles towards them. One pole is pulled towards the zero in direct manner. The pattern of two poles coming together, looping around a zero, then separating is very

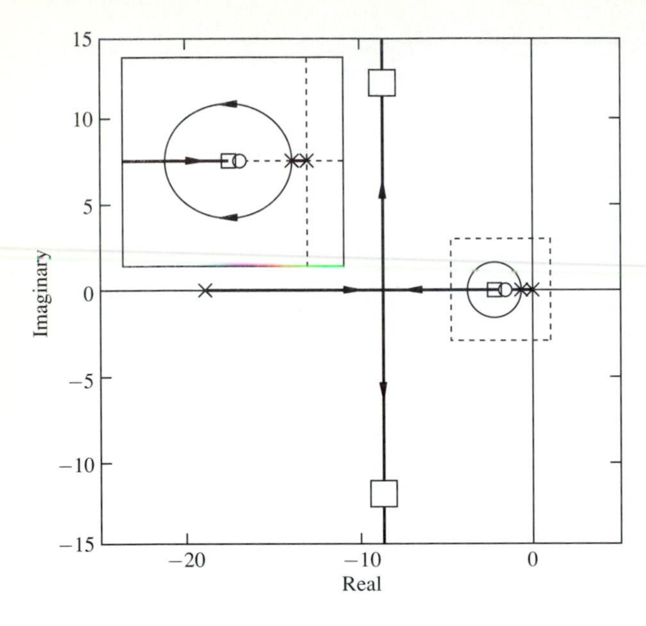

FIGURE 7.A-2
Example 7.A-2.

common. The pole-zero excess is now only two, which means the asymptotes make angles of 90° and −90° and do not cross into the right half-plane. The closed-loop system is stable for all positive values of K.

Example 7.A-3. The transfer function of Example 7.A-1 is a typical transfer function for a plant. The results of placing a zero in a compensator to achieve the transfer function of Example 7.A-2 appear promising. However, if the controller is to be realizable a pole must also be added. A controller with a zero inside of a pole on the negative real axis is called a lead compensator and is a standard building block of design considered in Chap. 8. Consider the loop gain

$$G_L(s) = \frac{K(s+2)}{s(s+1)(s+19)(s/20+1)}$$

The root locus for this transfer function appears in Fig. 7.A-3. Again, the poles for $K = 250$ are marked with boxes and again an inset shows the behavior near the origin.

This root locus shows that poles far away from the origin have little effect on the part of the root locus near the origin. The near-in root locus is similar to the near-in root locus of Example 7.A-2. The poles begin their loop around the zero. For larger values of K the root locus is different as the extra pole has more of an effect.

The root locus of Example 7.A-3 should be compared to the root locus of Example 7.A-1. While the pole-zero excess is three in both examples, the lead compensator has pulled the poles further into the left half-plane at each value of K. The closed-loop system is now stable for $K = 250$. At lower values of K the poles are much better placed.

These examples of typical root loci should give the reader who chooses not to cover Chap. 7 completely some insight into root locus behavior. The reader should also quickly review in Table 7.4-2 how Rules 4 and 9 change when the gain K is negative. Also, if the reader has a CAD package available, it might be interesting to read the CAD

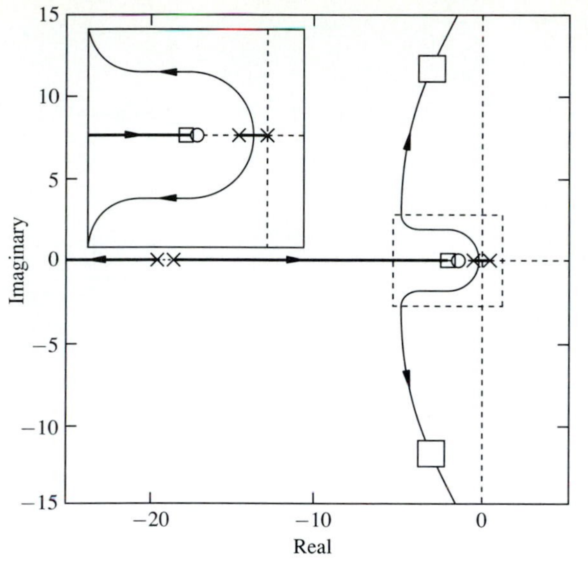

FIGURE 7.A-3
Example 7.A-3.

Notes at the end of Sec. 7.3 and try some problems. We suggest Problems 7.4, 5, 6, and 7 where the reader can use the CAD package to plot the root loci rather than sketching them.

CHAPTER 8

THE DESIGN OF CONTROL SYSTEMS

8.1 INTRODUCTION

In this chapter techniques that can be used to design feedback control systems are examined. In the previous chapters techniques were developed to analyze the behavior of control systems. It is now time to apply the insight gained by the analysis of these systems to the design of the systems. A major tool for design is a combination of the Nyquist and Bode plots.

Let's examine what we have learned that may be useful in design. In Chap. 3 it was established that many of the desirable properties of a closed-loop controller come from creating a large return difference transfer function. More specifically, a large loop gain produces good disturbance rejection and good sensitivity reduction. The Bode plots of the loop gain that were developed in Chap. 5 graphically display the size of the loop gain as a function of frequency. In Chap. 6 it was shown how to establish stability of the closed-loop system from the Nyquist plot of the loop gain transfer function.

The transient response of the closed-loop system to inputs and disturbances can be estimated by combining the results of Chaps. 4, 5, and 6. The closed-loop frequency response is attainable from the M-circles of the Nyquist plot of the loop gain as shown in Sec. 6.5. The closed-loop transfer function can be estimated from the frequency plot using the identification techniques of Sec. 5.5. Finally the transient response is estimated from the transfer function using the results of Chap. 4. This chain of analysis is not actually performed in detail for each design; rather, insights from general analyses create guidelines for design. The root locus techniques of Chap. 7 may be used to gain a different view of how transient response can be altered. Robustness results from Chap. 6 indicate that the loop gain must be allowed to become small at frequencies where modeling uncertainty becomes large.

These ideas are put together in Sec. 8.2 where an overall strategy for shaping the loop gain as a function of frequency is developed. In that section it is also shown how the loop gain is manipulated by adding elements to a series compensator. In Sec. 8.3 the process of introducing the commonly used building blocks of controller design is begun. The design of proportional compensators is examined in Sec. 8.3. The design of proportional-integral (PI) or, equivalently, lag compensators is examined in Sec. 8.4. The design of proportional-integral-derivative (PID) or, equivalently, lead-lag compensators is examined in Sec. 8.5. Design by canceling plant dynamics with notch filters is examined in Sec. 8.6. A controller design is demonstrated by examples of realistic controller designs of stable minimum phase plants in Sec. 8.7 and 8.8. In the design of Sec. 8.8 the question of active versus passive control of oscillatory dynamics is addressed. Techniques for stabilizing unstable open-loop plants are developed in Sec. 8.9. In Sec. 8.10 it is shown that plants containing right half-plane zeros contain natural limitations to achievable controller performance. In Sec. 8.11 it is shown that the mechanized approach of a pole placement technique can show how to attack difficult control problems but that well-placed closed-loop poles do not necessarily guarantee acceptable performance. Finally, in Sec. 8.12 the potential benefits of extra measurements in overcoming the limitations imposed by right half-plane zeros are displayed. First a state-variable approach is shown and then a technique for dealing with missing measurements is displayed. Detailed analysis of the state-variable design would lead into the analysis of multi-input, multi-output systems. This is left to another book and another course.[1]

8.1.1 CAD Notes

The CAD tools that are useful in design have been introduced during the previous chapters of the book. The student should use these tools wherever they help in understanding the design process. However, it is our suggestion that the student think through as much of the problem as possible and use the CAD system as a computational aid to eliminate dreary calculations. Almost always, the time spent to solve a problem can be minimized by thinking through the problem and gaining insight rather than making many disoriented computer iterations.

---

[1] The book by Doyle, Francis, and Tennenbaum is recommended as continuing the approach of this book. The book by Maciejowski is encyclopedic in delineating the varied approaches to the multi-input, multi-output systems. See the references listed at the end of this book.

8.2 GENERAL PRINCIPLES FOR DESIGNING SERIES COMPENSATORS USING FREQUENCY RESPONSE TECHNIQUES

In this chapter, we consider series compensators of the form given in Fig. 8.2-1 where $G_p(s)$ is a model of the plant and $G_C(s)$ is the compensator transfer function that is to be designed. The series compensator is easily translated into the now familiar G configuration by observing that

$$G(s) = G_C(s)G_p(s) \tag{8.2-1}$$

We have already spent a good deal of effort learning how to analyze control systems in the G configuration. In particular we have learned that, for these systems, much information is available concerning performance, stability, and robustness of the closed-loop system from the Bode and Nyquist plots of the loop gain $G(s)$. The series compensator is particularly convenient because we can start with the Bode plot of the plant model $G_P(s)$ and modify that plot as we add new pole and zero factors to $G(s)$ by including them in the compensator. As we learned in Chap. 5, new factors in $G(s)$ affect the Bode diagram of $G(s)$ by simply summing the values on the Bode diagram of the original $G(s)$ with the values on the Bode diagram of the new factors. In this way, we can build the Bode diagram of $G(s)$ into a desired shape. This process is sometimes referred to as the *loop shaping* method of design. We now demonstrate this process with an example. This example shows the thought process that goes into a controller design. You will learn the elements of this process in this chapter. For this example simply watch how a controller design evolves and notice the utility of the Bode plot and the ease with which this plot is modified by modifying the compensator.

Example 8.2-1. A plant model is given by

$$G_p(s) = \frac{1}{(s+1)(s+0.1)}$$

To obtain higher gain at low frequencies and a Type 1 system for total dc disturbance rejection and zero steady state tracking error, a design that includes an integrator in the loop is desirable. However, it is also desirable to negate the 90° phase lag of the integrator at higher frequencies. The designer sets the series compensator to be

$$G_{C1}(s) = \frac{s+0.1}{s}$$

It is seen in Sec. 8.4 that such a compensator is a standard building block called a *lag compensator* or a *PI compensator*. Here it is not important where $G_{C1}(s)$ comes from; it is only important to see how the addition of transfer functions in series affects the Bode plot of the evolving loop gain.

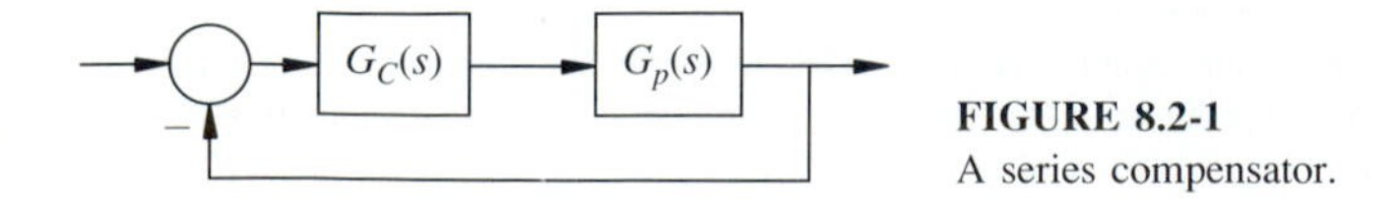

FIGURE 8.2-1
A series compensator.

The loop gain $G_1(s)$ is the product of the plant and the compensator.

$$G_1(s) = \frac{1}{s(s+1)}$$

Fig. 8.2-2 shows the Bode magnitude and phase plots of $G_P(s)$, $G_{C1}(s)$ and $G_1(s)$. Notice how the plots associated with $G_P(s)$, and $G_{C1}(s)$ simply add. (The logarithms or decibels in the magnitude plot add—not the magnitudes themselves).

If the designer decides to add to the overall gain of the loop gain transfer function, it can be done by multiplying the compensator transfer function by a constant, say 10.

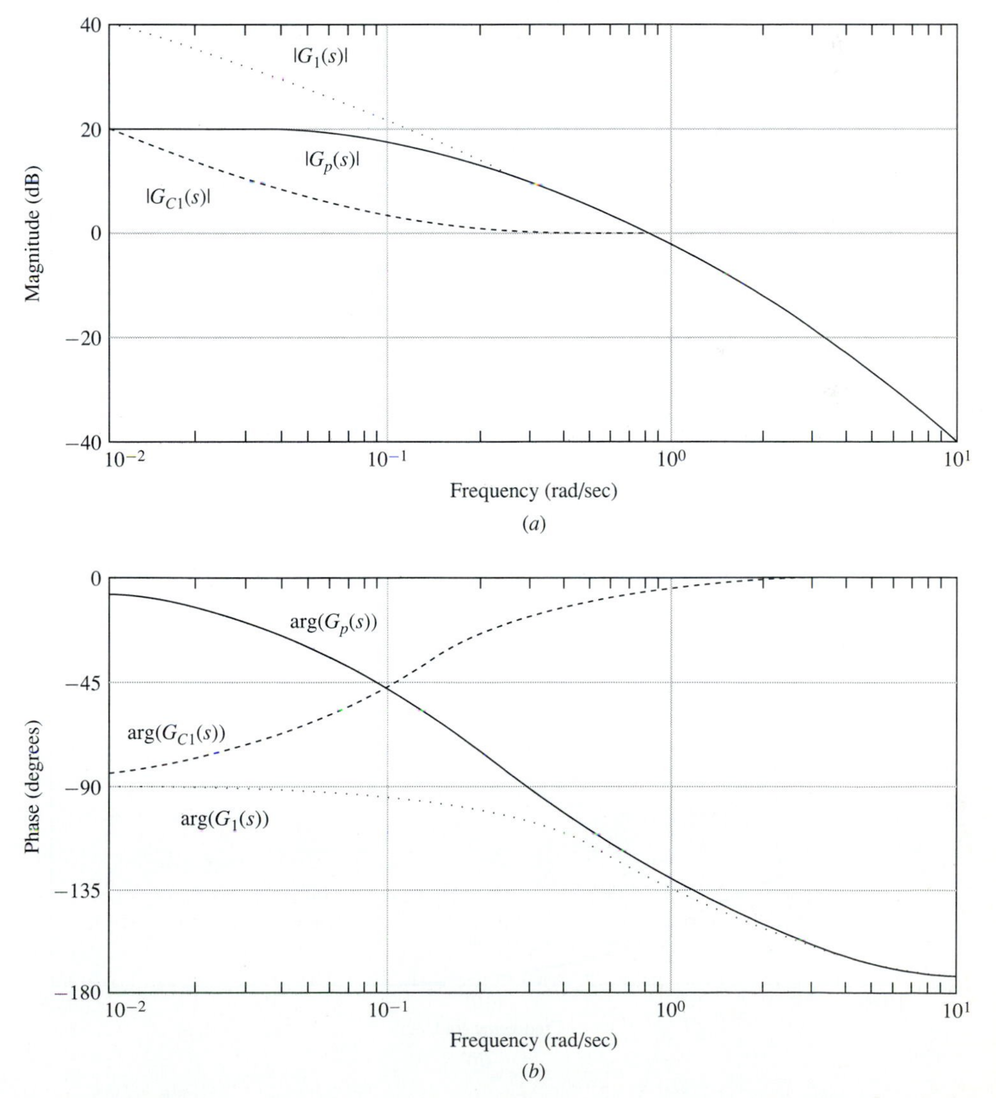

FIGURE 8.2-2
Bode plot for Example 8.2-1. (*a*) Magnitude; (*b*) phase.

The series compensator $G_{C2}(s)$ is given by

$$G_{C2}(s) = 10G_{C1}(s) = \frac{10(s+0.1)}{s}$$

The loop gain becomes

$$G_2(s) = G_{C2}(s)G_p(s) = 10G_1(s)$$

The Bode magnitude plot is shifted upward by a factor of 10, or equivalently, 20 log 10 = 20 dB, as shown in Fig. 8.2-3. The phase plot is unaltered. While the resulting design

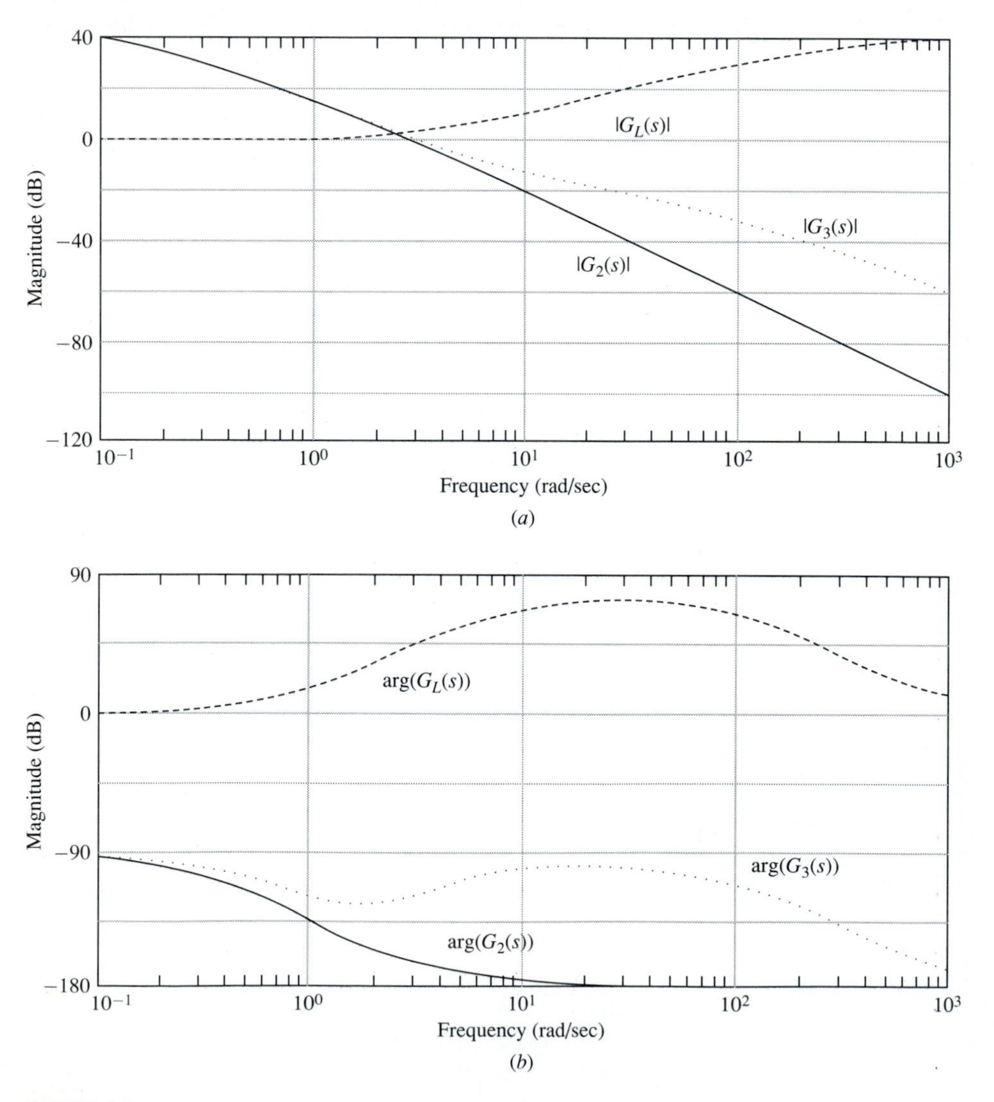

FIGURE 8.2-3
Bode plots for the revised design in Example 8.2-1. (*a*) Magnitude; (*b*) phase.

has adequate loop gain at low frequency, a look at the Nyquist plot of this loop gain indicates serious problems. If $G_{C2}(s)$ were used as a controller the resulting system would be highly oscillatory and would have so little robustness that unless the plant model were extremely precise the closed-loop system would likely be unstable. (Draw the Nyquist plot and verify these statements.)

The designer remedies this situation as you will learn to do in this chapter. Another pole and zero are added to the compensator in what we will come to call a lead compensator with transfer function

$$G_L(s) = \frac{100(s+3)}{s+300}$$

The new compensator $G_{C3}(s)$ becomes

$$G_{C3}(s) = G_L(s)G_{C2}(s) = \frac{1000(s+3)}{s(s+300)}$$

The new loop gain $G_3(s)$ becomes

$$G_3(s) = G_L(s)G_2(s)$$

Figure 8.2-3 displays the Bode magnitude and phase plots of $G_2(s)$, $G_L(s)$ and $G_3(s)$. Notice how the plots of $G_2(s)$ are changed into the plots of $G_3(s)$ by adding the plots of $G_2(s)$ and $G_L(s)$ point by point.

The Nyquist plot of $G_2(s)$ and $G_3(s)$ are given in Fig. 8.2-4. Notice how the compensator adds positive phase margin (phase lead) to the Nyquist plot around the critical point $s = -1$ to avoid approaching the critical point too closely. An analysis of the resulting loop gain $G_3(s)$ shows an adequate controller design. The designer has created an adequate design by modifying the loop gain one step at a time, adding new features with each new transfer function placed in series with the compensator of the previous step.

In previous chapters we learned how to analyze control systems. In Sec. 6.7 we saw that it is possible to define a function $p(j\omega)$ such that many aspects of good

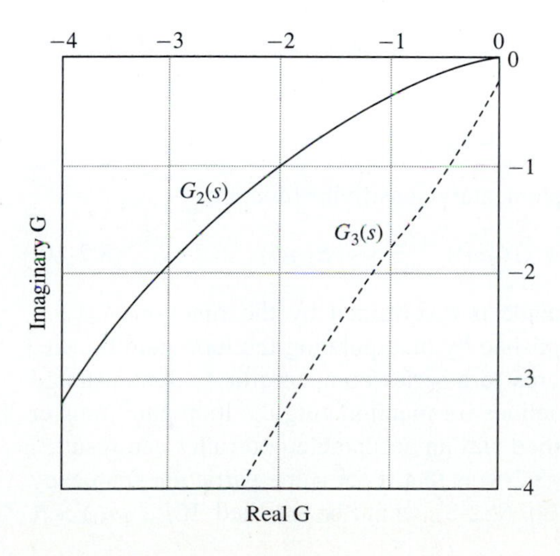

FIGURE 8.2-4
Nyquist plots for Example 8.2-1.

performance will be satisfied if Eq. (6.7-1) is satisfied. Equation (6.7-1) is repeated here as Eq. (8.2-2).

$$\left|1 + \tilde{G}(j\omega)\right| > p(j\omega) \tag{8.2-2}$$

The set of transfer functions represented by $\tilde{G}(j\omega)$ is the set of loop gains in the G configuration when there are perturbations present. Let a set of plant models be characterized by Eqs. (6.6-5) and (6.6-6), repeated here as Eqs. (8.2-3) and (8.2-4).

$$\tilde{G}_p(s) = G_p(s)(1 + L_m(s)) \tag{8.2-3}$$

$$|L_m(j\omega)| < l_m(j\omega) \tag{8.2-4}$$

Using a series compensator the set of loop gains is given by

$$\tilde{G}(s) = G_C(s)\tilde{G}_p(s) \tag{8.2-5}$$

The nominal loop gain is given by

$$G(s) = G_C(s)G_p(s) \tag{8.2-6}$$

and using Eqs. (8.2-3) through (8.2-6), we find that

$$\tilde{G}(s) = G(s)(1 + L_m(s)) \tag{8.2-7}$$

with Eq. (8.2-4) still satisfied. The controller is robustly stable and has the performance qualities specified by Eq. (8.2-2) if Eq. (6.7-18) is satisfied for all ω. Equation (6.7-18b) is repeated here as Eq. (8.2-8).

$$\left|\frac{1}{1 + G(j\omega)}\right| p(j\omega) + \left|\frac{G(j\omega)}{1 + G(j\omega)}\right| l_m(j\omega) < 1 \tag{8.2-8}$$

As discussed in Section 6.7, satisfying Eq. (8.2-8) requires a tradeoff between the magnitude of the nominal sensitivity function

$$S(j\omega) = (1 + G(j\omega))^{-1} \tag{8.2-9}$$

and the magnitude of the nominal complementary sensitivity function

$$T(j\omega) = G(j\omega)(1 + G(j\omega))^{-1} = 1 - S(j\omega) \tag{8.2-10}$$

The manner in which the tradeoff is made is determined by the functions $l_m(j\omega)$ and $p(j\omega)$. This tradeoff can be accomplished by manipulating the loop gain transfer function $G(j\omega)$ by adding poles and zeros to a series compensator as demonstrated in Example 8.2-1. We now develop guidelines for manipulating the loop gain transfer function so that Eq. (8.2-8) can be satisfied and an acceptable controller can result.

The first thing to notice from Eq. (8.2-8) is that if for some particular frequency $\omega_0, l_m(j\omega_0) > 1$ and $p(j\omega_0) > 1$, then Eq. (8.2-8) cannot be satisfied. If $l_m(j\omega_0) > 1$

and $p(j\omega_0) > 1$, then

$$\left|\frac{G(j\omega_0)}{1+G(j\omega_0)}\right| l_m(j\omega_0) + \left|\frac{1}{1+G(j\omega_0)}\right| p(j\omega_0)$$
$$> \left|\frac{G(j\omega_0)}{1+G(j\omega_0)}\right| + \left|\frac{1}{1+G(j\omega_0)}\right|$$
$$\geq \left|\frac{G(j\omega_0)}{1+G(j\omega_0)} + \frac{1}{1+G(j\omega_0)}\right| = 1$$

This fact can be interpreted as showing that aggressive performance objectives cannot be met for frequencies where the system is not well modeled. In general $l_m(j\omega)$, which represents the modeling error, is small for low frequencies and grows large at high frequencies. To accommodate this, $p(j\omega)$, which represents performance requirements, is made acceptably low at high frequencies where the model is uncertain. A large performance constraint is required and attainable for low frequency inputs where the model is well known. A typical plot of $l_m^{-1}(j\omega)$ and $p(j\omega)$ is shown in Fig. 8.2-5. The frequency axis splits into three regions:

low frequencies, where $p(j\omega)$ is large and $l_m(j\omega)$ is small ($l_m^{-1}(j\omega)$ is large)
high frequencies, where $p(j\omega)$ is small and $l_m(j\omega)$ is large ($l_m^{-1}(j\omega)$ is small)
transition frequencies, where $p(j\omega)$ and $l_m(j\omega)$ are both fairly close to one

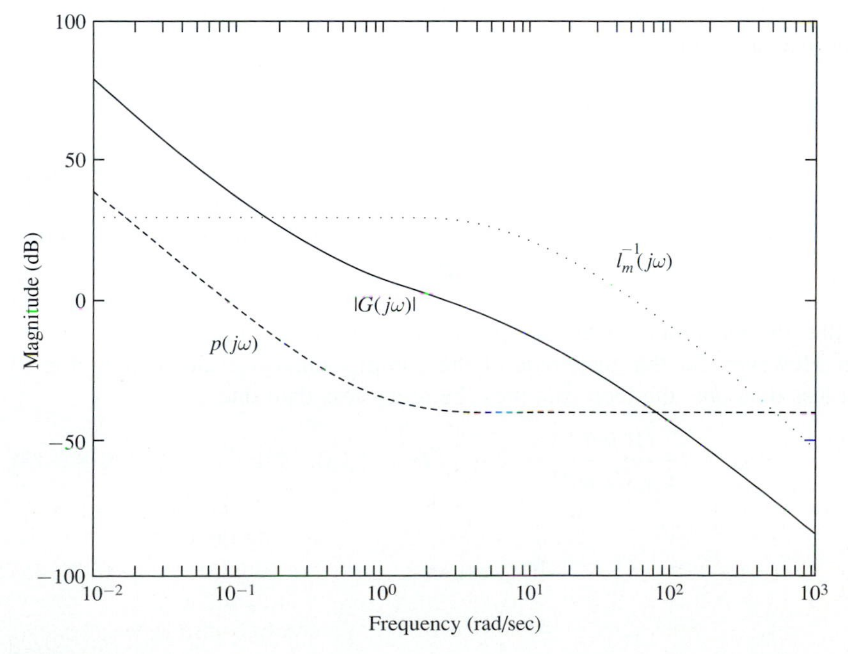

FIGURE 8.2-5
A typical acceptable design.

The requirements on $G(j\omega)$ so that Eq. (8.2-8) is satisfied can now be developed fairly easily for the high and low frequency sections while the requirements on $G(j\omega)$ for the transition frequency section is more difficult to determine.

At low frequencies $p(j\omega)$ is much larger than one. In this case Eq. (8.2-8) dictates that the magnitude of the sensitivity function $S(j\omega)$ must be much less than one. This situation can be accomplished if the magnitude of the loop gain $G(j\omega)$ is made much larger than one. When $|G(j\omega)|$ is much larger than one we can make approximations in Eq. (8.2-8). Specifically, if

$$|G(j\omega)| \gg 1 \tag{8.2-11a}$$

then

$$\left|\frac{1}{1+G(j\omega)}\right| \approx \frac{1}{|G(j\omega)|} \tag{8.2-11b}$$

and

$$\left|\frac{G(j\omega)}{1+G(j\omega)}\right| \approx 1 \tag{8.2-11c}$$

If the approximations are used in Eq. (8.2-8) we obtain the approximate requirement

$$\frac{1}{|G(j\omega)|}p(j\omega) < 1 - l_m(j\omega) \tag{8.2-11d}$$

Thus, the requirements of Eq. (8.2-8) are very close to being satisfied at low frequencies (where $p(j\omega) \gg 1$ and $l_m(j\omega) \ll 1$) if $|G(j\omega)|$ is made large enough, i.e., if the following inequality, obtained from Eq. (8.2-11*d*) is satisfied.

$$|G(j\omega)| > \frac{p(j\omega)}{1 - l_m(j\omega)} \tag{8.2-12}$$

Inequality (8.2-12) shows that at low frequencies, large loop gains are required to meet performance requirements. The nominal loop gain must be made slightly larger than the performance requirement dictates so that the performance is achieved in the face of the small amount of modeling error expected at low frequencies.

At high frequencies, $l_m(j\omega)$ is much larger than one. In this case, Eq. (8.2-8) dictates that the magnitude of the complementary sensitivity function be much less than one. However, for the magnitude of the complementary sensitivity function to be much less than one, the loop gain must be much less than one.

$$\left|\frac{G(j\omega)}{1+G(j\omega)}\right| \ll 1 \text{ implies } |G(j\omega)| \ll 1 \tag{8.2-13}$$

When

$$|G(j\omega)| \ll 1 \tag{8.2-14a}$$

then

$$\left|\frac{1}{1+G(j\omega)}\right| \approx 1 \tag{8.2-14b}$$

and

$$\left|\frac{G(j\omega)}{1+G(j\omega)}\right| \approx |G(j\omega)| \tag{8.2-14c}$$

Using these approximations in Eq. (8.2-8) leads to approximate requirement

$$|G(j\omega)|l_m(j\omega) + p(j\omega) < 1 \tag{8.2-14d}$$

The requirements of Eq. (8.2-8) are very nearly satisfied at high frequencies (where $p(j\omega) \ll 1$ and $l_m(j\omega) \gg 1$) if $|G(j\omega)|$ is made small enough, i.e., if the following inequality obtained from Eq. (8.2-14*d*) is satisfied.

$$|G(j\omega)| < \frac{1 - p(j\omega)}{l_m(j\omega)} \tag{8.2-15}$$

Comparing inequalities (8.2-15) and (6.6-11) shows that the main concern at high frequencies is to keep the loop gain small enough so that its magnitude cannot exceed one and cause instability in the face of large modeling errors and unknown phase characteristics. Thus $|G(j\omega)|$ should be less than $l_m^{-1}(j\omega)$ as required by Eq. (6.6-11) when $|G(j\omega)|$ is small. An additional reduction is dictated by the high frequency performance requirement. If, at some frequency, the perturbed loop gain is allowed to approach the -1 point on the Nyquist plot too closely, the system will demonstrate the unacceptable performance attributes of large oscillations, disturbance amplification, and high sensitivity at that frequency. Thus, we must allow for a safety margin and keep the magnitude of the *perturbed* loop gain below a threshold that is somewhat less than one. The nominal loop gain should be kept somewhat less than $l_m^{-1}(j\omega)$ to provide not only for robust stability but also for robust performance at high frequencies.

The strategy for manipulating the magnitude of the loop gain of a control design is now taking shape. Fig. 8.2-5 introduced previously shows a plot of $p(j\omega)$, $l_m^{-1}(j\omega)$ and $|G(j\omega)|$ for a possibly acceptable controller design. The loop gain is made larger than $p(j\omega)$ at low frequencies and smaller than $l_m^{-1}(j\omega)$ for high frequencies. There is a safety margin above $p(j\omega)$ at low frequencies to account for possible modeling errors and for the fact that approximations are used in attaining Eq. (8.2-12). There is a safety margin below $l_m^{-1}(j\omega)$ at high frequencies to ensure reasonable performance and to account for the approximations used in developing Eq. (8.2-15). Notice that the loop gain need not be above $p(j\omega)$ at *high* frequencies where $p(j\omega)$ is small or below $l_m^{-1}(j\omega)$ at *low* frequencies where $l_m^{-1}(j\omega)$ is small.

What qualities should $G(j\omega)$ have through the transition frequencies? This question is more difficult since the assumptions leading to the approximations Eq. (8.2-11) and Eq. (8.2-14) are not valid. We know that the $|G(j\omega)|$ must change from being large at low frequencies to being small at high frequencies. Somewhere in the transition band $|G(j\omega)|$ must pass through the point ω_c where

$$|G(j\omega_c)| = 1 \tag{8.2-16}$$

This frequency ω_c is defined in Sec. 6.5 as the crossover frequency of the control system. In Sec. 6.5, the phase margin ϕ_m of the control system is defined as the angle between the polar representation of $G(j\omega_c)$ and the negative real axis.

$$\phi_m = 180^\circ + \arg\left(G\left(j\omega_c\right)\right) \tag{8.2-17}$$

Much of what we wish to learn about how we should make $G(j\omega)$ behave in the transition frequencies can be seen by studying the control system around ω_c.

It is desirable to create a loop gain with as large a phase margin as possible. A small phase margin means that the Nyquist plot of the loop gain passes very close to the -1 point, and indicates two problems. The first problem is that the closed-loop system responds to inputs and disturbances with poorly damped oscillations at the same frequency as the crossover frequency if the phase margin is small. This fact is shown in Sec. 6.5 where we find how to determine the closed-loop behavior from the Nyquist plot of the loop gain. The second problem is that if ϕ_m is small, a small modeling error in the plant at the crossover frequency causes the Nyquist plot of the loop gain to pass extremely close to the -1 point or actually encircle the -1 point, causing instability. Thus, there is a nominal performance problem and a robustness problem associated with a small phase margin.

The combination of a nominal performance and robustness problem can be seen by examining Eq. (8.2-8) at ω_c. Since

$$|G(j\omega_c)| = 1 \tag{8.2-18a}$$

then

$$\left|\frac{G\left(j\omega_c\right)}{1+G\left(j\omega_c\right)}\right| = \frac{1}{\left|1+G\left(j\omega_c\right)\right|} \tag{8.2-18b}$$

so that Eq. (8.2-8) becomes

$$\left|\frac{1}{1+G\left(j\omega_c\right)}\right| \quad \left(l_m\left(j\omega_c\right)+p\left(j\omega_c\right)\right) < 1 \tag{8.2-18c}$$

At ω_c the magnitude of the sensitivity function that relates to performance and the magnitude of complementary sensitivity function that relates to robustness are both made small by making the magnitude of the return difference large. In fact Eq. (8.2-18c) and also Eq. (8.2-8) will both be satisfied at the crossover frequency if the magnitude of the return difference is made larger than the sum of $l_m(j\omega_c)$ and $p(j\omega_c)$.

$$\left|1+G\left(j\omega_c\right)\right| > l_m\left(j\omega_c\right)+p\left(j\omega_c\right) \tag{8.2-19}$$

In Sec. 6.5 it is shown that, at the crossover frequency, the magnitude of the return difference can be expressed as a function of the phase margin.

$$\left|1+G\left(j\omega_c\right)\right| = 2\left|\sin\left(\frac{\phi_m}{2}\right)\right| \tag{8.2-20}$$

The largest the return difference can be at the crossover frequency is 2. This maximum occurs if the phase margin is 180°, i.e., if $\arg(G(j\omega_c)) = 0$. Any phase lag in the loop gain is associated with a decrease in the return difference and an associated drop in robust performance measure.

As is seen in the remainder of this chapter, attaining a large phase margin is difficult. In practice, a phase margin of 60° is usually considered to be large. From Eq. (8.2-20) a 60° phase margin corresponds to a return difference equal to one at the crossover frequency. At frequencies higher than the crossover frequency, the return difference is usually slightly smaller than one. As ω gets large the loop gain approaches zero so that the return difference approaches one.

In general, $l_m(j\omega)$ grows as ω grows. Equation (8.2-19) dictates that, in general, the crossover frequency be chosen below the frequency where $l_m(j\omega)$ first becomes larger than one. Since at the crossover frequency $|G(j\omega_c)| = 1$ Eq. (8.2-19) shows that the crossover frequency is absolutely required to be at a frequency where $l_m(j\omega) < 2$. The crossover frequency ω_c marks the boundary between frequencies where the loop gain is large, producing enhanced performance, and frequencies where the loop gain is small, producing less desirable performance. Thus, the crossover frequency is often used to describe the bandwidth of a control system. The bandwidth of a control system is limited by the frequencies where an accurate model of the plant can be attained. If perfect modeling were attainable so that $l_m(j\omega)$ were always zero then any bandwidth could be attained. Such ideal situations, however, exist only in textbooks.

We have found that unmodeled dynamics dictate that the complementary sensitivity function and thus that the loop gain be rolled off to small values at high frequencies. The desire to enhance performance requires that the return difference and thus the loop gain be large at low frequencies. These two desires are fairly easily accomplished. The skill of the control designer comes in making appropriate trade-offs as the loop gain goes from large values to small values through the transition frequencies.

There are two conflicting desires in the transition region. We would like the cutoff between low frequencies and high frequencies to be as sharp as possible so that the low frequency region where performance is enhanced becomes as large as possible given the high frequency constraints imposed by modeling uncertainties. The second desire is to keep the amount of phase lag near crossover frequency as small as possible so that the loop gain avoids getting too close to the -1 point or creating extra encirclements.

The fact that these desires conflict can be seen by recalling from Chap. 5 that the phase plot and the slope of the magnitude plot of a transfer function are related. Assume that the loop gain transfer function of the system is minimum phase. (The situation is worse for nonminimum phase systems). Recall the discussion surrounding the Bode phase relationship of Eq. (5.4-1). If a system's magnitude plot has a slope of -40dB per decade for a decade on either side of the crossover frequency, then the phase shift at the crossover frequency is very close to $-180°$. Larger negative slopes in the magnitude plot correspond to greater phase lags. To achieve a reasonably large phase margin, the magnitude plot must slope gradually through the crossover frequency. If a phase margin of 60° is required the magnitude plot must have only a −20dB per decade slope for about two-thirds of a decade on either side of the crossover frequency. Although we would like to transition quickly from high loop gain at low frequencies to low loop gain at high frequencies, the need to maintain a reasonable degree of robust stability and performance throughout the transition region dictates that the transition occur more slowly.

We have now seen the overall objective of shaping the frequency response of the loop gain transfer function to achieve an acceptable level of controller performance and robustness. If a feasible performance specification $p(j\omega)$ and modeling uncertainty function $l_m(j\omega)$ are specified, an acceptable controller is one in which the sensitivity function and complementary sensitivity function satisfy Eq. (8.2-8).

It should be noted that it is somewhat artificial to summarize various performance specifications in the function $p(j\omega)$. The lessons of loop shaping introduced here do not actually depend upon the ability to produce such a $p(j\omega)$. The role of $p(j\omega)$ is mainly to express mathematically the generally true concept that the performance of a control system is made better if the return difference is made larger.

In the next section we examine the building blocks that have been used in classical control theory for years to develop series compensators to shape the loop gain of a control system.

Exercises 8.2

8.2-1. Turn the following performance specifications into a guideline $p(j\omega)$ so that if

$$|1 + \tilde{G}(j\omega)| > p(j\omega) \qquad \text{for all } \omega$$

then the performance specifications are likely to be met.

1. If there is an output disturbance with a frequency below $\omega = 0.01$ rad/sec, its effect on the output must be less than 0.001 times its original magnitude.
2. The phase margin should be greater than 30°.
3. The closed-loop response to a unit step input should have zero steady-state error. It should have less than 25 percent overshoot in the transient and it should settle to within 2 percent of its final value in 10 secs.

8.2-2. Sketch the Bode magnitude straight line approximation plots for the following set of loop gain transfer functions on the same plot. Do the same for the phase plot. Notice how adding additional terms in series changes the plots.

$$G_1(s) = \frac{1}{s+1}$$

$$G_2(s) = \frac{10}{(s+1)}$$

$$G_3(s) = \frac{10}{s(s+1)}$$

$$G_4(s) = \frac{10(s+3)}{(s+30)}\left(\frac{10}{s(s+1)}\right)$$

8.3 SERIES COMPENSATOR BUILDING BLOCKS: PROPORTIONAL CONTROL

Having determined some goals for shaping the loop gain transfer function in the previous section, we use this section to begin an investigation of the elements of a series compensator that can be used to achieve the desired shape.

The simplest element for series compensation is a constant gain. On the Bode plot a positive gain leaves the phase plot unchanged. A gain greater than one raises the magnitude plot while a gain less than one lowers the magnitude plot. Raising the magnitude plot helps to achieve the large loop gain desired at low frequencies but, at the same time, creates the undesirable effect of raising the loop gain at high frequencies. Since the Bode magnitude plots of plants generally slope down to the right, raising the gain also has the effect of increasing the crossover frequency of the loop gain. In frequency response design methods, a first setting of the gain may be used to set the bandwidth of the system near the limit allowed by modeling uncertainty. A standard root locus analysis shows the effect of a change of gain on the closed-loop pole positions; the closed-loop poles tend to move away from the origin as the gain is increased, manifesting the increasing bandwidth of the system.

Generally, as the gain of a control system is increased, the system's response becomes fast, then oscillatory and finally unstable. Let's examine the effect of increasing the gain on a typical plant model. We restrict ourselves to a series compensator with a gain that is constant across all frequencies. The simple constant gain compensator produces a control signal that is proportional to the error signal between the desired reference input and the plant output. Such a control scheme is often referred to as a *proportional compensator*. Let the plant model $G_p(s)$ be

$$G_p(s) = \frac{10}{(s+1)(s+10)^2} \tag{8.3-1}$$

Our purpose here is to explore the effects of basic control building blocks on nominal plant models; consequently, we consider any perturbations to the model only in a qualitative, not in a quantitative sense.

Plants of the general form shown in Eq. (8.3-1) are very common; they arise in many applications such as process control and control of mechanical systems. An automobile speed control system is a plant that can be modeled with Eq. (8.3-1). Most of us have experienced trying to control such a system manually where the human operator functions as the controller. The operator's actions can be modeled as a proportional compensator as the driver compares the measured speed on the speedometer with the desired speed and changes the throttle position accordingly. If the difference in the two speeds is large, the driver makes a large change in the throttle position thus producing a proportional control law. The transfer function of Eq. (8.3-1) models the response of the car's speed to a change in the throttle position.

As given in Eq. (8.3-1), the plant's time response to a step change in the throttle position is that of a dominantly first-order system with a time constant of 1 sec, which is slowed or delayed slightly by the presence of additional poles. The dominant pole can be thought of as arising from the car's attempt to gain speed against the dissipative

force of aerodynamic drag while the additional poles arise from the dynamics required for the motor to increase its output in response to the change in throttle position.

The constant in the numerator is dependent on the units chosen to represent the input and output. It changes depending on whether the speed is in miles per hour or kilometers per hour and whether the throttle position is measured in inches from the floor or some angular deflection. Once the units are chosen, however, the constant is set and is part of the plant description. We assume that sensors and actuators are represented as part of the plant transfer function so that for an automated controller the input is really a voltage proportional to the throttle position and the output is really a voltage proportional to the speed. The sensor and actuator dynamics are assumed to be fast enough (of wide enough bandwidth) so that they are constant across frequencies of interest and add no dynamics. (They then become high frequency unmodeled dynamics and could be addressed by plant perturbations in a more refined design.)

The effect on the closed-loop system response that arises from increasing the gain of a proportional compensator can best be seen using a root locus. The root locus for the proportional control system of Fig. 8.3-1 using the plant model of Eq. (8.3-1) is given in Fig. 8.3-2. The scenario described at the beginning of this example indeed occurs as K is increased. First the dominant pole moves out along the negative real axis, indicating a faster response. Then complex conjugates poles break away from the real axis and move towards the imaginary axis, indicating oscillatory responses of increasing frequency and decreasing damping. Finally, the closed-loop poles move across the imaginary axis, indicating an unstable response, specifically, an oscillatory response with growing amplitude.

The results should match the intuition we have developed from driving a car. If we overreact to small desired changes in speed by making large swings in the the throttle pedal position the result will be a herky-jerky ride marked by sudden accelerations and decelerations. We have probably all ridden with stressed out drivers whose gain is too high, and the resulting oscillation was probably quite annoying.

While we can see how the gain affects the closed-loop pole position on the root locus, we do not get much information from the root locus about how much of the possible benefits of feedback control are achieved. Remember from Sec. 3.2 that, as well as a good transient response, a good feedback control design achieves sensitivity reduction and disturbance rejection. Remember also that these latter goals are achieved by creating as large a return difference transfer function over as large a bandwidth as possible.

Observe Fig. 8.3-3 where the Bode and Nyquist plots are drawn for the loop gain of Fig. 8.3-1 for three different values of K. Only the portion of the Nyquist plot corresponding to the positive imaginary axis portion of the Nyquist D-contour is drawn. This is the usual practice in design as it prevents clutter on the diagram. The

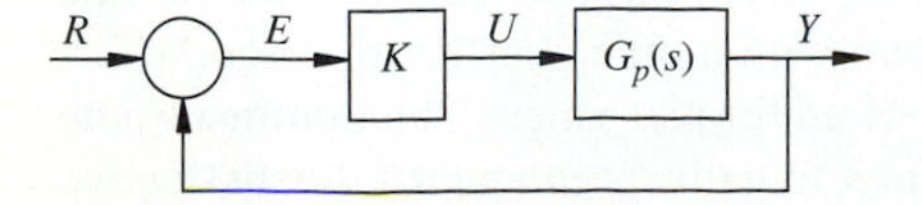

FIGURE 8.3-1
A proportional compensator system.

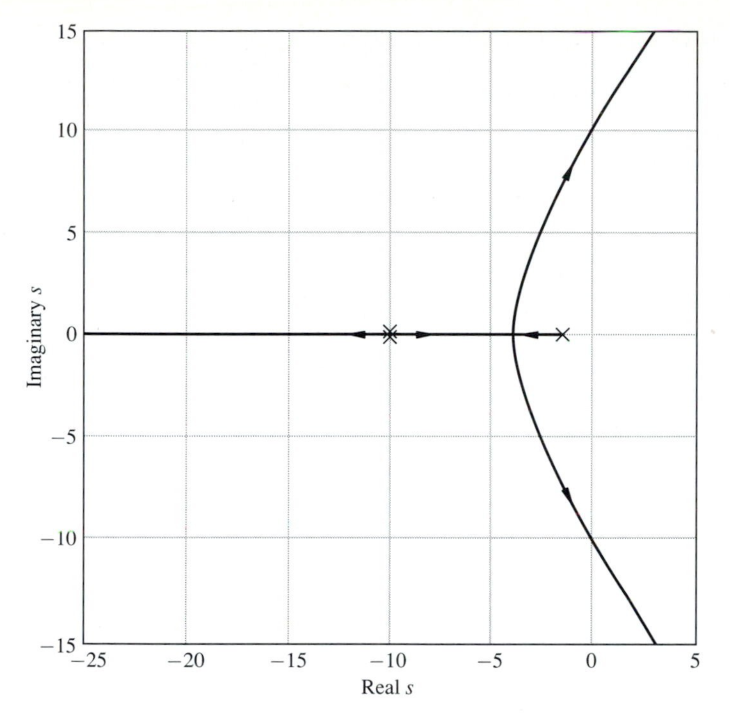

FIGURE 8.3-2
Root locus for proportional compensator.

designer usually thinks of altering this portion of the Nyquist plot with the design. The rest of the plot is usually completed only within the designer's mind.

Qualitative information about the transient response can be read off Nyquist and Bode plots. For $K = 20$, the (completed) Nyquist plot shows no encirclements of the origin and a stable closed-loop system. In addition, the Nyquist plot avoids the -1 point well at all frequencies, indicating no peak in the closed-loop frequency response and a non-oscillatory closed-loop time response. The speed of the closed-loop system response can be approximated by finding the bandwidth of the closed-loop frequency response. The closed-loop frequency response has magnitude of approximately two-thirds at low frequencies where the loop gain has magnitude two and little phase lag. As the loop gain transitions from values larger than one to values smaller than one, the closed-loop frequency response falls off from near unity to smaller values approximately equal to the loop gain. The bandwidth of a control system is usually defined to be up to the frequency where the loop gain crosses one in magnitude; this frequency is the crossover frequency ω_c. It can be seen from the Bode plots of Fig. 8.3-3*a* that the crossover frequency for $K = 20$ is $\omega_c = 1.5$ This indicates the speed of the step response of the closed-loop system. The closed-loop system will respond with a time constant of $T = 1/1.5 = 2/3$ sec.

Now, notice the trend on the Nyquist plot of Fig. 8.3-3*d* as K is increased. As K is increased, the entire plot expands radially. The -1 point is first approached closely,

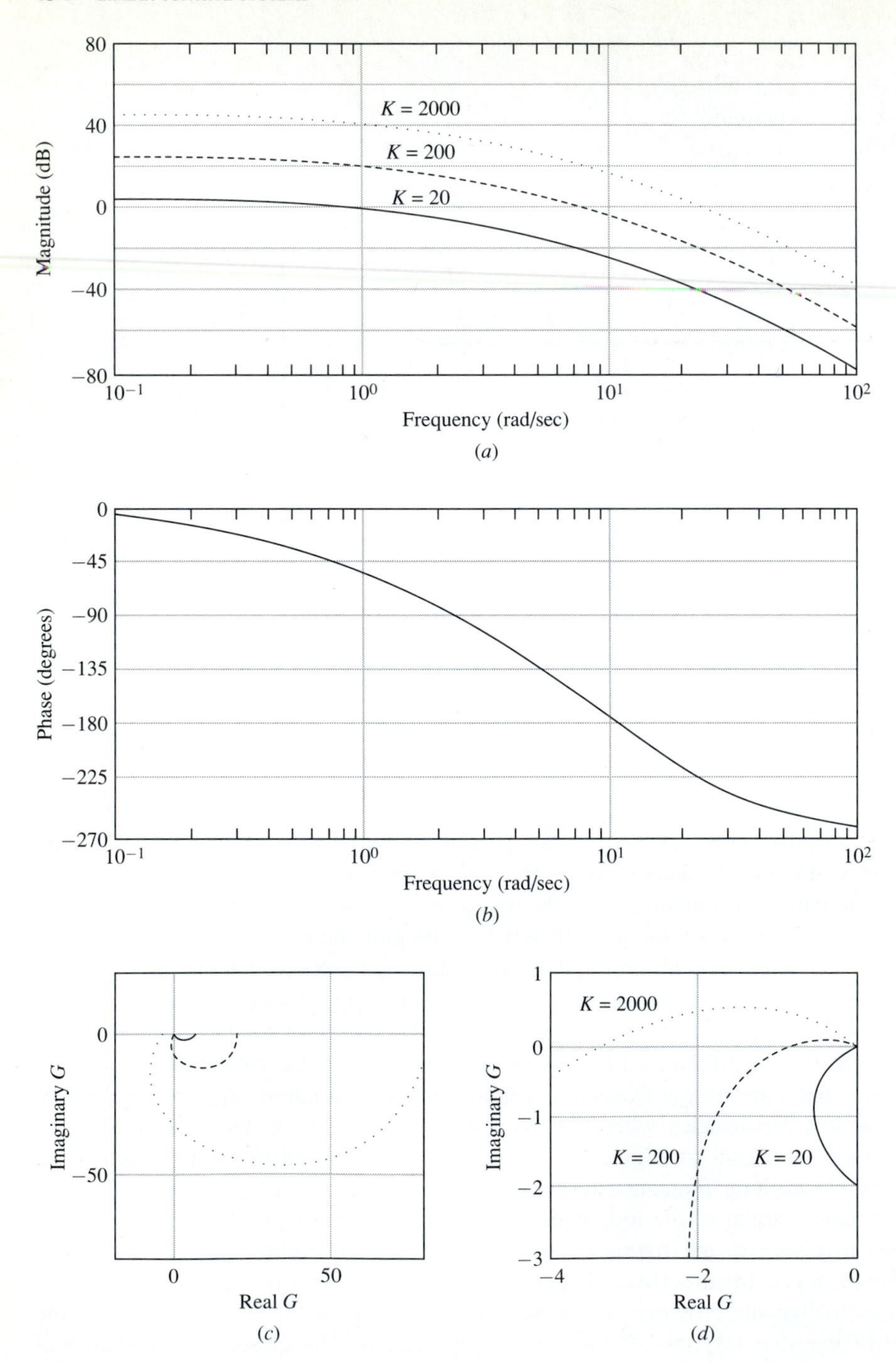

FIGURE 8.3-3

Proportional control loop with 20, 200, and 2,000. (*a*) Bode magnitude plot; (*b*) Bode phase plot; (*c*) Nyquist plot (wide scale); (*d*) Nyquist plot (expanded scale).

indicating a peak in the closed-loop frequency response (a large M_p peak value) and an oscillatory closed-loop time response. As K is increased further the -1 point is reached and encircled, indicating an unstable closed-loop system. This is consistent with the root locus analysis.

Look at the Bode plot as K is increased. The basic change in the Bode plot is simple. The magnitude plot is shifted up, as K is increased while the phase plot is unaltered. However, there are also more subtle changes. As the Bode magnitude plot is shifted up, the crossover frequency ω_c is shifted out. The larger crossover frequency indicates a larger closed-loop system bandwidth. The larger closed-loop system bandwidth indicates faster closed-loop time responses in following changes in reference inputs and in rejecting disturbances.

Although the Bode phase plot doesn't change when the constant gain of the loop gain transfer function increases, the shifting out of the crossover frequency on the magnitude plot means that the system phase margin is read off the Bode phase plot at a higher frequency. Since the phase plot is sloping down, this means that, in this typical situation, increasing the gain decreases the phase margin. Remember from Sec. 6.5 that for smooth Nyquist plots where the Nyquist plot most closely approaches the -1 point near the crossover frequency, the phase margin provides a good indicator of the degree of oscillation of the closed-loop step response. As the gain increases, the crossover frequency becomes higher, the phase margin decreases, the Nyquist plot passes closer to the -1 point, and the closed-loop transient response becomes more oscillatory. The phase margin going to zero corresponds to an infinite closed-loop frequency response at crossover frequency and two closed-loop poles on the imaginary axis at the crossover frequency ω_c. Such a system is unstable. Thus the point on the imaginary axis of the root locus plot where the poles cross into the right half-plane can be found by examining the open-loop Bode phase plot for the frequency where the phase equals 180°. This frequency is often called the *phase crossover* frequency. In the example this frequency is near 10 rad/sec. The gain margin of the system can be read off the Bode magnitude plot at the phase crossover frequency. In the example, when the gain is 20, the gain margin is 20 dB or a factor of 10. Obviously, when the gain is 200 there is almost zero gain margin and the system is at the edge of instability.

From the Bode plot one can also see fairly directly how well the controller can handle disturbance rejection and sensitivity reduction. Remember that a loop gain that is much greater than one indicates good disturbance rejection and sensitivity reduction. We can examine the limits on the disturbance rejection and sensitivity reduction that are possible using a proportional compensator in the example. The largest gain that produces a stable controller is $K = 200$. When $K = 200$ the loop gain is greater than 11 (22 dB) for all frequencies less than 0.9 rad/sec. This implies that for these frequencies the sensitivity function satisfies

$$\left|\frac{1}{1 + G(j\omega)}\right| \le \frac{1}{-1 + |G(j\omega)|} < 0.1 \text{ for } |G(j\omega)| > 11 \tag{8.3-2}$$

Thus with $K = 200$ the steady-state magnitudes of disturbances at frequencies less than 0.9 rad/sec are attenuated to less than 10 percent of their original magnitudes.

Also, the sensitivity reduction means that changes in the plant transfer functions at frequencies below 0.9 rad/sec map into changes in the closed-loop transfer function of only one-tenth the magnitude of the change in the loop gain. Of course, an actual proportional compensator would have to be set with a lower gain due to the oscillatory nature of the time response when $K = 200$. Systems with lower gain would have less steady-state disturbance rejection and sensitivity reduction.

Finally, we note that as K increases, the robustness of the closed-loop system to perturbations in the plant decreases. This is due to two factors: first, the decrease in phase margin with increasing K indicates that the nominal loop gain approaches the -1 point more closely as K increases. (This is easily seen on the Nyquist plot of Fig. 8.3-3*d*). Secondly, since the crossover frequency increases as K increases and plant models are relatively less precise at higher frequencies, the multiplicative perturbation near the critical crossover frequency becomes larger as K increases. Again, this is not because the perturbation as a function of frequency changes with K but because the critical crossover frequency becomes higher as K increases.

The question of performance and robustness can be addressed more directly by referring to Fig. 8.3-4. In Fig. 8.3-4*a*, the magnitude of the sensitivity function $(1 + G(j\omega))^{-1}$ is plotted for various values of K that result in stable closed-loop systems. In Fig. 8.3-4*b*, the magnitude of the complementary sensitivity function $G(j\omega)(1 + G(j\omega))^{-1}$ is plotted for various values of K that produce stable closed-loop systems. Recall the robust performance specification, Eq. (8.2-8), repeated here as Eq. (8.3-3).

$$\left|\frac{1}{1 + G(j\omega)}\right| p(j\omega) + \left|\frac{G(j\omega)}{1 + G(j\omega)}\right| l_m(j\omega) < 1 \tag{8.3-3}$$

The smaller the complementary sensitivity function is at any frequency, the larger the multiplicative perturbation that can be handled at that frequency. Fig. 8.3-4*b* shows that, as K increases, the peak in the complementary sensitivity function increases; also the peak moves to a higher frequency. This confirms the double effect of loss of robustness explained in the previous paragraph. The margin for plant mismodeling decreases and the worst case frequency increases to frequencies where more mismodeling is expected. Remember that the complementary sensitivity function is also the closed-loop frequency response function; thus the peaking that occurs with large K indicates oscillatory time responses as discussed previously.

The performance tradeoff displayed in Fig. 8.3-4*a* is even more interesting. As K increases, the sensitivity function decreases at most frequencies, indicating improved steady-state disturbance rejection and sensitivity reduction at these frequencies. At other frequencies (near the shifting crossover frequency) the sensitivity function peaks badly. This is due to the loop gain approaching -1 closely for larger values of K. At some point (upon which the designer and the customer must mutually decide) the disadvantages associated with the increase in sensitivity at some frequencies override the advantages associated with the decrease in sensitivity at most other frequencies.

With all the information available, how does one actually go about designing a proportional compensator, i.e., how does one choose K? When K is small, increasing K has only positive effects. The increasing loop gain provides decreasing sensitiv-

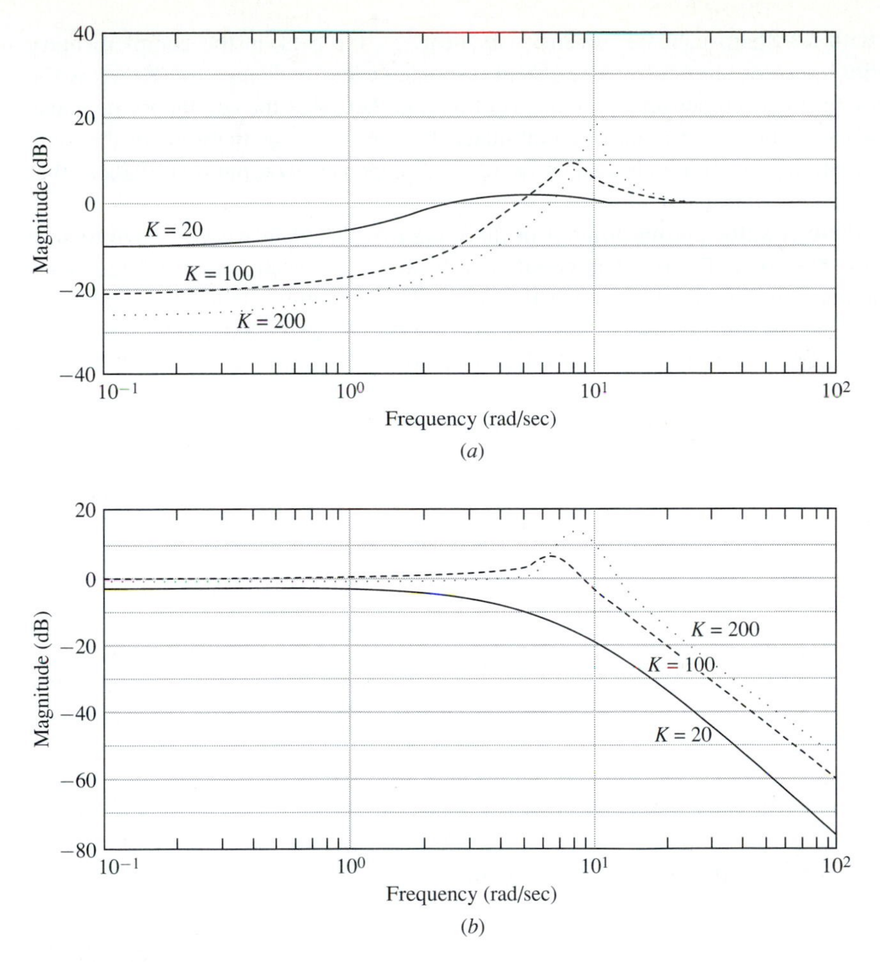

FIGURE 8.3-4
Sensitivity and complementary sensitivity functions for the proportional compensator with $K = 20$, 100, and 200. (*a*) Sensitivity; (*b*) complementary sensitivity.

ity (and improving disturbance rejection) at almost all frequencies with very little increase in sensitivity at the few frequencies where it does increase. The bandwidth becomes larger and the transient response becomes faster and remains non-oscillatory. Obviously, we'd like to keep increasing the gain. What stops us?

Clearly we must stop increasing the gain before the closed-loop system becomes unstable. In actuality something else always stops us short of this point. If the system were perfectly modeled and the noise associated with the output measurement were small, the first negative effect of further increasing K would be the simultaneous peaking in the sensitivity and complementary sensitivity functions near the crossover frequency. The peaking occurs from the increase in bandwidth accompanied by the decrease in phase margin. This means the Nyquist plot approaches -1 too closely.

Along with the sharp increase in sensitivity near ω_c, the peak in the complementary sensitivity indicates an oscillatory transient response. Clearly we must set K below the value where this becomes unacceptable. (Let us note here that the oscillatory response to reference inputs can possibly be eliminated by the use of a prefilter but the peak in the sensitivity function indicates an oscillatory response to output disturbances that cannot be easily corrected.)

If there is sufficient uncertainty in the model of the plant we will have to stop increasing K sooner. The problem could arise from larger uncertainties at frequencies much higher than the initial crossover frequency. Uncertainties due to resonant modes may be large enough to require keeping the complementary sensitivity very small at high frequencies. Remember the robust performance condition of Eq. (8.3-3). If $l_m(j\omega)$ is large enough it will limit the size of the complementary sensitivity function and thus limit the value of K. The need for robust performance is even more likely to limit the value of K because of the effect near the crossover frequency. The requirement of good performance for a set of plants is more stringent than the requirement of good performance for the nominal plant model. While the Nyquist plot of the nominal model for a particular K may be kept far enough away from the -1 point to avoid large peaks in the sensitivity and complementary sensitivity function, it requires a smaller K to assure that every possible plant within the perturbed set remains far enough away from the -1 point. Stated another way using Eq. (8.3-3), if $l_m(j\omega)$ is significant at some ω_0, the complementary sensitivity function must be kept small at ω_0 so that the second term in Eq. (8.3-3) can remain small. It may be that the control system bandwidth needs to be kept lower than it would in the nominal case if the modeling uncertainty becomes significantly large at frequencies at or below the bandwidth that could be achieved if there were no modeling error.

Let us review the situation of designing the value K of the proportional compensator using the control design strategy outlined in Sec. 8.2. Assume that performance requirements have been established and have been translated into a performance bound $p(j\omega)$ on the possible perturbed sensitivity function in a manner similar to that presented in Example 6.7-1. Assume also that along with the plant model of Eq. (8.3-1) there has been established a bound $l_m(j\omega)$ on the possible multiplicative plant perturbations. One can use the approximations of Eqs. (8.2-12) and (8.2-15) as a guide for selecting the loop gain. Assume that the magnitudes of $p(j\omega)$ and $l_m^{-1}(j\omega)$ are as plotted in Fig. 8.3-5. From Eqs. (8.2-12) we find that we would like to keep the loop gain above $p(j\omega)$ for the frequencies where $p(j\omega) > 1$ and below $l_m^{-1}(j\omega)$ for frequencies where $l_m^{-1}(j\omega) < 1$. In this case these objectives can be accomplished by setting $K = 200$. The loop gain with $K = 200$ is also plotted in Fig. 8.3-5.

Unfortunately if we set $K = 200$ the magnitude crosses unity while sloping -20 dB per decade (one pole of slope) for a significant length before crossover frequency but begins sloping -60 dB per decade (three poles of slope) shortly after crossover frequency. For this minimum phase system, these slopes indicate a phase very close to 180° at crossover and little phase margin. We have already seen that when a Nyquist plot passes close to the -1 point as the Nyquist plot corresponding

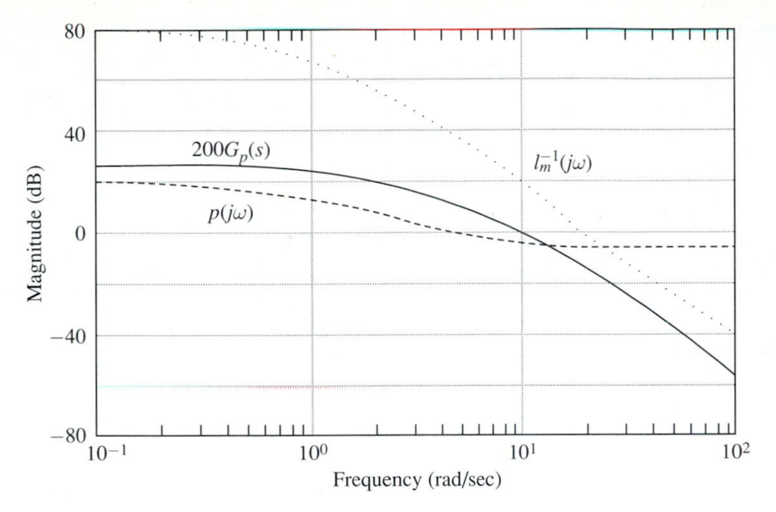

FIGURE 8.3-5
Loop gain for $K = 200$ with constraints.

to $K = 200$ does, the sensitivity function and the complementary sensitivity function both peak to large values near the crossover frequency. While the approximations of Eqs. (8.2-12) and (8.2-15) are satisfied, the actual constraint of Eq. (8.2-8) is violated. This can be seen by comparing the plot of the sensitivity function and the complementary sensitivity functions for $K = 200$ in Fig. 8.3-4 with the plots of $p(j\omega)$ and $l_m^{-1}(j\omega)$ in Fig. 8.3-5 at the crossover frequency. In this case, the factor limiting the loop gain is the behavior near crossover rather than the high frequency unmodeled dynamics. The loop gain must be lowered until an acceptable transient response is obtained.

We now realize that we cannot meet all of our objectives with a simple proportional compensator. Suppose that we decide on the following compromise. Let us find the largest value of K that will still produce a phase margin of 45°. In that way, an adequate transient response will likely be achieved. How do we find this value of K? Refer to the Bode plots for the loop gain with $K = 20$ in Figs. 8.3-3*a* and *b*. A 45° phase margin means that the loop gain's phase should be 135° at crossover. Remember that the phase plot of Fig. 8.3-3*b* does not change as K changes. The required crossover frequency of $\omega = 5$ rad/sec can be read from Fig. 8.3-3*b*. At $\omega = 5$ rad/sec, the magnitude plot for $K = 20$ passes through -10 dB. To make $\omega = 5$ rad/sec the new crossover frequency we must add 10 dB to the gain of $K = 20$, i.e., we must multiply the gain $K = 20$ by 3 to obtain $K = 60$.

The loop gain with $K = 60$ is plotted in Fig. 8.3-6 along with $p(j\omega)$ and $l_m^{-1}(j\omega)$. The crossover frequency is $\omega = 5$ rad/sec as predicted and, since the phase plot of Fig. 8.3-3*b* is still the loop gain phase plot, the phase margin is 45° as planned. Notice that for $K = 60$ the loop gain has a slope of -20 dB per decade (a one pole slope) for a decade preceding the crossover frequency and also for half a decade

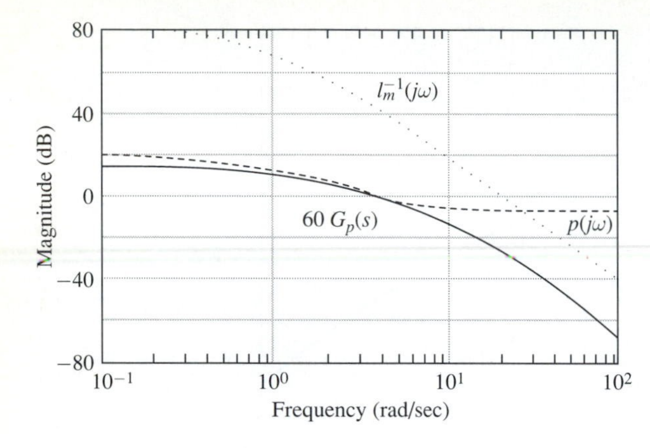

FIGURE 8.3-6
Loop gain for $K = 60$ with constraints.

after the crossover frequency before the slope of −60 dB per decade (the three pole slope) kicks in. This supports the known 45° phase margin. Evaluating the return difference for a 45° phase margin from Eq. (8.2-20), we see that the return difference has magnitude close to 0.75 at crossover so that the sensitivity function has a value of between 2 and 3 dB at crossover. Thus, the design violates the performance criteria near crossover frequency where the value of $p(j\omega)$ is 0 dB. Also, the low frequency performance specification is no longer met.

Sometimes a simple proportional compensator is adequate; usually, as in this case, it is not. In the next section we learn how to correct the low frequency part of the loop gain. Two sections hence, we learn how to correct the performance near crossover frequency. The use of these building blocks for creating more complex controllers is the subject of the remainder of this chapter.

It should be noted that formal specifications of $p(j\omega)$ and $l_m(j\omega)$ such as we used here are rarely available. However, the control designer must interface with the modeling team and higher level system designers to translate the information that is given into a form similar to $p(j\omega)$ and $l_m(j\omega)$ for design. The principles are generally true. The loop gain and therefore the controller bandwidth should be as large as possible to produce closed-loop advantages in speed of response, disturbance rejection, and sensitivity reduction. The size of the loop gain is limited either by the phase margin or crossover frequency indicating oscillatory responses, by the possible effects of model uncertainty at crossover and higher frequencies, or by the need to filter the effects of noisy sensors.

Exercises 8.3.

8.3-1. Find the closed-loop poles and the closed-loop step response for the control system using the plant given by Eq. (8.3-1) and a proportional compensator with $K = 60$. How well does the bandwidth obtained from the Bode plots predict the speed of response? Explain.

Answer: $s = -16.7 \quad s = -2.4 + j6 \quad s = -2.4 - j6$
$\omega_n = 6.5$, slightly higher than crossover of $\omega_c = 5.5$.
M_p occurs at ω_p, slightly higher than ω_c.

8.3-2. Given

$$G_P(s) = \frac{100}{(s+1)(s+10)^2}$$

$$l_m(s) = \frac{(s+1)^3}{10000}$$

If possible, find a proportional compensator so that the closed-loop system meets the following objectives:

1. The system is robustly stable to the perturbations encompassed by $l_m(s)$.
2. The magnitude of steady-state response to a disturbance at $\omega = 0.2$ is less than 20 percent of the value of the disturbance.
3. The transient response oscillates as little as possible given that the first two conditions are met.

Answer: $K = 5$

8.3-3. Let

$$G_P(s) = \frac{-(s-10)}{s(s+1)(s+10)}$$

(a) Find the value of K which produces a 45° phase margin. What is the crossover frequency?
(b) What is the crossover frequency and phase margin if K is doubled?
(c) What is the crossover frequency and phase margin if K is halved?

Answer: (a) $K = 0.8$, $\omega_c = 0.8$ rad/sec
(b) $\omega_c = 1.2$ $\phi_m = 30°$
(c) $\omega_c = 0.4$ $\phi_m = 60°$

8.4 SERIES COMPENSATOR BUILDING BLOCKS: LAG COMPENSATORS, PI COMPENSATORS

A method of increasing the loop gain transfer function at low frequencies while producing minimal changes at transition frequencies and high frequencies is provided by a lag compensator. A lag compensator is characterized by a pole and a zero. The pole is located either at the origin or very close to the origin. The zero is located somewhat further out the negative real axis, but still at a frequency lower than the crossover frequency. A lag transfer function is given by the equation

$$G_{\text{lag}}(s) = \frac{s+a}{s+b} \qquad \text{with} \qquad a > b \geq 0 \tag{8.4-1}$$

Often, b is set to 0 to provide an integrator in the loop, which increases the System Type and provides improved steady-state behavior, as discussed in Section 4.7. The actual Bode plots and the straight line approximations of a typical lag compensator are shown in Fig. 8.4-1. The transfer function that corresponds to this Bode plot is given by

$$G_{\text{lag}}(s) = \frac{s+1}{s} \tag{8.4-2}$$

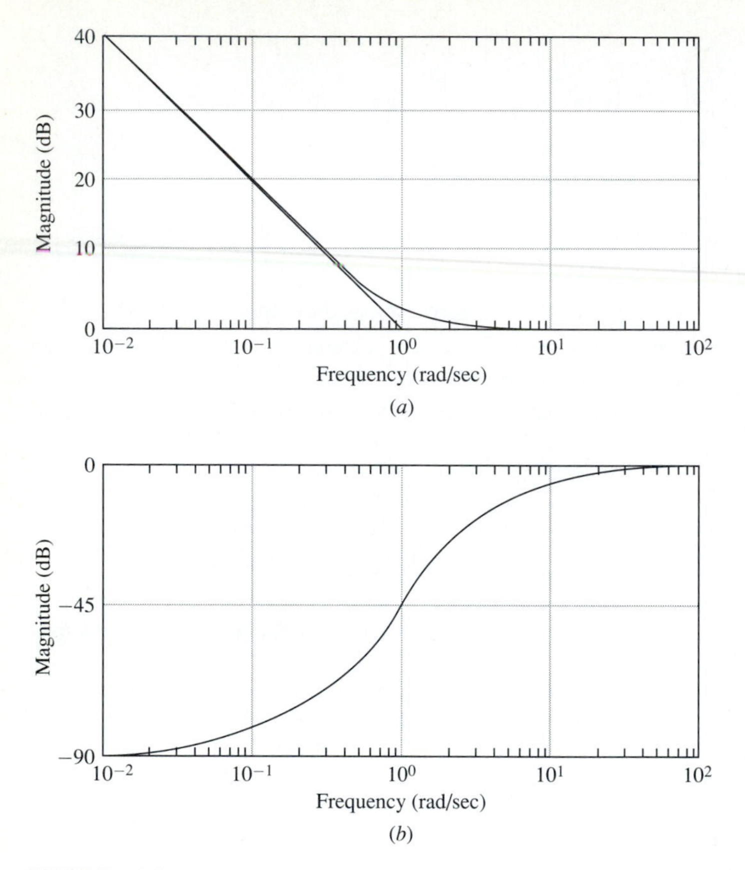

FIGURE 8.4-1
Bode plots of lag compensator. (*a*) Magnitude; (*b*) phase.

which is clearly of the form of Eq. (8.4-1). The lag compensator has high gain at low frequencies with a gain close to unity for frequencies above the value of the zero placement. When placed in series with a plant the lag compensator produces the desired effect of raising the magnitude of the loop gain at low frequencies while leaving the loop gain at the crossover frequency and beyond relatively untouched.

There is a price to be paid in achieving the benefits of a lag compensator. As its name suggests, the lag compensator introduces phase lag into the loop gain. Increased phase lag around the crossover frequency usually moves the Nyquist plot of a loop gain closer to the -1 point with an attendant loss of performance and robustness. As can be seen from Fig. 8.4-1 the amount of phase lag introduced by the lag compensator decreases as frequency increases beyond the frequency of the zero placement. To minimize the phase lag at crossover frequency, the zero of the lag compensator is placed as close to the origin as the need for low frequency loop gain allows. If the zero is placed a decade below the crossover frequency, the lag compensator reduces the phase margin by only 6°. A wider frequency gap results in a smaller impact on the phase margin.

The thought of adding an integrator in parallel with a proportional compensator provides another interpretation of a lag compensator. A lag compensator can be interpreted as a PI compensator, where *PI* represents Proportional-Integral. A diagram of a PI compensator is shown in Fig. 8.4-2. The effect of the integrator has an interesting intuitive explanation. The error between the reference input $r(t)$ and the plant output $y(t)$ is formed at the summing junction preceding the series compensator. If there were a constant steady-state error in response to a step input, a proportional compensator would result in a constant plant input signal and the error would remain. The output of the integrator, however, would grow in time if its input were constant. This would produce a large control signal, which then must reduce the error to zero if the plant input is to reach steady state. Such behavior is referred to as integral action and the PI compensator is said to have *reset capability* in that it can reset the output to the desired value in the presence of a constant offset disturbance.

The transfer function for a PI compensator is given as:

$$G_{PI}(s) = K_p + \frac{K_I}{s} = \frac{K_p s + K_I}{s} = \frac{K_p(s + K_I/K_p)}{s} \tag{8.4-3}$$

Comparing Eq. (8.4-3) with Eq. (8.4-2) we see that the PI compensator is the same as a lag compensator with $b = 0$ and $a = K_I/K_p$ in series with a proportional gain K_p. We have analyzed the effect of the integral action of a PI compensator from a different viewpoint when we discussed System Type and steady-state Error Constants in Sec. 4.7.

Let's examine the effect of a lag compensator placed in series with the proportional compensator designed for the automobile cruise control system in Sec. 8.3. Recall from Fig. 8.3-6 that a proportional compensator design with $K = 60$ achieves a phase margin of 45° at crossover frequency and acceptable high frequency attenuation, but fails to meet the desired performance specification at low frequency. Recall from the form of the lag compensator of Eq. (8.4-1) that its magnitude is one for frequencies higher than the zero position a. The lag compensator placed in series should raise the loop gain at low frequencies while leaving it relatively unaffected at crossover and high frequencies. Consider using the compensator

$$G_{\text{lag}}(s) = 60\frac{(s + a)}{s} \tag{8.4-4}$$

in series with the plant of Eq. (8.3-1).

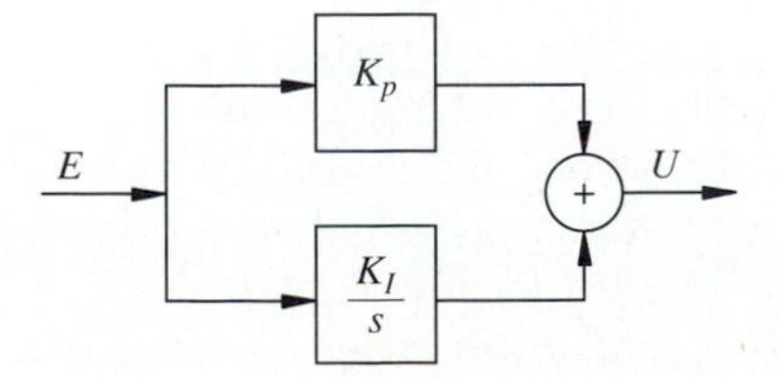

FIGURE 8.4-2
A PI series compensator.

The Bode plots for the resulting loop gains for $a = 0.5$, $a = 2.5$, and $a = 5.0$ are given in Fig. 8.4-3. The hypothetical performance specification $p(j\omega)$ and model uncertainty bound $l_m^{-1}(j\omega)$ from Fig. 8.3-6 are repeated in Fig. 8.4-3*a*. Notice that two phenomena occur as the design parameter a, the zero position, is increased. As a increases the loop gain at low frequencies increases and the range of frequencies affected also increases. At low frequencies the gain from the lag compensator is

$$\left|\frac{s+a}{s}\right|_{s=j\omega} \approx \frac{a}{\omega} \qquad \text{for} \qquad \omega \ll a \tag{8.4-5}$$

The second effect is that as a increases and the zero moves towards crossover frequency, the additional phase lag from the lag compensator at the crossover frequency erodes the phase margin of the proportional compensator.

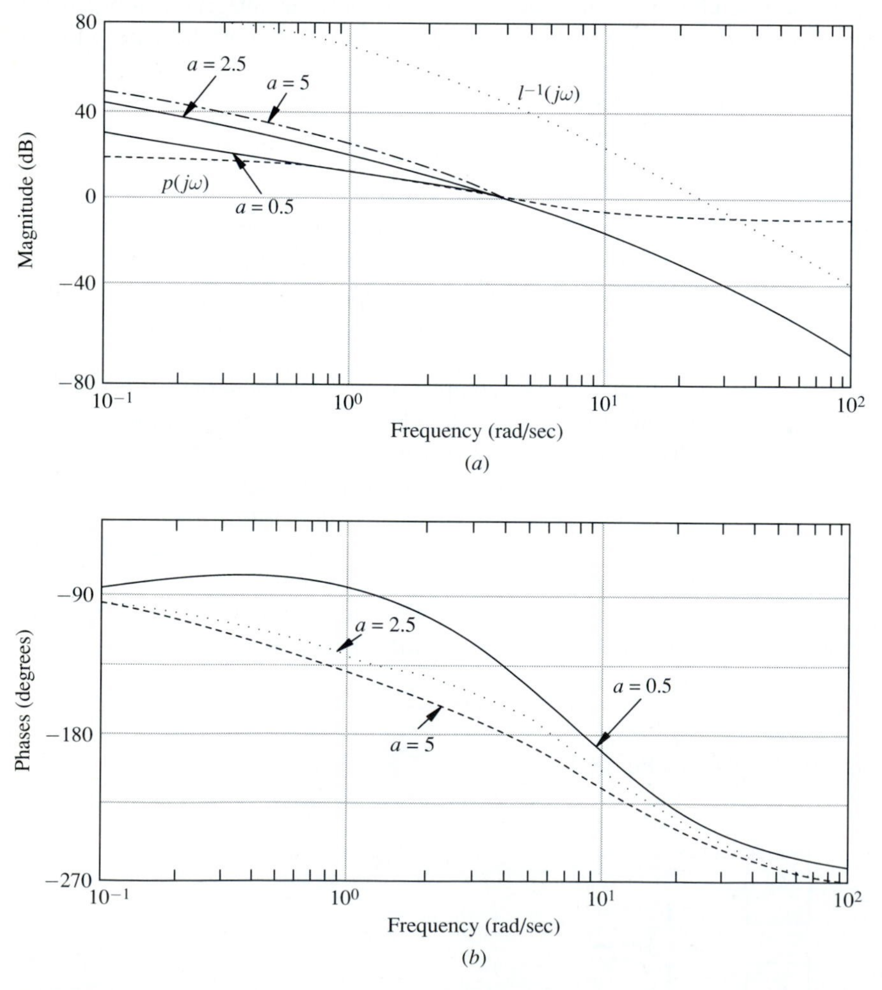

FIGURE 8.4-3
Bode plots for lag compensator control loops. (*a*) Magnitude; (*b*) phase.

When $a = 2.5 = 0.5\omega_c$ the phase margin has been reduced from 45° for the proportional compensator to 22° for the lag compensator. By the time $a = 5.0 = \omega_c$ almost all of the original 45° phase margin has been eroded by the lag compensator. The tradeoff in designing a lag compensator is to set the zero position for sufficient low frequency gain to achieve the performance required at low frequencies while leaving an adequate phase margin at crossover frequency. Note that, in the example, when $a = 0.5 = 0.1\omega_c$ the low frequency performance specification is just met while a phase margin of 39° is salvaged. (Remember that the performance specification curve plotted in Fig. 8.4-3 is only a guideline that the loop gain need exceed only when the magnitude of the loop gain is significantly greater than one.) We know that the resulting 39° phase margin is not adequate because the original 45° phase margin was inadequate. We will correct this in the next section. For now we need to check that we have indeed corrected the low frequency behavior.

Figure 8.4-4*a* shows the complementary sensitivity function for the control loop with $a = 0.5$. The complementary sensitivity function stays well below the inverse

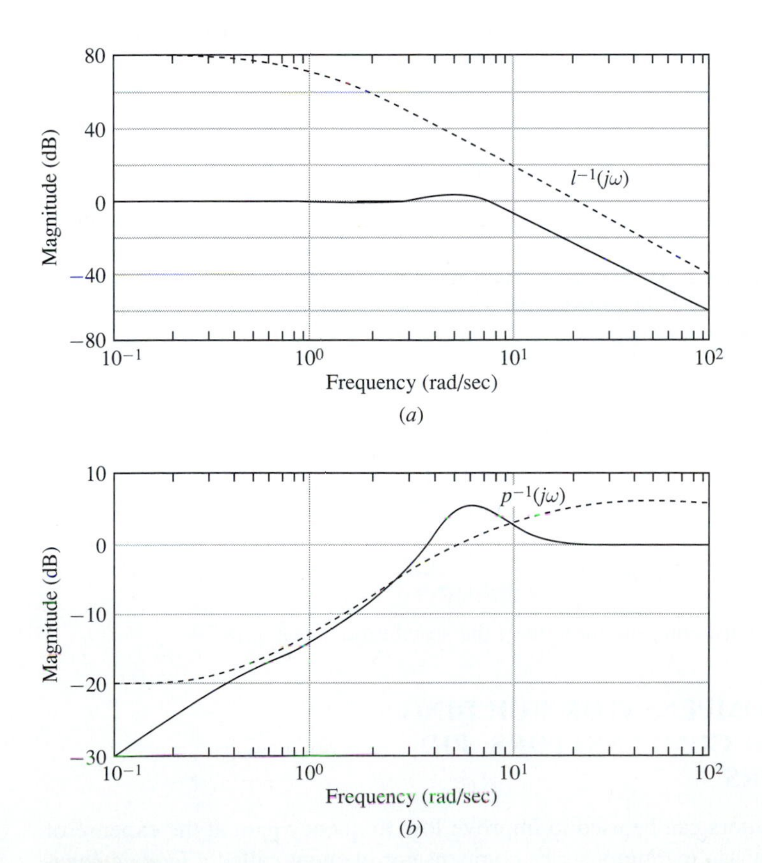

FIGURE 8.4-4
Sensitivity functions for lag control loop with $a = 0.5$. (*a*) Complementary sensitivity function; (*b*) sensitivity function.

of the multiplicative perturbation bound, rendering the design adequate in its robust stability property. Figure 8.4-4*b* shows the sensitivity function for the design. The sensitivity function is indeed sufficiently small at low frequencies. As expected, it rises above the inverse of the performance specification near the crossover frequency. Thus this design is unacceptable. At this point it is unclear how to fix this problem. If the gain is reduced or the zero position of the lag compensator is reduced the performance specification is violated at low frequencies. What is needed is a way of increasing the phase margin around crossover frequency. We learn how to achieve this in the next section, where we study lead compensators.

Exercises 8.4

8.4-1. Compare the closed-loop step responses of the proportional compensator $G_C(s) = 60$ and the lag compensator

$$G_{\text{lag}}(s) = \frac{60(s+0.5)}{s}$$

when used with the plant of Eq. (8.3-1). Next, compare the steady-state response of the two systems to an output disturbance $d(t) = 10 \sin 0.01t$.

8.4-2. Design a lag compensator series compensator for the plant

$$G_p(s) = \frac{100}{(s+1)(s+10)}$$

that does not effect the plant's high frequency magnitude response and has minimal effect on the phase margin. The lag compensator should make the plant-compensator combination into a Type 1 system so that it will follow a step input with no steady-state error. Sketch the Bode magnitude and phase plots of both the plant and combination plant-controller.

8.4-3. Let

$$l_m(s) = \frac{(s+0.1)^2}{100}$$

$$G_p(s) = \frac{10}{(s+0.1)(s+10)}$$

$$p(j\omega) = \frac{10}{(s+0.1)(s+1)}$$

Design a PI or lag compensator to meet the specifications of Eq. (8.2-8).

8.5 SERIES COMPENSATOR BUILDING BLOCKS: LEAD COMPENSATORS, PID COMPENSATORS

While lag compensators can be used to improve low frequency gain at the expense of slightly decreased phase margin, a series compensator element called a *lead compensator* can be used to increase a system's phase margin. A lead compensator provides a positive phase shift or phase lead over a limited frequency range. The equation for a lead compensator with unity dc gain is

$$G_{\text{lead}}(s) = \frac{1 + \dfrac{s}{c}}{1 + \dfrac{s}{d}} = \frac{d}{c}\left(\frac{s+c}{s+d}\right) \qquad 0 < c < d \tag{8.5-1}$$

The lead compensator may also be written with unity high frequency gain as

$$G_{\text{lead}}(s) = \frac{s+c}{s+d} = \frac{c}{d}\frac{\left(1 + \dfrac{s}{c}\right)}{\left(1 + \dfrac{s}{d}\right)} \qquad 0 < c < d \tag{8.5-2}$$

The lead compensator consists of a zero at $s = -c$ and a pole at $s = -d$ on the negative real axis with the zero closer to the origin than the pole. The Bode plots and straight line approximations of a lead compensator are shown in Fig 8.5-1. The transfer function plotted in Fig. 8.5-1 is

$$G_{\text{lead}}(s) = \frac{1 + \dfrac{s}{0.1}}{1 + s} \tag{8.5-3}$$

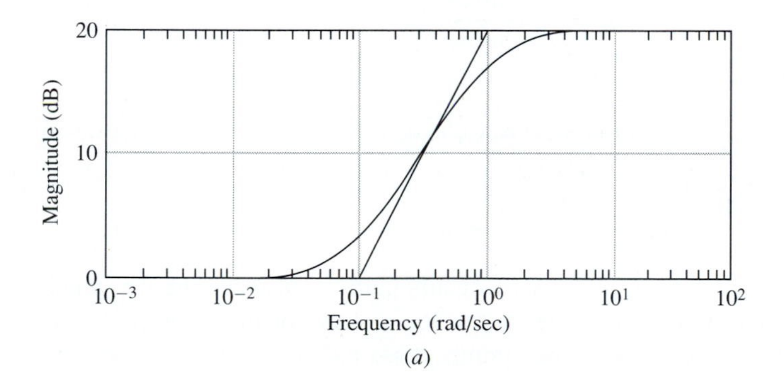

(*a*)

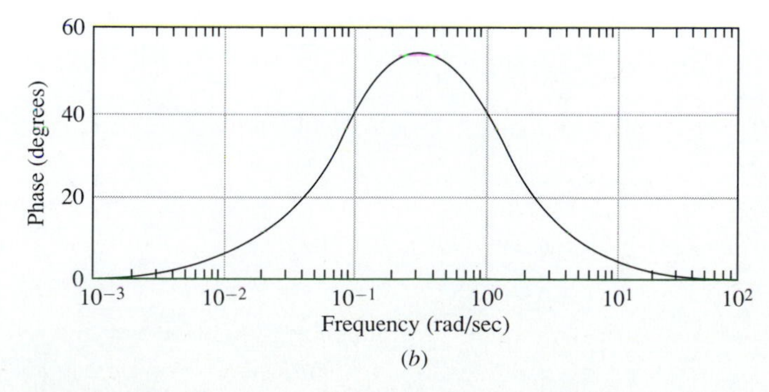

(*b*)

FIGURE 8.5-1
Bode plots for a lead compensator. (*a*) Magnitude; (*b*) phase.

The concept of the lead compensator is that the zero produces a positive slope in the Bode magnitude plot with its attendant phase lead. The pole is provided to create a proper and realizable transfer function; it returns the magnitude plot to its original slope. The need for this pole limits the frequency range over which phase lead is obtained and it also limits the maximum amount of phase lead obtained. In Eqs. (8.5-1) and (8.5-2), $d > c$ and we can define m as

$$m = \frac{d}{c}$$

In its usual application, a lead compensator is arranged so that its zero and pole straddle the crossover frequency of the resulting loop gain so that the maximum phase lead occurs at the resulting crossover frequency. This strategy provides for the greatest increase in phase margin, thereby increasing the return difference transfer function and improving performance through the transition frequencies. From Eq. (8.5-1) or Eq. (8.5-2), we can solve for the maximum amount of phase lead achieved, ϕ_l, and the frequency at which this maximum occurs, ω_l. The equation for ϕ_l is

$$\phi_l = \sin^{-1}\left(\frac{m-1}{m+1}\right), m = \frac{d}{c} \tag{8.5-4}$$

Solving for m in terms of ϕ_l, we find

$$m = \frac{1 + \sin\phi_l}{1 - \sin\phi_l} \tag{8.5-5}$$

The equation for ω_l is

$$\omega_l = \sqrt{cd} = c\sqrt{m} \tag{8.5-6}$$

Notice that

$$\log\omega_l = \frac{1}{2}(\log c + \log d) \tag{8.5-7}$$

so that the frequency of maximum phase shift occurs halfway between the frequency of the zero and the frequency of the pole on the log ω scale of the Bode plot. The relationship of Eq. (8.5-5) between the maximum phase attained and m can be seen in Table 8.5-1.

TABLE 8.5-1
The maximum phase lead of a lead compensator as a function of *m*

m	ϕ_l
3	30°
6	45°
10	55°
20	65°
100	78°

There is a side effect in using the lead compensator. As demonstrated in Fig. 8.5-1, the lead compensator of Eq. (8.5-1) produces a magnitude plot with gain that is $20 \log m$ dB larger at high frequencies than at low frequencies. This gain increase tends to shift the crossover frequency out to a higher frequency. Conversely, if the lead compensator is implemented in the form of Eq. (8.5-2) with unity high frequency gain the low frequencies are attenuated by $20 \log m$ dB and the crossover frequency is decreased.

In using the lead compensator, the effect of the lead compensator on the magnitude plot should be considered so that the maximum phase lead occurs at the new crossover frequency rather that the old crossover frequency. Here is the key observation. Since ω_l, the frequency of maximum phase lead, occurs halfway between the zero and pole positions on the Bode plot, the compensator achieves half of its gain (on a logarithmic basis) at this frequency. For the unity dc gain version of Eq. (8.5-1) the value of the magnitude of the lead compensator at ω_l is given as

$$20 \log |G_{\text{lead}}(j\omega_l)| = 10 \log m \text{ dB} \tag{8.5-8a}$$

or

$$|G_{\text{lead}}(j\omega_l)| = \sqrt{m} \tag{8.5-8b}$$

The new crossover frequency occurs at the frequency where the magnitude of the previous loop gain passes through $1/\sqrt{m}$ or $-10 \log m$ dB.

For the unity high frequency gain version of Eq. (8.5-2) the loop gain is partially attenuated at ω_l

$$20 \log |G_{\text{lead}}(j\omega_l)| = -10 \log m \text{ dB} \tag{8.5-9a}$$

or

$$|G_{\text{lead}}(j\omega_l)| = 1/\sqrt{m} \tag{8.5-9b}$$

The new crossover frequency in this case occurs at the frequency where the magnitude of the previous loop gain passes through $\sqrt{m}$ or $+10 \log m$ dB. These facts are used to properly place a lead compensator.

Let us examine the effects of a lead compensator in a loop gain transfer function and see how the previous observations can be used to properly place a lead compensator for maximal benefit, i.e., with maximal phase margin at the resulting crossover frequency. Let us first examine the use of the unity dc gain version of the lead compensator given by Eq. (8.5-1).

Assume a plant is given by

$$G_p(s) = \frac{1}{s(s+1)} \tag{8.5-10}$$

Such a plant is typical of position control systems where there is a first-order response between the input actuator and the velocity of the output. The velocity is integrated to produce the position output.

The Bode plots of the plant of Eq. (8.5-10) are given in Fig. 8.5-2. The closed-loop controller is nominally stable for all proportional compensators with negative feedback followed by positive gain. As the gain of the controller is increased the crossover frequency of the system is increased and the phase margin is eroded.

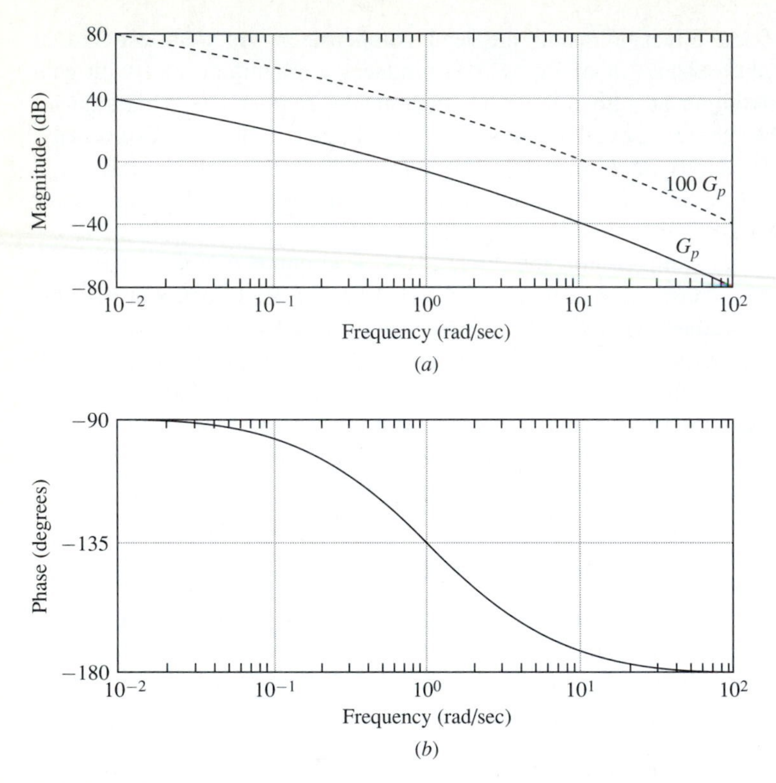

FIGURE 8.5-2
Bode plots of G_p and $100G_p$. (*a*) Magnitude; (*b*) phase.

Assume that a bandwidth of $\omega_c = 10$ is considered adequate speed. From the plant's Bode plots we can see that, at $\omega = 10$, the plant's magnitude is about 0.01 and its phase is 175°. Thus we conclude that, if a proportional compensator is used in the first controller design iteration, a gain of 100 is needed to set the crossover frequency at $\omega_c = 10$. The resulting loop gain Bode plot, also shown in Fig. 8.5-2, is large at low frequencies and small at high frequencies as generally desired, but it transitions from high to low frequencies too quickly. The large slope at crossover frequency around the Bode magnitude plot indicates a large amount of phase lag at crossover frequency and a small phase margin. The phase plot is unchanged by the proportional compensator so that the phase margin at the new crossover frequency of $\omega_c = 10$ is about 5°.

Now we attempt to modify the phase margin with the use of a lead compensator. Let us first examine adding a unity dc gain lead compensator of Eq. (8.5-1) to the loop gain $100G_P(s)$. Assume that we'd like to increase the phase at the new crossover frequency by 55°. Using either Table 8.5-1 or Eq. (8.5-5) we find that we must use $m = 10$. The only parameter left to choose is ω_l. After selecting m and ω_l, the c and d needed in Eq. (8.5-1) can be found using Eq. (8.5-6).

Choosing ω_l is somewhat tricky. Remember that the presence of the lead compensator changes the crossover frequency and we want the maximal phase lead which appears at ω_l to coincide with this new crossover frequency. Since the unity dc gain lead compensator adds $+10\log m$ dB of gain at ω_l we set ω_l equal to the frequency where the old loop gain passes through $-10\log m$ dB. The $10\log m$ dB gain produced by the lead compensator moves the gain at ω_l to 0 dB, i.e., ω_l becomes the new crossover frequency. In our case with m=10 we find the point where $100G_P(s)$ passes through -10 dB.

Figure 8.5-3 is a magnification of the magnitude plot of $100G_P(s)$ near the crossover frequency. From Fig. 8.5-3 we find that we should set ω_l around $\omega_l = 20$. (Closer calculation would put $\omega_l = 18$.) The resulting lead compensator with $m = 10$ and $\omega_l = 20$ is then calculated to be

$$G_{\text{lead1}}(s) = \frac{\dfrac{s}{6.3}+1}{\dfrac{s}{63}+1} = \frac{10s+63}{s+63} \tag{8.5-11}$$

The Bode plots of the proportionally compensated $100G_P(s)$ and the lead compensated loop gains are shown in Fig. 8.5-4. The crossover frequency and the maximal phase lead are indeed near $\omega_l = 20$ and the new phase margin is 58° (55° from the lead compensator and 3° from $100G_P(s)$ at $\omega_l = 20$). All the calculations work out as predicted.

The root locus for the loop gain with the lead compensator is a typical root locus for lead compensators and is shown in Fig. 8.5-5. The zero of the lead compensator at $s = -6.3$ "pulls" the closed-loop poles further into the left half-plane. The additional pole at $s = -63$ is chosen far enough out so that it has little effect on the dominant part of the root locus. The reason for this pole is to make the lead compensator physically realizable and to make the Bode magnitude plot of the compensated loop gain attenuate quickly enough. The chosen closed-loop poles are marked with stars.

From the root locus it appears that we could get better response by increasing the gain of the controller. Actually, for this plant, this observation is mathematically correct. In real applications, however, too large a loop gain would cause a loss of ro-

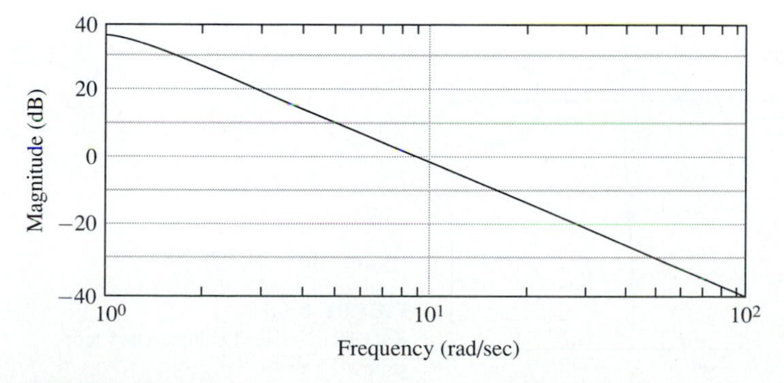

FIGURE 8.5-3
Magnification of the magnitude plot of Fig. 8.5-2 near crossover.

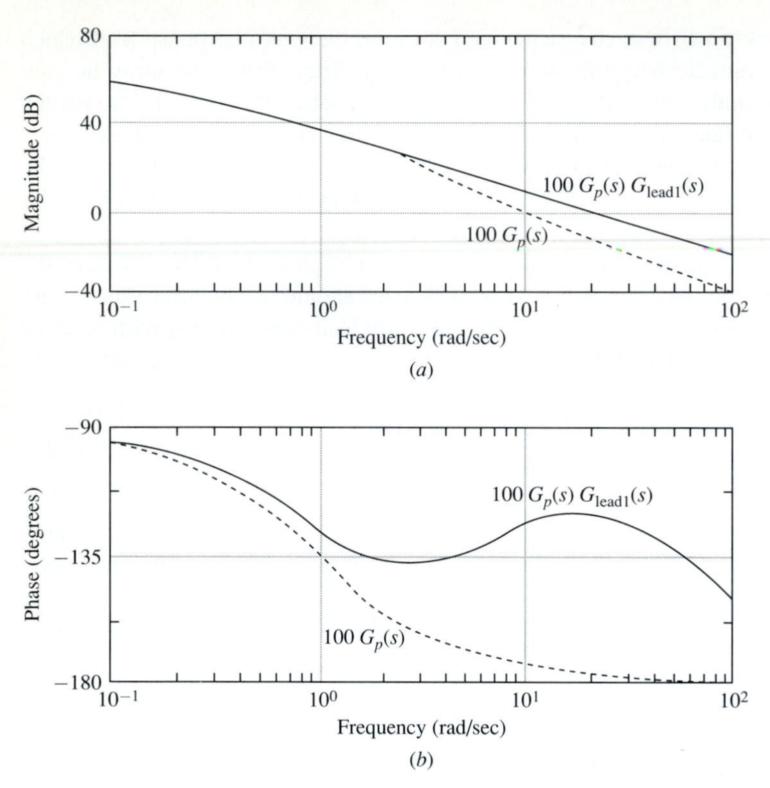

FIGURE 8.5-4
Bode plots of proportional and lead compensated loop gains. (*a*) Magnitude; (*b*) phase.

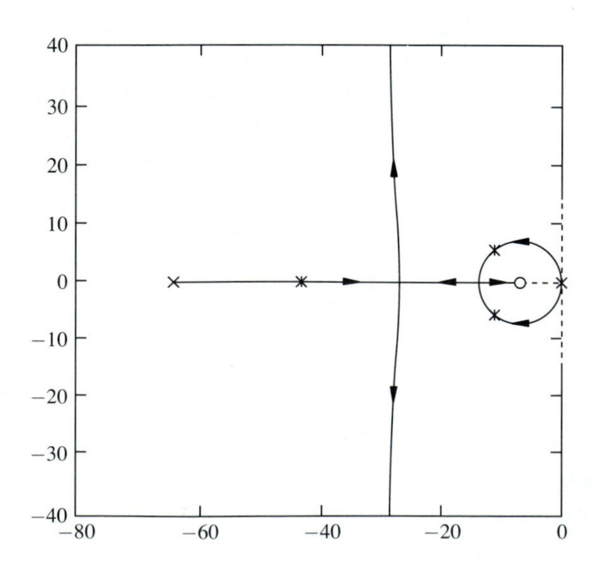

FIGURE 8.5-5
Root locus of lead compensated loop gain.

bustness properties. Since we wanted to concentrate on the lead design we chose not to complicate the problem statement by including a bound on unmodeled dynamics in the form a multiplicative perturbation. Instead we chose to state simply that the bandwidth of the system should be around 10 rad/sec. This is actually a more common way for model builders and control practitioners to describe model limititations. Because they know that bad things will happen if the system response is too fast, they express the model's limitations by putting a constraint on the bandwidth. It is the control designer's job to press the modeler on the exact nature of the uncertainties if the goal is to press the performance further. (It should be noted that often the control designer and the modeler are the same person. In that case, that person must develop a more refined understanding of the system and the requirements if the desire is to extend performance.)

Starting again from the loop gain, $100G_p(s)$, let us examine the effect of using a unity high frequency gain lead compensator given in Eq. (8.5-2). This time, the lead compensator attenuates the gain below and around the transition frequencies so that the unity high frequency gain lead compensator lowers the crossover frequency. For the unity high frequency gain lead compensator the gain at ω_l, the frequency of maximal phase lead, is $-10\log m$ dB (note the minus sign). To assure that ω_l coincides with the new crossover frequency, ω_l is chosen as the frequency where the previous loop gain magnitude crosses $+10\log m$ dB. (Notice the plus sign.) Suppose that this time, for a change, we intend to increase the phase margin by 45°. Now Eq. (8.5-5) dictates that $m = 6$. In this case we compute $10\log 6 = 7.8$ and from the Bode magnitude plot of Fig. 8.5-3, we select $\omega_l = 7$. The lead compensator is computed from Eq. (8.5-2) as

$$G_{\text{lead2}}(s) = \frac{s + 2.8}{s + 16.8} \tag{8.5-12}$$

The Bode plots of $100G_p(s)$ and $100G_p(s)G_{\text{lead2}}(s)$ are given in Fig. 8.5-6. Notice that the new crossover frequency is very near $\omega_l = 7$ as planned and that 45° phase lead has been added to the previous phase lead of 8° at this new crossover frequency. Notice also that the addition of the lead compensator reduces the system's bandwidth and lowers the low frequency gain but leaves the high frequency magnitude basically unaltered. The root locus of this lead compensator is similar to the root locus of the unity dc gain lead compensator since the root locus does not display the small difference in high frequency gain, low frequency gain and bandwidth as clearly as the Bode plot does.

8.5.1 PID and Lead-Lag Compensators

Very early in the design of control systems it was discovered that a great many systems can be adequately controlled with a simple series compensator. This is especially applicable to the control of chemical and manufacturing processes. This simple series compensator, called a *PID compensator*, consists of three terms

$$G_{\text{PID}}(s) = K_p + \frac{1}{T_I s} + \frac{T_D s}{\dfrac{s}{p} + 1} \tag{8.5-13}$$

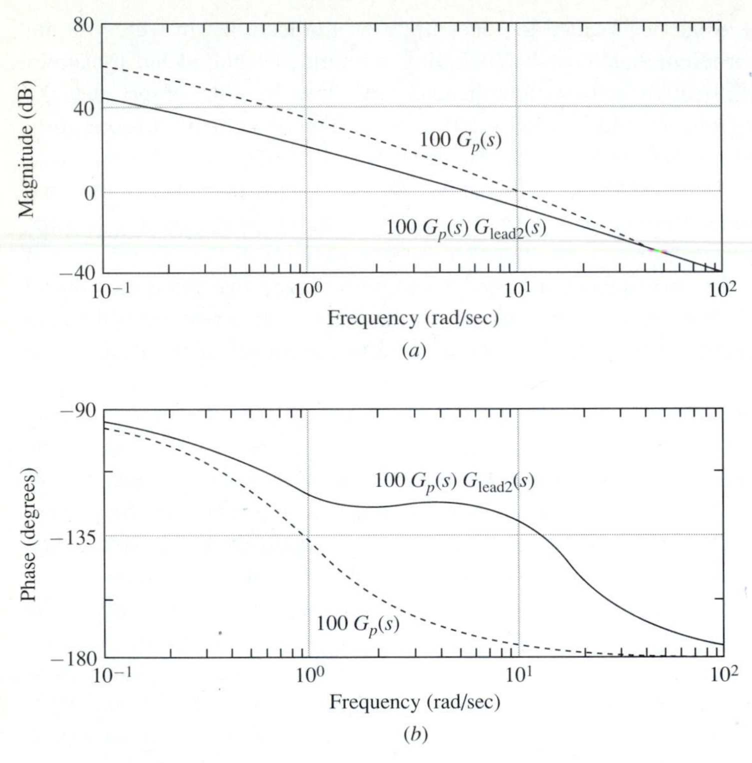

FIGURE 8.5-6
Bode plots for proportional and second lead compensated loop gains. (*a*) Magnitude; (*b*) phase.

The first two terms comprise the *PI* compensator, which was found in Sec. 8.4 to be equivalent to a lag compensator. The three parameters K_p, T_I, and T_D are adjusted to produce acceptable control performance. The first term is called the *proportional* term. The second term is called the *integral* or *reset* term. The third term is called the *derivative* term. The parameter p is generally chosen large. The pole controlled by p is present so that the compensator is realizable and the high frequency gain is limited. Intuitively, derivative action creates a control signal that is sensitive to changes in the error between the actual output and the desired output. Derivative action enables the system to react more quickly than a purely proportional compensator. By combining the terms in Eq. (8.5-13) we obtain

$$G_{\text{PID}}(s) = \frac{\left(T_D + \dfrac{K_p}{p}\right)s^2 + \left(K_p + \dfrac{1}{T_I P}\right)s + \dfrac{1}{T_I}}{s\left(1 + \dfrac{s}{p}\right)} \tag{8.5-14}$$

The PID series compensator contains two poles, one at the origin and one at a large negative value. The second-order numerator indicates that there are also two zeros. The zeros are usually located at intermediate frequencies between the two poles. By considering each zero as paired with a pole, we can see that the PID compensator is equivalent to the serial combination of a lag compensator and a lead compensator. The utility of this combination is fundamental to controller design. The overall gain is set to provide enough attenuation at high frequencies. The lead compensator, corresponding to the addition of the derivative term, provides for a non-oscillatory response by making the return difference larger at intermediary frequencies. The lag compensator, corresponding to the addition of the integral term, provides for adequate low frequency sensitivity reduction, disturbance rejection, and steady-state error properties by providing high loop gain at low frequencies.

Let us close this section by completing the design of the compensator for the automobile cruise control system discussed in Secs. 8.3 and 8.4.

Example 8.5-1. Recall from Fig. 8.4-4 that after placing a lag compensator (equivalently, a PI compensator) in front of the plant of Eq. (8.3-1) we are left with a design that meets the high frequency constraints on the complementary sensitivity function imposed by the multiplicative perturbation structure but does not meet the performance criterion. In particular, Fig. 8.4-4*b* shows that the sensitivity function stays just below its performance specification at low frequencies and that it exceeds its allowable values near crossover frequency. The problems can be solved by including a lead compensator in the design. Since we cannot afford to lower the low frequency gain much we choose the unity low frequency gain form of the lead compensator given by Eq. (8.5-1). This form increases the bandwidth of the system slightly and leaves the low frequency portion of the Bode plots almost unchanged.

We must be concerned with how much we increase the high frequency gain of the loop. Recall that the unity low frequency gain lead compensator increases the gain at high frequencies by $20 \log m$ dB. If the high frequency gain in Fig. 8.4-3*a* is increased by 20 dB the loop gain butts up against the model uncertainty bound, leaving no margin for robustness of performance. We choose $m = 8$, which raises the high frequency gain 18 dB. (This choice is not obvious; some educated guessing and some iteration of choices is part the design process. Of course, all such guesses are thoroughly tested through analysis and experimentation before implementation.)

The choice of $m = 8$ means that the gain of the system at the frequency of maximal phase lead is increased by $10 \log 8 = 9$ dB. Thus, ω_l is taken as the frequency where the working loop gain passes through -9 dB, in this case $\omega_l = 10$. The resulting lead compensator is given by

$$G_{L1}(s) = 8\frac{s + 3.5}{s + 28} \tag{8.5-15}$$

The loop gain without the lead compensator is shown in Fig. 8.5-7 by the solid line while the modified loop gain with the lead compensator is shown by the dashed line. The new crossover frequency is 10 rad/sec as expected. There is an increase of 51° of phase lead at this frequency as a properly placed lead compensator with $m = 8$ should attain.

The resulting design must be checked against the actual specifications. The results plotted in Fig 8.5-8*a* show that the complementary sensitivity function remains below its constraint as determined by the model bound. The results plotted in Fig. 8.5-8*b* show that the sensitivity function barely remains below its constraint as determined by the performance bound. Thus, assuming that the constraints are conservative enough,

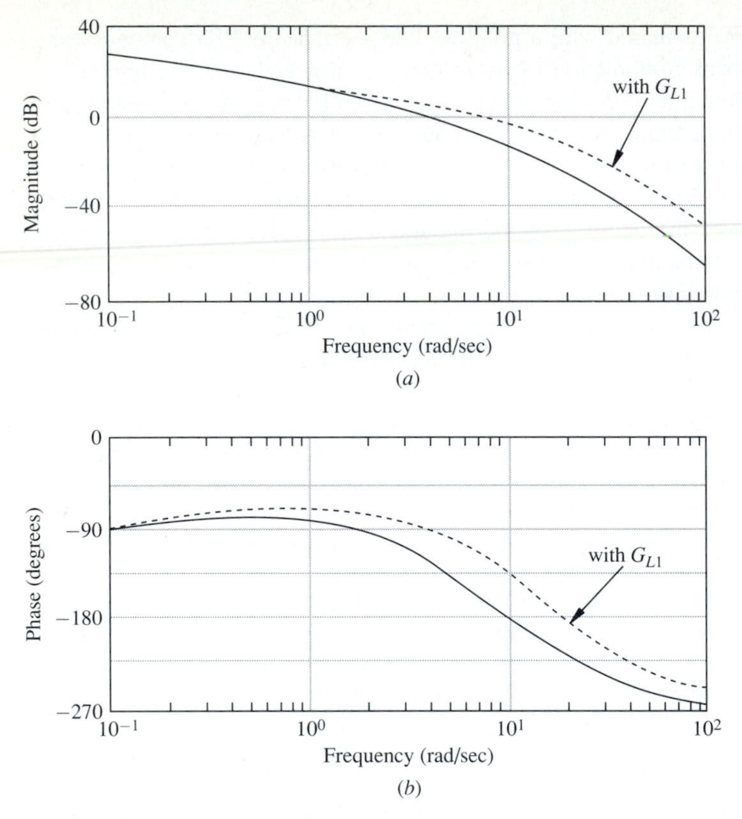

FIGURE 8.5-7
Bode plots of loop gains with and without G_{L1}. (*a*) Magnitude; (*b*) phase.

nominal performance and robust stability are achieved for this design. However, the small margin by which the nominal performance is met leads one to suspect that the robust performance condition of Eq. (8.2-8) is not met.

The results shown in Fig. 8.5-9 confirm that the robust performance condition is not met. Figure 8.5-9 contains a plot of 20 times the log of the quantity

$$\left| \frac{G(j\omega)}{1+G(j\omega)} \right| l_m(j\omega) + \left| \frac{1}{1+G(j\omega)} \right| p(j\omega) \tag{8.5-16}$$

To satisfy the robust performance condition of Eq. (8.2-8) this quantity should remain less that 0 dB for all frequencies. The current loop produces a peak of almost 2 dB around $\omega = 14$ or 15.

What do we do now? Unfortunately, we have run out of standard tricks to play. It is time to use some creativity in applying background knowledge, to use the computer to check out some guesses of what might work and to hope we get lucky. The following explanation is included as a reminder that, while the fundamentals of control design are well understood, any difficult design requires some creative effort. The creativity is greatly enhanced by a sound understanding of fundamental causes and effects and by logical reasoning. It is not clear that a novice control student should be able to adjust

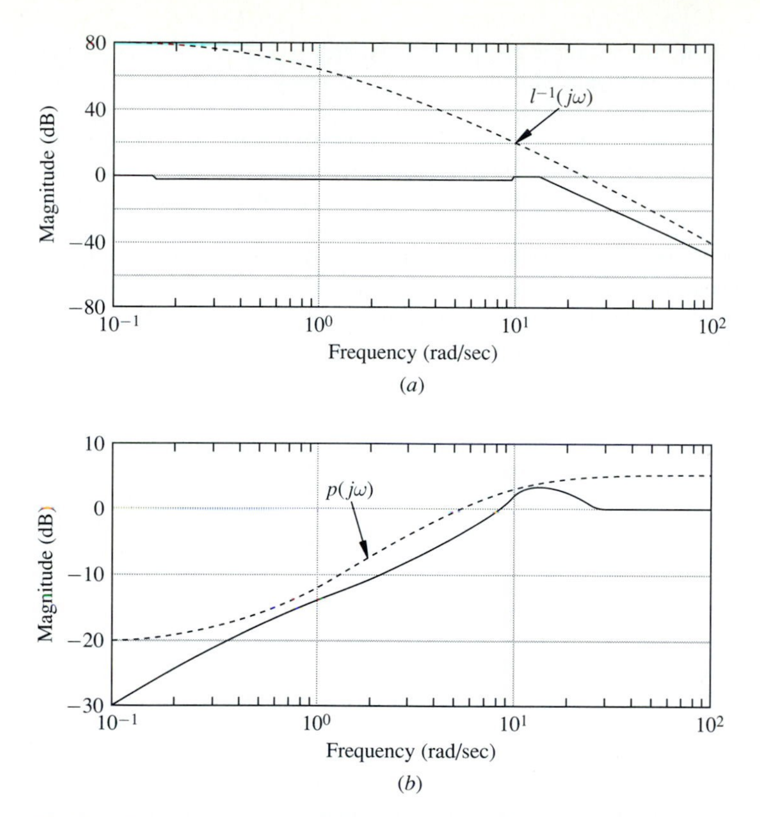

FIGURE 8.5-8
Sensitivity functions for loop gains with G_{L1}. (*a*) Complementary sensitivity functions; (*b*) sensitivity function.

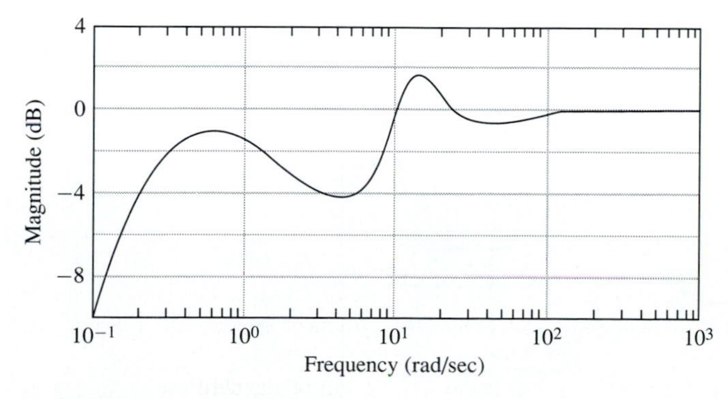

FIGURE 8.5-9
Robust performance measure using G_{L1}.

this design to make it work but a student should have the background by now to follow the reasoning in what follows. We believe that it should be instructive to see the kind of thinking required to fine tune a design in a logical manner.

The first helpful observation is that the problem area in Fig. 8.5-9 occurs not at the crossover frequency but slightly after the crossover frequency. It can be reasoned that the problem of an excessively small return difference might be corrected by increasing the phase lead at this critical frequency. The lead compensator that was tried first can be replaced by a redesigned lead compensator that achieves its maximal phase lead at the slightly higher frequency of $\omega = 14$. The value $m = 8$ is retained so that the very high and very low frequency performance is the same as it is for the first lead compensator. We emphasize that the new lead compensator is not just added to the loop but it *replaces* the original lead compensator. It represents a fine tuning away from the generally correct thought that maximal phase lead is desirable at crossover frequency to place the maximal phase lead where it is needed most *in this particular case*. The resulting lead compensator is computed to be

$$G_{L2}(s) = 8\frac{s+5}{s+40} \tag{8.5-17}$$

The resulting robust performance measure shown in Fig. 8.5-10 is encouraging. The peak has come down substantially. Now there is a real possibility of meeting the objective of keeping the robust performance measure below 0 dB. While the bound is still tight at high frequecies there is some room at low frequencies to trade off against the speed of response to further increase the phase margin near $\omega = 14$.

Consider the following. If the pole of the lead compensator is moved out along the negative real axis, more phase lead will be allowed to develop throughout the frequencies where the lead compensator has effect. The pole must be moved in such a way as to avoid changing the high frequency gain. The increased phase margin is offset by lowering the low frequency gains and increasing the sensitivity function at low frequency. Thus the pole should not be moved too far. Computer iterations show that this logic is sound.

As the parameter placing the pole is increased from 40, the peak in the robust performance measure of Fig. 8.5-10, which is located around $\omega = 14$, decreases while the local peak around $\omega = 0.7$ increases. After some trial and error the following redesigned lead compensator results.

$$G_{L3}(s) = 8\frac{s+5}{s+44} \tag{8.5-18}$$

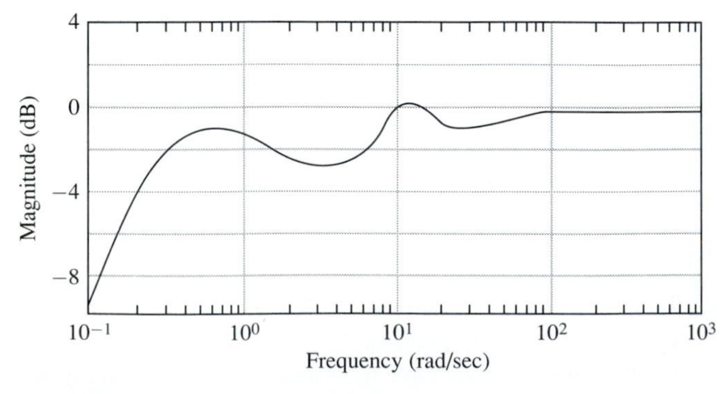

FIGURE 8.5-10
Robust performance measure using G_{L2}.

The plot of Fig. 8.5-11 shows the final robust performance measure for the plant of Eq. (8.3-1) in series with the lead-lag compensator given by

$$G_c(s) = 480\frac{(s+0.5)(s+5)}{s(s+44)} \tag{8.5-19}$$

The designed system meets the design criteria. It should be noted that in this example the design criteria were fairly difficult to meet, requiring some iteration in design beyond the straightforward application of standard building blocks. Often, the original requirements are easily met. Sometimes the original requirements cannot be met. If the original requirements cannot be met, the control designer must modify the design, while those who specify the requirements and those who build the models must also contribute their own modifications. If the requirements are loose it may be possible to take advantage of the additional margins available. If the requirements are unrealistic it must be determined where the expectations are too high, the bounding is too conservative, or the model is too inaccurate. Remember that the vast majority of effort in a control design is involved in creating a model, setting expectations, and validating the design. The actual design which lends itself to mathematical analysis and which is treated in this book is just one part of the overall design.

Exercises 8.5

8.5-1. Using the plant given by

$$G_p(s) = \frac{100}{(s+1)(s+10)}$$

design a lead compensator to increase the phase margin. Use a unity dc gain compensator with $m = d/c = 10$. Set ω_l equal to the original plant crossover frequency. Generate Bode magnitude and phase plots of the controller-plant combination. Then answer the following questions:

(*a*) Where in the frequency spectrum is performance lost due to the addition of the lead compensator?

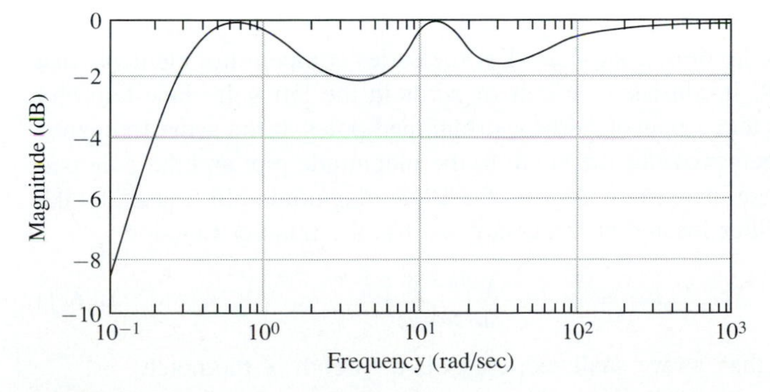

FIGURE 8.5-11
Robust performance measure for final design.

(*b*) Where does the crossover frequency move to when the lead compensator is added? Does this agree with the text?

(*c*) How much phase margin was actually added by the compensator? How does this compare to the maximum phase lead possible when $m = d/c = 10$?

8.5-2. Given the plant

$$G_p(s) = \frac{40}{(s+2)^2}$$

design a unity dc gain lead compensator so that the resulting crossover frequency has a phase margin of 50° and this new crossover frequency is located where the maximum phase lead of the lead compensator occurs.

8.6 SERIES COMPENSATOR BUILDING BLOCKS: HIGH FREQUENCY ROLL-OFF, NOTCH FILTERS, CANCELING PLANT DYNAMICS

Occasionally, additional control elements are needed to provide sufficient attenuation at high frequencies. This attenuation is often referred to as high frequency *roll-off*. It is most simply accomplished by adding poles in series with the previously designed compensator at a frequency higher than the crossover frequency. The poles cause the magnitude plot of the loop gain to roll off at a greater negative slope. However, care must be exercised as the additional poles also produce phase lag at frequencies less than the frequency position of the poles. Each additional pole produces 6° lag at a frequency one decade before the pole position and greater phase lag for higher frequencies. If poles for roll-off are placed too close to the crossover frequency, the system's phase margin decreases, performance suffers, and robustness at intermediate frequencies is adversely affected. This fact is in accord with the discussion in Sec. 8.2 where it was noted that the magnitude plot of the loop gain generally should be allowed to slope gradually through the area near the crossover frequency.

There is a series compensator element that produces a more sudden attenuation of the loop gain's magnitude plot than simple roll-off poles can produce. The element is called a *notch filter*. As its name suggests, the notch filter produces sharp attenuation over a small frequency range, creating a notch in the magnitude plot as seen in Fig. 8.6-1.

The notch filter is the most complicated of the series compensator elements that we address in this book. It consists of a pair of zeros in the left half-plane but very near the imaginary axis and a pair of fairly well-damped poles at the same frequency as the zeros. The zero pair provides the notch in the magnitude plot and the pole pair counteracts the long range increase in slope in the Bode magnitude plot caused by the zeros. A typical notch filter located at frequency ω_N has the transfer function

$$G_{\text{notch}}(s) = \frac{s^2 + 2\zeta_N\omega_N s + \omega_N^2}{s^2 + \omega_N s + \omega_N^2} \tag{8.6-1}$$

where ζ is a parameter that, as we shall see, controls the depth of the notch.

The Bode plots for a notch filter with $\omega_N = 1$ and $\zeta_N = 0.005$ are shown in Fig. 8.6-1. The straight line approximations of the magnitude plots for the numerator

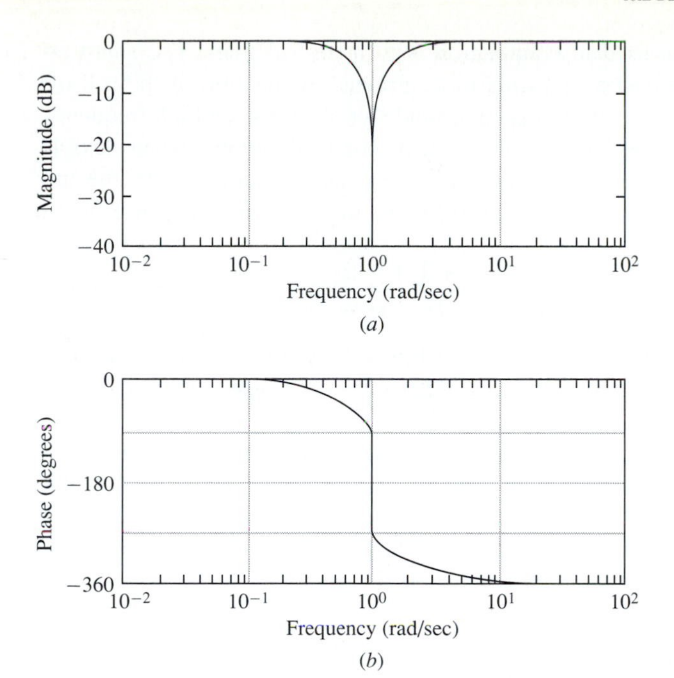

FIGURE 8.6-1
Bode plots of a notch filter.
(*a*) Magnitude; (*b*) phase.

and the denominator exactly offset each other. The damping of 0.5 for the pole pair has its actual magnitude plot following the straight line approximation for that contributor closely but the actual Bode magnitude plot for the zero pair has a notch in it of depth $-20 \log 2\zeta_N$. This creates the notch of the notch filter. With $\zeta_N = 0.005$ the depth of the notch is $20 \log(0.01) = -40$ dB. The poorly damped zero pair produces a 180° phase shift at the notch frequency. Notice that the notch filter also produces phase lag at frequencies less than ω_N. This phase lag can be traded off against the amount of attenuation by manipulating the damping constant of the second-order poles and zeros.

The plots of Fig. 8.6-1 also provide a convenient demonstration of what can be expected from computer plots. The most obvious surprise in the plot is that the phase seems to shift by a negative 180° as the frequency passes the zero pair. Of course we think of the left half-plane zero pair as producing a sudden 180° positive phase shift. There is really no difference if the shifts are sudden enough since phase values are equal when 360° is added or subtracted. The computer computes the phase shift and connects the points. Some programs are better than others about correcting the plot to match the way control theorists think about the plots. Notice also that the depth of the notch is not obvious from the magnitude plot of Fig. 8.6-1. When things are changing quickly the accuracy of the computer plots depends on how closely to the bottom of the notch the computer takes a sample point. We must know enough about what we expect to see in order to properly interpret the computer's output. (In particular, when working with computer aids we must be able to quickly ascertain whether the data have been entered correctly and the computer is correctly programmed.)

Notch filters are often used in conjunction with a roll-off filter. The sharp attenuation of the notch is passed over to the more gradual attenuation of the roll-off filter. In this way the loop gain can be shaped around sharply cornered high frequency constraints. Such constraints appear in practice in mechanical systems when flexible bending modes or other poorly damped oscillatory modes are lumped into the unmodeled dynamics. The Bode plots of Fig. 8.6-2 show the frequency response of the transfer function

$$G_{NR}(s) = \frac{\left(s^2 + 0.1s + 1^2\right)}{\left(s^2 + s + 1^2\right)} \frac{2}{(s+2)} \tag{8.6-2}$$

When using the notch filter in conjunction with the roll-off filter the phase lag that occurs before the magnitude notch must be carefully considered lest the phase margin of the system near the crossover frequency be adversely affected.

There is a second possible use of notch filters. Since the zeros produce a narrow notch in frequency, a notch filter can be used to counteract a narrow peak in the plant's magnitude plot caused by poorly damped poles. This technique amounts to a near pole-zero cancellation near the stability boundary of the imaginary axis. When using this technique, great care must be taken that the position of the poles is well known so that near perfect cancellation is achieved. In addition, such a cancellation has no effect on oscillatory responses to input disturbances as closed-loop poles still exist near the imaginary axis and the canceling zeros do not appear in the resulting closed-loop transfer function from the input disturbance to the plant output.

A few words about canceling dynamics are in order. It has been mentioned previously that the cancellation of poles and zeros in the right half-plane is not allowed.

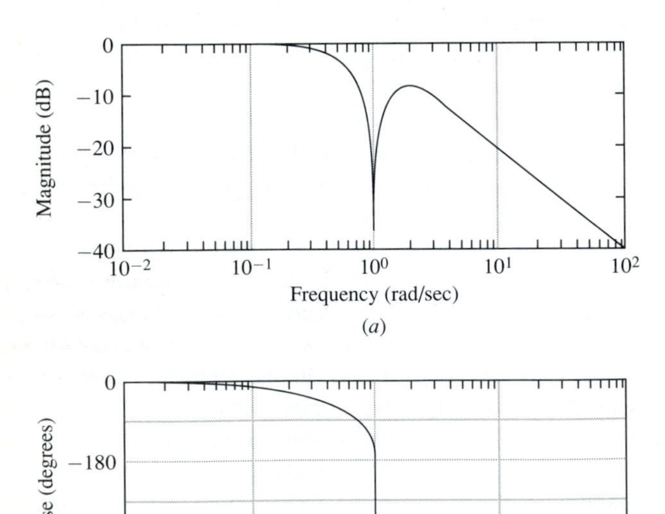

FIGURE 8.6-2
Bode plots of a notch filter with a roll-off pole. (*a*) Magnitude; (*b*) phase.

Let's examine this more closely. Look at the situation of Fig. 8.6-3, which shows an unstable plant pole at $s = 1$ canceled by a zero in a series compensator. (For this demonstration, we ignore the questions of improperness and physical unrealizability of the compensator.)

If the cancellation is perfect, the response to a reference input $R(s)$ is

$$\frac{Y(s)}{R(s)} = \frac{G_c(s)G_p(s)}{1 + G_c(s)G_p(s)} = \frac{\dfrac{1}{s+1}}{1 + \dfrac{1}{s+1}} = \frac{1}{s+2} \tag{8.6-3}$$

The response to an output disturbance $D_o(s)$ is

$$\frac{Y(s)}{D_o(s)} = \frac{1}{1 + G_c(s)G_p(s)} = \frac{1}{1 + \dfrac{1}{s+1}} = \frac{s+1}{s+2} \tag{8.6-4}$$

The two transfer functions are stable. However, the response to a disturbance at the plant input is

$$\frac{Y(s)}{D_i(s)} = \frac{G_p(s)}{1 + G_c(s)G_p(s)} = \frac{\dfrac{1}{(s+1)(s-1)}}{1 + \dfrac{1}{(s+1)}} = \frac{1}{(s+2)(s-1)} \tag{8.6-5}$$

This transfer function is unstable even though we assumed a perfect pole-zero cancellation. If the unstable plant pole were perfectly modeled and the controller zero perfectly implemented, the controller can make sure it doesn't pass signals that excite the unstable dynamics but it can do nothing about countering the effects of other signals or initial conditions that might excite the unstable dynamics. Thus, canceling unstable plant poles is not allowed.

A slightly different situation occurs in attempting to cancel right half-plane plant zeros. In Fig. 8.6-4, the plant zero at $s = 1$ is canceled with a controller pole at $s = 1$.

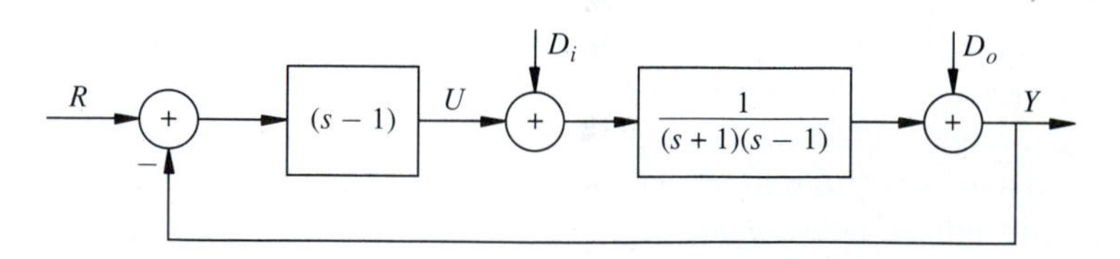

FIGURE 8.6-3
Canceling a right half-plane plant pole.

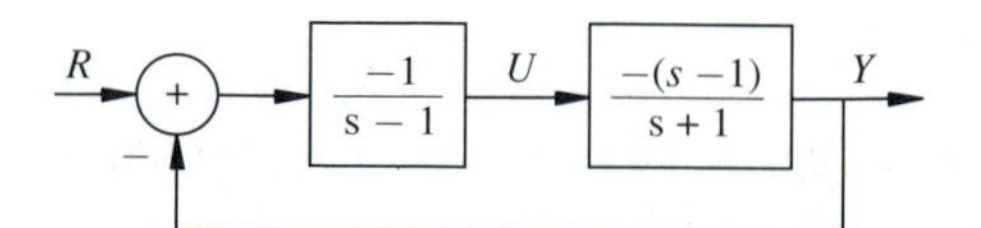

FIGURE 8.6-4
Canceling a right half-plane plant zero.

Again, the transfer function from R to Y is stable.

$$\frac{Y(s)}{R(s)} = \frac{G_c(s)G_p(s)}{1 + G_c(s)G_p(s)} = \frac{\dfrac{1}{s+1}}{1 + \dfrac{1}{s+1}} = \frac{1}{s+2}$$

Now the problem is that the transfer function from the reference input R to the plant input U is unstable.

$$\frac{U(s)}{R(s)} = \frac{G_c(s)}{1 + G_p(s)G_c(s)} = \frac{\dfrac{-1}{s-1}}{1 + \dfrac{1}{s+1}} = \frac{-1}{(s+1)(s-1)} \tag{8.6-6}$$

The plant input $u(t)$ grows without bound and surely saturates the plant's actuators.

So we see that even if perfect pole-zero cancellation could be attained, cancellation of right half-plane zeros or poles cannot be allowed. Of course, perfect cancellation is impossible and as a result the transfer functions just considered are unstable in practical situations. Unstable poles in any closed-loop transfer function cannot be allowed. Systems that yield stable closed-loop transfer functions from the input of any block to the output of any block are called *internally stable*. For the series compensator, if the closed-loop system is stable and there are no pole-zero cancellations between $G_c(s)$ and $G_p(s)$ the system is internally stable.

What about canceling left half-plane poles and zeros? Any time that something like a cancellation is forbidden for right half-plane poles or zeros, a thinking engineer must take warning against the same actions when considering poles or zeros in the left half-plane but near the $j\omega$ axis. We have seen that canceled poles and zeros do not disappear completely. Repeating the argument used above, it can be seen that if highly oscillatory plant poles are canceled by a compensator then the oscillatory plant poles still appear in the transfer function between an input disturbance and the output. If the dynamic modes associated with these poles are excited they produce poorly damped oscillations in the output. By repeating the argument used in considering the cancellation of right half-plane zeros, it can be seen that if left half-plane zeros near the $j\omega$ axis are canceled by a compensator these zeros appear as poorly damped poles in the transfer function between the reference input and the plant input. In this position the dynamic modes associated with these poles produce large oscillations in the plant input, which may saturate or stress the input actuators.

While canceling dynamics (poles or zeros) that are in the right half-plane is forbidden and canceling dynamics that are near the imaginary axis is not recommenced unless absolutely necessary, canceling well-behaved plant dynamics is perfectly legitimate. Indeed, one easy way to shape a loop gain into a desirable position when the plant contains well-behaved dynamics is to invert the plant transfer function, canceling all plant dynamics with the controller and to add the desired loop shape in series. This strategy of canceling all well-behaved plant poles and zeros is legitimate although it most probably leads to a controller with more dynamic elements than are absolutely necessary.

We have now been exposed to a number of building blocks for series compensator design. In the next section, we explore a realistic example that shows how to combine these building blocks in our overall control design strategy to produce satisfactory control designs.

Exercises 8.6

8.6-1. Roll-off poles are used to drop the loop gain under high frequency constraints in much the same way that lag compensators are used to raise the loop gain above low frequency constraints. Let

$$l_m(s) = \frac{(s+0.1)^2}{100}$$

$$G_p(s) = \frac{3(s+0.3)}{s^2}$$

$$p(s) = \frac{(s+0.5)}{s^2}$$

Find the phase margin for a unity proportional compensator, i.e., $G_c(s) = 1$. Design a controller using only roll-off poles so that the robust performance measure of Eq. (8.2-8) is met. Find the phase margin of the resulting loop gain. Explain how and why the phase margin changes between the two controllers.

8.6-2. Let

$$G_p(s) = \frac{100}{(s+1)\left(s^2+s+100\right)}$$

and

$$G_c(s) = \frac{10^4 K\left(s^2+s+100\right)}{(s+1000)^2}$$

Sketch the Bode plots of the loop gain when $K = 1$ and when $K = 100$. Find the response to a step disturbance at the plant input when $K = 1$ and $K = 100$. Explain the difference between the two situations.

8.7 A REALISTIC DESIGN EXAMPLE USING A LEAD-LAG COMPENSATOR[1]

In this section, the process of designing a control system is demonstrated by the use of a simple but very real design problem. The design problem concerns attitude stabilization of an experimental communication satellite called ATS-6 (Applications Technology Satellite 6). This spacecraft was launched in May of 1974 to perform communication experiments at geosynchronous (stationary) altitude. It performed its mission virtually flawlessly with control laws very similar to those we develop here.

---

[1]Much of this material relies on material in Stein and Sandel, *Classical and Modern Methods for Control System Design*, MIT, 1979. It is used with permission.

Detailed information about ATS-6 and its mission can be found in the May 27, 1974 issue of *Aviation Week*. Here we introduce just enough of the mission description to explain the control design problem. Physically, the satellite is a large, 30-foot-diameter, plated nylon mesh parabolic antenna. At its focus is a box of equipment called the earth viewing module (EVM), which contains the communication equipment and all attitude control, thermal control, and telemetry hardware. Power is supplied by two large solar panel drums extended on booms above the antenna. A sketch of the satellite is presented in Fig. 8.7-1*a*.

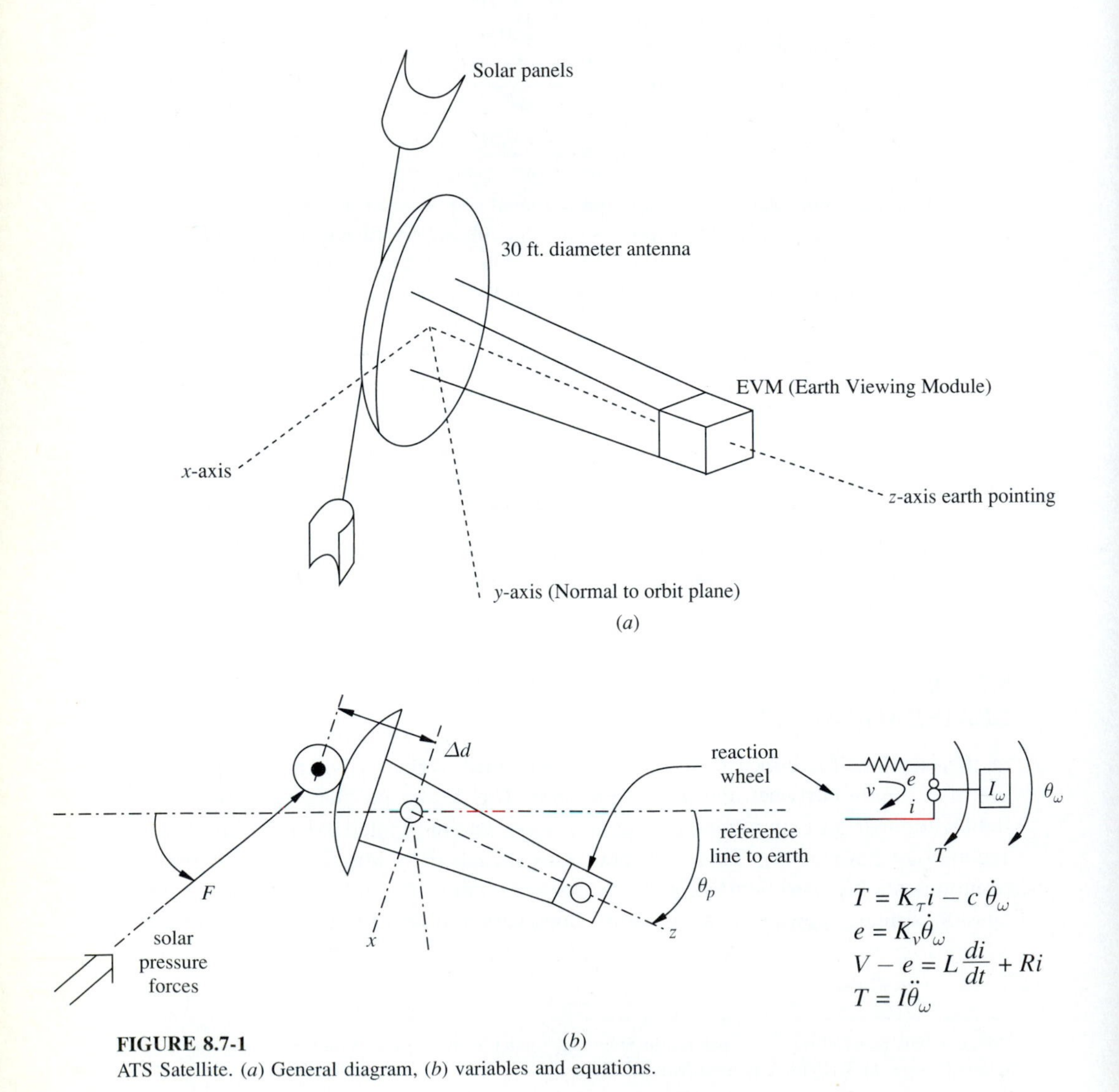

FIGURE 8.7-1
ATS Satellite. (*a*) General diagram, (*b*) variables and equations.

A satellite at geostationary altitude moves along its orbit at the same rate that the earth rotates about its axis. Hence the satellite is stationary with respect to points on earth. The controller's job is to point the communication antenna at selected receiving areas to within an absolute accuracy of 0.01 degree to 0.10 degree of arc, depending on the mission phase. The receiving areas include remote sites over Appalachia, the Rocky Mountains, and India. They also include special targets such as low-earth-orbit satellites (for data relay purposes such as the joint US-USSR Apollo-Soyuz space mission, where ATS-6 was the prime communications link), and aircraft within the earth's atmosphere (for air traffic control experiments). To sense where the antenna is pointing, the controller has two information sources. One is an infrared (IR) sensor on the EVM, which measures pointing angles with respect to the earth's infrared image. As seen from geostationary altitude, this image is a disc that the sensor attempts to keep centered in its field of view. The second information source is the communication system itself. Using wave interference techniques, the system can measure the antenna's pointing attitude directly to within approximately 0.01 degree of arc. The controller is expected to work equally well with either of these information sources. There is also a third sensor, which measures rotations about the line of sight to the earth by reference to the north star, Polaris. This sensor is not considered in this example.

We will consider the control design problem for one axis of motion for this satellite—the pitch axis or y axis. This axis is normal to the orbit plane and, hence, motion about it represents east-west pointing errors of the communication system. Such motions are uncoupled from other attitude motions if all motions are sufficiently small.

Referring to the sketch of Fig. 8.7-1*b* we see the variables of interest defined. The angle θ_p is the pointing angle to be controlled. The control is actuated by a reaction wheel, a device that produces torques by changing the speed of a rotating wheel through the use of a dc motor. The control variable V is the voltage driving the motor. As derived in Exercise 8.7-1, the transfer function between the control input V and the torque T produced by the motor is

$$\frac{T(s)}{V(s)} = \frac{s}{s + 0.01} \qquad (8.7\text{-}1)$$

Since there is essentially no friction in space the transfer function between the torque T and the angular position Θ_p is a simple double integrator (with appropriate scaling). The relationship between the control input V and the angular position Θ_p is

$$\frac{\Theta_p(s)}{V(s)} = \frac{1}{10^3 s(s + 0.01)} \qquad (8.7\text{-}2)$$

The angle of the satellite is not sensed instantaneously; there are dynamics associated with the sensors. The measured angle Θ_m is related to the actual angle Θ_p by the transfer function

$$\frac{\Theta_m(s)}{\Theta_p(s)} = \frac{10}{s + 10} \qquad (8.7\text{-}3)$$

As the sensor follows the pointing angle much faster than the pointing angle is expected to change, we consider the actual and the measured angles to be practically identical and consider the measured angle Θ_m as our variable both to be fed back and to be controlled. The plant transfer function becomes

$$\frac{\Theta_m(s)}{V(s)} = \frac{10}{10^3 s(s+0.01)(s+10)} \tag{8.7-4}$$

The pointing control is to be kept accurate in the face of torque disturbances caused by the pressure of solar radiation on the solar collector panels. The torques are cyclical in nature with the period equal to the 24 hours it takes the earth to rotate; the frequency is then equal to 7×10^{-5} rad/sec. The maximum torque is 10^{-5} foot-pounds. The maximal disturbance $d'(t)$ is

$$d'(t) = 10^{-5} \sin\left(7 \times 10^{-5} t\right) \tag{8.7-5}$$

As the disturbance is itself a torque on the spacecraft, it enters the plant at the same place as the control torque as shown in Fig. 8.7-2a. Since it is easier to treat disturbances as output disturbances, the disturbance is reflected to the output as in Fig. 8.7-2b. When there is no controller the transfer function from the disturbance $D'(s)$ to $\Theta_m(s)$ is

$$\frac{\Theta_m(s)}{D'(s)} = \frac{-(0.001)(10)}{s^2(s+10)} \tag{8.7-6}$$

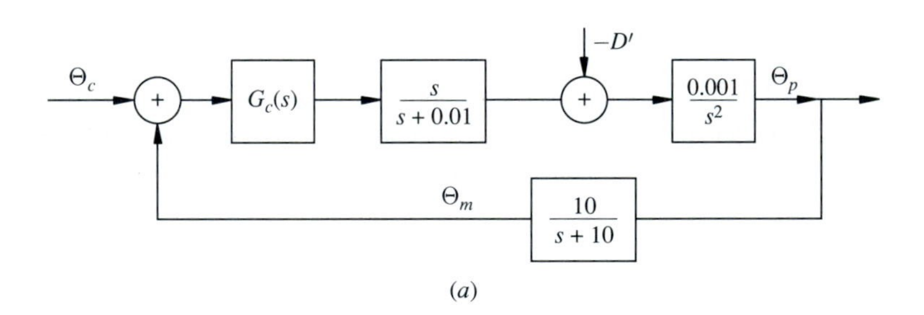

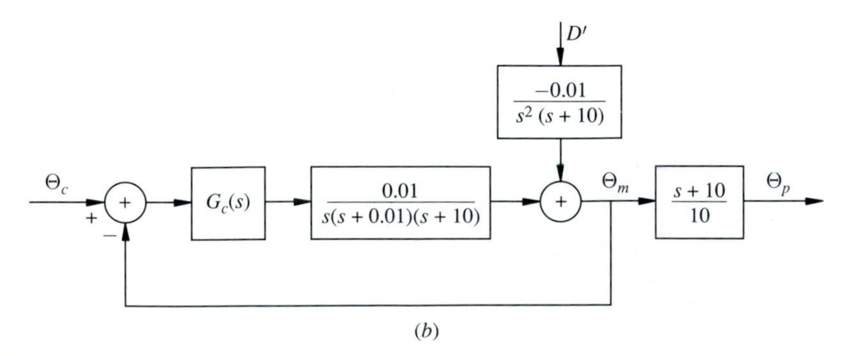

FIGURE 8.7-2
Block diagram of the plant. (a) Original; (b) revised.

The block diagram manipulations between Fig. 8.7-2*a* and Fig. 8.7-2*b* also demonstrate the decision to consider $\Theta_m(s)$ as the output to be controlled since it is available for feedback. The resulting loop is in the G configuration with the inverse of the sensor dynamics appended in series. Again, this filter can be ignored since it passes with little change the signals of frequency less than 10 rad/sec, and the bandwidth of the control loop should be much less than that.

The model is fairly accurate for low frequencies when the satellite is moved slowly enough to avoid exciting any bending modes or flexible resonances. Since the satellite material must be light to be launched economically, the satellite's structural members are flexible and oscillate if the satellite is torqued hard or fast enough to create bending. The fundamental bending mode comes from the beams connecting the EVM to the dish antenna. The oscillations are poorly damped due to the lack of atmosphere in space. The resonant frequency occurs at $\omega = 1$ rad/sec and produces a resonant peak on the magnitude plot of up to 30 dB along with a sudden phase shift of 180°. Since there are also bending modes at harmonics of 1 rad/sec and nonlinearities that become more important at higher frequencies, all these model inaccuracies can be covered by a magnitude-bounded multiplicative perturbation. The perturbation model is given as

$$\tilde{G}_p(s) = G_p(s)\,(1 + L_m(s)) \tag{8.7-7a}$$

$$|L_m(j\omega)| < l_m(j\omega) = \left| \frac{\left(\dfrac{j\omega}{-0.03} + 1\right)^2}{30} \right| = \frac{1 + \dfrac{\omega^2}{0.0009}}{30} \tag{8.7-7b}$$

There is one specific performance requirement. In spite of the disturbance, the pointing angle must be held to within 0.01° or 1.7×10^{-4} rad. From the block diagram we see that

$$\Theta_m(s) = \frac{\dfrac{-0.01}{s^2(s+10)}}{1 + G_c(s)G_p(s)} D'(s) \tag{8.7-8}$$

Take the magnitude of both sides of Eq. (8.7-8). By substituting the maximum disturbance and evaluating the expression at the frequency of the disturbance ω_e, the bound on the size of the pointing error can be transformed into a point on the performance bound of the return difference function.

$$|\Theta_m\,(j\omega_e)|_{\max} = \frac{\dfrac{0.01}{\omega_e^2\left(\omega_e^2 + 100\right)^{1/2}} 10^{-5}}{\left|1 + G_c\,(j\omega_e)\,G_p\,(j\omega_e)\right|} < 1.7 \times 10^{-4} \tag{8.7-9}$$

To meet this criterion the return difference must be large enough at $\omega_e = 7 \times 10^{-5}$ rad/sec. Specifically,

$$\left|1 + G_c\,(j\omega_e)\,G_p\,(j\omega_e)\right| \geq \frac{0.01\left(10^{-5}\right)}{\left(7 \times 10^{-5}\right)^2 10\left(1.7 \times 10^{-4}\right)} = 12{,}000 \tag{8.7-10}$$

The other performance requirements are general in nature. The return difference should never be much smaller than one so that the sensitivity remains low and the responses to disturbances are not too oscillatory or amplifying at any frequency. The bandwidth should be as large as the model accuracy allows, so that the response to any disturbance occurs as quickly as possible. The response to a step change in command input θ_c should be smooth with little or no overshoot. These requirements can be translated into guidelines for the shape of the loop as discussed in Sec. 6.7. Rather than establish a complete performance bounding function before designing the system we try to achieve good performance and then examine the resulting worst case performance. The one specific requirement we have is from Eq. (8.7-10).

$$p\,(j\omega_e) = 12{,}000 \tag{8.7-11}$$

Otherwise we expect $p(j\omega)$ to slope down to a value slightly less than one near crossover and then return to approach one asymptotically.

The guiding equation for robust performance is Eq. (8.2-8), repeated here as Eq. (8.7-12).

$$\left|\frac{G(j\omega)}{1+G(j\omega)}\right| l_m(j\omega) + \left|\frac{1}{1+G(j\omega)}\right| p(j\omega) < 1 \tag{8.7-12}$$

In Sec. 8.2 we made approximations and turned these into conditions on the loop gain for low frequencies and high frequencies as in Eqs. (8.2-12) and (8.2-15). Equation (8.2-12), repeated here as Eq. (8.7-13), is accurate at low frequencies when $l_m(j\omega) \ll 1$ and $p(j\omega) \gg 1$.

$$|G(j\omega)| > \frac{p(j\omega)}{1 - l_m(j\omega)} \tag{8.7-13}$$

The performance requirement given by Eq. (8.7-11) occurs at $\omega_e = 7 \times 10^{-5}$ rad/sec. At this frequency, $l_m(j\omega_e) = 1/30$. We extend the requirement of Eq. (8.7-13) for lower frequencies and mark it as an objective for the loop gain on the Bode plot of Fig. 8.7-3.

Equation (8.2-15), repeated here as Eq. (8.7-14), is accurate at high frequencies when $l_m(j\omega) \gg 1$ and $p(j\omega) \ll 1$.

$$|G(j\omega)| < \frac{1 - p(j\omega)}{l_m(j\omega)} \tag{8.7-14}$$

The approximation is fairly accurate when $l_m(j\omega) > 10$. The curve of $l_m^{-1}(j\omega)$ is sketched on Fig. 8.7-3 as an objective for the loop gain magnitude. The area where $l_m^{-1}(j\omega) < 0.1$ is marked off in a distinct way as the loop gain must stay below this objective. Although there is no specific performance requirement here we need to leave a safety margin so that not only stability but reasonable performance is maintained in the face of perturbations.

Figure 8.7-3 contains the Bode plots of the plant $G_p(s)$ and $10G_p(s)$. The effect of a simple proportional compensator is seen here. With the additional gain both the low and high frequency objectives for the loop gain are met. There are, however,

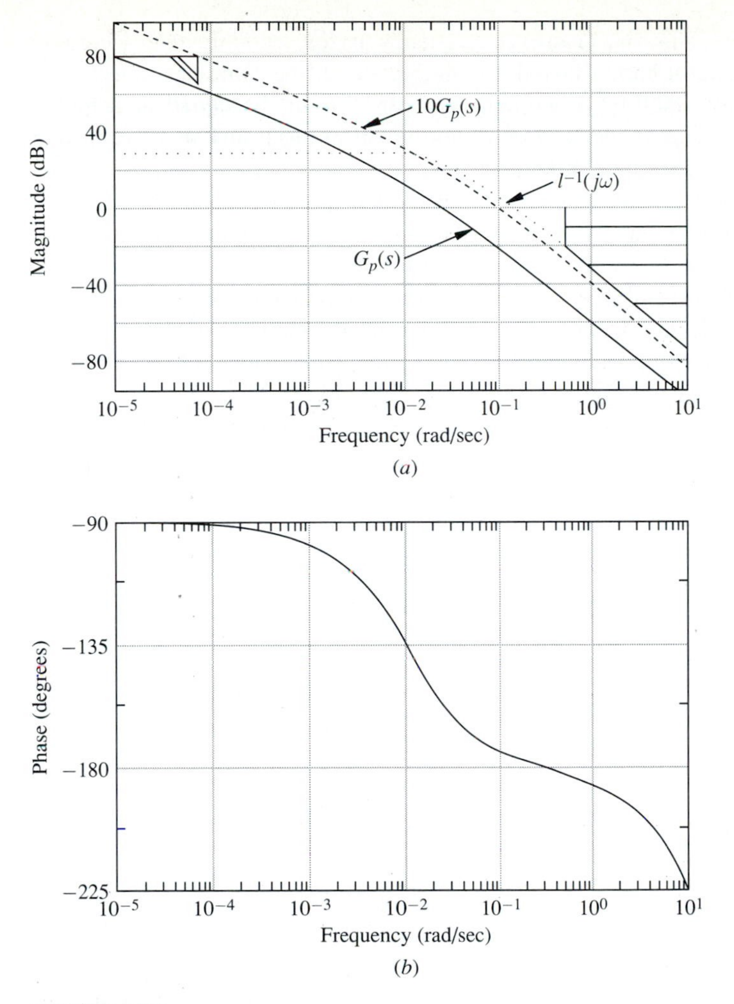

FIGURE 8.7-3
Bode plots of $G_p(s)$ and $10G_p(s)$. (*a*) Magnitude; (*b*) phase.

severe problems at intermediate frequencies. The crossover frequency is $\omega_c = 0.1$ and the phase margin is only 5°. This is unacceptably small since it implies

$$|1 + G\,(j\omega_c)| = 2\sin\frac{\phi_m}{2} = 0.09 \tag{8.7-15}$$

If the control design were stopped here, the closed-loop frequency response would contain a peak at ω_c with height $1/0.09 \approx 11$ or 21 dB. This would indicate closed-loop poles with a damping of approximately $\zeta = 0.05$. The step response of such a system would have overshoot and oscillation. What is worse is that the plant model could be inaccurate at $\omega_c = 0.1$ so that the actual Nyquist plot is closer to the -1 point than the nominal plot and the system would be even more oscillatory.

Notice, however, that the crossover frequency is low enough that, according to the multiplicative perturbation bound, the model should be quite accurate at this frequency and at least stability is assured. A phase margin this small at a higher frequency where the model is less accurate could result in instability if the controller were implemented on an actual satellite.

The phase margin can be improved by using a lead compensator. It is known from Sec. 8.5 that a lead compensator produces either high frequency amplification if Eq. (8.5-1) is used or low frequency attenuation if Eq. (8.5-2) is used. Referring to the loop gain plots of $10G_p(s)$ in Fig. 8.7-3 it can be seen that in either case the objectives are not met. We decide to use the unity high frequency gain form of Eq. (6.5-2) since the low frequency objective can be recovered by the later addition of a lag compensator. The fact that the high frequency constraint allows about three decades of frequency between the attainable crossover frequency and the low frequency constraint provides confidence that such a lag compensator can be built.

If 60° phase margin is to be achieved, the magnitude of the return difference at the crossover frequency must equal one. As this seems to be a reasonable objective, we try to get 55° of lead out of the lead compensator. Referring to Table 8.5-1, a value of $m = 10$ is needed for 55° phase lead at the frequency of maximum lead. The new crossover frequency coincides with ω_l if ω_l is chosen as the frequency where the old loop gain has a magnitude of $+10 \log m$ dB. From Fig. 8.7-3 we observe that the frequency where $10G_p(s)$ passes through $10 \log m = 10$ dB is $\omega = 0.06$. Therefore, we choose $m = 10$ and $\omega_l = 0.06$ for our lead compensator. The lead compensator is computed using Eqs. (8.5-6), and (8.5-2).

$$G_{\text{lead}}(s) = \frac{s + 0.019}{s + 0.19} \tag{8.7-16}$$

$$G_{c1}(s) = \frac{10(s + 0.019)}{s + 0.19} \tag{8.7-17}$$

The Bode plots of the old loop gain and the new loop gain are given in Fig. 8.7-4. As computed earlier, the new crossover frequency is $\omega_l = 0.06$ with 65° phase margin. The high frequency gain is unaffected while the response at low frequency has been attenuated. The next step is to append a lag compensator to the compensator to raise the low frequency gain while leaving the plots at and above crossover frequency nearly unaffected.

Before designing the needed lag compensator, let's examine the lead compensator from the root locus point of view. The root locus provides intuitively pleasing information about a system's transient response but little information about other performance and robustness qualities. The root locus of the plant with a lead compensator is shown in Fig. 8.7-5. The plant poles at $s = 0$ and $s = -0.01$ are shown along with the lead compensator's zero at $s = -0.019$ and pole at $s = -0.19$. The plant's pole at $s = -10$ is not shown as it moves very little. The basic concept of a lead compensator in the root locus setting is the placement of a zero in the left half-plane to the left of the dominant poles and the placement of a pole much further out in the left half-plane to provide realizability. The distant pole should have very little effect

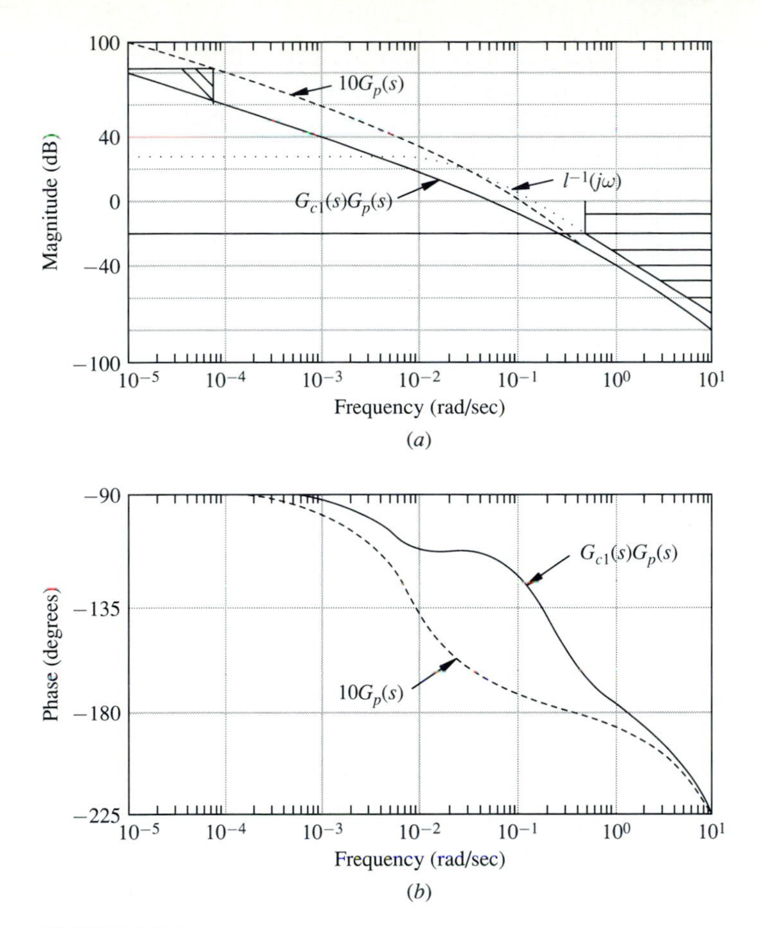

FIGURE 8.7-4
Bode plots of $10G_p(s)$ and $G_{c1}(s)G_p(s)$. (*a*) Magnitude; (*b*) phase.

on the root locus. The zero has the effect of pulling the dominant poles towards itself and further into the left half-plane. The result is a faster and less oscillatory response as the dominant poles are pulled away from the imaginary axis.

The closed-loop poles for the correct gain of the compensator of Eq. (6.7-17) are marked with stars on Fig. 8.7-5. Since the system is in the G configuration the root locus plane coincides with the closed-loop response plane. It is interesting to observe that there is a zero closer to the origin than the first closed-loop pole. We know from Chap. 4 that the step response of such a system will have some overshoot even though there are no complex pole pairs. This problem is easily corrected with a prefilter as will be seen later.

We now design the lag compensator to meet the low frequency performance constraint. The key part of the lag compensator design is to avoid causing problems at the crossover frequency. Since the phase margin of the lead compensated system is 65°, we can afford to place the zero of the lag compensator only a decade below the

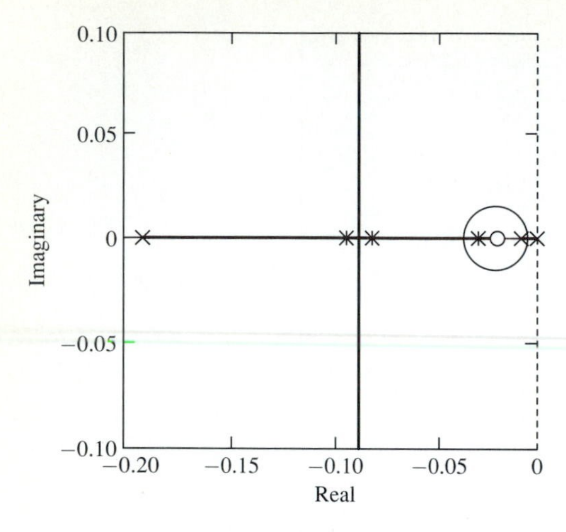

FIGURE 8.7-5
Root locus of $G_{c1}(s)G_p(s)$.

crossover frequency, creating 6° phase lag at the crossover frequency. The equation for the lag compensator is

$$G_{\text{lag}}(s) = \frac{s + 0.005}{s} \tag{8.7-18}$$

The compensator is now

$$G_{c2}(s) = \frac{10(s + 0.005)}{s} \frac{(s + 0.019)}{(s + 0.19)} \tag{8.7-19}$$

The Bode plots of the loop gains using compensators $G_{c1}(s)$ of Eq. (8.7-17) and $G_{c2}(s)$ of Eq. (8.7-19) are given in Fig. 8.7-6. It can be seen that the lag compensator raises the low frequency gain while leaving the frequencies higher than the crossover frequency unaffected.

The final loop gain using $G_{c2}(s)$ as a compensator appears to be acceptable. Both high and low frequency objectives are met with enough margin to expect adequately robust performance. The phase margin of 60° also indicates adequately robust performance.

It is interesting that, since the lag compensators are designed to have very little effect on a system's transient response, lag compensators have very little effect on closed-loop pole positions on a root locus plot. The root locus plot of the poles and zeros close to the origin for the plant with the lead-lag compensator $G_{c2}(s)$ is shown in Fig. 8.7-7. The insert in the figure gives a magnified view of the behavior of the poles and zeros that are very near the origin. The second pole at the origin and the low frequency zero from the lag compensator are so close to each other that they have very little effect on the root locus except for points very close to the origin. The resulting closed-loop pole near the low frequency zero from the lag compensator is effectively canceled in the closed-loop transfer function since the series compensator zeros appear in the closed-loop transfer function.

It is time now to visually check the sensitivity function and the complementary sensitivity function. In addition, the closed-loop step response should be checked.

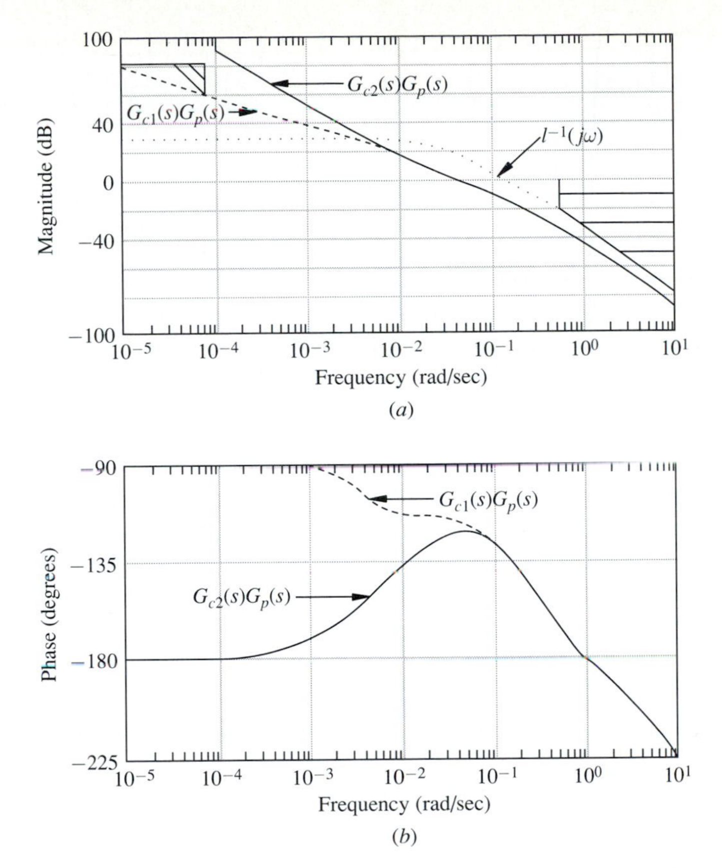

FIGURE 8.7-6
Bode plots of $G_{c2}(s)G_p(s)$. (*a*) Magnitude; (*b*) phase.

The plots of Fig. 8.7-8 show the sensitivity function and the complementary sensitivity function for the loop with the controller $G_{c2}(s)$. These plots can be used to check the robust performance condition of Eq. (8.2-8). First note that the specific low frequency disturbance rejection requirement is easily met as shown on the plot of the sensitivity function, which ascends from very small values at low frequency to a maximum value of about 2 dB at $\omega = 0.1$ and returns toward unity at high frequencies. The magnitude of the complementary sensitivity function is near unity for low frequencies, has a peak of about 1 dB at $\omega = 0.03$ and falls off rapidly, remaining at least 10 dB below the $l_m^{-1}(j\omega)$ curve. This 10 dB gap indicates that not only does the system remain stable in the face of all allowable perturbations but also maintains reasonable performance.

The sensitivity function that results when the system is confronted with the worst case allowable perturbation is obtained using Eq. (6.7-16). In Fig. 8.7-9 the nominal sensitivity function and the worst case sensitivity function are plotted. Even when the worst case perturbation is used the maximal sensitivity is about 4 dB = 1.6. This indicates that the system performs robustly for the mismodeling considered.

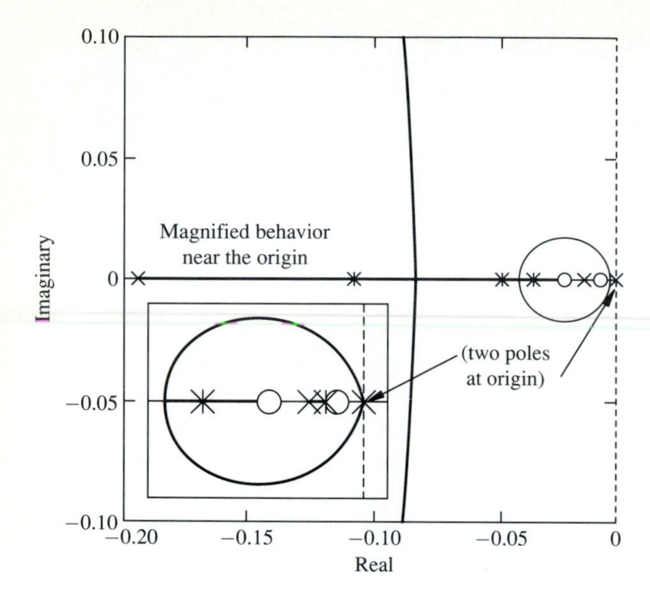

FIGURE 8.7-7
Root locus of $G_{c2}(s)G_p(s)$.

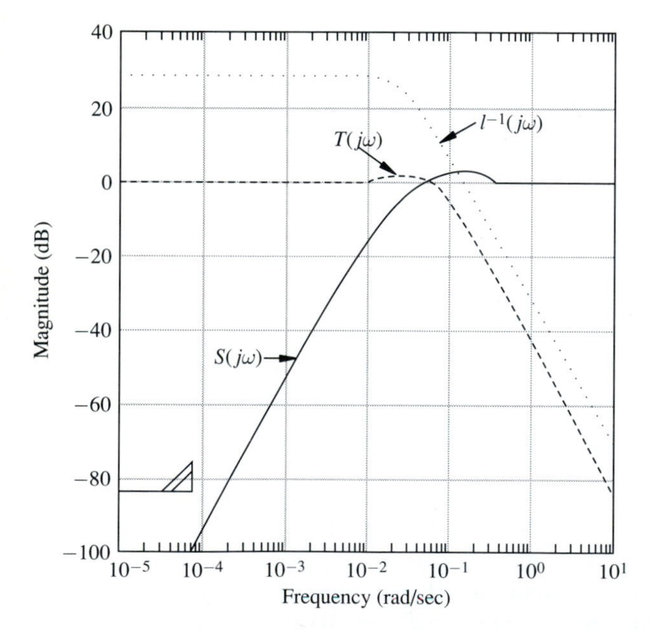

FIGURE 8.7-8
Sensitivity function $S(j\omega)$ and complementary sensitivity function $T(j\omega)$ plots for loop gain $G_{c2}(s)G_p(s)$.

The last measure of the design to check is the closed-loop response to a step change in command angle θ_c. This step response is plotted in Fig. 8.7-10. Despite no indications of a complex pole pair from the flat closed-loop frequency response curve (i.e., the complementary sensitivity function curve) of Fig. 8.7-9, the closed-loop step response displays some overshoot. The cause of the overshoot was discussed in examining the root locus of the lead compensator. The overshoot arises from the

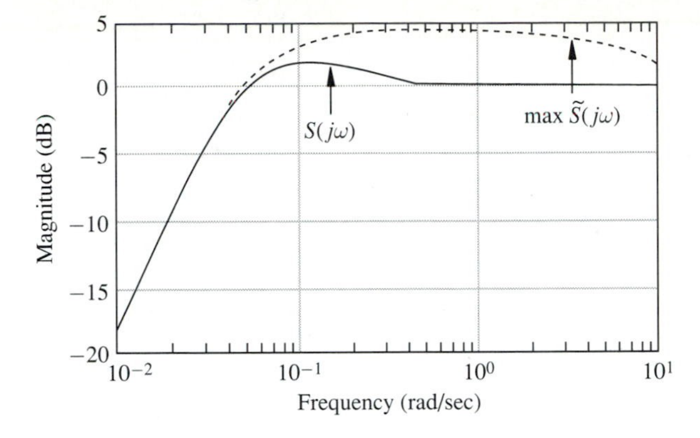

FIGURE 8.7-9
Nominal sensitivity function $S(j\omega)$ and worst case perturbed sensitivity function $\max \tilde{S}(j\omega)$.

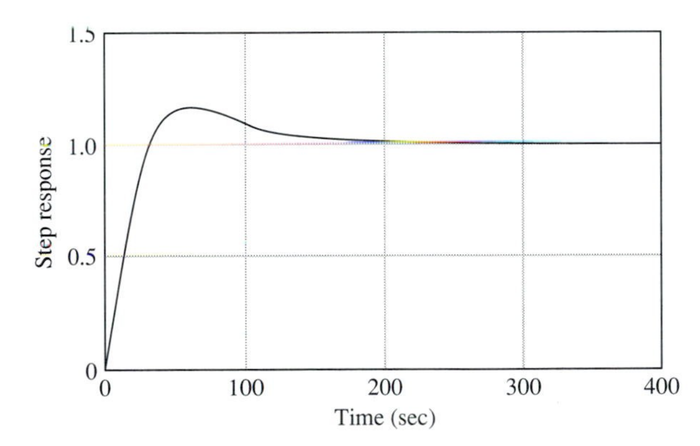

FIGURE 8.7-10
Step response of the closed-loop system.

fact that the zero of the lead compensator remains closer to the origin than does the nearest closed-loop pole. This often happens with lead compensators. The cause of the overshoot is more clearly shown by examining the closed-loop transfer function

$$M(s) = 0.1 \frac{(s + 0.005)(s + 0.019)}{(s + 0.00517)(s + 0.033)(s + 0.05)(s + 0.11)(s + 10)} \tag{8.7-20}$$

The system has a pole-zero pair from the lag compensator that essentially cancels at $s = -0.005$. The next element is a zero closer to the origin than is the first dominant pole. This is the cause of the overshoot. Since the zero is part of the compensator, it does not appear in the transfer function from either an input disturbance or an output disturbance. Since the overshoot affects only the closed-loop transient response, it is easily remedied with a prefilter that cancels the offending zero.

$$P(s) = \frac{0.019}{s + 0.019}$$

The resulting control configuration, given in Fig. 8.7-11, uses block diagram manipulation to separate the sensor dynamics from the plant. The final step response, in Fig. 8.7-12, shows smooth response. The time constant of approximately 20 sec agrees

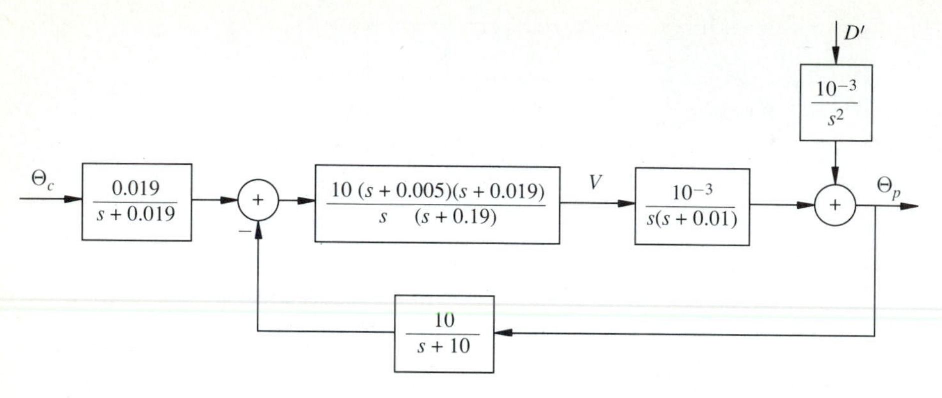

FIGURE 8.7-11
Final control configuration.

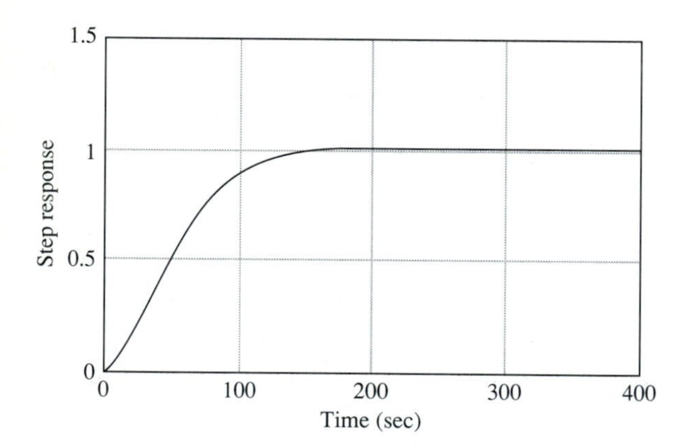

FIGURE 8.7-12
Step response of closed-loop system with prefilter.

with the loop gain bandwidth of 0.05 rad/sec, as shown in Fig. 8.7-6. The response can be made faster by further manipulation of the prefilter.

The final design of the ATS-6 was indeed quite similar to the design specified here. The control system was very successful in its mission.

Exercises 8.7

8.7-1. Using the equations given in Fig. 8.7-1*b* (*a*) Find the transfer function from V to T; (*b*) Find how this transfer function reduces for frequencies where $L\omega \ll R$. Hint: Use block diagrams.

Answers: (*a*) $\dfrac{K_\tau Is}{(Ls+R)(Is+C)+K_v K_\tau}$ (*b*) $\dfrac{K_\tau Is}{RIs+RC+K_v K_\tau}$

8.7-2. It is interesting to examine how the satellite control system might fare if it contained unmodeled dynamics that satisfy the magnitude bound $l_m(j\omega)$. According to the previous analysis, stability should be maintained and performance should be just partially degraded. Suppose that the ATS-6 satellite does indeed contain a lightly damped bending mode with a resonance at $\omega = 1$. Suppose that a better model for the plant is given by

$$G'_p(s) = G_p(s)G_u(s)$$

with $G_p(s)$ given by the transfer function in Eq. (8.7-3) and

$$G_u(s) = \frac{(s+1)^2}{\left(s^2 + 0.04s + 1\right)}$$

Solve for the particular $L_m(s)$ that results from this $G_u(s)$

$$L_m(s) = G_u(s) - 1$$

Plot the magnitude of $L_m(s)$ and show that this particular $L_m(s)$ satisfies the magnitude bound of Eq. (8.4-5*b*). Plot the Bode plot of the loop gain, $G_{c2}(s)G'_p(s)$ when the perturbed plant $G'_p(s)$ and the compensator $G_{c2}(s)$ are together in the loop. Plot the response to a step change in the reference input. Comment on the resulting frequency and time plots.

8.8 AN EXAMPLE USING ROLL-OFF AND A NOTCH FILTER TO CANCEL PLANT DYNAMICS

Often in the design of control systems for mechanical structures the plant model contains a high frequency, poorly damped pair of poles. If this pole pair can be modeled accurately enough it need not limit the bandwidth of the control system.

Let's create a hypothetical twist on the design problem of the previous section. Suppose that when the controller design was submitted, the customer felt that a larger bandwidth is needed so that the system could react more quickly to disturbances. (If the customer's only reason for wanting increased bandwidth were the desire to follow command inputs more closely, this problem could have been resolved by the use of a prefilter. However, prefilters have no effect on disturbance rejection properties.)

If we were using the plant model of Eq. (8.7-4) with the uncertainty of Eq. (8.7-7) we would have to tell our customer, that we could increase the bandwidth only by an insignificant amount. Referring to Fig. 8.7-6, we'd explain that the loop gain must be down significantly below −20 dB by $\omega = 0.7$ to account for modeling uncertainty. If we push the bandwidth out of its current position of $\omega = 0.06$ the gain would have to roll off more sharply. This increased slope would indicate additional phase lag, decreased phase margin and oscillatory transient responses to disturbances. If, however, the system could be modeled better so that the uncertainty wouldn't become large until higher frequencies were reached, then the bandwidth could be increased and still roll off soon enough to meet the high frequency uncertainty constraint.

Let's suppose that the customer responds by giving our modeling department a contract to determine a better model of the satellite. The modeling department returns the model

$$\tilde{G}_{p2}(s) = G_{p2}(s)(1 + L_m(s)) \qquad (8.8\text{-}1a)$$

$$G_{p2}(s) = \frac{10}{10^3 s(s+0.01)(s+10)} \qquad \frac{(s+1)^2}{s^2 + 0.04s + 1} \qquad (8.8\text{-}1b)$$

with

$$|L_m(j\omega)| \leq l_m(j\omega) = \frac{\left(1+\frac{\omega^2}{0.09}\right)(1+\omega^2)}{30\left(1+\frac{\omega^2}{100}\right)} \tag{8.8-1c}$$

Figure 8.8-1 contains the Bode plots of $10G_{p2}(s)$ and a magnitude plot of $l_m^{-1}(j\omega)$. The Bode plots show the loop gain meeting the high and low frequency constraints with a crossover frequency of $\omega_c = 0.1$ but a poor phase margin of 15°. If we try to add a lead compensator, either the low frequency constraint will be violated or the high frequency constraint will be approached too closely. We begin a set of new design iterations with the intent of meeting the low frequency constraint, keeping about 10 dB or more below the high frequency constraint, while achieving as high a crossover frequency as possible with a 60° phase margin.

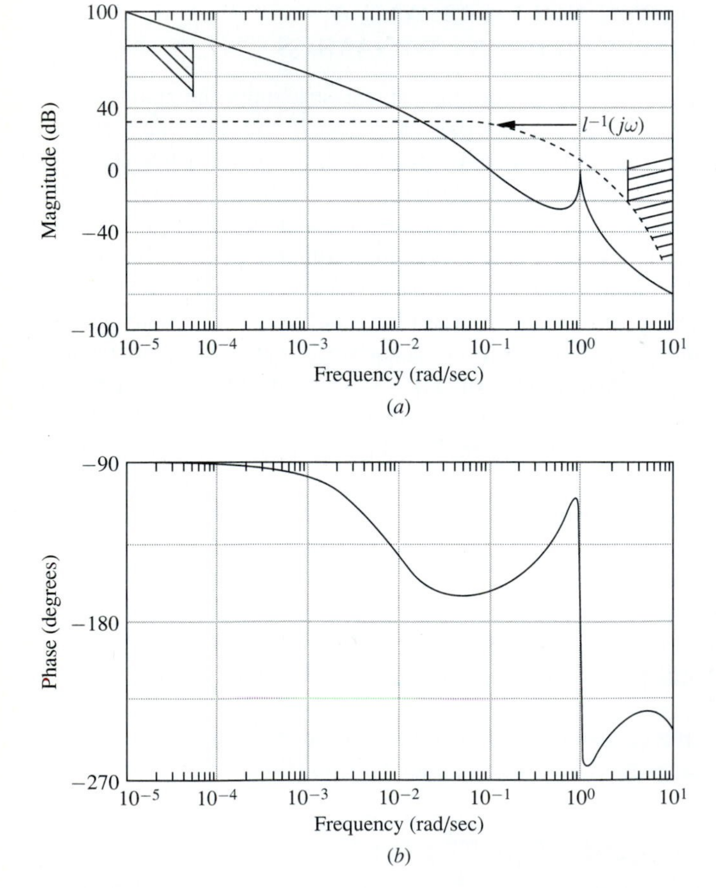

FIGURE 8.8-1
Bode plot of $10G_{p2}(s)$. (*a*) Magnitude; (*b*) phase.

Look what happens if the loop gain is raised by another factor of 10. The Bode magnitude plot of Fig. 8.8-1 is raised by 20 dB. While the first 0 dB crossover point is extended to $\omega = 0.4$, the peak at $\omega = 1$ is driven above the 0 dB line. A sketch of the Nyquist plot of $100G_{p2}(s)$ is shown in Fig. 8.8-2. The main plot is the image of the entire Nyquist D-contour including positive frequencies, negative frequencies and the small semicircle that runs counterclockwise around the pole at the origin. The sketch is not drawn to scale. The insert shows the computer-generated polar plot of $100G_{p2}(j\omega)$ for positive ω only as ω goes from $\omega = 0.01$ to infinity. The loop gain of $100G_{p2}(s)$ gives rise to an unstable closed-loop system as there are two clockwise encirclements of the -1 point caused by the peak in the Bode plot at $\omega = 1$.

One strategy that can be used to avoid this unpleasant situation is to remove the peak with a notch filter. The notch is to be centered at $\omega = 1$ where the peak is located. It should have a depth of -30 dB. (Refer to Sec. 8.6 where the parameters of a notch filter are discussed.) The depth of notch is given by $2\zeta_N$. We choose

$$2\zeta_N = 0.0316 = -30 \text{ dB}$$

$$\zeta_N = 0.0158$$

The notch filter is given by

$$G_{\text{notch}}(s) = \frac{s^2 + 0.0316s + 1}{s^2 + s + 1} \tag{8.8-2}$$

The Bode plots of the loop gain $100G_{p2}(s)G_{\text{notch}}(s)$ are given in Fig. 8.8-3. The peak in magnitude is removed along with the associated phase shift. Note that the elimination of the peak is possible only when the parameters of the oscillatory pole pair are precisely modeled. If the frequency of the notch is slightly off from the frequency of

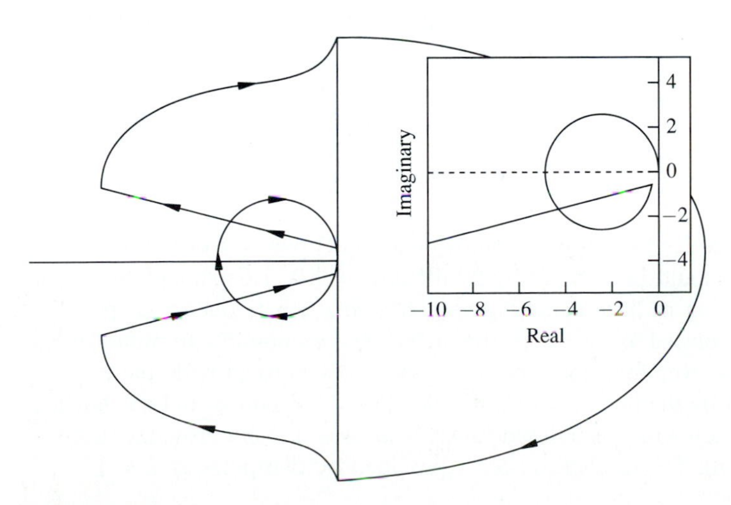

FIGURE 8.8-2
Nyquist of plot of $100G_{p2}(s)$.

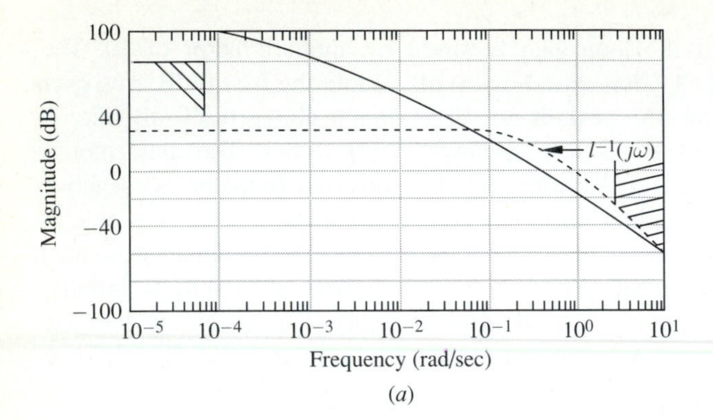

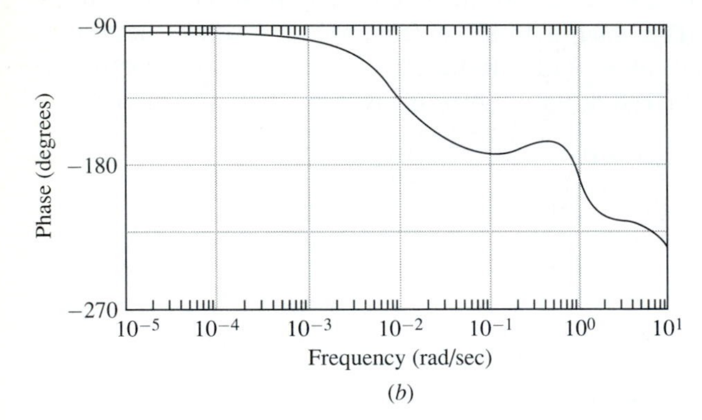

FIGURE 8.8-3
Bode plots of $100G_{p2}(s)G_{\text{notch}}(s)$. (*a*) Magnitude; (*b*) phase.

the oscillatory plant poles the resulting Bode plot would have a peak next to a notch and the closed-loop system would be unstable. The major effect of the notch filter is the near cancellation of the plant poles located at $s = -0.02 \pm j0.9996$ by the compensator zeros located at $s = -0.0158 \pm j0.9997$. The near cancellation makes the poles almost disappear from the loop gain. Of course, as discussed in Sec. 8.6, canceling oscillatory plant poles leaves the possibility of an oscillatory response to input disturbances in the closed-loop system.

Two problems remain in the loop gain of Fig. 8.8-3. High frequency attenuation is needed and the phase margin should be increased. The high frequency attenuation is addressed first using poles for roll-off. To provide a 10 dB pad on the high frequency constraint, the gain should be reduced by 20 dB at $\omega = 10$. This should be accomplished while producing as little phase lag near the crossover frequency as possible. The poles should be placed as far out the frequency axis as possible to minimize the phase lag at the crossover frequency; however, the poles must provide the required attenuation of 20 dB by the point $\omega = 10$. A pole at $\omega = 3$ produces only 6° lag near the crossover frequency which is a decade away near $\omega = 0.3$ and provides 10 dB of attenuation by $\omega = 10$. The needed roll-off is provided by two poles at $\omega = 3$.

$$G_{\text{roll}}(s) = \frac{9}{(s+3)^2} \tag{8.8-3}$$

The Bode plots of the new loop gain, $100G_{p2}(s)G_{\text{notch}}(s)G_{\text{roll}}(s)$, are shown in Fig. 8.8-4. The loop gain now meets the high frequency constraint with a 10 dB margin as specified.

The last step in the design is to use a lead compensator to increase the phase margin. Since it is desirable to maintain the margin from the high frequency constraint, some of the bandwidth gained previously is sacrificed and a unity high frequency gain lead compensator is used. The loop gain of Fig. 8.8-4 has about 5° phase margin. A lead compensator that produces 55° maximum phase lead is used. The information given in Sec. 8.4 shows that $m = 10$ is appropriate. Since the gain from the lead compensator at the frequency of maximum phase lead is $-10\log m = -10$ dB, ω_l is taken as the frequency where the loop gain crosses 10 dB. From Fig. 8.8-4, $\omega_l = 0.2$. The pole of the lead compensator needs to be at $s = -\omega_l\sqrt{m} = -0.2\sqrt{10} = -0.63$ and the zero needs to be at $s = -0.063$.

$$G_{\text{lead}}(s) = \frac{s + 0.063}{s + 0.63} \tag{8.8-4}$$

The total compensator is the product of the parts.

$$G_C(s) = \frac{900(s^2 + 0.0316s + 1)(s + 0.063)}{(s^2 + s + 1)(s + 0.63)(s + 3)^2} \tag{8.8-5}$$

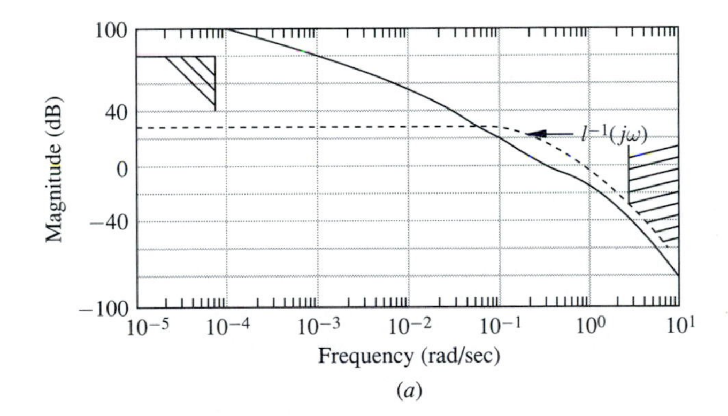

(a)

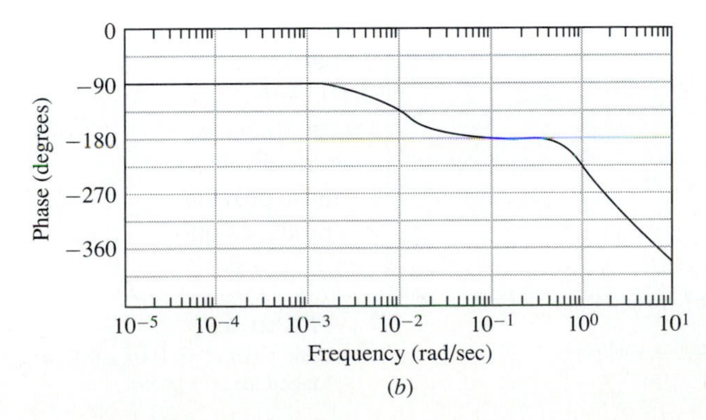

(b)

FIGURE 8.8-4
Bode plots of $100G_{p2}(s)G_{\text{notch}}(s)G_{\text{roll}}(s)$. (*a*) Magnitude; (*b*) phase.

The Bode plots of the total loop gain are shown in Fig. 8.8-5. The low frequency constraint is just met and the high frequency constraint is met with a 10 dB margin. The phase margin is 60° and the crossover frequency has been increased to $\omega_c = 0.2$ from the crossover frequency of $\omega_c = 0.06$, which was achieved in Sec. 8.7 with the inferior model.

Figure 8.8-6 displays the magnitude plots of the sensitivity function $S(j\omega)$ and the complementary sensitivity function $T(j\omega)$. The plot shows that the complementary sensitivity function stays at least 10 dB below $l_m^{-1}(j\omega)$ for all ω. This indicates that the system is stable for all plants within the set of plants described by Eq. (8.8-1). The 10 dB margin indicates that every plant in that set produces a loop gain that comes no closer than -10 dB $= 0.31$ to the -1 point on the Nyquist plot. Reasonable performance is thus maintained for all plants within the set. The sensitivity function shows that the low frequency performance specification is met. In addition the maximum nominal sensitivity is about 1.2. At frequencies lower than the crossover frequency the sensitivity function is much less than one and disturbances and the effects of changes in parameters are greatly attenuated. The design needs a prefilter to clean up the response to reference inputs just as the design of Sec. 8.7 did.

Can the design of the loop gain be improved? Yes, the low frequency gain could be raised with a lag compensator with little or no effect on the intermediate or high

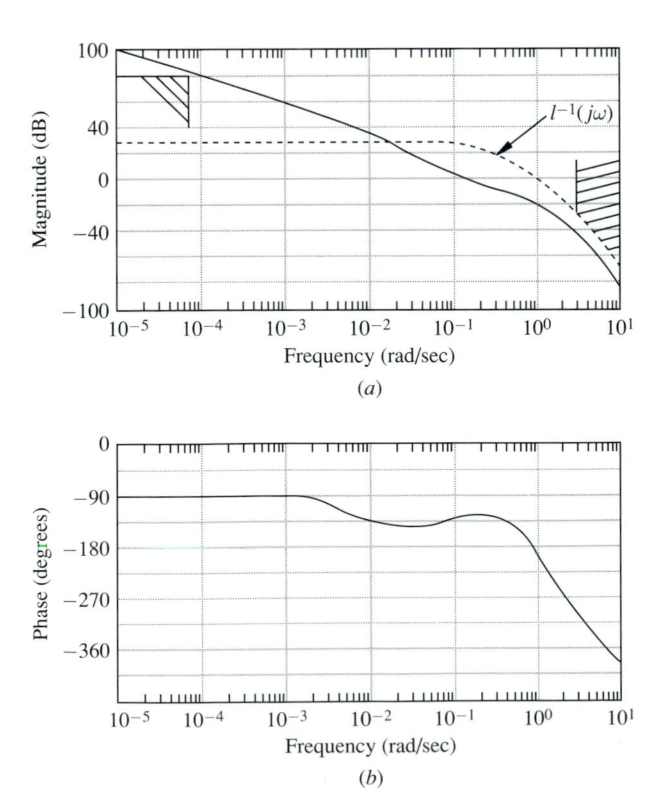

FIGURE 8.8-5
Bode plots of $G_c(s)G_{p2}(s)$. (*a*) Magnitude; (*b*) phase.

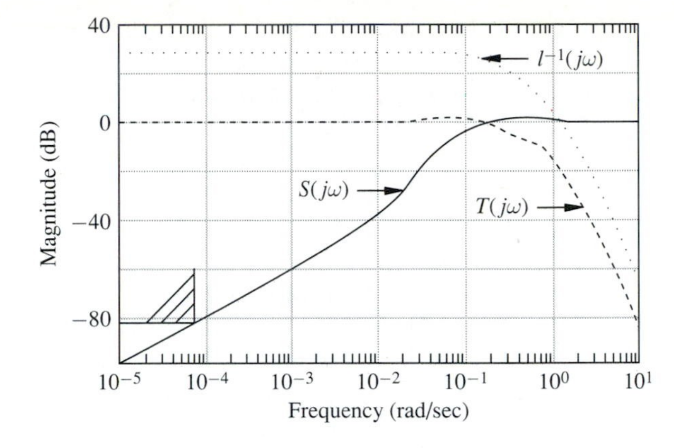

FIGURE 8.8-6
Plot of the sensitivity function $s(j\omega)$ and the complementary sensitivity function $T(j\omega)$.

frequency behavior. Can the bandwidth be further increased? The answer is probably yes. Almost any design measure can be improved somewhat with more careful and more complex designs, but a level of diminishing returns is always met. In this design, the bandwidth cannot be increased much. The loop gain must meet the high frequency constraint arising from the modeling uncertainty. If the bandwidth is pushed out much further the magnitude must slope down at a sharper angle. This indicates phase lag at lower frequencies and destroys the attainable phase margin. In general, the inability to model a plant well at high frequencies limits the bandwidth attainable in a control design. In a later section we see that delays in the plant dynamics may also limit the attainable bandwidth.

It is interesting to see how the new control design responds to disturbances. Let's compare the controller designed in this section with the controller designed in Sec. 8.4 when placed in a loop with the plant of Eq. 8.8-1b and excited with a step disturbance at the output, i.e., let $D_o(s)$ in Fig. 8.8-7 be a unit step. The two outputs are displayed in Fig. 8.8-8.

First, notice that the increased bandwidth of the design of this section has indeed resulted in a faster correction of the output from its disturbed value. Notice also that there is an oscillation at a rate of 1 rad/sec in the response of the Section 8.7 control system that doesn't exist in the response of the revised control system. The step disturbance contains some amount of signal of all frequencies including 1 rad/sec. The controller of Section 8.7 feeds back some of that frequency and excites the resonant mode in the plant. The zeros from the notch filter in the controller of this section block signals with a frequency of 1 rad/sec from reaching the plant and exciting the resonant mode. Another way to see this is to write the transfer function from the output disturbance $D_o(s)$ to the output $Y(s)$ for the controller of this section. The

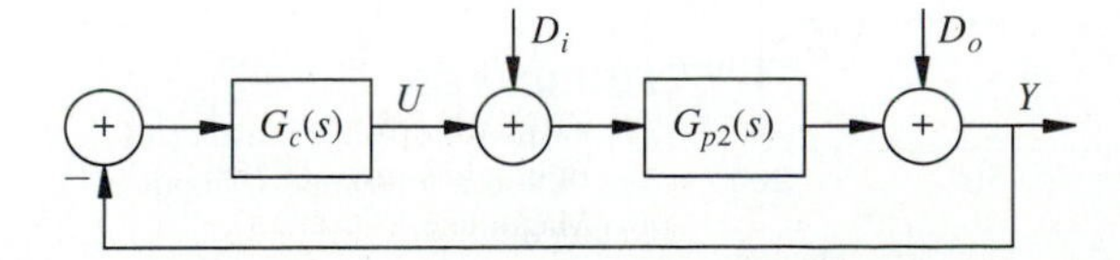

FIGURE 8.8-7
Loop with output disturbance.

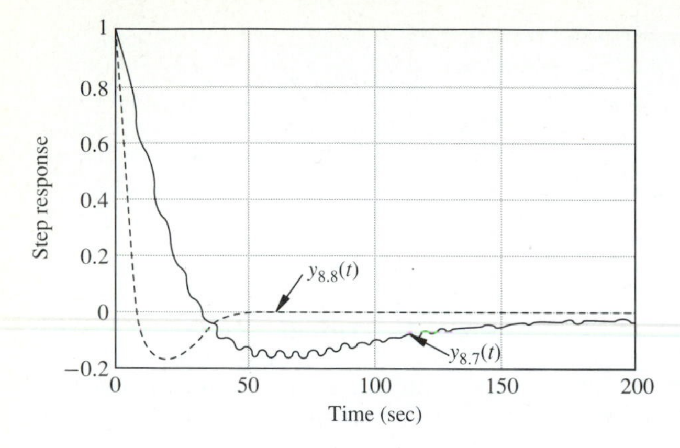

FIGURE 8.8-8
The responses of the control loop of Sec. 8.7, $y_{8.7}(t)$, and Sec. 8.8, $y_{8.8}(t)$, to a unit step output disturbance.

poles of this transfer function are the closed-loop poles, and the zeros of this transfer function are the poles of the loop gain transfer function.

$$\frac{Y(s)}{D_o(s)} = \frac{s(s+0.01)(s+10)(s^2+0.04s+1)}{\left((s+0.1)^2+(0.23)^2\right)(s+10)(s^2+0.0414s+1)} \cdot \frac{(s+3)^2(s+0.634)(s^2+s+1)}{(s^2+0.03s+9.09)(s+0.572)(s^2+0.82s+0.80)}$$

The dominant dynamics here are the zeros at $s=0$ and $s=-0.01$ and the pole pair at $s=0.1+j0.23$. All other poles and zeros almost cancel, including the resonant poles represented by $s^2+0.0414s+1$ and the zeros represented by $s^2+0.04s+1$. Therefore no oscillation appears in the output.

We should remember, however, that for the satellite system under study, disturbances are more likely to enter the system at the input to the plant. The disturbance comes in as a force on the satellite. It excites the plant dynamics. The plot of Fig. 8.8-9 shows the output in response to an impulse disturbance at the plant input, i.e., an impulse in $D_i(s)$ as shown in Figure 8.8-7. The impulse disturbance excites the bending

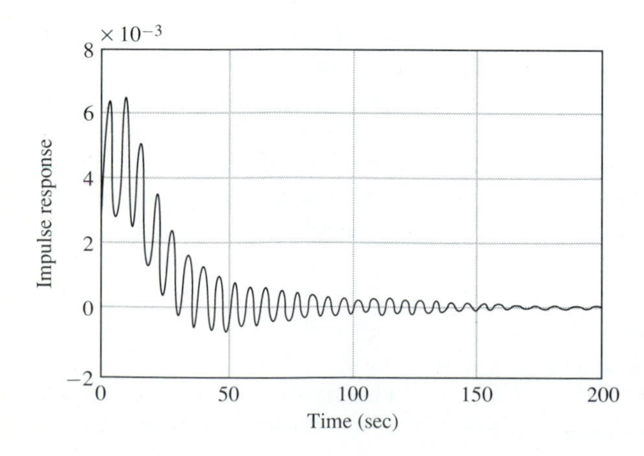

FIGURE 8.8-9
The response of the controller of Sec. 8.8 to a unit impulse input disturbance.

mode at $\omega = 1$. Since the loop gain does not have a large gain at this frequency, the controller cannot take much action to counteract the oscillations. They must play themselves out. The control that comes from canceling plant poles with zeros in the controller is a *passive* control of the oscillations. If a disturbance does not directly excite the oscillatory mode, the notch in the controller will ensure that the actions of the feedback control law also do not excite the oscillations. If the oscillations are directly excited, the controller does not allow enough signal into the plant input at the frequency of the oscillation to produce corrective action. This situation is in contrast to the *active* control that takes place with a large loop gain. At frequencies where the loop gain is large, if oscillations are excited, the oscillating signals are fed back around the loop and into the input correctly shifted in phase so that they can counteract the oscillations. In this way, the oscillation are eliminated very quickly. (The student may want to review the results of Exercise 8.6-2 in light of this discussion on active versus passive control.)

In the design in this section, large gain at the frequency of oscillation is not a feasible option since the uncertainty in the model limits the bandwidth. While canceling oscillatory plant dynamics is not something to be done carelessly, sometimes it is the best option available. In this case, the notch filter allows the bandwidth to be increased while restricting oscillations in response to reference inputs and output disturbances. No design can remove the oscillation from input disturbances since the loop gain at the resonant frequency $\omega = 1$ cannot be made larger while retaining the required robust stability properties.

Exercises 8.8

8.8-1. The goal of the roll-off poles of Eq. (8.8-3) is to provide 20 dB of attenuation by $\omega = 10$. It was decided to place two poles at $\omega = 3$. This produces 12° of phase lag at the original crossover frequency, $\omega = 0.3$.

(*a*) Where must a single roll-off pole be placed to provide 20 dB of attenuation at $\omega = 10$? How much phase lag is produced by this pole at $\omega = 0.3$?

(*b*) Where should a three pole roll-off be placed to produce 20 dB of attenuation at $\omega = 10$? How much phase lag do these poles produce at $\omega = 0.3$?

8.8-2. (*CAD Exercise*) Find the response of the controller designed in Sec. 8.7 when used with the plant of Eq. (8.8-1*b*) and excited with a unit impulse $D_i(s)$ at the input to the plant. Compare this response with that of Fig. 8.8-9 and explain the differences.

8.9 CONTROLLING UNSTABLE PLANTS

So far we have demonstrated control designs that improve the performance of systems containing only stable plants. One of the major advantages of designing closed-loop control systems is that an unstable plant can be stabilized and also meet performance criteria. In this section we consider control system designs starting with plants that contain poles in the right half-plane, i.e., unstable plants. It is interesting that the key technique used to stabilize loops with an unstable plant is a lead compensator—the same lead compensator used to achieve an improved margin of stability in loops with stable plants. Indeed, there is little difference between improving the performance of a stable loop gain and creating acceptable performance with an initially unstable loop gain except for the urgency inherent in stabilizing an otherwise unstable system.

Let's compare a controller placed in series with two different plants, one stable and one unstable. Consider the stable plant

$$G_{p1}(s) = \frac{100}{(s+0.1)(s+1)(s+10)} \tag{8.9-1}$$

and the unstable plant

$$G_{p2}(s) = \frac{100}{(s+0.1)(s-1)(s+10)} \tag{8.9-2}$$

If these plants are placed in feedback loops with constant gain compensators the two root loci of Fig. 8.9-1 result. The two loci are very similar except for the location of the $j\omega$ axis, which indicates that the loop with the stable plant remains stable for small gains while the loop with the unstable plant is unstable for all constant gain compensators.

We know that the zero of a lead compensator *pulls* the poles of the plant of Eq. (8.9-1) further into the left half-plane and allows a higher gain while retaining stability. The same effect occurs when a lead compensator is used in series with the unstable plant of Eq. (8.9-2). Consider the lead compensator

$$G_C(s) = \frac{K(s+1)}{s+10} \tag{8.9-3}$$

The root loci of the combination of this lead compensator in series with each of the plants is shown in Fig. 8.9-2. In Fig. 8.9-2*b*, it can be seen that for intermediate values of the gain K, the closed-loop system is stabilized.

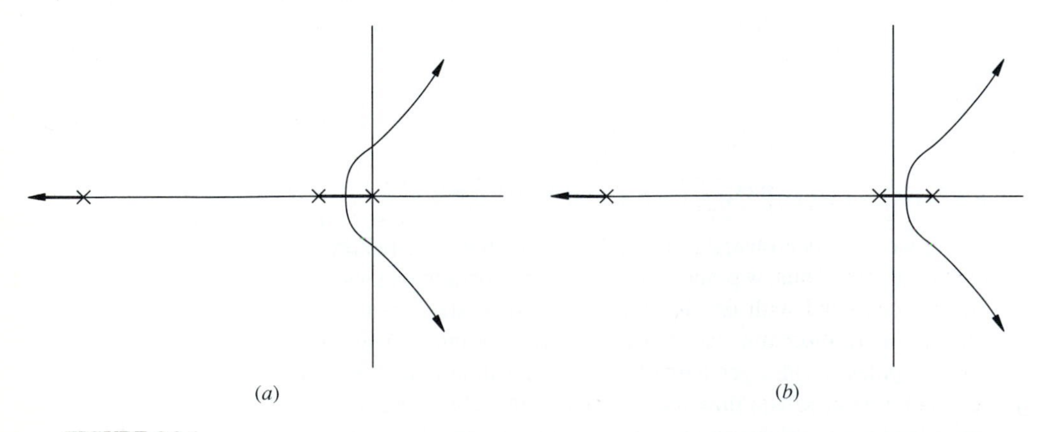

FIGURE 8.9-1
Root locus plot. (*a*) Stable plant; (*b*) unstable plant.

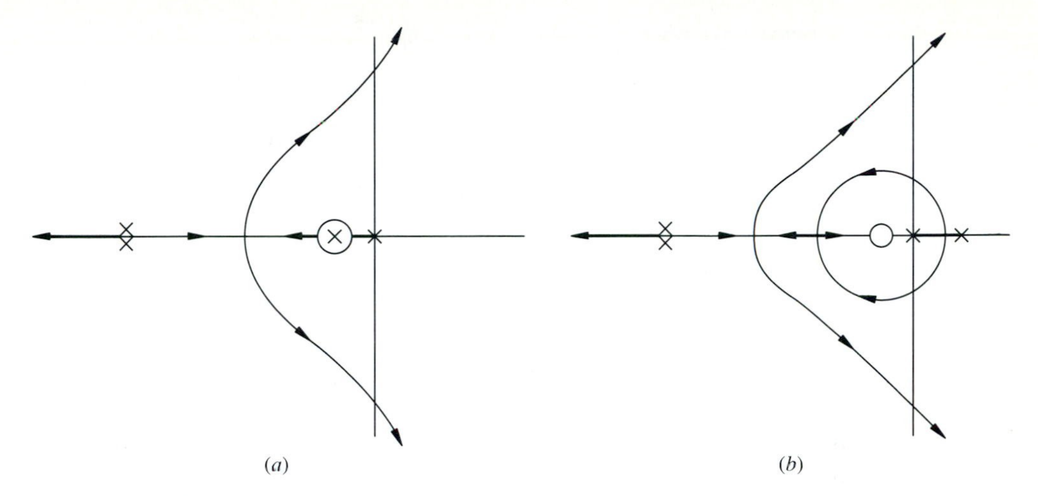

FIGURE 8.9-2
Plant with lead compensator. (*a*) Stable plant; (*b*) unstable plant.

To further determine the behavior of the system with the unstable plant, the Bode and Nyquist plots of the system must be examined. In the compensator of Eq. (8.9-3), let $K = 3$. We examine the Bode plots of the plant $G_{p2}(s)$ given in Eq. (8.9-2) and the combination of the plant $G_{p2}(s)$ with the series compensator $G_C(s)$ given by Eq. (8.9-3).

$$G_C(s)G_{p2}(s) = \frac{300(s+1)}{(s+0.1)(s-1)(s+10)^2} \tag{8.9-4}$$

The Bode plots are given in Fig. 8.9-3. The usual effect of the lead compensator can be seen in these Bode plots. From the Bode plot, the Nyquist plots can be drawn. The entire Nyquist plot of the unstable plant $G_{p2}(s)$ is sketched (not to scale) in Fig. 8.9-4. The Nyquist plot contains one clockwise encirclement of the -1 point. Since the plant has one right half-plane pole, the Nyquist analysis indicates that a unity feedback compensator in a closed-loop system with this plant produces two unstable closed-loop poles. This observation is in agreement with the root locus of Fig. 8.9-1*b*.

The entire Nyquist plot of $G_C(s)G_{p2}(s)$ is sketched (not to scale) in Fig. 8.9-5. The lead compensator has pulled the portion of the Nyquist plot corresponding to intermediate positive frequency values below the negative real axis on the Nyquist plot. This produces a single counterclockwise encirclement of the -1 point and, because there is one right half-plane loop gain pole, the closed-loop system is stable. Thus, we see with frequency domain analysis that an unstable plant can be stabilized with the addition of a lead compensator. We should note that further compensation may be placed in series with $G_C(s)G_{p2}(s)$ to further improve the performance and robustness of the system. Once the Nyquist plot has the correct number of encirclements, a further lead compensator can pull the plot further away from the -1 point, giving a better phase margin and a less oscillatory response.

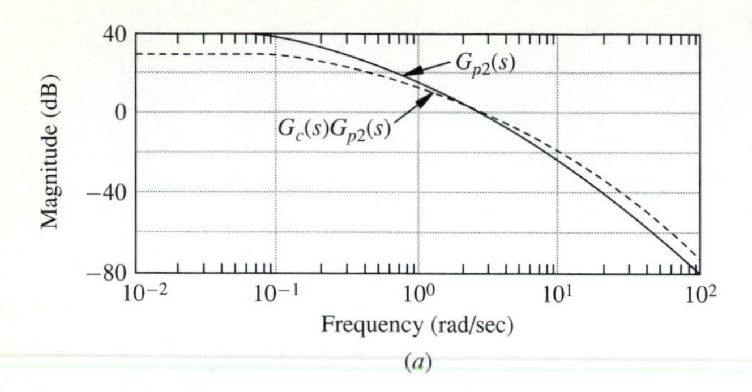

(a)

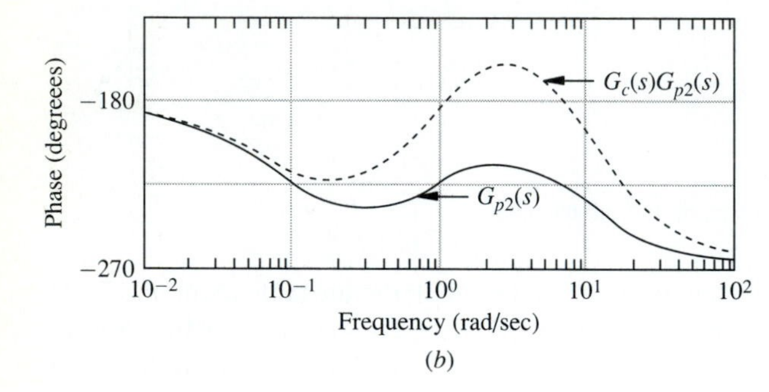

(b)

FIGURE 8.9-3
Bode plots $G_{p2}(s)$ and $G_C(s)G_{p2}(s)$. (a) Magnitude; (b) phase.

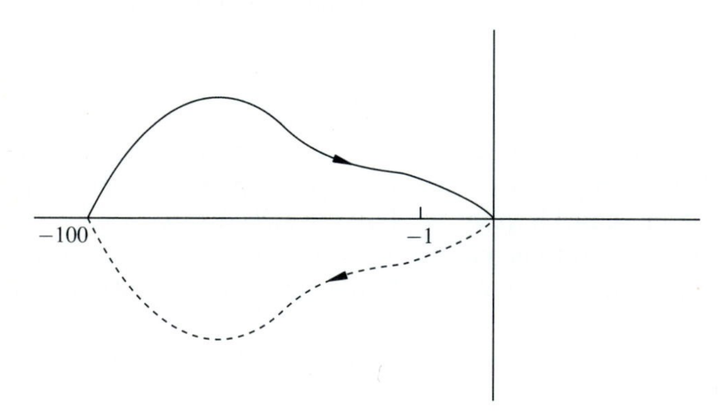

FIGURE 8.9-4
The Nyquist plot of $G_{p2}(s)$.

It is interesting to see the Nyquist plots before and after a plant with two unstable poles is stabilized. Let the plant be given by

$$G_{p3}(s) = \frac{10}{(s+0.1)(s-1)^2} \tag{8.9-5}$$

A sketch of the entire Nyquist plot (not to scale) of $G_{p3}(s)$ given by Eq. (8.9-5) is shown in Fig. 8.9-6. There are no encirclements of the −1 point, indicating two unstable closed-loop poles.

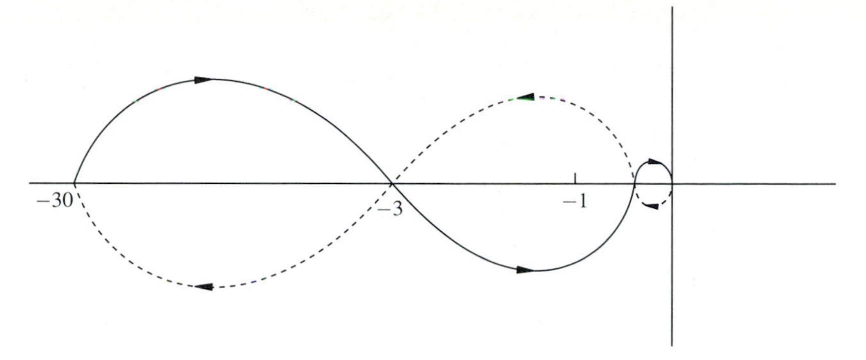

FIGURE 8.9-5
The Nyquist plot of $G_C(s)G_{p2}(s)$

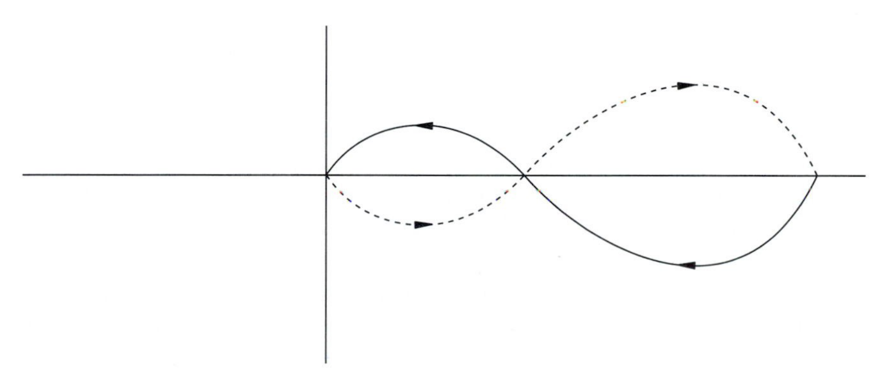

FIGURE 8.9-6
Nyquist plot of $G_{p3}(s)$.

It is not clear how one could possibly create two counterclockwise encirclements starting from this plot. However, if enough phase lead can be created, the Nyquist plot of Fig. 8.9-7 can result. This plot has the required two counterclockwise encirclements to create a stable closed-loop system. The transfer function used to sketch Fig. 8.9-7 is

$$G_{C2}(s)G_{p3}(s) = \frac{6(s+0.1)^2}{(s+10)^2}\frac{10}{(s+0.1)(s-1)^2} \tag{8.9-6}$$

The two lead compensators in series generate almost 180° phase lead near $\omega = 1$. This brings the Nyquist plot around the -1 point to create the counterclockwise encirclements.

We have seen that stable closed-loop systems can result from unstable plants by the introduction of phase lead in the loop gain. Two further observations can be made. First, since lead compensators produce only phase lead over a restricted frequency range, the gain of the system must be set so that the loop gain crosses over through unity magnitude when the phase lead is near its greatest value. In both

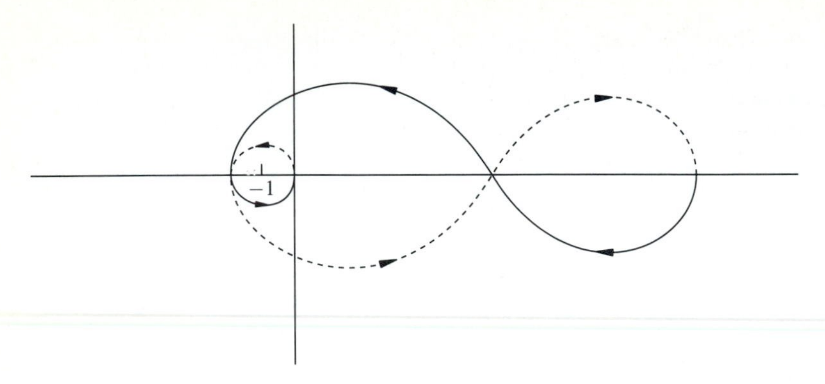

FIGURE 8.9-7
Nyquist plot of $G_{C2}(s)G_{p3}(s)$.

of the examples, $G_C(s)G_{p2}(s)$ and $G_{C2}(s)G_{p3}(s)$, the closed-loop system is unstable if too large or too small a gain is used. Such systems are sometimes referred to as *conditionally stable*.

Second, recall that on a Bode *phase* plot right half-plane poles produce the same effect as left half-plane zeros, i.e., they produce phase lead. In general, so much phase lead is needed to create counterclockwise encirclements on the Nyquist plot that both the lead of a series compensator and the lead produced by unstable poles themselves are needed near the crossover frequency.

There is a subtle indication here. Since the lead contribution from the unstable poles usually must be allowed to develop before the crossover frequency, it is a general rule of thumb that the bandwidth of control systems that stabilize unstable plants cannot be much less than the frequency associated with the highest frequency unstable pole. This relates to the concept of active control introduced in Sec. 8.8. Unstable plant poles must be actively, not passively, controlled. The range of frequencies where the return difference is greater than one is the area of active control. The area of active control should include the frequencies around any unstable pole.

The need for the bandwidth to be large enough to actively control the highest frequency unstable poles means that the frequency range where there is little modeling uncertainty must extend beyond the frequency of the highest frequency unstable pole. If such is the case the loop gain can still be rolled off to meet the magnitude constraint given by the unmodeled dynamics at high frequencies after attaining the necessary bandwidth to stabilize the loop.

Now that we have seen how right half-plane poles affect controller designs, it is natural to investigate how right half-plane zeros affect controller designs. This we do in the next section.

Exercise 8.9

8.9-1. Given the unstable plant

$$G_p(s) = \frac{100}{s(s-1)(s+10)}$$

design a stabilizing series compensator G_C which achieves 40° phase margin. Include along with the transfer function of your compensator the Bode plots of the plant and loop gain (i.e., combination plant and compensator) and answer the following questions:

(*a*) What is the bandwidth of the loop gain transfer function and how does it compare with the location of the unstable pole?

(*b*) What advantage is there in having the bandwidth near the point of maximum phase lead of a plant?

8.9-2. Using the plant of Exercise 8.9-1, design a stabilizing series compensator G_C such that the bandwidth of the loop gain is less than the unstable pole location (i.e., $G_C G_p$ crosses through 0 dB well below 1 rad/sec).

In this design, you cannot use much of the phase lead available from the pole. How does this constraint change your compensator?

8.10 CONTROLLING PLANTS WITH RIGHT HALF-PLANE ZEROS

The presence of right half-plane zeros in a plant places some natural limitations upon what performance can be achieved with a control system. In this section we explore some of these limitations. A zero in the right half-plane means that the plant is a non-minimum phase plant as defined in Chap. 5. A right half-plane zero is often referred to as a *non-minimum phase zero*.

Picture the root locus of a loop gain transfer function that has a zero in the right half-plane. We know that as the gain constant of a root locus is increased, the closed-loop poles move either towards the loop gain zeros or towards infinity. If the loop gain has a right half-plane zero, a closed-loop pole must approach that zero and create an unstable closed-loop pole if the gain constant is made too large. Thus even in this simple analysis, one can see that a right half-plane zero limits the control gain that can be used.

Another way of exploring the limitations imposed by a non-minimum phase zero in the plant is to realize that a right half-plane zero is the one element in a plant model that cannot be effectively eliminated in the closed-loop design, even under ideal conditions. We learned in Chap. 3 that a designer can place closed-loop poles in desired locations using either state-variable feedback or solving a Diophantine equation to create a series or feedback compensator. The only restrictions on the pole to be moved is that it must not be canceled by a zero in the plant. Also, we have seen that left half-plane zeros can be canceled either directly in the loop gain or by placing a closed-loop pole on top of the left half-plane zero. However, a right half-plane zero cannot be moved or canceled. It is the one plant element that must appear in both the loop gain and the closed-loop transfer function.

Since the final loop gain must be zero at the location of a right half-plane zero, the type of return difference transfer function and sensitivity function that can be achieved are restricted by the fact that each of these functions must equal one when evaluated at a right half-plane plant zero. Arguments from the mathematical theory of complex variables can be used with this fact to show that, when the plant contains a right half-plane zero, the sensitivity function evaluated along the $j\omega$ axis must meet certain conditions. These conditions indicate that if the sensitivity is made less than one at some frequencies it must become significantly larger than one at other frequencies. The closer the right half-plane zero is to the $j\omega$ axis, the more severe are the tradeoff restrictions. Also, if a right half-plane zero is close to a right half-plane pole, the tradeoff restrictions indicate that the sensitivity function must be much larger than one for some frequencies. Of course, many of the goals of designing closed-loop control systems can be translated into the objective of achieving a large return difference and thus a small sensitivity function. Right half-plane zeros are natural obstructions to this objective.

Let's examine how a non-minimum phase zero affects any attempts to design an appropriate loop gain for a control system. The objectives of the design effort were explained in Sec. 8.2. A designer would like a large loop gain at small frequencies with a sharp transition to a small loop gain at high frequencies. The sharpness of the transition from large loop gain to small loop gain is limited by the fact that large negative slopes on the magnitude plot are associated with large amounts of phase lag. Large amounts of phase lag cause problems with stability, performance and robustness near the crossover frequency. The control engineer is constantly trying to negate the effect of phase lag. We have seen the importance of phase lead compensators. In the previous section we saw that phase lead is even more important if unstable poles are present in the loop gain.

The effect of a non-minimum phase zero upon the Bode plot of a loop gain is a deadly combination of an increase in magnitude in conjunction with an increase of phase *lag*! Remember that in the transition frequency region, the control engineer is trying to sharply decrease the magnitude while minimizing the phase lag. The non-minimum phase zero works in direct opposition to the control designers objectives and it cannot be eliminated by cancellation. No wonder that plants with non-minimum phase zeros are the most difficult to control!

Notice that the effect of a non-minimum phase zero is minimal if that zero occurs at a frequency that is much higher than the crossover frequency. In the high frequency range, the magnitude of the loop gain is small enough to keep the sensitivity function near one, independent of the phase. Therefore, the extra phase lag from a non-minimum phase zero in this frequency range has little effect.

Let's look at an example which shows the difficulty presented by a non-minimum phase zero.

Example 8.10-1. Consider a plant described by the transfer function

$$G_{P1}(s) = G_{P0}(s)G_{AP}(s) \tag{8.10-1}$$

where

$$G_{P0}(s) = \frac{0.1}{s\left(\frac{s}{0.01}+1\right)\left(\frac{s}{10}+1\right)} \tag{8.10-2}$$

and

$$G_{AP}(s) = \frac{(-s+0.01)}{(s+0.01)} = \frac{\left(\frac{-s}{0.01}+1\right)}{\left(\frac{s}{0.01}+1\right)} \tag{8.10-3}$$

The transfer function $G_{P0}(s)$ is the same as the plant transfer function of the satellite given by Eq. (8.7-4), which was examined in Sec. 8.7. Let's examine $G_{AP}(s)$ and see how it affects the overall transfer function.

The transfer function $G_{AP}(s)$ contains a right half-plane zero with a pole located in the left half-plane at the mirror image of the zero. Notice that the pole and the zero in this arrangement have exactly equal and opposite effects on the Bode magnitude plot. Indeed, the magnitude plot of $G_{AP}(s)$ is equal to unity at all frequencies.

$$|G_{AP}(j\omega)|^2 = \left(\frac{0.01 - j\omega}{0.01 + j\omega}\right)\left(\frac{0.01 + j\omega}{0.01 - j\omega}\right) = 1 \tag{8.10-4}$$

It is said that $G_{AP}(s)$ is an *all-pass* transfer functions since, interpreted as a filter, it passes all frequencies equally. Since the magnitude of $G_{AP}(s)$ is unity for all frequencies, the magnitude plot of $G_{P1}(s)$ is the same as the magnitude plot of $G_{P0}(s)$ from Sec. 8.7.

In Fig. 8.10-1 the Bode plots of $10G_{P1}(s)$ and $10G_{P0}(s)$ are shown. Unlike the magnitude plot, the phase plot of $G_{P1}(s)$ is greatly affected by the all-pass section with its non-minimum phase zero. The left half-plane pole and the right half-plane zero of Eq. (8.10-3) each add about 45° of phase lag per decade to the phase plot in the area between $\omega = 0.001$ and $\omega = 0.1$. Thus, by the point $\omega = 0.1$ the plant $G_{P1}(s)$ has almost 180° more phase lag than the plant $G_{P0}(s)$. This can be seen in Fig. 8.10-1.

Recall from Sec. 8.7 that a major difficulty in the design of the satellite system was to achieve enough phase lead through the crossover frequency range so that a good phase

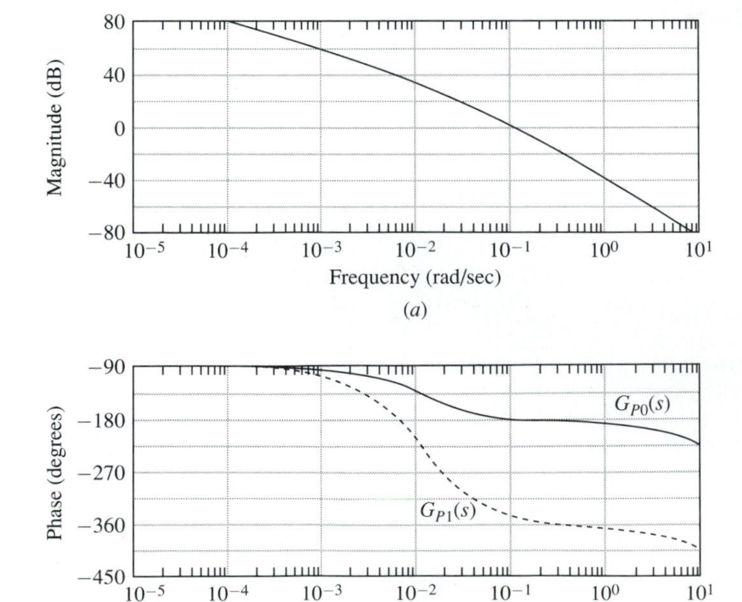

FIGURE 8.10-1
Bode plots of $10G_{P1}(s)$ and $10G_{P0}(s)$. (*a*) Magnitude; (*b*) phase.

margin and transient response result. With the plant $G_{P1}(s)$, this problem is exacerbated by the fact that almost 180° of additional phase lead is needed if the crossover frequency is to remain near $\omega = 0.1$.

Let's try to use a series compensator to negate the effects of the all-pass section of the plant. The pole at $s = -0.01$ can be directly canceled by a zero. Unfortunately, the zero at $s = 0.01$ cannot be canceled. However, its affect on the *phase* plot can be completely countered by adding another zero at $s = -0.01$. The magnitude plot is greatly affected. For the moment, let's ignore the fact that a compensator with two zeros and no poles is unrealizable. We can add more poles later. Let's examine the behavior of

$$G_{L1}(s) = G_{P1}(s)G_{C1}(s) \tag{8.10-5}$$

with

$$G_{C1}(s) = 0.01\left(\frac{s}{0.01} + 1\right)^2 \tag{8.10-6}$$

The Bode plots of $G_{L1}(s)$ and $10G_{P0}(s)$ are given in Fig. 8.10-2. The phase plot of $G_{P0}(s)$ has been recovered by $G_{L1}(s)$. Now, however, the magnitude plot is much flatter. (The difference between $G_{P0}(s)$ and $G_{L1}(s)$ is two zeros, the non-minimum phase zero at $s = +0.01$ and the zero at $s = -0.01$ which restores the phase plot.)

The magnitude of $G_{L1}(s)$ at $\omega = 0.1$ is 1.005 while the value of the magnitude of $G_{L1}(s)$ at $\omega = 10$ is 0.7. If this system is to meet the kind of performance constraints that the loop in Sec. 8.7 needed to meet, a number of poles need to be added well below

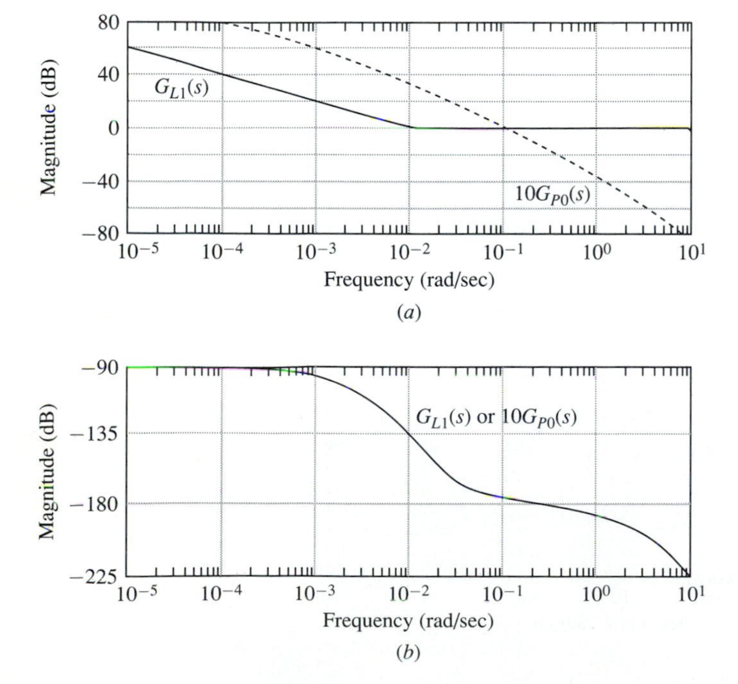

FIGURE 8.10-2
Bode plots of $G_{L1}(s)$ and $10G_{P0}(s)$. (*a*) Magnitude; (*b*) phase.

$\omega = 0.7$ in order to provide enough roll-off in the magnitude plot to meet the high frequency constraint. These poles will cause phase lag in the range $\omega = 0.01$ to $\omega = 0.1$. In addition, a multiple order phase lag is necessary to raise the magnitude over the low frequency performance constraint. This will further add to the phase lag near crossover. Consider the transfer function

$$G_{L2}(s) = G_{L1}(s)G_R(s)G_{lag}(s) \tag{8.10-7}$$

where

$$G_R(s) = \frac{1}{(10s+1)^2} \tag{8.10-8}$$

is used for roll-off, and

$$G_{\text{lag}}(s) = \frac{(s+0.001)^2(s+0.0003)}{s^3} \tag{8.10-9}$$

is a third-order lag compensator.

The Bode plots of $G_{L2}(s)$ are given in Fig. 8.10-3. Notice that the system now meets the magnitude plot objectives (which can be found on Fig. 8.7-3) with some margin for added robustness. Notice, however, that the magnitude plot near crossover is very flat and that the phase lag is greater than 180° and has a large slope at crossover. The phase lag at the crossover frequency cannot be decreased because the increase in the slope of the magnitude plot that accompanies a phase lead compensator causes the crossover

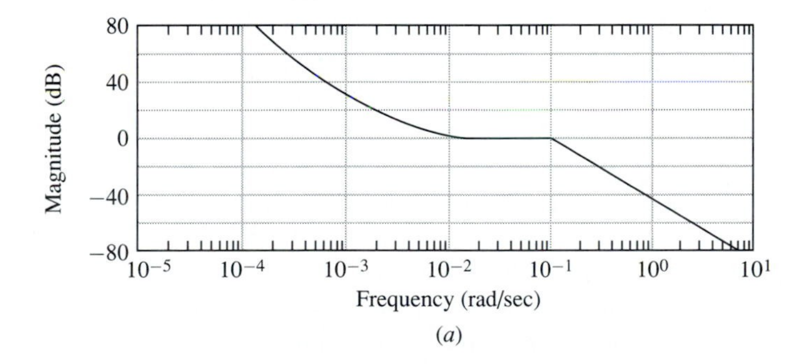

(*a*)

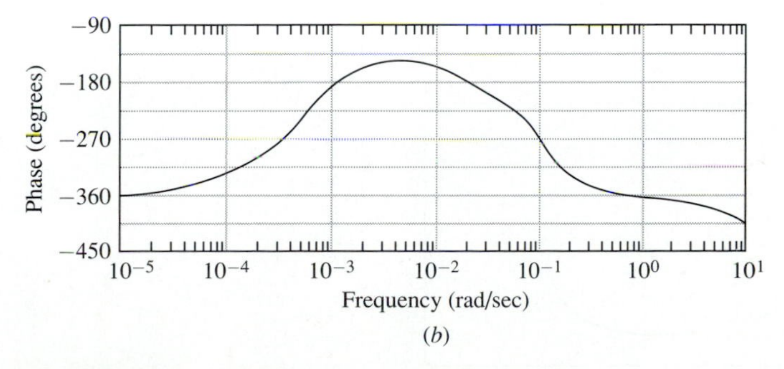

(*b*)

FIGURE 8.10-3
Bode plots of $G_{L2}(s)$. (*a*) Magnitude; (*b*) phase.

frequency to move out to the area where the phase lag is too large to be overcome. The bandwidth must be decreased. A reasonable phase margin can be achieved by decreasing the crossover frequency without the benefit of a lead compensator.

Simply allowing

$$G_{L3}(s) = \frac{G_{L2}(s)}{3} \tag{8.10-10}$$

lowers the magnitude plot about 10 dB. The final crossover frequency occurs at $\omega = 0.004$ with a phase margin of approximately 30°. A polar plot of $G_{L3}(s)$ for $\omega \geq 10^{-3}$ is shown in Fig. 8.10-4. (Use this plot and the transfer function $G_{L3}(j\omega)$ to draw the entire Nyquist plot of $G_{L3}(j\omega)$ and convince yourself that the closed-loop design is stable. Be careful how you treat the indentation around the *four* poles at the origin.) The plot of Fig. 8.10-4 indicates that the minimum value of the return difference for this design is 0.5 or -6 dB. This is somewhat more sensitive than is the design of Sec. 8.7, where the minimal return difference was -2 dB.

The key difference between the design here starting with a non-minimum phase plant and the design of Sec. 8.7 starting with a minimum phase plant is that the non-minimum phase zero in this design forces a reduction of bandwidth. Indeed, the bandwidth is reduced to $\omega_c = 0.004$. This is below the frequency of $\omega = 0.01$ associated with the non-minimum phase zero. It is fortunate that the low frequency performance specification in this problem is loose enough that it was possible to meet the specification even with the reduced bandwidth.

It is a general rule that non-minimum phase zeros restrict the bandwidth of a control system. The bandwidth is usually restricted to be less than the frequency of the lowest frequency non-minimum phase zero so that the phase lag from all the non-minimum phase zeros is absorbed when the magnitude is low. Couple this rule of thumb with the guideline of the last section which states that the bandwidth of a system is usually larger than the frequency associated with an unstable pole and you see that a right half-plane pole at a frequency higher than the right half-plane zero is a devastating combination. In such a situation something should be done to physically redesign the plant and eliminate either the pole or the zero.

It is important to understand how right half-plane zeros physically arise so that they can be eliminated or avoided if plant redesign is possible. There are two common physical situations that manifest themselves as right half-plane zeros in a

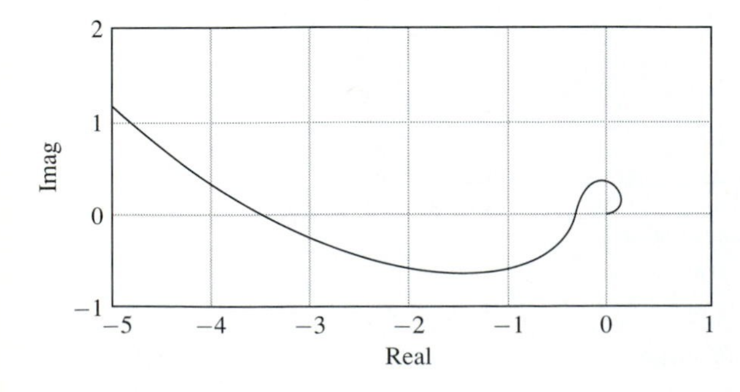

FIGURE 8.10-4
Polar plot of $G_{L3}(s)$.

system model. Right half-plane zeros arise due to time delays in a system and they arise when the measurement of important variables cannot or does not occur. We address time delays here and the importance of measurements in the final section of this chapter.

The impulse response for a system which consists of a time delay of T seconds is

$$h(t) = \delta(t - T) \tag{8.10-11}$$

The system function is given by

$$H(s) = e^{-sT} \tag{8.10-12}$$

Taking the magnitude and phase of the frequency response we get

$$|H(j\omega)| = 1 \qquad \text{for all } \omega$$
$$\arg(H(j\omega)) = -\omega T \tag{8.10-13}$$

A delay system is an all-pass system with a phase lag that increases linearly with frequency. A Bode phase plot of the system with a delay of one second is shown in Fig. 8.10-5. Notice that the phase lag, which increases linearly with ω, becomes an exponential curve on the semi-log Bode phase plot. One can see that it is extremely difficult to overcome this phase lag over a substantial frequency range as lead compensators produce a phase increase that is only linear on the Bode plot and this increase affects only a small frequency range.

The frequency response of the delay can be approximated over a finite frequency range by an all-pass network consisting of right half-plane zeros and mirror image left half-plane poles. The all-pass network matches the magnitude plot of a delay exactly since both systems have unity magnitude for all frequencies. The phase lag of the

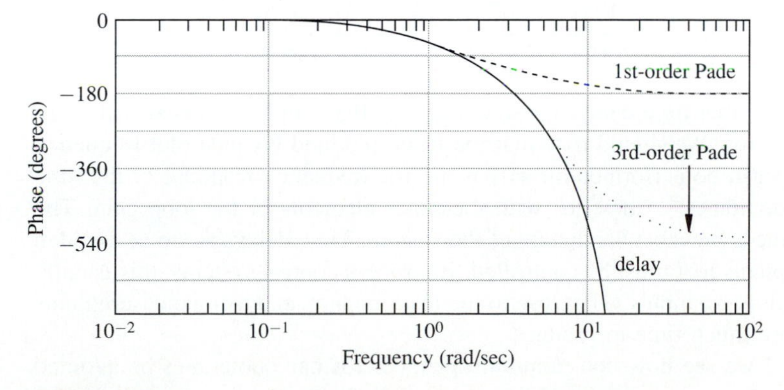

FIGURE 8.10-5
Bode phase plots of a time delay and approximations.

all-pass network using poles and zeros can match the phase lag of the delay over a limited frequency range. This frequency range can be extended by adding more poles and zeros. The total phase lag from each pole-zero combination is 180°. The phase lag from the delay increases without bound.

A reasonable position of the zeros and poles can be found by creating a series expansion for the two transfer functions and equating as many terms as possible. Such an approximation is called a *Padé approximation*. We assume that the number of zeros equals the number of poles and call the number of poles the order of the approximation. Let's demonstrate the technique on a first-order Padé approximation

$$e^{-sT} = 1 - sT + \frac{s^2T^2}{2} - \frac{s^3T^3}{6} + \cdots$$

$$\frac{a-s}{a+s} = 1 - \frac{2}{a}s + \frac{2}{a^2}s^2 + \cdots$$

Equating the first two terms gives $a = \frac{2}{T}$. Higher-order Padé approximations can match more terms and approximate the transfer function of a delay over a wider frequency range. The Bode phase plots of a first-order and a third-order Padé approximation of a one second time delay are shown along with the actual Bode phase plot of a one second time delay in Fig. 8.10-5. The first-order approximation is given by

$$G_{D1}(s) = \frac{-0.5s + 1}{0.5s + 1} \tag{8.10-14}$$

The third-order approximation is given by

$$G_{D3}(s) = \frac{-0.0083s^3 + 0.1s^2 - 0.5s + 1}{0.0083s^3 + 0.1s^2 + 0.5s + 1}$$

$$= \frac{-1\left((s - 3.67)^2 + (3.5)^2\right)(s - 4.64)}{\left((s + 3.67)^2 + (3.5)^2\right)(s + 4.64)} \tag{8.10-15}$$

We have seen that time delays cause phase lags that can be modeled with non-minimum phase zeros. We should note that the Bode plot and Nyquist plot frequency response analysis can be performed directly using the frequency response of the time delay transfer function e^{-sT} in series with the other elements of the loop gain. The phase lag limits the achievable bandwidth of the system. Time delays are to be avoided in the design of plants that must be controlled. In process control systems, this usually means placing valves on inputs very close to the reaction tank and using measurements that do not require much time to produce.

In Sec. 8.12 we see how non-minimum phase zeros can sometimes be avoided by taking extra measurements and using state-variable feedback.

8.10.1 CAD Notes

Bode plots of plants with time delays can be plotted directly with MATLAB by adding the phase shift associated with the time delay directly. The following example demonstrates this.

Example 8.10-2. Plot the Bode phase plot of the following plants.

$$G_{P1}(s) = \frac{1}{s + 0.1}$$

$$G_{P2}(s) = \frac{e^{-2s}}{s + 0.1}$$

```
% clear the graph
clg;
% set up the omega vector of frequencies
w=logspace(-2,1,100);
% enter the plant P1
ng=[1];dg=[1 .1];
[mg,pg]=bode(ng,dg,w);
% loop to compute the phase of P2 from the phase of P1
% the time delay causes an extra phase lag which is
% linear in w
% conversion from radians to degrees must be made
for ind=1:100
    pg2(ind)=pg(ind)-(w(ind)*2*(180/pi));
end
% place axis limits
v=[-2 1 -720 0];
axis(v);
% plot the two plots
p=[pg pg2'];
semilogx(w,p),xlabel('frequency (rad/sec)'),
ylabel('phase(degrees)');
% set up to place a grid on the plot
a=axis;
hold on
% plot horizontal line every 90 degrees
for m=a(3):90:a(4);
    x=[a(1) a(2)];
    y=[m,m];
    plot(x,y)
end
% plot vertical line every factor of ten
for x=a(1):1:a(2);
    x1=[x,x];
    y=[a(3) a(4)];
    plot(x1,y)
end
% turn hold off for next plot
hold off
```

The resulting plots appear in Fig. 8.10-6. Notice the grid lines placed in the program every 90° and the different scale used by MATLAB for tick marks.

Exercises 8.10

8.10-1. Given the non-minimum phase plant

$$G_P(s) = \frac{30(-s+1)}{(s+0.1)(s+10)}$$

(*a*) Design a series compensator G_C that achieves 20 degrees of phase margin. There are no other frequency domain requirements (i.e., bandwidth) so that the compensator can be as simple as possible while achieving as large a bandwidth as possible.

Include along with the transfer function of your compensator the Bode plots of the plant and loop gain (i.e., combination plant and compensator). What is the bandwidth of the loop gain transfer function and how does it compare with the location of the non-minimum phase zero?

(*b*) Try to design a controller with bandwidth larger than the right half-plane zero. Is it possible?

8.10-2. Consider the plant

$$G_P(s) = \frac{e^{-s}}{s}$$

and a set of possible controllers given by

$$G_C(s) = K\left(\frac{100(s+1)}{s+100}\right)^n$$

Sketch the Bode plots of the loop gains for $K = 1$ and $n = 1, 2, 3$. For each $n = 1, 2, 3$ set K for the maximum achievable bandwidth. What is the bandwidth achieved in each case? Are the diminishing returns mentioned in the text evident?

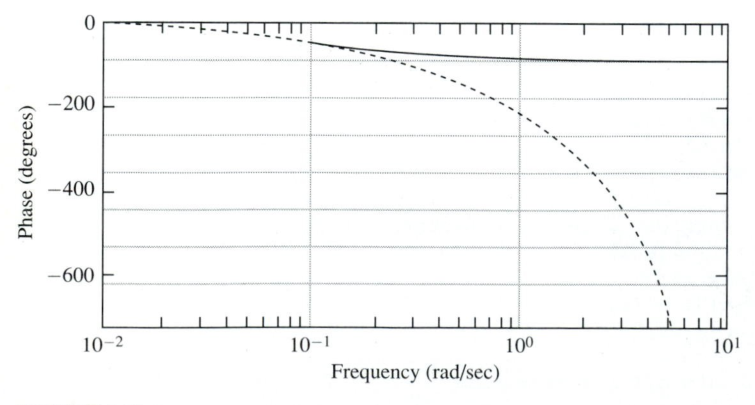

FIGURE 8.10-6
Phase plots with and without time delay.

8.11 POLE PLACEMENT CONTROL

In this section we examine the algorithms studied in Chap. 3 that allow arbitrary placement of the closed-loop poles. Such pole placement can be accomplished using transfer functions and the Diophantine equation or using state-variable feedback. We use a well-known example of a difficult control problem, the inverted pendulum on a cart. This problem is sometimes known as the broom balancing problem. In this section we examine a Diophantine equation approach and show that just because we can place closed-loop poles in a good position we don't necessarily have a good controller. However, a pole placement algorithm may provide a reasonable initial control design that can then be analyzed and improved using frequency domain techniques. In the next section we examine the state-variable approach.

Let's start by deriving a plant model for our example plant. We all have probably at some time tried to balance a broom upside down on our hands. Even with the fantastic processing power of the human brain, this is a difficult task. It becomes even more difficult if we attempt it blindfolded. The plant we examine describes a similar problem. Imagine a "broomstick" on a cart as depicted in Fig. 8.11-1. The dynamics are approximated by assuming that all the mass m of the broomstick is concentrated at the end of the broomstick located at a length l from the cart. The cart has mass M and we assume that, through a motor, we can control directly the force $u(t)$ on the cart. The variables of interest are the angle $\theta(t)$ that the broomstick makes with cart and the distance $y(t)$ that the cart has moved from some arbitrary zero position.

From the physics of the situation, we get the following equations. Applying $\boldsymbol{F} = m\boldsymbol{a}$ in the horizontal direction, we get

$$u(t) = M\frac{d^2y(t)}{dt^2} + m\frac{d^2(y(t) + l\sin\theta(t))}{dt^2} \tag{8.11-1}$$

The vertical force on the broomstick due to gravity is broken into two components. The component working along the shaft is opposed by the shaft and produces no net effect. The component perpendicular to the shaft is given by $mg\sin\theta(t)$. It produces a motion perpendicular to the shaft.

$$mg\sin\theta(t) = m\frac{d^2(y(t)\cos\theta(t) + l\theta(t))}{dt^2} \tag{8.11-2}$$

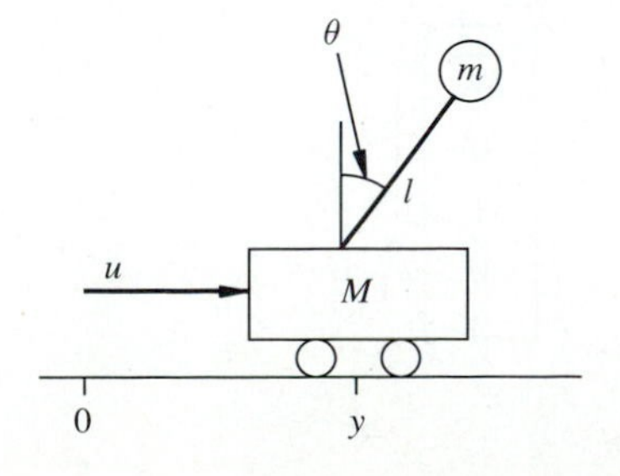

FIGURE 8.11-1
Broomstick on a cart.

The system of Eqs. (8.11-1) and (8.11-2) is nonlinear. These equations can be linearized for small $\theta(t)$ with the approximation

$$\sin\theta(t) \approx \theta(t) \qquad \text{and} \qquad \cos\theta(t) \approx 1 \tag{8.11-3}$$

The two plant equations become

$$u(t) = (M+m)\frac{d^2y(t)}{dt^2} + ml\frac{d^2\theta(t)}{dt^2} \tag{8.11-4}$$

$$mg\theta(t) = m\frac{d^2y(t)}{dt^2} + ml\frac{d^2\theta(t)}{dt^2} \tag{8.11-5}$$

Subtracting Eq. (8.11-5) from Eq. (8.11-4) to eliminate $d^2\theta(t)/dt^2$ produces

$$u(t) - mg\theta(t) = M\frac{d^2y(t)}{dt^2} \tag{8.11-6}$$

or

$$\frac{d^2y(t)}{dt^2} = \frac{-mg}{M}\theta(t) + \frac{1}{M}u(t) \tag{8.11-7}$$

Substituting Eq. (8.11-7) into Eq. (8.11-5) yields

$$mg\theta(t) = \frac{-m^2g}{M}\theta(t) + \frac{m}{M}u(t) + ml\frac{d^2\theta(t)}{dt^2} \tag{8.11-8}$$

or

$$\frac{d^2\theta(t)}{dt^2} = \left(\frac{g}{l} + \frac{mg}{Ml}\right)\theta(t) - \frac{1}{Ml}u(t) \tag{8.11-9}$$

Equations (8.11-7) and (8.11-9) can be turned into state-variable equations by defining a state-variable vector

$$\boldsymbol{x}(t) = \begin{bmatrix} x_1(t) \\ x_2(t) \\ x_3(t) \\ x_4(t) \end{bmatrix} = \begin{bmatrix} \dot{\theta}(t) \\ \theta(t) \\ \dot{y}(t) \\ y(t) \end{bmatrix} \tag{8.11-10}$$

where $\dot{\theta}(t) = d\theta(t)/dt$ and $\dot{y}(t) = dy(t)/dt$.

We then achieve the physical state-variable description

$$\dot{\boldsymbol{x}}(t) = \boldsymbol{A}\boldsymbol{x}(t) + \boldsymbol{b}u(t) \tag{$\boldsymbol{Ab}$}$$

$$y(t) = \boldsymbol{c}^T\boldsymbol{x}(t) \tag{$\boldsymbol{c}$}$$

$$\boldsymbol{A} = \begin{bmatrix} 0 & \dfrac{g}{l} + \dfrac{mg}{Ml} & 0 & 0 \\ 1 & 0 & 0 & 0 \\ 0 & \dfrac{-mg}{M} & 0 & 0 \\ 0 & 0 & 1 & 0 \end{bmatrix} \qquad \boldsymbol{b} = \begin{bmatrix} \dfrac{-1}{Ml} \\ 0 \\ \dfrac{1}{M} \\ 0 \end{bmatrix} \tag{8.11-11}$$

$$\boldsymbol{c}^T = [0\ 0\ 0\ 1]$$

Inserting hypothetical values for the constants we arrive at our example plant model

$$A = \begin{bmatrix} 0 & 11 & 0 & 0 \\ 1 & 0 & 0 & 0 \\ 0 & -1 & 0 & 0 \\ 0 & 0 & 1 & 0 \end{bmatrix} \quad b = \begin{bmatrix} -1 \\ 0 \\ 1 \\ 0 \end{bmatrix} \tag{8.11-12}$$

$$c^T = [0\ 0\ 0\ 1]$$

We have considered the output variable to be the position of the cart. We wish to move the cart from one position to another with the pendulum coming to rest in the vertical position. Notice that the output does not include a measurement of the angle $\theta(t)$. This is similar to balancing a broomstick blindfolded. The angle must be inferred from the forces on the cart. This is expected to be a difficult if not impossible task.

Let's find the resolvent matrix of our system and determine the plant poles.

$$(sI - A)^{-1} = \frac{1}{s^2(s^2-11)} \begin{bmatrix} s^3 & 11s^2 & 0 & 0 \\ s^2 & s^3 & 0 & 0 \\ -s & -s^2 & s(s^2-11) & 0 \\ -1 & -s & s^2-11 & s(s^2-11) \end{bmatrix} \tag{8.11-13}$$

There are four plant poles; two poles are at the origin, one is at $s = -\sqrt{11}$ and another is an unstable pole at $s = +\sqrt{11}$. It should not be surprising that the plant is unstable. If the state-variable vector is perturbed from its zero state and no further control action is supplied, the pendulum swings down. In the linear model, $\theta(t)$ grows until the linear model assumption of small $\theta(t)$ becomes invalid. However, if a stable closed-loop system can be created then any initial offset in the position of the cart or the angle of the pendulum is returned to zero and the cart returns to the zero position with the pendulum vertical. A step input then produces a constant offset in the position $y(t)$ and a vertical pendulum represented by $\theta(t) = 0$. Let's proceed by deriving the plant transfer function.

$$(sI - A)^{-1}b = \frac{1}{s^2(s^2-11)} \begin{bmatrix} -s^3 \\ -s^2 \\ s(s^2-10) \\ s^2-10 \end{bmatrix} \tag{8.11-14}$$

$$\frac{Y(s)}{U(s)} = c^T(sI - A)^{-1}b = \frac{s^2-10}{s^2(s^2-11)} \tag{8.11-15}$$

The expectations based on physical intuition that this plant would be very difficult to control are verified in the transfer function of the plant. It contains a non-minimum phase zero at a lower frequency than an unstable pole. This is the difficult combination mentioned in the previous section. It is especially difficult here because the zero is very near the pole. Remember, if the zero were to cancel the pole exactly, it would be impossible to move the pole. In a root locus analysis a zero very near the unstable pole means a large gain is required to wrench that pole away from the zero.

Let's proceed with a pole placement design using the transfer function and Diophantine equation approach. Ideally, the bandwidth of the system should be greater than that of the unstable pole at $s = \sqrt{11} \simeq 3.32$ but less that of the non-minimum phase zero at $s = \sqrt{10} \simeq 3.16$. Of course, this is not possible so we pick our closed-loop poles to have a compromise bandwidth of $\omega = 3.25$.

Recall from Sec. 3.2 that to have a realizable controller, the degree of the desired closed-loop polynomial $D_M(s)$ must equal $n + m$ when n is the degree of the plant denominator and $m \geq n - 1$.

$$\deg D_M(s) \geq n + n - 1 = 2(4) - 1 = 7 \tag{8.11-16}$$

The desired closed-loop polynomial must be at least seventh-order. We choose to use an eighth-order polynomial for convenience. $D_M(s)$ is chosen rather arbitrarily to have well-damped poles with a bandwidth of $\omega = 3.25$ and other poles at larger frequencies. Let

$$\begin{aligned} D_M s &= (s + 2.3 + j2.3)(s + 2.3 - j2.3)(s + 5 + j5)(s + 5 - j5)(s + 10)^4 \\ &= s^8 + 54.6s^7 + 1290.58s^6 + 17359s^5 + 146309s^4 \\ &\quad + 794960s^3 + 2726400s^2 + 5474000s + 5290000 \end{aligned} \tag{8.11-17}$$

Let

$$G_c(s) = \frac{b_3 s^3 + b_2 s^2 + b_1 s + b_0}{s^4 + a_3 s^3 + a_2 s^2 + a_1 s + a_0} \tag{8.11-18}$$

Then

$$\begin{aligned} &D_c(s)D_p(s) + N_c(s)N_p(s) = \\ &\quad s^8 + a_3 s^7 + (a_2 - 11)s^6 + (a_1 - 11a_3 + b_3)s^5 + (a_0 - 11a_2 + b_2)s^4 \\ &\quad + (-11a_1 + b_1 - 10b_3)s^3 + (-11a_0 + b_1 - 10b_3)s^3 + (-11a_0 + b_0 - 10b_2)s^2 \\ &\quad + (-10b_1)s + (-10b_0) \end{aligned} \tag{8.11-19}$$

Equating coefficients in Eq. (8.11-17) and Eq. (8.11-19) we can solve for $G_c(s)$

$$\begin{aligned} G_c(s) &= \frac{1539925.6s^3 + 5022290.18s^2 - 547400s - 529000}{s^4 + 54.6s^3 + 1301.58s^2 - 1521956s - 4861663.8} \\ &= \frac{1539926(s + 3.337)(s - 0.3609)(s + 0.2852)}{\left[(s + 74.44)^2 + (100.5)^2\right](s - 97.48)(s + 3.187)} \end{aligned} \tag{8.11-20}$$

The second equality in Eq. (8.11-20) is only approximate because in computing the roots some rounding occurs.

It is interesting to note that the series compensator of Eq. (8.11-20) is itself unstable, having a pole at $s = 97.48$. However, the closed-loop system is stable. The closed-loop transfer function is computed to be

$$\begin{aligned} M_c(s) &= \frac{1539926(s + 3.162)(s - 3.162)(s + 3.337)(s - 0.361)(s + 0.285)}{\left[(s + 2.30)^2 + (2.29)^2\right]\left[(s + 4.8)^2 + (4.91)^2\right]\left[(s + 9.78)^2 + (4.28)^2\right]} \\ &\quad \times \frac{1}{(s + 4.8)(s + 6.54)} \end{aligned} \tag{8.11-21}$$

Notice that the closed-loop poles are close to the desired positions but not exactly on the desired positions. The reason for this is not a problem with the theory of the pole placement algorithm but rather it is due to numerical inaccuracies in the computation. Such movement in the closed-loop poles due to small numerical inaccuracies may alert us to the fact that, in this case, even with feedback, the closed-loop system is highly sensitive to parameter inaccuracies. While the closed-loop poles of Eq. (8.11-21) are not exactly where we desire them, they are the poles of an apparently well-behaved closed-loop transfer function. Notice, however, that the closed-loop transfer function must contain all the zeros of both the plant, $G_p(s)$, and the series compensator, $G_c(s)$. In this case, there are two zeros much closer to the origin than the closest poles and the effects of these zeros greatly alter the dynamics. The transient response of this system is shown in Fig. 8.11-2. The response is very different from what we had in mind when we chose the closed-loop poles. It is unlikely that such a wild response would be acceptable. Indeed, except for all but the smallest inputs, the assumptions made about the linearity of the model and the abilities of a motor to provide the required force would be invalid.

Let's examine the frequency domain plots of the loop gain of this controller. In Fig. 8.11-3, the Bode magnitude and phase plots are given. The magnitude of the loop gain is near unity for almost two decades from $\omega = 1$ until $\omega = 100$. Also, the phase is close to 180° for these same frequencies. This clearly indicates trouble as the loop gain is near the critical point, $s = -1$, for all these frequencies. The phase lag of nearly 180° near the almost zero magnitude slope is due to the extra 180° additional phase lag that the non-minimum phase zeros provide over the phase lag of a minimum phase system with the same magnitude plot.

The Bode plots are magnified in Fig. 8.11-4 to show only the range $\omega = 1$ to $\omega = 100$. Notice the scale of the magnitude and phase axes. The magnitude plot crosses 0 dB three times while the phase plot crosses 180° three times. The polar plot for frequencies in the same range is shown in Fig. 8.11-5. Again, notice the scale of the plot. The crossings of the Bode plots occur in such a manner as to create a counterclockwise encirclement of the critical point on the polar plot. A second counterclockwise encirclement occurs when the negative frequencies are traced in the Nyquist D-contour. Two counterclockwise encirclements are needed for closed-loop stability since there are two right half-plane poles in the loop gain—one from the

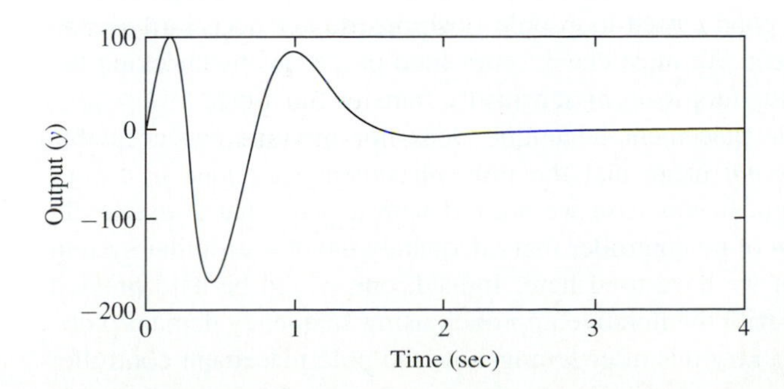

FIGURE 8.11-2
Unit step response of the closed-loop system.

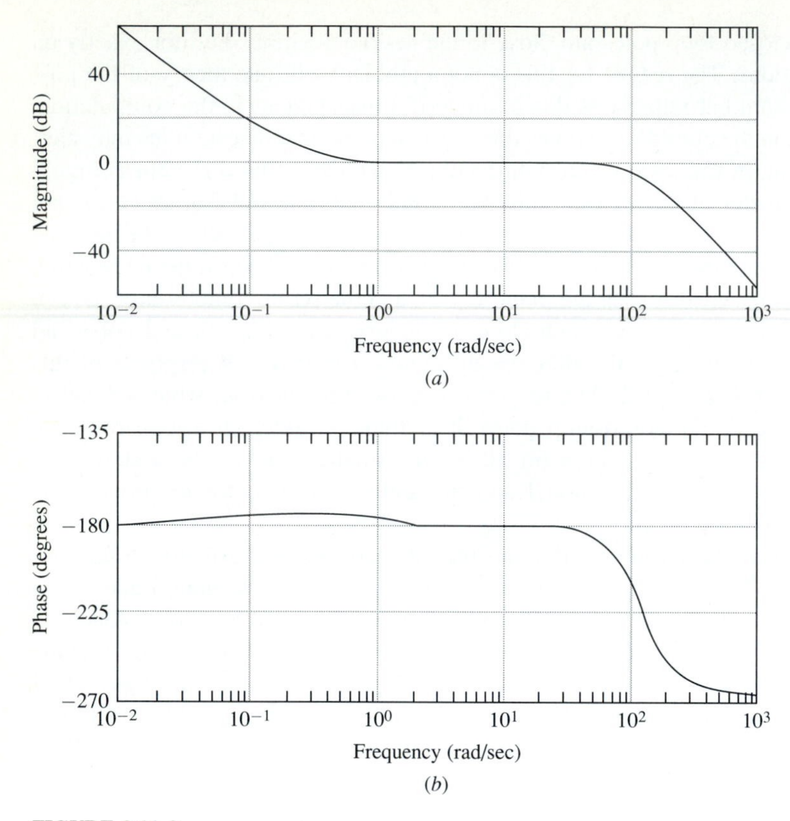

FIGURE 8.11-3
Bode plots of the loop gain. (*a*) Magnitude; (*b*) phase.

plant and one from the controller. The reader should verify that the Nyquist theorem for this loop gain does indeed show closed-loop stability.

Here is a case in which we have stability for the mathematical models involved but the loop gain passes so close to the critical point on all sides that if the mathematical model is less than perfect the system will be unstable. It has almost no robustness. One would not attempt to use this controller on a physical system. It could not be used despite the fact that mathematically the closed-loop poles are placed well into the left half-plane. Apparently, good closed-loop pole positions do not necessarily mean a well-behaved control system. We must check other loop properties by checking the frequency domain plots of the loop gain or sensitivity transfer functions.

The fact that the pole placement technique does not produce an acceptable controller in this case does *not* mean that the pole placement technique is a poor approach to controller design. In this case we started with a plant that is extremely difficult to control. We know of no controller that adequately handles a similar system using only the measurement we have used here. Indeed, one would be hard pressed to even stabilize the system with the iterative approach using frequency domain plots. It is an interesting exercise to try this after seeing how the pole placement controller

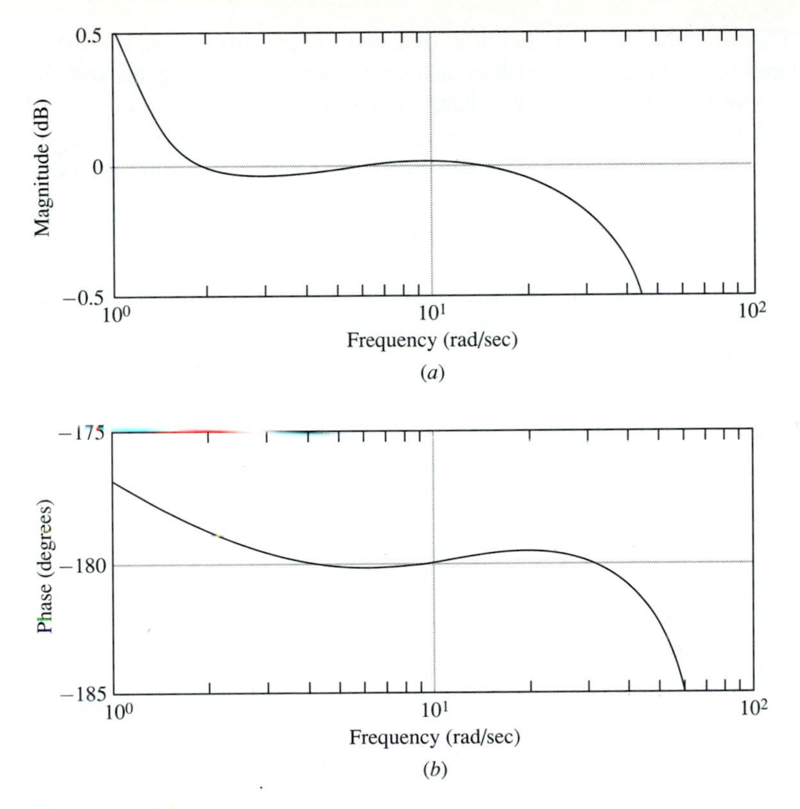

FIGURE 8.11-4
Magnification of Bode plots of the loop gain. (*a*) Magnitude; (*b*) phase.

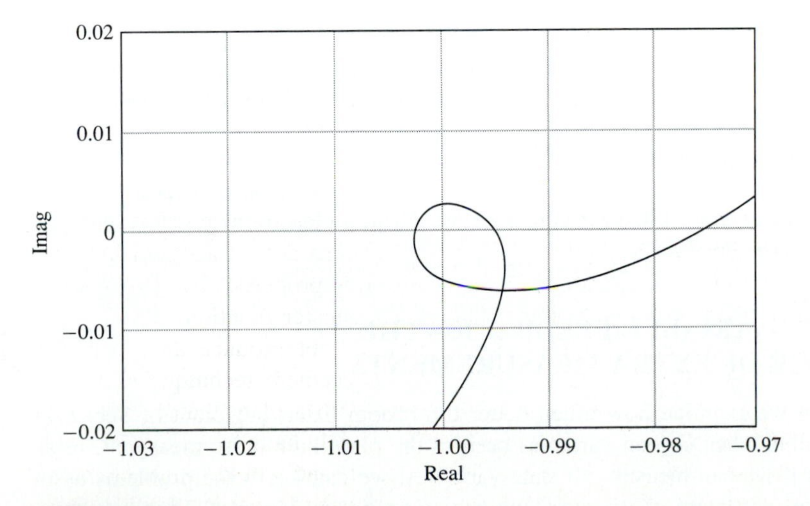

FIGURE 8.11-5
Polar plot of the loop gain.

achieves a stabilizing gain. One can also note that there is no way to use frequency domain techniques to improve the existing design as any lead or lag around the intermediate range of frequencies causes a change in the number of encirclements and an unstable closed-loop system. It would also require a clever designer to stabilize this system using the root locus technique. The root locus technique also suffers the same problem as the pole placement technique in that the method gives little indication of the properties of the loop gain, specifically robustness, sensitivity, and frequency response.

The pole placement technique using a series compensator is often a good way to get an initial design on difficult control problem. That initial design should then be analyzed using the frequency domain methods with Bode and Nyquist plots. If the design is inadequate it can possibly be improved by using series compensator techniques such as lead or lag compensators. If the initial design cannot be sufficiently improved, it may be necessary to alter the design of the plant itself by reducing time delays or adding measurements or control variables. In the next section we approach the broom balancing problem from a state-variable pole placement approach. We will see that if certain state-variables can be measured, this system can be adequately controlled.

Exercises 8.11

8.11-1. There is nothing in the theory of pole placement indicating that particular placements of poles won't work. For example, nothing in the theory prevents choosing closed-loop system poles at a higher frequency than right half-plane zeros even though we know from Bode-Nyquist ideas that such a design does not generally produce an acceptable loop gain.

Consider the plant from Exercise 8.10-1.

$$G_p(s) = \frac{30(-s+1)}{(s+0.1)(s+10)}$$

Design a pole placement controller with all the closed-loop poles placed at a frequency $\omega = 10$ or greater. Examine the Bode and Nyquist plots of the resulting loop gain and critique the design.

8.11-2. Now redesign a controller for the plant of Exercise 8.11-1, placing the dominant closed-loop pole at $s = -0.5$ and the rest of the closed-loop poles at a frequency greater than $\omega = 10$. Examine the Bode and Nyquist plots of the resulting loop gain and critique the design.

8.12 STATE-VARIABLE FEEDBACK—THE ADVANTAGE OF EXTRA MEASUREMENTS

In this section we examine how much better the broom balancing plant of Sec. 8.11 can be controlled when certain variables besides the plant output are measured. First, we show that if we can measure all state-variables, we can avoid the problems associated with non-minimum phase zeros and achieve a good controller. Then, we show how to adjust a state-variable controller if not all states are measurable. Of course, as less states are measurable, we lose the benefits of state-variable control. When we get to the point where only the plant output is measurable, the state-variable controller is equivalent to an output feedback controller. The final conclusion is that the problem associated with non-minimum phase zeros can be alleviated if certain specific additional measurements can be made.

Let's consider the same plant as in Sec. 8.11 but this time let's assume all the state-variables are available for measurement so that pole placement can be achieved by state-variable feedback. The plant is represented by the state Eqs. ($\boldsymbol{Ab}$) and ($\boldsymbol{c}$) with $\boldsymbol{A}$, $\boldsymbol{b}$, and $\mathbf{c}$ given by Eq. (8.11-12). With the state-variable approach we need only place the four poles associated with the eigenvalues of the state matrix $\boldsymbol{A}$. The closed-loop system has four poles given by the eigenvalues of $\boldsymbol{A} - \boldsymbol{bk}^T$. The appropriate feedback vector $\boldsymbol{k}$ is determined using the methods of Sec. 3.5. Let's place the closed-loop poles so that the closed-loop pole polynomial is given by

$$
\begin{aligned}
D_H(s) &= (s + 2.3 + j2.3)(s + 2.3 - j2.3)(s + 5 + j5)(s + 5 - j5) \\
&= s^4 + 14.6s^3 + 128.58s^2 + 335.8s + 529
\end{aligned}
\tag{8.12-1}
$$

The closed-loop poles chosen are the same as the four poles closest to the origin of the eight desired poles chosen in Sec. 8.11. The appropriate feedback gains in $\boldsymbol{k}$ are determined according to Eq. (3.5-21), repeated here as Eq. (8.12-2).

$$
\boldsymbol{k}^T = \frac{\boldsymbol{d}^T - \boldsymbol{a}^T}{K} \boldsymbol{C}_{A_p b_p} \boldsymbol{C}_{Ab}{}^{-1} \tag{8.12-2}
$$

In Equation (8.12-2), $\boldsymbol{C}_{Ab}$ is the controllability matrix associated with the physical state-variable representation of the system.

In this case

$$
\boldsymbol{C}_{Ab} = \begin{bmatrix} -1 & 0 & 11 & 0 \\ 0 & -1 & 0 & -11 \\ 1 & 0 & 1 & 0 \\ 0 & 1 & 0 & 1 \end{bmatrix} \tag{8.12-3}
$$

This matrix is invertible, indicating that the system is controllable and arbitrary pole placement is achievable.

The inverse is given by

$$
\boldsymbol{C}_{Ab}{}^{-1} = \begin{bmatrix} 0.1 & 0 & 1.1 & 0 \\ 0 & 0.1 & 0 & 1.1 \\ -0.1 & 0 & -0.1 & 0 \\ 0 & -0.1 & 0 & -0.1 \end{bmatrix} \tag{8.12-4}
$$

We also need the phase variable description of the system. The phase variable description of the system can be obtained directly from the plant transfer function

Eq. (8.11-15), repeated here as Eq. (8.12-5)

$$G_p(s) = \frac{s^2 - 10}{s^2(s^2 - 11)} = \frac{s^2 - 10}{s^4 - 11s^2} \tag{8.12-5}$$

$$\boldsymbol{A}_p = \begin{bmatrix} 0 & 1 & 0 & 0 \\ 0 & 0 & 1 & 0 \\ 0 & 0 & 0 & 1 \\ 0 & 0 & 11 & 0 \end{bmatrix} \qquad \boldsymbol{b}_p = \begin{bmatrix} 0 \\ 0 \\ 0 \\ 1 \end{bmatrix}$$

$$\boldsymbol{c_p}^T = [-10\ 0\ 0\ 0] \tag{8.12-6}$$

The vector $\boldsymbol{a}^T$ in Equation (8.12-2) is the negative of last row in $\boldsymbol{A}_p$

$$\boldsymbol{a}^T = [0\ 0\ -11\ 0] \tag{8.12-7}$$

The matrix $\boldsymbol{A_P b_P}$ in Eq. (8.12-2) is the controllability matrix associated with the phase variable description of Eq. (8.12-6)

$$\boldsymbol{C}_{A_p b_p} = \begin{bmatrix} 0 & 0 & 0 & 1 \\ 0 & 0 & 1 & 0 \\ 0 & 1 & 0 & 11 \\ 1 & 0 & 11 & 0 \end{bmatrix} \tag{8.12-8}$$

The vector $\boldsymbol{d}^T$ is composed of the coefficients of the desired pole polynomial of Eq. (8.12-1)

$$\boldsymbol{d}^T = [\,529 \quad 335.8 \quad 128.58 \quad 14.6\,] \tag{8.12-9}$$

Feeding all the appropriate information into Eq. (8.12-2), and letting the extra parameter $K = 1$, the feedback gains $\boldsymbol{k}^T$ are computed as

$$\boldsymbol{k}^T = [\,-48.18 \quad -192.48 \quad -33.58 \quad -52.9\,] \tag{8.12-10}$$

One can check the eigenvalues of $\boldsymbol{A} - \boldsymbol{bk}^T$ to ascertain that the closed-loop poles are placed as desired.

Now let us analyze the resulting controller. Look first at the elementary block diagram of the system given by Fig. 8.12-1. An output disturbance has been added to this diagram for later analysis. The controller is realized by measuring the four state-variables and feeding each through a constant gain. There is no problem with pole-zero cancellation or trying to realize an improper transfer function. The closed-loop transfer function is given by

$$\boldsymbol{M}_c(s) = \boldsymbol{c}^T(s\boldsymbol{I} - \boldsymbol{A} + \boldsymbol{bk}^T)^{-1}\mathbf{b} = \frac{s^2 - 11}{\left((s+2.3)^2 + (2.3)^2\right)\left((s+5)^2 + (5)^2\right)} \tag{8.12-11}$$

The plant zeros remain in the closed-loop transfer function as they must. The poles have been moved to the desired location. The system is not very sensitive to changes in the coefficients of the controller as was the output feedback system of the

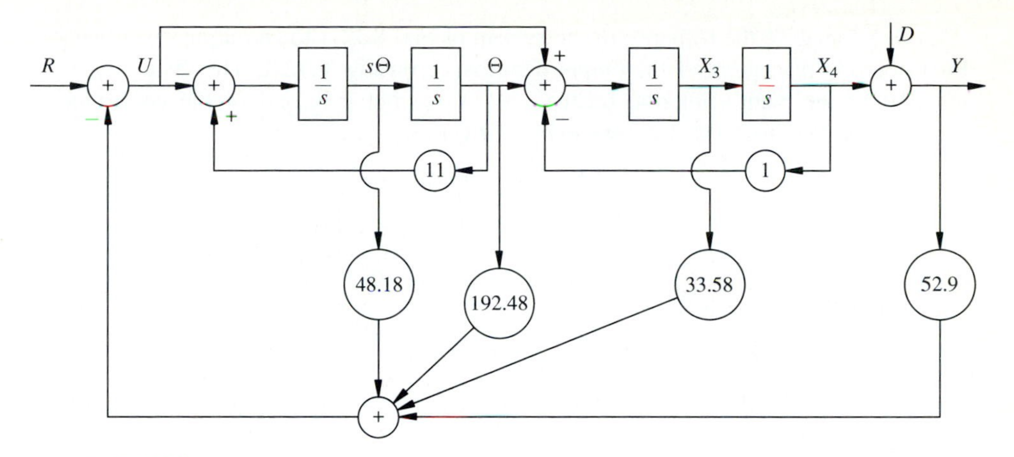

FIGURE 8.12-1
State-variable feedback controller.

last section. The closed-loop transfer function can also be solved by finding $H_{eq}(s)$ and using

$$M_c(s) = \frac{G_p(s)}{1 + G_p(s)H_{eq}(s)} \tag{8.12-12}$$

The transfer function $H_{eq}(s)$ is

$$H_{eq}(s) = \frac{\boldsymbol{k}^T(s\boldsymbol{I} - \boldsymbol{A})^{-1}\boldsymbol{b}}{\mathbf{c}^T(s\boldsymbol{I} - \boldsymbol{A})^{-1}\boldsymbol{b}} = \frac{14.6\left((s+1.6)^2 + (2.2)^2\right)(s + 4.85)}{\left(s + \sqrt{10}\right)\left(s - \sqrt{10}\right)} \tag{8.12-13}$$

Of course, we could not implement a controller by building a transfer function equal to $H_{eq}(s)$ and using it in the feedback path. The transfer $H_{eq}(s)$ is improper and, more seriously, has a right half-plane pole at the same location as a right half-plane zero of the plant. Implementing $H_{eq}(s)$ would create an illegal right half-plane pole-zero cancellation. The resulting system would not be internally stable and small reference inputs would result in unbounded plant inputs. The state-variable feedback of Fig. 8.12-1 is implementable and mathematically produces the same closed-loop transfer function as an output feedback controller using $H_{eq}(s)$ would. The $H_{eq}(s)$ is a mathematical abstraction that allows a different way of thinking about and analyzing the state-variable controller.

8.12.1 Disturbance Rejection

One must be careful in analyzing the reaction of a state-variable control system to a disturbance. The problem is that, depending on where the disturbance enters, only part of the controller affects the disturbance by feeding it back. The state-variable feedback system does not respond to an output disturbance in the same way that the $H_{eq}(s)$ abstraction does because $H_{eq}(s)$ contains parts of the controller and plant that do not directly act upon the disturbance.

Let's analyze the response of the system of Fig. 8.12-1 to an output disturbance. First note the placement of the output disturbance on Fig. 8.12-1. The disturbance is not part of the plant's internal feedback of $X_4(s)$ but it is part of the controller's feedback of $Y(s)$ to $U(s)$. First we express $Y(s)$ as

$$\begin{aligned} Y(s) &= \boldsymbol{c}^T \boldsymbol{X}(s) + D(s) \\ &= \boldsymbol{c}^T (s\boldsymbol{I} - \boldsymbol{A})^{-1} \boldsymbol{b} U(s) + D(s) \end{aligned} \tag{8.12-14}$$

The tricky part is to correctly factor $D(s)$ into the equation that describes how the plant input $U(s)$ is formed.

$$\begin{aligned} U(s) &= -\boldsymbol{k}^T \boldsymbol{X}(s) - 52.9\ D(s) \\ &= -\mathbf{k}^T (s\boldsymbol{I} - \boldsymbol{A})^{-1} \boldsymbol{b} U(s) - 52.9\ D(s) \end{aligned} \tag{8.12-15}$$

Notice that only the term $k_4 = 52.9$ multiplies the $D(s)$ term in the feedback. The other feedback gains come before the disturbance. Equation (8.12-15) can be rewritten as

$$U(s) = \frac{-52.9}{1 + \boldsymbol{k}^T (s\boldsymbol{I} - \boldsymbol{A})^{-1} \boldsymbol{b}} D(s) \tag{8.12-16}$$

Substituting Eq. (8.12-16) into Eq. (8.12-14), we get

$$Y(s) = \left(1 - \frac{52.9 \boldsymbol{c}^T (s\boldsymbol{I} - \boldsymbol{A})^{-1} \boldsymbol{b}}{1 + \boldsymbol{k}^T (s\boldsymbol{I} - \boldsymbol{A})^{-1} \boldsymbol{b}}\right) D(s) \tag{8.12-17}$$

Recalling

$$G_p(s) = \boldsymbol{c}^T (s\boldsymbol{I} - \boldsymbol{A})^{-1} \boldsymbol{b}$$

and

$$G_p(s) H_{\text{eq}}(s) = \boldsymbol{k}^T (s\boldsymbol{I} - \boldsymbol{A})^{-1} \boldsymbol{b}$$

We can rewrite Eq. (8.12-17) in terms of $G_p(s)$ and $H_{\text{eq}}(s)$ as

$$\begin{aligned} Y(s) &= \left(1 - \frac{52.9 G_p(s)}{1 + G_p(s) H_{\text{eq}}(s)}\right) D(s) \\ &= \frac{1 + G_p(s)(H_{\text{eq}}(s) - 52.9)}{1 + G_p(s) H_{\text{eq}}(s)} \end{aligned} \tag{8.12-18}$$

This expression is more complicated than the usual sensitivity function. When more than one variable is measured and fed back, the plant is no longer a single-input, single-output system but it must be considered as a single-input, multi-output system. The analysis of such systems with respect to sensitivity and robustness is somewhat more complicated and beyond the scope of this book. The principle of large loop gain being good for sensitivity reduction still holds but must be interpreted properly. In Eq. (8.12-18), if the loop gain is large and if most of its size comes from the last loop that works on the disturbance, the reaction to the output disturbance is small.

Notice that, in Eq. (8.12-18), the amount of disturbance rejection depends on how closely $H_{\text{eq}}(s)$ matches the outside loop feedback constant, 52.9. If $H_{\text{eq}}(s)$ is near

52.9 then the response to a disturbance is approximately

$$\frac{D(s)}{1 + G_p(s)H_{\text{eq}}(s)}$$

If 52.9 is small compared to $H_{\text{eq}}(s)$, the response to a disturbance is approximately $D(s)$ itself, i.e., there is almost no disturbance attenuation. Thus we see that when using state-variable feedback the attenuation of a disturbance depends not as much on the size of the overall loop gain but more on the size of the gain that surrounds the disturbance.

We have seen that, if we are able to measure all state-variables for feedback, then the problems of right half-plane zeros can be avoided. While poles arise from the internal dynamics of the plant, zeros arise from the interaction of the dynamics with the plant inputs and outputs. If we can create more measurements or control inputs we can eliminate bothersome plant zeros. This allows the successful control of unusually difficult plants that cannot be adequately controlled using only output feedback.

8.12.2 Inaccessible State Variables

Can some of the advantages of state-variable feedback still be achieved without having to measure *all* of the state-variables? The answer is yes, if certain needed variables are measurable. Let's look at the system of Fig. 8.12-1 and assume that y and θ are measurable but their derivatives $\dot{y}$ and $\dot{\theta}$ are not. By a simple block-diagram manipulation the system of Fig. 8.12-1 can be turned into the system of Fig. 8.12-2 with

$$H_{01}(s) = 4.81s + 17.04 \tag{8.12-19}$$

$$H_{02}(s) = 33.5s + 52.9 \tag{8.12-20}$$

These two transfer functions are improper and thus unrealizable. They do not, however, try to cancel right half-plane poles or zeros. The transfer functions can

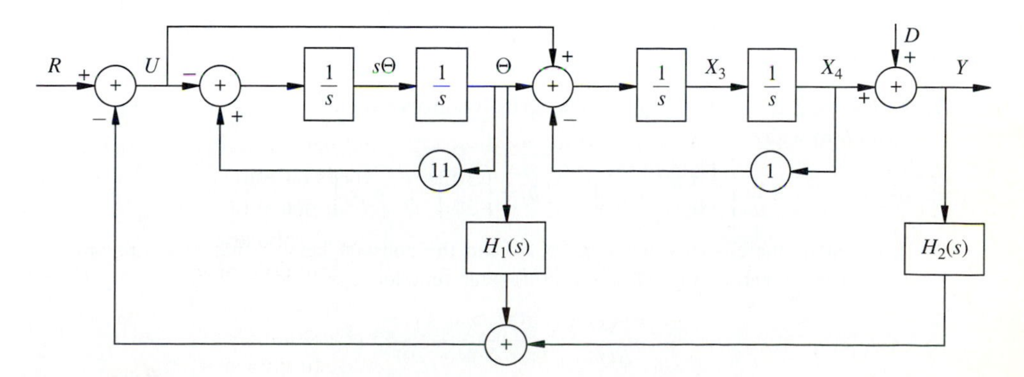

FIGURE 8.12-2
Controller with $Y(s)$ and $\theta(s)$ measured.

be made realizable by adding a denominator with a pole in the left half-plane at a frequency high enough that the extra pole does not slow the system down even after it is moved to its closed-loop position. If the poles are fast enough, the systems will react almost exactly like those of Eqs. (8.12-19) and (8.12-20) for the slower signals fed into them. Since the eventual bandwidth of the closed loop system is about $\omega = 3$, a reasonable placement for the extra poles is $\omega = 30$. The transfer functions

$$H_1(s) = \frac{4.81s + 17.04}{\dfrac{s}{30} + 1} \tag{8.12-21}$$

$$H_2(s) = \frac{33.5s + 52.9}{\dfrac{s}{30} + 1} \tag{8.12-22}$$

are realizable and produce almost the same effect as the state-variable controller.

The key extra measurement in controlling this plant is θ, the angle, of the pendulum. It makes sense physically that measuring this variable facilitates control. It supplies the measurement provided visually in the broom balancing problem. Being able to measure θ is like removing the blindfold. If we try to perform block-diagram manipulation to eliminate the feedback from θ, thus turning the system into an output feedback control system, the feedback transfer function which results is the $H_{eq}(s)$ from Eq. (8.12-13), which cannot be used because of right half-plane pole-zero cancellations. Thus, the measurement of θ is indeed the key to controlling this system.

We have seen that additional measurements can make plants easier to control. Plants can always be controlled with reasonable sensitivity if all state-variables are measured, but often all variables are not needed. If the key variables are identified the added expense of achieving extra measurements can be applied only when needed. Whenever extra measurements are used the system can no longer be simply analyzed for sensitivity and robustness properties using the single-input, single-output techniques of this book. Research on multi-input, multi-output techniques is ongoing. Much has been learned about these systems and many schools teach graduate courses in multivariable control systems.

Exercise 8.12

8.12-1. The objective is to design a state-space controller for the plant given by Eqs. ($\boldsymbol{Ab}$) and ($\boldsymbol{c}$) with

$$\boldsymbol{A} = \begin{bmatrix} -0.1 & 0 \\ 0 & -10 \end{bmatrix} \qquad \boldsymbol{b} = \begin{bmatrix} 30 \\ 30 \end{bmatrix} \qquad \boldsymbol{c} = \begin{bmatrix} 1/9 \\ -10/9 \end{bmatrix}$$

Calculate the elements in the $\boldsymbol{k}$ vector and the constant gain K that, in combination with this plant, gives a closed-loop transfer function

$$\frac{Y(s)}{R(s)} = \frac{75(-s+1)}{(s+4)(s+20)}$$

(i.e., place the closed loop poles at -4 and -20). For this control design, find the transfer function H_{eq}. Is H_{eq} realizable? Why? Generate the Bode plot of the loop gain, $G_p H_{eq}$. What is the bandwidth and phase margin of this design? How does H_{eq} appear to increase the bandwidth while maintaining some phase margin?

8.13 CONCLUSIONS

In this chapter we have completed our study of the basic principles of control theory by applying the insights we have gained in modeling and analyzing control systems to the problem of designing control systems. First a strategy for controller design by manipulation of the loop gain transfer function with series compensators was developed. After stability of the nominal control system is assured, performance objectives are pursued.

At low frequencies a large loop gain corresponding to a large return difference and small sensitivity function is sought to provide low sensitivity to plant perturbations, good low frequency disturbance rejection and reasonable reference input following for low frequency inputs. At high frequencies a small loop gain corresponding to a small complementary sensitivity function or closed-loop transfer function is necessary for the design to be robust in the face of high frequency model uncertainty. A small loop gain is also advantageous for minimizing the effect of sensor noise and limiting the control action at high frequencies. In the transition frequencies where the loop gain goes from large to small, a large return difference is desired so that the system has an adequate transient response and enjoys a large degree of robustness. The crossover frequency should be as large as high frequency constraints allow and the phase margin should be as large as a fast transition from high loop gain to small loop gain allows.

There are a number of building blocks for shaping the loop gain of a system and the more fundamental of these were explored. It was found that unstable plants can be stabilized using the same techniques as are used to improve the performance of stable plants. In particular, the control system designer is constantly trying to achieve as much phase lead and as large a bandwidth as conditions allow. The design is often limited by model uncertainty, which limits the bandwidth, and often by right half-plane or non-minimum phase zeros, which create extra phase lag. Non-minimum phase zeros can arise from time delays within the plant or from insufficient measurements.

Pole placement techniques can produce reasonable starting points for controller designs involving difficult plants. They do not, however, overcome problems associated with non-minimum phase zeros or modeling uncertainty. The loop gain, sensitivity function, and complementary sensitivity function associated with a pole placement design must be analyzed to assess the design. Sometimes problems with non-minimum phase zeros can be overcome with the use of extra plant measurements. The extreme is to measure all state-variables and perform state-variable feedback. If not all state-variables can be measured, then some can be accounted for with block-diagram manipulations. At some point, however, a set of required measurements is reached.

In attempting to analyze the sensitivity, disturbance rejection properties and robustness properties of a state feedback controller, one comes to the realization that the techniques developed in this book do not directly apply. Since the controller's position in the loop is physically distributed it is difficult to define a single loop gain.

Indeed, we have relied on the assumption that the systems we have been analyzing deal with a single input and a single output. A controller with more than one measurement must be analyzed as a system with more than one output. The techniques developed in this book can be generalized to include multi-input and multi-output systems (often called *multivariable* systems). We leave this generalization to the next course of study. Let us remark, however, that the major difference is that, instead of dealing with transfer functions, multivariable systems can be described with matrices of transfer functions. Many problems of multivariable control arise from two difficult properties of these matrices—matrix multiplication does not commute and matrices can have both large and small elements at the same frequency. After the details concerning these problems with matrices are worked out the principles explored in this book remain.

There are many other topics of interest in a continuing study of control theory. Optimal control techniques have been applied in both the time and frequency domain. Researchers are just beginning to understand the control of nonlinear plants and the use of nonlinear controllers. On line plant identification and adaptive control techniques that allow modeling and control to occur simultaneously are in need of more study. We hope that the principles presented in this book provide a solid foundation for those who go on to design working controllers and those who will continue forging new results in the theory of control systems.

PROBLEMS

8.1. (*a*) Why does a control system designer want the loop gain high at low frequencies? Why does the designer want the loop gain low at high frequencies? Write a short, one sentence answer for each reason. You should mention at least six reasons total.

(*b*) Using ideas from Bode plots, explain why increasing the loop gain of a controller generally makes the closed-loop system respond faster.

8.2. For a design problem a disturbance exists at the input to the plant as shown in Fig. P8.2. The maximum magnitude of the disturbance is $A = 100$. The design can not tolerate a sinusoidal output from this disturbance with a magnitude greater than 0.01. If at $\omega = 0.1$, $|G_p| = 10$ find the constraint on $|1 + G_c G_p|$ at $\omega = 0.1$. What is the constraint on $|G_c G_p|$ at $\omega = 0.1$?

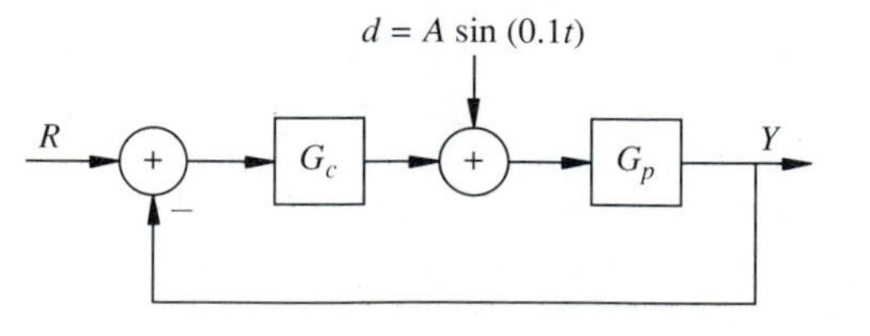

FIGURE P8.2
Problem 8.2.

8.3. A plant model set is described by

$$\tilde{G}_p(s) = G_p(s)\,(1 + L_m(s))$$

and

$$|L_m(j\omega)| < l_m(j\omega)$$

where

$$G_p(s) = \frac{100}{s^2}$$

and

$$l_m(s) = \left| \frac{\left(s + 10^2\right)}{3000} \right|$$

Design a controller $G_c(s)$ so that the closed-loop system is stable for all plants in $\tilde{G}_p(s)$ and has reasonable performance for all plants in $\tilde{G}_p(s)$. This loop should also achieve 60° of phase margin at crossover frequency and a loop gain of greater than 90 dB for $\omega < 0.1$ when the nominal plant $G_p(s)$ is used. The crossover frequency is not specified.

8.4. The following parts of the problem refer to Fig. P8.4.

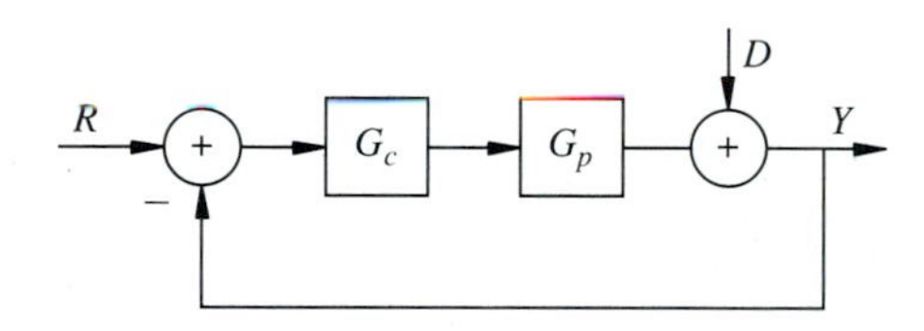

FIGURE P8.4
Problem 8.4.

(*a*) Sketch the Bode plots of the plant

$$G_p(s) = \frac{0.1(s+1)}{s^2 + 0.1s + 1}$$

(*b*) Assume that $\tilde{G}(s) = G_p(s)\,(1 + L_m(s))$.
with $|L_m(j\omega)| < \left|0.003(j\omega + 1)^2\right|$
and that it is required that $\left|\dfrac{Y(j\omega)}{D(j\omega)}\right| < 0.1$ for $\omega < 1$.
where $D(j\omega)$ is an output disturbance.
Design a cascade compensator $G_c(s)$ so that

(i) The closed-loop system is stable.
(ii) The closed-loop system remains stable for all $L_m(j\omega)$ satisfying the stated bound.
(iii) The nominal system has at least a 30° phase margin.
(iv) The disturbance rejection requirement is met.

8.5. A control system is to be designed with reference to Fig. P8.5.

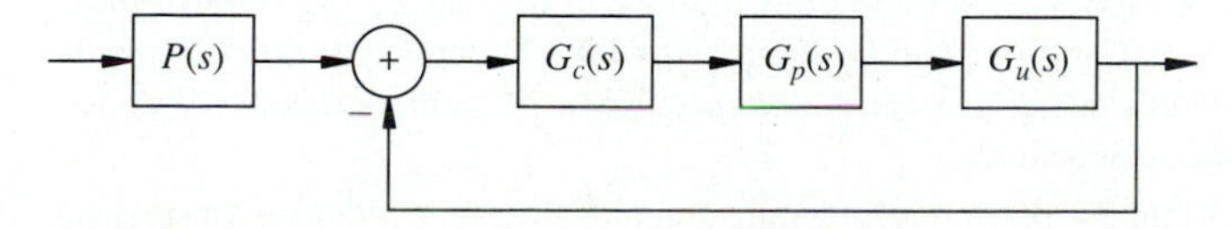

FIGURE P8.5
Problem 8.5.

where $G(s)$ is the nominal plant model.

$G_u(s)$ represents dynamics that are missing from the nominal plant model.

$G_c(s)$ and $P(s)$ are the controller and prefilter you are to design. (You may not need $P(s)$.)

Let

$$G_p(s) = \frac{1}{(s+1)^2}$$

Design a controller to meet the following specifications:

1. If there is an output disturbance with a frequency below $\omega = 0.01$ rad/sec, its effect on the output must be less than 0.001 times its original magnitude.
2. The nominal phase margin should be greater than 30°.
3. The closed-loop response of the nominal system to a unit step input should have zero steady-state error. It should have less than 25 percent overshoot in the transient and it should settle to within 5 percent of its final value in 10 sec.
4. The closed-loop system should be stable in the presence of either of the following representations of unmodeled dynamics.

$$G_{u1}(s) = \frac{100}{s^2 + 0.1s + 100}$$

$$G_{u2}(s) = \frac{10{,}000}{s^2 + 0.01s + 10{,}000}$$

8.6. Unstable poles and non-minimum phase zeros limit the ability to attain good control performance even when there is no modeling error in the plant.

(*a*) Consider the unstable plant whose transfer function is

$$G_{p1}(s) = \frac{100}{s-10}$$

Sketch the Bode phase and magnitude plot and the Nyquist plot for $G_{p1}(s)$. Is there a simple controller that stabilizes this plant? If there is, describe one such controller. If there is not, explain why?

(*b*) Now assume that there is a non-minimum phase zero in the plant so that

$$G_{p2}(s) = \frac{100\left(\dfrac{-s}{z}+1\right)}{(s-5)(s+20)}$$

If z is large, $G_{p2}(s)$ has a right half-plane zero far from the origin. How will the job of stabilizing $G_{p2}(s)$ compare to the job of stabilizing $G_{p1}(s)$ if z is greater than 1000? Will it be much harder, harder, the same, easier or much easier? Explain your answer.

(*c*) If z is small, the zero is now close to the origin. Sketch the Bode magnitude and phase plots for $G_{p2}(s)$ when $z = 1$. Add these plots to those of (*a*). Sketch the Nyquist plot for $G_{p2}(s)$ with $z = 1$. Can this plant be stabilized with a simple controller? Explain how or why not. (Don't design the controller specifically.) Compare this situation to that of (*a*) and write conclusions.

8.7. We know from Chap. 3 that the poles of $H_{eq}(s)$ are zeros of the plant. We also know that it is acceptable to cancel plant zeros if the zeros are in the left half-plane and away from the $j\omega$ axis. This implies that a state-variable pole placement controller can be turned into a realizable transfer function feedback controller if it is acceptable to cancel all the zeros of the plant.

Given the plant in state-variable form with

$$A = \begin{bmatrix} 1 & -1 & 2 \\ -52 & 12 & -26 \\ -53 & 13 & -27 \end{bmatrix} \qquad b = \begin{bmatrix} 0 \\ 1 \\ 1 \end{bmatrix} \qquad c = \begin{bmatrix} 9 \\ 16 \\ -15 \end{bmatrix}$$

(*a*) Find the state-variable feedback controller that places the closed-loop poles at $s = -10$, $s = -15 + j15$ and $s = -15 - j15$.

(*b*) Find the $H_{\text{eq}}(s)$ associated with this controller.

(*c*) Add enough distant poles to $H_{\text{eq}}(s)$ to make it into $H(s)$, a realizable transfer function that reacts similarly to $H_{\text{eq}}(s)$.

(*d*) Find the closed-loop poles using the new $H(s)$ in feedback with the plant.

8.8. (*CAD Problem*) Ziegler-Nichols tuning laws for PID compensators

It is quite fortuitous that the beneficial properties of feedback control are often obtained when a favorable transient response is obtained. The following conditions lead to a favorable transient response in a simple series compensator; they also are favorable properties to produce a loop gain with low sensitivity:

1. An integrator in the loop indicates zero steady-state error for constant reference inputs. (An integrator in the loop indicates an infinite loop gain at dc, which indicates complete steady-state rejection of constant disturbances.)
2. A fast transient response indicates a large bandwidth. (A large bandwidth indicates that the sensitivity function is kept small for a large range of frequencies.)
3. A non-oscillatory transient response indicates large phase margin at crossover. (A large phase margin indicates that the sensitivity function does not have a large peak near the crossover frequency.)
4. A first order step response with zero steady-state error indicates that the closed-loop system's magnitude response is near unity for low frequencies, which indicates a large loop gain at low frequencies. (A large loop gain at low frequencies indicates low sensitivity at low frequencies.)

A large number of control loops operating today, especially in the process control industry, are not designed by mathematical manipulations on a model. Instead, a simple PID compensator is implemented and the three parameters associated with the PID compensator are *tuned* to provide an acceptable transient response. The disturbance rejection and sensitivity reduction benefits are achieved from the structure of the controller that includes an integrator. The results listed above indicate that, for a simple controller, a good transient response leads to good loop properties.

In 1942 two control practitioners, Ziegler and Nichols, produced a prescription for tuning P, PI, and PID compensators. These tuning rules bear their names today. The prescription that Ziegler and Nichols produced is summarized here.

Set up a proportional compensator, i.e., let $G_c(s) = K_p$. Starting from very small K_p increase K_p until the output of the system in response to a unit step input becomes a sinusoid whose amplitude does not grow with time but also does not decay with time. Call the value of K that produces this condition of borderline instability $K_{\max}$. Measure the period of the oscillation and call this value T_p. Set up a P, PI, or PID compensator according to the following equation and Table P8.1.

$$G_c(s) = K\left(1 + \frac{1}{T_1 s} + \frac{T_D s}{1 + (s/10T_D)}\right)$$

The pole at $10T_D$ in the derivative compensator term is there only to make the compensator realizable. It has little effect and is largely ignored in this analysis.

TABLE P8.1

Ziegler-Nichols PID tuning rules

Controller type	K	T_I	T_D
P	$0.5K_{max}$	∞	0
PI	$0.45K_{max}$	$Tp/1.2$	0
PID	$0.6K_{max}$	$Tp/2$	$Tp/8$

The Ziegler-Nichols rules work well on plants whose step response closely matches that of a simple first-order system with a delay. Since time can be scaled, the important parameter of such a system is the speed of the pole in relation to the length of the delay. Therefore, we can analyze an entire group of plants of this type by assuming a one time unit delay and moving the pole relative to the delay length. Let

$$G_p(s) = \frac{a}{(s+a)}e^{-s}$$

(*a*) A PID compensator is equivalent to a lead-lag compensator. Without regard to the Ziegler-Nichols rules, design a lead-lag compensator for the plant, $G_p(s)$, with $a = 0.01$. Make the bandwidth as large as you can without letting the sensitivity function become greater than 1.5.

Now let's analyze the Ziegler-Nichols proportional control laws for this plant.

(*b*) Plot the Bode plots for $G_p(s)$ with $a = 0.01$, 0.1, 1,and 10. From the Bode plots determine the K_{max} and T_p that result from the Ziegler-Nichols experiment for each value of a. For each value of a, find the phase margin that results from the Ziegler-Nichols proportional compensator. What can you say about the expected transient responses for the Ziegler-Nichols proportional compensators for various values of a? If the dc gain of $G_p(s)$ were to double while the pole position and the delay remained the same, what happens to the gains K of the Ziegler-Nichols *proportional* compensators? What happens to the loop gains that result? What is the relation between T_p and the crossover frequency ω_c when $a = 0.01$ and $a = 0.1$? Can you argue that there is a general approximate relationship between T_p and ω_c for the Ziegler-Nichols *proportional* compensators when a is small compared to the delay?

In (*b*) you should have convinced yourself that the Ziegler-Nichols rules result in a reasonable loop gain transfer function as long as the plant is well modeled by the $G_p(s)$ given above. This is true for all values of the pole position, the delay and the dc gain of the plant. The Ziegler-Nichols PI compensators behave similarly to the proportional compensators near crossover frequency but add much low frequency gain with the integral action.

(*c*) Using the values of K_{max} and T_p found in (*b*), plot the Bode plots of prescribed Ziegler-Nichols PI compensators for each of the four values of the pole position parameter a. The zero of the PI compensator is set based upon T_p. For each value of a find the relationship between the zero of the PI compensator and the crossover frequency? When a is small how does this result relate to the relationship between T_p and ω_c found in (*b*)? For each a how much phase lag does the addition of the integral term cause at the ω_c associated with the proportional compensator? How is this additional phase lag counteracted?

(*d*) Using the values of $K_{\max}$ and T_p found in (*b*), plot the Bode plots of prescribed Ziegler-Nichols PID compensators for each of the four values of the pole position parameter a. Why can K be increased for the PID compensators over its value for the proportional compensators? How does the Ziegler-Nichols PID compensator for $a = 0.01$ compare with your lead-lag design of (*a*)? If a CAD tool is available plot the unit step responses for each of the Ziegler-Nichols PID compensators. To do this more easily replace the unit delay by the third-order Padé approximation for a unit delay given in Eq. (8.10-15).

8.9. (*CAD Problem*) Control of airplane phugoid mode

The transfer function between the differential stick position which commands the elevator position $\Delta_e(s)$, and the differential pitch angle $\Theta(s)$ is developed in the appendix to Chap. 2 and Problem 4.11. We consider this to be our plant.

$$G_P(s) = \frac{\Theta(s)}{\Delta_e(s)} = \frac{-0.5632s^2 - 0.3123s - 0.0102}{s^4 + 1.0480s^3 + 0.7862s^2 + 0.0193s + 0.0120}$$

Since aircraft must be designed to be light the dynamics are subject to various bending modes. These dynamics are too complicated to model precisely but their effect can be summarized with a bound on the magnitude of all possible multiplicative perturbations. Let

$$l_m(j\omega) = 0.01(s+1)^2$$

Design a series compensator for this plant that meets the following design criteria:

1. The nominal loop gain stays at least 20 dB below $l_m^{-1}(j\omega)$ for all values of ω where $l_m^{-1}(j\omega) < 0.1$.
2. Output disturbances of frequencies less than $\omega = 0.1$ are rejected to less than 10 percent of their original magnitude in steady-state. Output disturbances of frequencies less than $\omega = 0.01$ are rejected to less than 0.1 percent of their original magnitude in steady-state. Both these objectives are still met when the worst case multiplicative perturbation is present.
3. When the worst case multiplicative perturbation is present, the perturbed sensitivity function remains less than 1.3 for all frequencies.
4. The bandwidth of the system is as large as possible given the other requirements are met.

Observe and comment upon the transient response of your design to a negative step change in elevator position. Design a prefilter to remove any overshoot in the transient response while maintaining a similar speed of response. Observe and comment upon the transient response of your design to a unit step disturbance at the plant input. Comment upon the utility of the prefilter in this situation. Is the oscillatory pole pair of the phugoid actively or passively controlled? Compare this with the control of the oscillatory pair in Sec. 8.8.

(Optional) To observe the effect of your feedback loop in suppressing the phugoid oscillations during a climbing maneuver repeat Problem 4.12 using the controller designed during this problem.

8.10. (*CAD Problem*) Control of an exothermic chemical reactor

In Example 2.7-3 and Problem 3.11 the transfer function for the linearized dynamics of an exothermic chemical reactor was derived. The transfer function relates the percentage change in the flow of liquid in the cooling jacket to the percentage change in temperature inside the reaction tank. The transfer function is given by

$$\frac{Y(s)}{U(s)} = \frac{\dfrac{\delta T(s)}{\overline{T}}}{\dfrac{\delta F_J(s)}{\overline{F}_J}} = \frac{-3.36s - 5.7}{s^3 + 18.3s^2 - 43.78s - 119.8}$$

Because the time unit used in deriving the transfer function is hours, all results must be interpreted in units of radians per hour rather than the usual radians per second. Assume that the most significant dynamics that are not captured by the transfer function are the time delays involved in mixing in the tank, measurement of the temperature, and the change in the flow of the liquids after the valve positions are changed. Assume that the total of the time delays in the transfer functions is less than 20 sec or $1/180$ of an hour.

(*a*) Use Eq. (5.6-21) to establish a bound on the multiplicative perturbation arising from a delay of $1/180$ of an hour.

(*b*) Design a series compensator transfer function to work in this setting. Your compensator should contain integral action and as large a bandwidth as possible. It should be robustly stable with respect to the multiplicative perturbations bounded in (*a*) and the maximum over frequency of the worst case perturbed sensitivity function should be as small as possible. The transient response should be as fast and as smooth as possible with little or no oscillations or overshoot.

(*c*) After you have settled on a design, discuss its strengths and weaknesses. Explain how some aspects of the design might be improved and at what expenses to other aspects these improvements might occur.

APPENDIX A

THE LAPLACE TRANSFORM—A SUMMARY

This appendix is designed to serve as a brief review of the properties of the Laplace transform methods that are needed in this book. As such, it does not contain a complete treatment of all phases of the Laplace transform nor does it serve as an introduction to that subject. The reader who is not familiar with Laplace transforms is directed to one of the many available textbooks for an introductory treatment.[1]

Definitions. The Laplace transform of a time function $f(t)$, written as $F(s) = \mathcal{L}[f(t)]$, is defined by the following integral operation:

$$F(s) = \mathcal{L}[f(t)] = \int_0^\infty f(t)e^{-st}\,dt \tag{A-1}$$

[1]See, for example, the books by Oppenheim, Willsky, and Young, the book by Papoulis or the book by McGillem and Cooper in the references.

In dealing with the Laplace transform definiton of Eq. (A-1), we assume that $f(t)$ is zero for $t < 0$. In order for $\mathcal{L}[f(t)]$ to exist, it is necessary and sufficient that $f(t)$ be *sectionally continuous* and *exponential order*.

A function $f(t)$ is sectionally continuous if it possesses at most a finite number of finite discontinuities in any finite interval. A function $f(t)$ is of exponential order if $|f(t)| \leq Me^{-\alpha t}$ for some numbers M and α and all values of t larger than some value. In effect, this last statement simply means that $f(t)e^{-\alpha t}$ approaches zero as t approaches infinity; all functions of the form $f(t) = t^n e^{\lambda t} \sin \omega t$ are of exponential order.

The Laplace transform possesses a unique inverse transform $f(t) = \mathcal{L}^{-1}[F(s)]$ given by the expression

$$f(t) = \mathcal{L}^{-1}[F(s)] = \frac{1}{2\pi j} \int_{\sigma - j\infty}^{\sigma + j\infty} F(s)e^{st}\, ds \tag{A-2}$$

Here σ is chosen to be greater than the real part of any of the poles of $f(s)$, thus ensuring the existence of an inverse. In practice, rather than use Eq. (A-2), one normally obtains a partial-fraction expansion[1] of $f(s)$ and makes use of a table of Laplace transforms such as the one given in Appendix B.

Transform properties. Some of the more common properties of the Laplace transform are given below. The proofs for these have been kept simple; the reader is again directed to the available textbooks for a more complete delineation of the properties of the Laplace transform and rigorous proofs of these properties.

1. Superposition. For any constants a_1 and a_2 and any two functions $f_1(t)$ and $f_2(t)$ that possess Laplace transforms,

$$\begin{aligned} \mathcal{L}[a_1 f_1(t) + a_2 f_2(t)] &= a_1 \mathcal{L}[f_1(t)] + a_2 \mathcal{L}[f_2(t)] \\ &= a_1 F_1(s) + a_2 F_2(s) \end{aligned} \tag{A-3}$$

Proof. By the use of the definition of Eq. (A-1), $\mathcal{L}[a_1 f_1(t) + a_2 f_2(t)]$ is given by

$$\begin{aligned} \mathcal{L}[a_1 f_1(t) + a_2 f_2(t)] &= \int_0^\infty [a_1 f_1(t) + a_2 f_2(t)] e^{-st}\, dt \\ &= a_1 \int_0^\infty f_1(t) e^{-st}\, dt + a_2 \int_0^\infty f_2(t) e^{-st}\, dt \end{aligned}$$

The last two integrals may be recognized as the Laplace transforms of $f_1(t)$ and $f_2(t)$, respectively, so that we have established that

$$\mathcal{L}[a_1 f_1(t) + a_2 f_2(t)] = a_1 F_1(s) + a_2 F_2(s)$$

[1]See Sec. 4.2 for a detailed discussion of partial-fraction expansion.

In simple terms, this first property is nothing more than a statement that the Laplace transform is a linear operator.

2. Derivatives. The Laplace transform of the time derivative of a function $f(t)$ is given by s times the Laplace transform of $f(t)$ minus the initial value of $f(t)$; that is,

$$\mathcal{L}\left[\frac{df(t)}{dt}\right] = \mathcal{L}[f'(t)] = sF(s) - f(t)\big|_{t=0} \tag{A-4}$$

where $F(s) = \mathcal{L}[f(t)]$.
Proof. By definition, $\mathcal{L}[f'(t)]$ is given by

$$\mathcal{L}[f'(t) = \int_0^\infty f'(t)e^{-st}dt \tag{A-5}$$

After applying the integration-by-parts formula

$$\int_0^\infty u\,dv = uv|_0^\infty - \int_0^\infty v\,du$$

to Eq. (A-5) with $u = e^{-st}$ and $dv = f'(t)dt$, we obtain

$$\begin{aligned}\mathcal{L}[f'(t)] &= f(t)e^{-st}\Big|_{t=0}^{t=\infty} + s\int_0^\infty f(t)e^{-st}dt \\ &= -f(t)|_{t=0} + sF(s)\end{aligned}$$

In the last step we have made use of the fact that $f(t)$ is of exponential order and have set $\lim_{t\to\infty} f(t)e^{-st} = 0$.

By continuing the above procedure k times, one can easily establish the general rule of derivatives as

$$\mathcal{L}\left[\frac{d^k f(t)}{dt^k}\right] = s^k F(s) - s^{k-1}f(t)\big|_{t=0} - s^{k-2}\frac{df(t)}{dt}\bigg|_{t=0} \cdots - \frac{d^{k-1}f(t)}{dt}\bigg|_{t=0} \tag{A-6}$$

3. Integration. The Laplace transform of the integral of a function $f(t)$ is given by $1/s$ times the Laplace transform of $f(t)$ if $f(t)|_{t=0}$ is assumed to be finite; that is

$$\mathcal{L}\left[\int_0^t f(\tau)d\tau\right] = \frac{1}{s}F(s) \tag{A-7}$$

Proof. By the use of the definition, $\mathcal{L}\left[\int_0^t f(\tau)d\tau\right]$ becomes

$$\mathcal{L}\left[\int_0^t f(\tau)d\tau\right] = \int_0^\infty e^{-st}\left[\int_0^t f(\tau)d\tau\right]dt$$

By applying integration by parts to this expression with $u = \int_0^t f(\tau)d\tau$ and $dv = e^{-st}dt$, we obtain

$$\mathcal{L}\left[\int_0^t f(\tau)d\tau\right] = -\frac{1}{s}e^{-st}\int_0^t f(\tau)d\tau\bigg|_0^\infty + \frac{1}{s}\int_0^\infty f(t)e^{-st}dt$$

$$= \frac{1}{s}F(s)$$

4. Real translation. The Laplace transform of a time function delayed by time T is given by e^{-sT} times the Laplace transform of $f(t)$; that is,

$$\mathcal{L}[f(t-T)u(t-T)] = e^{-sT}F(s) \tag{A-8}$$

where $u(t)$ is the unit step function.
Proof. By definition, $\mathcal{L}[f(t-T)\mu(t-T)]$ is given by

$$\mathcal{L}[f(t-T)\mu(t-T)] = \int_0^\infty f(t-T)\mu(t-T)e^{-st}dt$$

$$= \int_T^\infty f(t-T)e^{-st}dt \tag{A-9}$$

Now let $\tau = t - T$ so that Eq. (A-9) becomes

$$\mathcal{L}[f(t-T)\mu(t-T)] = \int_0^\infty f(\tau)e^{-s(\tau+T)}d\tau$$

$$= e^{-st}\int_0^\infty f(\tau)e^{-st}d\tau = e^{-sT}F(s)$$

5. Complex translation. The Laplace transform of $e^{-\alpha t}$ times $f(t)$ is equal to the Laplace transform of $f(t)$ with s replaced by $s+\alpha$; that is

$$\mathcal{L}[e^{-\alpha t}f(t)] = F(s+\alpha) \tag{A-10}$$

Proof. By definition, $\mathcal{L}[e^{-\alpha t}f(t)]$ is

$$\mathcal{L}[e^{-\alpha t}f(t)] = \int_0^\infty e^{-\alpha t}f(t)e^{-st}dt$$

$$= \int_0^\infty f(t)e^{-(s+\alpha)t}dt = F(s+\alpha)$$

6. Initial value. If the initial value of $f(t)$ is finite, then $f(t)|_{t=0}$ is equal to the limit of $sF(s)$ as s approaches infinity; that is,

$$f(t)|_{t=0} = \lim_{s\to\infty} sF(s) \tag{A-11}$$

Note that this is an *equality*, not a transform pair.
Proof. Let us consider the Laplace transform of $f'(t)$ given by

$$\mathcal{L}[f'(t)] = sF(s) - f(t)|_{t=0} = \int_0^\infty f'(t)e^{-st}dt \tag{A-12}$$

Taking the limit as s approaches ∞ on both sides of Eq. (A-12) we obtain

$$\lim_{s\to\infty} sF(s) - f(t)\Big|_{t=0} = \lim_{s\to\infty}\int_0^\infty f'(t)e^{-st}dt$$

Now, assuming that the limit and integration can be interchanged, we obtain

$$\lim_{s\to\infty} sF(s) - f(t)\Big|_{t=0} = \int_0^\infty f'(t)\lim_{s\to\infty} e^{-st}dt = 0$$

7. Final value. If both limits exist, the limit of $f(t)$ as t approaches ∞ is equal to the limit of $sF(s)$ as s approaches zero; that is,

$$\lim_{t\to\infty} f(t) = \lim_{s\to 0} sf(s) \tag{A-13}$$

Proof. Let us begin with Eq. (A-12) and in this case take the limit as s approaches zero on both sides so that we obtain

$$\lim_{s\to 0} sF(s) - f(t)\Big|_{t=0} = \lim_{s\to 0}\int_0^\infty f'(t)e^{-st}dt$$

Once again interchanging the limit and integration operations, we obtain

$$\begin{aligned}\lim_{s\to 0} sF(s) - f(t)\Big|_{t=0} &= \int_0^\infty f'(t)\lim_{s\to 0} e^{-st}dt\\ &= \int_0^\infty f'(t)dt = f(t)\Big|_0^\infty\\ &= \lim_{t\to\infty} f(t) - f(t)\Big|_{t=0}\end{aligned}$$

Therefore we have established that

$$\lim_{s\to 0} sF(s) = \lim_{t\to\infty} f(t)$$

One must be very careful in applying the above results to ensure that both limits exist. For example, the Laplace transform of $f(t) = e^t$ is $F(s) = 1/(s-1)$ and

$$\lim_{s\to 0} sF(s) = \lim_{s\to 0}\frac{s}{s-1} = 0$$

Since $\lim_{t\to\infty} f(t) = \infty$, the result above is meaningless. In terms of the stability development of Chap. 6, we know that both limits exist if $F(s)$ has all its poles in the interior of the left half s plane.

8. Convolution. The inverse Laplace transform of the product of two transformed functions is given by

$$\begin{aligned}\mathcal{L}^{-1}[F_1(s)F_2(s)] &= \int_0^\infty f_1(\tau)f_2(t-\tau)d\tau\\ &= \int_0^\infty f_1(t-\tau)f_2(\tau)d\tau \qquad \text{(A-14)}\end{aligned}$$

Proof. The proof of this result is part of the standard texts of system theory and is not repeated here.

APPENDIX B

LAPLACE TRANSFORM TABLE

	$F(s)$	$f(t) \quad 0 \leq t$
1.	1	$\delta(t)$ unit impulse at $t = 0$
2.	$\frac{1}{s}$	1 or $\mu(t)$ unit step at $t = 0$
3.	$\frac{1}{s^2}$	$t\mu(t)$ ramp function
4.	$\frac{1}{s^n}$	$\frac{1}{(n-1)!}t^{n-1}$ n is a positive integer
5.	$\frac{1}{s}e^{-as}$	$\mu(t-a)$ unit step starting at $t = a$
6.	$\frac{1}{s}(1 - e^{-as})$	$\mu(t) - \mu(t-a)$ rectangular pulse
7.	$\frac{1}{s+a}$	e^{-at} exponential decay
8.	$\frac{1}{(s+a)^n}$	$\frac{1}{(n-1)!}t^{n-1}e^{-st}$ n is a positive integer
9.	$\frac{1}{s(s+a)}$	$\frac{1}{a}(1 - e^{-at})$
10.	$\frac{\omega}{s^2 + \omega^2}$	$\sin \omega t$

	$F(s)$	$f(t) \quad 0 \le t$
11.	$\dfrac{s}{s^2+\omega^2}$	$\cos\omega t$
12.	$\dfrac{s+\alpha}{s^2+\omega^2}$	$\dfrac{\sqrt{\alpha^2+\omega^2}}{\omega}\sin(\omega t+\phi) \qquad \phi=\arctan\dfrac{\omega}{\alpha}$
13.	$\dfrac{s\sin\theta+\omega\cos\theta}{s^2+\omega^2}$	$\sin(\omega t+\theta)$
14.	$\dfrac{1}{s(s^2+\omega^2)}$	$\dfrac{1}{\omega^2}(1-\cos\omega t)$
15.	$\dfrac{s+\alpha}{s(s^2+\omega^2)}$	$\dfrac{\alpha}{\omega^2}-\dfrac{\sqrt{\alpha^2+\omega^2}}{\omega^2}\cos(\omega t+\phi) \qquad \phi=\arctan\dfrac{\omega}{\alpha}$
16.	$\dfrac{1}{(s+a)^2+b^2}$	$\dfrac{1}{b}e^{-at}\sin bt$
17.	$\dfrac{1}{s^2+2\zeta\omega_n s+\omega_n^2}$	$\dfrac{1}{\omega_n\sqrt{1-\zeta^2}}e^{-\zeta\omega_n t}\sin\omega_n\sqrt{1-\zeta^2}t$
18.	$\dfrac{\lvert R\rvert e^{j\phi_R}}{s+\alpha-j\beta}+\dfrac{\lvert R\rvert e^{-j\phi_R}}{s+\alpha+j\beta}$	$2\lvert R\rvert e^{-\alpha t}\cos(\beta t+\phi_R)u(t)$
19.	$\dfrac{s+a}{(s+a)^2+b^2}$	$e^{-at}\cos bt$
20.	$\dfrac{s+\alpha}{(s+a)^2+b^2}$	$\dfrac{\sqrt{(\alpha-a)^2+b^2}}{b}e^{-st}\sin(bt+\phi)$
		$\phi=\arctan\dfrac{b}{\alpha-a}$
21.	$\dfrac{1}{s[(s+a)^2+b^2]}$	$\dfrac{1}{a^2+b^2}+\dfrac{1}{b\sqrt{a^2+b^2}}e^{-at}\sin(bt-\phi)$
		$\phi=\arctan\dfrac{b}{-a}$
22.	$\dfrac{1}{s(s^2+2\zeta\omega_n s+\omega_n^2)}$	$\dfrac{1}{\omega_n^2}-\dfrac{1}{\omega_n^2\sqrt{1-\zeta^2}}e^{-\zeta\omega_n t}\sin(\omega_n\sqrt{1-\zeta^2}t+\phi)$
		$\phi=\arccos\zeta$
23.	$\dfrac{s+\alpha}{s[(s+a)^2+b^2]}$	$\dfrac{\alpha}{a^2+b^2}+\dfrac{1}{b}\sqrt{\dfrac{(\alpha-a)^2+b^2}{a^2+b^2}}e^{-at}\sin(bt+\phi)$
		$\phi=\arctan\dfrac{b}{\alpha-a}-\arctan\dfrac{b}{-a}$

APPENDIX C

MATRIX INVERSION, EIGENVALUES, AND EIGENVECTORS

This appendix considers some needed properties concerning the inversion of matrices and the use of the eigenvalues and eigenvectors of a matrix.

C.1 INVERSION

Division in not defined for matrices; it is replaced by an operation known as *matrix inversion* for *square* matrices. The square matrix $\boldsymbol{A}^{-1}$ is defined as the inverse of the square matrix $\boldsymbol{A}$ if

$$\boldsymbol{A}^{-1}\boldsymbol{A} = \boldsymbol{A}\boldsymbol{A}^{-1} = \boldsymbol{I} \tag{C-1}$$

where $\boldsymbol{I}$ is the identity matrix. Not all square matrices have an inverse, since in order to have an inverse the determinant of the matrix must be nonzero. If the determinant is nonzero, the matrix possesses an inverse and is said to be *nonsingular*. If, on the other hand, the determinant is zero, the matrix has no inverse and is *singular*.

For nonsingular matrices, the inverse may be determined by the use of the adjoint matrix. The adjoint matrix, adj($\boldsymbol{A}$), is related to the matrix $\boldsymbol{A}$ by means of the cofactors of the elements of $\boldsymbol{A}$. The *cofactor*, cof(a_{ij}), of the element a_{ij} is $(-1)^{i+j}$ times the determinant of the matrix formed by deleting the ith row and the jth column of $\boldsymbol{A}$. The cofactors are often referred to as *signed* minors since they differ from the minor of the elements only by the $(-1)^{i+j}$ factor. The adjoint of the matrix $\boldsymbol{A}$ is then defined as the transpose of the cofactor matrix, that is, the matrix whose elements are the cofactors of A. Thus the adjoint of A is

$$\operatorname{adj}(A) = \left[\operatorname{cof}\left(a_{ij}\right)\right]^T = \begin{bmatrix} \operatorname{cof}(a_{11}) & \operatorname{cof}(a_{21}) & \cdots & \operatorname{cof}(a_{n1}) \\ \operatorname{cof}(a_{12}) & \operatorname{cof}(a_{22}) & \cdots & \operatorname{cof}(a_{n2}) \\ \vdots & \vdots & \ddots & \vdots \\ \operatorname{cof}(a_{1n}) & \operatorname{cof}(a_{2n}) & \cdots & \operatorname{cof}(a_{nn}) \end{bmatrix} \tag{C-2}$$

In terms of the adjoint matrix, the inverse of A is given by

$$A^{-1} = \frac{\operatorname{adj}(A)}{\det(A)} \tag{C-3}$$

Here we see why the determinant of A must be nonzero, since otherwise the above expression would involve division by zero.

Example C-1. To illustrate the above procedure, let us determine the inverse of the matrix.

$$A = \begin{bmatrix} 1 & 0 & 2 \\ 0 & 3 & 5 \\ 0 & 2 & 4 \end{bmatrix}$$

In order to check if the matrix is nonsingular, the determinant may be formed by expanding in terms of the minors of the first column.

$$\det(A) = 1(12 - 10) + 0 + 0 = 2 \neq 0$$

and A is nonsingular and therefore possesses an inverse.

The next step is to determine the cofactor matrix which is given by

$$\left[\operatorname{cof}\left(a_{ij}\right)\right] = \begin{bmatrix} 2 & 0 & 0 \\ 4 & 4 & -2 \\ -6 & -5 & 3 \end{bmatrix}$$

Here, for example, the cofactor of a_{23} is formed by eliminating the second row and the third column from the given matrix and multiplying the determinant of the resulting 2×2 matrix by $(-1)^{2+3}$. Thus $\operatorname{cof}(a_{23})$ is formed as

$$[\operatorname{cof}(a_{23})] = (-1)^{2+3} \det \begin{bmatrix} 1 & 0 & 2 \\ 0 & 3 & 5 \\ 0 & 2 & 4 \end{bmatrix} = (-1)^{2+3} \det \begin{bmatrix} 1 & 0 \\ 0 & 2 \end{bmatrix} = -2$$

The other cofactors are found by a similar process.

The adjoint matrix is easily found by transposing the cofactor matrix,

$$\operatorname{adj}(A) = \left[\operatorname{cof}\left(a_{ij}\right)\right]^T = \begin{bmatrix} 2 & 4 & -6 \\ 0 & 4 & -5 \\ 0 & -2 & 3 \end{bmatrix}$$

and the inverse of A is computed by means of Eq. (C-3):

$$A^{-1} = \frac{\operatorname{adj}(A)}{\det(A)} = \begin{bmatrix} 1 & 2 & -3 \\ 0 & 2 & -2.5 \\ 0 & -1 & 1.5 \end{bmatrix}$$

It is suggested that the reader verify that $A^{-1}A = AA^{-1} = I$.

C.2 EIGENVALUES AND EIGENVECTORS

The eigenvalues of an $n \times n$ matrix $\boldsymbol{A}$ are the n complex values, λ_i, $i = 1, 2, \ldots, n$, which solve the equation

$$\det(\lambda_i \boldsymbol{I} - \boldsymbol{A}) = 0 \tag{C-4}$$

We assume the eigenvalues are distinct. The matrix $\lambda_i \boldsymbol{I} - \boldsymbol{A}$ where λ_i solves the equation above is said to be singular and since singular matrices are non-invertible there exist vectors $\boldsymbol{v}_i$ and $\boldsymbol{w}_i$ such that

$$(\lambda_i \boldsymbol{I} - \boldsymbol{A})\, \boldsymbol{v}_i = 0 \quad i = 1, 2, \ldots, n \tag{C-5}$$

and

$$\boldsymbol{w}_j^T \left(\lambda_j \boldsymbol{I} - \boldsymbol{A}\right) = 0 \quad j = 1, 2, \ldots, n \tag{C-6}$$

Equivalently Eqs. (C-5) and (C-6) can be expressed as

$$\boldsymbol{A}\boldsymbol{v}_i = \lambda_i \boldsymbol{v}_i \quad i = 1, 2, \ldots, n \tag{C-7}$$

and

$$\boldsymbol{w}_j^T \boldsymbol{A} = \lambda_j \boldsymbol{w}_j^T \quad j = 1, 2, \ldots, n \tag{C-8}$$

The vectors $\boldsymbol{v}_i$ are call *right eigenvectors* and the vectors $\boldsymbol{w}_j$ are called *left eigenvectors*. Eigenvectors can always be scaled and remain eigenvectors as can be seen from Eqs. (C-7) and (C-8). By pre-multiplying Eq. (C-7) by $\boldsymbol{w}_j^T$ and post-multiplying Eq. (C-8) by $\boldsymbol{v}_i$, the following result is achieved.

$$\boldsymbol{w}_j^T \boldsymbol{A}\boldsymbol{v}_i = \lambda_i \boldsymbol{w}_j^T \boldsymbol{v}_i \tag{C-9}$$

$$\boldsymbol{w}_j^T \boldsymbol{A}\boldsymbol{v}_i = \lambda_j \boldsymbol{w}_j^T \boldsymbol{v}_i \tag{C-10}$$

Therefore, we have

$$\left(\lambda_i - \lambda_j\right) \boldsymbol{w}_j^T \boldsymbol{v}_i = 0$$

Since the eigenvalues are assumed to be distinct we have

$$\boldsymbol{w}_j^T \boldsymbol{v}_i = 0 \qquad \text{when} \qquad i \neq j \tag{C-11}$$

Let Λ be a diagonal matrix whose diagonal elements are the n eigenvalues of $\boldsymbol{A}$.

$$\boldsymbol{\Lambda} = \begin{bmatrix} \lambda_i & 0 & 0 & \ldots & 0 \\ 0 & \lambda_2 & 0 & \ldots & 0 \\ \vdots & \vdots & \vdots & \ddots & \vdots \\ 0 & 0 & 0 & \ldots & \lambda_n \end{bmatrix}$$

Let $\boldsymbol{V}$ be a matrix whose columns are the n right eigenvectors of $\boldsymbol{A}$.

$$\boldsymbol{V} = [\boldsymbol{v}_1 \ \boldsymbol{v}_2 \ \ldots \ \boldsymbol{v}_n] \tag{C-12}$$

Let $\boldsymbol{W}$ be a matrix whose rows are the transposes of the n left eigenvectors of $\boldsymbol{A}$.

$$\boldsymbol{W} = \begin{bmatrix} \boldsymbol{w}_1^T \\ \boldsymbol{w}_2^T \\ \vdots \\ \boldsymbol{w}_n^T \end{bmatrix} \tag{C-13}$$

With these definitions the n equations of Eq. (C-7) can be written as

$$\boldsymbol{AV} = \boldsymbol{V\Lambda} \tag{C-14}$$

and the n equations of Eq. (C-8) can be written as

$$\boldsymbol{WA} = \boldsymbol{\Lambda W} \tag{C-15}$$

Using the property of Eq. (C-11) and the fact that eigenvectors can be scaled, we obtain

$$\boldsymbol{WV} = \boldsymbol{I} \tag{C-16}$$

Using Eq. (C-16) in Eqs. (C-14) and (C-15) we obtain

$$\boldsymbol{A} = \boldsymbol{V\Lambda V}^{-1} = \boldsymbol{V\Lambda W} \tag{C-17}$$

and

$$\boldsymbol{A} = \boldsymbol{W}^{-1}\boldsymbol{\Lambda W} = \boldsymbol{V\Lambda W} \tag{C-18}$$

The expressions above allow us to obtain a useful representation of the resolvent matrix of $\boldsymbol{A}$.

$$\begin{aligned}(s\boldsymbol{I} - \boldsymbol{A})^{-1} &= \left(s\boldsymbol{W}^{-1}\boldsymbol{W} - \boldsymbol{W}^{-1}\boldsymbol{\Lambda W}\right)^{-1} = \left(\boldsymbol{W}^{-1}(s\boldsymbol{I} - \boldsymbol{\Lambda})\boldsymbol{W}\right)^{-1} \\ &= \boldsymbol{W}^{-1}(s\boldsymbol{I} - \boldsymbol{\Lambda})^{-1}\boldsymbol{W} \\ &= \boldsymbol{V}(s\boldsymbol{I} - \boldsymbol{\Lambda})^{-1}\boldsymbol{W}\end{aligned} \tag{C-19}$$

Now, since $\boldsymbol{\Lambda}$ is a diagonal,

$$(s\boldsymbol{I} - \boldsymbol{\Lambda})^{-1} = \begin{bmatrix} \frac{1}{s-\lambda_1} & 0 & \dots & 0 \\ 0 & \frac{1}{s-\lambda_2} & \dots & 0 \\ \vdots & \vdots & \ddots & \vdots \\ 0 & 0 & \dots & \frac{1}{s-\lambda_n} \end{bmatrix}$$

and, using the definitions of $\boldsymbol{V}$ and $\boldsymbol{W}$

$$(s\boldsymbol{I} - \boldsymbol{A})^{-1} = \sum_{i=1}^{n} \frac{\boldsymbol{v}_i \boldsymbol{w}_i^T}{s - \lambda_i} \tag{C-20}$$

Note that each term in the sum is itself a matrix.

APPENDIX D

COMPUTER AIDED DESIGN (CAD) TOOLS FOR CONTROL SYSTEMS: INTRODUCTION TO MATLAB®

There are a number of computer programs available today which greatly ease the burden of computing the quantities and producing the graphics required by the control system designer. These programs go under the general heading of Computer Aided Design (CAD) tools for control systems. Each of these products has various strengths and weaknesses. Each will probably be improved and made obsolete in the future. (The hope is that the programs will become obsolete before this book is obsolete.) In any event, it is not the purpose here to claim superiority of one program over another, nor is it the intention here to require the purchase of a particular program in

MATLAB is a registered trademark of The MathWorks, Inc.

order to make full use of this book. Computers should be used judiciously in learning control theory; insight is gained by doing the sketches by hand when learning the subject. That said, we have used computer aids extensively in the course of this book to demonstrate what aids are generally available and to provide some examples that are more true to life but are messy when worked by hand. Computer plots also provide attractive graphs for demonstrations.

This book uses the computer program MATLAB®, by MathWorks, Inc. We have also used the supplemental Control Toolbox, also produced by MathWorks. We have used these programs on Macintosh computers. This gave an added benefit of being able to easily alter the markings on the plots using standard graphics packages.

The sections in the text marked *CAD Notes* give examples of MATLAB programs that demonstrate the usefulness of CAD tools in general and the operation of one tool in particular. Other tools work similarly and have similar capabilities. MATLAB was chosen as the example program because it was made accessible to the author and interfaced well with the author's Macintosh system. The MATLAB examples should also prove useful to those who have access to other CAD tools.

MATLAB, like most programs, provides the ability to run a list of commands stored in what is called a *macro*. (MATLAB uses the term *M-files*.) Typical macros are included in the text. They are there to demonstrate typical capabilities of computer aided design programs, not to teach the student how to run MATLAB. (MathWorks has spent time producing manuals to provide that function.) What follows is a short introduction to some features of MATLAB so that the examples make sense.

The basic data structures of MATLAB are matrices and, in special cases, vectors. MATLAB can be run interactively or by M-files. If a line in MATLAB is not terminated with a semicolon, MATLAB will display the resulting quantity on the left-hand side of the equal sign after executing the line. This feature in used in the book to show how the program is responding. Comment lines are made in MATLAB by preceding the comment with a `%` symbol. User-supplied lines are presented in *italics* and usually follow the MATLAB prompt, >>. The value of a variable can be obtained by simply typing its name. Variables are erased or cleared using the `clear` command. MATLAB has a number of ways of entering, displaying, and manipulating matrices and vectors. Some of those ways are shown here by example.

```
>> v=[1 3 2 5]
v =
   1   3   2   5
>> w=v'
w =
   1
   3
   2
   5
>> x=v(4)
x =
   5
```

```
>>a=[1 5 3 8
 2 7 4 6
 3 4 3 9]

a =
   1  5  3  8
   2  7  4  6
   3  4  3  9
>>x=a(2,4)
x =
   6
>>b=a'
b =
   1  2  3
   5  7  4
   3  4  3
   8  6  9
```

The operator : is a special wild card operator in MATLAB

```
>>v=b(2,:)
v =
   5  7  4
>>w=b(:,1)
w =
   1
   5
   3
   8
```

The operator : is also used in *for* loops.

```
>>clear v
>>for i=0:2:10
 v(i/2+1)=i;
 end
>>v
v =
   0  2  4  6  8  10
```

The operator : can also be used to build vectors easily. The preceding vector could be built by

```
>>w=[0:2:10]
w =
   0  2  4  6  8  10
```

The operators `+`, `-`, `*`, `/` are interpreted on matrices where appropriate even to the point of a/b meaning ab^{-1} if possible. Matrices of zeros are formed using `zeros(m,n)`; matrices of ones are formed by using `ones(m,n)` and identity matrices are formed using `eye(n)`. In these functions square matrices need only one argument while vectors are formed by setting the appropriate argument to 1. The program handles complex numbers simply. The symbols i and j are set equal to the square root of -1 unless changed. Complex numbers are entered by writing, for example, `x=3+4*j`. The program is powerful in producing plots with simple instructions. It should be clear how this is working in the examples.

To reiterate, the examples of this book are there to demonstrate how programs *comparable to* MATLAB work. We have made no attempt to fully exercise MATLAB's capabilities and, while we are familiar with a number of computer aided control design programs, we make no claim to expertise in the intricacies of any of these programs or their relative merits.

REFERENCES

Anderson, B. D. O., and J. B. Moore: *Linear Optimal Control*, Prentice-Hall, Englewood Cliffs, N.J., 1971.

Åström, K. J.: *Computer Controlled Systems*, Prentice-Hall, Englewood Cliffs, N.J., 1984.

Åström, K. J.: "Frequency Domain Properties of Otto Smith Regulators," *Int. J. Control*, 26(2):307–314, 1977.

Åström, K. J., and A. Ostberg, "A Teaching Laboratory for Process Control," *IEEE Control Systems*, pp. 37–42, October 1986.

Athans, M., and P. L. Falb: *Optimal Control*, McGraw-Hill, New York, 1966.

Bell, R. F., et al.: "Head Positioning in a Large Disk Drive," *Hewlett-Packard Journal*, pp. 14–20, January 1984.

Bernard, J.: "Laser Jet Printer," *Computers and Electronics*, pp. 36–39, July 1984.

Black, H. S.: "Inventing the Negative Feedback Amplifier," *IEEE Spectrum*, pp. 55–60, December 1977.

Black, H. S.: "Stabilized Feedback Amplifiers," Bell Systems Tech. J., 1934.

Blakelock, J. H.: *Automatic Control of Aircraft and Missiles*, John Wiley & Sons, New York, 1965.

Bode, H. W.: "Feedback—The History of an Idea," in *Selected Papers on Mathematical Trends in Control Theory*, Dover, New York, pp. 106–123, 1969.

Bode, H. W.: *Network Analysis and Feedback Amplifier Design*, Van Nostrand, New York, 1945.

Bode, H. W.: "Relations Between Attenuation and Phase in Feedback Amplifier Design," *Bell System Tech. J.*, pp. 421–454, July 1940. Also in Thaler, G. S., ed.: *Automatic Control: Classical Linear Theory*, Dowden, Hutchinson and Ross, Inc., Stroudsburg, Pa., pp. 145–178, 1974.

Brasch, F. M., and J. B. Pearson, "Pole Placement Using Dynamic Compensators," *IEEE Trans. Automatic Control*, vol. AC-15, 1970, pp. 34–43.

Brockett, R. W.: "Poles, Zeros, and Feedback: State Space Interpretation," *IEEE Trans. Automatic Control*, vol. AC-10, April 1965, pp. 129–135.

Brockett, R. W., and J. L. Wilems: "Frequency Domain Stability Criteria—Part I," *IEEE Trans. Automatic Control*, vol. AC-10, July 1965, pp. 255–261.

Brockett, R. W., and J. L. Wilems: "Frequency Domain Stability Criteria—Part II," *IEEE Trans. Automatic Control*, Vol. AC-10, October 1965, pp. 407–413.

Churchill, R. V., J. W. Brown, and R. F. Verhey: *Complex Variables and Applications*, McGraw-Hill, New York, 1976.

Cruz, J. B., Jr.: *Feedback Systems*, McGraw-Hill, New York, 1972.

D'Azzo, J. J., and C. H. Houpis: *Feedback Control System Analysis and Synthesis*, 3rd ed., McGraw-Hill, New York, 1988.

DeLaCierva, J.: "Rate Servo Keeps TV Picture Clear," *Control Engineering*, p. 112, May 1965.

Donner, M. D.: *Real-Time Control of Walking*, Birkhauser Books, Boston, 1987.

Dorato, P.: "Robust Control: A Historical Review," *IEEE Control Systems*, pp. 44–46, April 1987.

Dorf, R. C.: *Encyclopedia of Robotics*, John Wiley & Sons, New York, 1988.

Dorf, R. C., *Modern Control Systems*, Fifth ed., Addison-Wesley, Reading, Mass., 1989.

Doyle, J. C., B. A. Francis, and A. R. Tannenbaum, *Feedback Control Theory*, Macmillan, New York, 1992.

Doyle, J. C., and G. Stein: "Multivariable Feedback Design: Concepts for a Classical/Modern Synthesis," *IEEE Trans. Automatic Control.*, AC-26 (1):4–16, February 1981.

Drela, M., and J. S. Langford: "Human Powered Flight," *Scientific American*, pp. 114–151, November 1985.

Etkin, B.: *Dynamics of Atmospheric Flight*, John Wiley & Sons, New York, 1959.

Evans, W. R., *Control System Dynamics*, McGraw-Hill, New York, 1954.

Evans, W. R.: "Control System Synthesis by Root Locus Method," *Trans. AIEE*, 69:66–69, 1950.

Evans, W. R.: "Graphical Analysis of Control Systems," *Trans. AIEE* 67:547–551, 1948.

Farquharson, F. B.: "Aerodynamic Stability of Suspension Bridges, With Special Reference to the Tacoma Narrows Bridge," *Bulletin 116, Part I*, The Engineering Experiment Station, University of Washington, 1950.

Francis, B. A.: "A Course in H_∞ Control Theory," *Lecture Notes in Control and Information Sciences*, 88, Springer-Verlag, 1987.

Francis, B.A., Zames, G.: "On H^∞-optimal sensitivity theory for siso feedback systems," *IEEE Trans. Automatic Control*, AC-29:9-16, 1984.

Franco, S.: *Design with Operational Amplifiers and Analog Integrated Circuits*, McGraw-Hill, New York, 1988.

Franklin, G. F., J. D. Powell, and M. L. Workman: *Digital Control of Dynamic Systems*, Second ed., Addison-Wesley, Reading, Mass., 1990.

Freudenberg, J. S., Looze, D. P.: "Right half-plane poles and zeros and design trade-offs in feedback systems," *IEEE Trans. Automatic Control*, AC-30:555-565, 1985.

Freudenberg, J. S., Looze, D. P.: "Frequency Domain Properties of Scalar and Multivariable Feedback Systems," *Lecture Notes in Control and Information Sciences*, 104, Springer-Verlag, 1988.

Gantmacher, F. R.: *The Theory of Matrices*, vols. I and II, Chelsea Publishing Co., New York, 1959.

Garbow, B. S., F. M. Boyle, J. J. Dongarra, and C. B. Moler, *Matrix Eigensystem Routines—EISPACK Guide Extension*, lecture notes in Computer Science 51, Springer-Verlag, Berlin, 1977.

Gilbert, E. G.: "Controllability and Observability in Multivariable Control Systems," *J. SIAM Control*, vol. 1, pp. 128–151, 1963.

Haber, R.: "Flight Simulation," *Scientific American*, pp. 96–103, July 1986.

Hall, A. C.: "Application of Circuit Theory to the Design of Servomechanisms," *J. Franklin Inst.*, 1946.

Hanselmann, H., and W. Moritz: "High Bandwidth Control of the Head-Positioning Mechanism in a Winchester Disk Drive," *IEEE Control Systems*, pp. 15–19, October 1987.

Horowitz, I. M.: *Synthesis of Feedback Systems*, Academic, New York, 1963.

Horowitz, I. M., and U. Shaked: "Superiority of Transfer Function over State-Variable Methods in Linear, Time Invariant Feedback System Design," *IEEE Trans. Automatic Control*, vol. AC-20, pp. 84–97, 1975.

Hurwitz, A.: "On the Conditions Under Which an Equation Has Only Roots With Negative Real Parts," *Mathematische Annalen*, 46, pp. 273–284, 1895. Also in *Selected Papers on Mathematical Trends in Control Theory*, Dover, New York, pp. 70–82, 1964.

Iso, H., et al.: "Instrumentation and Control for a Refining Process in Steelmaking," *IEEE Control Systems*, pp. 3-8, October 1987.

Jacobsen, S. C., et al.: "Design of Tactile Sensing Systems for Dextrous Manipulators," *IEEE Control Systems*, pp. 3–8, February 1988.

James, H. M., N. B. Nichols, and R. S. Phillips: *Theory of Servomechanisms*, Radiation Lab. Series, vol. 25, McGraw-Hill, New York, 1947.

Jurgen, R.: "Detroit '88 Driver Friendly Innovations," *IEEE Spectrum*, pp. 53–56, December 1987.

Kailath, T.: *Linear Systems*, Prentice-Hall, Englewood Cliffs, N.J., 1980.

Kalman, R. E.: "Mathematical Description of Linear Dynamical Systems," *J. SIAM*, vol. II, no. 1, ser. A, pp. 151–192, 1963.

Kalman, R. E.: "A New Approach to Linear Filtering and Prediction Problems," *ASME J. Basic Eng.*, 82:34–45, 1960.

Kalman, R. E.: "When is a Linear Control System Optimal?," *ASME J. Basic Eng.*, 86:51–60, March 1964.

Kalman, R. E., Y. C. Ho, and K. S. Narendra: "Controllability of Linear Dynamical Systems," *Contribution to Differential Equations*, vol. 1, no. 2, pp. 189–213, 1962.

Kuo, B. C.: *Automatic Control Systems*, Fourth ed., Prentice-Hall, Englewood Cliffs, N.J., 1982.

Kwakernaak, H., and R. Sivan: *Linear Optimal Control Systems*, Wiley-Interscience, New York, 1972.

Laub, A. J., A. Linnemann, and M. Wette: "Algorithms and Software for Pole Assignment by State Feedback," *Proc. 2nd Symp. CACSD*, March 1985.

Luenberger, D. G.: "Canonical Forms for Linear Multivariable Systems," *IEEE Trans. Automatic Control*, vol. AC-12, pp. 290–293, June 1967.

MacDuffe, C. C.: *Theory of Equations*, John Wiley & Sons, Inc., New York, pp. 29–104, 1954.

Maciejowski, J. M.: *Multivariable Feedback Design*, Addison Wesley, Reading, Mass., 1989.

Mar, B. W., and O. A. Bakken: "Applying Classical Control Theory to Energy-Economics Modeling," *Management Science*, pp. 81–91, January 1984.

MATLAB® User's Guide, The MathWorks, Inc., South Natick, MA, 1989.

Maxwell, J. C.: "On Governors," *Proc. of the Royal Society of London*, 16, 1868; in *Selected Papers on Mathematical Trends in Control Theory*. Dover, New York, pp. 270–283, 1964.

Mayr, O.: "Adam Smith and the Concept of the Feedback System," *Technology and Culture*, 12, 1, pp. 1–22, January 1971.

Mayr, O.: *The Origins of Feedback Control*, MIT Press, Cambridge, Mass., 1970.

Mayr, O.: "The Origins of Feedback Control," *Scientific American*, 223, 4, pp. 110-118, October 1970.

"The Mechanical Heart," *Mechanical Engineering*, pp. 77–81, September 1984.

McGillem, C. D. and G. R. Cooper: *Continuous and Discrete Signal and System Analysis*, 2nd ed., Holt, Rinehart, and Winston, New York, 1984.

Minami, S.: "Optical Scanner Design Leads to Laser Printer," *Laser Focus*, pp. 98–99, October 1987.

Misra, P., and R. V. Patel: "Numerical Algorithms for Eigenvalue Assignment by Constant and Dynamic Output Feedback," *IEEE Trans. Automatic Control*, AC-34(6): 577–588, 1989.

Mitchell, J. R.: "Comments on Bode Compensator Design," *IEEE Trans. Automatic Control*, pp. 869–870, October 1977.

Monzingo, R. A.: "On Approximating the Step Response of a Third-Order Linear System by a Second-Order Linear System," *IEEE Trans. Automatic Control*, vol. AC-13, p. 739, December 1968.

Moretti, P. M., and L. V. Divone: "Modern Windmills," *Scientific American*, pp. 110–118, June 1986.

Newton, G., L. Gould, and J. Kaiser: *Analytical Design of Linear Feedback Controls*, John Wiley & Sons, New York, 1957.

Nyquist, H.: "Regeneration Theory," *Bell Sys. Tech J.*, 11:126–147, 1932.

Ogata, K.: *Modern Control Engineering*, Prentice-Hall, Englewood Cliffs, N.J., 1970.

Ogata, K.: *State Space Analysis of Control Systems*, Prentice-Hall, Inc., Englewood Cliffs, N.J., 1967.

Oppenheim, A. V., A. S. Willsky, and I. T. Young: *Signals and Systems*, Prentice-Hall, Englewood Cliffs, N.J., 1983.

Papoulis, A.: *Circuits and Systems: A Modern Approach*, Holt, Rinehart and Winston, New York, 1980.

Parr, E. A.: *Industrial Control Handbook*, Industrial Press, New York, 1988.

Phillips, C. L.: "Analytical Bode Design of Controllers," *IEEE Transactions on Education*, vol. E-28, no. 1, pp. 43–44, February 1985.

Pottle, C.: "On the Partial Fraction Expansion of a Rational Function with Multiple Poles by Digital Computer," *IEEE Trans. Circuit Theory*, vol. CT-11, pp. 161–162, March 1964.

Powell, B. K., et al.: "Modeling and Analysis of an Inherently Fuel Injected Engine Idle Speed Control Loop," *Proceedings of the 1987 American Control Conference*, pp. 1543–1547.

Rivard, J. G.: "The Automobile in 1997," *IEEE Spectrum*, pp. 67–71, October 1987.

Rohrer, R.: *Circuit Analysis: An Introduction to the State Variable Approach*, McGraw-Hill, New York, 1970.

Rosenbrock, H. H.: *Computer-Aided Control System Design*, Academic, New York, 1974.

Rosenbrock, H. H.: *State Space and Multivariable Theory*, John Wiley & Sons, New York, 1970.

Routh, E. J.: *Dynamics of a System of Rigid Bodies*, Macmillan & Co., London, 1905.

Safonov, M. G., A. J. Laub, and G. Hartmann: "Feedback Properties of Multivariable Systems: The Role and Use of Return Difference Matrix," *IEEE Trans. Automatic Control*, AC-26, 47–65, 1981.

Schultz, D. G., and J. L. Melsa: *State Functions and Linear Control Systems*, McGraw-Hill, New York, 1967.

Shahinpoor, M.: *A Robot Engineering Textbook*, Harper and Row, New York, 1987.

Shung, J. B., et al., "Feedback Control and Simulation of a Wheelchair," *Trans. ASME, J. of Dynamic Systems*, pp. 96–100, June 1983.

Silverman, L. M.: "Transformation of Time-Variable Systems to Canonical (Phase-Variable) Form," *IEEE Trans. Automatic Control*, vol. AC-11, pp. 300–303, April 1966.

Sinha, N. K., and B. Kuszta: *Modeling and Identification of Dynamic Systems*, Van Nostrand Reinhold Co., New York, 1983.

Smith, H. W., and E. J. Davison: "Design of Industrial Regulators," *Proc. IEE* (London), vol. 119, pp. 1210–1216, August 1972.

Smith, O. J. M.: *Feedback Control Systems*, McGraw-Hill, New York, 1958.

Sridhar, B.: "Design of a Precision Pointing Control System for the Space Infrared Telecsope Facility," *IEEE Control Systems*, pp. 28–34, February 1986.

Stein, G., and N. R. Sandell, Jr.: *Classical and Modern Methods for Control System Design*, MIT, Cambridge, MA, 1979.

Strang, G.: *Linear Algebra and Its Applications*, Third ed., Harcourt Brace Jovanovich, New York, 1988.

Thompson, P. M.: *User's Guide to Program CC, Version 3*, Systems Technology, Inc., Hawthorne, Calif., March 1985.

Truxal, J. G.: *Control System Synthesis*, McGraw-Hill, New York, 1955.

Vidyasagar, M.: *Nonlinear Systems Analysis*, Prentice-Hall, Englewood Cliffs, N.J., 1978.

Vidyasagar, M.: "System Theory and Robotics," *IEEE Control Systems*, pp. 16–17, April 1987.

Wakeland, W. R., "Bode Compensator Design," *IEEE Trans. Automatic Control*, pp. 771–773, October 1976.

Watts, J. W.: "Control of an Inverted Pendulum," *Proceedings of the American Society of Engineering Education*, pp. 706–709, 1984.

Wiberg, D. W.: *Theory and Problems of State Space and Linear Systems* (Schaum's Outline Series), McGraw-Hill, New York, 1971.

Wiener, N.: *The Extrapolation, Interpolation and Smoothing of Stationary Time Series*, John Wiley & Sons, New York, 1949.

Willems, J. C.: "Least Squares Optimal Control and the Algebraic Riccati Equation," *IEEE Trans. Automatic Control*, vol. AC-16, December 1971.

Willems, J. C., and S. K. Mitter: "Controllability, Observability, Pole Allocation, and State Reconstruction," *IEEE Trans. Automatic Control*, vol. AC-16, pp. 582–95, December 1971.

Zadeh, L. A., and C. A. Desoer: *Linear System Theory*, McGraw-Hill, New York, 1963.

Zames, G.: "Feedback and optimal sensitivity: model reference transformations, multiplicative seminorms, and approximate inverses," *IEEE Trans. Automatic Control*, AC-26:319-338, 1981.

Zames, G.: "On the Input-Output Stability of Time-Varying Nonlinear Feedback Systems—Part I: Conditions Derived Using Concepts of Loop Gain. Conicity and Positivity," *IEEE Trans. Automatic Control*, AC-11:465–476, 1966.

Zames, G.: "On the Input-Output Stability of Time-Varying Nonlinear Feedback Systems—Part II: Conditions Involving Circles in the Frequency Plane and Sector Nonlinearities," *IEEE Trans. Automatic Control*, AC-11:228–238, 1966.

Zames, G., and B. A. Francis: "Feedback, minimax sensitivity, and optimal robustness," *IEEE Trans. Automatic Control*, AC-28:585-601, 1983.

Zames, G., and D. Bensoussan: "Multivariable Feedback Sensitivity and Optimal Robustness," *IEEE Trans. Automatic Control*, vol. AC-28, pp. 1030–1035, 1983.

Ziegler, J. G., and N. B. Nichols; "Optimum Settings for Automatic Controllers," *Trans. ASME*, 64:759–768, 1942.

Ziegler, J. G., and N. B. Nichols; "Process Lags in Automatic Control Circuits," *Trans. ASME*, 65(5):433–444, July 1943.

INDEX ENTRIES FOR PHYSICALLY ORIENTED PHENOMENA

Author's note: In this book much emphasis has been placed on including examples and discussions that are relevant in the modeling and control of physical systems. Here is a guide to the index entries to which the reader may refer in order to find these physically oriented phenomena.

INDEX